HCS12 MICROCONTROLLER AND EMBEDDED SYSTEMS

Using Assembly and C with CodeWarrior

Muhammad Ali Mazidi
Danny Causey

Upper Saddle River, New Jersey
Columbus, Ohio

Library of Congress Cataloging in Publication Data

Mazidi, Muhammad Ali.
HCS12 microcontroller and embedded systems: using Assembly and C with CodeWarrior / Muhammad Ali Mazidi, Danny Causey.
p. cm.
ISBN 978-0-13-607229-4 (alk. paper)
1.Programmable controllers. I. Causey, Danny. II. Title.

TJ223.P76M3782009
629.8'95--dc22

2008040719

Editor in Chief: Vernon Anthony
Acquistions Editor: Wyatt Morris
Editorial Assistant: Christopher Reed
Project Manager: Rex Davidson
Senior Operations Supervisor: Pat Tonneman
Operations Specialist: Laura Weaver
Art Director: Candace Rowley
Cover Designer: Diane Lorenzo
Cover Image: Don Bayley/iStockphoto
Director of Marketing: David Gesell
Marketing Assistant: Les Roberts

This book was set in Times Roman by Muhammd Ali Mazidi and Danny Causey and was printed and bound by R. R. Donnelly & Sons Company. The cover was printed by Demand Production Center.

Pearson Education Ltd., London
Pearson Education Singapore Pte. Ltd.
Pearson Education Canada, Ltd.
Pearson Education—Japan
Pearson Education Australia Pty. Limited
Pearson Education North Asia Ltd.
Pearson Educación de Mexico, S.A. de C.V.
Pearson Education Malaysia Pte. Ltd.

10 9 8 7 6 5 4 3 2 1

ISBN-13: 978-0-13-607229-4
ISBN-10: 0-13-607229-1

Trademark Information and Acknowledgments

Regard man as a mine rich in gems of inestimable value. Education can, alone, cause it to reveal its treasures, and enable mankind to benefit therefrom.

Baha'u'llah

BRIEF CONTENTS

CHAPTERS

0:	Introduction to Computing	1
1:	The HCS12/9S12 Microcontroller: History and Features	43
2:	HCS12 Architecture and Assembly Language Programming	59
3:	Branch, Call, and Time Delay Loop	105
4:	HCS12/9S12 I/O Port Programming	133
5:	Arithmetic, Logic Instructions, and Programs	151
6:	Advanced Addressing Modes, Look-up Table, Macros, and Modules	195
7:	HCS12 Programming in C	225
8:	HCS12 Hardware Connection, BDM, and S19 Hex File	257
9:	HCS12 Timer Programming in Assembly and C	277
10:	HCS12 Serial Port Programming in Assembly and C	327
11:	Interrupt Programming in Assembly and C	359
12:	LCD and Keyboard Interfacing	407
13:	ADC, DAC, and Sensor Interfacing	433
14:	Accessing Flash and EEPROM, and Page Switching	471
15:	Relay, Optoisolator, and Stepper Motor Interfacing with HCS12	507
16:	SPI Protocol and RTC Interfacing with HCS12	525
17:	PWM and DC Motor Control	555

APPENDICES

A:	HCS12 Instructions Explained	583
B:	AsmIDE, ImageCraft C Compiler, and D-BUG12	633
C:	IC Interfacing, System Design Issues, and Wire Wrapping	643
D:	Flowcharts and Pseudocode	661
E:	HCS12 Primer for x86 Programmers	667
F:	ASCII Codes	668
G:	Assemblers, Development Resources, and Suppliers	670
H:	Data Sheets	672

CONTENTS

CHAPTER 0: INTRODUCTION TO COMPUTING **1**
SECTION 0.1: NUMBERING AND CODING SYSTEMS 2
SECTION 0.2: DIGITAL PRIMER 9
SECTION 0.3: SEMICONDUCTOR MEMORY 13
SECTION 0.4: HARVARD AND von NEUMANN CPU ARCHITECTURES 29
SECTION 0.5: RISC ARCHITECTURE 33

CHAPTER 1: THE HCS12/9S12 MICROCONTROLLER: HISTORY AND FEATURES **43**
SECTION 1.1: MICROCONTROLLERS AND EMBEDDED PROCESSORS 44
SECTION 1.2: OVERVIEW OF THE CPU12 AND CPU08 48

CHAPTER 2: HCS12 ARCHITECTURE AND ASSEMBLY LANGUAGE PROGRAMMING **59**
SECTION 2.1: INSIDE THE HCS12 60
SECTION 2.2: THE HCS12 MEMORY MAP 64
SECTION 2.3: THE HCS12 ADDRESSING MODES 70
SECTION 2.4: HCS12 CONDITION CODE REGISTER 77
SECTION 2.5: HCS12 DATA FORMAT AND DIRECTIVES 81
SECTION 2.6: INTRODUCTION TO HCS12 ASSEMBLY LANGUAGE 86
SECTION 2.7: ASSEMBLING AND LINKING AN HCS12 PROGRAM 88
SECTION 2.8: STACK AND DATA TRANSFER INSTRUCTIONS 92

CHAPTER 3: BRANCH, CALL, AND TIME DELAY LOOP **105**
SECTION 3.1: BRANCH INSTRUCTIONS AND LOOPING 106
SECTION 3.2: JSR AND CALL INSTRUCTIONS 117
SECTION 3.3: PROGRAMS USING INDEXED ADDRESSING MODE 126

CHAPTER 4: HCS12/9S12 I/O PORT PROGRAMMING **133**
SECTION 4.1: I/O PORT PROGRAMMING IN HCS12 134
SECTION 4.2: I/O BIT MANIPULATION PROGRAMMING 142

CHAPTER 5: ARITHMETIC, LOGIC INSTRUCTIONS, AND PROGRAMS **151**
SECTION 5.1: ARITHMETIC INSTRUCTIONS AND PROGRAMS 152
SECTION 5.2: LOGIC INSTRUCTIONS AND PROGRAMS 162
SECTION 5.3: COMPARE INSTRUCTION AND PROGRAMS 168
SECTION 5.4: ROTATE, SHIFT INSTRUCTIONS, AND DATA SERIALIZATION 173
SECTION 5.5: BCD AND ASCII CONVERSION 180
SECTION 5.6: SIGNED NUMBER CONCEPTS AND ARITHMETIC OPERATIONS 182

CHAPTER 6: ADVANCED ADDRESSING MODES, LOOK-UP TABLE, MACROS, AND MODULES **195**
SECTION 6.1: ADVANCED INDEXED ADDRESSING MODE 196
SECTION 6.2: ACCESSING LOOK-UP TABLE IN FLASH 201
SECTION 6.3: CHECKSUM AND ASCII SUBROUTINES 206
SECTION 6.4: MACROS AND MODULES 212

CHAPTER 7: HCS12 PROGRAMMING IN C **225**
SECTION 7.1: DATA TYPES AND TIME DELAYS IN C 226
SECTION 7.2: LOGIC OPERATIONS IN C 234
SECTION 7.3: DATA CONVERSION PROGRAMS IN C 237
SECTION 7.4: I/O BIT MANIPULATION AND DATA SERIALIZATION IN C 243
SECTION 7.5: PROGRAM ROM ALLOCATION IN C 248

CHAPTER 8: HCS12 HARDWARE CONNECTION, BDM, AND S19 HEX FILE **257**
SECTION 8.1: HCS12 PIN CONNECTION AND BDM 258
SECTION 8.2: SETTING THE PLL FREQUENCY IN THE HCS12 267
SECTION 8.3: EXPLAINING THE S19 FILE FOR THE HCS12 270

CHAPTER 9: HCS12 TIMER PROGRAMMING IN ASSEMBLY AND C **277**
SECTION 9.1: FREE-RUNNING TIMER AND OUTPUT COMPARE FUNCTION 278
SECTION 9.2: INPUT CAPTURE PROGRAMMING 290
SECTION 9.3: PULSE ACCUMULATOR AND EVENT COUNTER PROGRAMMING 296
SECTION 9.4: HCS12 TIMER PROGRAMMING IN C 304

CHAPTER 10: HCS12 SERIAL PORT PROGRAMMING IN ASSEMBLY AND C **327**
SECTION 10.1: BASICS OF SERIAL COMMUNICATION 328
SECTION 10.2: HCS12 CONNECTION TO RS232 335
SECTION 10.3: HCS12 SERIAL PORT PROGRAMMING IN ASSEMBLY 337
SECTION 10.4: HCS12 SERIAL PORT PROGRAMMING IN C 351

CHAPTER 11: INTERRUPT PROGRAMMING IN ASSEMBLY AND C **359**
SECTION 11.1: HCS12 INTERRUPTS 360
SECTION 11.2: PROGRAMMING TIMER INTERRUPTS 366
SECTION 11.3: PROGRAMMING EXTERNAL HARDWARE INTERRUPTS 373
SECTION 11.4: PROGRAMMING THE SERIAL COMMUNICATION INTERRUPT 379
SECTION 11.5: INTERRUPT PRIORITY IN THE HCS12 383
SECTION 11.6: INTERRUPT PROGRAMMING IN C 388
SECTION 11.7: PROGRAMMING THE REAL TIME INTERRUPT 398

CHAPTER 12: LCD AND KEYBOARD INTERFACING **407**
SECTION 12.1: LCD INTERFACING 408
SECTION 12.2: KEYBOARD INTERFACING 421

CHAPTER 13: ADC, DAC, AND SENSOR INTERFACING **433**
SECTION 13.1: ADC CHARACTERISTICS 434
SECTION 13.2: ATD PROGRAMMING IN THE HCS12 439
SECTION 13.3: SENSOR INTERFACING AND SIGNAL CONDITIONING 458
SECTION 13.4: DAC INTERFACING 462

CHAPTER 14: ACCESSING FLASH AND EEPROM, AND PAGE SWITCHING **471**
SECTION 14.1: PAGE SWITCHING OF FLASH MEMORY IN HCS12 472
SECTION 14.2: ERASING AND WRITING TO FLASH 478
SECTION 14.3: WRITING TO EEPROM IN THE HCS12 490
SECTION 14.4: CLOCK SPEED FOR FLASH AND EEPROM 498

CHAPTER 15: RELAY, OPTOISOLATOR, AND STEPPER MOTOR INTERFACING WITH HCS12 **507**
SECTION 15.1: RELAYS AND OPTOISOLATORS 508
SECTION 15.2: STEPPER MOTOR INTERFACING 514

CHAPTER 16: SPI PROTOCOL AND RTC INTERFACING WITH HCS12 **525**
SECTION 16.1: SPI BUS PROTOCOL 526
SECTION 16.2: SPI MODULES IN THE HCS12 529
SECTION 16.3: DS1306 RTC INTERFACING WITH HCS12 536
SECTION 16.4: DS1306 RTC PROGRAMMING IN C 544
SECTION 16.5: ALARM AND INTERRUPT FEATURES OF THE DS1306 546

CHAPTER 17: PWM AND DC MOTOR CONTROL **555**
SECTION 17.1: DC MOTOR INTERFACING AND PWM 556
SECTION 17.2: PROGRAMMING PWM IN HCS12 566

APPENDIX A: HCS12 INSTRUCTIONS EXPLAINED 583

APPENDIX B: AsmIDE, ImageCraft C COMPILER, AND D-BUG12 633

APPENDIX C: IC INTERFACING, SYSTEM DESIGN ISSUES, AND WIRE WRAPPING 643

APPENDIX D: FLOWCHARTS AND PSEUDOCODE 661

APPENDIX E: HCS12 PRIMER FOR x86 PROGRAMMERS 667

APPENDIX F: ASCII CODES 668

APPENDIX G: ASSEMBLERS, DEVELOPMENT RESOURCES, AND SUPPLIERS 670

APPENDIX H: DATA SHEETS 672

INDEX 729

This book is dedicated
to the memory of Dr. F. Samandari and Mr. Y. Astani
for their inspring examples of dedication and devotion to the cause of unification of our planet.
– Muhammad Ali Mazidi

I dedicate my part to my family
for the patience and support they provided during the production of this book.
Their love helped me to see this book through.
– Danny Causey

PREFACE

Products using microprocessors generally fall into two categories. The first category uses high-performance microprocessors such as the Pentium in applications where system performance is critical. We have an entire book dedicated to this topic, *The 80x86 IBM PC and Compatible Computers, Volumes I and II*, from Prentice Hall. In the second category of applications, performance is secondary; issues of cost, space, power, and rapid development are more critical than raw processing power. The microprocessor for this category is often called a microcontroller.

This book is for the second category of applications. The HCS12 is a widely used microcontroller. There are many reasons for this, including the legacy of once popular chip called 68HC11. This book is intended for use in college-level courses teaching microcontrollers and embedded systems. It not only establishes a foundation of Assembly language programming, but also provides a comprehensive treatment of HCS12 interfacing for engineering students. From this background, the design and interfacing of microcontroller-based embedded systems can be explored. This book can also be used by practicing technicians, hardware engineers, computer scientists, and hobbyists. It is an ideal source for those wanting to move away from 68HC11 to a more powerful chip.

Prerequisites

Readers should have had an introductory digital course. Knowledge of Assembly language would be helpful but is not necessary. Although this book is written for those with no background in Assembly language programming, students with prior Assembly language experience will be able to gain a mastery of HCS12 architecture very rapidly and start on their projects right away. For the HCS12 C programming sections of the book, a basic knowledge of C programming is required. We use the CodeWarrior compiler IDE from Freescale throughout the book. The CodeWarrior compiler is available for free from the Freescale website (www.freescale.com). We encourage you to use the CodeWarrior or some other IDE to simulate and run the programs in this book.

Overview

A systematic, step-by-step approach is used to cover various aspects of HCS12 C and Assembly language programming and interfacing. Many examples and sample programs are given to clarify the concepts and provide students with an opportunity to learn by doing. Review questions are provided at the end of each section to reinforce the main points of the section.

Chapter 0 covers number systems (binary, decimal, and hex), and provides an introduction to basic logic gates and computer memory. This chapter is designed especially for students, such as mechanical engineering students, who have not taken a digital logic course or those who need to refresh their memory on these topics.

Chapter 1 discusses the history of the HCS12/9S12 and features of the original family members such as the 68HC11. It also provides a list of various members of the HCS12 family.

Chapter 2 discusses the internal architecture of the HCS12 and explains the use of a HCS12 assembler to create ready-to-run programs. It also explores the stack and the flag register.

In Chapter 3 the topics of loop, jump, and call instructions are discussed, with many programming examples.

Chapter 4 is dedicated to the discussion of I/O ports. This allows students who are working on a project to start experimenting with HCS12 I/O interfacing and start the hardware project as soon as possible.

Chapter 5 is dedicated to arithmetic, logic instructions, and programs.

Chapter 6 covers the HCS12 advanced addressing modes and explains how to access the data stored in the look-up table, as well as how to do macros and modules.

The C programming of the HCS12 is covered in Chapter 7. We use the CodeWarrior compiler from Freescale for this and other C programs of the HCS12 family throughout the book. The CodeWarrior is available for free from the www.freescale.com website.

In Chapter 8 we discuss the hardware connection of the HCS12 chip.

Chapter 9 describes the HCS12 timers and how to use them as event counters.

Chapter 10 is dedicated to serial data communication of the HCS12 and its interfacing to the RS232. It also shows HCS12 communication with COM ports of the x86 IBM PC and compatible computers.

Chapter 11 provides a detailed discussion of HCS12 interrupts with many examples on how to write interrupt handler programs.

Chapter 12 shows HCS12 interfacing with real-world devices such as LCDs and keyboards.

Chapter 13 shows HCS12 interfacing with real-world devices such as DAC chips, ADC chips, and sensors.

In Chapter 14 we cover how to use HCS12 Flash and EEPROM memories for data storage and explains how to do page switching in Flash memory.

Chapter 15 covers the basic interfacing of the HCS12 chip to relays, optoisolators, and stepper motors.

Chapter 16 shows how to connect and program the DS1306 real-time clock chip using the SPI bus protocol.

Finally, Chapter 17 shows PWM and basic interfacing to DC motors.

The appendices have been designed to provide all reference material required for the topics covered in the book. Appendix A describes each HCS12 instruction in detail, with examples. Appendix B covers AsmIDE, the ImageCraft C compiler, and DBUG-12. Appendix C examines IC interfacing and logic families, as well as HCS12 I/O port interfacing and fan-out. Make sure you study this section before connecting the HCS12 to an external device. In Appendix D, the use of flowcharts and pseudocode is explored. Appendix E is for students familiar with x86 architectures who need to make a rapid transition to HCS12 architecture. Appendix F provides the table of ASCII characters. Appendix G lists resources for assembler shareware, HCS12 trainers, and electronics parts. Appendix H contains data sheets for the HCS12/9S12 chip.

Lab Manual

The lab manual covers some very basic labs and can be found at the **www.MicroDigitalEd.com** website. The more advanced and rigorous lab assignments are left up to the instructors depending on the course objectives, class level, and whether the course is graduate or undergraduate. The support materials for this and other books by the authors can be found on this website, too.

Solutions Manual/PowerPoint® Slides

The end-of-chapter problems cover some very basic concepts. The more challenging and rigorous homework assignments are left up to the instructors depending on the course objectives, class level, and whether the course is graduate or undergraduate. The solutions manual was produced with the help of Mr. Doug Gannon. The solutions manual and PowerPoint® slides for the drawings are available online for instructors only.

Online Instructor Resources

To access supplementary materials online, instructors need to request an instructor access code. Go to **www.prenhall.com**, click the **Instructor Resource Center** link, and then click **Register Today** for an instructor access code. Within 48 hours after registering you will receive a confirming e-mail including an instructor access code. Once you have received your code, go to the site and log on for full instructions on downloading the materials you wish to use.

Acknowledgments

This book is the result of the dedication and encouragement of many individuals. Our sincere and heartfelt appreciation goes to all of them.

First, we would like to thank Professor Richard Henderson of Devry University for his thorough reading of the chapters and suggestions. With his vast and deep knowledge of embedded systems, it is an honor to have him as a reviewer. We would also like to thank Professor Norm Grossman for his enthusiasm and support. His encouragement meant a great deal to us in writing this book.

Thanks to the reviewers of this edition:

Rick Henderson, DeVry University – Kansas City, MO;
Alireza Kavianpour, DeVry University – Pomona, CA;
Mehrdad Nourani, University of Texas – Dallas, TX;
Ken Johns, DeVry University – Sherman Oaks, CA;
Sarmad Naimi, BIHE University;
Faramarz Mortezaie, DeVry University – Fremont; and
Sepehr Naimi, BIHE University.

Numerous students found errors or made suggestions in improving this book. We would like to thank all of them for their enthusiasm and support. Those students are: Eris Flores, Doug Gannon, Patrick Holt, Ronnie Matus, Daniel Milligan, Tyrone Harris, Simon Njoki, Alberto Anchondo, and Ahmad Al-Othman.

Finally, we would like to thank the people at Prentice Hall, in particular our editor Wyatt Morris, who continues to support and encourage our writing, and our project manager Rex Davidson, who made the book a reality. We were lucky to get the best copy editors in the world, Janice Mazidi and Bret Workman. Thank you both for your fantastic job, as usual.

We enjoyed writing this book, and hope you enjoy reading it and using it for your courses and projects. Please let us know if you have any suggestions or find any errors.

Assemblers/Compiler

The CodeWarrior can be downloaded from the following website:
http://www.freescale.com

The ICCV7 C compiler for HCS12 can be downloaded from the following website:
http://www.imagecraft.com

The GNU C compiler (gcc) and AsmIDE assembler for HCS12 can be downloaded from the following website:
http://www.ericengler.com

The MiniIDE assembler for HCS12 can be downloaded from the following website:
http://www.mgtek.com/miniide

The tutorials for all the above assemblers/compilers and HCS12 Trainer boards can be found on the following website:
http://www.MicroDigitalEd.com

ABOUT THE AUTHORS

Muhammad Ali Mazidi went to Tabriz University and holds Master's degrees from both Southern Methodist University and the University of Texas at Dallas. He is currently a.b.d. on his Ph.D. in the Electrical Engineering Department of Southern Methodist University. He is co-author of some widely used textbooks, including *The 80x86 IBM PC and Compatible Computers, The 8051 Microcontroller and Embedded Systems,* and *The PIC Microcontroller and Embedded Systems,* also available from Prentice Hall. He teaches microprocessor-based system design at DeVry University in Dallas, Texas. He is the founder of MicroDigitalEd.com.

Danny Causey is a U.S. Army veteran who has served in Germany and Iraq. He is a graduate of the CET department of DeVry University. His areas of interest include networking, game development, and microcontroller and FPGA embedded system design. He is a partner in MicroDigitalEd.com.

The authors can be contacted at the following e-mail addresses if you have any comments or suggestions, or if you find any errors.

mdebooks@yahoo.com
mmazidi@microdigitaled.com
dcausey@microdigitaled.com

CHAPTER 0

INTRODUCTION TO COMPUTING

OBJECTIVES

Upon completion of this chapter, you will be able to:

>> Convert any number from base 2, base 10, or base 16 to any of the other two bases
>> Describe the logical operations AND, OR, NOT, XOR, NAND, and NOR
>> Use logic gates to diagram simple circuits
>> Explain the difference between a bit, a nibble, a byte, and a word
>> Give precise mathematical definitions of the terms *kilobyte*, *megabyte*, *gigabyte*, and *terabyte*
>> Describe the purpose of the major components of a computer system
>> Contrast and compare various types of semiconductor memories in terms of their capacity, organization, and access time
>> Describe the relationship between the number of memory locations on a chip, the number of data pins, and the chip's memory capacity
>> Contrast and compare PROM, EPROM, UV-EPROM, EEPROM, Flash memory EPROM, and mask ROM memories
>> Contrast and compare SRAM, NV-RAM, and DRAM memories
>> List the steps a CPU follows in memory address decoding
>> List the three types of buses found in computers and describe the purpose of each type of bus
>> Describe the role of the CPU in computer systems
>> List the major components of the CPU and describe the purpose of each
>> Understand the RISC and Harvard architectures

To understand the software and hardware of a microcontroller-based system, one must first master some very basic concepts underlying computer architecture. In this chapter (which in the tradition of digital computers is called Chapter 0), the fundamentals of numbering and coding systems are presented in Section 0.1. In Section 0.2, an overview of logic gates is given. The semiconductor memory and memory interfacing are discussed in Section 0.3. In Section 0.4, the von Neumann and Harvard CPU (Central Processing Unit) architectures are discussed. Finally, in the last section we give a brief history of of RISC architecture. Although some readers may have an adequate background in many of the topics of this chapter, it is recommended that the material be reviewed, however briefly.

SECTION 0.1: NUMBERING AND CODING SYSTEMS

Whereas human beings use base 10 (*decimal*) arithmetic, computers use the base 2 (*binary*) system. In this section we explain how to convert from the decimal system to the binary system, and vice versa. The convenient representation of binary numbers, called *hexadecimal,* also is covered. Finally, the binary format of the alphanumeric code, called *ASCII*, is explored.

Decimal and binary number systems

Although there has been speculation that the origin of the base 10 system is based on the fact that human beings have 10 fingers, there is absolutely no speculation about the reason behind the use of the binary system in computers. The binary system is used in computers because 1 and 0 represent the two voltage levels of on and off. Whereas in base 10 there are 10 distinct symbols, 0, 1, 2, ..., 9, in base 2 there are only two, 0 and 1, with which to generate numbers. Base 10 contains digits 0 through 9; binary contains digits 0 and 1 only. These two binary digits, 0 and 1, are commonly referred to as *bits*.

Converting from decimal to binary

One method of converting from decimal to binary is to divide the decimal number by 2 repeatedly, keeping track of the remainders. This process continues until the quotient becomes zero. The remainders are then written in reverse order to obtain the binary number. This is demonstrated in Example 0-1.

Example 0-1

Convert 25_{10} to binary.

Solution:

```
          Quotient  Remainder
25/2  =   12        1     LSB (least significant bit)
12/2  =   6         0
6/2   =   3         0
3/2   =   1         1
1/2   =   0         1     MSB (most significant bit)
```

Therefore, $25_{10} = 11001_2$.

Converting from binary to decimal

To convert from binary to decimal, it is important to understand the concept of weight associated with each digit position. First, as an analogy, recall the weight of numbers in the base 10 system, as shown in the diagram. By the same token, each digit position of a number in base 2 has a weight associated with it:

740683_{10}	=	
3×10^0	=	3
8×10^1	=	80
6×10^2	=	600
0×10^3	=	0000
4×10^4	=	40000
7×10^5	=	700000
		740683

110101_2 =				***Decimal***	***Binary***
1×2^0	=	1×1	=	1	1
0×2^1	=	0×2	=	0	00
1×2^2	=	1×4	=	4	100
0×2^3	=	0×8	=	0	0000
1×2^4	=	1×16	=	16	10000
1×2^5	=	1×32	=	32	100000
				53	110101

Knowing the weight of each bit in a binary number makes it simple to add them together to get the number's decimal equivalent, as shown in Example 0-2.

Example 0-2

Convert 11001_2 to decimal.

Solution:

Weight:	16	8	4	2	1
Digits:	1	1	0	0	1
Sum:	16 +	8 +	0 +	0 +	1 = 25_{10}

Knowing the weight associated with each binary bit position allows one to convert a decimal number to binary directly instead of going through the process of repeated division. This is shown in Example 0-3.

Example 0-3

Use the concept of weight to convert 39_{10} to binary.

Solution:

Weight:	32	16	8	4	2	1
	1	0	0	1	1	1
	32 +	0 +	0 +	4 +	2 +	1 = 39

Therefore, $39_{10} = 100111_2$.

Hexadecimal system

Base 16, or the *hexadecimal* system as it is called in computer literature, is used as a convenient representation of binary numbers. For example, it is much easier for a human being to represent a string of 0s and 1s such as 100010010110 as its hexadecimal equivalent of 896H. The binary system has two digits, 0 and 1. The base 10 system has ten digits, 0 through 9. The hexadecimal (base 16) system has 16 digits. In base 16, the first ten digits, 0 to 9, are the same as in decimal, and for the remaining six digits, the letters A, B, C, D, E, and F are used. Table 0-1 shows the equivalent binary, decimal, and hexadecimal representations for 0 to 15.

Table 0-1: Base 16 Number System

Decimal	Binary	Hex
0	0000	0
1	0001	1
2	0010	2
3	0011	3
4	0100	4
5	0101	5
6	0110	6
7	0111	7
8	1000	8
9	1001	9
10	1010	A
11	1011	B
12	1100	C
13	1101	D
14	1110	E
15	1111	F

Converting between binary and hex

To represent a binary number as its equivalent hexadecimal number, start from the right and group 4 bits at a time, replacing each 4-bit binary number with its hex equivalent shown in Table 0-1. To convert from hex to binary, each hex digit is replaced with its 4-bit binary equivalent. See Examples 0-4 and 0-5.

Example 0-4

Represent binary 100111110101 in hex.

Solution:

First the number is grouped into sets of 4 bits: 1001 1111 0101.
Then each group of 4 bits is replaced with its hex equivalent:

```
1001   1111   0101
   9      F      5
```

Therefore, 100111110101_2 = 9F5 hexadecimal.

Example 0-5

Convert hex 29B to binary.

Solution:

```
                 2      9      B
29B    =      0010   1001   1011
```

Dropping the leading zeros gives 1010011011.

Converting from decimal to hex

Converting from decimal to hex could be approached in two ways:

1. Convert to binary first and then convert to hex. Example 0-6 shows this method of converting decimal to hex.
2. Convert directly from decimal to hex by repeated division, keeping track of the remainders. Experimenting with this method is left to the reader.

Example 0-6

(a) Convert 45_{10} to hex.

32	16	8	4	2	1	
1	0	1	1	0	1	First, convert to binary. 32 + 8 + 4 + 1 = 45

$45_{10} = 0010\ 1101_2 = 2D$ hex

(b) Convert 629_{10} to hex.

512	256	128	64	32	16	8	4	2	1
1	0	0	1	1	1	0	1	0	1

$629_{10} = (512 + 64 + 32 + 16 + 4 + 1) = 0010\ 0111\ 0101_2 = 275$ hex

(c) Convert 1714_{10} to hex.

1024	512	256	128	64	32	16	8	4	2	1
1	1	0	1	0	1	1	0	0	1	0

$1714_{10} = (1024 + 512 + 128 + 32 + 16 + 2) = 0110\ 1011\ 0010_2 = 6B2$ hex

Converting from hex to decimal

Conversion from hex to decimal can also be approached in two ways:

1. Convert from hex to binary and then to decimal. Example 0-7 demonstrates this method of converting from hex to decimal.
2. Convert directly from hex to decimal by summing the weights of all the digits.

Example 0-7

Convert the following hexadecimal numbers to decimal.

(a) $6B2_{16} = 0110\ 1011\ 0010_2$

1024	512	256	128	64	32	16	8	4	2	1
1	1	0	1	0	1	1	0	0	1	0

$1024 + 512 + 128 + 32 + 16 + 2 = 1714_{10}$

(b) $9F2D_{16} = 1001\ 1111\ 0010\ 1101_2$

32768	16384	8192	4096	2048	1024	512	256	128	64	32	16	8	4	2	1
1	0	0	1	1	1	1	1	0	0	1	0	1	1	0	1

$32768 + 4096 + 2048 + 1024 + 512 + 256 + 32 + 8 + 4 + 1 = 40{,}749_{10}$

Table 0-2: Counting in Bases

Decimal	Binary	Hex
0	00000	0
1	00001	1
2	00010	2
3	00011	3
4	00100	4
5	00101	5
6	00110	6
7	00111	7
8	01000	8
9	01001	9
10	01010	A
11	01011	B
12	01100	C
13	01101	D
14	01110	E
15	01111	F
16	10000	10
17	10001	11
18	10010	12
19	10011	13
20	10100	14
21	10101	15
22	10110	16
23	10111	17
24	11000	18
25	11001	19
26	11010	1A
27	11011	1B
28	11100	1C
29	11101	1D
30	11110	1E
31	11111	1F

Counting in bases 10, 2, and 16

To show the relationship between all three bases, in Table 0-2 we show the sequence of numbers from 0 to 31 in decimal, along with the equivalent binary and hex numbers. Notice in each base that when one more is added to the highest digit, that digit becomes zero and a 1 is carried to the next-highest digit position. For example, in decimal, 9 + 1 = 0 with a carry to the next-highest position. In binary, 1 + 1 = 0 with a carry; similarly, in hex, F + 1 = 0 with a carry.

Addition of binary and hex numbers

The addition of binary numbers is a very straightforward process. Table 0-3 shows the addition of two bits. The discussion of subtraction of binary numbers is bypassed since all computers use the addition process to implement subtraction. Although computers have adder circuitry, there is no separate circuitry for subtracters. Instead, adders are used in conjunction with *2's complement* circuitry to perform subtraction. In other words, to implement "$x - y$", the computer takes the 2's complement of y and adds it to x. The concept of 2's complement is reviewed next. Example 0-8 shows the addition of binary numbers.

Table 0-3: Binary Addition

A + B	Carry	Sum
0 + 0	0	0
0 + 1	0	1
1 + 0	0	1
1 + 1	1	0

Example 0-8

Add the following binary numbers. Check against their decimal equivalents.

Solution:

	Binary	*Decimal*
	1101	13
+	1001	9
	10110	22

2's complement

To get the 2's complement of a binary number, invert all the bits and then add 1 to the result. Inverting the bits is simply a matter of changing all 0s to 1s and 1s to 0s. This is called the *1's complement*. See Example 0-9.

Example 0-9

Take the 2's complement of 10011101.

Solution:

	10011101	binary number
	01100010	1's complement
+	1	
	01100011	2's complement

Addition and subtraction of hex numbers

In studying issues related to software and hardware of computers, it is often necessary to add or subtract hex numbers. Mastery of these techniques is essential. Hex addition and subtraction are discussed separately below.

Addition of hex numbers

This section describes the process of adding hex numbers. Starting with the least significant digits, the digits are added together. If the result is less than 16, write that digit as the sum for that position. If it is greater than 16, subtract 16 from it to get the digit and carry 1 to the next digit. The best way to explain this is by example, as shown in Example 0-10.

Example 0-10

Perform hex addition: 23D9 + 94BE.

Solution:

	23D9	LSD:	9 + 14 = 23	23 – 16 = 7 with a carry
+	94BE		1 + 13 + 11 = 25	25 – 16 = 9 with a carry
	B897		1 + 3 + 4 = 8	
		MSD:	2 + 9 = B	

Subtraction of hex numbers

In subtracting two hex numbers, if the second digit is greater than the first, borrow 16 from the preceding digit. See Example 0-11.

Example 0-11

Perform hex subtraction: 59F – 2B8.

Solution:

	59F	LSD:	8 from 15 = 7
–	2B8		11 from 25 (9 + 16) = 14 (E)
	2E7		2 from 4 (5 – 1) = 2

ASCII code

Hex	*Symbol*	*Hex*	*Symbol*
41	A	61	a
42	B	62	b
43	C	63	c
44	D	64	d
...	...	...	...
59	Y	79	y
5A	Z	7A	z

Figure 0-1. Selected ASCII Codes

The discussion so far has revolved around the representation of number systems. Because all information in the computer must be represented by 0s and 1s, binary patterns must be assigned to letters and other characters. In the 1960s a standard representation called *ASCII* (American Standard Code for Information Interchange) was established. The ASCII (pronounced "ask-E") code assigns binary patterns for numbers 0 to 9, all the letters of the English alphabet, both uppercase (capital) and lowercase, and many control codes and punctuation marks. The great advantage of this system is that it is used by most computers, so that information can be shared among computers. The ASCII system uses a total of 7 bits to represent each code. For example, 100 0001 is assigned to the uppercase letter "A" and 110 0001 is for the lowercase "a". Often, a zero is placed in the most-significant bit position to make it an 8-bit code. Figure 0-1 shows selected ASCII codes. A complete list of ASCII codes is given in Appendix F. The use of ASCII is not only standard for keyboards used in the United States and many other countries but also provides a standard for printing and displaying characters by output devices such as printers and monitors.

Notice that the pattern of ASCII codes was designed to allow for easy manipulation of ASCII data. For example, digits 0 through 9 are represented by ASCII codes 30 through 39. This enables a program to easily convert ASCII to decimal by masking off (changing to zero) the "3" in the upper nibble. Also notice that there is a relationship between the uppercase and lowercase letters. The uppercase letters are represented by ASCII codes 41 through 5A while lowercase letters are represented by codes 61 through 7A. Looking at the binary code, the only bit that is different between the uppercase "A" and lowercase "a" is bit 5. Therefore, conversion between uppercase and lowercase is as simple as changing bit 5 of the ASCII code.

Review Questions

1. Why do computers use the binary number system instead of the decimal system?
2. Convert 34_{10} to binary and hex.
3. Convert 110101_2 to hex and decimal.
4. Perform binary addition: 101100 + 101.
5. Convert 101100_2 to its 2's complement representation.
6. Add 36BH + F6H.
7. Subtract 36BH – F6H.
8. Write "80x86 CPUs" in its ASCII code (in hex form).

SECTION 0.2: DIGITAL PRIMER

This section gives an overview of digital logic and design. First, we cover binary logic operations, then we show gates that perform these functions. Next, logic gates are put together to form simple digital circuits. Finally, we cover some logic devices commonly found in microcontroller interfacing.

Binary logic

As mentioned earlier, computers use the binary number system because the two voltage levels can be represented as the two digits 0 and 1. Signals in digital electronics have two distinct voltage levels. For example, a system may define 0 V as logic 0 and +5 V as logic 1. Figure 0-2 shows this system with the built-in tolerances for variations in the voltage. A valid digital signal in this example should be within either of the two shaded areas.

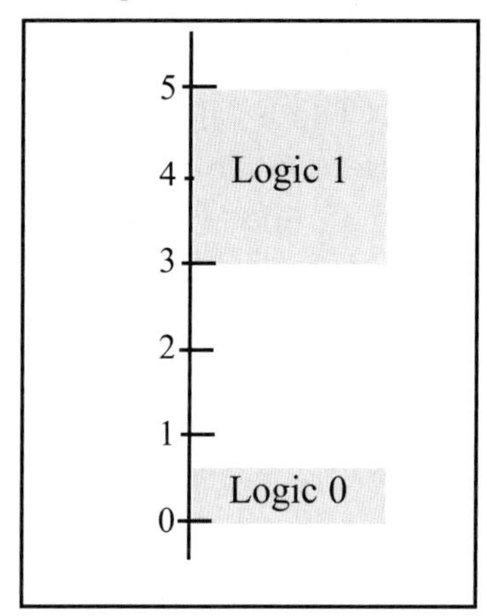

Figure 0-2. Binary Signals

Logic gates

Binary logic gates are simple circuits that take one or more input signals and send out one output signal. Several of these gates are defined below.

Logical AND Function

Inputs	Output
X Y	**X AND Y**
0 0	0
0 1	0
1 0	0
1 1	1

X
Y
X AND Y

AND gate

The AND gate takes two or more inputs and performs a logic AND on them. See the truth table and diagram of the AND gate. Notice that if both inputs to the AND gate are 1, the output will be 1. Any other combination of inputs will give a 0 output. The example shows two inputs, *x* and *y*. Multiple outputs are also possible for logic gates. In the case of AND, if all inputs are 1, the output is 1. If any input is 0, the output is 0.

Logical OR Function

Inputs	Output
X Y	**X OR Y**
0 0	0
0 1	1
1 0	1
1 1	1

X
Y
X OR Y

OR gate

The OR logic function will output a 1 if one or more inputs is 1. If all inputs are 0, then and only then will the output be 0.

Tri-state buffer

A buffer gate does not change the logic level of the input. It is used to isolate or amplify the signal.

Buffer

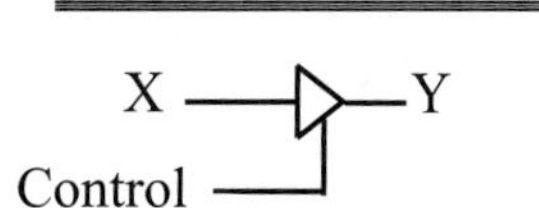

Inverter

The inverter, also called NOT, outputs the value opposite to that input to the gate. That is, a 1 input will give a 0 output, while a 0 input will give a 1 output.

Logical Inverter

Input	Output
X	NOT X
0	1
1	0

X —▷o— NOT X

XOR gate

The XOR gate performs an exclusive-OR operation on the inputs. Exclusive-OR produces a 1 output if one (but only one) input is 1. If both operands are 0, the output is 0. Likewise, if both operands are 1, the output is also 0. Notice from the XOR truth table, that whenever the two inputs are the same, the output is 0. This function can be used to compare two bits to see if they are the same.

Logical XOR Function

Inputs	Output
X Y	X XOR Y
0 0	0
0 1	1
1 0	1
1 1	0

X, Y — X XOR Y

NAND and NOR gates

The NAND gate functions like an AND gate with an inverter on the output. It produces a 0 output when all inputs are 1; otherwise, it produces a 1 output. The NOR gate functions like an OR gate with an inverter on the output. It produces a 1 if all inputs are 0; otherwise, it produces a 0. NAND and NOR gates are used extensively in digital design because they are easy and inexpensive to fabricate. Any circuit that can be designed with AND, OR, XOR, and INVERTER gates can be implemented using only NAND and NOR gates. A simple example of this is given below. Notice in NAND, that if any input is 0, the output is 1. Notice in NOR, that if any input is 1, the output is 0.

Logical NAND Function

Inputs	Output
X Y	X NAND Y
0 0	1
0 1	1
1 0	1
1 1	0

X, Y — X NAND Y

Logic design using gates

Next we will show a simple logic design to add two binary digits. If we add two binary digits there are four possible outcomes:

	Carry	*Sum*
0 + 0 =	0	0
0 + 1 =	0	1
1 + 0 =	0	1
1 + 1 =	1	0

Logical NOR Function

Inputs	Output
X Y	X NOR Y
0 0	1
0 1	0
1 0	0
1 1	0

X, Y — X NOR Y

Notice that when we add 1 + 1 we get 0 with a carry to the next higher place. We will need to determine the sum and the carry for this design. Notice that the sum column above matches the output for the XOR function, and that the carry column matches the output for the AND function. Figure 0-3(a) shows a simple adder implemented with XOR and AND gates. Figure 0-3(b) shows the same logic circuit implemented with AND and OR gates and inverters.

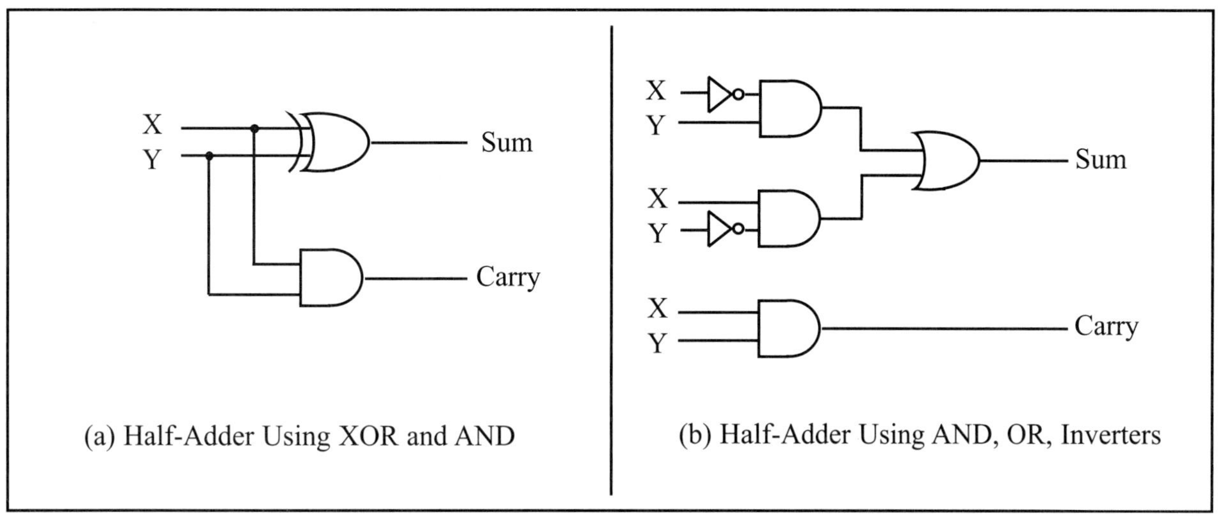

Figure 0-3. Two Implementations of a Half-Adder

Figure 0-4 shows a block diagram of a half-adder. Two half-adders can be combined to form an adder that can add three input digits. This is called a full-adder. Figure 0-5 shows the logic diagram of a full-adder, along with a block diagram that masks the details of the circuit. Figure 0-6 shows a 3-bit adder using three full-adders.

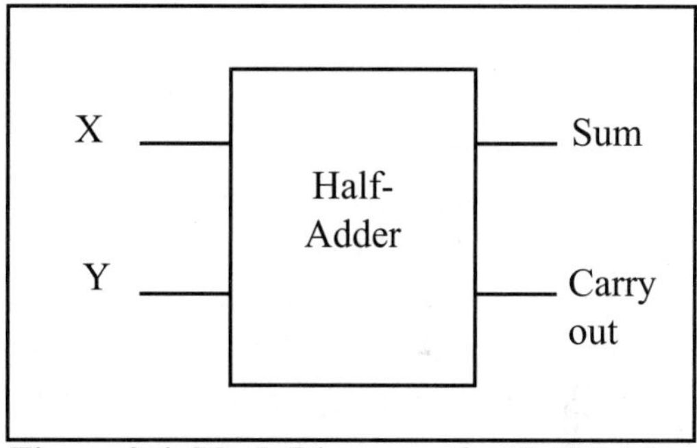

Figure 0-4. Block Diagram of a Half-Adder

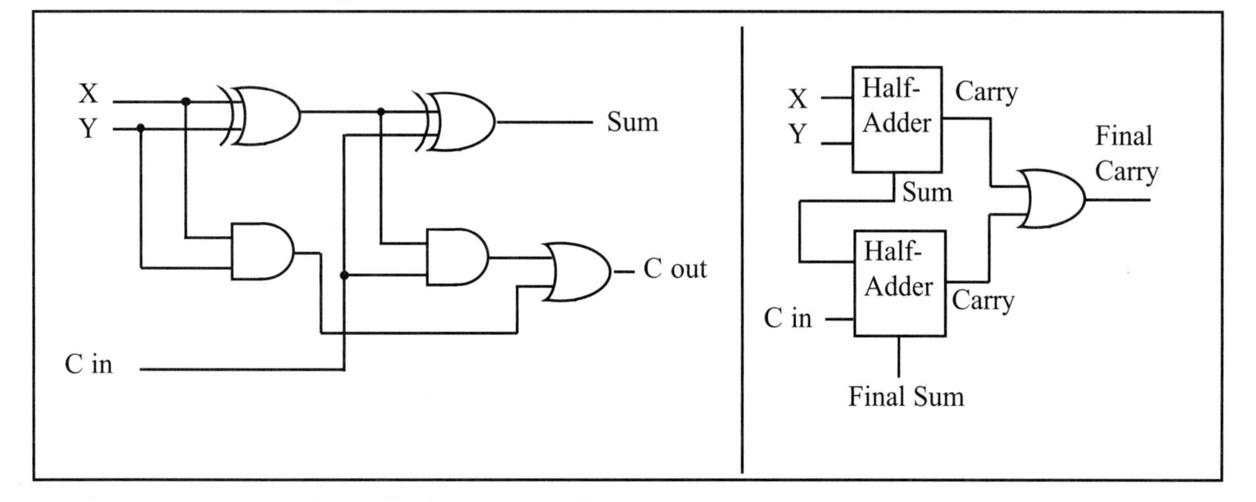

Figure 0-5. Full-Adder Built from a Half-Adder

Decoders

Another example of the application of logic gates is the decoder. Decoders are widely used for address decoding in computer design. Figure 0-7 shows decoders for 9 (1001 binary) and 5 (0101) using inverters and AND gates.

Flip-flops

A widely used component in digital systems is the flip-flop. Frequently, flip-flops are used to store data. Figure 0-8 shows the logic diagram, block diagram, and truth table for a flip-flop.

The D flip-flop is widely used to latch data. Notice from the truth table that a D-FF grabs the data at the input as the clock is activated. A D-FF holds the data as long as the power is on.

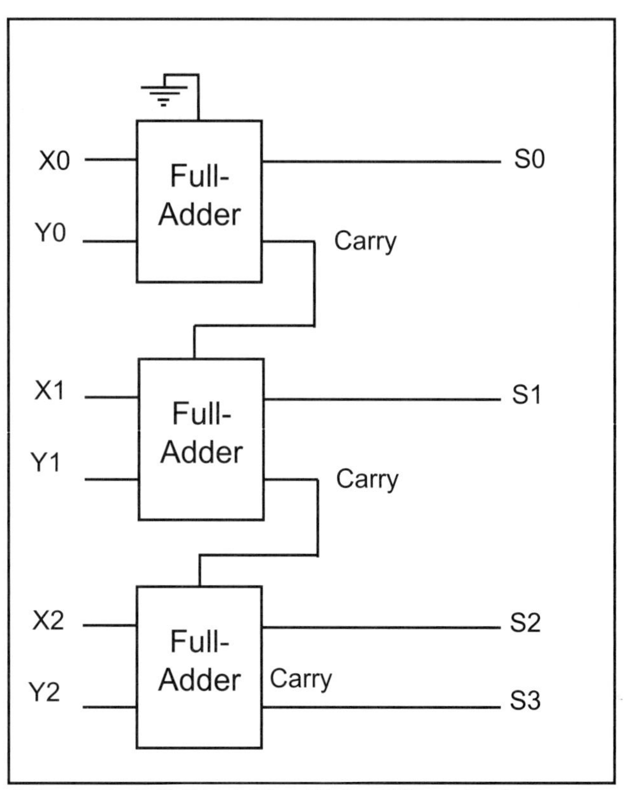

Figure 0-6. 3-Bit Adder Using Three Full-Adders

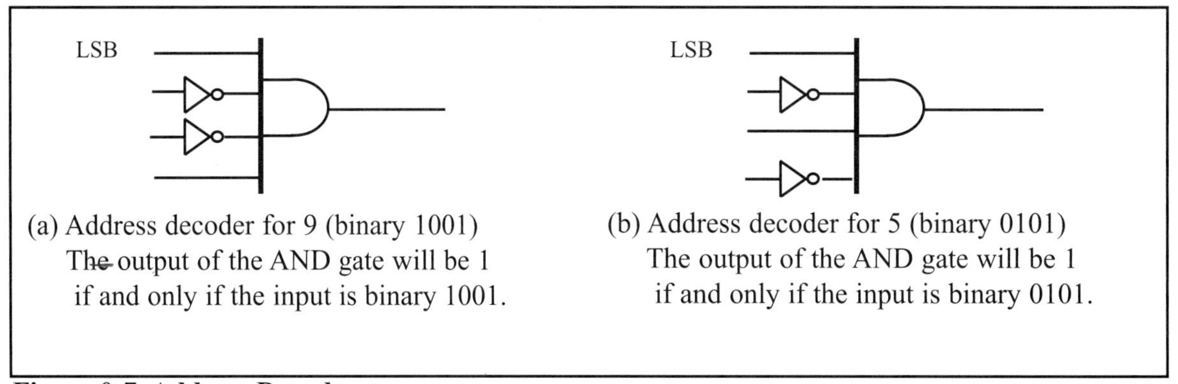

LSB
(a) Address decoder for 9 (binary 1001)
The output of the AND gate will be 1
if and only if the input is binary 1001.
LSB
(b) Address decoder for 5 (binary 0101)
The output of the AND gate will be 1
if and only if the input is binary 0101.

Figure 0-7. Address Decoders

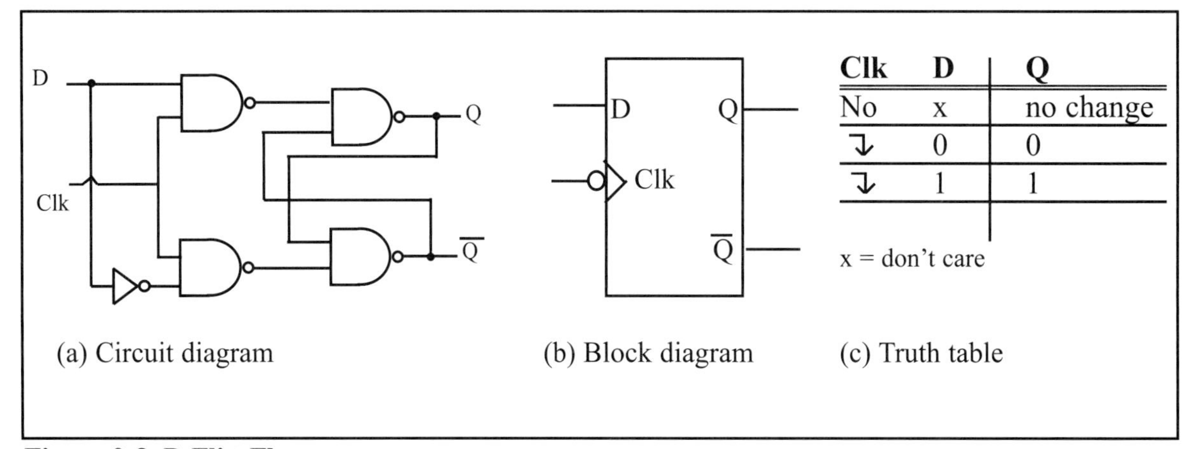

Clk	D	Q
No	x	no change
↴	0	0
↴	1	1

Figure 0-8. D Flip-Flops

Review Questions

1. The logical operation _____ gives a 1 output when all inputs are 1.
2. The logical operation _____ gives a 1 output when one or more of its inputs is 1.
3. The logical operation _____ is often used to compare two inputs to determine whether they have the same value.
4. A _____ gate does not change the logic level of the input.
5. Name a common use for flip-flops.
6. An address ______ is used to identify a predetermined binary address.

SECTION 0.3: SEMICONDUCTOR MEMORY

In this section we discuss various types of semiconductor memories and their characteristics such as capacity, organization, and access time. We will also show how the memory is connected to the CPU. Before we embark on the subject of memory, it will be helpful to give an overview of computer organization and review some terminology that is widely used in computer literature.

Some important terminology

```
Bit                         0
Nibble                   0000
Byte                0000 0000
Word 0000 0000 0000 0000
```

Recall from the discussion above that a *bit* is a binary digit that can have the value 0 or 1. A *byte* is defined as 8 bits. A *nibble* is half a byte, or 4 bits. A *word* is two bytes, or 16 bits. The display is intended to show the relative size of these units. Of course, they could all be composed of any combination of zeros and ones.

A *kilobyte* is 2^{10} bytes, which is 1,024 bytes. The abbreviation K is often used to represent kilobytes. A *megabyte*, or *meg* as some call it, is 2^{20} bytes. That is a little over 1 million bytes; it is exactly 1,048,576 bytes. Moving rapidly up the scale in size, a *gigabyte* is 2^{30} bytes (over 1 billion), and a *terabyte* is 2^{40} bytes (over 1 trillion). As an example of how some of these terms are used, suppose that a given computer has 16 megabytes of memory. That would be 16×2^{20}, or $2^4 \times 2^{20}$, which is 2^{24}. Therefore 16 megabytes is 2^{24} bytes.

Two types of memory commonly used in microcomputers are *RAM*, which stands for "random access memory" (sometimes called *read/write memory*), and *ROM*, which stands for "read-only memory." RAM is used by the computer for temporary storage of programs that it is running. That data is lost when the computer is turned off. For this reason, RAM is sometimes called *volatile memory*. ROM contains programs and information essential to the operation of the computer. The information in ROM is permanent, cannot be changed by the user, and is not lost when the power is turned off. Therefore, ROM is called *nonvolatile memory*.

Internal organization of computers

The internal working of every computer can be broken down into three parts: CPU (central processing unit), memory, and I/O (input/output) devices.

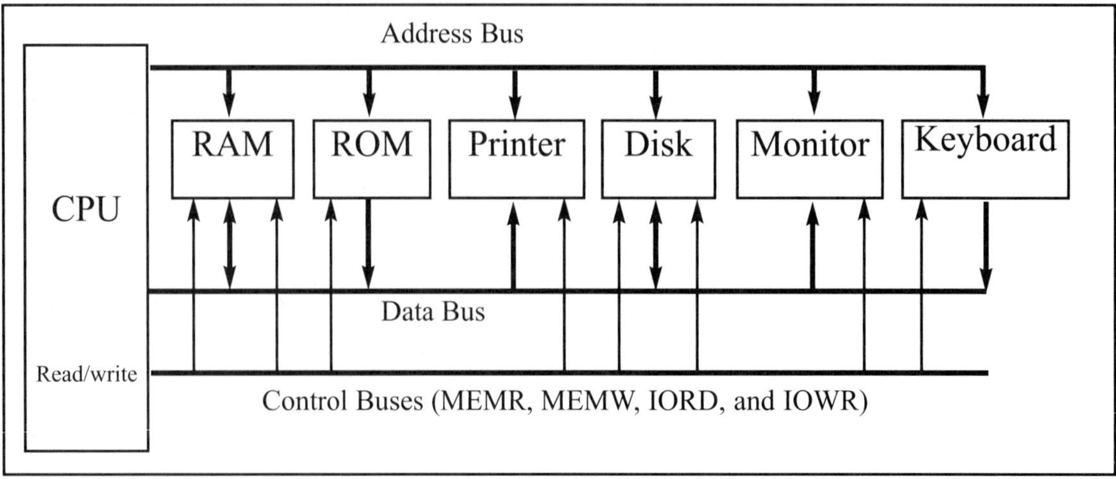

Figure 0-9. Internal Organization of a Computer

Figure 0-9 shows a block diagram of the internal organization of a computer. The function of the CPU is to execute (process) information stored in memory. The function of I/O devices such as the keyboard and video monitor is to provide a means of communicating with the CPU. The CPU is connected to memory and I/O through a group of wires called a *bus*. The bus inside a computer allows signals to carry information from place to place just as a street allows cars to carry people from place to place. In every computer there are three types of buses: address bus, data bus, and control bus.

For a device (memory or I/O) to be recognized by the CPU, it must be assigned an address. The address assigned to a given device must be unique; no two devices are allowed to have the same address. The CPU puts the address (in binary, of course) on the address bus, and the decoding circuitry finds the device. Then the CPU uses the data bus either to get data from that device or to send data to it. Control buses are used to provide read or write signals to the device to indicate if the CPU is asking for information or sending information. Of the three buses, the address bus and the data bus determine the capability of a given CPU.

More about the data bus

Because data buses are used to carry information in and out of a CPU, the more data buses available, the better the CPU. If one thinks of data buses as highway lanes, it is clear that more lanes provide a better pathway between the CPU and its external devices (such as printers, RAM, ROM, etc.; see Figure 0-9). By the same token, that increase in the number of lanes increases the cost of construction. More data buses mean a more expensive CPU and computer. The average size of data buses in CPUs varies between 8 and 64 bits. Early personal computers such as Apple 2 used an 8-bit data bus, while supercomputers such as Cray used a 64-bit data bus. Data buses are bidirectional, because the CPU must use them either to receive or to send data. The processing power of a computer is related to the size of its buses, because an 8-bit bus can send out 1 byte a time, but a 16-bit bus can send out 2 bytes at a time, which is twice as fast.

More about the address bus

Because the address bus is used to identify the devices and memory connected to the CPU, the more address bits available, the larger the number of devices that can be addressed. In other words, the number of address bits for a CPU determines the number of locations with which it can communicate. The number of locations is always equal to 2^x, where x is the number of address lines, regardless of the size of the data bus. For example, a CPU with 16 address lines can provide a total of 65,536 (2^{16}) or 64K of addressable memory. Each location can have a maximum of 1 byte of data. This is because all general-purpose microprocessor CPUs are what is called *byte addressable*. As another example, the IBM PC AT uses a CPU with 24 address lines and 16 data lines. Thus, the total accessible memory is 16 megabytes (2^{24} = 16 megabytes). In this example there would be 2^{24} locations, and because each location is one byte, there would be 16 megabytes of memory. The address bus is a *unidirectional* bus, which means that the CPU uses the address bus only to send out addresses. To summarize: The total number of memory locations addressable by a given CPU is always equal to 2^x where x is the number of address bits, regardless of the size of the data bus.

The CPU and its relation to RAM and ROM

For the CPU to process information, the data must be stored in RAM or ROM. The function of ROM in computers is to provide information that is fixed and permanent. This is information such as tables for character patterns to be displayed on the video monitor, or programs that are essential to the working of the computer, such as programs for testing and finding the total amount of RAM installed on the system, or for displaying information on the video monitor. In contrast, RAM stores temporary information that can change with time, such as various versions of the operating system and application packages such as word processing or tax calculation packages. These programs are loaded from the hard drive into RAM to be processed by the CPU. The CPU cannot get the information from the disk directly because the disk is too slow. In other words, the CPU first seeks the information to be processed from RAM (or ROM). Only if the data is not there does the CPU seek it from a mass storage device such as a disk, and then it transfers the information to RAM. For this reason, RAM and ROM are sometimes referred to as *primary memory* and disks are called *secondary memory*. Next, we discuss various types of semiconductor memories and their characteristics such as capacity, organization, and access time.

Memory capacity

The number of bits that a semiconductor memory chip can store is called chip *capacity*. It can be in units of Kbits (kilobits), Mbits (megabits), and so on. This must be distinguished from the storage capacity of computer systems. While the memory capacity of a memory IC chip is always given in bits, the memory capacity of a computer system is given in bytes. For example, an article in a technical journal may state that the 128M chip has become popular. In that case, it is understood, although it is not mentioned, that 128M means 128 megabits since the article is referring to an IC memory chip. However, if an advertisement states that a computer comes with 128M memory, it is understood that 128M means 128 megabytes since it is referring to a computer system.

Memory organization

Memory chips are organized into a number of locations within the IC (integrated circuit). Each location can hold 1 bit, 4 bits, 8 bits, or even 16 bits, depending on how it is designed internally. The number of bits that each location within the memory chip can hold is always equal to the number of data pins on the chip. How many locations exist inside a memory chip? That depends on the number of address pins. The number of locations within a memory IC always equals 2 to the power of the number of address pins. Therefore, the total number of bits that a memory chip can store is equal to the number of locations times the number of data bits per location. To summarize:

1. A memory chip contains 2^x locations, where x is the number of address pins.
2. Each location contains y bits, where y is the number of data pins on the chip.
3. The entire chip will contain $2^x \times y$ bits, where x is the number of address pins and y is the number of data pins on the chip.

Speed

One of the most important characteristics of a memory chip is the speed at which its data can be accessed. To access the data, the address is presented to the address pins, the READ pin is activated, and after a certain amount of time has elapsed, the data shows up at the data pins. The shorter this elapsed time, the better, and consequently, the more expensive the memory chip. The speed of the memory chip is commonly referred to as its *access time*. The access time of memory chips varies from a few nanoseconds to hundreds of nanoseconds, depending on the IC technology used in the design and fabrication process.

The three important memory characteristics of capacity, organization, and access time will be explored extensively in this chapter. Table 0-4 serves as a reference for the calculation of memory organization. Examples 0-12 and 0-13 demonstrate these concepts.

Table 0-4: Powers of 2

x	2^x
10	1K
11	2K
12	4K
13	8K
14	16K
15	32K
16	64K
17	128K
18	256K
19	512K
20	1M
21	2M
22	4M
23	8M
24	16M
25	32M
26	64M
27	128M

ROM (read-only memory)

ROM is a type of memory that does not lose its contents when the power is turned off. For this reason, ROM is also called *nonvolatile* memory. There are different types of read-only memory, such as PROM, EPROM, EEPROM, Flash EPROM, and mask ROM. Each is explained next.

PROM (programmable ROM) and OTP

PROM refers to the kind of ROM that the user can burn information into. In other words, PROM is a user-programmable memory. For every bit of storage in the PROM, there exists a fuse. PROM is programmed by blowing the fuses. If the information burned into PROM is wrong, that PROM must be discarded since its internal fuses are blown permanently. For this reason, PROM is also referred to

Example 0-12

A given memory chip has 12 address pins and 4 data pins. Find:
(a) the organization, and (b) the capacity.

Solution:

(a) This memory chip has 4,096 locations ($2^{12} = 4{,}096$), and each location can hold 4 bits of data. This gives an organization of $4{,}096 \times 4$, often represented as 4Kx4.
(b) The capacity is equal to 16K bits since there is a total of 4K locations and each location can hold 4 bits of data.

Example 0-13

A 512K memory chip has 8 pins for data. Find:
(a) the organization, and (b) the number of address pins for this memory chip.

Solution:

(a) A memory chip with 8 data pins means that each location within the chip can hold 8 bits of data. To find the number of locations within this memory chip, divide the capacity by the number of data pins. 512K/8 = 64K; therefore, the organization for this memory chip is 64Kx8.
(b) The chip has 16 address lines since 2^{16} = 64K.

as OTP (one-time programmable). The programming of ROM, also called *burning* ROM, requires special equipment called a *ROM burner* or *ROM programmer*.

EPROM (erasable programmable ROM) and UV-EPROM

EPROM was invented to allow making changes in the contents of PROM after it is burned. In EPROM, one can program the memory chip and erase it thousands of times. This is especially necessary during development of the prototype of a microprocessor-based project. A widely used EPROM is called UV-EPROM, where UV stands for ultraviolet. The only problem with UV-EPROM is that erasing its contents can take up to 20 minutes. All UV-EPROM chips have a window through which the programmer can shine ultraviolet (UV) radiation to erase its contents. For this reason, EPROM is also referred to as UV-erasable EPROM or simply UV-EPROM. Figure 0-10 shows the pins for UV-EPROM chips.

To program a UV-EPROM chip, the following steps must be taken:

1. Its contents must be erased. To erase a chip, remove it from its socket on the system board and place it in EPROM erasure equipment to expose it to UV radiation for 15 to 20 minutes.
2. Program the chip. To program a UV-EPROM chip, place it in the ROM burner (programmer). To burn code or data into EPROM, the ROM burner uses a programming voltage of 12.5 volts or higher, depending on the EPROM type. This voltage is referred to as V_{PP} in the UV-EPROM data sheet.

3. Place the chip back into its socket on the system board.

As can be seen from the above steps, not only is there an EPROM programmer (burner), but there is also separate EPROM erasure equipment. The main problem, and indeed the major disadvantage of UV-EPROM, is that it cannot be erased and programmed while it is on the system board. To find a solution to this problem, EEPROM was invented.

Notice the patterns of the IC numbers in Table 0-5. For example, part number 27128-25 refers to a UV-EPROM that has a capacity of 128K bits and access time of 250 nanoseconds. The capacity of the memory chip is indicated in the part number and the access time is given with a zero dropped. See Example 0-14. In part numbers, C refers to CMOS technology. Notice that 27XX always refers to UV-EPROM chips. For a comprehensive list of available memory chips see the JAMECO (jameco.com) or JDR (jdr.com) catalogs.

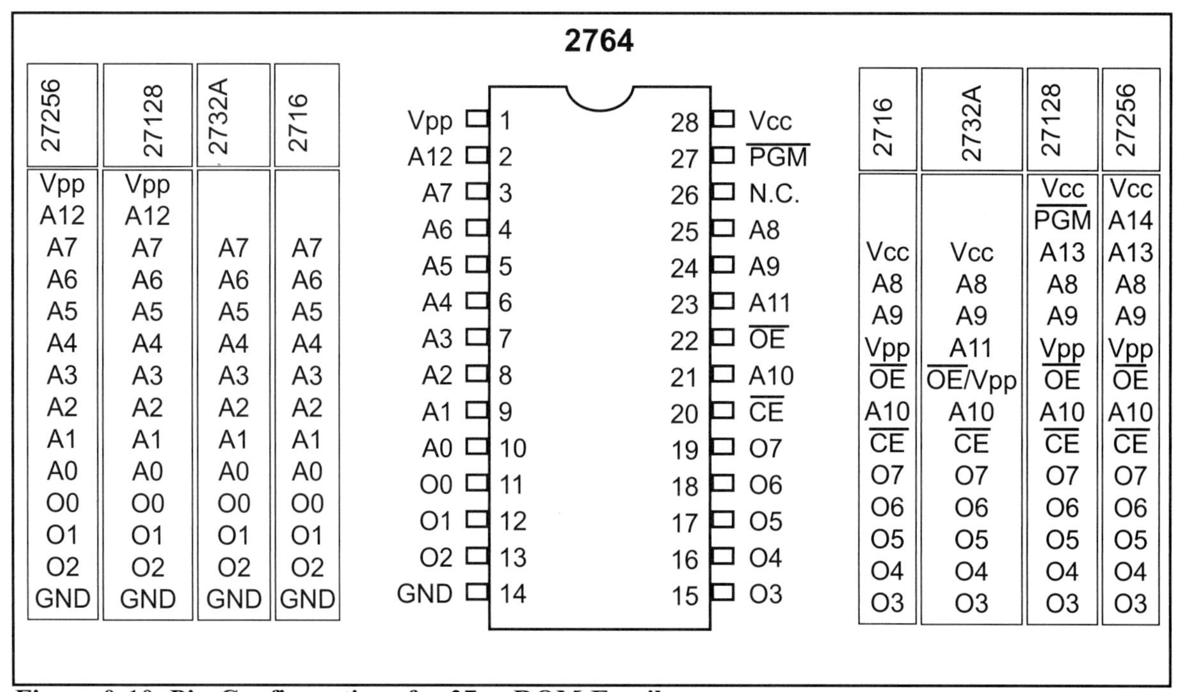

Figure 0-10. Pin Configurations for 27xx ROM Family

Example 0-14

For ROM chip 27128, find the number of data and address pins.

Solution:

The 27128 has a capacity of 128K bits. It has 16Kx8 organization (all ROMs have 8 data pins), which indicates that there are 8 pins for data and 14 pins for address (2^{14} = 16K).

Table 0-5: Some UV-EPROM Chips

Part #	Capacity	Org.	Access	Pins	V_{PP}
2716	16K	2Kx8	450 ns	24	25 V
2732	32K	4Kx8	450 ns	24	25 V
2732A-20	32K	4Kx8	200 ns	24	21 V
27C32-1	32K	4Kx8	450 ns	24	12.5 V CMOS
2764-20	64K	8Kx8	200 ns	28	21 V
2764A-20	64K	8Kx8	200 ns	28	12.5 V
27C64-12	64K	8Kx8	120 ns	28	12.5 V CMOS
27128-25	128K	16Kx8	250 ns	28	21 V
27C128-12	128K	16Kx8	120 ns	28	12.5 V CMOS
27256-25	256K	32Kx8	250 ns	28	12.5 V
27C256-15	256K	32Kx8	150 ns	28	12.5 V CMOS
27512-25	512K	64Kx8	250 ns	28	12.5 V
27C512-15	512K	64Kx8	150 ns	28	12.5 V CMOS
27C010-15	1024K	128Kx8	150 ns	32	12.5 V CMOS
27C020-15	2048K	256Kx8	150 ns	32	12.5 V CMOS
27C040-15	4096K	512Kx8	150 ns	32	12.5 V CMOS

EEPROM (electrically erasable programmable ROM)

EEPROM has several advantages over EPROM, such as the fact that its method of erasure is electrical and therefore instant, as opposed to the 20-minute erasure time required for UV-EPROM. In addition, in EEPROM one can select which byte is to be erased, in contrast to UV-EPROM, in which the entire contents of ROM are erased. However, the main advantage of EEPROM is that one can program and erase its contents while it is still in the system board. It does not require physical removal of the memory chip from its socket. In other words, unlike UV-EPROM, EEPROM does not require an external erasure and programming device. To utilize EEPROM fully, the designer must incorporate the circuitry to program the EEPROM into the system board. In general, the cost per bit for EEPROM is much higher than for UV-EPROM.

Flash memory EPROM

Since the early 1990s, Flash EPROM has become a popular user-programmable memory chip, and for good reasons. First, the erasure of the entire contents takes less than a second, or one might say in a flash, hence its name, Flash memory. In addition, the erasure method is electrical, and for this reason it is sometimes referred to as Flash EEPROM. To avoid confusion, it is commonly called *Flash memory*. The major difference between EEPROM and Flash memory is that when Flash memory's contents are erased, the entire device is erased, in contrast to EEPROM, where one can erase a desired byte. In Flash memories the contents are divided into blocks and the erasure can be done block by block. Unlike EEPROM, Flash memory has no byte erasure option. Because Flash memory can be programmed while it is in its socket on the system board, it is widely used to upgrade the BIOS ROM of the PC. Some designers believe that Flash memory will replace the hard disk as a mass storage medium. This would increase the performance of

Table 0-6: Some EEPROM and Flash Chips

EEPROMs

Part No.	Capacity	Org.	Speed	Pins	V_{PP}
2816A-25	16K	2Kx8	250 ns	24	5 V
2864A	64K	8Kx8	250 ns	28	5 V
28C64A-25	64K	8Kx8	250 ns	28	5 V CMOS
28C256-15	256K	32Kx8	150 ns	28	5 V
28C256-25	256K	32Kx8	250 ns	28	5 V CMOS

Flash

Part No.	Capacity	Org.	Speed	Pins	V_{PP}
28F256-20	256K	32Kx8	200 ns	32	12 V CMOS
28F010-15	1024K	128Kx8	150 ns	32	12 V CMOS
28F020-15	2048K	256Kx8	150 ns	32	12 V CMOS

the computer tremendously, since Flash memory is semiconductor memory with access time in the range of 100 ns compared with disk access time in the range of tens of milliseconds. For this to happen, Flash memory's program/erase cycles must become infinite, just like hard disks. Program/erase cycle refers to the number of times that a chip can be erased and reprogrammed before it becomes unusable. At this time, the program/erase cycle is 100,000 for Flash and EEPROM, 1,000 for UV-EPROM, and infinite for RAM and disks. See Table 0-6 for some sample chips.

Mask ROM

Mask ROM refers to a kind of ROM in which the contents are programmed by the IC manufacturer. In other words, it is not a user-programmable ROM. The term *mask* is used in IC fabrication. Since the process is costly, mask ROM is used when the needed volume is high (hundreds of thousands) and it is absolutely certain that the contents will not change. It is common practice to use UV-EPROM or Flash for the development phase of a project, and only after the code/data have been finalized is the mask version of the product ordered. The main advantage of mask ROM is its cost, since it is significantly cheaper than other kinds of ROM, but if an error is found in the data/code, the entire batch must be thrown away. Many manufacturers of 8051 microcontrollers support the mask ROM version of the 8051. It must be noted that all ROM memories have 8 bits for data pins; therefore, the organization is x8.

RAM (random access memory)

RAM memory is called *volatile* memory since cutting off the power to the IC results in the loss of data. Sometimes RAM is also referred to as RAWM (read and write memory), in contrast to ROM, which cannot be written to. There are three types of RAM: static RAM (SRAM), NV-RAM (nonvolatile RAM), and dynamic RAM (DRAM). Each is explained separately.

SRAM (static RAM)

Storage cells in static RAM memory are made of flip-flops and therefore do not require refreshing in order to keep their data. This is in contrast to DRAM, discussed below. The problem with the use of flip-flops for storage cells is that each cell requires at least 6 transistors to build, and the cell holds only 1 bit of data. In recent years, the cells have been made of 4 transistors, which still is too many. The use of 4-transistor cells plus the use of CMOS technology has given birth to a high-capacity SRAM, but its capacity is far below DRAM. Figure 0-11 shows the pin diagram for an SRAM chip.

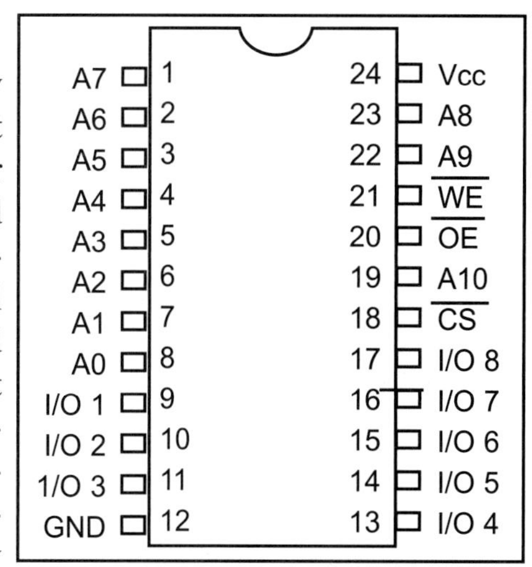

Figure 0-11. 2Kx8 SRAM Pins

The following is a description of the 6116 SRAM pins:

A0–A10 are for address inputs, where 11 address lines gives 2^{11} = 2K.
I/O0–I/O7 are for data I/O, where 8-bit data lines gives an organization of 2Kx8.
WE (write enable) is for writing data into SRAM (active LOW).
OE (output enable) is for reading data out of SRAM (active LOW)
CS (chip select) is used to select the memory chip.

The functional diagram for the 6116 SRAM is given in Figure 0-12.

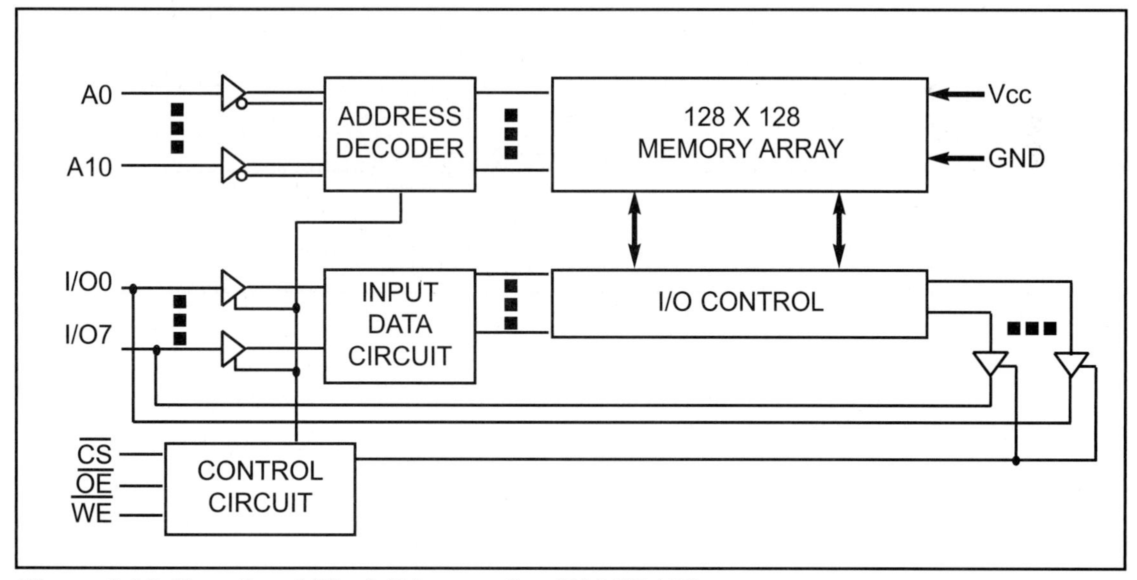

Figure 0-12. Functional Block Diagram for 6116 SRAM

Figure 0-13 shows the following steps to write data into SRAM.

1. Provide the addresses to pins A0–A11.
2. Activate the CS pin.
3. Make WE = 0 while RD = 1.

4. Provide the data to pins I/O0–I/O7.
5. Make WE = 1 and data will be written into SRAM on the positive edge of the WE signal.

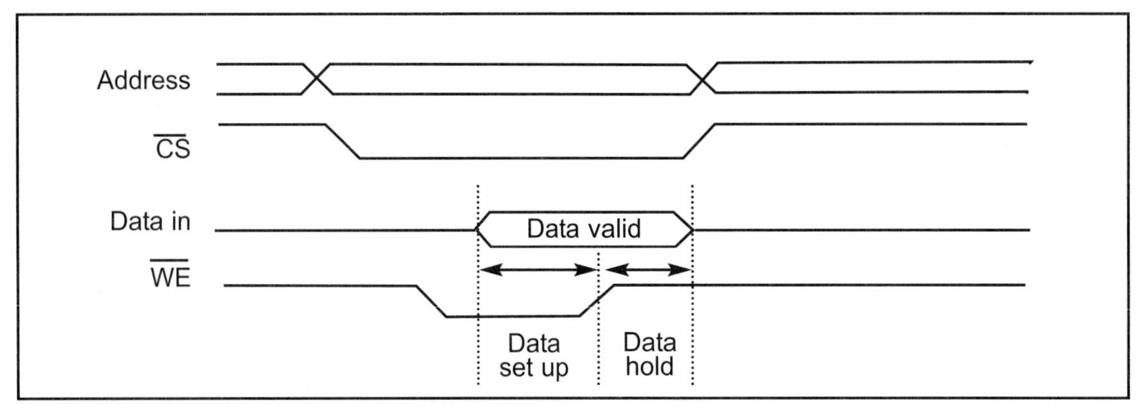

Figure 0-13. Memory Write Timing for SRAM

The following are steps to read data from SRAM. See Figure 0-14.

1. Provide the addresses to pins A0–A11. This is the start of the access time (t_{AA}).
2. Activate the CS pin.
3. While WE = 1, a high-to-low pulse on the OE pin will read the data out of the chip.

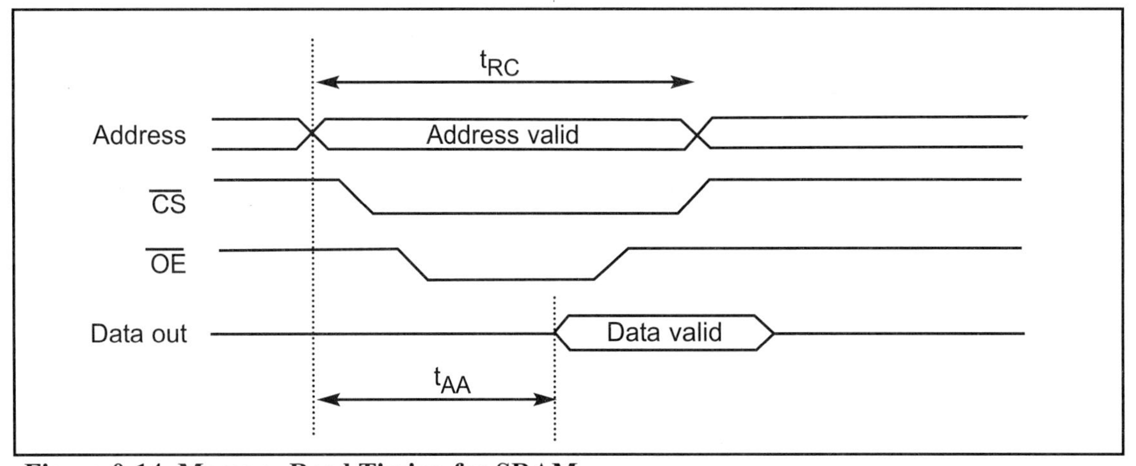

Figure 0-14. Memory Read Timing for SRAM

NV-RAM (nonvolatile RAM)

Whereas SRAM is volatile, there is a new type of nonvolatile RAM called *NV-RAM*. Like other RAMs, it allows the CPU to read and write to it, but when the power is turned off the contents are not lost. NV-RAM combines the best of RAM and ROM: the read and write ability of RAM, plus the nonvolatility of ROM. To retain its contents, every NV-RAM chip internally is made of the following components:

1. It uses extremely power-efficient (very low power consumption) SRAM cells built out of CMOS.

(a) data bus (b) address bus

9. If an address bus for a given computer has 16 lines, what is the maximum amount of memory it can access?
10. The speed of semiconductor memory is in the range of
    (a) microseconds (b) milliseconds
    (c) nanoseconds (d) picoseconds
11. Find the organization and chip capacity for each ROM with the indicated number of address and data pins.
    (a) 14 address, 8 data (b) 16 address, 8 data (c) 12 address, 8 data
12. Find the organization and chip capacity for each RAM with the indicated number of address and data pins.
    (a) 11 address, 1 data SRAM (b) 13 address, 4 data SRAM
    (c) 17 address, 8 data SRAM (d) 8 address, 4 data DRAM
    (e) 9 address, 1 data DRAM (f) 9 address, 4 data DRAM
13. Find the capacity and number of pins set aside for address and data for memory chips with the following organizations.
    (a) 16Kx4 SRAM (b) 32Kx8 EPROM (c) 1Mx1 DRAM
    (d) 256Kx4 SRAM (e) 64Kx8 EEPROM (f) 1Mx4 DRAM
14. Which of the following is (are) volatile memory?
    (a) EEPROM (b) SRAM (c) DRAM (d) Flash
15. A given memory block uses addresses 4000H–7FFFH. How many K bytes is this memory block?
16. The 74138 is a(n) _____ to _____ decoder.
17. In the 74138 give the status of G2A and G2B for the chip to be enabled.
18. In the 74138 give the status of G1 for the chip to be enabled.
19. In Example 0-16, what is the range of addresses assigned to Y5?

SECTION 0.4: HARVARD AND von NEUMANN CPU ARCHITECTURES

In this section we will examine the inside of a CPU. Then, we will compare the Harvard and von Neumann architectures.

Inside the CPU

A program stored in memory provides instructions to the CPU to perform an action. See Figure 0-19. The action can simply be adding data such as payroll numbers or controlling a machine such as a robot. The function of the CPU is to fetch these instructions from memory and execute them. To perform the actions of fetch and execute, all CPUs are equipped with resources such as the following:

1. Foremost among the resources at the disposal of the CPU are a number of *registers*. The CPU uses registers to store information temporarily. The information could be two values to be processed, or the address of the value needed to be fetched from memory. Registers inside the CPU can be 8-bit, 16-bit, 32-bit, or even 64-bit registers, depending on the CPU. In general, the more and bigger the registers, the better the CPU. The disadvantage of more and bigger registers is the increased cost of such a CPU.

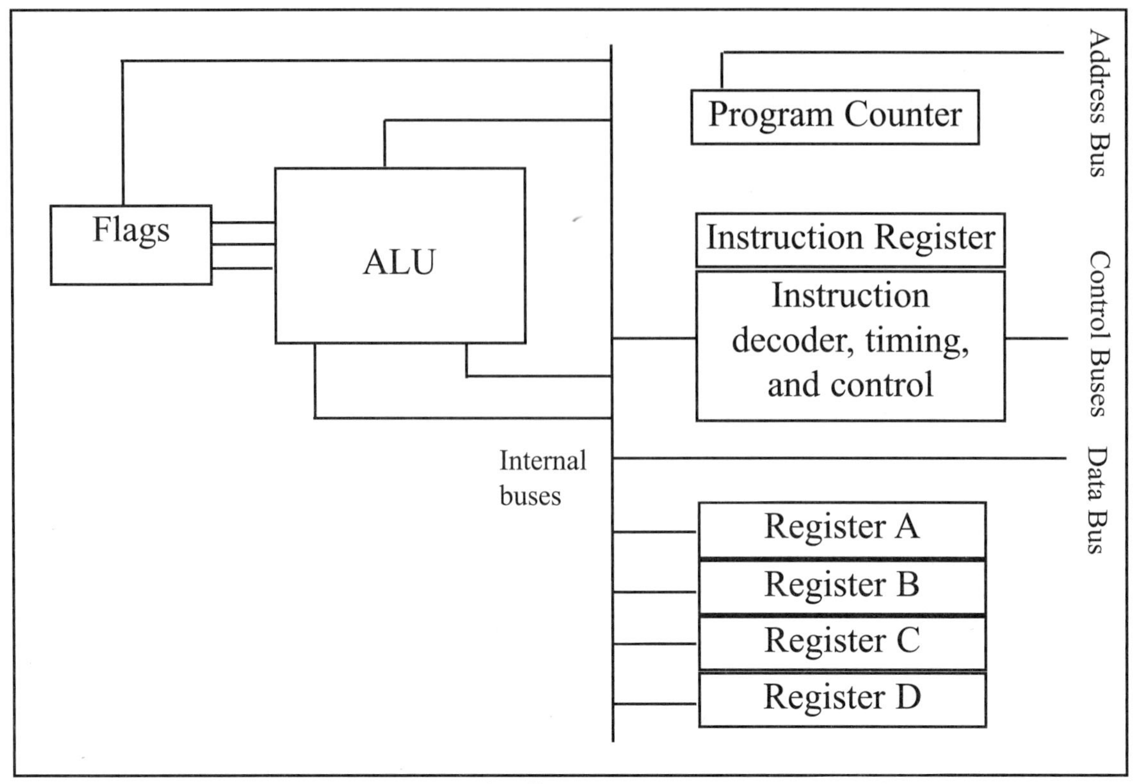

Figure 0-19. Internal Block Diagram of a CPU

2. The CPU also has what is called the *ALU* (arithmetic/logic unit). The ALU section of the CPU is responsible for performing arithmetic functions such as add, subtract, multiply, and divide, and logic functions such as AND, OR, and NOT.

3. Every CPU has what is called a *program counter*. The function of the program counter is to point to the address of the next instruction to be executed. As each instruction is executed, the program counter is incremented to point to the address of the next instruction to be executed. The contents of the program counter are placed on the address bus to find and fetch the desired instruction. In the x86 microprocessor, the program counter is a register called *IP*, or the instruction pointer.
4. The function of the *instruction decoder* is to interpret the instruction fetched into the CPU. One can think of the instruction decoder as a kind of dictionary, storing the meaning of each instruction and what steps the CPU should take upon receiving a given instruction. Just as a dictionary requires more pages the more words it defines, a CPU capable of understanding more instructions requires more transistors to design.

Internal workings of CPUs

To demonstrate some of the concepts discussed above, a step-by-step analysis of the process a CPU would go through to add three numbers is given next. Assume that an imaginary CPU has registers called A, B, C, and D. It has an 8-bit data bus and a 16-bit address bus. Therefore, the CPU can access memory from addresses 0000 to FFFFH (for a total of 10000H locations). The action to be performed by the CPU is to put hexadecimal value 21 into register A, and then add to register A values 42H and 12H. Assume that the code for the CPU to move a value to register A is 1011 0000 (B0H) and the code for adding a value to register A is 0000 0100 (04H). The necessary steps and the code to perform them are as follows.

Action	*Code*	*Data*
Move value 21H into register A	B0H	21H
Add value 42H to register A	04H	42H
Add value 12H to register A	04H	12H

If the program to perform the actions listed above is stored in memory locations starting at 1400H, the following would represent the contents for each memory address location:

Memory address	*Contents of memory address*
1400	(B0)code for moving a value to register A
1401	(21)value to be moved
1402	(04)code for adding a value to register A
1403	(42)value to be added
1404	(04)code for adding a value to register A
1405	(12)value to be added
1406	(F4)code for halt

The actions performed by the CPU to run the program above would be as follows:

1. The CPU's program counter can have a value between 0000 and FFFFH. The program counter must be set to the value 1400H, indicating the address of the first instruction code to be executed. After the program counter has been loaded with the address of the first instruction, the CPU is ready to execute.
2. The CPU puts 1400H on the address bus and sends it out. The memory circuitry finds the location while the CPU activates the READ signal, indicating to memory that it wants the byte at location 1400H. This causes the content of memory location 1400H, which is B0, to be put on the data bus and brought into the CPU.
3. The CPU decodes the instruction B0 with the help of its instruction decoder dictionary. When it finds the definition for that instruction it knows it must bring into register A of the CPU the byte in the next memory location. Therefore, it commands its controller circuitry to do exactly that. When it brings in value 21H from memory location 1401, it makes sure that the doors of all registers are closed except register A. Therefore, when value 21H comes into the CPU it will go directly into register A. After completing one instruction, the program counter points to the address of the next instruction to be executed, which in this case is 1402H. Address 1402 is sent out on the address bus to fetch the next instruction.
4. From memory location 1402H the CPU fetches code 04H. After decoding, the CPU knows that it must add the byte sitting at the next address (1403) to the contents of register A. After the CPU brings the value (in this case, 42H) into register A, it provides the contents of register A along with this value to the

ALU to perform the addition. It then takes the result of the addition from the ALU's output and puts it in register A. Meanwhile the program counter becomes 1404, the address of the next instruction.

5. Address 1404H is put on the address bus and the code is fetched into the CPU, decoded, and executed. This code again is adding a value to register A. The program counter is updated to 1406H.
6. Finally, the contents of address 1406 (HALT code) are fetched in and executed. The HALT instruction tells the CPU to stop incrementing the program counter and asking for the next instruction. Without the HALT, the CPU would continue updating the program counter and fetching instructions.

Now suppose that address 1403H contained the value 04 instead of 42H. How would the CPU distinguish between data 04 to be added and code 04? Remember that code 04 for this CPU means "move the next value into register A". Therefore, the CPU will not try to decode the next value. It simply moves the contents of the following memory location into register A, regardless of its value.

Harvard and von Neumann architectures

Every microprocessor must have memory space to store program (code) and data. While code provides instructions to the CPU, the data provides the information to be processed. The CPU uses buses (wire traces) to access the code ROM and data RAM memory spaces. The early computers used the same bus for accessing both the code and data. Such an architecture is commonly referred to as *von Neumann* (Princeton) architecture. That means for von Neumann computers, the process of accessing code or data could cause each to get in the other's way and slow down the processing speed of the CPU, because each had to wait for the other to finish fetching. To speed up the process of program execution, some CPUs use what is called *Harvard architecture*. The Harvard architecture has separate buses for the code and data memory. See Figure 0-20. That means that we need four sets of buses: (1) a set of data buses for carrying data into and out of the CPU, (2) a set of address buses for accessing the data, (3) a set of data buses for carrying code into the CPU, and (4) an address bus for accessing the code. See Figure 0-20. This is easy to implement inside an IC chip such as a microcontroller where both ROM code and data RAM are internal (on-chip) and distances are on the micron and millimeter scale. But to implement Harvard architecture for systems such as x86 IBM PC-type computers is very expensive because the RAM and ROM that hold code and data are external to the CPU. Separate wire traces for data and code on the motherboard will make the board large and expensive. For example, a Pentium microprocessor with a 64-bit data bus and a 32-bit address bus will need about 100 wire traces on the motherboard if it is von Neumann architecture (96 for address and data, plus a few others for control signals of read and write and so on). But the number of wire traces will double to 200 if we use Harvard architecture. Harvard architecture will also necessitate a large number of pins coming out of the microprocessor itself. For this reason Harvard architecture is not implemented in the world of PCs and workstations. This is also the reason that microcontrollers such as PIC use Harvard architecture internally, but they still use von Neumann architecture if they need external memory for code and data space. The von Neumann

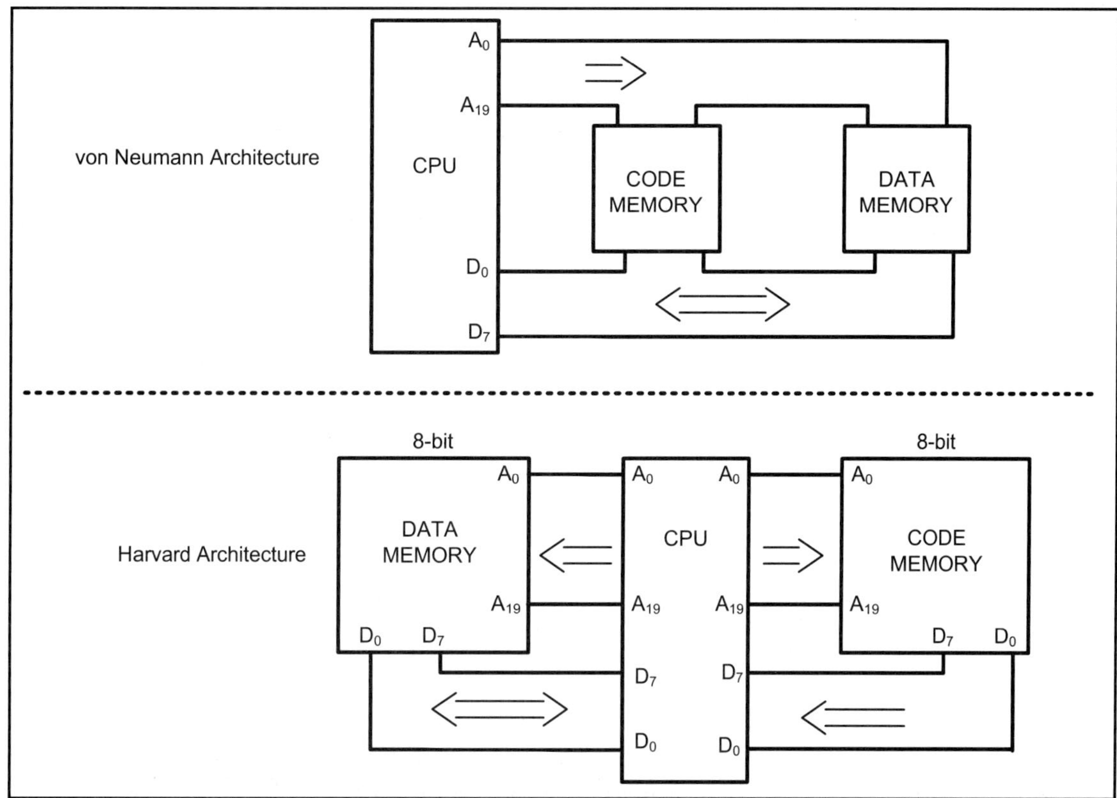

Figure 0-20. von Neumann vs. Harvard Architecture

architecture was developed at Princeton University, while the Harvard architecture was the work of Harvard University.

Review Questions

1. What does "ALU" stand for? What is its purpose?
2. How are registers used in computer systems?
3. What is the purpose of the program counter?
4. What is the purpose of the instruction decoder?
5. True or false. Harvard architecture uses the same address and data buses to fetch both code and data.

SECTION 0.5: RISC ARCHITECTURE

In this section we will examine the merits of the RISC architecture. If you wish, you can skip this section and come back to it after you have studied chapters 1 and 2.

Ways to increase the CPU power

There are three ways available to microprocessor designers to increase the processing power of the CPU:

1. Increase the clock frequency of the chip. One drawback of this method is that the higher the frequency, the more power and heat dissipation. Power and heat dissipation is especially a problem for handheld devices.

2. Use Harvard architecture by increasing the number of buses to bring more information (code and data) into the CPU to be processed. While in the case of x86 and other general-purpose microprocessors this architecture is very expensive and unrealistic, in today's single-chip computers (microcontrollers) this is not a problem.
3. Change the internal architecture of the CPU and use what is called *RISC* architecture.

Next, we discuss the merits of RISC architecture.

RISC architecture

In the early 1980s, a controversy broke out in the computer design community, but unlike most controversies, it did not go away. Since the 1960s, in all mainframes and minicomputers, designers put as many instructions as they could think of into the CPU. Some of these instructions performed complex tasks. An example is adding data memory locations and storing the sum into memory. Naturally, microprocessor designers followed the lead of minicomputer and mainframe designers. Because these microprocessors used such a large number of instructions and many of them performed highly complex activities, they came to be known as CISC (complex instruction set computer). According to several studies in the 1970s, many of these complex instructions etched into the brain of the CPU were never used by programmers and compilers. The huge cost of implementing a large number of instructions (some of them complex) into the microprocessor, plus the fact that a good portion of the transistors on the chip are used by the instruction decoder, made some designers think of simplifying and reducing the number of instructions. As this concept developed, the resulting processors came to be known as RISC (reduced instruction set computer).

Features of RISC

The following are some of the features of RISC.

Feature 1

RISC processors have a fixed instruction size. In a CISC microcontroller such as the HCS12, instructions can be 1, 2, or even 3 bytes. For example, look at the following instructions in the HCS12:

```
CLRA                    ;Clear Accumulator, a 1-byte  instruction
ADDA #mybyte            ;Add mybyte to Accumulator,
                        ;a 2-byte instruction
JMP  target_address     ;Jump, a 3-byte instruction
```

This variable instruction size makes the task of the instruction decoder very difficult because the size of the incoming instruction is never known. In a RISC architecture, the size of all instructions is fixed. Therefore, the CPU can decode the instructions quickly. This is like a bricklayer working with bricks of the same size as opposed to using bricks of variable sizes. Of course, it is much more efficient to use bricks of the same size.

Feature 2

One of the major characteristics of RISC architecture is a large number of registers. All RISC architectures have at least 32 registers. Of these 32 registers, only a few are assigned to a dedicated function. One advantage of a large number of registers is that it avoids the need for a large stack to store parameters. Although a stack can be implemented on a RISC processor, it is not as essential as in CISC because so many registers are available. In the HCS12 microcontrollers the use of a 512-byte RAM space satisfies this RISC feature.

Feature 3

RISC processors have a small instruction set. RISC processors have only the basic instructions such as ADD, SUB, MUL, LOAD, STORE, AND, OR, EXOR, CALL, JUMP, and so on. The limited number of instructions is one of the criticisms leveled at the RISC processor because it makes the job of Assembly language programmers much more tedious and difficult compared to CISC Assembly language programming. This is one reason that RISC is used more commonly in high-level language environments such as the C programming language rather than in Assembly language environments. It is interesting to note that some defenders of CISC have called it "complete instruction set computer" instead of "complex instruction set computer" because it has a complete set of every kind of instruction. How many of these instructions are used and how often is another matter. The limited number of instructions in RISC leads to programs that are large. Although these programs can use more memory, this is not a problem because memory is cheap. Before the advent of semiconductor memory in the 1960s, however, CISC designers had to pack as much action as possible into a single instruction to get the maximum bang for their buck.

Feature 4

At this point, one might ask, with all the difficulties associated with RISC programming, what is the gain? The most important characteristic of the RISC processor is that more than 95% of instructions are executed with only one clock cycle, in contrast to CISC instructions. Even some of the 5% of the RISC instructions that are executed with two clock cycles can be executed with one clock cycle by juggling instructions around (code scheduling). Code scheduling is most often the job of the compiler.

Feature 5

Because CISC has such a large number of instructions, each with so many different addressing modes, microinstructions (microcode) are used to implement them. The implementation of microinstructions inside the CPU takes more than 40–60% of transistors in many CISC processors. In the case of RISC, however, due to the small set of instructions, they are implemented using the hardwire method. Hardwiring of RISC instructions takes no more than 10% of the transistors.

Feature 6

RISC uses load/store architecture. In CISC microprocessors, data can be manipulated while it is still in memory. For example, in instructions such as "ADD Reg, Memory", the microprocessor must bring the contents of the external memory location into the CPU, add it to the contents of the register, then move the result back to the external memory location. The problem is there might be a delay in accessing the data from external memory. Then the whole process would be stalled, preventing other instructions from proceeding in the pipelinc. In RISC, designers did away with these kinds of instructions. In RISC, instructions can only load data from external memory into registers or store data in registers into external memory locations. There is no direct way of doing arithmetic and logic operations between a register and the contents of external memory locations. All these instructions must be performed by first bringing both operands into the registers inside the CPU, then performing the arithmetic or logic operation, and then sending the result back to memory. This idea was first implemented by the Cray 1 supercomputer in 1976 and is commonly referred to as load/store architecture.

In concluding this discussion of RISC processors, it is interesting to note that RISC technology was explored by the scientists in IBM in the mid-1970s, but it was David Patterson of the University of California at Berkeley who in 1980 brought the merits of RISC concepts to the attention of computer scientists. It must also be noted that in recent years CISC processors such as the Pentium have used some of the RISC features in their design. This was the only way they could enhance the processing power of the x86 processors and stay competitive. Of course, they had to use lots of transistors to do the job, because they had to deal with all the CISC instructions of the 8086/286/386 processors and the legacy software of DOS.

Review Questions

1. What do RISC and CISC stand for?
2. True or false. Instructions such as "ADD memory, memory" do not exist in a RISC CPU.
3. True or false. While CISC instructions are of variable sizes, RISC instructions are all the same size.
4. Which of the following operations do not exist for the ADD instruction in RISC?
   (a) register to register (b) immediate to register (c) memory to memory

SUMMARY

The binary number system represents all numbers with a combination of the two binary digits, 0 and 1. The use of the binary system is necessary in digital computers because only two states can be represented: on or off. Any binary number can be coded directly into its hexadecimal equivalent for the convenience of humans. Converting from binary/hex to decimal, and vice versa, is a straightforward process that becomes easy with practice. The ASCII code is a binary code used to represent alphanumeric data internally in the computer. It is frequently used in peripheral devices for input and/or output.

The AND, OR, and inverter logic gates are the basic building blocks of simple circuits. NAND, NOR, and XOR gates are also used to implement circuit design. Diagrams of half-adders and full-adders were given as examples of the use of logic gates for circuit design. Decoders are used to detect certain addresses. Flip-flops are used to latch in data until other circuits are ready for it.

The major components of any computer system are the CPU, memory, and I/O devices. "Memory" refers to temporary or permanent storage of data. In most systems, memory can be accessed as bytes or words. The terms *kilobyte*, *megabyte*, *gigabyte*, and *terabyte* are used to refer to large numbers of bytes. There are two main types of memory in computer systems: RAM and ROM. RAM (random access memory) is used for temporary storage of programs and data. ROM (read-only memory) is used for permanent storage of programs and data that the computer system must have in order to function. All components of the computer system are under the control of the CPU. Peripheral devices such as I/O (input/output) devices allow the CPU to communicate with humans or other computer systems. There are three types of buses in computers: address, control, and data. The address bus is used by the CPU to locate a device or a memory location. Control buses are used by the CPU to direct acitons of other devices. Data buses are used to send information back and forth between the CPU and other devices.

This chapter also provided an overview of semiconductor memories. Types of memories were compared in terms of their capacity, organization, and access time. ROM (read-only memory) is nonvolatile memory typically used to store programs in embedded systems. The relative advantages of various types of ROM were described in this chapter, including PROM, EPROM, UV-EPROM, EEPROM, Flash memory EPROM, and mask ROM.

Address decoding techniques using simple logic gates, decoders, and programmable logic were covered.

The computer organization and the internals of the CPU were covered. The relative advantages of Harvard and RISC architectures were also discussed.

PROBLEMS

SECTION 0.1: NUMBERING AND CODING SYSTEMS

1. Convert the following decimal numbers to binary:
   (a) 12 (b) 123 (c) 63 (d) 128 (e) 1,000
2. Convert the following binary numbers to decimal:
   (a) 100100 (b) 1000001 (c) 11101 (d) 1010 (e) 00100010
3. Convert the values in Problem 2 to hexadecimal.
4. Convert the following hex numbers to binary and decimal:
   (a) 2B9H (b) F44H (c) 912H (d) 2BH (e) FFFFH
5. Convert the values in Problem 1 to hex.
6. Find the 2's complement of the following binary numbers:
   (a) 1001010 (b) 111001 (c) 10000010 (d) 111110001

7. Add the following hex values:
   (a) 2CH + 3FH (b) F34H + 5D6H (c) 20000H + 12FFH
   (d) FFFFH + 2222H
8. Perform hex subtraction for the following:
   (a) 24FH – 129H (b) FE9H – 5CCH (c) 2FFFFH – FFFFFH
   (d) 9FF25H – 4DD99H
9. Show the ASCII codes for numbers 0, 1, 2, 3, ..., 9 in both hex and binary.
10. Show the ASCII code (in hex) for the following string:
    "U.S.A. is a country" CR,LF
    "in North America" CR,LF
    (CR is carriage return, LF is line feed)

SECTION 0.2: DIGITAL PRIMER

11. Draw a 3-input OR gate using a 2-input OR gate.
12. Show the truth table for a 3-input OR gate.
13. Draw a 3-input AND gate using a 2-input AND gate.
14. Show the truth table for a 3-input AND gate.
15. Design a 3-input XOR gate with a 2-input XOR gate. Show the truth table for a 3-input XOR.
16. List the truth table for a 3-input NAND.
17. List the truth table for a 3-input NOR.
18. Show the decoder for binary 1100.
19. Show the decoder for binary 11011.
20. List the truth table for a D-FF.

SECTION 0.3: SEMICONDUCTOR MEMORY

21. Answer the following:
    (a) How many nibbles are 16 bits?
    (b) How many bytes are 32 bits?
    (c) If a word is defined as 16 bits, how many words is a 64-bit data item?
    (d) What is the exact value (in decimal) of 1 meg?
    (e) How many K is 1 meg?
    (f) What is the exact value (in decimal) of 1 gigabyte?
    (g) How many K is 1 gigabyte?
    (h) How many megs is 1 gigabyte?
    (i) If a given computer has a total of 8 megabytes of memory, how many bytes (in decimal) is this? How many kilobytes is this?
22. A given mass storage device such as a hard disk can store 2 gigabytes of information. Assuming that each page of text has 25 rows and each row has 80 columns of ASCII characters (each character = 1 byte), approximately how many pages of information can this disk store?
23. In a given byte-addressable computer, memory locations 10000H to 9FFFFH are available for user programs. The first location is 10000H and the last location is 9FFFFH. Calculate the following:
    (a) The total number of bytes available (in decimal)
    (b) The total number of kilobytes (in decimal)

24. A given computer has a 32-bit data bus. What is the largest number that can be carried into the CPU at a time?
25. Below are listed several computers with their data bus widths. For each computer, list the maximum value that can be brought into the CPU at a time (in both hex and decimal).
    (a) Apple 2 with an 8-bit data bus
    (b) x86 PC with a 16-bit data bus
    (c) x86 PC with a 32-bit data bus
    (d) Cray supercomputer with a 64-bit data bus
26. Find the total amount of memory, in the units requested, for each of the following CPUs, given the size of the address buses:
    (a) 16-bit address bus (in K)
    (b) 24-bit address bus (in megs)
    (c) 32-bit address bus (in megabytes and gigabytes)
    (d) 48-bit address bus (in megabytes, gigabytes, and terabytes)
27. Regarding the data bus and address bus, which is unidirectional and which is bidirectional?
28. What is the difference in capacity between a 4M memory chip and 4M of computer memory?
29. True or false. The more address pins, the more memory locations are inside the chip. (Assume that the number of data pins is fixed.)
30. True or false. The more data pins, the more each location inside the chip will hold.
31. True or false. The more data pins, the higher the capacity of the memory chip.
32. True or false. The more data pins and address pins, the greater the capacity of the memory chip.
33. The speed of a memory chip is referred to as its _______________.
34. True or false. The price of memory chips varies according to capacity and speed.
35. The main advantage of EEPROM over UV-EPROM is _______________.
36. True or false. SRAM has a larger cell size than DRAM.
37. Which of the following, EPROM, DRAM, or SRAM, must be refreshed periodically?
38. Which memory is used for PC cache?
39. Which of the following, SRAM, UV-EPROM, NV-RAM, or DRAM, is volatile memory?
40. RAS and CAS are associated with which memory?
    (a) EPROM (b) SRAM (c) DRAM (d) all of the above
41. Which memory needs an external multiplexer?
    (a) EPROM (b) SRAM (c) DRAM (d) all of the above
42. Find the organization and capacity of memory chips with the following pins.
    (a) EEPROM A0–A14, D0–D7 (b) UV-EPROM A0–A12, D0–D7
    (c) SRAM A0–A11, D0–D7 (d) SRAM A0–A12, D0–D7
    (e) DRAM A0–A10, D0 (f) SRAM A0–A12, D0
    (g) EEPROM A0–A11, D0–D7 (h) UV-EPROM A0–A10, D0–D7
    (i) DRAM A0–A8, D0–D3 (j) DRAM A0–A7, D0–D7

43. Find the capacity, address, and data pins for the following memory organizations.

    (a) 16Kx8 ROM (b) 32Kx8 ROM
    (c) 64Kx8 SRAM (d) 256Kx8 EEPROM
    (e) 64Kx8 ROM (f) 64Kx4 DRAM
    (g) 1Mx8 SRAM (h) 4Mx4 DRAM
    (i) 64Kx8 NV-RAM
44. Find the address range of the memory design in the diagram.
45. Using NAND gates and inverters, design decoding circuitry for the address range 2000H–2FFFH.
46. Find the address range for Y0, Y3, and Y6 of the 74LS138 for the diagrammed design.

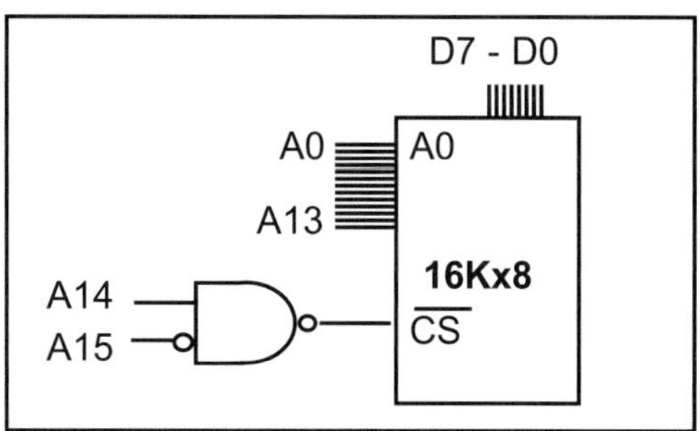

Diagram for Problem 44

47. Using the 74138, design the memory decoding circuitry in which the memory block controlled by Y0 is in the range 0000H to 1FFFH. Indicate the size of the memory block controlled by each Y.
48. Find the address range for Y3, Y6, and Y7 in Problem 47.
49. Using the 74138, design memory decoding circuitry in which the memory block controlled by Y0 is in the 0000H to 3FFFH space. Indicate the size of the memory block controlled by each Y.

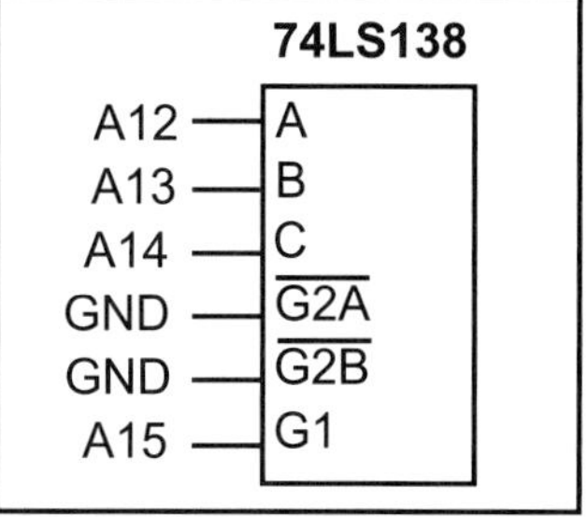

Diagram for Problem 46

50. Find the address range for Y1, Y2, and Y3 in Problem 49.

SECTION 0.4: HARVARD AND von NEUMANN CPU ARCHITECTURES

51. Which register of the CPU holds the address of the instruction to be fetched?
52. Which section of the CPU is responsible for performing addition?
53. List the three bus types present in every CPU.

SECTION 0.5: RISC ARCHITECTURE

54. What do RISC and CISC stand for?
55. In ______ (RISC, CISC) architecture we can have 1-, 2-, 3-, or 4-byte instructions.
56. In ______ (RISC, CISC) architecture, instructions are fixed in size.
57. In ______ (RISC, CISC) architecture, instructions are mostly executed in one or two cycles.
58. In ______ (RISC, CISC) architecture, we can have an instruction to ADD a register to external memory.

ANSWERS TO REVIEW QUESTIONS

SECTION 0.1: NUMBERING AND CODING SYSTEMS

1. Computers use the binary system because each bit can have one of two voltage levels: on and off.
2. $34_{10} = 100010_2 = 22_{16}$
3. $110101_2 = 35_{16} = 53_{10}$
4. 1110001
5. 010100
6. 461
7. 275
8. 38 30 78 38 36 20 43 50 55 73

SECTION 0.2: DIGITAL PRIMER

1. AND
2. OR
3. XOR
4. Buffer
5. Storing data
6. Decoder

SECTION 0.3: SEMICONDUCTOR MEMORY

1. 24,576
2. Random access memory; it is used for temporary storage of programs that the CPU is running, such as the operating system, word processing programs, etc.
3. Read-only memory; it is used for permanent programs such as those that control the keyboard, etc.
4. The contents of RAM are lost when the computer is powered off.
5. The CPU, memory, and I/O devices
6. Central processing unit; it can be considered the "brain" of the computer; it executes the programs and controls all other devices in the computer.
7. The address bus carries the location (address) needed by the CPU; the data bus carries information in and out of the CPU; the control bus is used by the CPU to send signals controlling I/O devices.
8. (a) bidirectional (b) unidirectional
9. 64K, or 65,536 bytes
10. c
11. (a) 16Kx8, 128K bits (b) 64Kx8, 512K (c) 4Kx8, 32K
12. (a) 2Kx1, 2K bits (b) 8Kx4, 32K (c) 128Kx8, 1M
    (d) 64Kx4, 256K (e) 256Kx1, 256K (f) 256Kx4, 1M
13. (a) 64K bits, 14 address, and 4 data (b) 256K, 15 address, and 8 data
    (c) 1M, 10 address, and 1 data (d) 1M, 18 address, and 4 data
    (e) 512K, 16 address, and 8 data (f) 4M, 10 address, and 4 data
14. b, c
15. 16K bytes
16. 3, 8
17. Both must be low.
18. G1 must be high.
19. 5000H–5FFFH

SECTION 0.4: HARVARD AND von NEUMANN CPU ARCHITECTURES

1. Arithmetic/logic unit; it performs all arithmetic and logic operations.
2. They are used for temporary storage of information.
3. It holds the address of the next instruction to be executed.
4. It tells the CPU what actions to perform for each instruction.
5. True

SECTION 0.5: RISC ARCHITECTURE

1. CISC stands for complex instruction set computer; RISC is reduced instruction set computer.
2. True
3. True
4. (c)

CHAPTER 1

THE HCS12/9S12 MICROCONTROLLER: HISTORY AND FEATURES

OBJECTIVES

Upon completion of this chapter, you will be able to:

>> Compare and contrast microprocessors and microcontrollers
>> Describe the advantages of microcontrollers for some applications
>> Explain the concept of embedded systems
>> Discuss criteria for considering a microcontroller
>> Explain the variations of speed, packaging, memory, and cost per unit and how these affect choosing a microcontroller
>> Explain the evolution of 68xx microcontrollers
>> Compare and contrast the various members of the HCS12 family
>> Give the history and features of HCS12/9S12 microcontrollers

SECTION 1.1: MICROCONTROLLERS AND EMBEDDED PROCESSORS

In this section we discuss the need for microcontrollers and contrast them with general-purpose microprocessors such as the Pentium and other x86 microprocessors. We also look at the role of microcontrollers in the embedded market. In addition, we provide some criteria on how to choose a microcontroller.

Microcontroller versus general-purpose microprocessor

What is the difference between a microprocessor and microcontroller? By microprocessor we mean the general-purpose microprocessors such as Intel's x86 family (8086, 80286, 80386, 80486, and the Pentium) or Motorola's PowerPC family. These microprocessors contain no RAM, no ROM, and no I/O ports on the chip itself. For this reason, they are commonly referred to as *general-purpose microprocessors*. See Figure 1-1.

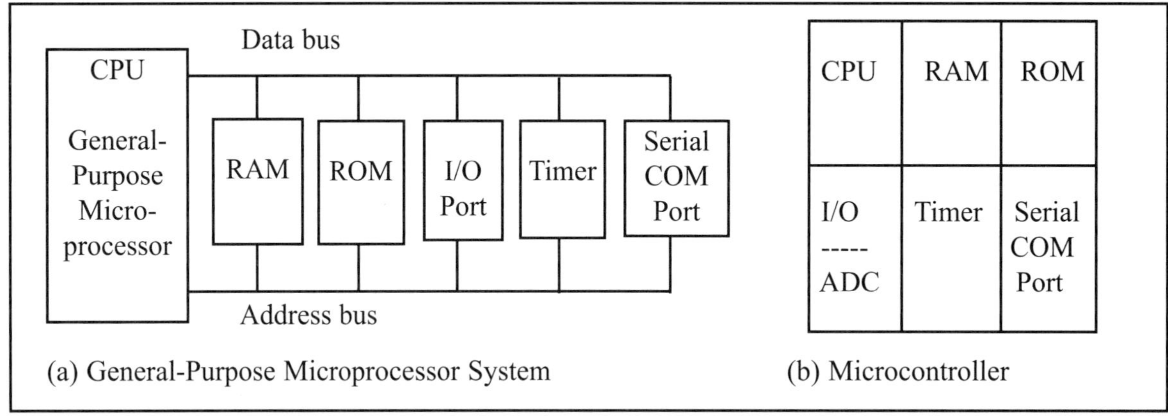

Figure 1-1. Microprocessor System Contrasted with Microcontroller System

A system designer using a general-purpose microprocessor such as the Pentium or the PowerPC must add RAM, ROM, I/O ports, and timers externally to make them functional. Although the addition of external RAM, ROM, and I/O ports makes these systems bulkier and much more expensive, they have the advantage of versatility, enabling the designer to decide on the amount of RAM, ROM, and I/O ports needed to fit the task at hand. This is not the case with microcontrollers. A microcontroller has a CPU (a microprocessor) in addition to a fixed amount of RAM, ROM, I/O ports, and a timer all on a single chip. In other words, the processor, RAM, ROM, I/O ports, and timer are all embedded together on one chip. The fixed amount of on-chip ROM, RAM, and number of I/O ports in microcontrollers makes them ideal for many applications in which cost and space are critical. In many applications, for example a TV remote control, there is no need for the computing power of a 486 or even an 8086 microprocessor. In many applications, the space used, the power consumed, and the price per unit are much more critical considerations than the computing power. These applications most often require some I/O operations to read signals and turn on and off certain bits. For this reason some call these processors IBP, "itty-bitty processors." (See "Good Things in Small Packages Are Generating Big Product Opportunities" by Rick Grehan, BYTE magazine, September 1994 (http://www.byte.com) for an excellent discussion of microcontrollers.)

It is interesting to note that some microcontroller manufacturers have gone as far as integrating an ADC (analog-to-digital converter) and other peripherals into the microcontroller.

Home
Appliances
Intercom
Telephones
Security systems
Garage door openers
Answering machines
Fax machines
Home computers
TVs
Cable TV tuner
VCR
Camcorder
Remote controls
Video games
Cellular phones
Musical instruments
Sewing machines
Lighting control
Paging
Camera
Pinball machines
Toys
Exercise equipment
Office
Telephones
Computers
Security systems
Fax machine
Microwave
Copier
Laser printer
Color printer
Paging
Auto
Trip computer
Engine control
Air bag
ABS
Instrumentation
Security system
Transmission control
Entertainment
Climate control
Cellular phone
Keyless entry

Table 1-1: Some Embedded Products Using Microcontrollers

Microcontrollers for embedded systems

In the literature discussing microprocessors, we often see the term *embedded system*. Microprocessors and microcontrollers are widely used in embedded system products. An embedded product is controlled by its own internal microprocessor (or microcontroller) as opposed to an external controller. Typically, in an embedded system, the microcontroller's ROM is burned with a purpose for specific functions needed for the system. A printer is an example of an embedded system because the processor inside performs one task only; namely, getting the data and printing it. Contrast this with a Pentium-based PC (or any x86 IBM-compatible PC), which can be used for any number of applications such as word processor, print-server, bank teller terminal, video game player, network server, or Internet terminal. A PC can also load and run software for a variety of applications. Of course, the reason a PC can perform a myriad of tasks is that it has RAM memory and an operating system that loads the application software into RAM and lets the CPU run it. In an embedded system, typically only one application software is burned into ROM. An x86 PC contains or is connected to various embedded products such as the keyboard, printer, modem, disk controller, sound card, CD-ROM driver, mouse, and so on. Each one of these peripherals has a microcontroller inside it that performs only one task. For example, inside every mouse a microcontroller performs the task of finding the mouse's position and sending it to the PC. Table 1-1 lists some embedded products.

x86 PC embedded applications

Although microcontrollers are the preferred choice for many embedded systems, sometimes a microcontroller is inadequate for the task. For this reason, in recent years many manufacturers of general-purpose microprocessors such as Intel, Freescale Semiconductor (formerly Motorola), and AMD (Advanced Micro Devices, Inc.) have targeted their microprocessor for the high end of the embedded market. Intel and AMD push their x86 processors for both the embedded and desktop PC markets. In the early 1990s, Apple computer began using the PowerPC microprocessors (604, 603, 620, etc.) in place of the 680x0 for the Macintosh. In 2006, Apple began to use the x86 processors for the design of the next generation of the Macintosh. The PowerPC microprocessor is a joint venture between IBM and Motorola, and is targeted for the high end of the embedded market. It must be noted that when a company targets a general-purpose microprocessor for the embedded market it optimizes the processor used for embedded systems. For this reason these processors are often called *high-end embedded processors*. Another chip widely used in the high end of the embedded system design is the ARM microprocessor. Very often the terms *embedded processor* and *microcontroller* are used interchangeably.

One of the most critical needs of an embedded system is to decrease power consumption and space. This can be achieved by integrating more functions into the CPU chip. All the embedded processors based on the x86 and PowerPC 6xx have low power consumption in addition to some forms of I/O, COM port, and ROM, all on a single chip. In high-performance embedded processors, the trend is to integrate more and more functions on the CPU chip and let the designer decide which features to use. This trend is invading PC system design as well. Normally, in designing the PC motherboard we need a CPU plus a chipset containing I/O, a cache controller, a Flash ROM containing BIOS, and finally a secondary cache memory. New designs are emerging in industry. For example, many companies have a chip that contains the entire CPU and all the supporting logic and memory, except for DRAM. In other words, we have the entire computer on a single chip.

Currently, because of Linux, MS-DOS, and Windows standardization, many embedded systems use x86 PCs. In many cases, using x86 PCs for the high-end embedded applications not only saves money but also shortens development time because a vast library of software already exists for the Linux, DOS, and Windows platforms. The fact that Windows and Linux are widely used and well-understood platforms means that developing a Windows-based or Linux-based embedded product reduces the cost and shortens the development time considerably.

Choosing a microcontroller

There are five major 8-bit microcontrollers. They are: Freescale Semiconductor's (formerly Motorola) 68HC08/68HC11/12, Intel's 8051, Atmel's AVR, Zilog's Z8, and PIC from Microchip Technology. Each microcontroller has a unique instruction set and register set; therefore, they are not compatible with each other. Programs written for one will not run on the others. There are also 16-bit and 32-bit microcontrollers made by various chip makers. With all these different microcontrollers, what criteria do designers consider in choosing one? The three criteria for choosing microcontrollers are as follows: (1) meeting the computing needs of the task at hand efficiently and cost effectively; (2) availability of software and hardware development tools such as compilers, assemblers, debuggers, and emulators; and (3) wide availability and reliable sources of the microcontroller. Next, we elaborate on each of the above criteria.

Criteria for choosing a microcontroller

1. The first and foremost criterion in choosing a microcontroller is that it must meet the task at hand efficiently and cost effectively. In analyzing the needs of a microcontroller-based project, we must first see whether an 8-bit, 16-bit, or 32-bit microcontroller can best handle the computing needs of the task most effectively. Among other considerations in this category are:

(a) Speed. What is the highest speed that the microcontroller supports?
(b) Packaging. Does it come in a 40-pin DIP (dual inline package) or a QFP (quad flat package), or some other packaging format? This is important in terms of space, assembling, and prototyping the end product.
(c) Power consumption. This is especially critical for battery-powered products.
(d) The amount of RAM and ROM on the chip.
(e) The number of I/O pins and the timer on the chip.
(f) Ease of upgrade to higher-performance or lower-power-consumption versions.
(g) Cost per unit. This is important in terms of the final cost of the product in which a microcontroller is used. For example, some microcontrollers cost 50 cents per unit when purchased 100,000 units at a time.

2. The second criterion in choosing a microcontroller is how easy it is to develop products around it. Key considerations include the availability of an assembler, debugger, a code-efficient C language compiler, emulator, tools, technical support, and both in-house and outside expertise. In many cases, third-party vendor (i.e., a supplier other than the chip manufacturer) support for the chip is as good as, if not better than, support from the chip manufacturer.
3. The third criterion in choosing a microcontroller is its ready availability in needed quantities both now and in the future. For some designers this is even more important than the first two criteria. Currently, of the leading 8-bit microcontrollers, the 8051 family has the largest number of diversified (multiple source) suppliers. (Supplier means a producer besides the originator of the microcontroller.) In the case of the 8051, which was originated by Intel, over fifty companies also currently produce the 8051.

Note that Freescale Semiconductor (Motorola), Atmel, Zilog, and Microchip Technology have all dedicated massive resources to ensure wide and timely availability of their products because their products are stable, mature, and single sourced. In recent years, companies have begun to sell *Field-Programmable Gate Array* (FPGA) and *Application-Specific Integrated Circuit* (ASIC) libraries for the different microcontrollers.

Mechatronics and microcontrollers

The microcontroller is playing a major role in an emerging field called *mechatronics*. Here is an excellent summary of what the field of mechatronics is all about, taken from the web site of Newcastle University (http://mechatronics2004.newcastle.edu.au/mech2004), which holds a major conference every year on this subject:

"Many technical processes and products in the area of mechanical and electrical engineering show an increasing integration of mechanics with electronics and information processing. This integration is between the components (hardware) and the information-driven functions (software), resulting in integrated systems called mechatronic systems.

The development of mechatronic systems involves finding an optimal balance between the basic mechanical structure, sensor and actuator implementation, automatic digital information processing and overall control, and this synergy results in innovative solutions. The practice of mechatronics requires multi disciplinary expertise across a range of disciplines, such as: mechanical engineering, electronics, information technology, and decision making theories."

Review Questions

1. True or false. Microcontrollers are normally less expensive than microprocessors.
2. When comparing a system board based on a microcontroller and a general-purpose microprocessor, which one is cheaper?
3. A microcontroller normally has which of the following devices on-chip?
   (a) RAM (b) ROM (c) I/O (d) all of the above
4. A general-purpose microprocessor normally needs which of the following devices to be attached to it?
   (a) RAM (b) ROM (c) I/O (d) all of the above
5. An embedded system is also called a dedicated system. Why?
6. What does the term *embedded system* mean?
7. Why does having multiple sources of a given product matter?

SECTION 1.2: OVERVIEW OF THE CPU12 AND CPU08

In this section, we first look at the history of the 68xx family of microprocessors and then examine the CPU08 and CPU12 families in more detail.

A brief history of the 68xx microprocessors

In 1974, Motorola Corporation introduced an 8-bit microprocessor called the 6800. This was only 3 years after Intel had introduced the 4004 microprocessor. In the 1980s, the 8085 from Intel and 68xx from Motorola dominated the 8-bit microprocessor market. Eventually both companies introduced microcontrollers. Intel used a whole new architecture called 8051, which was not based on 8085. Since the introduction of the first 6800 microprocessor, Motorola has marketed wide variety of 68xx products such as the 6801, 6803, 6805, 6808, 6809, and 6811. In the 1980s, microcontrollers had small amounts of data RAM, a few Kbytes of on-chip ROM for the program, a couple of timers, and many pins for I/O ports, all on a single chip with 40 pins. (See Figure 1-2.) For many years, the greatly improved version of 6811 called 68HC11 (high-density CMOS) was one of the most widely used 8-bit microcontrollers. This changed in the late 1990s with the introduction of PIC microcontrollers from Microchip. In the late 1990s, Motorola faced a major competition in the 8-bit market mainly from the Microchip's PIC microcontroller. In recent years, Motorola has introduced an array of 8-bit microcontrollers too numerous to list here. They include the HC08, HCS08, and RS08 families. Motorola refers to 8-bit architecture as CPU08. These microcontrollers are all 8-bit processors, meaning that the CPU can work on only 8 bits of data at a time. To increase the performance of 68HC11, Motorola introduced the 16-bit microcontroller called *68HC12* in 1996. Motorola uses the term CPU12 to refer to the 16-bit architecture of the 68HC12. According to Motorola "the CPU12 instructions are a superset of the 68HC11 instruction set", which means "code written for an 68HC11 can be reassembled and run on a CPU12 with

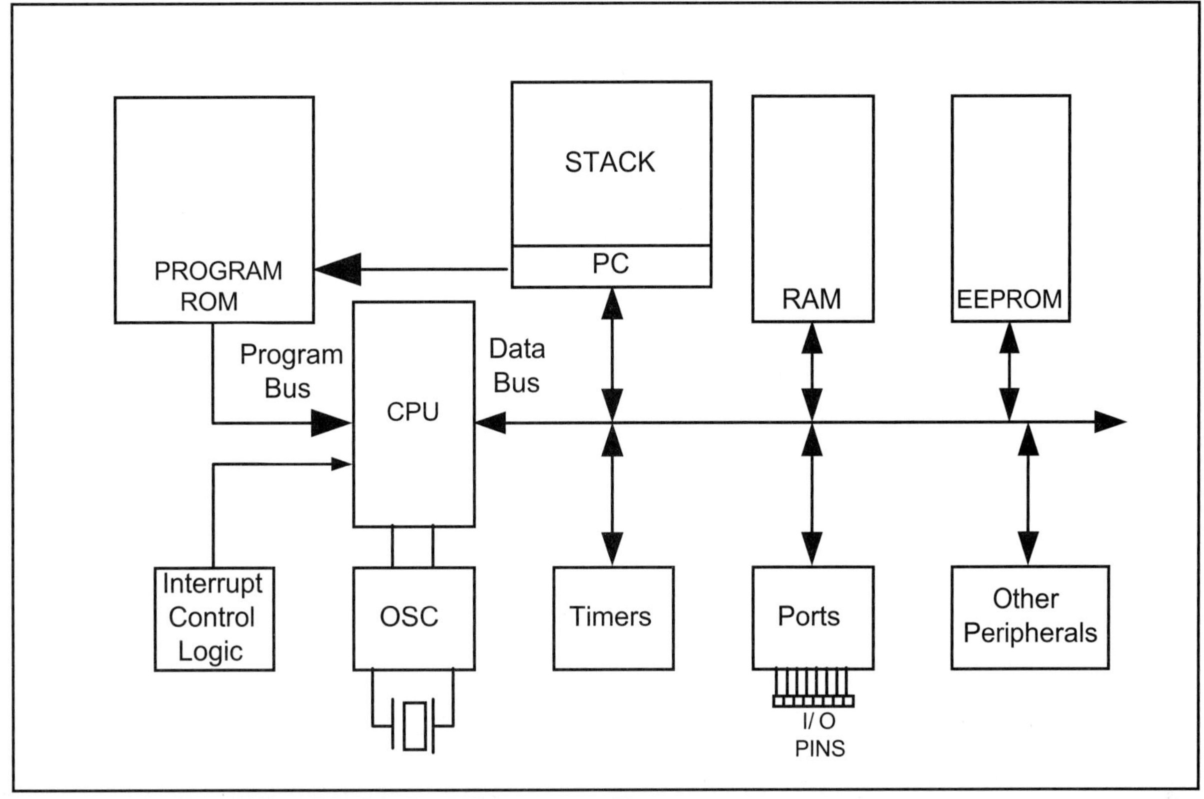

Figure 1-2. Simplified View of a Microcontroller

no changes." Again it must be noted that the CPU08 (68HC08, HC11, HCS08, and RS08) uses 8-bit wide ALU, while the CPU12 has 16-bit wide ALU with many new instructions. In other words, one can reassemble a program written for the CPU08 and run it on a CPU12, but not the other way around. For those who have mastered one family, understanding the other family is easy and straightforward since the CPU08 is a subset of the CPU12. See Figures 1-3 and 1-4.

From Motorola to Freescale

In 2004, the semiconductor division of Motorola became a separate and independent entity called Freescale Semiconductor. Since then the Freescale semiconductor has put massive efforts and resources into revamping the microcontroller products, hoping to recapture the ground lost to companies such as Microchip. Nowhere is this effort more evident than the new roadmap for the CPU08 line of products. According to Freescale, the goal is to allow "easy migration between Freescale's 8-bit microcontroller family and the higher-performance 32-bit ColdFire devices. Designers can develop new applications using the same software and hardware development tools for both 8-bit and 32-bit MCUs, and as products mature, they can more easily scale to the next-generation offerings." In the CPU08 line of products, the RS08 series with less than 16K bytes of Flash memory allows an inexpensive entry into the ultra-low end of the Controller Continuum. Again according to Freescale "the Controller Continuum offers stepwise compatibility across our portfolio of consumer and industrial microcontrollers (MCUs), with a common set of peripherals, tools and software to provide unbounded flexibility." See Figure 1-5.

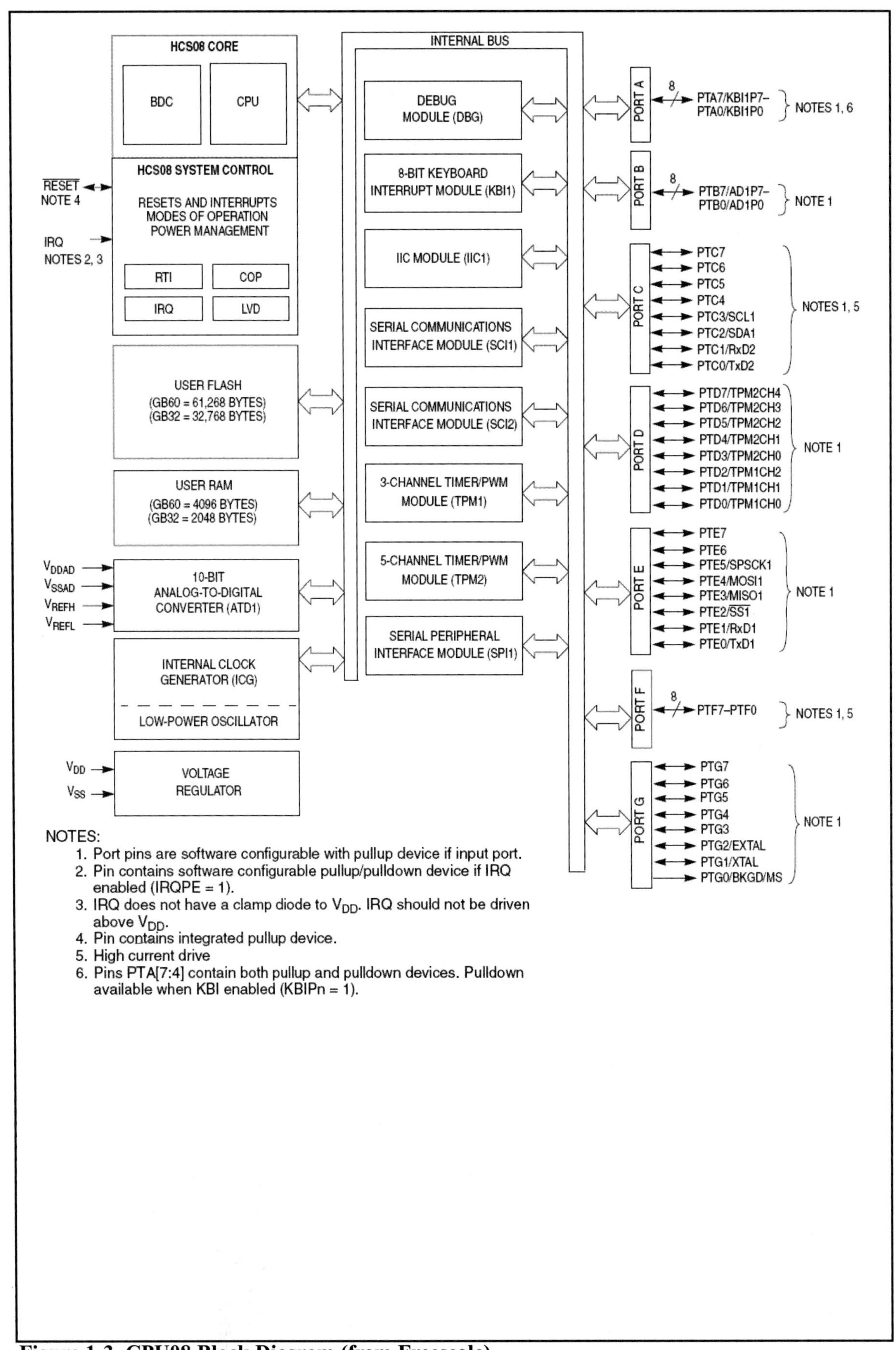

Figure 1-3. CPU08 Block Diagram (from Freescale)

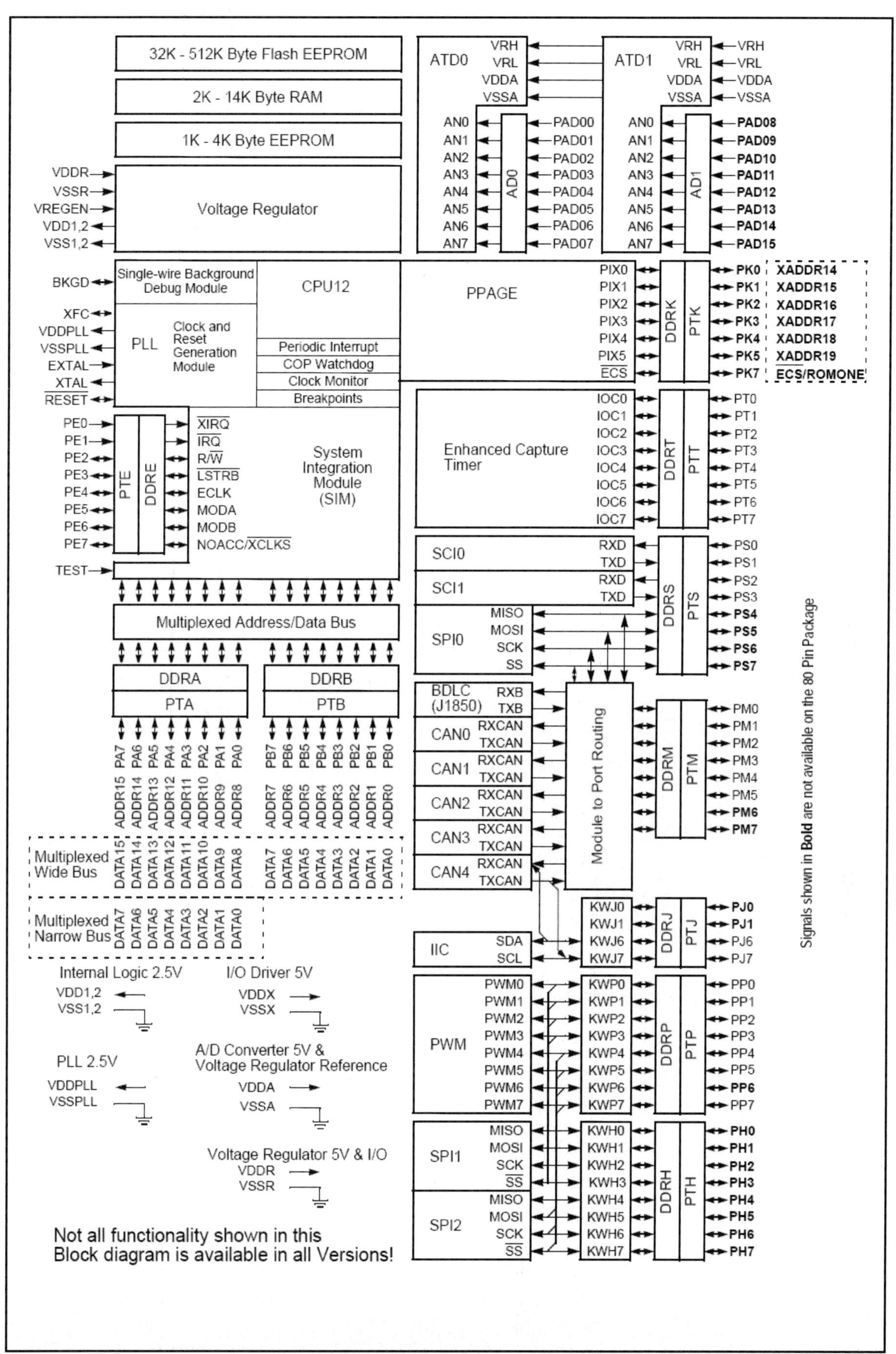

Figure 1-4. CPU12 Block Diagram (from Freescale)

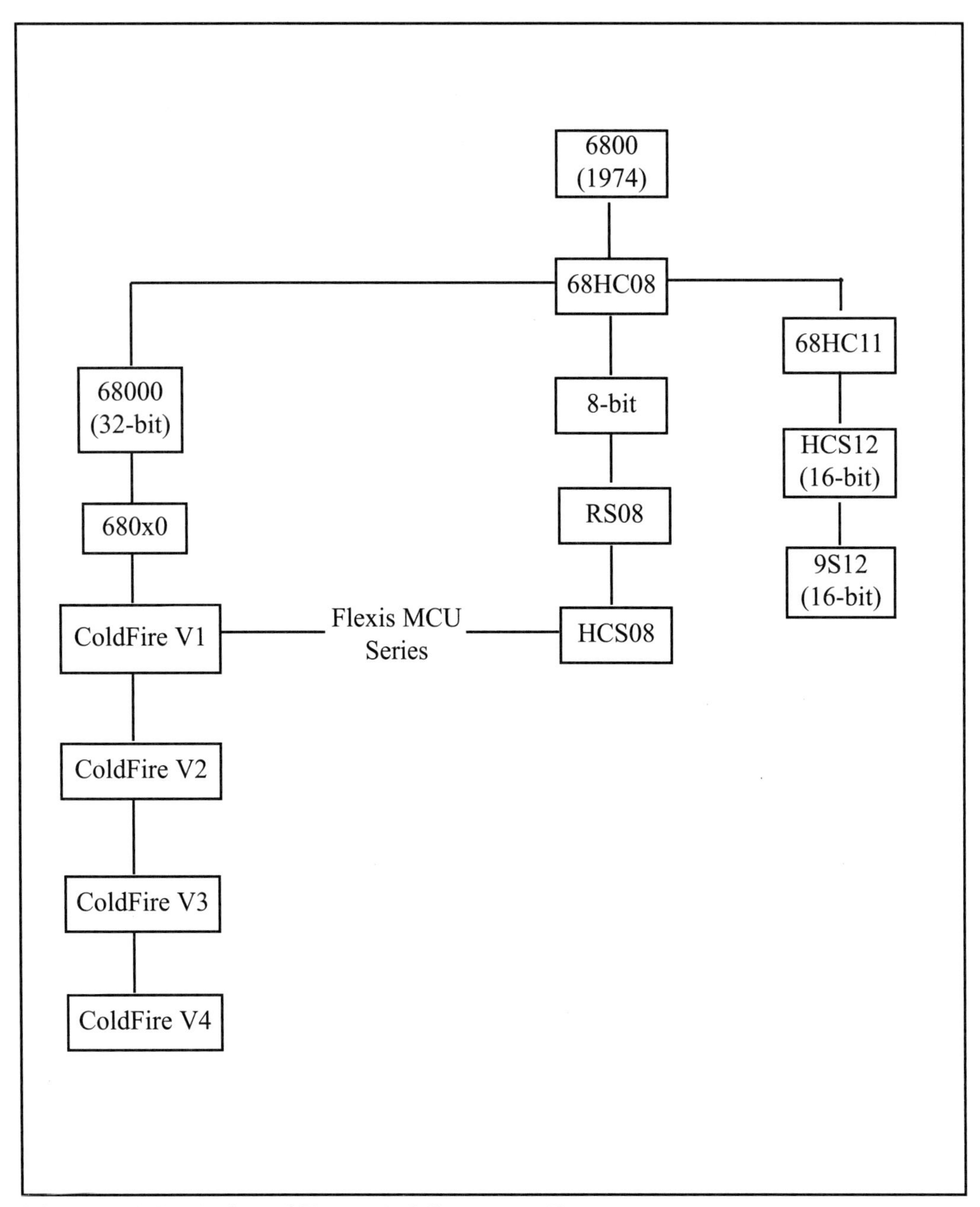

Figure 1-5. Evolution of Freescale Microcontrollers

HCS and RS chips

In the 1980s, advances in chip design led the semiconductor companies to move away from NMOS to more power-efficient CMOS technology. That is where we get the label HC (high-density CMOS) in the Freescale products. With the rising popularity of the handheld devices and battery-powered systems in the 1990s, the challenge was to increase the performance and the features of microcontrollers while keeping the power consumption low. The HCS08/HCS12 and RS08/S12 series have been the answers to these challenges. In recent years, Freescale has added many new features such as CAN, I^2C, SPI, BDM, and fuzzy logic to their microcontrollers, making them highly competitive products. See the Freescale website http://www.freescale.com for more information.

CPU08 and CPU12 features

The CPU08 and CPU12 come with some standard features such as on-chip program (code) ROM, data RAM, timers, ADC, and USART (Universal Synchronous Asynchronous Receiver Transmitter) and I/O ports. Although the size of the program ROM, data RAM, data EEPROM, and I/O ports varies among the family members, they all have peripherals such as timers, ADC (analog-to-digital converter), and USART. See Figures 1-3 and 1-4. Due to the importance of these peripherals, we have dedicated an entire chapter to each of them. The details of the RAM/ROM memory, peripherals, and I/O features of the CPU08/12 are given in future chapters. Next, we discuss the ROM memory type in microcontrollers.

Microcontroller program ROM type

In microcontrollers, the ROM is used to store programs and for that reason it is called *program* or *code* ROM. Although microcontrollers can have up to a few megabytes of program (code) ROM space, not all family members come with that much ROM installed. The program ROM size can vary from a few kilobytes to several megabytes at the time of this writing, depending on the family member. The program ROM is available in different memory types, such as Flash, OTP, and masked, all of which have different part numbers. A discussion of the various types of ROM was given in Chapter 0. See Chapter 0 to review these important memory technologies. Next, we discuss briefly the program ROM type for each microcontroller family.

Microcontroller with UV-EPROM

Some of the microcontrollers used UV-EPROM for on-chip program ROM. To use these kinds of chips for development required access to a PROM burner, as well as a UV-EPROM eraser to erase the contents of ROM. The window on the UV-EPROM chip allowed the UV light to erase the ROM. The problem with the UV-EPROM is that it takes around 20 minutes to erase the chip before it can be programmed again. This led companies to introduce a Flash version of their microcontroller. At this time Flash has replaced the UV-EPROM altogether. In the 1970s and 1980s, the UV-EPROM version used the number 7 in the part number to indicate that the on-chip ROM is UV-EPROM.

Microcontrollers with Flash

Today's microcontroller chips have on-chip program ROM in the form of Flash memory. Sometimes, the Flash version uses the letter F or number 9 in the part number to indicate that the on-chip ROM is Flash. The Flash version is ideal for fast development because Flash memory can be erased in seconds compared to the 20 minutes or more needed for the UV-EPROM version. For this reason, Flash has been used in place of the UV-EPROM to eliminate the waiting time needed to erase the chip, thereby speeding up the development time. To use Flash chips in the development of microcontroller-based systems requires a Flash programmer; however, a ROM eraser is not needed, because Flash is an EEPROM (electrically erasable PROM). Depending on the way the Flash memory is designed, you must erase the entire contents of a block of ROM in order to program it again. This erasing of Flash is done by the ROM programmer itself, so a separate eraser is not needed.

OTP version of the microcontroller

OTP (one-time-programmable) versions of the microcontroller are also available from manufacturers where the letter C (or number 7) indicates the OTP ROM, while the letter F (or number 9) is for the Flash. The Flash version is typically used for product development. When a product is designed and absolutely finalized, the OTP version of the chip is used for mass production because it is cheaper than Flash in terms of price per unit. The problem with the OTP is that you cannot reprogram it if you want to modify your program. Freescale uses the number 7 in the part number to indicate that the on-chip ROM is OTP.

Masked version of microcontroller

Many corporations provide a service in which you can send in your program and they will burn the program into the chip during the fabrication process of the chip. This chip is commonly referred to as *masked*, which is one of the stages of IC fabrication. A masked chip is the cheapest of all types, if the unit numbers are high enough. This is because there is a minimum order for the masked version of the microcontrollers. Freescale uses the number 3 in the part number to indicate that the on-chip ROM is of the masked memory type. See Figure 1-6.

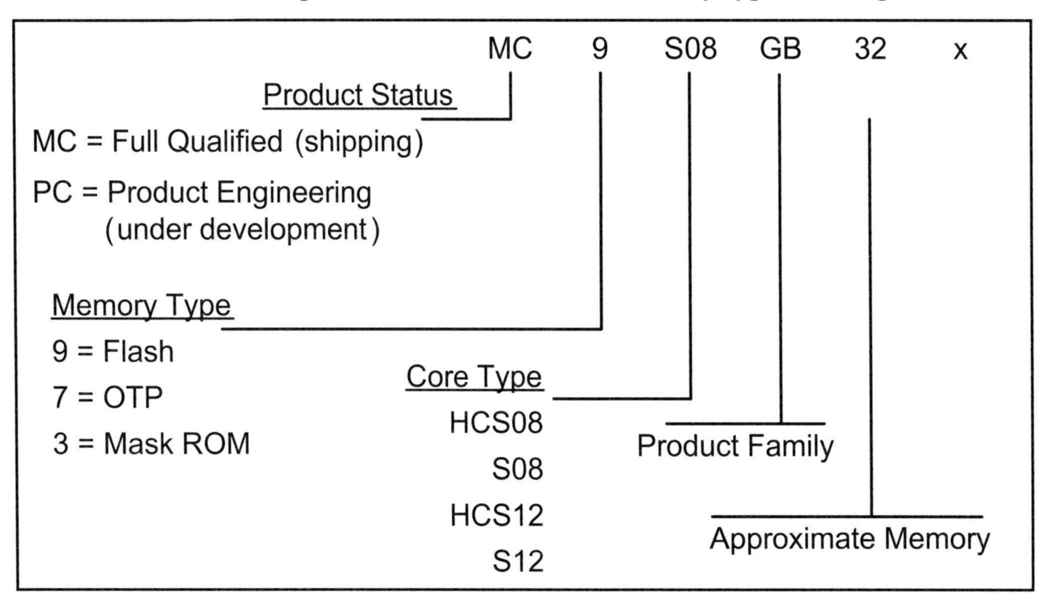

Figure 1-6. Freescale Chip Identification Scheme

Microcontroller peripherals

Nowadays the vast majority of microcontrollers, including the CPU08 and CPU12 family members, come with ADC (analog-to-digital converter), timers, and USART (Universal Synchronous Asynchronous Receiver Transmitter) as standard peripherals. The vast majority of ADCs used in microcontrollers are either 8-bit or 10-bit and the number of ADC channels in each chip varies from none to

16, depending on the package. We will examine ADC in Chapter 13. The microcontrollers can have up to as many as 8 timers besides a watchdog timer. We will examine timers in Chapter 9. The USART peripheral allows us to connect the microcontroller-based system to serial ports such as the COM port of the IBM PC, as we will see in Chapter 10. Many of the newer generations of microcontrollers come with the I^2C and CAN bus as well. See Tables 1-2 and 1-3.

Microcontroller data RAM and EEPROM

While ROM is used to store program (code), the RAM space is for data storage. The data RAM size for the CPU08/12 varies from a few hundred bytes to several kilobytes. As we will see in the next chapter, the data RAM space has two components: General-Purpose RAM (GPR) and Special Function Registers (SFRs). The SFRs are registers associated with the peripherals such as ADC, timers, and serial ports (UART) present on the chip. The number of bytes set aside for SFRs depends on the number of peripherals supported by a given microcontroller chip. The RAM GPR space is used for read/write scratch pad and general-purpose registers. Many of the microcontrollers also have a small amount of EEPROM to store critical data that does not need to be changed very often. While every microcontroller must have some data RAM for scratch pad, the EEPROM is optional, so not all microcontrollers come with EEPROM. EEPROM is used mainly for storage of critical data, as we will see in Chapter 14.

Table 1-2: Some Members of the CPU08 Family (http://www.freescale.com)

Part Num	Code ROM	Data RAM	SCI UART	ADC Channels	Timers
MC9S08AW60	60 KB	2 KB	2	16	8
MC9S08AW32	32 KB	2 KB	2	16	8
MC9S08AW16	16 KB	1 KB	2	16	6
MC9S08GB60A	60 KB	4 KB	1	8	8
MC9S08GB32A	32 KB	2 KB	1	8	8
MC9S08QG4	4 KB	256 B	1	8	2

Notes:
1. All code ROM memories are Flash.

Table 1-3: Some Members of the CPU12 Family (http://www.freescale.com)

Part Num	Code ROM	Data RAM	EEPROM	SCI UART	ADC Channels	I/O pins
MC9S12DP512	512 KB	14 KB	4 KB	2	16	91
MC9S12DT256	256 KB	12 KB	4 KB	2	16	91
MC9S12DT128	128 KB	8 KB	2 KB	2	16	91
MC9S12D64	64 KB	4 KB	1 KB	2	8	59
MC9S12D32	32 KB	2 KB	1 KB	2	8	59

Notes:
1. All code ROM memories are Flash.
2. All the chips have on-chip timers, PWM, CAN, and I^2C peripherals.

Microcontroller I/O pins

Microcontrollers can have from 8 to 100 pins dedicated for I/O. The number of I/O pins depends on the number of pins in the package itself and varies widely among the family members of a given microcontroller such as CPU08 and CPU16. We will study I/O pins and programming in Chapter 4.

BDM (background debug mode)

Freescale microcontrollers come with a powerful feature called BDM (background debug mode). The BDM allows us to examine the contents of the CPU such as RAM, ROM, and registers as we trace and debug the program. It uses both hardware and software to access the internal contents of the CPU. In Chapter 8, we will discuss the BDM feature in the HCS12 chips.

Trainers

In Chapter 8, we will discuss the pin connection of the HCS12 chip. Although one can design a trainer (a development board) for the HCS12, we recommend using the trainers made by companies such as Axiom and Wytec. See Appendix G for a list of companies that produce HCS12 trainers. The www.MicroDigitalEd.com website provides tutorials on how to use these trainers to test the programs used in this book.

Other microcontrollers

There are many other popular 8-bit microcontrollers besides the HCS08 chip. Among them are the 8051, PIC18, AVR, and Z8. Besides Intel, a number of other companies make the 8051 family, as seen in Table 1-4. The AVR is made by Atmel Corp. Microchip makes the PIC18F/PIC16F and many of its variations. Zilog produces the Z8 microcontroller. For a comprehensive treatment of the 8051 and PIC microcontrollers, see "The 8051 Microcontroller and Embedded Systems" by Mazidi, et al. and "PIC Microcontroller and Embedded Systems," also by Mazidi, et al.

Table 1-4: Some of the Companies that Produce Widely Used 8-bit Microcontrollers

Company	Web Site	Architecture
Freescale	http://www.freescale.com	RS08/HCS08/HC11
Microchip	http://www.microchip.com	PIC16xxx/18xxx
Intel	http://www.intel.com/design/mcs51	8051
Atmel	http://www.atmel.com	AVR and 8051
Philips/Signetics	http://www.semiconductors.philips.com	8051
Zilog	http://www.zilog.com	Z8 and Z80
Dallas Semi/Maxim	http://www.maxim-ic.com	8051

See http://www.microcontroller.com for a complete list.

Review Questions

1. True or false. The CPU08 is a superset of the CPU12.
2. What is the main difference between the HCS12 and HCS08 microcontrollers?
3. Give the size of RAM in each of the following:
   (a) MC9S12DP512 (b) MC9S12DT256
4. Give the size of the on-chip program ROM in each of the following:
   (a) MC9S12DP512 (b) MC9S12DT256
5. The HCS12 is a(n) _______-bit microprocessor.

Some useful websites:

1) For CPU12, CPU08, and CodeWarrior documents:

http://www.freescale.com

2) For tutorials on using the trainers and assembler/compilers:

http://www.MicroDigitalEd.com

PROBLEMS

SECTION 1.1: MICROCONTROLLERS AND EMBEDDED PROCESSORS

1. True or false. A general-purpose microprocessor has on-chip ROM.
2. True or false. Generally, a microcontroller has on-chip ROM.
3. True or false. A microcontroller has on-chip I/O ports.
4. True or false. A microcontroller has a fixed amount of RAM on the chip.
5. What components are usually put together with the microcontroller onto a single chip?
6. Intel's Pentium chips used in Windows PCs need external ______ and _____ chips to store data and code.
7. List three embedded products attached to a PC.
8. Why would someone want to use an x86 as an embedded processor?
9. Give the name and the manufacturer of some of the most widely used 8-bit microcontrollers.
10. In Problem 9, which one has the most manufacture sources?
11. In a battery-based embedded product, what is the most important factor in choosing a microcontroller?
12. In an embedded controller with on-chip ROM, why does the size of the ROM matter?
13. In choosing a microcontroller, how important is it to have multiple sources for that chip?

14. What does the term "third-party support" mean?
15. Suppose that a microcontroller architecture has both 8-bit and 16-bit versions. Which of the following statements is true?
    (a) The 8-bit software will run on the 16-bit system.
    (b) The 16-bit software will run on the 8-bit system.

SECTION 1.2: OVERVIEW OF THE CPU12 AND CPU08

16. The MC9S12DP512 has _____ bytes of on-chip program ROM.
17. The MC9S12D64 has _____ bytes of on-chip program ROM.
18. The MC9S12DT128 has ____ bytes of on-chip data RAM.
19. The MC9S12D64 has ____ bytes of on-chip data RAM.
20. The MC9S12DP512 has circuitry to support _____ serial ports.
21. The MC9S12DP512 on-chip program ROM is of the ______ type.
22. The MC9SDT64 on-chip program ROM is of the _____ type.
23. Give the amount of data RAM for the following chips:
    (a) MC9S12DP512 (b) MC9S12DT256 (c) MC9S12D32
24. Of the HCS12 family, which memory type is the most cost effective if you are using a million of them in an embedded product?

ANSWERS TO REVIEW QUESTIONS

SECTION 1.1: MICROCONTROLLERS AND EMBEDDED PROCESSORS

1. True
2. A microcontroller-based system
3. (d)
4. (d)
5. It is dedicated because it is dedicated to doing one type of job.
6. Embedded system means that the application and processor are combined into a single system.
7. Having multiple sources for a given part means you are not hostage to one supplier. More importantly, competition among suppliers brings about lower cost for that product.

SECTION 1.2: OVERVIEW OF THE CPU12 AND CPU08

1. False
2. HCS08 has an 8-bit ALU while the ALU in HCS12 is 16-bit.
3. (a) 14 KB
   (b) 12 KB
4. (a) 512 KB
   (b) 256 KB
5. 16

CHAPTER 2

HCS12 ARCHITECTURE AND ASSEMBLY LANGUAGE PROGRAMMING

OBJECTIVES

Upon completion of this chapter, you will be able to:

>> List the registers of the HCS12 microcontroller
>> Manipulate data using the registers and load and add instructions
>> Code simple HCS12 Assembly language instructions
>> Assemble and run an HCS12 program
>> Describe the sequence of events that occur upon HCS12 power-up
>> Examine programs in ROM code of the HCS12
>> Explain the ROM memory map of the HCS12
>> Detail the execution of HCS12 Assembly language instructions
>> Describe HCS12 addressing modes
>> Describe HCS12 directives for Assembly language programming
>> Explain the purpose of the CCR (condition code register) register
>> Discuss memory space allocation in the HCS12
>> Diagram the use of the stack in the HCS12

SECTION 2.1: INSIDE THE HCS12

In this section we examine the major registers of the HCS12 and show their use with the simple instructions dealing with load and add.

Registers

In the CPU, registers are used to store information temporarily. That information could be a byte of data to be processed, or an address pointing to the data to be fetched. The HCS12 registers are either 8-bit or 16-bit. The 8 bits of a register are shown in Figure 2-1 from the MSB (most significant bit) D7 to the LSB (least significant bit) D0. Since there are a large number of registers in the HCS12, we will concentrate here on some of the widely used general-purpose registers and cover special registers in future chapters.

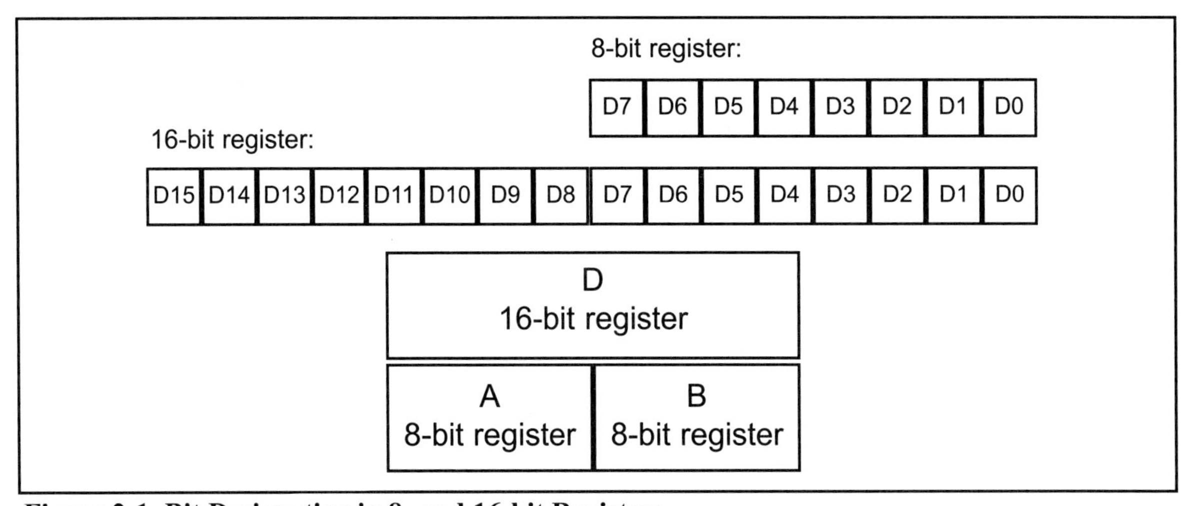

Figure 2-1. Bit Designation in 8- and 16-bit Registers

The most widely used registers of the HCS12 are A, B, X, Y, SP (stack pointer), CCR (condition code register), and PC (program counter). All of the above registers are 16-bits, except A, B, and CCR. See Figure 2-2. The A and B registers are referred to as accumulators and are widely used by the arithmetic and logic instructions. Combining registers A and B will give us a 16-bit register called D, which is used for the 16-bit arithmetic and logic operations. To understand the use of these registers, we will show them in the context of many simple instructions dealing with load and add.

Load instructions

Simply stated, the LDAA instruction loads a value into accumulator A. It has the following format:

```
LDAA source     ;copy source to accumulator A.
```

This instruction tells the CPU to load the source operand into the accumulator A. The source can be an immediate value, or a value held by a memory location, as we will see in Section 2.3. For example, the instruction "`LDAA,#$35`" loads the constant value of 35 (in hex) to register A. After this instruction is executed, register A will have the value 35 in hex. Notice the "#" in the instruction. This signifies that it is an immediate value. The importance of the # sign will be discussed soon. Also notice the $ sign. This signifies that the value is in hex. We also have the LDAB instruction, which loads a value into accumulator B. The following program loads registers A and B with the values 55H and 21H, respectively, then adds the value in register B to the value in register A.

Figure 2-2. CPU12 Registers

```
LDAA #$55       ;load value 55H into Accumulator A
LDAB #$21       ;load value 21H into Accumulator B
                ;(now A=55H and B=21H)
ABA             ;Add contents of B to A
                ;now A=76H and B=21H)
```

When programming the HCS12 microcontroller, the following points should be noted:

1. Values can be loaded into any of the registers A, B, D, X, Y, or SP. However, the value must be preceded with a pound sign (#) to indicate that it is an immediate value. This is shown next.

```
LDAA #$23       ;load 23H into A (A=23H)
LDAB #$12       ;load 12H into B (B=12H)
LDX  #$1F34     ;load 1F34H into X reg (X=1F34H)
LDY  #$2BC      ;load 2BCH into Y reg (Y=2BCH)
LDD  #$3C9F     ;load 3C9FH into D reg (D=3C9FH)
LDAA #$F9       ;load F9H into A (A=F9H)
LDAA #12        ;load 12 decimal (0CH)
                ;into reg. A (A=0CH)
```

Notice in the instruction "`LDAA #$F9`" that a 0 is not needed (used) between the $ and F to indicate that F is a hex number and not a letter. In other words, `$` tells the assembler that a hex number follows.

2. If hex values are used, we must use the $ sign to indicate that. For example, in "`LDAA #$9F`" the result will be A = 9F in hex; that is, A = 10011111 in binary. We can use the % sign to indicate the value is in binary. For example,

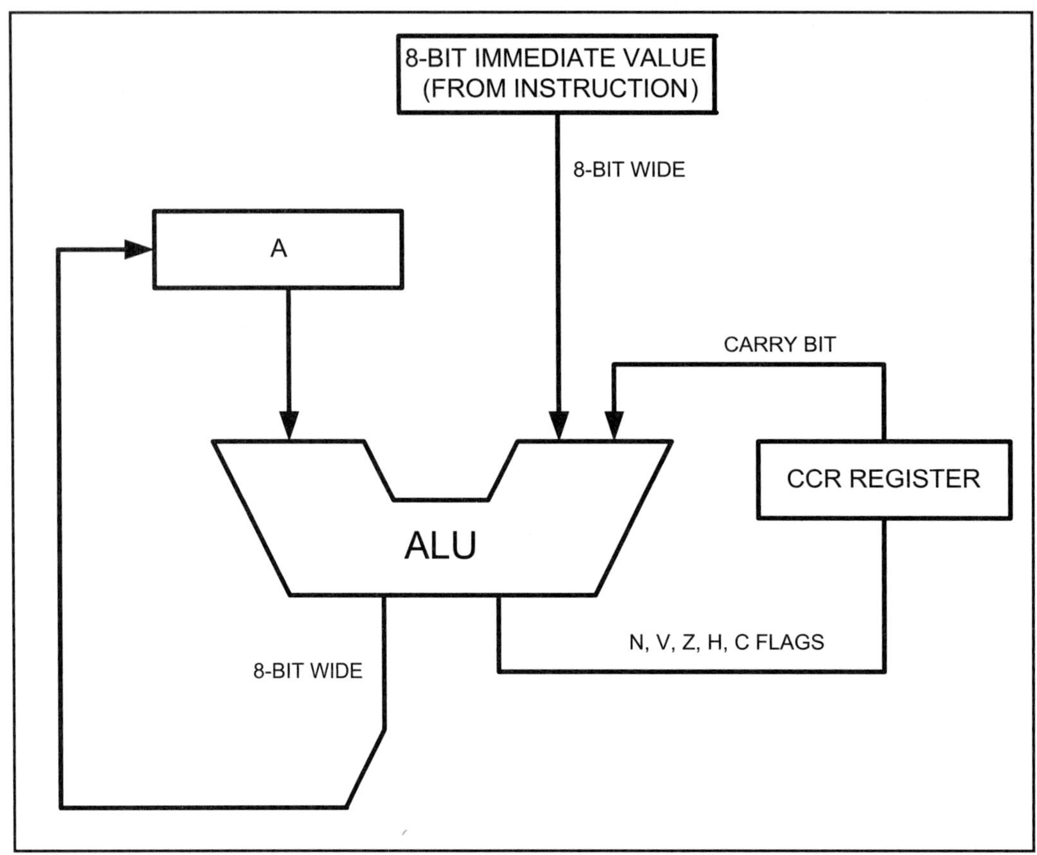

Figure 2-3. HCS12 ALU Using Immediate Value

in "`LDAA #%10011111`" the result will be A = 10011111 in binary; that is, A = 9F in hex. If we use nothing, it indicates the value is in decimal. For example, in "`LDAA #255`" the result will be A = 11111111 in binary; that is, A = FF in hex. This is summarized as follows:

Symbol	value is in
	decimal
%	binary
$	hexadecimal

3. If values 0 to F are moved into an 8-bit register, the rest of the bits are assumed to be all zeros. For example, in "`LDAA #$5`" the result will be A = 05; that is, A = 00000101 in binary.

4. Moving a value that is too large into a register will cause an error.

```
LDAA #$7F2     ;ILLEGAL: $7F2  > 8 bits (FFH)
LDAB #456      ;ILLEGAL: 456 > 255 decimal (FFH)
LDX  #$7FFF2 ;ILLEGAL: $7FFF2  > 16 bits (FFFFH)
LDY  #75456 ;ILLEGAL:75456 > 65535 deci (FFFFH)
```

5. A value to be loaded into a register must be preceded with a pound sign (#). Otherwise the command will load from a memory location. For example "`LDAA $17`" means to move into register A the byte held in memory location 17H, which could have any value. In order to load the value 17H into the accumulator we must write "`LDAA #$17`" with the # preceding the number.

Notice that the absence of the pound sign will not cause an error by the assembler since it is a valid instruction. However, the result would not be what the programmer intended. This is a common error for beginning programmers in the HCS12.

ADDA and ADDB instructions

The ADDA and ADDB instructions have the following format:

```
ADDA source ;ADD source operand to the accu A
ADDB source ;ADD source operand to the accu B
```

The ADDA instruction tells the CPU to add the source byte to register A and put the result in the destination register A. To add two numbers such as 25H and 34H, each can be moved to a register and then added together:

```
LDAA #$25       ;load 25H into A
LDAB #$34       ;load 34H into B
ABA         ;add B to accumulator A (A = A + B)
```

Executing the program above results in A = 59H (25H + 34H = 59H) and B = 34H. Notice that the content of B does not change. There are always many ways to write the same program. One question that might come to mind after looking at the program above, is whether it is necessary to move both data items into registers before adding them together. The answer is no, it is not necessary. See Figure 2-3. Look at the following variation of the same program:

```
LDAA #$25   ;load one operand into A (A=25H)
ADDA #$34   ;add the second operand 34H to A
```

Another variation is:

```
LDAB #$25   ;load one operand into B (B=25H)
ADDB #$34   ;add the second operand 34H to B
```

In the above cases, while one register contained one value, the second value followed the instruction as an operand. This is called an *immediate* operand. See Figure 2-3. The examples shown so far for the ADD instruction indicate that the source operand can be either a register or immediate data, but the destination must always be register A or B (the accumulators). In other words, an instruction such as "`ADD #$25,#$34`" is invalid since an accumulator, register A or B, must be involved in any 8-bit arithmetic operation.

There are several 16-bit registers in the HCS12: PC (program counter), D, X, Y, and SP (stack pointer). The importance and use of the program counter are covered in Section 2.2. The X and Y registers are used in accessing data using index addressing mode, which is discussed in Section 2.3. The use of the stack pointer is discussed in Section 2.8.

Review Questions

1. Write the instructions to move value 34H into register A and value 3FH into register B, then add them together.

2. Write the instructions to add the values 16H and CDH. Place the result in register A.
3. True or false. No value can be moved directly into registers A and B.
4. What is the largest hex value that can be moved into an 8-bit register? What is the decimal equivalent of the hex value?
5. What is the largest hex value that can be moved into the Y register? What is the decimal equivalent of the hex value?

SECTION 2.2: THE HCS12 MEMORY MAP

In this section we discuss the memory map for various HCS12 family members.

Memory space allocation in the HCS12

The HCS12 has 64K bytes of directly accessible memory space. This memory space has addresses 0000 to FFFFH. The 64K bytes of memory space is divided into four sections. They are as follows:

1. ***Register space***. This area is dedicated to I/O and special function registers (SFRs) such as timers, serial communication, ADC, and so on. The function of each SFR is fixed by the CPU designer at the time of design because it is used for I/O (Input/Output) port registers and peripherals. The I/O registers and SFRs are 8-bit registers. The number of locations set aside for I/O registers and SFRs depends on the pin numbers and peripheral functions supported by that chip. That number can vary from chip to chip even among members of the same family. Some chips have as few as 64 bytes. In the HCS12 chips, the registers start at address location 0000. The last location assigned to registers varies from chip to chip since the more peripherals and I/O pins we have, the more locations are assigned to them. See Table 2-1 and Figure 2-4. We will study the use of I/O registers and peripheral SFRs in future chapters as we cover the I/O and peripherals.
2. ***Data RAM space***. A RAM space ranging from 128 bytes to several kilobytes is set aside for data storage. The data RAM space is used for data variables and stack and is accessed by the microcontroller instructions. The data RAM space is read/write memory used by the CPU for storage of data variables, scratch pad, and stack. The HCS12 microcontrollers' data RAM size ranges from 256 bytes to several thousand bytes depending on the chip. Even within the same family, the size of the data RAM space varies from chip to chip. A larger data RAM size means more difficulties in managing these RAM locations if you use Assembly language programming. In today's high-performance microcontroller, however, with over a thousand bytes of data RAM, the job of managing them is handled by the C compilers. Indeed, the C compilers are the very reason we need a large data RAM since it makes it easier for C compilers to store parameters and allows them to perform their jobs much faster. The amount and the location of the RAM space varies from chip to chip in the Freescale chips. A section of the data RAM space is used by stack as we will see in Section 2.8. Notice that the data RAM has a byte-size width. In HCS12 we have only A and B for general-purpose registers. In this chapter and future

Table 2-1: Partial List of MCHC912B32 Register Addresses

Symbol	Function	Address (in Hex)
PORT A	PORTA	00
PORTB	PORTB	01
DDRA	Data Direction register for PORTA	02
DDRB	Data Direction register for PORTB	03
PORTE	PORTE	08
DDRE	Data Direction register for PORTE	09
PORTP	PORTP	56
DDRP	Data Direction register for PORTP	57
PORTAD	Analog Input	6F
TCNT	16-bit Timer	84
PORTT	PORTT	AE
DDRT	Data Direction register for PORTT	AF
SC0BD	Serial Port baud rate generator	C0
SC0CR1	Serial Port control register	C2
SC0CR2	Serial Port control register	C3
SC0SR1	Serial Port status register	C4

chapters we will see how to use the RAM locations as general-purpose registers.

3. ***EEPROM space***. A block of memory from 128 bytes to several thousand bytes is set aside for EEPROM memory. The amount and the location of the EEPROM space varies from chip to chip in the Freescale chips. Although in some applications the EEPROM is used for program code storage, it is used most often for saving critical data. We will study the use of EEPROM in Chapter 14.
4. ***Code ROM space***. A block of memory from a few kilobytes to several hundred kilobytes is set aside for program

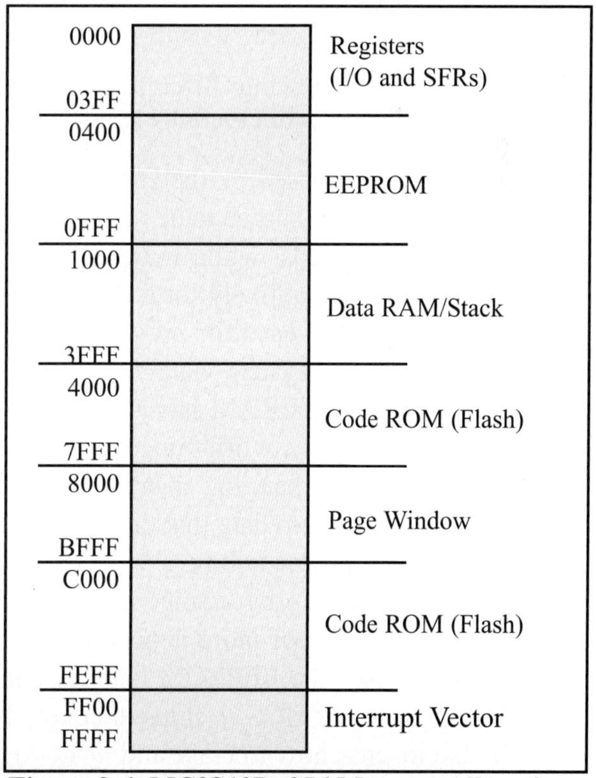

0000–03FF	Registers (I/O and SFRs)
0400–0FFF	EEPROM
1000–3FFF	Data RAM/Stack
4000–7FFF	Code ROM (Flash)
8000–BFFF	Page Window
C000–FEFF	Code ROM (Flash)
FF00–FFFF	Interrupt Vector

Figure 2-4. MC9S12Dx256 Memory Allocation

space. The program space is used for the program code. In today's microcontroller chips the code ROM space is flash memory. The amount and the location of the code ROM space varies from chip to chip in the Freescale products. See Table 2-2 and Examples 2-1 and 2-2. The flash memory of code ROM is under the control of the PC (program counter). The code ROM memory can also be used for storage of static fixed data such as ASCII data strings. See Chapter 14. In Chapter 14 we will see how to access the program ROM memory of greater than 64K bytes by using page switching.

From Tables 2-1 and 2-3, notice that the register locations vary from chip to chip. See Chapter 4 for additional information on the I/O ports and special function registers associated with them.

Table 2-2: HCS12 Memory Allocation

	Registers (I/O & SFRs)	**EEPROM**	**RAM** (Data & Stack)	**Code ROM**
MC9S12Dx64	1 KB	1 KB	4 KB	64 KB Flash
	000–3FF	**800–FFF**	**3000–3FFF**	**4000–FFFF**
MC9S12Dx128	1 KB	2 KB	8 KB	128 K Flash
	000–3FF	**800–FFF**	**2000–3FFF**	**4000–FFFF**
MC9S12Dx256	1 KB	4 KB	12 KB	256 KB Flash
	000–3FF	**400–FFF**	**1000–3FFF**	**4000–FFFF**
MC9S12DP512	1 KB	4 KB	14 KB	512 KB Flash
	000–3FF	**400–7FF**	**0800–3FFF**	**4000–FFFF**

Notes: (1) The page window address of $8000 through $BFFF is used to access the code ROM beyond 64 KB. See Chapter 14. (2) Upon reset, some memory addresses are overlapped. One can remap them to different address locations after reset.

Notice the following differences among the Flash ROM, data RAM, and EEPROM memories in microcontrollers:

a) The data RAM is used by the CPU for data variables and stack, whereas the EEPROMs are considered to be memory that one can also add externally to the chip. In other words, while many microcontroller chips have no EEPROM memory, it is very unlikely for a microcontroller to have no data RAM.
b) The Flash ROM is used for program code, while the EEPROM is used most often for critical system data that must not be lost if power is cut off. Remember that data RAM is volatile memory and its contents are lost if the power to the chip is cut off. Since volatile data RAM is used for dynamic variables (constantly changing data) and stack, we need EEPROM memory to secure critical system data that does not change very often and will not be lost in the event of power failure.
c) The Flash ROM is programmed and erased in block size. The block size is 8, 16, 32, or 64 bytes or more depending on the chip. That is not the case with EEPROM, since the EEPROM is byte programmable and erasable. The EEPROM memory of HCS12 chips is covered in Chapter 14. In Chapter 14, we will also discuss how to erase and write to flash memory in block sizes.

Table 2-3: HCS912DP512 Special Function Register (SFR) Addresses

Symbol	Function	Address (in Hex)
PORT A	PORTA	00
PORTB	PORTB	01
DDRA	Data Direction register for PORTA	02
DDRB	Data Direction register for PORTB	03
PORTE	PORTE	08
DDRE	Data Direction register for PORTE	09
	Timer	40–7F
ATD0	Analog-to-Digital Converter 0	80–9F
	PWM	A0–C7
	Asynch. Serial Port 0	C8–CF
	Asynch. Serial Port 1	D0–D7
	SPI (first one)	F0–F7
	SPI (second one)	F8–FF
	Flash Memory Control Registers	100–10F
ATD1	Analog-to-Digital Converter 1	120–13F

Example 2-1

Verify the memory address for each of the following sections.

(a) Registers: 1 KB (000–3FF)
(b) RAM: 12 KB (1000–3FFF)
(c) Flash ROM: 32 KB (8000–FFFF)

Solution:

(a) With 1,024 bytes of register space, we have 0000–3FFH. Since we are converting 1,024 to hex, we get 400H, or 000–3FF.

(b) With 12K bytes we have 12,288 bytes (12 × 1,024 = 12,288). Converting 12,288 to hex, we get 3000H (0000–2FFF); therefore, the memory space for RAM is 1000 to 3FFFH since the starting location is 1000H (1000 + 2FFF = 3FFF).

(c) With 32K bytes of on-chip ROM memory space, we have 32,768 bytes (32 × 1,024 = 32,768), which gives 0000–7FFFH. Since we are converting 32,768 to hex, we get 8000H. Now, if the first location is 8000H, the last location will be FFFF since 8000 + 7FFF = FFFFH.

In many HCS12 chips, such as MC9S12Dx512, when adding register, EEPROM, and RAM spaces we get more than 64K bytes. This is due to the fact that some of the addresses are overlapped. We need to remap the addresses if we want to access the full memory space for each section.

Example 2-2

HCS12 divides the 64 KB of memory space into four equal sections. Calculate the space and the amount of memory given to each section.

Solution:

1) With address space of 0000 to 3FFF, we have 3FFF–0000 = 3FFF bytes. Converting 3FFF to decimal, we get 16,383 + 1, which is equal to 16K bytes.
2) With address space of 4000 to 7FFF, we have 7FFF–4000 = 3FFF bytes. Converting 3FFF to decimal, we get 16,383 + 1, which is equal to 16K bytes.
3) With address space of 8000 to BFFF, we have BFFF–8000 = 3FFF bytes. Converting 3FFF to decimal, we get 16,383 + 1, which is equal to 16K bytes.
4) With address space of C000 to FFFF, we have FFFF–C000 = 3FFF bytes. Converting 3FFF to decimal, we get 16,383 + 1, which is equal to 16K bytes.

See Figure 2-5.

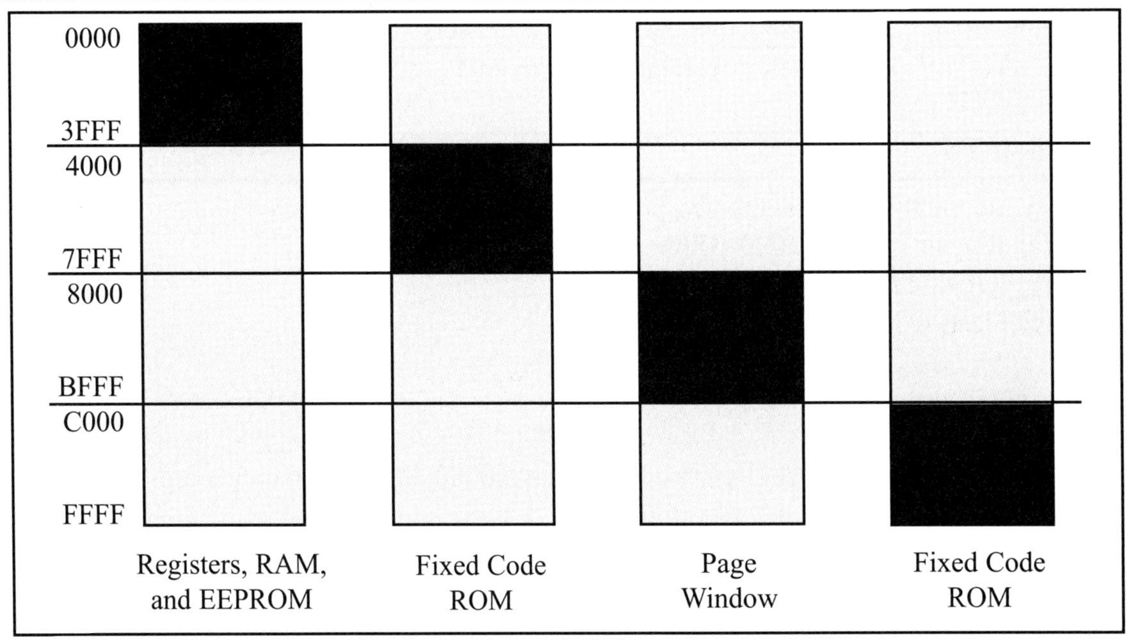

Figure 2-5. HCS12 Address Range for Example 2-2

Accessing memory beyond 64 KB

One of the most important registers in the microcontroller is the PC (program counter). The program counter points to the address of the next instruction to be executed. As the CPU fetches the opcode from the program ROM, the program counter is incremented to point to the next instruction. The program counter in the HCS12 is 16 bits wide. This means that the HCS12 can access address space 0000 to FFFFH, a total of 64K bytes since 2^{16} = 64 KB. However, no member of the HCS12 family has the entire 64K bytes dedicated to on-chip ROM, since a portion of the 64 KB space is also used by the RAM and EEPROM for the data storage, as we saw earlier. For the HCS12 chips such as MC9S12DP512 with 512K bytes of on-chip program ROM, we must use page switching to access the memory beyond 64 KB. See Figure 2-6. In Chapter 14 we will show how this is done.

Where the HCS12 wakes up when it is powered up

One question that we must ask about any microcontroller (or microprocessor) is: At what address does the CPU wake up when power is appled to it? Each microprocessor is different. In the case of the HCS12 family, the microcontroller wakes up at memory address $FFFE when it is powered up. By powering up we mean applying power to the V_{CC} pin or activating the RESET pin as discussed in Chapter 8. In other words, when the HCS12 is powered up, the PC (program counter) has the value of $FFFF:$FFFE in it. This means that it expects the address of the first instruction to be stored at ROM address $FFFE. In the HCS12 chips, locations $FFFE and $FFFF belong to the interrupt vector table. In some HCS12-based trainers the interrupt vector table locations are redirected so that users do not overwrite the bootloader's address with their own program's address.

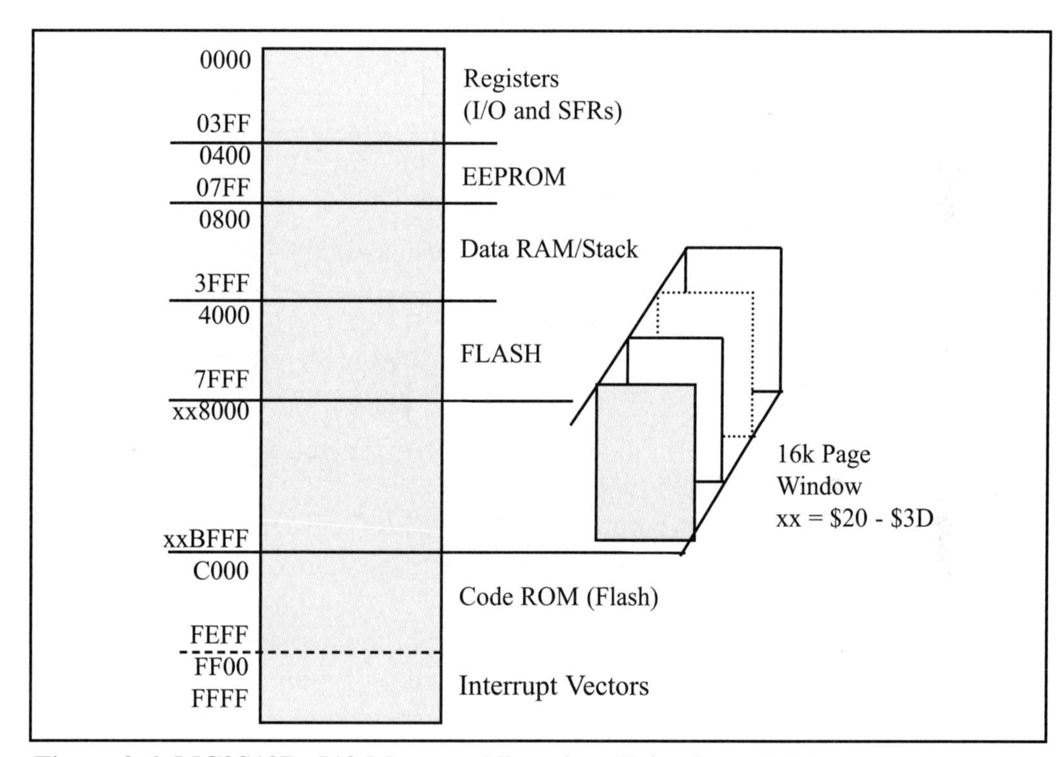

Figure 2-6. MC9S12Dx512 Memory Allocation (Using Page Window)

Single-chip and expanded modes in HCS12

HCS12 can work in several modes. The most widely used modes are single-chip and expanded. In the single-chip mode, all the on-chip memories are visible to the programmer and their addresses are set to fixed locations upon reset. See Figure 2-7. In the expanded mode, one can connect the HCS12 chip to external memories of RAM and ROM and remap the on-chip memories to different address locations. In some development boards the HCS12 uses the external RAM for code development since the on-chip Flash ROM has a limited number of program/erase cycles. Flash memory with a limited number of program/erase cycles is easily worn out after being programmed and erased so many times. In such cases the use of external RAM for code development is common since RAM has an unlimited number of program/erase cycles.

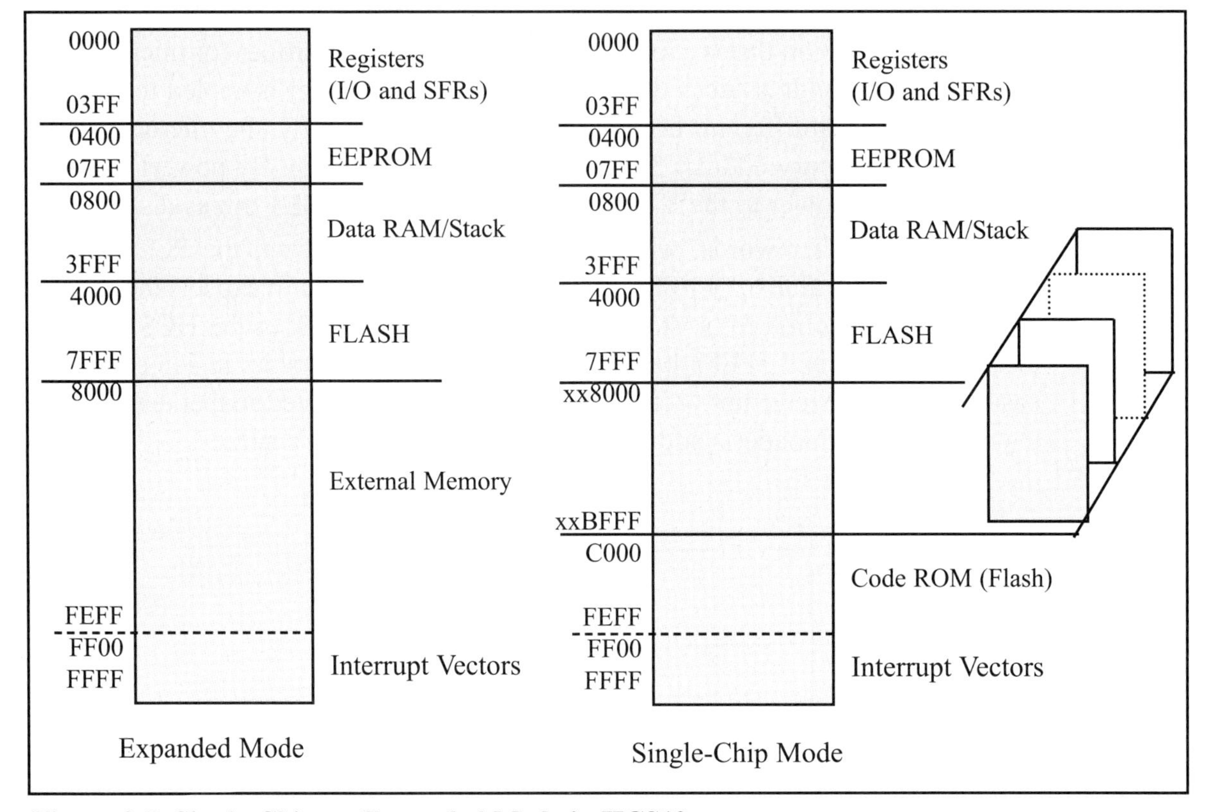

Figure 2-7. Single-Chip vs. Expanded Mode in HCS12

Review Questions

1. True or false. Data space in the HCS12 is SRAM memory, whereas program (code) space is of the ROM type.
2. True or false. The general-purpose RAM and SFRs registers are nonvolatile.
3. True or false. The larger the size of RAM space, the more difficult it is to manage.
4. True or false. Registers A and B are not a part of the register address space.
5. The vast majority of registers in the HCS12 are _____-bit.
6. The register space in the HCS12 always starts at address location______.
7. True or false. The register space in HCS12 ends at the same address for every HCS12 family member.

SECTION 2.3: THE HCS12 ADDRESSING MODES

The CPU can access data operands in various ways. The data could be in a register, in memory, or be provided as an immediate value. These various ways of accessing operand data are called *addressing modes*. In this section we discuss the HCS12 addressing modes in the context of some examples.

The various addressing modes of a microprocessor are determined when it is designed, and therefore cannot be changed by the programmer. The HCS12 provides six distinct addressing modes. They are as follows:

(1) inherent
(2) immediate
(3) extended
(4) direct
(5) relative
(6) indexed

Inherent / register addressing mode

Inherent addressing mode involves no operand or it uses a register to hold the operand. The following are examples of inherent/register addressing modes:

```
NOP         ;No Operation. waste a clock cycle
STOP        ;stop the program counter
ABA         ;add the contents of B to contents of A
CLRA        ;clear register A
INCB        ;increment register B
INX         ;increment register X
```

Immediate addressing mode

In immediate addressing mode, the source operand is a constant. As the name implies, when the instruction is assembled, the operand comes immediately after the opcode. Notice that the immediate data must be preceded by the pound sign, "#". This addressing mode can be used to load information into any of the registers, including the stack pointer (SP) register. See the examples below.

```
LDAA #$25        ;load 25 hex value into A
LDAB #62         ;load the decimal value 62 into B
LDD  #$9FF2      ;load 9FF2H into register D
LDS  #$7FF       ;SP=7FFH
LDX  #$559D      ;X=559DH
```

Although the D register is 16-bit, it can also be accessed as two 8-bit registers, A and B, where A is the high byte and B is the low byte. Look at the following code.

```
LDD  #$2550
```

is the same as:

```
LDAB #$50
LDAA #$25
```

Also notice that the following would produce an error since the value is larger than 16 bits.

```
LDD  #$68975    ;illegal!! value > 65535 (FFFFH)
```

In the first two addressing modes, the operands are either inside one of the registers or tagged along with the instruction itself. In most programs, the data to be processed is often in some memory location of RAM. There are many ways to access this data. Next we describe these different methods.

Extended addressing mode

In extended addressing mode, the data is in a RAM memory location whose address is known, and this address is given as a part of the instruction. Contrast this with immediate addressing mode, in which the operand itself is provided with the instruction. The "#" sign distinguishes between the two modes. See the examples below, and note the absence of the "#" sign.

```
LDAA $800        ;load A from RAM location 800H
LDAB $801        ;load B from RAM location 801H
```

The above examples should reinforce the importance of the "#" sign in HCS12 instructions. See the following code.

```
LDAA #$25 ;load #25 into A (A=25H)
LDAA $801 ;load RAM location 801H
          ;into A (A=99)
```

Address	Data
800	..
801	99
802	..
803	..
804	..

STAA instruction

The STAA instruction tells the CPU to move (in reality, copy) the source, register A, to a destination in the RAM. After this instruction is executed, the location pointing to the register or RAM will have the same value as register A. The STAA stores the contents of A into any location in the RAM region using extended addressing mode. The following program will put 99H into locations 800–804 of the RAM region:

Address	Data
800	99
801	99
802	99
803	99
804	99

```
    LDAA #$99 ;A = 99H
    STAA $800 ;store (copy) A contents
to location 800h
    STAA $801 ;store (copy) A contents
to location 801h
    STAA $802
    STAA $803
    STAA $804
```

The following program will a) first put value 33H into RAM locations 1200, 1201, and 1202, b) then add them to accumulator A, and c) store the result in RAM location 1204:

Address	Data
1200	33
1201	33
1202	33
1203	xx
1204	99

```
LDAA #$33        ;A = 33H
STAA $1200;move(copy) A contents to location 1200H
STAA $1201;move(copy) A contents to location 1201H
STAA $1202;move(copy) A contents to location 1202H
CLRA             ;clear register A (A=00)
```

```
ADDA $1200;add A and loc 1200H, result in A (A = 33H)
ADDA $1201;add A and loc 1201H, result in A (A = 66H)
ADDA $1202;add A and loc 1202H, result in A (A = 99H)
STAA $1204      ;store A into RAM location 1204H
```

Example 2-3

State the contents of RAM locations after the following program:

```
LDAA #$99        ;load A with value 99H
STAA $812
LDAA #$85        ;load A with value 85H
STAA $813
LDAA #$3F        ;load A with value 3FH
STAA $814
LDAA #$63        ;load A with value 63H
STAA $815
LDAA #$12        ;load A with value 12H
STAA $816
```

Solution:

After the execution of STAA $812 RAM location $812 has value $99;
after the execution of STAA $813 RAM location $813 has value $85;
after the execution of STAA $814 RAM location $814 has value $3F;
after the execution of STAA $815 RAM location $815 has value $63;
and so on, as shown in the chart.

Address	Data
0812	99
0813	85
0814	3F
0815	63
0816	12

Little endian vs. big endian war

Previous examples used 8-bit or 1-byte data. In this case the bytes are stored one after another in memory. What happens when 16-bit data is used? For example:

```
LDX   #$95F3     ;load 95F3H into X
STX   $525 ;copy the contents of X to address 525H
```

Address	Data
524	xx
525	95
526	F3
527	xx
528	xx

The "STX $525" instruction stores the 2-byte value in register X in memory locations 525H and 526H. In cases like this, the high byte goes to the low memory location and the low byte goes to the high memory address. In the example above, memory location 525H contains 35H and memory location 526H contains F3H (525H = 35H, 526H = F3H).

This convention is called big endian versus little endian. The origin of the terms *big endian* and *little endian* is from a *Gulliver's Travels* story about how an egg should be opened: from the little end or the big end. In the big endian method, the high byte goes to the low address and the low byte to the high address, whereas in the little endian method, the high byte goes to the high address and the low byte to the low address. See Examples 2-4 and 2-5. All Intel microprocessors use the little endian convention. Freescale microprocessors use big endian. This difference might seem as trivial as whether to break an egg from the big end or little end, but it is a nuisance in converting software from one camp to be run on a computer of the other camp.

Example 2-4

Assume memory locations with the following contents: 826H = ($48) and 827H = ($22). Show the contents of register D in the instruction "LDD $826".

Solution:

According to the big endian convention used in all Freescale microprocessors, register B should contain the value from the high address 827H and register A the value from the address 826H, giving A = 48H and B = 22H.

Address	Data
825	xx
826	48
827	22
828	xx

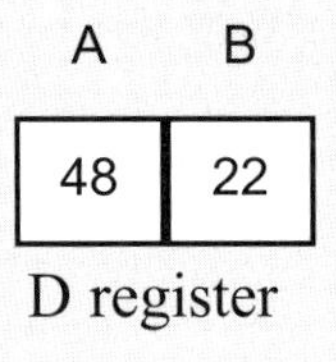

Example 2-5

Show the contents of RAM locations for the code:

```
LDY #$9245     ;Y reg. has 9245H
STY $804       ;store Y reg. values in loc 804H and 805H.
```

Solution:

According to the big endian convention, location 804H should contain the value from the high byte and location 805H the value from the low byte.

High Low

92 45

Y register

Address	Data
803	xx
804	92
805	45
807	xx

Direct addressing mode

Direct addressing mode is a special case of extended addressing mode. In the extended addressing mode, the address of RAM is 16-bit, which means we can go anywhere in the 64K bytes address space of HCS12. In the direct ddressing mode, the address is 8-bit, which limits us to locations 00 to FFH. The fact that the special function registers are located at addresses of 00–FF allows us to use the direct addressing mode to access them. The SFR, can be accessed by their names (which is much easier) or by their addresses. For example, PORTA has address 00, and PORTB has been designated the address 01, as was shown in Table 2-2.

Notice how the following pairs of instructions mean the same thing.

```
STAA $00        ;is the same as
STAA PORTA;which means place register A into PORTA
```

The following program first loads the A register with value 55H, then moves this value to ports A and B:

```
LDAA #$55       ;A = 55H
STAA PORTA      ;copy A to Port A (Port A = 55H)
STAA PORTB      ;copy A to Port B (Port B = 55H)
```

PORTA and PORTB are part of the special function registers in the register area. They can be connected to the I/O pins of the HCS12 microcontroller as we will see in Chapter 4.

Relative addressing mode

The relative addressing mode is used exclusively by branch (jump) instructions. In the relative addressing mode, the address of the target location is relative to the current value of program counter (PC). Chapter 3 covers this topic with some examples.

Indexed addressing mode

In the indexed addressing mode, index registers X and Y, as well as a displacement value, are used to calculate what is called the *effective address* (EA) of the operand data. We can use indexed addressing mode to access data stored in the RAM or the ROM sections. Since X and Y are 16-bit registers, this allows us access to the entire 64K bytes of memory space. This is similar to register-indirect addressing mode in other microprocessors except we can also add a fixed offset value to the X or Y pointer. Adding the offset value to pointers X and Y gives us an effective address. Examine the following code:

Address	Data
800	55
801	55
802	55
803	55
804	xx

```
LDX  #$800      ;load 800H into X
LDAA #$55       ;A = 55H
STAA 0,X  ;store contents of reg A into RAM loc X+0
STAA 1,X  ;store contents of reg A into RAM loc X+1
STAA 2,X  ;store contents of reg A into RAM loc X+2
STAA 3,X  ;store contents of reg A into RAM loc X+3
```

The above code is another version of an earlier program that used extended addressing mode. In the above code, the EA (effective address) = X + fixed offset value. Another variation of the above program is as follows:

```
LDX  #$800      ;load 800H into X
LDAA #$55
STAA 0,X  ;store contents of reg A into RAM loc 800+0
INX       ;increment X (X=801)
STAA 0,X  ;store contents of reg A into RAM loc 801+0
INX       ;increment X (X=802)
STAA 0,X  ;store contents of reg A into RAM loc 802+0
```

```
INX         ;increment X (X=803)
STAA 0,X    ;store contents of reg A into RAM loc 803+0
INX         ;increment X (X=804)
STAA 0,X    ;store contents of reg A into RAM loc 804+0
```

We can write the above code more efficiently using a loop. The loop version of the above program is shown in Chapter 3. Besides X and Y, we can also use the SP (stack pointer) and PC (program counter) registers as indexed registers. The indexed addressing mode has many powerful variations and we will discuss them in Chapter 6 along with some applications. Table 2-4 shows load and store instructions.

Table 2-4: Load and Store Instructions

Instruction	Operation
LDAA	Load A
LDAB	Load B
LDD	Load D
LDS	Load SP
LDX	Load X
LDY	Load Y
STAA	Store A
STAB	Store B
STD	Store D
STS	Store SP
STX	Store X
STY	Store Y

Instruction size of the HCS12

The instructions of HCS12 can be anywhere from 1 to 5 bytes of machine code. For example, the "NOP" is a 1-byte instruction since there is no operand and all we have is the opcode "01" for the instruction. The "LDAA #$25" is a 2-byte instruction since the opcode "$86" is the first byte and the immediate value of $25 is the second byte. Instructions using the extended addressing mode are 3-byte instructions since one byte is used for opcode and 2 bytes are for the address of the RAM location. Examine the following instructions and their sizes:

```
ABA             ;1-byte instruction
LDAA #$25       ;2-byte instruction
LDAA $750       ;3-byte instruction
LDD  #$29F2     ;3-byte instruction
```

Next we explore the instruction formation for a few of the instructions we have used in this chapter. This should give you some insights into the instructions of the HCS1218.

"LDAA #imm_val" instruction formation

The "LDAA #imm_val" is a 2-byte instruction. Of the 2 bytes, the first byte is set aside for the opcode and the second byte is used for a value in the range from 00 to FFH. This is shown below.

$0 \leq i \leq FF$

"LDAA ram_locl" instruction formation

The "LDAA ram_locl" is a 3-byte instruction. Of the 3 bytes, the first byte is set aside for the opcode and the other 2 bytes are used for the location of the memory from which the value will be loaded.

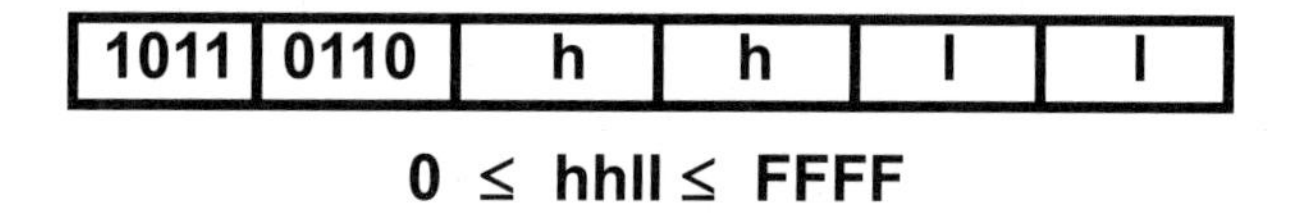

"STAA ram_loc" instruction formation

The "STAA ram_loc" is a 3-byte instruction. Of the 3 bytes, the first byte is set aside for the opcode and the other 2 bytes are used for the location of the RAM.

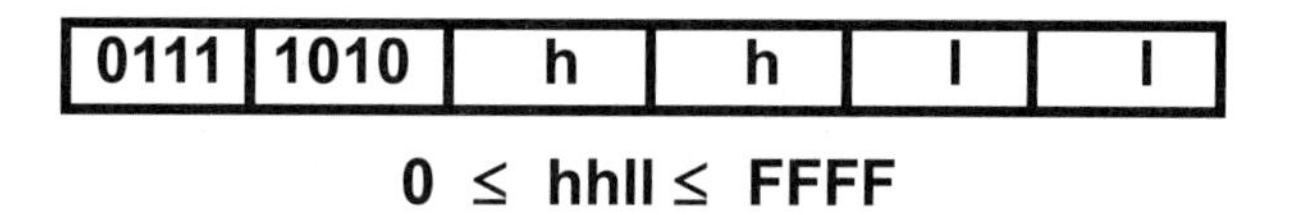

The above examples show that the HCS12 is not a RISC processor since all the instructions have the same size in the RISC architecture. See Chapter 0.

Review Questions

1. Can the programmer of a microcontroller make up new addressing modes?
2. Show the instruction to load FF (hex) into register B.
3. Why is the following invalid? "LDAA #$5F3"
4. True or false. The D register is a 16-bit register that is also accessible in low-byte and high-byte formats.
5. Is the X register also available in low-byte and high-byte formats?
6. Write instructions to add the values 16H and CDH. Place the result in location 200H of the RAM.
7. True or false. Instructions using direct addressing mode are 3-byte instructions.
8. True or false. Instructions using extended addressing mode are 3-byte instructions.
9. Which registers can be used for indexed addressing mode?
10. Give the size of registers X and Y.

SECTION 2.4: HCS12 CONDITION CODE REGISTER

Like any other microprocessor, the HCS12 has a flag register to indicate arithmetic conditions such as the carry bit. The flag register in the HCS12 is called the *condition code register* (CCR). In this section we discuss the various bits of this register and provide some examples of how it is altered.

CCR (condition code register) register

The condition code register (CCR) is an 8-bit register. It is also referred to as the *flag register*. Five of the flags are called *conditional flags*, meaning that they indicate some conditions that result after an instruction is executed. These five are C (carry), H (half carry), Z (zero), N (negative), and V (overflow). Two of the bits, I and X, are set aside for the interrupt masking and one bit, the S bit, is for the stop option.

The following is a brief explanation of five of the flag bits of the CCR register. The impact of instructions on these registers is then discussed.

C, the carry flag

This flag is set whenever there is a carry out from the D7 bit. This flag bit is affected after an 8-bit or 16-bit addition or subtraction.

H, the half-byte carry flag

If there is a carry from D3 to D4 (half-byte) during an arithmetic operation such as add or subtract, this bit is set; otherwise, it is cleared. This flag is used by instructions that perform BCD (binary coded decimal) arithmetic. In some microprocessors this is called the *AC* (Auxiliary Carry) flag. See Chapter 5 for more information.

Z, the zero flag

The zero flag reflects the result of an arithmetic or logic operation. If the result is zero, then Z = 1. Therefore, Z = 0 if the result is not zero. See Chapter 3 to see how to use the Z flag for looping.

V, the overflow flag

This flag is set whenever the result of a signed number operation is too large, causing the high-order bit to overflow into the sign bit. In general, the carry flag is used to detect errors in unsigned arithmetic operations while the overflow flag is used to detect errors in signed arithmetic operations. The V and N flag bits are used for the signed number arithmetic operations and are discussed in Chapter 5.

S	X	H	I	N	Z	V	C

C Carry flag
V Overflow flag
Z Zero flag
N Negative flag
H Half-byte carry flag

I Interrupt mask bit
X Interrupt mask bit
S Stop bit

Figure 2-8. Bits of the CCR Register

N, the negative flag

Binary representation of 8-bit signed numbers uses bit D7 as the sign bit. The negative flag reflects the result of an arithmetic operation. If the D7 bit of the result is zero, then N = 0 and the result is positive. If the D7 bit is one, then N = 1 and the result is negative. The negative and V flag bits are used for the signed number arithmetic operations and are discussed in Chapter 5.

ADD instructions and CCR

Next we examine the impact of the ADD instruction on the flag bits C, H, and Z of the CCR register. Some examples should clarify their status. Although the flag bits affected by the ADD instruction are C (carry flag), Z (zero flag), H (half-byte carry flag), and V (overflow flag), we will focus on flags C, H, and Z for now. A discussion of the overflow flag is given in Chapter 5, since it relates only to signed number arithmetic. The use of the various flag bits in programming is discussed in future chapters in the context of many applications.

See Examples 2-6 through 2-8 for the impact on selected flag bits as a result of the ADD instruction.

Example 2-6

Show the status of the C, H, N, and Z flags after the addition of 38H and 2FH in the following instructions.

```
LDAA #$38
ADDA #$2F ;after the addition A=67H, C=0
```

Solution:

```
   38      00111000
 + 2F      00101111
   67      01100111
```

C = 0 since there is no carry beyond the D7 bit.
H = 1 since there is a carry from the D3 to the D4 bit.
N = 0 since the D7 bit = 0.
Z = 0 since the accumulator has a value other than zero in it.

Example 2-7

Show the status of the C, H, N, and Z flags after the addition of 9CH and 64H in the following instructions.

```
LDAA #$9C
ADDA #$64         ;after addition A=00 and C=1
```

Solution:

```
   9C      10011100
 + 64      01100100
  100      00000000
```

C = 1 since there is a carry beyond the D7 bit.
H = 1 since there is a carry from the D3 to the D4 bit.
N = 0 since the D7 bit = 0.
Z = 1 since the accumulator has a zero value in it.

Example 2-8

Show the status of the C, H, S, and Z flags after the addition of 88H and 73H in the following instructions.

```
        LDAA #$88
        ADDA #$73 ;after the addition A=FBH,C=0
```

Solution:

```
     88        10001000
   + 73        01110011
     FB        11111011
```

C = 0 since there is no carry beyond the D7 bit.
H = 0 since there is no carry from the D3 to the D4 bit.
N = 1 since the D7 bit = 1.
Z = 0 since the result in the accumulator is not zero.

Not all instructions affect the flags

Some instructions affect all five flag bits C, H, Z, V, and N (e.g., ADDA). But some instructions affect no flag bits at all. The XGDX instruction is in this category. Some instructions affect only some such as N, Z, and V. Table 2-5 shows some of the instructions and the flag bits affected by them. Appendix H provides a complete list of all the instructions and their associated flag bits.

Table 2-5: Status Flag Bits for Some of the Instructions

Instruction	H	N	Z	V	C
ABA	X	X	X	X	X
ADDA	X	X	X	X	X
ADDB	X	X	X	X	X
ANDA			X	X	0
CLRA		0	1	0	0
COMA		1	X	0	X
DAA		X	X	?	X
DECA		X	X	X	
INCA		X	X	X	
ORAB		X	X	X	
XGDX					

Note: X can be 0 or 1. Blank means it is not affected. ? means not known. See Chapter 5 for how to use these instructions.

Flag bits and decision making

Because status flags are also called conditional flags, there are instructions that will make a conditional jump (branch) based on the status of the flag bits. Table 2-6 provides the list. Chapter 3 will discuss the conditional branch instructions and how they are used.

Table 2-6: HCS12 Branch (Jump) Instructions Using Flag Bits

Instruction	Action
BCS	Branch if C = 1
BCC	Branch if C = 0
BEQ	Branch if Z = 1
BNE	Branch if Z = 0
BMI	Branch if N = 1
BPL	Branch if N = 0
BVS	Branch if V = 1
BVC	Branch if V = 0

Review Questions

1. The flag register in the HCS12 is called the ________.
2. What is the size of the CCR register in the HCS12?
3. True or false. Some instructions affect no flag bits.
4. Find the C, Z, and H flag bits for the following code.
```
LDAA #$9F
ADDA #$61
```
5. Find the C, Z, and H flag bits for the following code.
```
LDAA #$82
ADDA #$22
```
6. Find the C, Z, and H flag bits for the following code:
```
LDAA #$67
ADDA #$99
```

SECTION 2.5: HCS12 DATA FORMAT AND DIRECTIVES

In this section we look at some widely used data formats and directives supported by the HCS12 assembler.

Data format representation

There are four ways to represent a byte of data in the HCS12 assembler. The numbers can be in hex, binary, decimal, or ASCII formats. The following are examples of how each works.

Hex numbers

There are two ways to show hex numbers:

1. In Assembly, we use $ right before the number like this: LDAA #$99
2. In C, we put 0x (or 0X) in front of the number like this: PORTB = 0x99

Here are a few lines of code that use the hex format:

```
LDAA #$25       ;A = 25H
ADDA #$11       ;A = 25H + 11H = 36H
ADDA #$12       ;A = 36H + 12H = 48H
ADDA #$2A       ;A = 48H + 2AH = 72H
```

Binary numbers

The % sign is used to represent binary numbers in an HCS12 assembler. It is as follows:

```
LDAA #%00100101      ;A = 25H
ADDA #%00010001      ;A = 25H + 11H = 36H
LDAA #%10011001      ;A = 10011001 or 99 in hex
```

Decimal numbers

There is only one way to represent decimal numbers in an HCS12 assembler. It is as follows:

```
LDAA #12             ;A = 00001100 or 0C in hex
```

Here are some examples of how to use it:

```
LDAA #37  ;A = 25H (37 in decimal is 25 in hex)
ADDA #17  ;A = 37 + 17 = 54 where 54 in dec is 36H
```

ASCII character

To represent ASCII data in an HCS12 assembler we use single quotes for single ASCII characters and double quotes for a string. Here are some examples:

```
LDAA #'9' ;A = 39H, which is hex number for ASCII '9'
LDAA #'1' ;A = 31H. 31 hex is for ASCII '1'
LDAB #'Z' ;B = 5AH, which is hex number for letter Z
```

To define ASCII strings (more than one character), we use the DC (define constant) directive. We will look at DC usage in future chapters.

Assembler directives

While instructions tell the CPU what to do, directives (also called *pseudo-instructions*) give directions to the assembler. For example, the LDAA and ADDA instructions are commands to the CPU, but EQU, ORG, and END are directives to the assembler. The following sections present some more widely used directives of the HCS12 and explain how they are used.

EQU (equate)

This is used to define a constant value or a fixed address. The EQU directive does not set aside storage for a data item, but associates a constant number with a data or an address label so that when the label appears in the program, its constant will be substituted for the label. The following uses EQU for the counter constant, and then the constant is used to load the A register:

```
COUNT     EQU  $25
          ...       ....
  LDAA #COUNT    ;A = 25H
```

When executing the above instruction "LDAA #COUNT", register A will be loaded with the value 25H. What is the advantage of using EQU? Assume that a constant (a fixed value) is used throughout the program, and the programmer wants to change its value everywhere. By using EQU, the programmer can change it once and have the assembler change all of its occurrences throughout the program, rather than searching the entire program trying to find every occurrence.

SET

This directive is used to define a constant value or a fixed address. In this regard, the SET and EQU directives are identical. The only difference is the value assigned by the SET directive may be reassigned later.

Using EQU for fixed data assignment

To get more practice using EQU to assign fixed data, examine the following:

```
                ;in hexadecimal
DATA1     EQU   $39                    ;8-bit hex data
DATA2     EQU   $96F2                  ;16-bit hex data

                ;in binary
DATA3     EQU   %00110101              ;binary (35 in hex)
DATA4     EQU   %11110111              ;binary (F7 in hex)
DATA5     EQU   %1001011110001000 ;binary (9788 in hex)

                ;in decimal
DATA6     EQU   28          ;decimal numbers (1C in hex)
DATA7     EQU   2500        ;16-bit number in decimal

                ;in ASCII
DATA8     EQU   '2'         ;ASCII characters
DATA9     EQU   'F'         ;ASCII char
```

We can use FCC (Form Constant Character) to allocate ROM memory code locations for fixed data such as ASCII strings. See Chapter 6 for more examples. There is also the DC.W (define constant word) directive, which allows us to allocate 2 bytes of memory locations.

Using EQU for RAM address assignment

Another common usage of EQU is for address assignment of RAM. Examine the following rewrite of an earlier example using EQU:

```
MYCOUNT   EQU   5
MYREG     EQU   $1200
     LDAA #MYCOUNT
     STAA MYREG      ;now location 1200H has value 5
```

The following code will first put value 33H into RAM locations 1200, 1201, and 1202, then add them together using the accumulator:

```
MYVAL     EQU   $33   ;MYVAL = 33H
MYRAM     EQU   $1200;assign RAM address 1200H to MYRAM
     LDAA #MYVAL            ;A = 33H
```

```
STAA MYRAM;move(copy) A contents to location 1200H
STAA MYRAM+1;move(copy) A contents to location 1201H
STAA MYRAM+2;move(copy) A contents to location 1202H
CLRA                 ;clear register A (A=00)
ADDA MYRAM;add A and loc 200H, result in A (A=33H)
ADDA MYRAM+1;add A and loc 201H, result in A (A=66H)
ADDA MYRAM+2;add A and loc 201, result in A (A=99H)
```

This is especially helpful when the address needs to be changed in order to use a different HCS12 chip for a given project. It is much easier to refer to a name than a number when accessing RAM address locations.

The following program will put 99H into locations 800 through 804 of the RAM region:

```
VAL_1      EQU  $99  ;VAL_1 = 99H
MYBUF      EQU  $800 ;assign RAM address 800H to MYBUF

     LDAA #VAL_1    ;A = 99H
     STAA MYBUF;store (copy) A contents to location 800h
     STAA MYBUF+1;store (copy) A contents to location 801h
     STAA MYBUF+2;store (copy) A contents to location 802h
     STAA MYBUF+3;store (copy) A contents to location 803h
     STAA MYBUF+4;store (copy) A contents to location 804h
```

The above method is widely used for defining the general-purpose registers since HCS12 has only registers A and B as general-purpose registers.

ORG (origin) directive

The ORG directive is used to indicate the beginning of the address space. It can be used for both code and data. The number that comes after ORG must be in hex.

END directive

Another important pseudocode is the END directive. This indicates to the assembler the end of the source (asm) file. The END directive is the last line of the HCS12 program, meaning that anything after the END directive in the source code is ignored by the assembler. In the next section we will see how the ORG and END directives are used.

FCB directive

The FCB (form constant byte) directive is used to define byte size variables. We can also use DC.B (define constant byte) and FCB to do the same thing. Chapter 6 shows how we use this directive to place table elements in Flash code ROM. Compare the following:

```
     ORG $850
MYDAT1      DC.B  $99,$55,$AA
     ORG $860
MYDAT2      FCB   $33,$66,$99,$CC
```

FCC directive

The FCC (form constant character) directive is used to define ASCII string characters such as a message displayed on the LCD. Chapter 6 shows how to use this directive.

```
        ORG   $8550
MYMESS      FCC   "Hello World"
```

DC.B directive (define constant byte)

The DC.B directive allocates a byte size value for reading only.

```
        ORG   $1000
MYVALUE     DC.B 5      ;MYVALUE = 5
```

DC.W directive (define constant word)

The DC.W directive allocates a word size value for reading only.

```
        ORG   $1000
MYWORD      DC.W $200 ;MYWORD = 200H
```

DS.B directive (define space byte)

The DS.B directive is used to reserve a number of bytes of RAM for variables in the program.

```
        ORG   $1000
MYARRAY     DS.B 5      ;set aside 5 bytes of RAM
COUNTER     DS.B 1      ;set aside location for counter
```

DS.W directive (define space word)

The DS.W directive is used to reserve a number of words of RAM for variables in the program.

```
        ORG   $1000
MYARRAY     DS.W 5      ;set aside 10 bytes or 5 words of RAM
COUNTER     DS.W 1 ;set aside locations for a 16-bit counter
```

Rules for labels in Assembly language

By choosing label names that are meaningful, a programmer can make a program much easier to read and maintain. There are several rules that names must follow. First, each label name must be unique. The names used for labels in Assembly language programming consist of alphabetic letters in both upper and lower case, the digits 0 through 9, and the special characters question mark (?), period (.), at (@), underline (_), and dollar sign ($). Every assembler has some reserved words that must not be used as labels in the program. Foremost among the reserved words are the mnemonics for the instructions. For example, "ABA" and "STOP" are reserved because they are instruction mnemonics. In addition to the mnemonics there are some other reserved words. Check your assembler for the list of reserved words.

Review Questions

1. Show how to represent decimal 99 in (a) hex, (b) decimal, and (c) binary formats in the HCS12 assembler.
2. What is the advantage in using the EQU directive to define a constant value?
3. Show the hex number value used by the following directives:
   (a) `ASC_DATA EQU '4'` (b) `MY_DATA EQU %00011111`
4. Give the value in register A for the following:
```
MYCOUNT    EQU  15
LDAA #MYCOUNT
```
5. Give the value in RAM location 220H for the following:
```
MYCOUNT    EQU  $95
MYREG      EQU  $220
LDAA       #MYCOUNT
STAA       MYREG
```

SECTION 2.6: INTRODUCTION TO HCS12 ASSEMBLY LANGUAGE

In this section we discuss Assembly language format and define some widely used terminology associated with Assembly language programming. While the CPU can work only in binary, it can do so at a very high speed. It is quite tedious and slow for humans, however, to deal with 0s and 1s in order to program the computer. A program that consists of 0s and 1s is called *machine language.* In the early days of the computer, programmers coded programs in machine language. Although the hexadecimal system was used as a more efficient way to represent binary numbers, the process of working in machine code was still cumbersome for humans. Eventually, Assembly languages were developed, which provided mnemonics for the machine code instructions, plus other features that made programming faster and less prone to error. The term *mnemonic* is frequently used in computer science and engineering literature to refer to codes and abbreviations that are relatively easy to remember. Assembly language programs must be translated into machine code by a program called an *assembler*. Assembly language is referred to as a *low-level language* because it deals directly with the internal structure of the CPU. To program in Assembly language, the programmer must know all the registers of the CPU and the size of each, as well as other details.

Today, one can use many different programming languages, such as BASIC, C, C++, Java, and numerous others. These languages are called *high-level languages* because the programmer does not have to be concerned with the internal details of the CPU. Whereas an *assembler* is used to translate an Assembly language program into machine code (sometimes also called *object code* or opcode for operation code), high-level languages are translated into machine code by a program called a *compiler*. For instance, to write a program in C, one must use a C compiler to translate the program into machine language. Next we look at HCS12 Assembly language format.

Structure of Assembly language

An Assembly language program consists of, among other things, a series of lines of Assembly language instructions. An Assembly language instruction consists of a mnemonic, optionally followed by one or two operands. The operands are the data items being manipulated, and the mnemonics are the commands to the CPU, telling it what to do with those items.

An Assembly language program (see Figure 2-9) is a series of statements, or lines, which are either Assembly language instructions such as ADDA and LDAA, or statements called *directives*. While instructions tell the CPU what to do, directives (also called *pseudo-instructions*) give directions to the assembler. For example, in Figure 2-9 while the LDAA and ADDA instructions are commands to the CPU, ORG and END are directives to the assembler. The ORG directive tells the assembler to place the opcode at memory location 0 while END indicates the end of the source code to the assembler. In other words, one directive is for the start of the program and the other for the end of the program.

```
     ORG  $8000      ;start (origin) at location 8000H
     LDAA #$25       ;load 25H into A (A=25H)
     ADDA #$12       ;add to A value 12H
                     ;now A = A + 12H = 25H+12H= 37H)
     ADDA #14        ;add to A value 14
                     ;now A = A + 14 = 37H+0EH=45H)
     ADDA #%00101000 ;add to A value 28H(101000 bin)
                     ;now A = A + 28H = 45H+28H=6DH)
HERE JMP HERE        ;stay in this loop
     END             ;end of asm source file
```

Figure 2-9. Sample of an Assembly Language Program

An Assembly language instruction consists of four fields:

```
[ label]   mnemonic  [ operands]   [ ;comment]
```

Brackets indicate that a field is optional and not all lines have them. Brackets should not be typed in. Regarding the above format, the following points should be noted:

1. The label field allows the program to refer to a line of code by name. The label field cannot exceed a certain number of characters. Check your assembler for the rule.
2. The Assembly language mnemonic (instruction) and operand(s) fields together perform the real work of the program and accomplish the tasks for which the program was written. In Assembly language statements such as

   ```
   LDAA #$55
   ADDA #$67
   ```

   ADDA and LDAA are the mnemonics that produce opcodes; the "$55" and "$67" are the operands. Instead of a mnemonic and an operand, these two fields could contain assembler pseudo-instructions, or directives. Remember that directives do not generate any machine code (opcode) and are used only by the assembler, as opposed to instructions that are translated into machine code (opcode) for the CPU to execute. In Figure 2-9 the ORG (origin) and END commands are examples of directives. More of these pseudo-instructions are discussed in future chapters.

3. The comment field begins with a semicolon comment indicator ";". Comments may be at the end of a line or on a line by themselves. The assembler ignores comments, but they are indispensable to programmers. Although comments are optional, they should be used to describe the program in a way that makes it easier for someone else to read and understand.
4. Notice the label "`HERE`" in the label field in Figure 2-9. In the JMP the HCS12 is told to stay in this loop indefinitely. If your system has a monitor program you do not need this line and it should be deleted from your program. In Section 2.7 we will see how to create a ready-to-run program.

Review Questions

1. What is the purpose of pseudo-instructions?
2. _____________ are translated by the assembler into machine code, whereas _________________ are not.
3. True or false. Assembly language is a high-level language.
4. Which of the following instructions produces opcode? List all that do.
   (a) LDAA #$25 (b) ADDA #12 (c) ORG $2000 (d) JMP HERE
5. Pseudo-instructions are also called ___________.
6. True or false. Assembler directives are not used by the CPU itself. They are simply a guide to the assembler.
7. In Question 4, which instuction is an assembler directive?

SECTION 2.7: ASSEMBLING AND LINKING AN HCS12 PROGRAM

Now that the basic form of an Assembly language program has been given, the next question is: How it is created, assembled, and made ready to run? The steps to create an executable Assembly language program (see Figure 2-10) are as follows:

1. First we use a text editor to type in a program similar to Figure 2-11. In the case of the HCS12 microcontrollers, we can use a number of assemblers available for free. They come with a text editor, assembler, linker, and much more, all in one software package. Many editors or word processors are also available that can be used to create or edit the program. A widely used editor is the Notepad in Windows, which comes with all Microsoft operating systems. Notice that the editor must be able to produce an ASCII file. For assemblers, the file names follow the usual DOS conventions, but the source file has the extension "asm". The "asm" extension for the source file is used by an assembler in the next step.

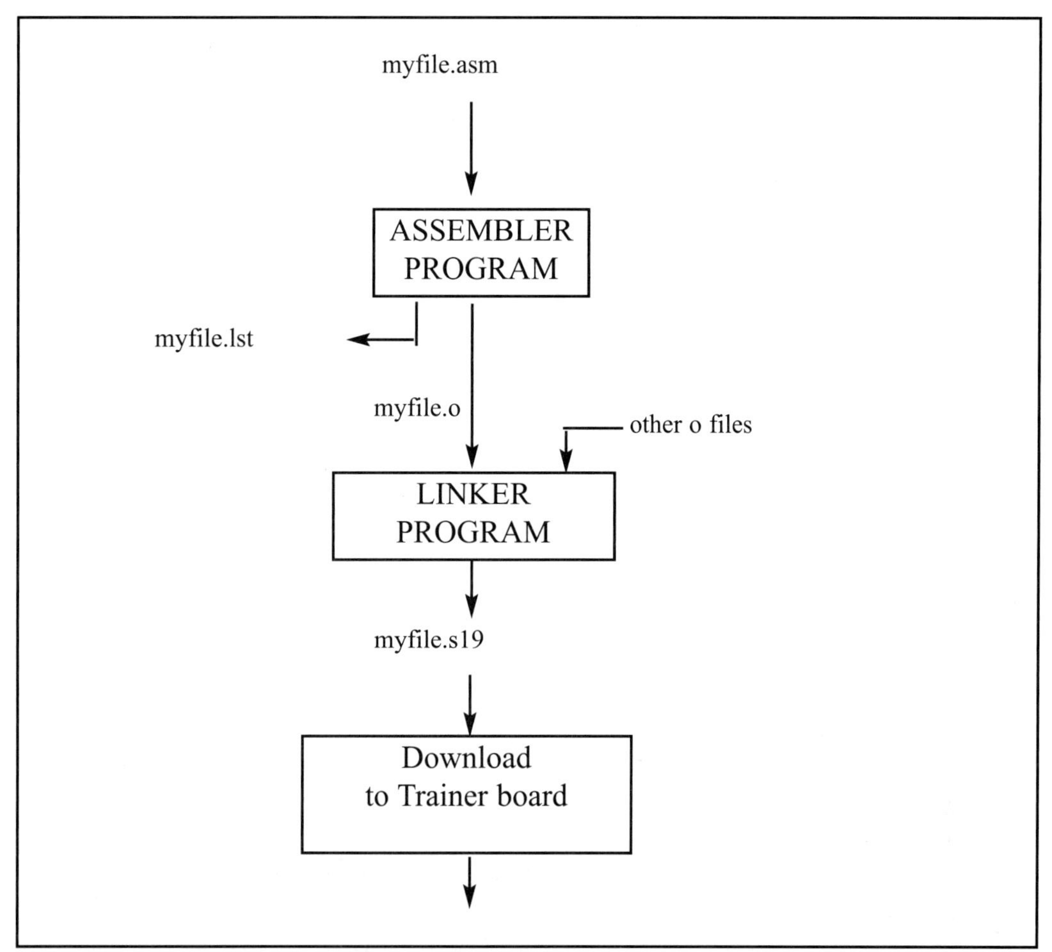

Figure 2-10. Steps to Create a Program

2. The "asm" source file containing the program code created in step 1 is fed to the HCS12 assembler. The assembler converts the instructions into machine code. The assembler will produce an object file. The extension for the object file is "o". The extension for the list file, which contains any syntax errors and their line numbers, is "lst". The lst file can be viewed with any text editor.
3. Assemblers require a third step called *linking*. The link program takes one or more object files and produces a hex file and map. The hex file has the extension "s19". After a successful link, the hex file is ready to be burned into the HCS12's program ROM or downloaded into the HCS12 Trainers. See Chapter 8 for more details.

Many of the Windows-based IDE assemblers combine steps 2 and 3 into one step after the program has been typed.

More about asm and object files

The asm file is also called the *source* file and must have the "asm" extension. As mentioned earlier, this file can be created with a text editor such as Windows Notepad. Many assemblers come with a text editor. The assembler converts the asm file's Assembly language instructions into machine language and provides the o (object) file. The HCS12 assembler produces the object and lst files. The object file, as mentioned earlier, has an "o" as its extension. In modular pro-

```
       ;HCS12 Assembly Language Program To Add Some Data.
       ;store SUM in RAM location 210H.

SUM    EQU    $210           ;RAM loc 210H for SUM

       ORG    $8000          ;start at address 8000H
       LDAA #$25             ;A = 25H
       ADDA #$34             ;add 34H to A (A=59H)
       ADDA #$11             ;add 11H to A (A= 6AH)
       ADDA #18              ;A = A + 12H = 7CH
       ADDA #%00011100       ;A = A + 1CH = 98H
       STAA SUM              ;save the SUM in loc 210H
HERE   JMP HERE              ;stay here forever
       END                   ;end of asm source file
```

Figure 2-11. Sample of an Assembly Language Program

gramming, we use the linker to link many object files together to create a ready-to-burn hex file as we will see in Chapter 6. But before we can link a program to create a ready-to-run program, we must make sure that it is error free. The HCS12 assembler provides us the list file with the extension of "lst", which is the file we examine to see the nature of syntax errors. The linker will not link the program until all the syntax errors are fixed. We can print the list file or use Notepad to examine the nature of the errors. Then we go back to the asm file and correct all the errors before we assemble it again. A sample of an lst file is shown in Figure 2-12.

```
as12, an absolute assembler for Motorola MCU's, version 1.2h

;HCS12 Assembly Language Program To Add Some Data.
;store SUM in RAM location 210H.

0210                SUM   EQU $210           ;RAM loc 210H for SUM
8000                      ORG $8000          ;start at address 8000H
8000 86 25                LDAA #$25          ;A = 25H
8002 8b 34                ADDA #$34          ;add 34H to A (A=59H)
8004 8b 11                ADDA #$11          ;add 11H to A (A=6AH)
8006 8b 12                ADDA #18           ;A = A + 12H = 7CH
8008 8b 1c                ADDA #%00011100     ;A = A + 1CH = 98H
800a 7a 02 10             STAA SUM            ;save the SUM in loc 210H
800d 06 80 0d       HERE  JMP HERE            ;stay here forever
                          END
```

Figure 2-12. List File for Figure 2-11

Placing code in program ROM

To get a better understanding of the role of the program counter in fetching and executing a program, we examine the action of the program counter as each instruction is fetched and executed. First, we examine once more the list file of the sample program and show how the code is placed in the ROM of the HCS12 chip. See Table 2-7. As you can see, the opcode and operand for each instruction are listed on the left side of the list file.

Table 2-7: Machine Code for Figure 2-11

ROM Address	Machine Language	Assembly Language
8000	8625	LDAA #$25
8002	8B34	ADDA #$34
8004	8B11	ADDA #$11
8006	8B12	ADDA #18
8008	8B1C	ADDA #%00011100
800A	7A0210	STAA SUM
800D	06800D	HERE JMP HERE

After the program is downloaded (burned) into flash ROM of a HCS12 chip, the opcode and operand are placed in ROM memory locations starting at 8000 as shown in Table 2-8.

Table 2-8: ROM Contents

Address	Code
8000	86
8001	25
8002	8B
8003	34
8004	8B
8005	11
8006	8B
8007	12
8008	8B
8009	1C
800A	7A
800B	02
800C	10
800D	06
800E	80
800F	0D

The list shows that address 8000 contains 86, which is the opcode for loading an immediate value into register A, and address 8001 contains the operand (in this case $25) to be moved to register A. Therefore, the instruction "LDAA #$25" has a machine code of "8625", where 86 is the opcode and 25 is the operand. Similarly, the machine code "8B34" is located in memory locations 8002 and 8003 and represents the opcode and the operand for the instruction "ADDA #$34". In the same way, machine code "8B11" is located in memory locations 8004 and 8005 and represents the opcode and the operand for the instruction "ADDA #$11". The memory location 8006 has the opcode of 8B, which is the opcode for the instruction "ADDA $18" and memory location 8007 has the content 12, which is the operand for the decimal 18 in the "ADDA #18" instruction. The opcode for instruction "ADDA %00011100" is located at address 8008 and the operand 1C at address 8009. The opcode for instruction "STAA SUM" is located at address 800A and its operand, 0210, at addresses 800B and 800C. The opcode for "JMP HERE" and its target address are located in locations 800D, 800E, and 800F. While most of the instructions in this program are 2-byte instructions, the STAA and JMP instructions are 3-byte instructions.

Executing a program byte by byte

Assuming that the above program is downloaded into the ROM of an HCS12 chip, the following is a step-by-step description of the actions taken by the HCS12:

1. After the user sets PC (program counter) to $8000 the HCS12 starts to fetch the first opcode from location $8000. In the case of the above program the first opcode is $86, which is the code for loading an operand into register A. Upon executing the opcode, the CPU places the value of $25 in A. Now one instruction is finished. Then the program counter is incremented to point to $8002 (PC = 8002), which contains opcode $8B, the opcode for the instruction "ADDA #$34".

2. Upon executing the opcode $8B, the value $34 is added to A. Then the program counter is incremented to $8004.
3. ROM location $8004 has the opcode for instruction "ADDA #$11". This instruction is executed and now PC = $8006. Notice that all the above instructions are 2-byte instructions; that is, each one takes two memory locations.
4. This process goes on until all the instructions up to "STAA SUM" are fetched and executed. Notice that the STAA instruction is a 3-byte instruction.
5. Now PC = $800D points to the next instruction, which is "JMP HERE". This is a 3-byte instruction. It takes ROM addresses of $800D, $800E, and $800F. After the execution of this instruction, PC = $800D. This keeps the program in an infinite loop. If your HCS12 Trainer has a monitor program you do not have to use the JMP instruction, and the program will go back to the monitor program. The fact that the program counter points at the next instruction to be executed explains why some microprocessors (notably the x86) call the program counter the *instruction pointer*.

Review Questions

1. True or false. The HCS12 assembler/editor and Windows Notepad text editor all produce an ASCII file.
2. True or false. The extension for the source file is "asm".
3. Which of the following files can be produced by the text editor?
   (a) myprog.asm (b) myprog.obj (c) myprog.s19 (d) myprog.lst
4. Which of the following files is produced by an assembler?
   (a) myprog.asm (b) myprog.obj (c) myprog.s19 (d) myprog.lst
5. Which of the following files lists syntax errors?
   (a) myprog.asm (b) myprog.obj (c) myprog.s19 (d) myprog.lst

SECTION 2.8: STACK AND DATA TRANSFER INSTRUCTIONS

The HCS12 microcontroller has many bytes of RAM for scratch pad and stack. In this section we discuss the use of RAM for stack.

Stack in the HCS12

The stack is a section of RAM used by the CPU to store information temporarily. This information could be data or an address. The CPU needs this storage area since there are only a limited number of registers.

How stacks are accessed in the HCS12

Since the stack is a section of RAM, there must be registers inside the CPU to point to it. The register used to access the stack is called the *SP* (stack pointer) register. See Figure 2-11. The stack pointer in the HCS12 is 16 bits wide, which means that it can take values of 0000 to $FFFF. The operation of storing a CPU register in the stack is called a *PUSH*, and operation of pulling the contents off the stack back into a CPU register is called a *PULL*. In other words, a register is pushed onto the stack to save it and pulled off the stack to retrieve it. The job of SP is very critical when push and pull actions are performed. To see how the stack works, let's look at the PUSH and PULL instructions.

Pushing onto the stack

In the HCS12 the stack pointer (SP) points to the last-used location of the stack. As we push data onto the stack, the stack pointer (SP) is decremented. Notice that this is the same as many other microprocessors, notably x86 processors in which the SP register is also decremented when data is pushed onto the stack. Examining Example 2-9, we see that as PSHA is executed, the contents of the register are saved on the stack and SP is decremented by 1. Notice that for the PSHY instruction, SP is decremented by 2 since Y is a 16-bit register.

Example 2-9

Show the stack and stack pointer for the following.

```
LDS   #$1200      ;SP=$1200
LDY   #$F395      ;Y=$F395
LDAA #$25         ;A=$25
LDAB #$12         ;B=$12
PSHA              ;Push Reg. A onto stack
PSHB              ;Push Reg. B onto stack
PSHY              ;Push Reg. Y onto stack
```

Solution:

		After PSHA		After PSHB		After PSHY	
11FC		11FC		11FC		11FC	F3
11FD		11FD		11FD		11FD	95
11FE		11FE		11FE	12	11FE	12
11FF		11FF	25	11FF	25	11FF	25
Start SP = 1200		SP = 11FF		SP = 11FE		SP = 11FC	

Pulling from the stack

Pulling the contents of the stack back into a given register is the opposite process of pushing. With every pull, the top byte of the stack is copied to the register specified by the instruction and the stack pointer is incremented. Examples 2-10 and 2-11 demonstrate the PULL instructions.

Stack and scratch pad conflict

Recall from our earlier discussion that the stack pointer register points to the last-used location of the stack. As data is pushed onto the stack, SP is decremented. Conversely, it is incremented as data is pulled off the stack into the registers. The reason that SP is decremented after the push is to make sure that the stack is growing toward the beginning of RAM address, from upper addresses to lower addresses. This decrementing of the stack pointer for push instructions also ensures that the stack will not reach the last location of RAM, and consequently run out of space for the stack. When programming the HCS12, the programmer sets the SP to the last location of RAM before stacks are used. See Figure 2-13.

Example 2-10

Examining the stack below, show the contents of the registers and SP after execution of the following instructions. All values are in hex.

```
PULA        ;PULL stack into A
PULB        ;PULL stack into B
PULX
```

Solution:

Address	Start	After PULA	After PULB	After PULX
12FC	54			
12FD	F9	F9		
12FE	76	76	76	
12FF	6C	6C	6C	
	Start SP = 12FC	SP = 12FD and A = 54	SP = 12FE and B = F9	SP = 1300 and X = 766C

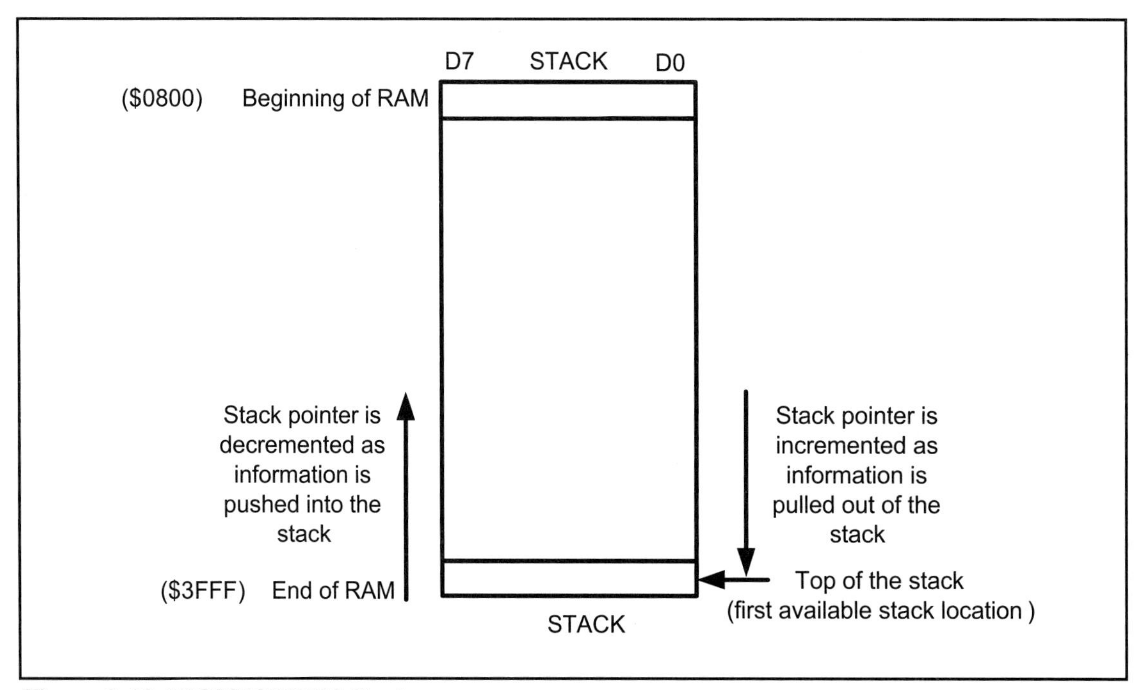

Figure 2-13. MC9S12DP512 Stack

Example 2-11

Show the stack and stack pointer for the following instructions being used for HCS12.

```
LDS   #$4000 ;make RAM location $3FFF,first stack Loc
LDAA #$25
LDAB #$12
ABA
PSHA
PSHB
ABA
PSHA
```

Solution:

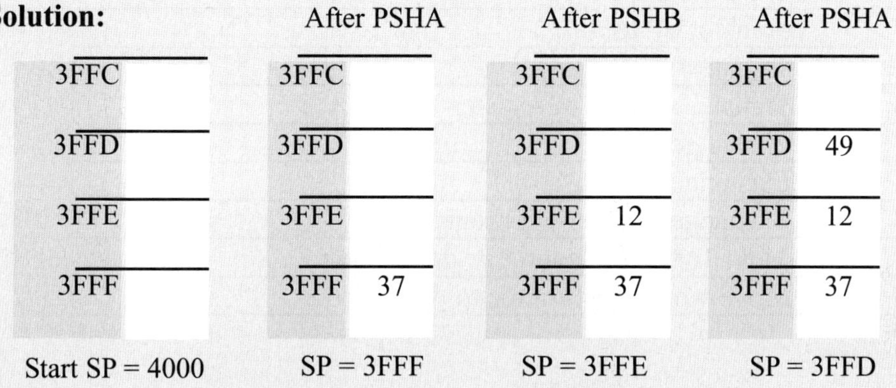

		After PSHA	After PSHB	After PSHA
3FFC				
3FFD				49
3FFE			12	12
3FFF		37	37	37
	Start SP = 4000	SP = 3FFF	SP = 3FFE	SP = 3FFD

Call operation and the stack

In addition to using the stack to save registers, the CPU also uses the stack to save the address of the instruction just below the call instructions such as JSR and CALL. This is how the CPU knows where to resume when it returns from the called subroutine. More information on this will be given in Chapter 3 when we discuss the JSR and CALL instructions. Table 2-9 shows instructions dealing with the stack. Chapter 3 will show how to use them.

Table 2-9: Stack Instructions

Instruction	Operation
PSHA	Push A
PSHB	Push B
PSHC	Push C
PSHD	Push D
PSHX	Push X
PSHY	Push Y
PULA	Pull A
PULB	Pull B
PULC	Pull D
PULD	Pull SP
PULX	Pull X
PULY	Pull Y

Data transfer instructions

There are some useful data transfer and exchange instructions in HCS12. See Table 2-10. These instructions allow us to move data around the CPU and from memory location to memory location. We use many of them in future chapters.

Table 2-10: Data Transfer and Exchange Instructions

Instruction	Operation
TAB	Transfer A to B
TBA	Transfer B to A
TAP	Transfer A to CCR
TPA	Transfer CCR to A
TSX	Transfer SP to X
TXS	Transfer X to SP
TSY	Transfer SP to Y
TYS	Transfer Y to SP
TFR	Transfer register to register
EGDX	Exchange D with X
EGDY	Exchange D with Y
EXG	Exchange register with register
MOVB	Move byte from source to destination
MOVW	Move word from source to destination

Coming from other microprocessors to the HCS12

If you have a background in programming other microprocessors/microcontrollers, making the transition from these devices to the HCS12 can be easy. The RAM area of the HCS12 can be viewed as a large number of registers, except they do not have names as in other processors. We can assign any register names we want, however, as long as we do not use any of the reserved names used by SFRs (X, Y, PC, SP, and so on). Here is an example if we are used to the 8051 or some other RISC processor:

```
R0   EQU  $800
R1   EQU  $801
R2   EQU  $802
R3   EQU  $803

...  ....      ..

     LDAA #$55
     STAA R0
     LDAA #$99
     STAA R1
     LDAA #$72
     STAA R3
```

Or look at the following for the x86:

```
BL    EQU   $800
BH    EQU   $801
CL    EQU   $802
CH    EQU   $803
DL    EQU   $804

...   ....        ..

      LDAA #$55
      STAA BL
      LDAA #$99
      STAA BH
      LDAA #$72
```

Viewing registers and memory with a simulator

Many assemblers and C compilers come with a simulator. Simulators allow us to view the contents of registers and memory after executing each instruction (single-stepping). We strongly recommend that you use a simulator to single-step some of the programs in this chapter and future chapters. Single-stepping a program with a simulator gives us a deeper understanding of microcontroller architecture, in addition to the fact that we can use it to find errors in our programs. Figure 2-14 shows a screen-shot of CodeWarrior from Freescale and Figure 2-15 shows the CodeWarrior Assembly language programming shell.

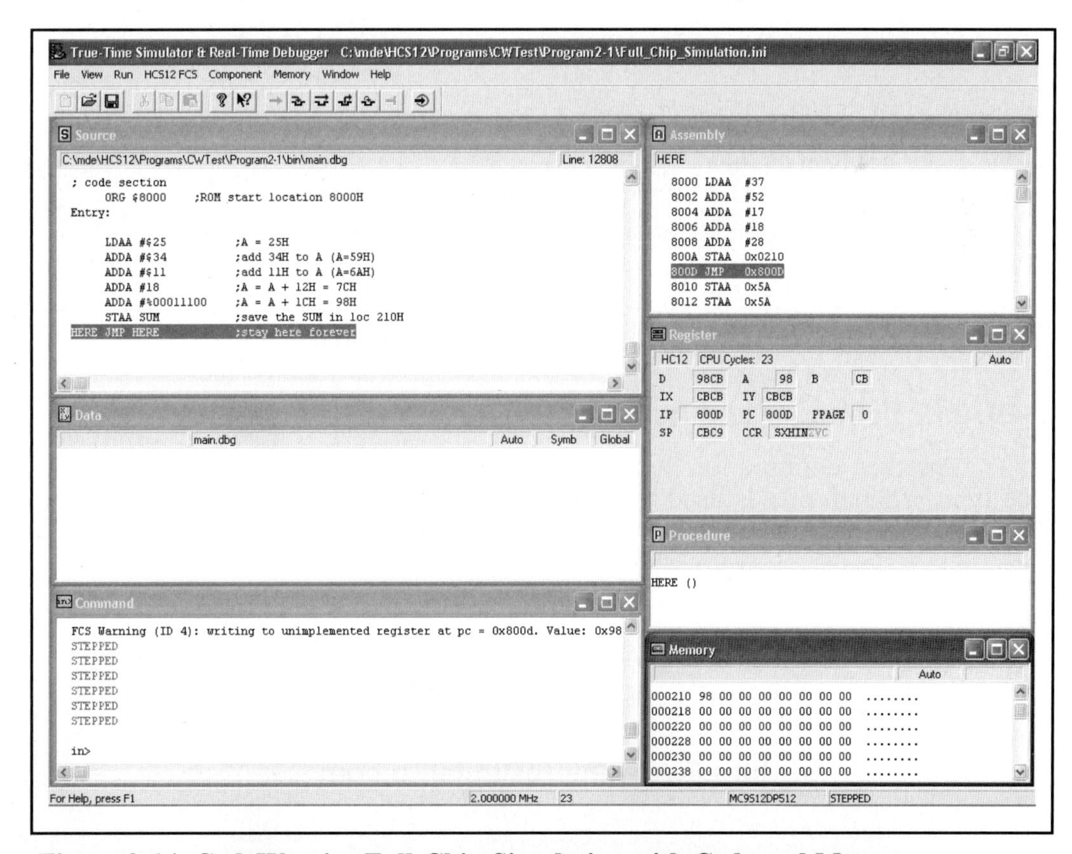

Figure 2-14. CodeWarrior Full-Chip Simulation with Code and Memory

```
     XDEF Entry           ; export 'Entry' symbol
     ABSENTRY Entry       ; for absolute assembly
;include derivative specific macros
     INCLUDE 'mc9s12dp512.inc'
ROMStart     EQU  $4000
;absolute address to place my code/constant data
    ...
    ...

;variable/data section
     ORG RAMStart
;Insert here your data definition.
Counter      DS.W 1
    ...
    ...

; code section
     ORG   ROMStart
Entry:
     LDS   #RAMEnd+1 ; initialize the stack pointer

    ...
    ; your code here
    ...

HERE:        BRA HERE         ; stay here

;*************************************
;*        Interrupt Vectors
;*************************************
             ORG   $FFFE
             DC.W  Entry              ; Reset Vector
```

Figure 2-15. CodeWarrior Assembly Language Progam Shell

D-BUG12 Debugger and AsmIDE

Another tool used for HCS12 is the D-BUG12 debugger. It is a command-line type program and is used with the AsmIDE and MiniIDE assemblers. See Appendix B for more.

See the following website for a tutorial on using CodeWarrior, AsmIDE, MiniIDE, and D-BUG12:

http://www.MicroDigitalEd.com

Review Questions

1. What is the size of the SP register?
2. With each PSHA instruction, the stack pointer register, SP, is _______________ (incremented, decremented) by 1.
3. With each PULA instruction, the SP is __________ (incremented, decremented) by 1.
4. True or false. We use EEPROM for stack.
5. True or false. On power-up, we set the address of the stack as the first location of RAM.

PROBLEMS

SECTION 2.1: INSIDE THE HCS12

1. HCS12 is a(n) _____-bit microcontroller.
2. Register A is _____ bits wide.
3. The immediate value in "LDAA #value" is _____ bits wide.
4. The largest number that can be loaded into A is _____ in hex.
5. To load A with the value 65H, the pound sign is ______________ (not necessary, optional, necessary) in the instruction LDAA "#$65".
6. What is the result of the following code and where is it kept?
```
LDAA #$15
ADDA #$13
```
7. Which of the following instructions is (are) illegal?
   (a) `LDAA #500` (b) `LDAA #50` (c) `LDAA #00`
   (d) `LDAA #$255` (e) `LDAA #$25` (f) `LDAA #$F5`
   (g) `ADDA mybyte,#$50`
8. Which of the following instructions is (are) illegal?
   (a) `ADDA #$300` (b) `ADDA #$50` (c) `ADDA #$500`
   (d) `ADDA #$255` (e) `ADDA #12` (f) `ADDA #$F5`
   (g) `ADDA #$25`
9. What is the result of the following code and where is it kept?
```
LDAA #$25
ADDA #$1F
```
10. What is the result of the following code and where is it kept?
```
LDAA #$15
ADDA #$EA
```
11. The largest number that K can take for the instruction "`ADDA #$K`" is ____ in hex.
12. True or false. We have many A and B registers in the HCS12.

SECTION 2.2: THE HCS12 MEMORY MAP

13. HCS12 has _____________ bytes of address space.
14. True or false. Register space in HCS12 starts at address 0000.
15. True or false. The registers are part of the RAM memory space.

16. True or false. The general-purpose registers such as A, B, and X are not part of the register space address.
17. True or false. All members of the HCS12 family have the same amount of RAM.
18. True or false. If we add together the RAM and ROM amounts of HCS12 we should get the total space for the 64K bytes.
19. Give the RAM size for the following HCS12 chips:
    (a) MC9S12DP512 (b) MC68HC912B32
20. What is the difference between the EEPROM and data RAM space in the HCS12?
21. Can we have a HCS12 chip with no EEPROM?
22. Can we have a HCS12 chip with no RAM?
23. In HCS9xxxx, the program ROM is of _______type memory.
24. Give the amount of on-chip flash memory in the MC68HC912B32.
25. What is the starting address of the registers section in HCS12?
26. Give the address map for MC9S12DP512.

SECTION 2.3: THE HCS12 ADDRESSING MODES

27. What is the starting address for the RAM section of MC9S12DP512?
28. Show a simple code to load values 30H and 97H into locations 805H and 806H, respectively.
29. Show a simple code to load value 55H into locations 300H through 308H.
30. Show a simple code to load value 5FH into Port B.
31. Which of the following are an invalid use of immediate addressing mode?
    (a) LDAA #$24 (b) LDAA $30 (c) LDAA #$60
32. Identify the addressing mode for each of the following:
    (a) STAA PORTB (b) LDAA #$50 (c) STAA $800
    (d) LDAA #0 (e) CLRA (f) ABA
33. Indicate the size of instruction for each of the following:
    (a) STAA PORTB (b) LDAA #$50 (c) STAA $200
    (d) LDAA #0 (e) CLRA (f) ABA
34. What is the starting address for registers (SFRs)?
35. In accessing the SFR registers, we must use ______ addressing mode.
36. What addressing mode does the following instruction use? "`LDAA #$F0`"
37. What addressing mode does the following instruction use? "`STAA PORTA`"
38. What addressing mode does the following instruction use? "`LDAA $805`"
39. "`CLRA $800`" is a(n) _______ (valid, invalid) instruction.
40. The byte addresses assigned to the 256 bytes of lower data RAM are _____ to _____.
41. The indexed addressing mode uses registers _____.
42. True or false. The indexed addressing mode uses the X and Y registers as pointers to memory locations.

SECTION 2.4: HCS12 CONDITION CODE REGISTER

43. The condition register is a(n) ______ -bit register.

44. Which bits of the condition register are used for the C and H flag bits, respectively?
45. Which bits of the condition register are used for the V and N flag bits, respectively?
46. In the "ADDA #imm_val" instruction, when is C raised?
47. In the "ADDA #imm_val" instruction, when is H raised?
48. What is the status of the C and Z flags after the following code?

```
LDAA #$FF
ADDA #1
```

49. Find the C flag value after each of the following codes:

```
(a) LDAA #$54        (b) LDAA #00        (c) LDAA #$FF
    ADDA #$0C4           ADDA #$FF           ADDA #05
```

50. Write a simple program in which the value 55H is added 3 times.

SECTION 2.5: HCS12 DATA FORMAT AND DIRECTIVES

51. State the value (in hex) used for each of the following data:

```
MYDAT_1    EQU  55
MYDAT_2    EQU  98
MYDAT_3    EQU  'G'
MYDAT_4    EQU  $50
MYDAT_5    EQU  200
MYDAT_6    EQU  'A'
MYDAT_7    EQU  $AA
MYDAT_8    EQU  255
MYDAT_9    EQU  %10010000
MYDAT_10   EQU  %01111110
MYDAT_11   EQU  10
MYDAT_12   EQU  15
```

52. State the value (in hex) for each of the following data:

```
DAT_1      EQU  22
DAT_2      EQU  $56
DAT_3      EQU  %10011001
DAT_4      EQU  32
DAT_5      EQU  $F6
DAT_6      EQU  %1111011
```

53. Show a simple code to (a) load value 11H into locations $1300–$1305, and (b) add them together and place the result in A as the values are added. Use EQU to assign the names R0–R5 to locations $1300–$1305.

SECTION 2.6: INTRODUCTION TO HCS12 ASSEMBLY LANGUAGE
and
SECTION 2.7: ASSEMBLING AND LINKING AN HCS12 PROGRAM

54. Assembly language is a________ (low, high)-level language while C is a ________ (low, high)-level language.

55. Of C and Assembly language, which is more efficient in terms of code generation (i.e., the amount of ROM space it uses)?
56. Which program produces the o file?
57. True or false. The source file has the extension "asm".
58. Which file provides the listing of error messages?
59. True or false. The source code file can be a non-ASCII file.
60. True or false. Every source file must have ORG and END directives.
61. Do the ORG and END directives produce opcodes?
62. Why are the ORG and END directives also called pseudocode?
63. True or false. The ORG and END directives appear in the "lst" file.
64. True or false. The linker produces the file with the extension "asm".
65. True or false. The linker produces the file with the extension "s19".
66. The file with the _____ extension is downloaded into HCS12 ROM.
67. Give three file extensions produced by HCS12 you have used for this chapter.

SECTION 2.8: STACK AND DATA TRANSFER INSTRUCTIONS

68. In the HCS12, the stack pointer (SP) points to (the last used, next available) location of the stack.
69. With each PSHY instruction, the stack pointer register, SP, is _______________ (incremented, decremented) by ______.
70. With each PULY instruction, the SP is __________ (incremented, decremented) by _____.
71. True or false. We use a section of flash memory for stack.
72. True or false. On power-up, we set the highest address of RAM as the first location of the stack.

ANSWERS TO REVIEW QUESTIONS

SECTION 2.1: INSIDE THE HCS12

1. LDAA #$34
   LDAB #$3F
   ABA
2. LDAA #$16
   ADDA #$CD
3. False
4. FF hex and 255 in decimal
5. FFFF hex and 65,535 in decimal

SECTION 2.2: THE HCS12 MEMORY MAP

1. True
2. True
3. True
4. False. There is plenty of space for both.
5. 8
6. 0000
7. False

SECTION 2.3: THE HCS12 ADDRESSING MODES

1. No
2. LDAB #$99
3. Too large for register A
4. True
5. No
6. LDAA #$16
   ADDA #$FD
   STAA $200
7. False
8. True
9. X and Y
10. Both are 16-bit.

SECTION 2.4: HCS12 CONDITION CONTROL REGISTER

1. Status register
2. 8 bits
3. True
4.

Hex	binary	
9F	1001 1111	
+ 61	+ 0110 0001	
100	10000 0000	This leads to C = 1, H = 1, and Z = 1.

5.

Hex	binary	
82	1000 0010	
+ 22	+ 0010 0010	
A4	1010 0100	This leads to C = 0, H = 0, and Z = 0.

6.

Hex	binary	
67	0110 0111	
+ 99	+ 1001 1001	
100	10000 0000	This leads to C = 1, H = 1, and Z = 1.

SECTION 2.5: HCS12 DATA FORMAT AND DIRECTIVES

1. DATA1 EQU $63
   DATA2 EQU 99
   DATA3 EQU %10011001
2. If the value is to be changed later, it can be done once in one place instead of at every occurrence.
3. (a) 34H (b) 1FH
4. A = 0FH
5. Value of location $220 = ($95)

SECTION 2.6: INTRODUCTION TO HCS12 ASSEMBLY LANGUAGE

1. The real work is performed by instructions such as MOV and ADD. Pseudo-instructions, also called *assembly directives*, instruct the assembler in doing its job.
2. The instruction mnemonics, pseudo-instructions
3. False
4. All except (c)
5. Assembler directives

6. True
7. (c)

SECTION 2.7: ASSEMBLING AND LINKING AN HCS12 PROGRAM

1. True
2. True
3. (a)
4. (b) through (d)
5. (d)

SECTION 2.8: STACK AND DATA TRANSFER INSTRUCTIONS

1. 16 bits
2. Decremented
3. Incremented
4. False
5. False

CHAPTER 3

BRANCH, CALL, AND TIME DELAY LOOP

OBJECTIVES

Upon completion of this chapter, you will be able to:

>> Code HCS12 Assembly language instructions to create loops
>> Code HCS12 Assembly language conditional branch instructions
>> Explain conditions that determine each conditional branch instruction
>> Code instructions for unconditional jumps
>> Calculate target addresses for conditional branch instructions
>> Code HCS12 subroutines
>> Describe the use of stack calling subroutines
>> Discuss pipelining and instruction queue in the HCS12
>> Discuss the relative and indexed addressing modes
>> Write a loop program using indexed addressing mode

SECTION 3.1: BRANCH INSTRUCTIONS AND LOOPING

In this section we first discuss how to perform a looping action in HCS12 and then explain the branch (jump) instructions, both conditional and unconditional.

Looping in HCS12

Repeating a sequence of instructions or an operation a certain number of times is called a *loop*. The loop is one of most widely used programming techniques. In the HCS12, there are several ways to repeat an operation many times. One way is to repeat the operation over and over until it is finished, as shown below:

```
CLRA          ;A = 0
ADDA #3       ;add value 3 to A
ADDA #3       ;add value 3 to A(A = 6)
ADDA #3       ;add value 3 to A(A = 9)
ADDA #3       ;add value 3 to A(A = 0CH)
ADDA #3       ;add value 3 to A(A = 0FH)
```

In the above program, we add 3 to register A total of 5 times. That makes $5 \times 3 = 15 = 0FH$. One problem with the above program is that too much code space would be needed to increase the number of repetitions to 50 or 100. A much better way is to use a loop. There are two ways to do a loop in HCS12. Next, we describe each method.

Using instruction BNE for looping

The BNE (branch if not equal) instruction is widely used for looping in the HCS12 family. It uses the zero flag in the status register. The BNE instruction is used as follows:

```
      LDAB #250   ;load the counter
BACK  .........   ;start of the loop
      .........   ;body of the loop
      .........   ;body of the loop
      DECB        ;decrement B, Z = 1 if B reg = 0
      BNE BACK    ;branch to BACK if Z = 0
```

In the last two instructions, the B register is decremented; if it is not zero, it branches (jumps) back to the target address referred to by the label. Prior to the start of the loop, the B reg is loaded with the counter value for the number of repetitions. Notice that the BNE instruction refers to the Z flag of the status register affected by the previous instruction, DECB. This is shown in Example 3-1. Figure 3-1 shows the flowchart for the DECB instruction. Study the flowchart structure in Appendix D to get familiar with the symbols. The flowchart is a widely used method to represent a sequence of actions pictorially. Its usage for program design is recommended very strongly.

Example 3-1

Write a program to (a) clear register A, then (b) add 3 to A ten times.
Use the zero flag and BNE with DECB.

Solution:

```
;this program adds value 3 to register A ten times

COUNT EQU 10      ;use 10 for counter

      LDAB #COUNT         ;load the counter
      CLRA                ;A = 0
AGAIN ADDA #3             ;add 03 to A (A = sum)
      DECB                ;decrement counter
      BNE AGAIN           ;repeat until COUNT = 0
      STAA $200           ;send sum to LOC 200H
```

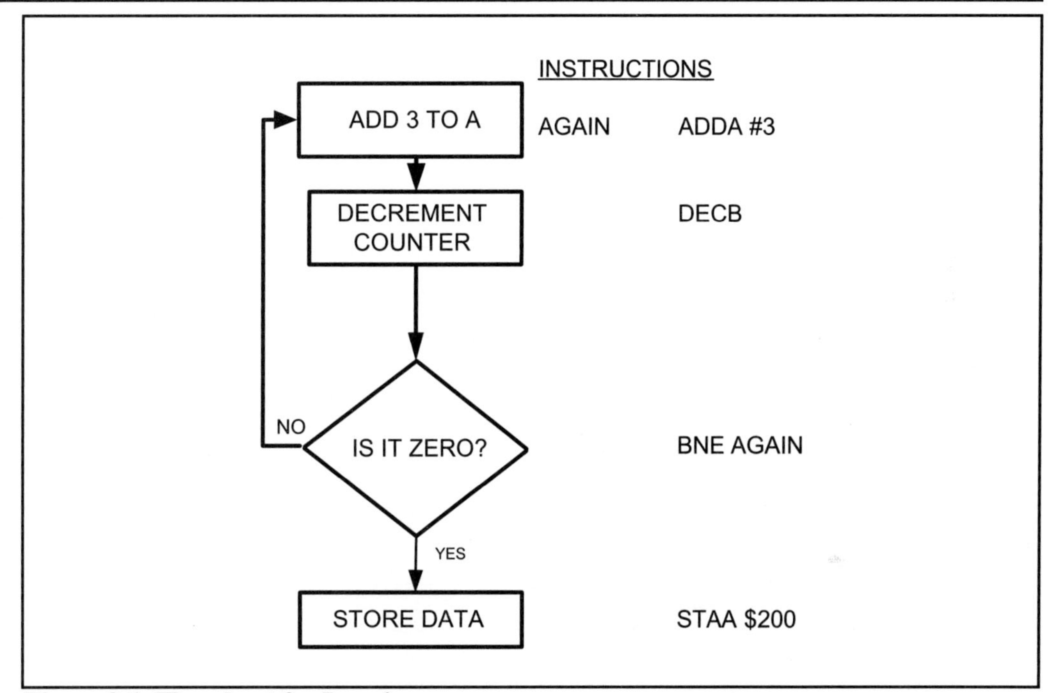

Figure 3-1. Flowchart for Looping

Using RAM locations for the count register

Sometimes we need many registers for counters. In the program in Example 3-2, RAM location $1225 is used as a counter instead of register B. The counter is first set to 15. In each iteration, the DEC instruction decrements the RAM location and sets the flag bits accordingly. If the RAM location is not zero ($Z \neq 0$), it jumps to the target address associated with the label "AGAIN". This looping action continues until the RAM location COUNT becomes zero. After variable COUNT becomes zero ($Z = 0$), the counter falls through the loop and executes the instruction immediately below it, in this case "STAA MYRAM". See Figure 3-2.

Notice, in the "DEC COUNT" instruction, that RAM location $1225 is used as a register to hold the count as it decrements instead of the B register. Since we have limited number of general-purpose registers we use RAM locations as registers.

Example 3-2

Write a program to (a) clear register A, and (b) add 3 to register A fifteen times and place the result in RAM location $1250. Use a RAM location for holding the counter.

Solution:

```
;this program adds value 3 to A fifteen times
COUNT   EQU $1225          ;use loc 1225H for counter
MYRAM   EQU $1250          ;use loc 1250H for sum
        LDAA  #15          ;A = 15 (decimal) for counter
        STAA  COUNT        ;load the counter
        CLRA               ;A = 0
AGAIN   ADDA  #3           ;add 03 to A REG (A = sum)
        DEC   COUNT        ;decrement counter
        BNE   AGAIN        ;repeat until count becomes 0
        STAA  MYRAM        ;store sum
```

Notice that the DEC instruction will decrement the counter (location $1225), which has 15 in it. It becomes 14. Because it is not zero, it will execute the "BNE AGAIN" instruction. The "BNE AGAIN" goes back to the start of the loop. It goes on like that until the counter becomes zero. Upon the counter becoming zero, it goes to the STAA, which gets it out of the loop, and stores the result.

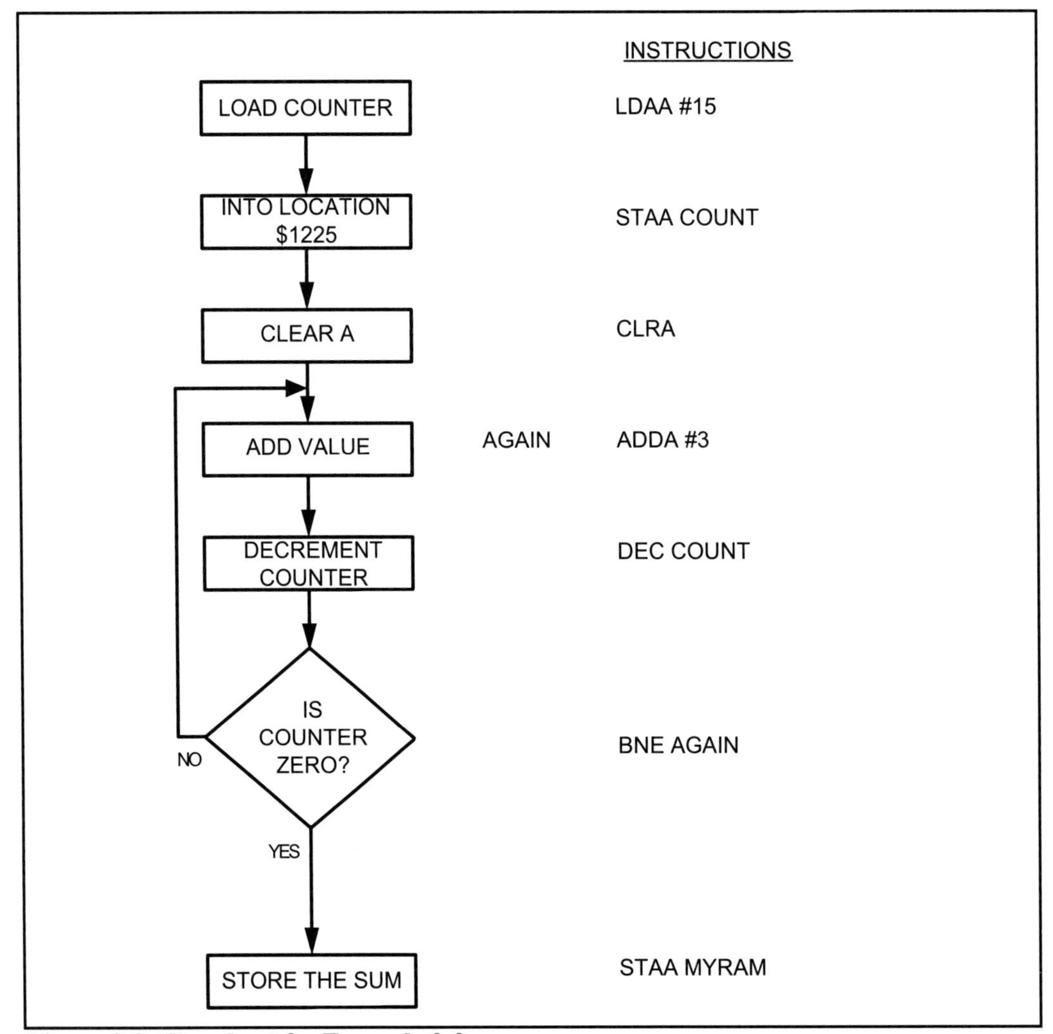

Figure 3-2. Flowchart for Example 3-2

Example 3-3

What is the maximum number of times that the loop in Example 3-2 can be repeated?

Solution:

Because RAM location COUNT is an 8-bit register, it can hold a maximum of FFH (255 decimal); therefore, the loop can be repeated a maximum of 255 times. See Example 3-4 to see how to overcome this limitation.

Loop inside a loop

As shown in Example 3-2, the maximum count is 255. What happens if we want to repeat an action more times than 255? To do that, we use a loop inside a loop, which is called a *nested loop*. In a nested loop, we use two registers to hold the count. See Example 3-4.

Example 3-4

Write a program to (a) load the PORTB register with the value 55H, and (b) complement PORTB 700 times.

Solution:

Because 700 is larger than 255 (the maximum capacity of any register), we use two registers to hold the count. The following code shows how to use RAM locations 1225H and 1226H as a register for counters.

```
      R1 EQU $1225
      R2 EQU $1226
      COUNT_1 EQU 10
      COUNT_2 EQU 70
      LDAA #$FF
      STAA DDRB           ;make PORTB an output. See Chapter 4.
      LDAA #$55           ;A = 55H
      STAA PORTB          ;PORTB = 55H
      LDAA #COUNT_1       ;A = 10, outer loop count value
      STAA R1             ;load 10 into loc 1225H (outer loop count)
LOP_1 LDAA #COUNT_2       ;A = 70, inner loop count value
      STAA R2             ;load 70 into loc 1226H
LOP_2 COM PORTB           ;complement Port B
      DEC R2              ;dec RAM loc 1226 (inner loop)
      BNE LOP_2           ;repeat it 70 times
      DEC R1              ;dec RAM loc 1225 (outer loop)
      BNE LOP_1           ;repeat it 10 times
```

In this program, RAM location $1226 is used to keep the inner loop count. In the instruction "`BNE LOP_2`", whenever location 1226H becomes 0 it falls through and "`DEC R1`" is executed. This instruction forces the CPU to load the inner count with 70 if R2 is not zero, and the inner loop starts again. This process will continue until location 1225H becomes zero and the outer loop is finished.

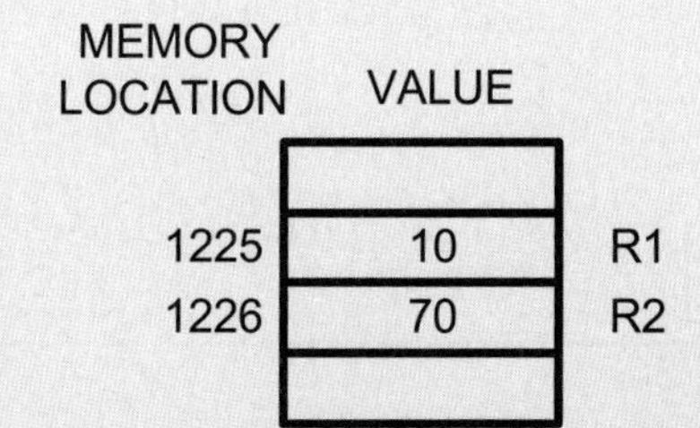

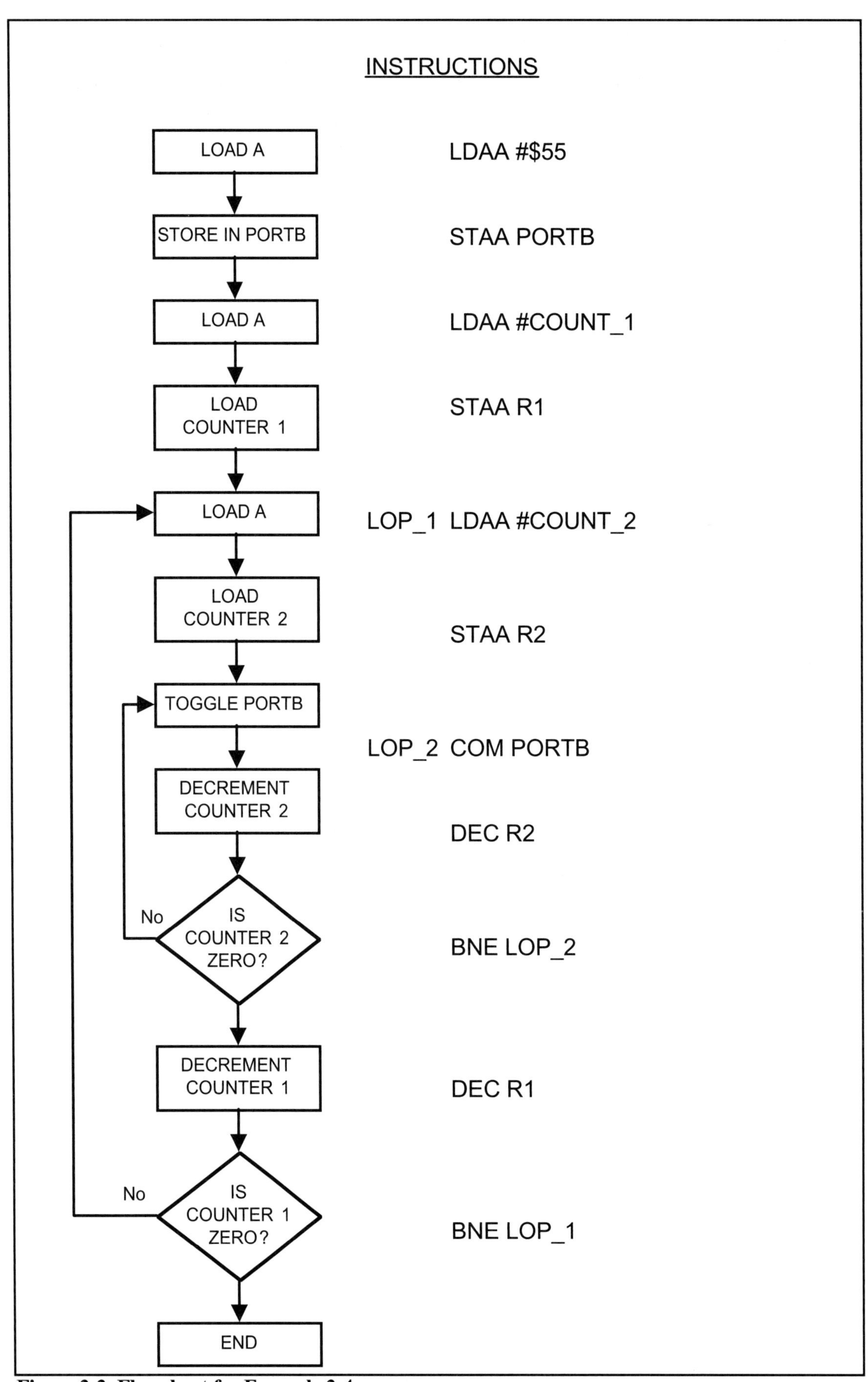

Figure 3-3. Flowchart for Example 3-4

Looping 100,000 times

Because two registers give us a maximum value of 65,025 ($255 \times 255 = 65{,}025$), we can use three registers to get up to more than 16 million (2^{24}) iterations. Program 3-1 complements PORTB 100,000 times:

```
;Program 3-1
R1 EQU $1201                     ;assign RAM loc for the R1-R2
R2 EQU $1202
R3 EQU $1203
COUNT_1 EQU 100   ;fixed value for 100,000 times
COUNT_2 EQU 100
COUNT_3 EQU 10

      LDAA   #$FF
      STAA   DDRB         ;make PORTB an output
      LDAA   #$55
      STAA   PORTB
      LDAA   #COUNT_3
      STAA   R3
LOP_3 LDAA   #COUNT_2
      STAA   R2
LOP_2 LDAA   #COUNT_1
      STAA   R1
LOP_1 COM    PORTB
      DEC    R1
      BNE    LOP_1
      DEC    R2
      BNE    LOP_2
      DEC    R3
      BNE    LOP_3
```

A simpler way would be to use the MOVB instruction to load the registers. See end of this chapter for more on this. This is shown in Program 3-2.

```
;Program 3-2
R1 EQU $1201                     ;assign RAM loc for the R1-R2
R2 EQU $1202
R3 EQU $1203
COUNT_1 EQU 100   ;fixed value for 100,000 times
COUNT_2 EQU 100
COUNT_3 EQU 10

      MOVB   #$FF,DDRB    ;make PORTB an output
      MOVB   #$55,PORTB
      MOVB   #COUNT_3,R3
LOP_3 MOVB   #COUNT_2,R2
LOP_2 MOVB   #COUNT_1,R1
LOP_1 COM    PORTB
      DEC    R1
      BNE    LOP_1
      DEC    R2
      BNE    LOP_2
      DEC    R3
      BNE    LOP_3
```

Other conditional jumps

Conditional branches for the HCS12 are summarized in Table 3-1. More details of each instruction are provided in Appendix A. In Table 3-1, notice that all of the instructions, such as BEQ (Branch if Z = 1) and BCS (Branch if C = 1), jump only if a certain condition is met. Next, we examine some conditional branch instructions with examples.

Table 3-1: HCS12 Short Branch Instructions Using Flag Bits

Instruction	Action
BCS	Branch if C = 1
BCC	Branch if C = 0
BEQ	Branch if Z = 1
BNE	Branch if Z = 0
BMI	Branch if N = 1
BPL	Branch if N = 0
BVS	Branch if V = 1
BVC	Branch if V = 0

BEQ (Branch if Z = 1)

In this instruction, the Z flag is checked. If it is high, the CPU jumps to the target address. For example, look at the following code.

```
OVER  LDAA PORTB        ;read Port B and put it in A
      BEQ OVER          ;jump if A is zero
```

In this program, if PORTB is zero, the CPU jumps to the label OVER. It stays there until PORTB has a value other than zero. Notice that the BEQ instruction can be used to see whether any RAM location has a zero. More importantly, you don't have to perform an arithmetic instruction such as decrement to use the BEQ instruction. See Example 3-5.

Example 3-5

Write a program to determine if RAM location $830 contains the value 0. If so, put 55H in it.

Solution:

```
MYLOC EQU $830
      LDAA   MYLOC            ;copy MYLOC to reg A
      BNE    NEXT             ;branch if MYLOC is not zero
      LDAA   #$55
      STAA   MYLOC            ;put 0x55 if MYLOC has zero value
NEXT  ...
```

BCC (branch if no carry, branch if CY = 0)

In this instruction, the carry flag bit in the condition code register is used to make the decision whether to jump. In executing "`BCC label`", the processor looks at the carry flag to see if it is raised (C = 1). If it is not, the CPU starts to fetch and execute instructions from the address of the label. If C = 1, it will not branch but will execute the next instruction below BCC. Study Example 3-6 to see how BCC is used to add numbers together when the sum is higher than FFH. Note that there is also a "`BCS label`" instruction. In the BCS instruction, if C = 1 it jumps to the target address. We will give more examples of these instructions in the context of some applications in Chapter 5.

The other conditional branch instructions in Table 3-1 are discussed in Chapter 5 when arithmetic operations with signed numbers are discussed.

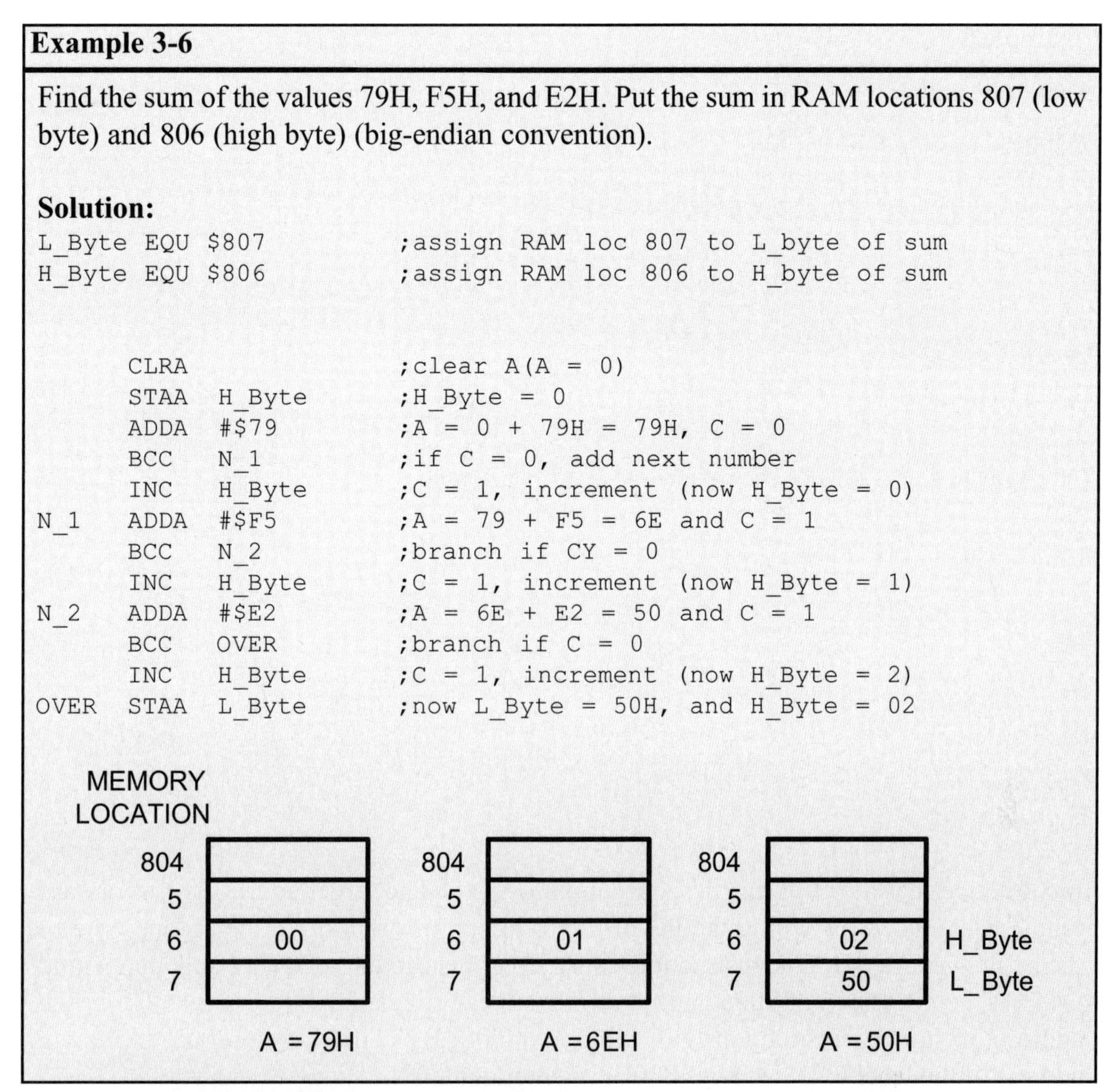

Example 3-6

Find the sum of the values 79H, F5H, and E2H. Put the sum in RAM locations 807 (low byte) and 806 (high byte) (big-endian convention).

Solution:

```
L_Byte EQU $807           ;assign RAM loc 807 to L_byte of sum
H_Byte EQU $806           ;assign RAM loc 806 to H_byte of sum


       CLRA               ;clear A(A = 0)
       STAA   H_Byte      ;H_Byte = 0
       ADDA   #$79        ;A = 0 + 79H = 79H, C = 0
       BCC    N_1         ;if C = 0, add next number
       INC    H_Byte      ;C = 1, increment (now H_Byte = 0)
N_1    ADDA   #$F5        ;A = 79 + F5 = 6E and C = 1
       BCC    N_2         ;branch if CY = 0
       INC    H_Byte      ;C = 1, increment (now H_Byte = 1)
N_2    ADDA   #$E2        ;A = 6E + E2 = 50 and C = 1
       BCC    OVER        ;branch if C = 0
       INC    H_Byte      ;C = 1, increment (now H_Byte = 2)
OVER   STAA   L_Byte      ;now L_Byte = 50H, and H_Byte = 02
```


Short conditional branches

It must be noted that we have both short and long conditional branches. In short branch jumps the address of the target must be within 256 bytes of the contents of the program counter (PC). This concept is discussed next.

Calculating the short branch address

Conditional branches such as BCC, BEQ, and BNE are short branches due to the fact that they are all 2-byte instructions. In these instructions the first byte is the opcode and the second byte is the relative address. The target address is relative to the value of the program counter. If the second byte is positive, the jump is forward. If the second byte is negative, the jump is backwards. The second byte can be a value from –127 to +128. To calculate the target address, we add the second byte of the instruction to the PC of the next instruction (target address = 2nd byte of instruction + PC of next instruction). See Example 3-7. We do the same thing for the backward branch, although the second byte is negative. That is, we add a negative number to the PC value of the next instruction. See Example 3-8.

Example 3-7

Using the following list file of Example 3-6, verify the jump forward address calculation.

```
as12, an absolute assembler for Motorola MCU's, version 1.2h

0807           L_Byte EQU $807        ;assign RAM loc 807 to L_byte of sum
0806           H_Byte EQU $806        ;assign RAM loc 806 to H_byte of sum

8000                  ORG $8000
8000 87               CLRA            ;clear A(A = 0)
8001 7a 08 06         STAA H_Byte     ;H_Byte = 0
8004 8b 79            ADDA #$79       ;A = 0 + 79H = 79H, C = 0
8006 24 03            BCC N_1         ;if C = 0, add next number
8008 72 08 06         INC H_Byte      ;C = 1, increment (now H_Byte = 0)
800b 8b f5     N_1    ADDA #$F5       ;A = 79 + F5 = 6E and C = 1
800d 24 03            BCC N_2         ;branch if CY = 0
800f 72 08 06         INC H_Byte      ;C = 1, increment (now H_Byte = 1)
8012 8b e2     N_2    ADDA #$E2       ;A = 6E + E2 = 50 and C = 1
8014 24 03            BCC OVER        ;branch if C = 0
8016 72 08 06         INC H_Byte      ;C = 1, increment (now H_Byte = 2)
8019 7a 08 07         OVER STAA L_Byte ;now L_Byte = 50H, and H_Byte = 02

Executed: Tue Jul 31 17:49:38 2007
Total cycles: 33, Total bytes: 28
Total errors: 0, Total warnings: 0
```

Solution:

First, notice that the BCC instruction jumps forward. The target address for a forward jump is calculated by adding the PC of the following instruction to the second byte of the branch instruction. The instruction "`BCC N_1`" has an opcode of 24 and an operand of 03 at the addresses of 8006 and 8007. The 03 is the relative address, relative to the address of the next instruction INC, which is 8008. By adding 03 to 8008, the target address of the label N_1, which is 800B, is generated. In the same way, the "`BCC N_2`" and "`BCC OVER`" instructions jump forward because their relative address values are positive.

Long conditional branches

In long branch instructions the address of the target must be within 32K bytes of the contents of the program counter (PC). Conditional long branches such as LBNC, LBZ, and LBNZ are long branches due to the fact that they are all 4-byte instructions. In these instructions the first 2 bytes are the opcode and the second 2 bytes are the relative address. The target address is relative to the value of the program counter. If the second 2-byte block is positive, the jump is forward. If the second 2-byte block is negative, the jump is backwards. The second 2-byte block can be a value from –32,768 to +32,767. To calculate the target address, we add the second 2-byte block of the instruction to the PC of the next instruction.

Table 3-2: HCS12 Long Branch Instructions Using Flag Bits

Instruction	Action
LBCS	Long Branch if C = 1
LBCC	Long Branch if C = 0
LBEQ	Long Branch if Z = 1
LBNE	Long Branch if Z = 0
LBMI	Long Branch if N = 1
LBPL	Long Branch if N = 0
LBVS	Long Branch if V = 1
LBVC	Long Branch if V = 0

Example 3-8

Verify the calculation of backward jumps for the listing of Example 3-2, shown below.

```
       0000 0225  COUNT EQU $225     ;use loc 225H for counter
       0000 0250  MYRAM EQU $250
                        ORG $8000
008000 860F             LDAA #15     ;A = 15 (decimal) for counter
008002 7A02 25          STAA COUNT   ;load the counter
008005 87               CLRA         ;A = 0
008006 8B03       AGAIN ADDA #3      ;add 03 to A reg (A = sum)
008008 7302 25          DEC COUNT    ;decrement counter
00800B 26F9             BNE AGAIN    ;repeat until count becomes 0
00800D 7A02 50          STAA MYRAM   ;store sum
```

Solution:

In the program list, "`BNE AGAIN`" has opcode 26 and relative address F9H. The F9H gives us –7. When the relative address of –7 is added to 800DH, the address of the instruction below the byte, we have –7 + 800DH = 8006H. Notice that 8006 is the address of the label AGAIN. F9H is a negative number, which means it will branch backward. For further discussion of the addition of negative numbers, see Chapter 5.

Unconditional branch instruction

The unconditional branch is a jump in which control is transferred unconditionally to the target location. In the HCS12 there are two unconditional branches: JMP (jump) and BRA (branch). Deciding which one to use depends on the target address. Each instruction is explained next.

JMP

JMP is an unconditional jump that can go to any memory location in the 64K address space of the HCS12. It is a 3-byte instruction in which 1 byte is used for the opcode, and the other 2 bytes are for the address of the target location. The 16-bit target address allows a jump to 64 KB of memory locations from 0000 to FFFFH. See Figure 3-4.

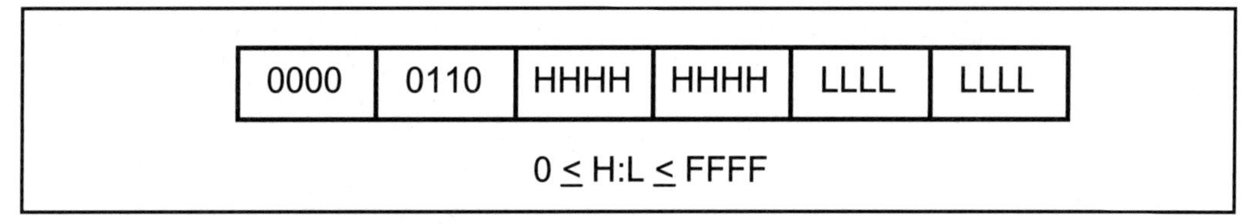

Figure 3-4. Jump Instruction Address Range

Remember that although the program counter in the HCS12 is 16-bit (thereby giving a ROM address space of 64 KB), not all HCS12 family members have that much on-chip program ROM. Some of the HCS12 family members have only 4K to 32K of on-chip Flash ROM for program space; consequently, every byte is precious. For this reason there is also a BRA (branch) instruction, which is a 2-byte instruction as opposed to the 3-byte JMP instruction. This can save some bytes of memory in many applications where ROM memory space is in short supply. BRA is discussed next.

BRA (branch always)

In this 2-byte instruction, the first byte is the opcode and the second byte is the relative address of the target location. The relative address range of 00–FFH is divided into forward and backward jumps; that is, within –128 to +127 bytes of memory relative to the address of PC (program counter) for the next instruction. If the jump is forward, the target address is positive. If the jump is backward, the target address is negative. In this regard, BRA is like the conditional branch instructions. See Figure 3-5.

Notice that this is a 2-byte instruction, and is preferred over the JMP because it takes less ROM space. Chapter 5 examines signed numbers.

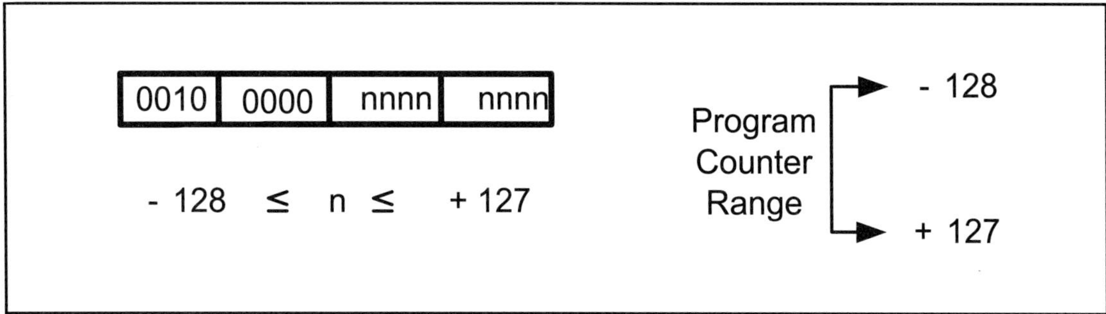

Figure 3-5. BRA (Branch Unconditionally) Instruction Address Range

JMP to itself using $ sign

In cases where there is no monitor program, we use the JMP to itself in order to keep the microcontroller busy. A simple way of doing that is to use the $ sign. That means in place of this:

```
HERE        JMP         HERE
```

we can use the following:

```
            JMP   $
```

This will also work for the BRA instruction, as shown below:

```
OVER        BRA   OVER
```

which is the same as:

```
            BRA   $             ;$ means same line
```

Review Questions

1. The mnemonic BNE stands for _______.
2. True or false. "`BNE  BACK`" makes its decision based on the last instruction affecting the Z flag.
3. "`BNE  HERE`" is a ___ -byte instruction.
4. In "`BEQ  NEXT`", which flag is checked?
5. BEQ is a(n) ___ -byte instruction.

Example 3-10

Analyze the stack for the JSR instructions in the following program.

Solution:

When the first JSR is executed, the address of the instruction "`LDAA #$AA`" is saved (pushed) on the stack. The last instruction of the called subroutine must be an RTS (return from subroutine) instruction, which directs the CPU to pull the contents of the top location of the stack into the PC and resume executing at address 800A. The diagrams show the stack frame after the JSR and RTS instructions.

```
0808                 MYREG  EQU $808      ;use location 808 as counter
8000                        ORG $8000
8000 cf 40 00               LDS #$4000    ;initialize the stack pointer
8003 86 55           BACK   LDAA #$55     ;load A with 55H
8005 5a 01                  STAA PORTB    ;send 55H to port B
8007 16 83 00               JSR DELAY     ;time delay
800a 86 aa                  LDAA #$AA     ;load A with AA (in hex)
800c 5a 01                  STAA PORTB    ;send AAH to port B
800e 16 83 00               JSR DELAY
8011 20 f0                  BRA BACK      ;keep doing this indefinitely
                            ;-------      this is the delay subroutine
8300                        ORG $8300     ;put time delay at address 8300H
8300 86 ff           DELAY  LDAA #$FF     ;A = 255, the counter
8302 7a 08 08               STAA MYREG
8305 a7              AGAIN  NOP           ;no operation wastes clock cycles
8306 a7                     NOP
8307 73 08 08               DEC MYREG
830a 26 f9                  BNE AGAIN     ;repeat until MYREG becomes 0
830c 3d                     RTS           ;return to caller
                            END           ;end of asm file
```

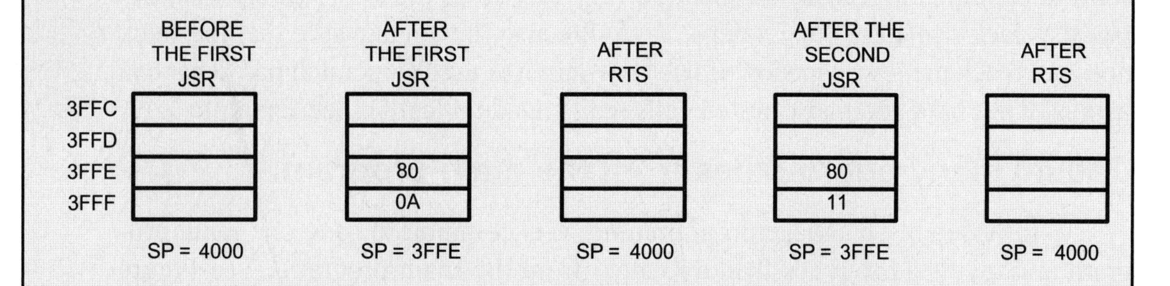

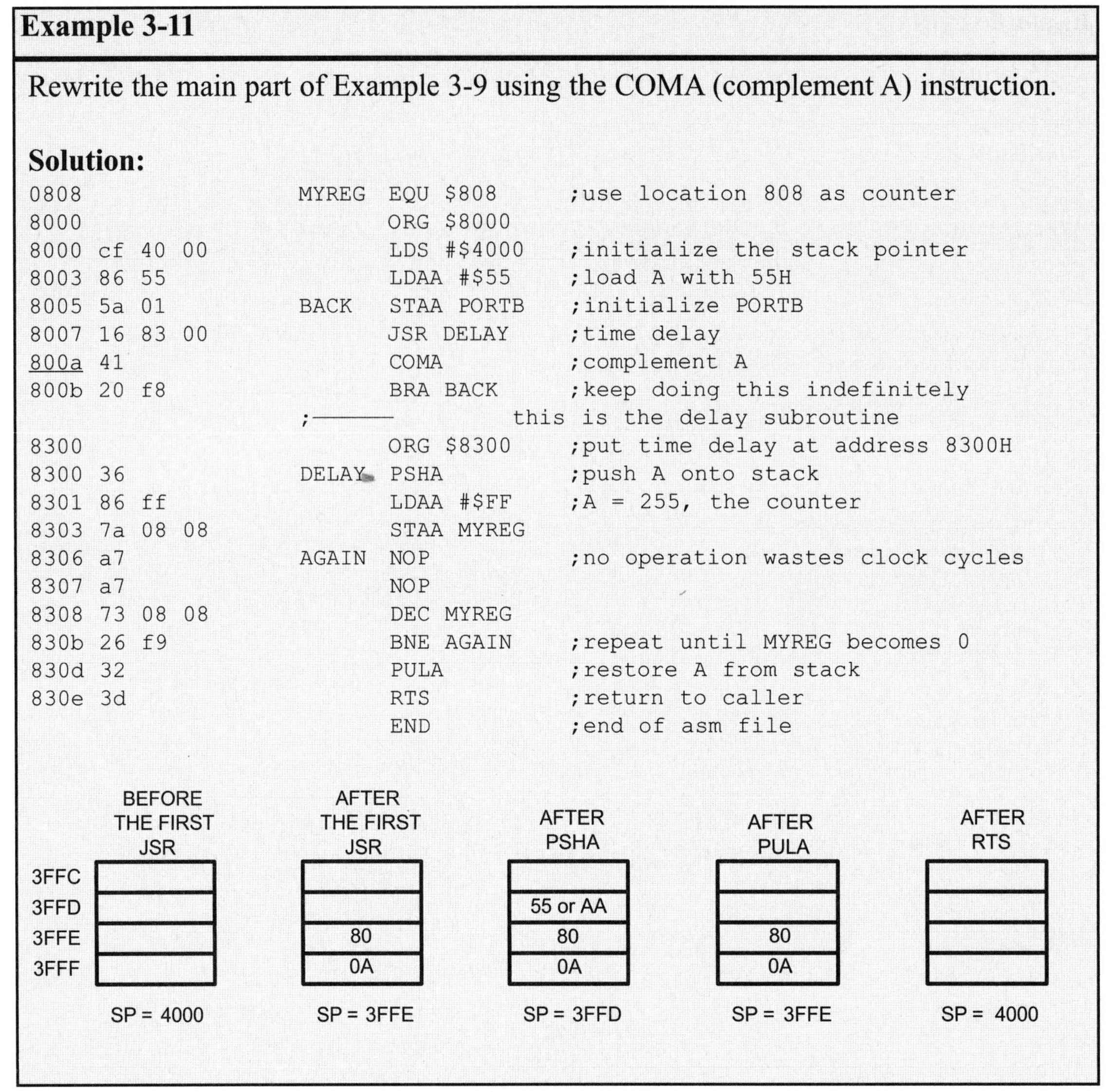

Example 3-11

Rewrite the main part of Example 3-9 using the COMA (complement A) instruction.

Solution:

```
0808                MYREG  EQU $808      ;use location 808 as counter
8000                       ORG $8000
8000 cf 40 00              LDS #$4000    ;initialize the stack pointer
8003 86 55                 LDAA #$55     ;load A with 55H
8005 5a 01          BACK   STAA PORTB    ;initialize PORTB
8007 16 83 00              JSR DELAY     ;time delay
800a 41                    COMA          ;complement A
800b 20 f8                 BRA BACK      ;keep doing this indefinitely
                    ;-------        this is the delay subroutine
8300                       ORG $8300     ;put time delay at address 8300H
8300 36             DELAY  PSHA          ;push A onto stack
8301 86 ff                 LDAA #$FF     ;A = 255, the counter
8303 7a 08 08              STAA MYREG
8306 a7             AGAIN  NOP           ;no operation wastes clock cycles
8307 a7                    NOP
8308 73 08 08              DEC MYREG
830b 26 f9                 BNE AGAIN     ;repeat until MYREG becomes 0
830d 32                    PULA          ;restore A from stack
830e 3d                    RTS           ;return to caller
                           END           ;end of asm file
```

	BEFORE THE FIRST JSR	AFTER THE FIRST JSR	AFTER PSHA	AFTER PULA	AFTER RTS
3FFC					
3FFD			55 or AA		
3FFE		80	80	80	
3FFF		0A	0A	0A	
	SP = 4000	SP = 3FFE	SP = 3FFD	SP = 3FFE	SP = 4000

The stack usage and its lower limit

In HCS12, the stack is used most often for calls and interrupts. We can also use it for storing some parameters. We must remember that upon calling a subroutine, the stack keeps track of where the CPU should return after completing the subroutine. For this reason, we must be very careful in the way that we manipulate the stack contents. See Example 3-11. We also must make sure that the stack does not reach the lower section of RAM, where it is used by scratch pad. That can happen if we have too many nested calls (calls inside of calls). See Example 3-12.

Calling many subroutines from the main program

In Assembly language programming, it is common to have one main program and many subroutines that are called from the main program. (See Figure 3-8.) This allows you to make each subroutine into a separate module. Each module can be tested separately and then brought together with the main program. More importantly, in a large program the modules can be assigned to different programmers in order to shorten development time. See Chapter 6 for a discussion of modules.

Example 3-12

Write a program to count up from 00 to FFH and send the count to Port B. Use one call subroutine for sending the data to Port B and another one for time delay. Put a time delay in between each issuing of data to Port B.

Solution:

```
LOC    OBJECT CODE  LINE   SOURCE TEXT
  VALUE
                           ORG $8000
008000 CF40 00             LDS #$4000    ;initialize the stack pointer
008003 8600                LDAA #$0      ;load A with 0H
008005 1683 00      BACK   JSR DISPLAY   ;time delay
008008 20FB                BRA BACK      ;keep doing this indefinitely
                    :-------------------------------------
                           ORG $8300
008300 42           DISPLAY INCA
008301 5A01                STAA PORTB
008303 1684 00             JSR DELAY
008306 3D                  RTS

                    ;----     this is the delay subroutine
008400 C6FF         DELAY  LDAB #$FF
008402 A7           OVER   NOP
008403 A7                  NOP
008404 A7                  NOP
008405 53                  DECB
008406 26FA                BNE OVER
008408 3D                  RTS
```

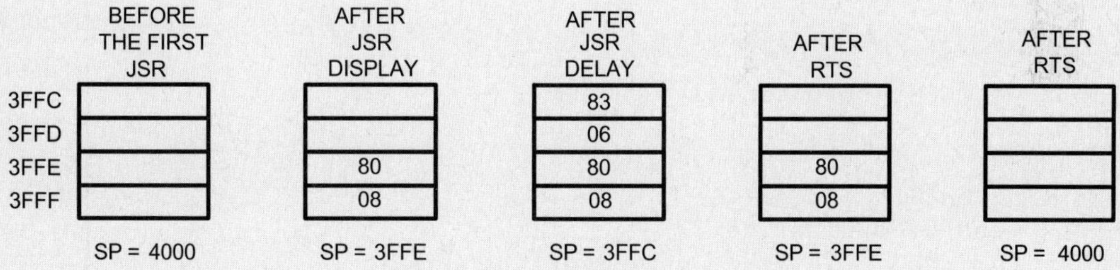

	BEFORE THE FIRST JSR	AFTER JSR DISPLAY	AFTER JSR DELAY	AFTER RTS	AFTER RTS
3FFC			83		
3FFD			06		
3FFE		80	80	80	
3FFF		08	08	08	
	SP = 4000	SP = 3FFE	SP = 3FFC	SP = 3FFE	SP = 4000

It needs to be emphasized that in using JSR, the target address of the subroutine can be anywhere within the 64 KB memory space of the HCS12. To go beyond the 64 KB space we use the CALL instruction, which is explained next.

CALL

In many variations of the HCS12 marketed by Freescale Corporation, on-chip ROM is as high as 512K bytes. In such cases, the use of CALL instead of JSR allows us to call a subroutine in program ROM residing outside the 64 KB space. To access the Flash ROM space beyond the 64 KB, we must use the page switching with the help of the PPAGE register. CALL is a 4-byte instruction in contrast to JSR, which is 3 bytes. See Example 3-13. Because CALL is a 4-byte instruction, the target address of the subroutine can be within 512K bytes of expanded memory since 1 byte of the instruction is used for the PPAGE value. There is also a difference between JSR and CALL in terms of saving the program counter on the stack or the function of the return instruction. The difference is that JSR uses 2 bytes of stack, while CALL uses 3 bytes of stack since it must save the PPAGE register too. We also must use the RTC (return from call) instruction instead of RTS (return from subroutine). The RTC pulls the PPAGE value from stack in addition to the program counter. Chapter 14 discusses page switching for flash ROM of 64 KB and beyond.

Delay calculation for the HCS12

In creating a time delay using Assembly language instructions, one must be

```
;MAIN program calling subroutines
            INCLUDE 'mc9s12dp512.inc'
            ORG    $8000
            LDS    #$4000      ;initialize the Stack Pointer
            ....               ;some other instructions
MAIN        JSR    SUBR_1
            ....
            JSR    SUBR_2
            JSR    SUBR_3

HERE        BRA    HERE        ;stay here
;———————————end of MAIN
;
SUBR_1      ....
            ....
            RTS   ;return from subroutine
;———————————end of subroutine 1
;
SUBR_2      ....
            ....
            RTS                ;return from subroutine
;———————————end of subroutine 2
SUBR_3      ....
            ....
            RTS
;———————————end of subroutine 3
            END                ;end of the asm file
```

Figure 3-8. HCS12 Assembly Main Program That Calls Subroutines

Example 3-13

A developer is using the HCS12 microcontroller chip for a product. This chip has only 32K of on-chip flash ROM. Which of the instructions, CALL or JSR, is more useful in programming this chip?

Solution:

The JSR instruction is more useful because it is a 3-byte instruction. It saves one byte each time the JSR instruction is used. More importantly, we use CALL to access memory beyond 64K bytes.

Table 3-3: HCS12 Call Instructions

Instruction	Action	Instruction size
JSR	Branch (call) to an address within 64 KB	3-byte
BSR	Branch (call) to an address in the range of 256 bytes	2-byte
RTS	Return from subroutine (used with JSR and BSR)	1-byte
CALL	Call an address beyond 64 KB	4-byte
RTC	Return from subroutine in expanded memory (used with CALL)	1-byte

mindful of two factors that can affect the accuracy of the delay:

1. The crystal frequency: The frequency of the crystal oscillator connected to the XTAL and EXTAL pins is one factor in the time delay calculation. The duration of the clock period for the instruction execution is a function of crystal frequency and PLL. See Chapter 8.
2. The HCS12 internal design: Since the 1970s, both the field of IC technology and the architectural design of microprocessors have seen great advancements. Advances in both IC technology and CPU design in the 1980s and 1990s have made the execution of an instruction in a single clock cycle a common feature of many microcontrollers. Indeed, one way to increase performance without losing code compatibility with the older generation of a given family is to reduce the number of clock cycles it takes to execute an instruction. One might wonder how microprocessors are able to execute an instruction in one clock cycle. There are three ways to do that: (a) Use Harvard architecture to get the maximum amount of code and data into the CPU, (b) use RISC architecture features such as fixed-size instructions, or (c) use pipelining to overlap fetching and execution of instructions. The HCS12 uses pipelining extensively to speed up the execution of instructions since it could not use RISC architecture due to compatibility issues with the older generations of 68HC11 and CPU08. Next, we discuss pipelining.

Pipelining

In early microprocessors such as the 6800, the CPU could either fetch or execute. It could not do both at the same time. In other words, the CPU had to fetch an instruction from memory and execute it, and then fetch the next instruction and execute it, and so on. The idea of pipelining in its simplest form is to allow the CPU to fetch and execute at the same time, as shown in Figure 3-9.

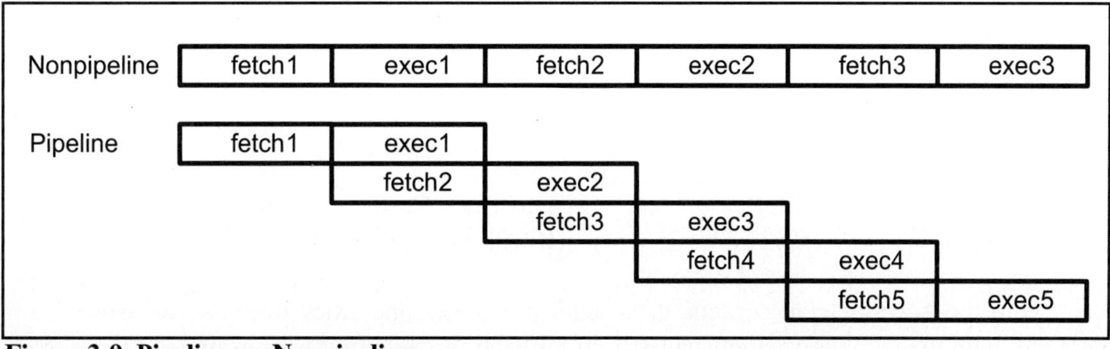

Figure 3-9. Pipeline vs. Nonpipeline

Instruction queue and branch penalty

The overlapping of fetch and execution of the instruction is widely used in today's microcontrollers such as HCS12. For the concept of pipelining to work, we need a buffer or queue in which instructions are prefetched and ready to be executed. In some circumstances, the CPU must flush out the queue. For example, when a branch instruction is executed, the CPU starts to fetch codes from the new memory location, and the code in the queue that was fetched previously is discarded. In this case, the execution unit must wait until the fetch unit fetches the new instruction. This is called a *branch penalty.* The penalty is extra clock cycles to fetch the instruction from the target location instead of executing the instruction right below the branch. Remember that the instruction below the branch has already been fetched, is sitting in the queue, and is next in line to be executed when the CPU branches to a different address. This means that while many HCS12 instructions take only one clock cycle, some instructions take two or more clock cycles. The latter are JMP, BRA, CALL, and all the conditional branch instructions such as BNE, BCS, and so on. The conditional branch instruction, however, can take only one clock cycle if it does not jump. For example, the BNE will jump if Z = 0 and that takes three clock cycles. If Z = 1, however, it falls through and it takes only one clock cycle. Another major problem in HCS12 is that for a given instruction, the different addressing modes demand different numbers of clock cycles for execution.

Instruction cycle time in HCS12 / 9S12 datasheet

Let's explain the instruction timing shown in the HCS12 manual. In the early generation of the 68HC08/11, the data sheet provided the exact number of clock cycles used to execute an instruction. This was easy since it did not use any pipelining. Since 1990s, pipelining has been used to increase the performance of the CPU. This is also the case with the HCS12 and 9S12 chips, since pipelining is used to get more performance out of an old architecture and instruction set. In the pipelined CPU architecture, the number of clocks used in executing an instruction is less clear-cut than in the nonpipelined architecture. Next, we examine how to interpret the instruction clock cycles for the HCS12 instructions. In the HCS12 manual we see the following for the ADDA instruction:

Instruction	Addressing Mode	Access Detail	Clock Cycles
ADDA	IMM	p	1
	DIR	rPf	3
	EXT	rPo	3
	IDX	rPf	3
	IDX1	rPo	3
	IDX2	frPP	4
	[D,IDX]	fIfrPf	6
	[IDX2]	fIPrPf	6

In the Access Detail column, each letter represents one clock cycle, so we can come up with the number of clock cycles used for each instruction. The meanings of the various letters in the Access Detail column are shown in the HCS12 datasheet and are provided in Appendix H of this book. See Examples 3-14 and 3-15.

From these discussions we conclude that the use of instructions in generating time delay is not the most reliable method. To get more accurate software-generated time delay we use timers, as described in Chapter 9. Meanwhile, to get an accurate time delay for a given HCS12 microcontroller, we must use an oscilloscope to measure the exact time delay.

Example 3-14

Estimate the length of the delay of the loop part if the crystal frequency is 5 MHz:

Solution:

```
                                       Clock Cycle
MYREG EQU    #$1208         ;use location $1208 as counter

DELAY     LDAA #200                    1
          STAA     MYREG               3

AGAIN     NOP                          1
          NOP                          1
          DEC      MYREG               4
          BNE      AGAIN               3
```

HCS12 uses 1/2 of the crystal oscillator frequency for CPU (instruction) cycle time. That means 5 MHz / 2 = 2.5 MHz is used for CPU cycle time. Now, the CPU cycle time = 1 / 2.5 MHz = 0.4 μs. Therefore, we have a time delay of $(200 \times 9) \times 0.4\ \mu s$ = 720 μs. Notice that BNE takes three instruction cycles if it jumps back, and takes only one when falling through the loop. The above delay was verified using CodeWarrior.

Review Questions

1. How wide is the stack in the HCS12?
2. True or false. In the HCS12, control can be transferred anywhere within the 64 KB of code space by using the JSR instruction.
3. The CALL instruction is a(n) ___ -byte instruction.
4. True or false. In the HCS12, control can be transferred anywhere within the 4MB of code space by using the CALL instruction.
5. With each JSR instruction, the stack pointer register, SP, is _______________ (incremented, decremented) by ________(1, 2).
6. With each RTS instruction, the SP is __________ (incremented, decremented) by _______ (1, 2).
7. True or false. On power-up, we must use the LDSP (load stack pointer) instruction before we call any subroutine.
8. How deep is the size of the stack in the HCS12?
9. The JSR instruction is a(n) ___ -byte instruction.
10. ______ (JSR, CALL) takes more ROM space.

Example 3-15

Find the time delay for the following subroutine, assuming a crystal frequency of 4 MHz.

```
MYREG EQU   $1150
                                        ;Clock Cycle

DELAY    LDAA      #200                    1
         STAA      MYREG                   3

AGAIN    NOP                               1
         NOP                               1
         NOP                               1
         NOP                               1
         NOP                               1
         NOP                               1
         NOP                               1
         NOP                               1
         NOP                               1
         NOP                               1
         NOP                               1
         NOP                               1
         DEC       MYREG                   4
         BNE       AGAIN                   3
```

Solution:

The CPU (instruction) cycle time = 1/2 MHz = 0.5 μs. The time delay inside the AGAIN loop is $(200 \times 19) \times 0.5\ \mu s = 1{,}900\ \mu s$. The above delay was verified using both CodeWarrior and oscilloscope.

SECTION 3.3: PROGRAMS USING INDEXED ADDRESSING MODE

In this section we provide some examples of using indexed addressing mode.

Clearing RAM buffer

The following shows how to clear RAM locations using indexed addressing mode:

```
LDX #$800     ;load 800H into X
CLRA          ;A = 0
STAA 0,X      ;clear RAM loc 800+0
INX           ;increment X (X=801)
STAA 0,X      ;clear RAM loc 801+0
INX           ;increment X (X=802)
STAA 0,X      ;clear RAM loc 802+0
INX           ;increment X (X=803)
STAA 0,X
```

The following program is the loop version of the above program clearing 50 RAM locations starting at address 800H:

```
R1      EQU $860
        LDX #$800     ;set the pointer X
        LDAA #50      ;set the counter
        STAA R1       ;counter = 50
        CLRA
OVER    STAA   0,X    ;clear RAM location pointed to by X and X+1
        INX           ;increment X
        DEC R1        ;decrement counter
        BNE    OVER   ;repeat
```

The above program cleared RAM locations one byte at a time. The following program clears the RAM locations 2 bytes at a time:

```
R1      EQU $860
        LDX #$800     ;set the pointer X
        LDAA #25      ;set the counter
        STAA R1
        LDD    #0     ;D=0
OVER    STD    0,X    ;clear RAM location pointed to by X and X+1
        INX           ;increment X
        INX           ;twice
        DEC R1        ;decrement counter
        BNE    OVER   ;repeat
```

Moving a block of data

The following program transfers a block of 50 bytes of data from RAM locations starting at 900H to RAM locations starting at 1100H:

```
R1      EQU $810
        LDX #$900     ;set the pointer X
        LDY #$1100    ;set the pointer Y
        LDAA #50      ;set the counter
        STAA R1
OVER    LDAA 0,X      ;load a byte from loc pointed to by X
        STAA 0,Y      ;store in loc pointed to by Y
        INX           ;point to location
        INY
        DEC R1        ;decrement counter
        BNE OVER
```

The above program moved data one byte at a time. The following program moves data 2 bytes at a time:

```
R1      EQU $810
        LDX #$900     ;set the pointer X
        LDY #$1100    ;set the pointer Y
        LDAA #25      ;set the counter
        STAA R1
OVER    LDD 0,X       ;load a word from loc pointed to by X
        STD 0,Y       ;store in loc pointed to by Y
        INX           ;point to location
        INX
        INY
        INY
        DEC R1        ;decrement counter
        BNE OVER
```

Using MOV instruction

There are two MOV instructions in HCS12: MOVB (move byte) and MOVW (move word). The MOVB instruction allows us to move a byte to a memory location. The source operand can be an immediate value or another memory location. The following program uses MOVB to toggle PORTB continuously.

```
BACK   MOVB #$55,PORTB    ;move 55H to PORTB
       JSR  DELAY
       MOVB #$AA,PORTB    ;move AAH to PORTB
       JSR DELAY
       BRA BACK
```

Compare the above program with an earlier version of the program in Example 3-8 to see how the MOVB replaces the two instructions of LDAA and STAA. The following program adds 3 to the accumulator 15 times. It is a rewrite of Example 3-3 using the MOVB to load the counter.

```
COUNT EQU $825            ;use loc 825H for counter
MYRAM EQU $850

      MOVB #15,COUNT      ;load the counter
      CLRA                ;A = 0
AGAIN ADDA #3             ;add 03 to A reg (A = sum)
      DEC COUNT           ;decrement counter
      BNE AGAIN           ;repeat until count becomes 0
      STAA MYRAM          ;store sum
```

The following is a repeat of an earlier version of the program moving a block of data using the MOVB instruction.

```
R1     EQU $810
       LDX #$900    ;set the pointer X
       LDY #$1100   ;set the pointer Y
       LDAA #50     ;set the counter
       STAA R1
OVER   MOVB 0,X, 0,Y ;move a byte from source to destination
       INX          ;point to next location
       INY
       DEC R1       ;decrement counter
       BNE   OVER   ;repeat
```

The following is a repeat of the above program transferring a word (2 bytes) at a time:

```
R1     EQU $810
       LDX #$900    ;set the pointer X
       LDY #$1100   ;set the pointer Y
       LDAA #25     ;set the counter
       STAA R1
OVER   MOVW 0,X, 0,Y ;move a word from source to destination
       INX          ;point to next location
       INX
       INY
       INY
       DEC R1       ;decrement counter
       BNE   OVER   ;repeat
```

Review Questions

1. What is the difference between the MOVB and MOVW instructions?
2. True or false. In using the indexed addressing mode, we can use the D register as a pointer.
3. Explain the action performed by the "LDD $2000" instruction.
4. Explain the action performed by the "STD $2000" instruction.

PROBLEMS

SECTION 3.1: BRANCH INSTRUCTIONS AND LOOPING

1. In the HCS12, looping action with the instruction "`BNE target`" is limited to ____ iterations.
2. If a conditional branch is not taken, what is the next instruction to be executed?
3. In calculating the target address for a branch, a displacement is added to the contents of register _______.
4. The mnemonic BRA stands for __________ and it is a(n) ___-byte instruction.
5. The JMP instruction is a(n) ___-byte instruction.
6. What is the advantage of using BRA over JMP?
7. True or false. The target of a BNE can be anywhere in the 64 KB address space.
8. True or false. All HCS12 branch instructions can branch to anywhere in the 64 KB address space.
9. Which of the following instructions are 2-byte instructions.
   (a) BEQ (b) BCC (c) JMP (d) BRA
10. What is the advantage of using BNE over LBNZ?
11. True or false. All conditional branches are 2-byte instructions.
12. Show code for a nested loop to perform an action 1,000 times.
13. Show code for a nested loop to perform an action 100,000 times.
14. Find the number of times the following loop is performed:

```
        REGA   EQU    $800
        REGA   EQU    $801

        LDAA          #200
        STAA          REGA
BACK    LDAA          #100
        STAA          REGB
HERE    DEC           REGB
        BNE           HERE
        DEC           REGA
        BNE           BACK
```

15. The target address of a BNE is backward if the second byte of opcode is _______ (negative, positive).
16. The target address of a BNE is forward if the second byte of opcode is ________ (negative, positive).

SECTION 3.2: JSR AND CALL INSTRUCTIONS

17. JSR is a(n) ___-byte instruction.
18. CALL is a(n) ___-byte instruction.
19. True or false. The JSR target address can be anywhere in the 64-KB address space.
20. True or false. The CALL target address can be anywhere in the 1-MB address space.
21. When JSR is executed, how many locations of the stack are used?
22. When CALL is executed, how many locations of the stack are used?
23. Explain the differences between RTS and RTC.
24. Describe the action associated with the RTS instruction.
25. Give the size of the stack pointer in HCS12.
26. In HCS12, which address is pushed onto the stack and what happens to the stack pointer when a JSR instruction is executed?

SECTION 3.3: PROGRAMS USING INDEXED ADDRESSING MODE

27. Write two versions of a program to transfer a block of data one byte at a time. Use a) the LDAA and STAA instructions, and b) the MOVB instruction. Compare the size of the two programs. Which one takes fewer ROM locations?
28. Write two versions of a program to transfer a block of data one word at a time. Use a) the LDD and STD instructions, and b) the MOVW instruction. Compare the size of the two programs. Which one takes fewer ROM locations?

ANSWERS TO REVIEW QUESTIONS

SECTION 3.1: BRANCH INSTRUCTIONS AND LOOPING

1. Branch if not equal
2. True
3. 2
4. Z flag
5. 3

SECTION 3.2: JSR AND CALL INSTRUCTIONS

1. 8-bit
2. True
3. 3
4. True
5. Decremented, 2
6. Incremented, 2
7. True
8. As deep as the RAM size
9. 3
10. CALL

SECTION 3.3: PROGRAMS USING INDEXED ADDRESSING MODE

1. The MOVB instruction moves one byte from source to destination, while MOVW moves two bytes.
2. False
3. The content of location 2000H is moved to register A and the content of location 2001H is moved to register B.
4. The content of register A is moved to location 2000H and the content of register B is moved to location 2001H.

CHAPTER 4

HCS12/9S12
I/O PORT
PROGRAMMING

OBJECTIVES

Upon completion of this chapter, you will be able to:

>> List all the ports of the HCS12
>> Describe the dual role of HCS12 pins
>> Code Assembly language to use the ports for input or output
>> Explain the dual role of HCS12 ports
>> Code HCS12 instructions for I/O handling
>> Code I/O bit-manipulation programs for the HCS12
>> Explain the bit-addressability of HCS12 ports

SECTION 4.1: I/O PORT PROGRAMMING IN HCS12

In the HCS12 family, there are many ports for I/O operations, depending on which family member you choose. Examine Figure 4-1 for the MC9S12D. The ports are designated as PORTA, PORTB, PORTE, PORTK, and so on. The rest of the pins are designated as V_{dd} (V_{cc}), V_{ss} (GND), XTAL, EXTAL, TEST, RESET, and so on. They are discussed in Chapter 8.

I/O port pins and their functions

The number of ports in the HCS12 family varies depending on the number of pins on the chip, the peripherals it supports, and package type. The MC9S12D family members come in either 80-pin QFP (quad flat pack) or 112-pin LQFP (low-profile quad pack) packages. The 80-pin QFP package ports are designated as PA, PB, PE, PJ, PM, PP, PS, PT, and PAD. The MC9S12DP512 comes in a 112-pin LQFP package only. Its ports are designated as PA, PB, PE, PH, PJ, PK, PM, PP, PS, PT, and PAD. See Table 4-1. In addition to being used for simple I/O, each port has some other functions such as ADC, timers, interrupts, and serial communication pins. Figures 4-1 and 4-2 show the alternate functions of the HCS9S12 pins. We will study all these alternate functions in future chapters. In this chapter we focus on the simple I/O function of the HCS12 family. Notice that not all ports have 8 pins. For example, in the 80-pin package, Port M has 6 pins. Each port has two registers associated with it, as shown in Figure 4-3. They are designated as PORTx and DDRx (data direction register). For example, for Port B we have PORTB and DDRB. Next, we describe the role of DDR in accessing the pins as simple input/output.

Table 4-1: Number of Ports in MC9S12D Family Members

		80-pin QFP	**112-pin LQFP**
Ports	**(Port's name)**	**MC9S12Dx256 MC9S12Dx128 MC9S12D64 MC9S12D32**	**MC9S12DP512 MC9S12Dx256 MC9S12Dx128 MC9S12D64**
Port A	PORTA	PA7–PA0	PA7–PA0
Port B	PORTB	PB7–PB0	PB7–PB0
Port E	PORTE	PE7–PE0	PE7–PE0
Port H	PTH		PH7–PH0
Port J	PTJ	PJ7–PJ6	PJ7, PJ6, PJ1, PJ0
Port K	PORTK		PK7, PK5–PK0
Port M	PTM	PM5–PM0	PM7–PM0
Port P	PTP	PP7, PP5–PP0	PP7–PP0
Port S	PTS	PS3–PS0	PS7–PS0
Port T	PTT	PT7–PT0	PT7–PT0
PAD (A/D converter)			

Note: 1) PE0 and PE1 are used for input only.
2) Many ports use the PT name instead of PORT.

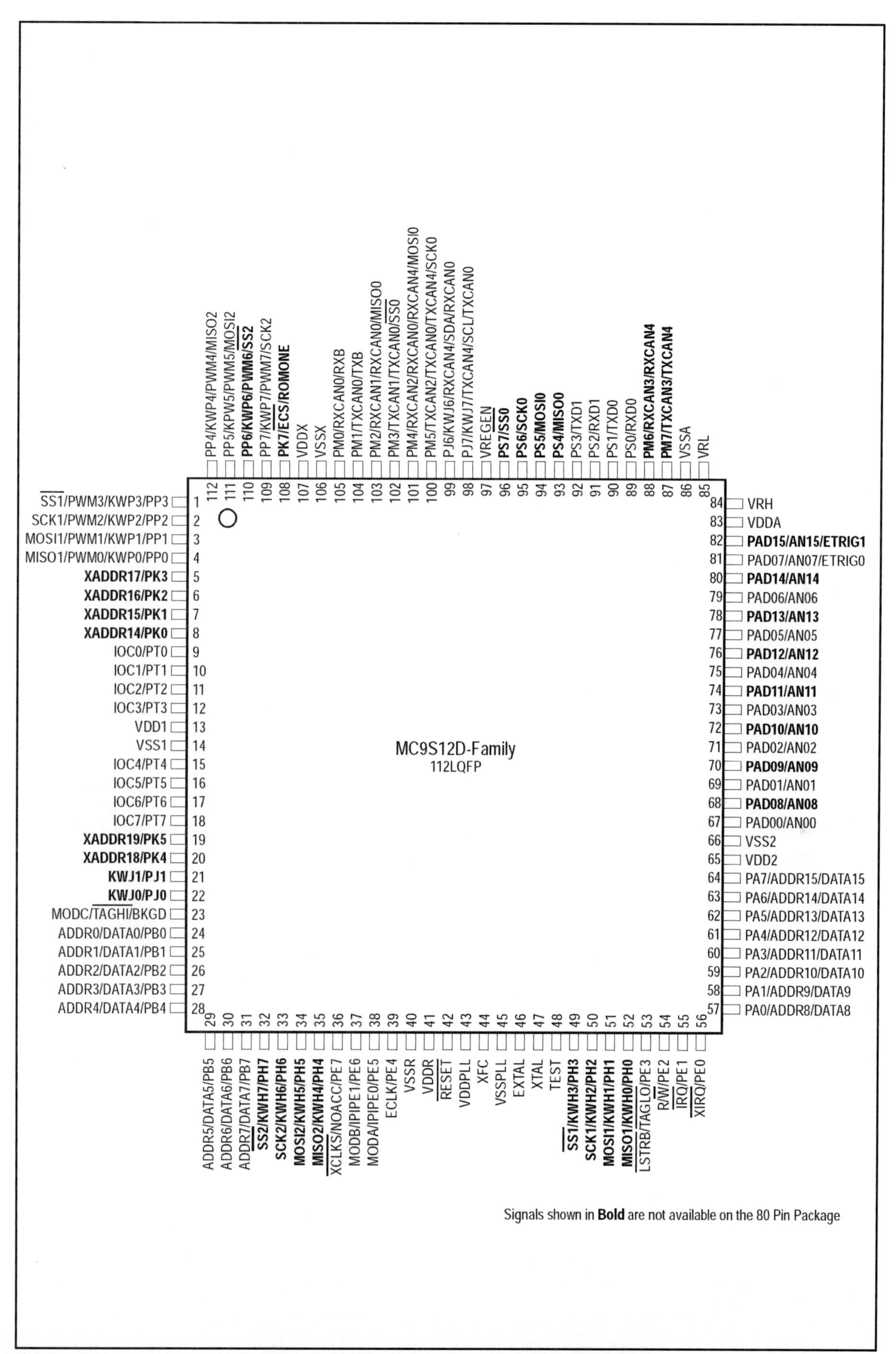

Figure 4-1. MC9S12D LQFP 112-Pin Diagram (from Freescale)

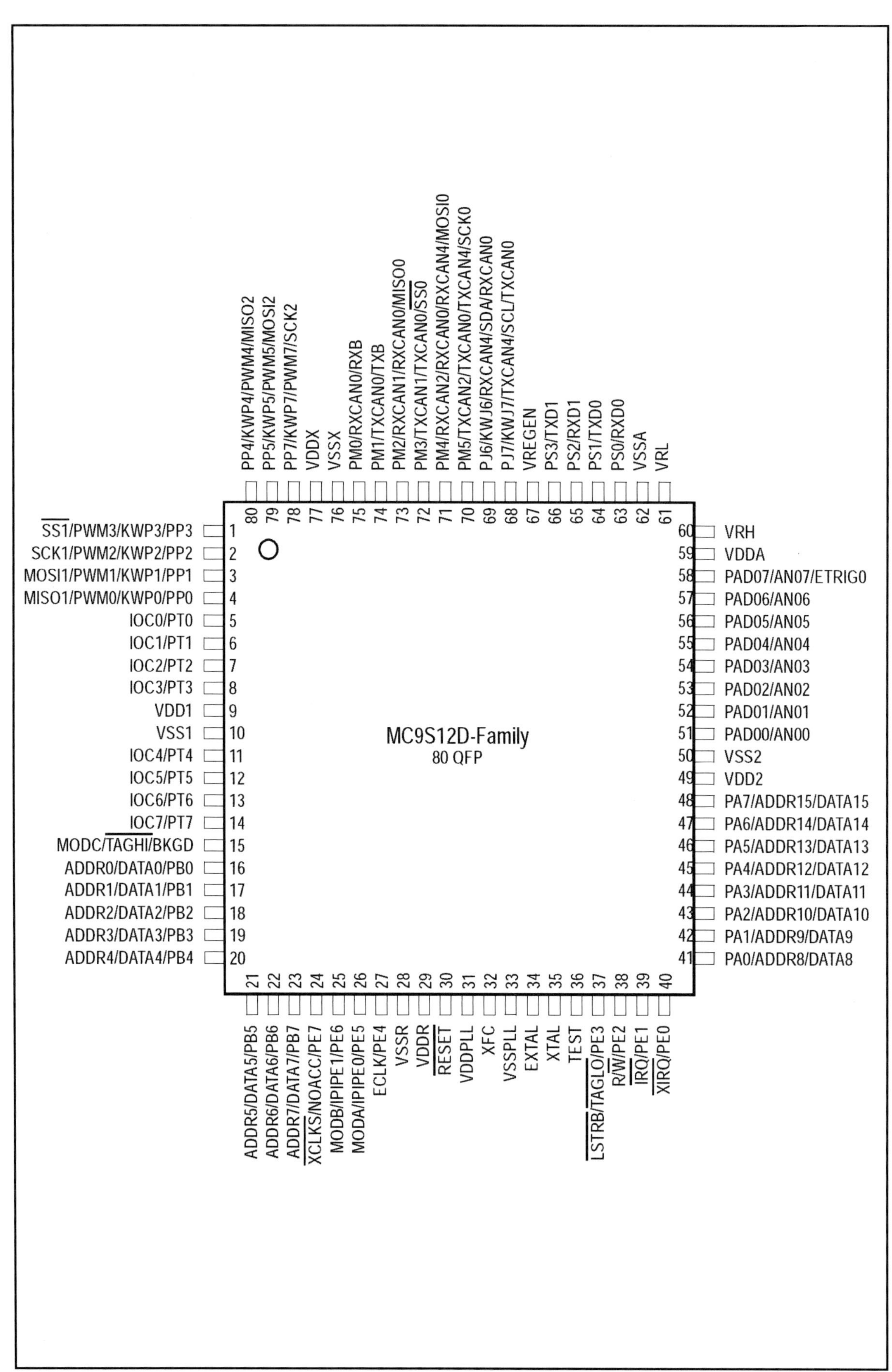

Figure 4-2. MC9S12D QFP 80-Pin Diagram (from Freescale)

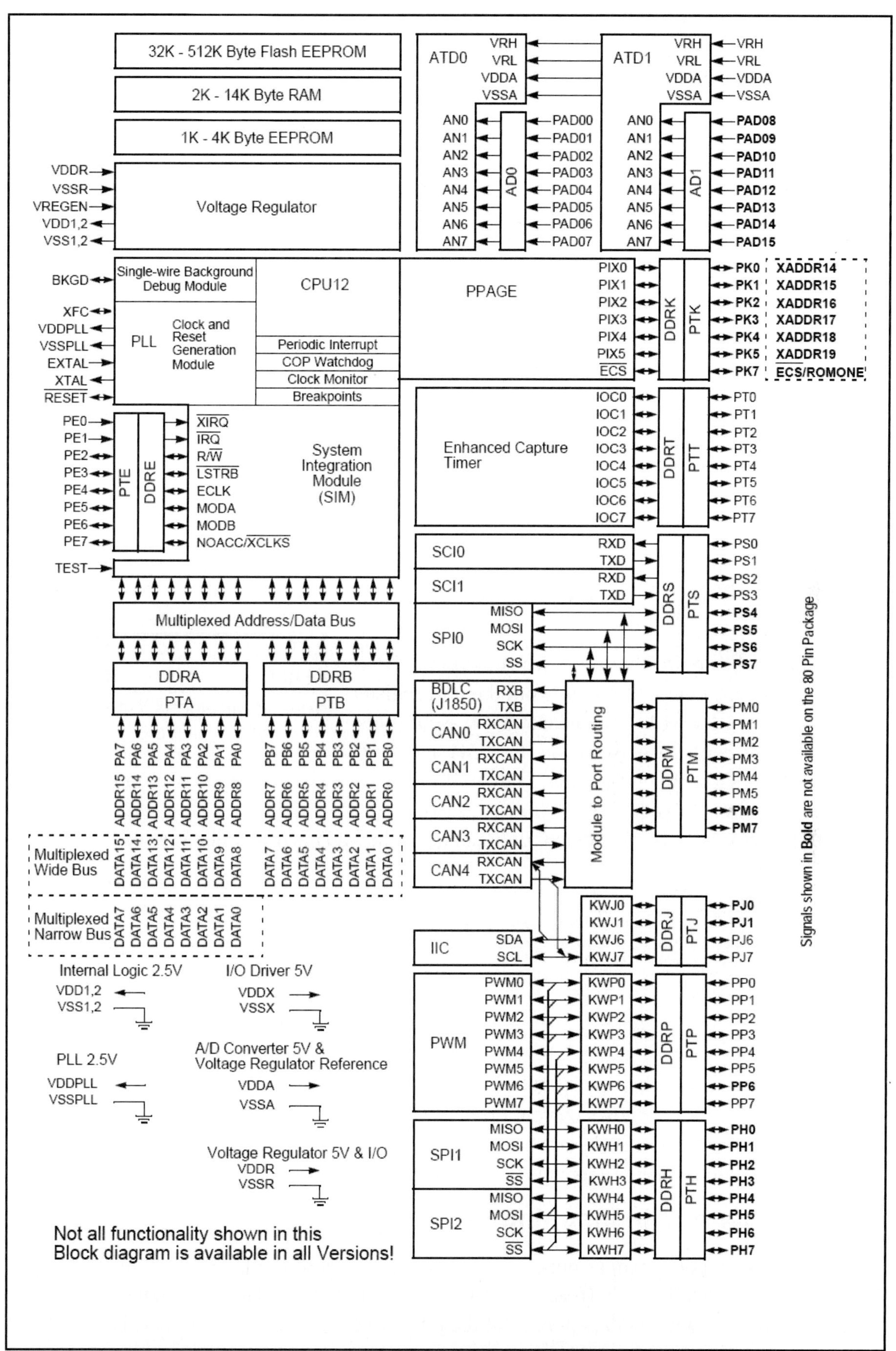

Figure 4-3. CPU12 Block Diagram (from Freescale)

DDR register role in outputting data

Many of the ports in the 9S12 can be used for input or output. The DDR register is used solely for the purpose of making a given port an input or output port. For example, to make a port an output, we write 1s to the DDRx register. In other words, to output data to any of the pins of Port B, we must first put 1s into the DDRB register to make it an output port, and then send the data to Port B itself.

The following code will toggle all 8 bits of Port B forever with some time delay in between "on" and "off" states:

```
      LDAA   #$FF   ;A = FFH
      STAA   DDRB   ;make Port B an output port
L1    LDAA   #$55   ;A = 55H
      STAA   PORTB  ;put 55H on port B pins
      JSR    DELAY
      LDAA   #$AA   ;A = AAH
      STAA   PORTB  ;put AAH on port B pins
      JSR    DELAY
      BRA    L1
```

It must be noted that unless we activate the DDR bit (set it to 1), the data will not go from the port register to the pins of the HCS12. This means that if we remove the first two lines of the above code, the 55H and AAH values will not show up on the pins. They will be sitting in the SFR of Port B inside the CPU.

Table 4-2: Register Addresses for MC9S12DP512 Ports

Name	Address (hex)
PORTA	0000
DDRA	0002
PORTB	0001
DDRB	0003
PORTE	0008
DDRE	0009
PTH	0260
DDRH	0262
PTJ	0268
DDRJ	026A
PORTK	0032
DDRK	0033
PTM	0250
DDRM	0252
PTP	0258
DDRP	025A
PTS	0248
DDRS	024A
PTT	0240
DDRT	0242

Note: Many ports use the PT name instead of PORT. Upon reset all ports are configured as input.

Another version of the previous program using the MOVB instruction is as follows:

```
      MOVB   #$FF,DDRB    ;make Port B an output port
L1    MOVB   #$55,PORTB   ;put 55H on port B pins
      JSR    DELAY
      MOVB   #$AA,PORTB   ;put AAH on port B pins
      JSR    DELAY
      BRA    L1
```

It must be noted that 55H (01010101) when complemented becomes AAH (10101010). Appendix C shows current driving capability for I/O pin logic levels.

Note that upon reset, all ports have value 00H in their DDRx registers. This means all ports are configured as inputs. See Example 4-1.

DDR register role in inputting data

To make a port an input port, we must first put 0s into the DDRx register for that port, and then bring in (read) the data present at the pins. The code in part (b) of Example 4-1 will get data present at the pins of port H and send it to port B indefinitely, after adding the value 5 to it:

Example 4-1

Write and test a program to a) toggle all the bits of PORTA and PORTB and b) get a byte of data from PTH and send it to PORTB.

Solution:

a)

```
R1      EQU $1107
R2      EQU $1108
        LDS    #$8FF ;set the stack pointer
        LDAA   #$FF          ;A = FF
        STAA   DDRA          ;make Port A an output port
        STAA   DDRB          ;make Port B an output port
        LDAA   #$55          ;A = 55h
L2      STAA   PORTA         ;put 55h on port A pins
        STAA   PORTB
        JSR    DELAY
        COMA                 ;complement reg A
        BRA    L2

;-----------DELAY
DELAY
        PSHA
        LDAA #200
        STAA R1
D1      LDAA #250
        STAA R2
D2      NOP
        NOP
        NOP
        DEC  R2
        BNE  D2
        DEC  R1
        BNE  D1
        PULA
        RTS
        END
```

b)

```
        LDAA   #%11111111   ;A = 11111111 (binary)
        STAA   DDRB         ;Port B an output port (1 for Out)
        LDAA   #%00000000   ;A = 00000000 (binary)
        STAA   DDRH         ;Port H an input port (0 for input)
L2      LDAA   PTH          ;move data from PTH pins to reg A
        ADDA   #5           ;add some value to it
        STAA   PORTB        ;send it to Port B
        BRA    L2           ;continue forever
```

Another version of the program is as follows:

```
        MOVB   #$FF,DDRB    ;make Port B an output port
        MOVB   #00,DDRH     ;make Port H an input port
L2      LDAA   PTH          ;move data from PTH pins to reg A
        ADDA   #5           ;add some value to it
        STAA   PORTB        ;send it to Port B
        BRA    L2           ;continue forever
```

Again, it must be noted that unless we activate the DDRx bits (by putting 0s there), the data will not be brought into the A register from the pins of Port H. For the current driving capability of the pins, see Appendix C.

Port A

Port A occupies a total of 8 pins (PA0–PA7). To use the pins of Port A as both input and output ports, each bit must be connected externally to the pin by enabling the bits of the DDRA register. For example, the following code will continuously send out to Port A the alternating values of 55H and AAH:

```
        ;toggle all bits of PORTA

        MOVB   #$FF,DDRA    ;make Port A an output port
L1      MOVB   #$55,PORTA   ;put 55H on port A pins
        JSR    DELAY
        MOVB   #$AA,PORTA   ;put AAH on port A pins
        JSR    DELAY
        BRA    L1
```

Port A as input

In order to make all the bits of Port A an input, DDRA must be programmed by writing 0 to all the bits. In the following code, Port A is configured first as an input port by writing all 0s to register DDRA, and then data is received from Port A and saved in some RAM location:

```
        MYREG EQU    0X820 ;save it here

        MOVB   #%00000000, DDRA  ;make Port A an input port
        LDAA   PORTA             ;move from PORTA pins to reg A
        STAA   MYREG             ;save it in RAM loc of MYREG
```

Port B

Port B occupies a total of 8 pins (PB0–PB7). To use the pins of Port B as I/O, the bits of register DDRB must be activated. For example, the following code will continuously send out the alternating values of 55H and AAH to Port B:

```
        ;toggle all bits of PORTB

        MOVB   #$FF,DDRB    ;make Port B an output port
L1      MOVB   #$55,PORTB   ;put 55H on port B pins
        JSR    DELAY
        MOVB   #$AA,PORTB   ;put AAH on port B pins
        JSR    DELAY
        BRA    L1
```

Port B as input

In order to make all the bits of Port B an input, DDRB must be programmed by writing 0 to all the bits. In the following code, Port B is configured first as an input port by writing all 0s to register DDRB, and then data is received from Port B and sent to Port A.

```
MOVB   #%00000000,DDRB    ;make Port B an input port
MOVB   #$FF,DDRA    ;make Port A an output port
LDAA   PORTB        ;move from PORTB pins to reg A
STAA   PORTA        ;send it to PORTA
```

Dual role of Ports A and B

In expanded mode, the HCS12 uses Port A and Port B for address/data multiplexing. This allows the connection of HCS12 to external RAM and ROM. The alternate functions of the pins for Ports A and B are shown in Tables 4-3 and 4-4, respectively.

Table 4-3: Port A Alternate Functions (in Expanded Mode)

Bit	Function
PA0	ADD8/DATA8
PA1	ADD9/DATA9
PA2	ADD10/DATA10
PA3	ADD11/DATA11
PA4	ADD12/DATA12
PA5	ADD13/DATA13
PA6	ADD14/DATA14
PA7	ADD15/DATA15

Table 4-4: Port B Alternate Functions (in Expanded Mode)

Bit	Function
PB0	ADD0/DATA0
PB1	ADD1/DATA1
PB2	ADD20/DAT2
PB3	ADD3/DATA3
PB4	ADD4/DATA4
PB5	ADD5/DATA5
PB6	ADD6/DATA6
PB7	ADD7/DATA7

Port E

Port E occupies a total of 8 pins (PE0–PE7). To use the pins of Port E as both input and output ports, each bit must be connected to the external pin by enabling the bits of register DDRE. In Port E, we can use only the six pins of PE2–PE7 for output. The PE0 and PE1 pins are configured for input permanently. For example, the following code will continuously send out the alternating values of 55H and AAH to Port E:

```
       ;toggle all bits of PORTE

       MOVB   #$FF,DDRE    ;make Port E an output port
L1     MOVB   #$55,PORTE   ;toggle PORTE pins PE2-PE7
                           ;PE0 and PE1 will not change
       JSR    DELAY
       MOVB   #$AA,PORTE   ;put AAH on port E pins
       JSR    DELAY
       BRA    L1
```

Since only pins PE5–PE0 can be used for output, we will not see any toggling of pins PB0 and PB1.

Port E as input

In order to make all the bits of Port E an input, DDRE must be programmed by writing 0 to all the bits. In the following code, Port E is configured first as an input port by writing all 0s to register DDRE, and then data is received from Port E and saved in some RAM location:

```
MYREG EQU    $850   ;save it here

MOVB  #%00000000,DDRE    ;make Port E an input port
LDAA  PORTE         ;move from PORTE pins to reg A
STAA  MYREG         ;save it in RAM loc of MYREG
```

Alternate functions of Port E

Many of the pins of Port E are used in expanded mode, as shown in Figure 4-1. Pins PE0 and PE1 are used for hardware interrupt, as we will see in Chapter 11.

Dual role of Ports H through T

Although we can use ports H through T for simple I/O programming, they have more important alternate functions, as shown in Figures 4-1 and 4-2. For example, the PS0 and PS1 pins are used for the RxD and TxD signals of serial communication. We will show how to use these alternate functions in future chapters.

Analog-to-digital converter pins

HCS12 chips can have one or two 8-channel ADC modules. In the 80-pin (QFP) HCS12 chip, we have one module, and the 8 channels are designated as the PAD00–PAD07 pins. The 122-pin LQFP chip has two modules of ADCs and the pins are designated as PAD00–PAD07 and PAD10–PAD17. See Figures 4-1 and 4-2. We will study ADC in Chapter 13. Note that we can use the ADC pins for I/O operations if they are not used by the ADC. The only difference is that they do not have DDR registers.

Review Questions

1. True or false. All of the HCS12 ports have 8 pins.
2. Which package type gives us more ports?
3. True or false. PORTK does exist in the 80-pin HCS12 chip.
4. Code a simple program to send 99H to Port A.
5. To make Port B an output port, we must place _______ in register _______.

SECTION 4.2: I/O BIT MANIPULATION PROGRAMMING

In this section we further examine the HCS12 I/O instructions. We pay special attention to I/O bit manipulation because it is a powerful and widely used feature of the HCS12 family.

I/O ports and bit-addressability

Sometimes we need to access only 1 or 2 bits of the port instead of the entire 8 bits. A powerful feature of HCS12 I/O ports is their capability to access individual bits of the port without altering the rest of the bits in that port. For all

HCS12 ports, we can access either all 8 bits or any group of bits without altering the rest. Table 4-5 lists the bit-oriented instructions for the HCS12. Although the instructions in Table 4-5 can be used for any RAM location, I/O port operations use them most often. We will see the use of these instructions throughout future chapters.

Table 4-5: Bit-Oriented Instructions for HCS12

Instruction	Function
BSET mem,mask-byte	Bit Set RAM loc
BCLR mem,mask-byte	Bit Clear RAM
BRCLR	Branch if bits clear
BRSET	Branch if bits set

Next we describe all these instructions and examine their usage.

BSET (bit set)

To set HIGH certain bits of a given location, we use the syntax "`BSET mem_loc, mask-byte`" where mem_loc is any location in the RAM (or register) and mask_byte has the desired bit number as 1. Although the bit-oriented instructions can be used for manipulation of bits D0–D7 of any RAM location, they are mostly used for I/O ports in embedded systems. For example, "`BSET PORTB, %00100000`" sets HIGH bit 5 of Port B. We can use this instruction to set HIGH more than one bit. For example, "`BSET PORTB, %00000011`" sets HIGH bits D0 and D1 of Port B. See Example 4-2.

Example 4-2

An LED is connected to each pin of Port B. Write a program to turn on each LED from pin D0 to pin D7. Call a delay module before turning on the next LED.

Solution:

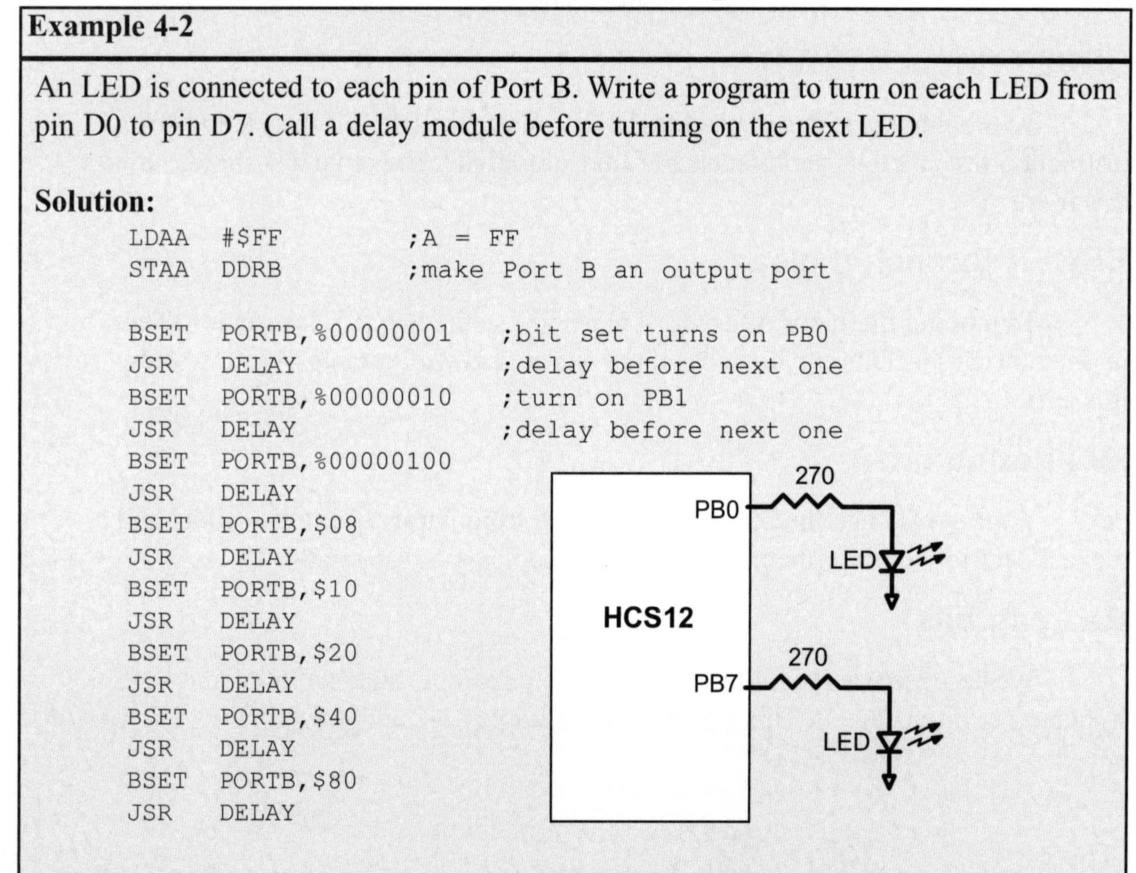

```
        LDAA   #$FF             ;A = FF
        STAA   DDRB             ;make Port B an output port

        BSET   PORTB,%00000001   ;bit set turns on PB0
        JSR    DELAY             ;delay before next one
        BSET   PORTB,%00000010   ;turn on PB1
        JSR    DELAY             ;delay before next one
        BSET   PORTB,%00000100
        JSR    DELAY
        BSET   PORTB,$08
        JSR    DELAY
        BSET   PORTB,$10
        JSR    DELAY
        BSET   PORTB,$20
        JSR    DELAY
        BSET   PORTB,$40
        JSR    DELAY
        BSET   PORTB,$80
        JSR    DELAY
```

BCLR (bit clear)

To clear bits of a given location, we use the "`BCLR mem_loc, mask_byte`" instruction. For example, "`BCLR PORTB, %00000001`" clears (makes LOW) bit 0 of Port B. We can use this instruction to set LOW more than one bit. For example, "`BCLR PORTB, %00001100`" clears bits 2 and 3 of Port B. Remember that for I/O ports, we must activate the appropriate bit in the DDRx register if we want the pin to reflect the changes. For example, the following code toggles pins PB2 and PB0 continuously:

```
      BSET  DDRB,%00000101    ;make PB2 and PB0 an output pin
AGAIN BSET  PORTB,%00000101   ;make PB2 and PB0 high
      JSR   DELAY
      BCLR  PORTB,%00000101   ;bit clear (PB2 = PB7 = low)
      JSR   DELAY
      BRA   AGAIN
```

See Example 4-3.

Checking an input pin

To make decisions based on the status of given bits of the port, we use the instructions BRCLR (branch if clear) and BRSET (branch if set). These bit-oriented instructions are widely used for I/O operations. They allow you to monitor pins and make a decision depending on whether they are 0 or 1. Again it must be noted that the instructions BRCLR and BRSET can be used for any bits in the memory space, including the I/O ports A, B, E, and so on.

BRSET (branch if set)

To monitor the status of bits for HIGH, we use the BRSET instruction. This instruction tests the bits and branches if they are HIGH. Example 4-4 shows how it is used.

BRCLR (branch if clear)

To monitor the status of bits for LOW, we use the BRCLR instruction. This instruction tests the bits and branches if the bits are LOW. Example 4-4 shows how it is used.

Monitoring bits

We can also use the bit test instructions to monitor the status of bits and make a decision to perform an action. See Examples 4-5, 4-6, and 4-7.

Reading bits

We can also use the bit test instructions to read the status of a bit and send it to another bit or save it. This is shown in Examples 4-7 and 4-8.

Example 4-3

Write the following programs:
(a) Create a square wave of 50% duty cycle on bit 0 of Port A.
(b) Create a square wave of 66% duty cycle on bit 3 of Port A.

Solution:

(a) The 50% duty cycle means that the "on" and "off" states (or the high and low portions of the pulse) have the same length. Therefore, we toggle PA0 with the same time delay between each state.

```
      BSET  DDRA,%000000001   ;set DDRA bit for PA0 = out
HERE  BSET  PORTA,%00000001   ;set to HIGH PA0 (PA0 = 1)
      JSR   DELAY             ;call the delay subroutine
      BCLR  PORTA,%00000001   ;PA0 = 0
      JSR   DELAY
      BRA   HERE        ;keep doing it
```

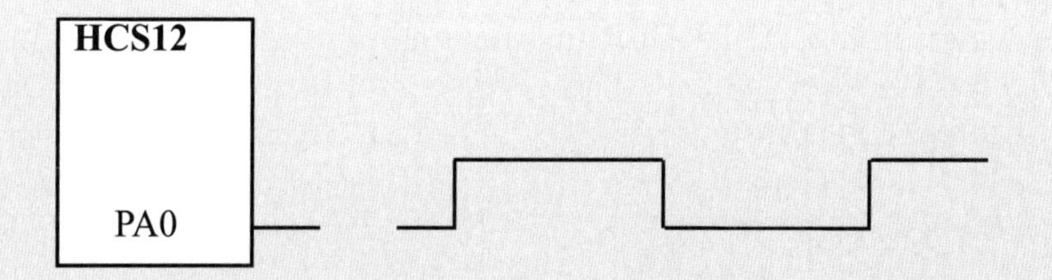

(b) A 66% duty cycle means that the "on" state is twice the "off" state.

```
      BSET  DDRA,%00001000    ;set DDRA bit for PA3 output
BACK  BSET  PORTA,%00001000   ;PA3 = 1
      JSR   DELAY             ;call the delay subroutine
      JSR   DELAY             ;twice for 66% part
      BCLR  PORTA,%00001000   ;PA3 = 0
      JSR   DELAY             ;call delay once for 33% part
      BRA   BACK              ;keep doing it
```

HCS12

PA3

CAUTION

We strongly recommend that you study Section C.2 (Appendix C) before connecting any external hardware to your HCS12 system. Failure to use the proper connection to port pins can damage the ports of your HCS12 chip.

Example 4-4

Write a program to perform the following:
(a) Keep monitoring the PB2 bit until it becomes HIGH;
(b) When PB2 becomes HIGH, write value 45H to Port A, and also send a HIGH-to-LOW pulse to PE3.

Solution:

```
      BCLR  DDRB,%00000100              ;make PB2 an input
      LDAA  #$FF
      STAA  DDRA                        ;make PORTA an output port
      BSET  DDRE,%00001000              ;make PE3 an output
      LDAA  #$45                        ;A = 45H
AGAIN BRSET PORTB,%00000100,OVER        ;branch if PB2 is HIGH
      BRA   AGAIN                       ;keep checking if LOW
OVER  STAA  PORTA                       ;issue reg A to Port A
      BSET  PORTE,%00001000             ;bit set PE3 (H-to-L)
      BCLR  PORTE,%00000000             ;bit clear PE3 (L)
```

Another way is to use the BRCLR instruction.

```
      BCLR  DDRB,%00000100              ;make PB2 an input
      LDAA  #$FF
      STAA  DDRA                        ;make PORTA an output port
      BSET  DDRE,%00001000              ;make PE3 an output
      LDAA  #$45                        ;A = 45H
AGAIN BRCLR PORTB,%00000100,AGAIN       ;keep checking if low
      STAA  PORTA                       ;issue reg A to Port A
      BSET  PORTE,%00001000             ;bit set PE3 (H-to-L)
      BCLR  PORTE,%00001000             ;bit clear PE3 (L)
```

Example 4-5

Assume that bit PB3 is an input and represents the condition of a door alarm. If it goes LOW, it means that the door is open. Monitor the bit continuously. Whenever it goes LOW, send a HIGH-to-LOW pulse to port PE5 to turn on a buzzer.

Solution:

```
      BCLR  DDRB,%00001000              ;make PB3 an input
      BSET  DDRE,%00100000              ;make PE5 an output
HERE  BRSET PORTB,%00010000,HERE        ;keep monitoring PB3 for LOW
      BSET  PORTE,%00100000             ;make PE5 HIGH
      BCLR  PORTE,%00100000             ;make PE5 LOW for H-to-L
      BRA   HERE
```

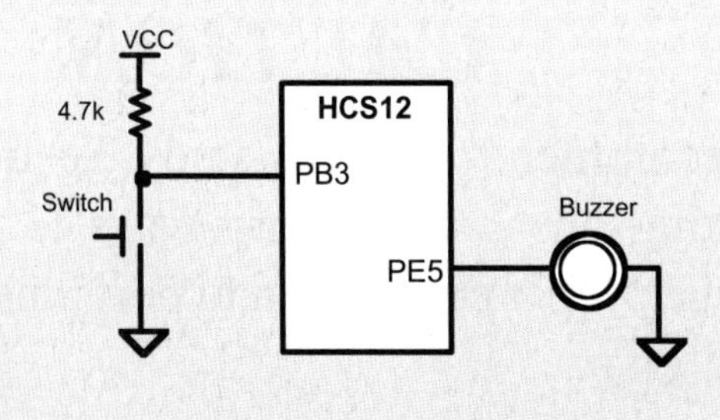

Example 4-6

A switch is connected to pin PB2. Write a program to check the status of SW and perform the following:
(a) If SW = 0, send the letter 'N' to PORTA.
(b) If SW = 1, send the letter 'Y' to PORTA.

Solution:

```
      BCLR  DDRB,%00000100           ;make PB2 an input
      LDAA  #$FF
      STAA  DDRA                     ;make PORTA an output port
AGAIN BRSET PORTB,%00000100,OVER     ;branch if PB2 is HIGH
      LDAA  #'N'
      STAA  PORTA                    ;send 'N' to Port A
      BRA   AGAIN
OVER  LDAA  #'Y'
      STAA  PORTA                    ;send 'Y' to Port A
      BRA   AGAIN
```

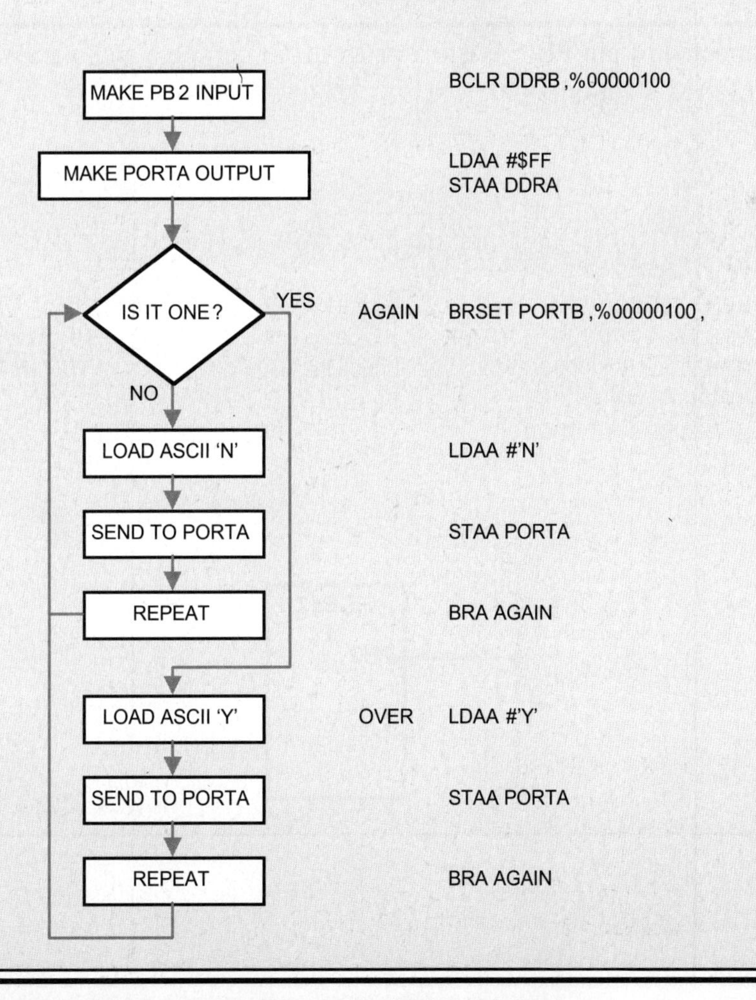

Example 4-7

A switch is connected to pin PB0 and an LED to pin PB7. Write a program to get the status of SW and send it to the LED. (When PB0 goes HIGH, i.e. a switch opens, an led lights.)

Solution:

```
        BCLR   DDRB,%00000001             ;make PB0 an input
        BSET   DDRB,%10000000             ;make PB7 an output
BACK    BRSET  PORTB,%00000001,OVER       ;keep monitoring PB0 for HIGH
        BCLR   PORTB,%10000000            ;make PB7 LOW
        BRA    BACK
OVER    BSET   PORTB,%10000000            ;make PB7 HIGH
        BRA    BACK
```

Example 4-8

A switch is connected to pin PB0. Write a program to get the status of SW and save it in the D0 bit of RAM location $820.

Solution:

```
MYBITREG EQU $820         ;set aside loc $820 as reg

        BCLR   DDRB,%00000001             ;make PB0 an input
BACK    BRSET  PORTB,%00000001,OVER       ;keep monitoring PB0 for HIGH
        BCLR   MYBITREG,%00000001         ;make D0 LOW
        BRA    BACK
OVER    BSET   MYBITREG,%00000001         ;make D0 HIGH
        BRA    BACK
```

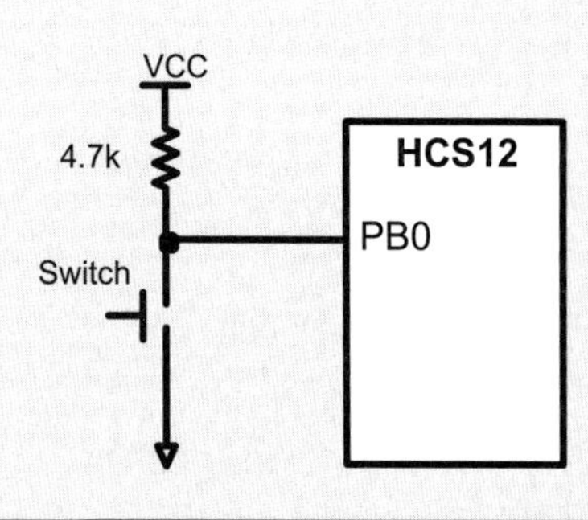

Review Questions

1. True or false. The instruction "`BSET PORTB,%000000010`" makes pin PB1 HIGH while leaving other pins of PORTB unchanged, if bit 1 of the DDRB bits is configured for output.
2. Show one way to toggle pin PB7 continuously using HCS12 instructions.
3. Using the instruction "`BRSET PORTA, %00100000,OVER`" assumes that bit PA5 is an _________ (input, output) pin.
4. True or false. We can use a bit instruction to test more than one bit.
5. True or false. This is a valid instruction "`BSET PORTB, %00110000`"

PROBLEMS

SECTION 4.1: I/O PORT PROGRAMMING IN HCS12

1. The MC9S12Dxx family with the LQFP package has ____pins.
2. The MC9S12Dxx family with the QFP package has ____pins.
3. In the MC9S12Dxx family, how many pins are assigned to Port A?
4. In the MC9S12Dxx family, how many pins are assigned to Port B?
5. In the MC9S12Dxx family, how many pins are assigned to Port E?
6. In the MC9S12Dxx family with the LQFP package, how many pins are assigned to Port H?
7. In the MC9S12Dxx family, how many pins are assigned to Port J?
8. Upon reset, all the bits of ports are configured as _____ (input, output).
9. For the HCS12, which register must be programmed in order for the port to be used as simple I/O?
10. Explain the role of DDRx in I/O operations.
11. Write a program to get 8-bit data from PORTA and send it to ports PORTB and PORTH.
12. Write a program to get 8-bit data from PORTB and send it to ports PORTA and PORTH.
13. Which pins are for RxD1 and TxD1 in the 80-pin QFP package?
14. Give the address locations in the register area assigned to Ports A–E and their DDR registers for the HCS12.
15. Write a program to toggle all the bits of PORTA and PORTB continuously (a) using AAH and 55H (b) using the complement instruction.

SECTION 4.2: I/O BIT MANIPULATION PROGRAMMING

16. Which ports of the HCS12 are bit-addressable?
17. What is the advantage of bit-addressability for HCS12 ports?
18. When "BSET PORTB, %00000001" is executed, which bit is set?
19. Is the instruction "`COM PORTB`" a valid instruction?
20. Write a program to toggle PB2 and PB5 continuously without disturbing the rest of the bits.
21. Write a program to toggle PB3, PB7, and PB5 continuously without disturbing the rest of the bits.
22. Write a program to monitor bit PH3. When it is HIGH, send 55H to PORTB.

ANSWERS TO REVIEW QUESTIONS

SECTION 4.1: I/O PORT PROGRAMMING IN HCS12

1. False
2. LQFP
3. False
4.
```
MOVB DDRA,$FF
LDAA #$99
STAA PORTA
```
5. $FF, DDRA

SECTION 4.2: I/O BIT MANIPULATION PROGRAMMING

1. True
2.
```
     BSET  DDRB,%100000000   ;set DDRB bit for PB7 = out
HERE BSET  PORTB,%10000000   ;set to HIGH PB7 (PB7 = 1)
     JSR   DELAY             ;call the delay subroutine
     BCLR  PORTA,%10000000   ;PB7 = 0
     JSR   DELAY
     BRA   HERE        ;keep doing it
```
3. Input
4. True
5. True

CHAPTER 5

ARITHMETIC, LOGIC INSTRUCTIONS, AND PROGRAMS

OBJECTIVES

Upon completion of this chapter, you will be able to:

- >> Define the range of numbers possible in the HCS12 for unsigned data
- >> Code addition and subtraction instructions for unsigned data
- >> Perform addition of BCD data
- >> Code HCS12 unsigned data multiplication instructions
- >> Code HCS12 unsigned data division instructions
- >> Code HCS12 Assembly language logic instructions AND, OR, and EX-OR
- >> Code HCS12 Assembly language for shift and rotate instructions
- >> Use HCS12 logic instructions for bit manipulation
- >> Use the compare instruction for program control
- >> Code HCS12 rotate instructions and data serialization
- >> Explain the BCD (binary coded decimal) system of data representation
- >> Contrast and compare packed and unpacked BCD data
- >> Code HCS12 programs for ASCII and BCD data conversion
- >> Define the range of numbers possible in HCS12 for signed data

SECTION 5.1: ARITHMETIC INSTRUCTIONS AND PROGRAMS

Unsigned numbers are defined as data in which all the bits are used to represent data, and no bits are set aside for the positive or negative sign. This means that the operand can be between 00 and $FF (0 to 255 decimal) for 8-bit data.

Addition of unsigned numbers

In order to add numbers together in the HCS12, the A register is involved. One form of the ADD instruction is

```
ADDA   Operand   ;A = A + Operand
```

The sum is stored in register A. The instruction could change any of the C, H, Z, N, or V bits of the status register. The effect of the ADDA instruction on N and V is discussed in Section 5.6. Look at Example 5-1.

Example 5-1

Show how the flag register is affected by the following instructions.

```
      LDAA #$F5              ;A = F5 hex
      ADDA #$B               ;A = F5 + 0B = 00 and C = 1
```

Solution:

```
     $F5          1111 0101
   + $0B        + 0000 1011
     100H         0000 0000
```

After the addition, register A contains 00 and the flags are as follows:
C = 1 because there is a carry out from D7. Z = 1 because the result in A is zero.
H = 1 because there is a carry from D3 to D4.

ADDA instruction and addition of individual bytes

Instruction "ADDA memory" allows the addition of A and individual bytes residing in RAM locations. Notice that register A must be involved because memory-to-memory arithmetic operations are never allowed in HCS12 Assembly language. To calculate the sum of any number of operands, the carry flag should be checked after the addition of each operand. Example 5-2 uses a RAM location to accumulate carries as the operands are added to A.

ADCA instruction and addition of 16-bit numbers

When adding two 16-bit data operands, we need to be concerned with the propagation of a carry from the lower byte to the higher byte. This is called *multi-byte addition* to distinguish it from the addition of individual bytes. The instruction ADCA (add with carry to A) is used on such occasions.

Example 5-2

Assume that file register RAM locations \$840–\$843 have the hex values shown below. Write a program to find the sum of the values. At the end of the program, location 807 of the RAM should contain the low byte and location 806 the high byte of the sum, which complies with the big endian convention.

Address	Data
840	7D
841	EB
842	C5
843	5B

Solution:

```
L_Byte EQU  $807          ;assign RAM location 807 to L_byte of sum
H_Byte EQU  $806          ;assign RAM location 806 to H_byte of sum

      CLRA                ;clear A (A = 0)
      STAA   H_Byte       ;H_Byte = 0
      ADDA   $840         ;A = 0 + 7DH = 7DH , C = 0
      BCC    N_1          ;branch if C = 0
      INC    H_Byte       ;increment (now H_Byte = 0)
N_1   ADDA   $841         ;A = 7D + EB = 68H and C = 1
      BCC    N_2          ;
      INC    H_Byte       ;C = 1, increment (now H_Byte = 1)
N_2   ADDA   $842         ;A = 68 + C5 = 2D and C = 1
      BCC    N_3          ;
      INC    H_Byte       ;C = 1, increment (now H_Byte = 2)
N_3   ADDA   $843         ;A = 2D + 5B = 88H and C = 0
      BCC    N_4          ;
      INC    H_Byte       ;(H_Byte = 2)
N_4   STAA   L_Byte       ;now L_Byte = 88H
```

Address	Data
805	xx
806	02
807	88
808	xx

At the end, the RAM location 807 = (88), and location 806 = (02) because 7D + EB + C5 + 5B = 288H. We can use the register indexed addressing mode to do this program much more efficiently with a loop.

```
L_Byte EQU  $807          ;assign RAM location 807 to L_byte of sum
H_Byte EQU  $806          ;assign RAM location 806 to H_byte of sum
MYCOUNT EQU $800          ;assign RAM location for counter

      LDAA   #4           ;the counter value
      STAA   MYCOUNT      ;load the counter
      CLRA                ;clear A (A = 0)
      STAA   H_Byte       ;H_Byte = 0
      LDX    #$840        ;set the pointer
BACK  ADDA   0,X          ;add byte
      BCC    OVER         ;bypass if C=0
      INC    H_Byte       ;add one for carry
OVER  INX                 ;next item
      DEC    MYCOUNT      ;dec counter
      BNE    BACK         ;do for all the bytes
      STAA   L_Byte       ;store the result
```

For example, look at the addition of \$3CE7 + \$3B8D, as shown next.

```
      1
      3C E7
+     3B 8D
      78 74
```

When the first byte is added, there is a carry (E7 + 8D = 74, CY = 1). The carry is propagated to the higher byte, which results in 3C + 3B + 1 = 78 (all in hex). Example 5-3 shows the above steps in a HCS12 program. We can also use the 16-bit add instruction to do the same thing since HCS12 supports the 16-bit arithmetic operations.

Example 5-3

Write a program to add two 16-bit numbers. The numbers are \$3CE7 and \$3B8D. Assume that RAM location 807 = (8D) and location 806 = (3B). Place the sum in RAM locations 808 and 809; location 809 should have the lower byte.

```
   3CE7
 + 3B8D
--------------------
   7874
```

Address	Data
806	3B
807	8D
808	78
809	74

Solution:

```
LDAA #$E7    ;load the low byte now (A = $E7)
ADDA $807    ;A = E7 + 8D = 74 and CY = 1
STAA $809
LDAA #$3C    ;load the high byte (A = $3C)
ADCA $806    ;A = A + RAM_loc + carry, adding the upper byte
STAA $808    ;with carry from lower byte
             ;A = 3C + 3B + 1 = $78 (all in hex)
```

Notice the use of ADDA for the lower byte and ADCA for the higher byte. We can do the 16-bit addition using the D register.

```
LDD    #$3CE7        ;D = 3CE7 hex
ADDD   $806          ;D = 3CE7 + 3B8D = 7478
STD    $808
```

Multibyte addition

In addition of multibyte numbers, the ADCA is much more useful than the ADDD instruction. Examine the addition of the following multibyte numbers:

```
      56F398C2
+     2B45F387
```

Address	Data
810	2B
811	45
812	F3
813	87

Address	Data
850	56
851	F3
852	98
853	C2

The following program adds two multibyte numbers:

```
MYCOUNT       EQU    $800
      LDAA   #4
      STAA   MYCOUNT       ;load the counter
      CLRA                 ;clear A (A = 0)
      LDX    #$813         ;set the pointer
      LDY    #$853         ;set the pointer
      CLC                  ;clear carry flag
BACK  LDAA   0,X           ;load the byte
      ADCA   0,Y           ;add the byte with carry if there is one
      STAA   0,Y           ;store the result
      DEX                  ;decrement pointer
      DEY                  ;next item
      DEC    MYCOUNT       ;dec counter
      BNE    BACK          ;do for all the bytes
```

BCD (binary coded decimal) number system

BCD stands for *binary coded decimal*. BCD is needed because in everyday life we use the digits 0 to 9 for numbers, not binary or hex numbers. Binary representation of 0 to 9 is called BCD (see Figure 5-1). In computer literature, one encounters two terms for BCD numbers: (1) unpacked BCD, and (2) packed BCD. We describe each one next.

Digit	BCD
0	0000
1	0001
2	0010
3	0011
4	0100
5	0101
6	0110
7	0111
8	1000
9	1001

Figure 5-1. BCD Code

Unpacked BCD

In unpacked BCD, the lower 4 bits of the number represent the BCD number, and the rest of the bits are 0. Example: "0000 1001" and "0000 0101" are unpacked BCD for 9 and 5, respectively. Unpacked BCD requires 1 byte of memory, or an 8-bit register, to contain it.

Packed BCD

In packed BCD, a single byte has two BCD numbers in it: one in the lower 4 bits, and one in the upper 4 bits. For example, "0101 1001" is packed BCD for 59H. Only 1 byte of memory is needed to store the packed BCD operands. Thus one reason to use packed BCD is that it is twice as efficient in storing data.

There is a problem with adding BCD numbers, which must be corrected. The problem is that after packed BCD numbers are added, the result is no longer BCD.

Look at the following.

```
LDAA #$17
ADDA #$28
```

Adding these two numbers gives 0011 1111B (3FH), which is not BCD! A BCD number can only have digits from 0000 to 1001 (or 0 to 9). In other words, adding two BCD numbers must give a BCD result. The result above should have been 17 + 28 = 45 (0100 0101). To correct this problem, the programmer must add 6 (0110) to the low digit: 3F + 06 = 45H. The same problem could have happened in the upper digit (for example, in 52H + 87H = D9H). Again, 6 must be added to the upper digit (D9H + 60H = 139H) to ensure that the result is BCD (52 + 87 = 139). This problem is so pervasive that most microprocessors such as the HCS12 have an instruction to deal with it. The HCS12 instruction "DAA" is designed to correct the BCD addition problem. This is discussed next.

DAA instruction

The DAA (decimal adjust accumulator) instruction in the HCS12 is provided to correct the aforementioned problem associated with BCD addition. The mnemonic "DAA" works only with an operand in the A register. The DAA instruction will add 6 to the lower nibble or higher nibble if needed; otherwise, it will leave the result alone. The following example will clarify these points.

```
LDAA   #$47   ;A = $47 first BCD operand
ADDA   #$25   ;hex(binary) addition (A = $6C)
DAA           ;adjust for BCD addition (A = $72)
```

After the program is executed, register A will contain $72 (47 + 25 = 72). Note that the "DAA" instruction works only on register A.

Summary of DAA action

After any instruction,

1. If the lower nibble (4 bits) is greater than 9, or if H = 1, add 0110 to the lower 4 bits.
2. If the upper nibble is greater than 9, or if C = 1, add 0110 to the upper 4 bits.

In reality there is no use for the H (half carry) flag bit other than for BCD addition and correction.

```
CLRA          ;A = 0
ADDA #$09     ;A = $09
ADDA #$08     ;A = 0x11, H = 1
DAA           ;A = $17 (9 + 8 = 17)
```

As another example, examine the case of adding $55 and $77. This will result in $CC, which is incorrect as far as BCD is concerned.

```
    Hex             BCD
     57             0101 0111
   + 77           + 0111 0111
     CE             1100 1110
   + 66           + 0110 0110
    134           1 0011 0100    Note C = 1
```

Note that the HCS12 does require the use of arithmetic instructions such as ADD, ADCA prior to execution of the "DAA" instruction.

Examine Example 5-4.

Subtraction of unsigned numbers

In the HCS12, there are two different instructions for subtraction: SUB and

Example 5-4

Assume that 4 BCD data items are stored in RAM locations starting at 840H, as shown below. Write a program to find the sum of all the numbers. The result must be in BCD.

Solution:

```
L_Byte      EQU   $807   ;assign RAM loc 807 to L_Byte of sum
H_Byte      EQU   $806   ;assign RAM loc 806 to H_Byte of sum
MYCOUNT     EQU   $800   ;assign RAM location for counter

       LDAA  #4          ;set the counter
       STAA  MYCOUNT
       CLRA              ;clear A (A = 0)
       STAA  H_Byte      ;H_Byte = 0
       LDX   #$840       ;set the pointer
BACK   ADDA  0,X         ;add byte
       DAA
       BCC   OVER        ;bypass if C=0
       INC   H_Byte      ;add one for carry
OVER   INX               ;next item
       DEC   MYCOUNT     ;dec counter
       BNE   BACK        ;do for all the bytes
       STAA  L_Byte      ;store the result
```

Address	Data
840	71
841	88
842	69
843	97

Address	Data
805	xx
806	03
807	25
808	xx

After this code executes, RAM location 806 = (03), and 807 = (25) because 71 + 88 + 69 + 97 = 325H.
This is the BCD version of Example 5-2.

SBC (subtract with borrow). Notice that we use the C (carry) flag for the borrow. We now will examine each of these commands.

SUBA Operand (A = A – Operand)

In subtraction, the HCS12 microcontrollers (indeed, all modern CPUs) use the 2's complement method. Although every CPU contains adder circuitry, it would be too cumbersome (and take too many transistors) to design separate subtracter circuitry. For this reason, the HCS12 uses adder circuitry to perform the subtraction command. Assuming that the HCS12 is executing a simple subtract instruction and that C = 0 prior to the execution of the instruction, one can summarize the steps of the hardware of the CPU in executing the SUBA instruction for unsigned numbers as follows:

1. Take the 2's complement of the source operand.
2. Add it to the A.
3. Invert the carry.

These three steps are performed for every SUB instruction by the internal hardware of the CPU, regardless of the source of the operands, provided that the addressing mode is supported. It is after these steps that the result is obtained and the flags are set. Example 5-5 illustrates the two steps.

NEG (2's complement) instruction

After the execution of SUBA, if C = 0, the result is positive; if C = 1, the

Example 5-5

Show the steps involved in the following.

```
LDAA #$3F           ;load 3FH into A (A = 3FH)
SUBA #$23           ;A = A - 23
```

Solution:

```
   A     = 3F    0011 1111        0011 1111
 - value = 23    0010 0011    +   1101 1101 (2's complement)
           1C                   1 0001 1100
                              C = 0 (result is positive)
```

The flag would be set as follows: C = 0. The programmer must look at the C flag to determine if the result is positive or negative.

result is negative and the destination has the 2's complement of the result. Normally, the result is left in 2's complement, but the NEG (negate, which is 2's complement) instruction can be used to change it. See Example 5-6.

Example 5-6

Write a program to subtract 4C – 6E.

Solution:

```
      MYREG EQU $820
      LDAA  #$6E        ;load A (A = 6EH)
      STAA  MYREG       ;MYREG = 6EH
      LDAA  #$4C        ;A = 4CH
      SUBA  MYREG       ;A = A - MYREG, 4C - 6E = DE, C = 1
      BCC   NEXT        ;if C = 0, jump to NEXT target
      NEGA              ;take 2's complement of A
NEXT  STAA  MYREG       ;save the result in MYREG
```

The following are the steps after the SUBA instruction:

```
  4C    0100 1100                0100 1100
 -6E    0110 1110   2's comp =   1001 0010
 -22                             1101 1110
```

After SUBA, we have C = 1, and the result is negative, in 2's complement. Then it falls through and NEG will be executed. The NEG instruction will take the 2's complement, and we have MYREG = 22H, the hex mangnitude of the negative result.

SBCA (A = A – Operand – Borrow) subtract with borrow

This instruction is used for multibyte numbers and will take care of the borrow of the lower byte. The C flag holds the borrow, so if C = 1 prior to executing the SBCA instruction, it also subtracts 1 from the result. See Example 5-7. Example 5-7 also shows how we can use the 16-bit instruction of SUBD to do the same thing. Table 5-1 shows some of the arithmetic instructions.

Example 5-7

Write a program to subtract two 16-bit numbers. The numbers are $2762 – $1296. Assume RAM location $806 = (12) and location $807 = (96). Place the difference in RAM locations $806 and $807; location $807 should have the lower byte.

```
  2762
– 1296
  14CC
```

Address	Data
806	12
807	96

Solution:

```
LDAA   #$62  ;load the low byte (A = 62H)
SUBA   $807 ;A = A - RAMLoc = 62 - 96 = CCH, C = borrow = 0
STAA   $807  ;store the low byte
LDAA   #$27  ;load the high byte (A = 27H)
SBCA   $806  ;A = A - RAMLoc - b, sub byte with the borrow
             ;A = 27 - 12 - 1 = 14H
STAA   $806
```

After the SUBA, location $806 has = $62 – $96 = $CC and the carry flag is set to 1, indicating there is a borrow. Because C = 1, when SBCA is executed the RAM location $806 has = 27H – 12H – 1 = 14H. Therefore, we have $2762 – $1296 = $14CC. Using the 16-bit subtraction instruction, we get the following:

```
LDD    #$2762        ;load the D reg
SUBD   $806          ;D = D - RAMLoc 806 and 807
STDD   $806          ;store the result
```

Table 5-1: Some Widely Used HCS12 Arithmetic Instructions

Instruction	Action
ADDA	Add Operand to Reg A (A = A + 8-bit Value)
ADCA	Add Operand to Reg A (A = A + 8-bit Value + Carry)
ADDD	Add Operand to Reg D (D = D + 16-bit value)
SUBA	Subtract Operand from Reg A (A = A – 8-bit Value)
SBCA	Subtract Operand and carry from reg A (A = A – 8-bit Value – C)
SUBD	Subtract 16-bit Operand from Reg D (D = A – 16-bit Value)
DAA	Decimal Adjust Addition

Multiplication of unsigned numbers

The HCS12 supports both byte-by-byte and word-by-word multiplications. The bytes and words are assumed to be unsigned data. The syntax for byte-by-byte multiplication is as follows:

```
MUL          ;A × B and 16-bit is result is in reg D
```

In byte-by-byte multiplication, one of the operands must be in the A register, and the second operand must be in register B. After multiplication, the result is in register D; the lower byte is in B, and the upper byte is in A. See Table 5-2.

Table 5-2: Unsigned Multiplication Summary for MUL and EMUL Instructions

Inst	Multiplication	Operand1	Operand2	Result
MUL	Byte × Byte	A	B	D (A = high byte, B = low byte)
EMUL	Word × Word	D	Y	Y:D (Y = high word, D = low word)

Note: Multiplication of operands larger than 16-bit takes some manipulation.

The following example multiplies 25H by 65H.

```
LDAA #$25            ;load 25H to A (A = 25H)
LDAB  #$65           ;B = 65H
MUL                  ;25H * 65H = E99 is in D where
                     ;A = 0EH and B = 99H
```

In word-by-word multiplication, one of the operands must be in the D register, and the second operand must be in register Y. After multiplication, the result is in the registers D and Y; the lower word is in D, and the upper word is in Y. The instruction for the word-by-word operation is EMUL. See Table 5-2. The following example multiplies 500 by 700.

```
LDD #500             ;load 500 to D (D = 500)
LDY #700             ;Y = 700
EMUL                 ;500 x 700 = 350,000 = $55730 = Y:D where
                     ;Y = 0005 and D = $5730
```

Division of unsigned numbers

In the division of unsigned numbers, the HCS12 supports both word/word and doubleword/word. See Table 5-3. The syntax for word/word is as follows.

```
IDIV         ;D/X
```

Table 5-3: Unsigned Division Summary

Instruc.	Division	Numerator	Denominator	Quotient	Remainder
IDIV	word / word	D	X	X	D
EDIV	doubleword/word	Y:D	X	Y	D

(If denominator = 0, then C = 1 indicating an error.)
Word = 16-bit and doubleword = 32-bit

When dividing a word/word, the numerator must be in register D and the denominator must be in X. After the IDIV instruction is performed, the quotient is in X and the remainder is in D. See the following example:

```
LDD    #255          ;D=250
LDX    #10           ;X=10
IDIV                 ;D/X, 255/10 = 25 and remainder = 5
                     ;X=25 and D=5
```

When dividing a 32-bit/16-bit, the numerator must be in registers Y and D and the denominator must be in X. After the EDIV instruction is performed, the quotient is in Y and the remainder is in D. See the following example for dividing 9,502,500/1,000.

```
LDD    #$FF24       ;D = FF24 (9,502,500 = 90FF24 in hex)
LDY    #$90         ;the upper word
LDX    #1000        ;X=1000
EDIV                ;Y:D/X, 9502500/1000 = 9502 (251E in hex)
                    ;is in Y and D = 500 (1F4 in hex) is for
                    ;remainder
```

An application for division

Sometimes an ADC (analog-to-digital converter) is connected to a sensor and the ADC represents some quantity such as temperature or pressure. The 8-bit ADC provides data in hex in the range of 00–FFH. This hex data must be converted to decimal. We do that by dividing it by 10 repeatedly, saving the remainders, as shown in Examples 5-8 and 5-9.

Example 5-8

Assume that we have value FD (hex). Write a program to convert it to decimal. Save the digits in locations $822, $823, and $824, where the least-significant digit is in $824.

Solution:

```
;HCS12 Assembly Language Program for division (Byte/Byte)

RMND_L      EQU    $824
RMND_M      EQU    $823
RMND_H      EQU    $822
MYNUM       EQU    $FD           ;FDH = 253 in decimal
MYDEN       EQU    10            ;253/10
            ORG    $8000         ;start at address 8000
            LDD    #MYNUM        ;load numerator
            LDX    #MYDEN        ;X = 10, the denominator
            IDIV                 ;D/X = 253/10 = 25, RMND =3
            STAB RMND_L          ;save the remainder (always < 10)
            XGDX                 ;exchange D and X for new numerator
            LDX   #MYDEN         ;X = 10, the denominator
            IDIV                 ;D/X
            STAB RMND_M          ;save the next digit
            XGDX                 ;exchange D and X
            STAB RMND_H          ;save the final digit
```

To convert a single decimal digit to ASCII format, we OR it with $30, as shown in Section 5.2.

Example 5-9

Analyze the program in Example 5-8 for a numerator of 253.

Solution:

To convert a binary (hex) value to decimal, we divide it by 10 repeatedly until the quotient is less than 10. After each division the remainder is saved. In the case of an 8-bit binary, such as $FD, we have 253 decimal, as shown below.

```
              Quotient      Remainder
253/10  =     25            3 (low digit)
25/10   =     2             5 (middle digit)
                            2 (high digit)
```

Therefore, we have $FD = 253. In order to display this data, it must be converted to ASCII, which is described in a later section in this chapter.

Review Questions

1. In multiplication of two bytes in the HCS1218, we can place one byte in register ________and the other one in ________.
2. In unsigned byte-by-byte multiplication, the product will be placed in register __________.
3. Is "`MULA VALUE`" a valid HCS12 instruction? Explain your answer.
4. In HCS12, the largest two numbers that can be multiplied are ________ and ________.
5. True or false. The DAA instruction works on A only.
6. Is "`DAA D`" a valid HCS1218 instruction? Explain your answer.
7. The instruction "`ADDA`" places the sum in __________.
8. The instruction "`ADCA Imm_value`" places the sum in _____________.
9. Find the value of the H and C flags in each of the following.

```
(a) LDAA #$4F          (b)    LDAA #$9C
    ADDA #$B1                 ADDA #$63
```

10. Show how the CPU would subtract $05 from $43.
11. If C = 1 and A = $95 prior to the execution of "`SBCA #$4F`", what will be the content of A after the subtraction?

SECTION 5.2: LOGIC INSTRUCTIONS AND PROGRAMS

Apart from I/O and arithmetic instructions, logic instructions are some of most widely used instructions. In this section we cover Boolean logic instructions such as AND, OR, and Exclusive-OR in HCS12. See Table 5-4.

Table 5-4: Some Widely Used Boolean Logic Instructions in HCS12

Instruction	Action
ANDA	AND A with memory (A = A ANDed with memory location)
ORAA	OR A with memory (A = A ORed with memory location)
EORA	EX-OR A with memory (A = A EX-ORed with memory location)
COMA	1's Complement (INVERT) A (A = NOT A)

AND

```
ANDA Operand   ;A = A AND Operand
```

This instruction will perform a logical AND on the two operands and place the result in A. The operand can be an immediate value or in the RAM location. The AND instruction will affect the Z and N flags. N is D7 of the result, and Z = 1 if the result is zero. The AND instruction is often used to mask (set to 0) certain bits of an operand. See Example 5-10.

Logical AND Function

Inputs		Output
X	Y	X AND Y
0	0	0
0	1	0
1	0	0
1	1	1

X, Y → X AND Y

Example 5-10

Show the results of the following.

```
LDAA   #$35          ;A = 35H
ANDA   #$0F          ;A = A ANDed with 0FH (now A = 05)
```

Solution:

```
$35   0 0 1 1 0 1 0 1
$0F   0 0 0 0 1 1 1 1
$05   0 0 0 0 0 1 0 1    ;$35 ANDed with $0F = $05, Z = 0, N = 0
```

OR

```
ORAA   Operand      ;A = A ORed with
Operand
```

This instruction will perform a logical OR on the two operands and place the result in A. The operand can be an immediate value or in any RAM location. The OR instruction will affect the Z and N flags. N is D7 of the result and Z = 1 if the result is zero. See Example 5-11.

Logical OR Function

Inputs		Output
X	Y	X OR Y
0	0	0
0	1	1
1	0	1
1	1	1

X, Y → X OR Y

Example 5-11

Show the results of the following:

```
LDAA #$04                 ;A = 04
ORAA #$30                 ;now A = $34
```

Solution:

(a)

```
$04      0000 0100
$30      0011 0000
$34      0011 0100        04 ORed with 30 = $34, Z = 0 and N = 0
```

EX-OR

```
EORA Operand        ;A = A EX-ORed with Operand
```

This instruction will perform a logical EX-OR on the two operands and place the result in A. The operand can be an immediate value or in any RAM location. The EX-OR instruction will affect the Z and N flags. N is D7 of the result and Z = 1 if the result is zero.

EX-OR can also be used to see if two registers have the same value. "EORA RAM_Loc" will EX-OR the A register and a RAM location, and put the result in A. If both registers have the same value, 00 is placed in A. Then we can use the BEQ instruction to make a decision based on the result. See Examples 5-12 and 5-13.

Logical XOR Function

Inputs		Output
A	**B**	**A XOR B**
0	0	0
0	1	1
1	0	1
1	1	0

A, B — A XOR B

Example 5-12

Show the results of the following:

```
LDAA #$54
EORA #$78
```

Solution:

```
$54     0 1 0 1 0 1 0 0
$78     0 1 1 1 1 0 0 0
$2C     0 0 1 0 1 1 0 0 $54 EX-ORed with $78 = $2C, Z = 0, N = 0
```

Example 5-13

The EX-OR instruction can be used to test the contents of a register by EX-ORing it with a known value. In the following code, we show how EX-ORing value $45 with itself will raise the Z flag:

```
OVER   LDAA   PORTB         ;get a byte from PORTB into A
       EORA   PORTB
       BNE    OVER          ;branch if not zero (i.e. if PORTB
                            ;value has changed)
```

Solution:

```
$45         01000101
$45         01000101
00          00000000
```

EX-ORing a number with itself sets it to zero with Z = 1. We can use the BNE instruction to make the decision. EX-ORing with any other number will result in a nonzero value.

Another application of EX-OR is to toggle the bits of an operand.

```
LDAA #$FF           ;A = $FF
EORA PORTB          ;EX-OR PORTB with 1111 1111 will
STAA PORTB  ;change all the bits of Port B to opposite
```

COM (complement)

This instruction complements the contents of a file register. The complement action changes the 0s to 1s and the 1s to 0s. This is also called *1's complement*.

```
      LDAA  #$FF
      STAA  DDRB   ;Port B = Output
      LDAA  #$55
OVER  COMA  ;complement A
      STAA  PORTB ;keep toggling PORTB
      BRA   OVER
```

Logical Inverter

Input	Output
X	**NOT X**
0	1
1	0

X —▷o— NOT X

NEG (negate)

This instruction takes the 2's complement of a value in A or RAM location. See Example 5-14.

Example 5-14

Find the 2's complement of the value $85.

Solution:

```
LDAA  #$85                $85 = 1000 0101
NEGA                      1's = 0111 1010
                                     + 1
                   2's comp     0111 1011 = 7BH
```

BCLR instruction

In Chapter 4 we showed how to use the BCLR (bit clear) instruction to clear specific bits of an I/O port. We can also use it for any RAM location in HCS12. In the "BCLR Memloc,mask_byte" instruction, the mask byte has the bits that need to be cleared. For example, "BCLR MYRAM,%11000000" will clear bits D6 and D7 of the MYRAM location. In reality, the BCLR instruction ANDs the inverted mask byte with the contents of the memory location and places the result back into that location. Look at the following case:

```
LDAA #$FF               ;A = $FF
STAA PORTB              ;PORTB = 1111 1111
BCLR PORTB,%11000000    ;clear bits 6 and 7 of PORTB
```

```
 1100 0000    mask byte
 0011 1111    mask byte inverted and ANDed with PORTB
 1111 1111
---------------
 0011 1111    Only bits 6 and 7 are cleared. The rest are unchanged.
```

Also see Example 5-15.

Example 5-15

Assume that RAM location 806 has value $9F. Clear bits 0, 2, and 4 of the RAM location, while leaving the rest of the bits unchanged.

Address	Data
806	9F

Solution:

```
        DATA1 EQU    $806
              BCLR   DATA1, %00010101

Inverted mask        11101010
DATA1                10011111 ANDed
                     10001010   8A
```

Address	Data
806	8A

BSET instruction

We use the BSET (bit set) instruction to set to HIGH specific bits of an I/O port or a RAM location. In the "BSET Memloc,maskbyte" instruction, the mask byte has the bits that need to be set high. For example, "BSET MYRAM,%00000011" will set bits D0 and D1 of the MYRAM location. In reality, the BSET instruction ORs the mask byte with the contents of the memory location and places the result back into that location. Look at the following case:

```
LDAA #$00               ;A = 00
STAA PORTB              ;PORTB = 0000 0000
BSET PORTB,%0000011     ;set high bits 0 and 1 of PORTB
```

```
0000 0011     mask byte ORed with PORTB
0000 0000     Port B
---------------
0000 0011     Only bits 6 and 7 are set HIGH. The rest are unchanged.
```

Also see Examples 5-16 and 5-17.

Example 5-16

Assume RAM location 806 has value 8A. Make bits 4, 5, 6, and 7 of the RAM location HIGH, while leaving the rest of the bits unchanged.

Address	Data
806	8A

Solution:

```
        DATA2 EQU    $806
              BSET   DATA2, %11110000

mask byte            11110000
DATA2                10001010 ORed
                     11111010   FA
```

Address	Data
806	FA

Example 5-17

Assume that Port B bit PB2 is used to control an outdoor light, and bit PB5 to control a light inside a building. Show how to turn "on" the outdoor light and turn "off" the inside one.

Solution:

```
      BSET  DDRB,%00100100          ;make PB2 and PB5 output
      BSET  PORTB,%00000100         ;PB2 = 1
      BCLR  PORTB,%00100000         ;PB5 = 0
```

BRSET instruction

We use the BRSET (branch if bit set) instruction to branch if certain bits of the memory location are HIGH. The mask_byte in "BRSET RAM_loc, mask_byte, target_addr" indicates the bits we are checking. For example, "BRSET PORTA, %10000000, OVER" will jump to location OVER if bit D7 is HIGH. In reality, the BRSET instruction ANDs the mask byte with the inverted contents of memory location and jumps only if all the bits are zeros. See Example 5-18.

Example 5-18

Assume that Port B bit PB7 is used to monitor the status of a switch. Write a simple program to toggle the PA0 bit whenever the SW goes high. Analyze the cases of PB7 = 1 and PB7 = 0.

Solution:

```
      BCLR  DDRB,%10000000          ;make PB7 and input
      BSET  DDRA,%00000001          ;make PA0 an output
BACK  BRSET PORTB,%10000000,TOG     ;branch if PB7 is HIGH
      BRA   BACK                    ;go back if LOW
TOG   BCLR  PORTA,%00000001         ;PA0 = 0 for toggle
      BSET  PORTA,%00000001         ;PA0 = 1 for toggle
      BRA   BACK                    ;keep doing it.

Case 1)PB7=1
      Mask byte          10000000
      Inverted PORTB     0xxxxxxx ANDed
                         00000000 since it is zero, it will branch
Case 2)PB7=0
      Mask byte          10000000
      Inverted PORTB     1xxxxxxx ANDed
                         10000000  it will not branch
```

BRCLR instruction

We use the BRCLR (branch if bit clear) instruction to branch if certain bits of memory location is LOW. The mask_byte in "BRCLR RAM_loc, mask_byte, target_addr" indicates the bits we are checking. For example, "BRCLR PORTA, %00000001, OVER" will jump to location OVER if bit D0 is LOW. In reality, the

BRCLR instruction ANDs the mask byte with the contents of the memory location and jumps only if all the bits are zeros. See Example 5-19.

Example 5-19

Assume that Port B bit PB2 is used to monitor the status of a switch. Write a simple program to toggle the PA7 bit whenever the SW goes low. Analyze the cases of PB2 = 1 and PB2 = 0.

Solution:

```
        BCLR   DDRB,%00000100             ;make PB2 an input
        BSET   DDRA,%10000000             ;make PA7 an output
BACK    BRCLR  PORTB,%000000100,TOG       ;branch if PB2 is LOW
        BRA    BACK                       ;go back since it is HIGH
TOG     BCLR   PORTA,%10000000            ;PA7 = 0 for toggle
        BSET   PORTA,%10000000            ;PA7 = 1 for toggle
        BRA    BACK                       ;keep doing it

Case 1)PB2=1
        Mask byte            00000100
        PORTB                xxxxx1xx ANDed
                             00000100  it will not branch
Case 2)PB2=0
        Mask byte            00000100
        PORTB                xxxxx0xx ANDed
                             00000000  since it is zero,it will branch
```

Review Questions

1. Find the content of register A after the following code in each case:
   (a) `LDAA #$37` `ANDA #$CA` (b) `MLDAA #$37` `ORAA #$CA` (c) `LDAA #$37` `EORA #$CA`
2. To mask certain bits of register A, we must AND it with ______.
3. To set certain bits of register A to 1, we must OR it with _____.
4. EX-ORing an operand with itself results in _______.
5. Find the contents of register A after execution of the following code:

```
CLRA
ORAA   #$99
EORA   #$FF
```

SECTION 5.3: COMPARE INSTRUCTION AND PROGRAMS

In this section we will study the compare instruction with some examples.

Compare instructions

The HCS1218 has many instructions for the compare operation, as shown in Table 5-5. In the case of an unsigned value, these instructions compare the value with the contents of a register, and sets the Z and C flags based on whether the value is greater than, equal to, or less than register. The value can be an immediate value or in a RAM location. The compare instruction is really a subtraction, except that the values of the operands do not change. It must be emphasized again that in compare instructions, the operands are not affected, regardless of the result of the comparison. See Examples 5-20 and 5-21.

Table 5-5: HCS12 Compare Instructions

		C = 0	C = 1	Z = 1
CBA	Compare A with B	A > B	A < B	A = B
CMPA	Compare A with a value	A > Value	A < Value	A = Value
CMPB	Compare B with a value	B > Value	B < Value	B = Value
CPD	Compare D with 16-bit value	D > Value	D < Value	D = Value

Example 5-20

Write code to determine if data on PORTB contains the value 99H. If so, write letter 'Y' to PORTA; otherwise, make PORTA = 'N'.

Solution:

```
        BCLR    DDRB,%11111111    ;PORTB = input
        BSET    DDRA,%11111111    ;PORTA = output
        LDAA    #'N'          ;A = 'N' (ASCII)
        STAA    PORTA         ;PORTA = 'N'
        LDAA    #$99          ;A = 99H
BACK    CMPA    PORTB         ;Is PORTB = 99?
        BNE     BACK          ;No. (Z = 0), keep checking
        LDAA    #'Y'          ;Yes, (Z = 1)
        STAA    PORTA         ;PORTA = 'Y'
```

Example 5-21

Write a program to find the greater of the two values 27 and 54, and place it in RAM location $820.

Solution:

```
        VAL_1 EQU    54
        VAL_2 EQU    27
        GREG  EQU    $820

        LDAA    #VAL_1        ;A = 54
        CMPA    #VAL_2        ;compare VAL_1 and VAL_2
        BCS     OVER          ;jump if VAL_1 < VAL_2 (C = 1)
        STAA    GREG ;place the greater in GREG since C = 0 (A > Value)
OVER    ....
```

Using BHI and BLO instructions for unsigned numbers

Although BCS (branch if carry set) and BCC (branch if carry clear) check the carry flag and can be used after a compare instruction, it is recommended that BHI (branch if higher) and BLO (branch if lower) be used for two reasons. One reason is that many assemblers will assemble BCC as BHI, and BCS as BLO, which may be confusing to beginning programmers. Another reason is that "branch if higher" and "branch if lower" are easier to understand than "branch if carry is set" and "branch if carry clear," since it is more immediately apparent that one number is larger than another, than whether a carry would be generated if the two numbers were subtracted. The BLO and BHI instructions are widely used in comparison of unsigned numbers. See Example 5-22.

Example 5-22

Assume that Port B is an input port connected to a pressure sensor. Write a program to read the sensor value and test it for the value 75. According to the test results, place the value into the locations indicated by the following.

If P = 75	then A = 75
If P > 75	then GREG = P
If P < 75	then LREG = P

Solution:

```
LREG   EQU   $820
GREG   EQU   $821
       BCLR  DDRB,%11111111         ;PORTB = input
       LDAA  #75                ;A = 75 decimal
       CMPA  PORTB         ;compare PORTB with 75
       BEQ   EXIT          ;branch if it is equal (Z = 1)
       LDAA  PORTB         ;not equal since Z = 0. Get the P value
       BLO   LOWER         ;save the lower (C = 1).
       STAA  GREG          ;save the greater (C = 0)
       BRA   EXIT
LOWER  STAA  LREG          ;save the less value
EXIT   .....               ;it must be equal, A = 75
```

Finding the highest number

Program 5-1 searches through 5 unsigned data items such as grades to find the highest grade. The program has a variable called "Highest" that holds the highest grade found so far. One by one, the grades are compared to Highest. If any of them is higher, that value is placed in Highest. This continues until all data items are checked. Figure 5-2 shows the flowchart and pseudocode for Program 5-1. A REPEAT-UNTIL structure was chosen in the program design. This design could be used to code the program in many different languages.

Assume that we have five grades with the following values: 69, 87, 96, 45, and 75. Find the highest one and save it in RAM location $810.

```
;Program 5-1: finding the highest value
            ORG $840
GRADE       dc.b  69,87,96,45,75 ;we use dc.b to define constant byte
MYCOUNT     EQU   $800
HIGHEST     EQU   $810
;------------------
            LDAA  #5          ;set up loop counter
            STAA  MYCOUNT
            LDX   #GRADE      ;X points to GRADE data
            CLRA              ;A holds highest grade found so far
AGAIN:      CMPA  0,X         ;compare next grade to highest
            BHI   NEXT        ;jump if A still highest
            LDAA  0,X         ;else A holds new highest
NEXT:       INX               ;point to next temp
            DEC   MYCOUNT
            BNE   AGAIN       ;continue search
            STAA  HIGHEST     ;store highest grade
```

Program 5-1. Finding the Highest Number

Program 5-1, as coded in Assembly language, uses the CMPA instruction to search through 5 bytes of data to find the highest grade. The program uses register A to hold the highest grade found so far. A is given the initial value of 0. A loop is used to compare each of the 5 bytes with the value in A. If register A contains a higher value, the loop continues to check the next byte. If A is smaller than the byte being checked, the contents of A are replaced by that byte and the loop continues.

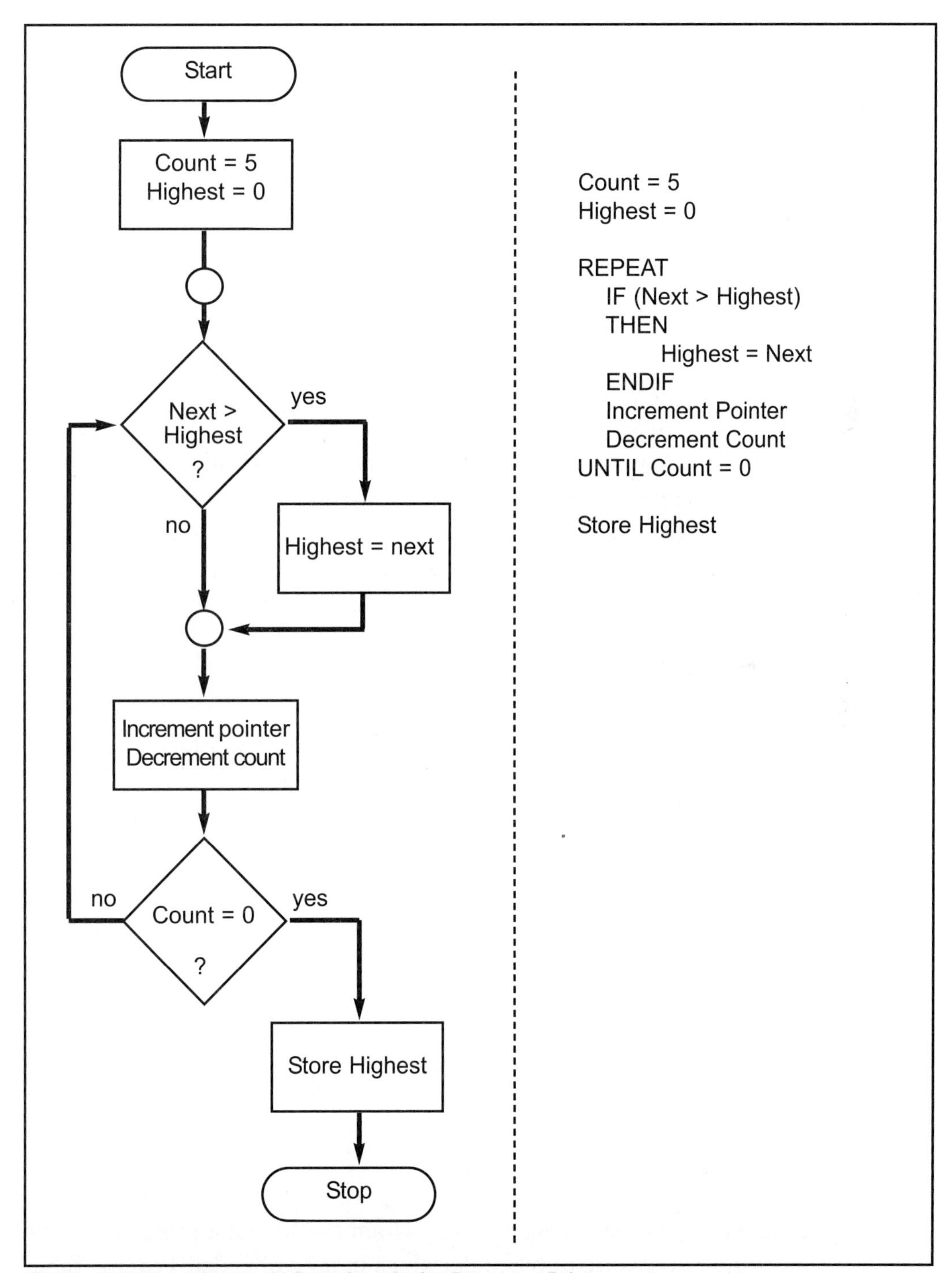

Figure 5-2. Flowchart and Pseudocode for Program 5-1

Lowercase to uppercase conversion

Program 5-2 uses the CMPA instruction to determine if an ASCII character is uppercase or lowercase. Note that small and capital letters in ASCII have the following values:

Letter	Hex	Binary	Letter	Hex	Binary
A	41	0100 0001	a	61	0110 0001
B	42	0100 0010	b	62	0110 0010
C	43	0100 0011	c	63	0110 0011
...	...	...	...	...	...
Y	59	0101 1001	y	79	0111 1001
Z	5A	0101 1010	z	7A	0111 1010

As can be seen, there is a relationship between the pattern of lowercase and uppercase letters, as shown below for A and a:

```
A     0100  0001  41H
a     0110  0001  61H
```

The only bit that changes is D5. To change from lowercase to uppercase, D5 must be masked. Program 5-2 first detects if the letter is in lowercase, and if it is, it is ANDed with 1101 1111 (binary) = $DF. Otherwise, it is simply left alone. To determine if it is a lowercase letter, it is compared with $61 and $7A to see if it is in the range a to z. Anything above or below this range should be left alone.

```
;Program 5-2. Lowercase to uppercase conversion
            ORG    $850
DATA1       DC.B   'mY NAME is jOe'
            ORG    $875
DATA2       DS.B   14
MYCOUNT     EQU    $800
;------------------
            LDAA   #14          ;set up loop counter
            STAA   MYCOUNT
            LDX    #DATA1       ;X points to original data
            LDY    #DATA2       ;Y points to uppercase data
BACK:       LDAA   0,X          ;get next character
            CMPA   #$61         ;if less than 'a'
            BLO    OVER         ;then no need to convert
            CMPA   #$7A         ;if greater than 'z'
            BHI    OVER         ;then no need to convert
            ANDA   #%11011111   ;mask D5 to convert to uppercase
OVER:       STAA   0,Y          ;store uppercase character
            INX                 ;increment pointer to original
            INY                 ;increment pointer to uppercase data
            DEC    MYCOUNT
            BNE    BACK         ;continue looping if MYCOUNT > 0
```

Program 5-2. Converting from Lowercase to Uppercase

One can use the concept covered in Program 5-2 to test a range of values such as temperature or pressure to make decisions in an embedded system. Table 5-6 shows the branch instructions used for comparison of unsigned numbers. The comparison of signed numbers will be discussed in the last section of this chapter.

Table 5-6: Branch Instructions for Unsigned Comparison

Instruction		Action	Relation
BHI	Branch if higher	Branch if C + Z = 0	A > Value
BHS	Branch if higher or same	Branch if C = 0	A >= Value
BLO	Branch if lower	Branch if C = 1	A < Value
BLS	Branch if lower or same	Branch if C + Z = 0	A <= Value

Review Questions

1. True or false. The CMPA instruction alters the contents of its operands.
2. What value must MYREG have in order for the BEQ instruction to fall through?

```
          LDAA #$99
     BACK CMPA        MYREG
          BEQ  BACK
```

3. True or false. The CMPA affects all flags.
4. What value must MYREG have in order for the BLO instruction to fall through?

```
          LDAA        #$99
     BACK CMPA        MYREG
          BLO  BACK
```

5. True or false. The BCC instruction is the same as BHS.
6. What value must MYREG have in order for the BHI instruction to fall through?

```
          LDAA        #$99
     BACK CMPA        MYREG
          BHI  BACK
```

SECTION 5.4: ROTATE, SHIFT INSTRUCTIONS, AND DATA SERIALIZATION

In many applications there is a need to perform a bitwise rotation or shift of an operand. In the HCS12 instructions such as ROR and LSL are designed specifically for the purpose of rotate and shift. See Table 5-7. We explore the rotate and shift instructions next.

Table 5-7: Some Widely Used Rotate and Shift Instructions

Instruction	Action
LSL	Logical shift left memory
LSLA	Logical shift left A
LSR	Logical shift right memory
LSRA	Logical shift right A
ROL	Rotate left memory
ROLA	Rotate left A
ROR	Rotate right memory
RORA	Rotate right A

Rotating right through carry

```
RORA        ;rotate Reg A right through carry
```

In RORA, as bits are rotated from left to right, the carry flag enters the MSB and the LSB exits to the carry flag. In other words, in RORA the C is moved to the MSB, and the LSB is moved to C. In reality, the carry flag acts as if it is part of the register, making it a 9-bit register. See the following code:

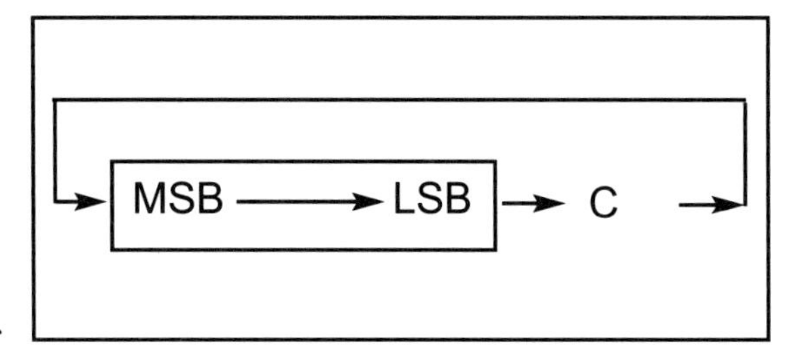


```
CLC                 ;make C = 0
LDAA    #$26        ;A = 0010 0110    C = 0
RORA                ;A = 0001 0011    C = 0
RORA                ;A = 0000 1001    C = 1
RORA                ;A = 1000 0100    C = 1
RORA                ;A = 1100 0010    C = 0
```

We can also rotate right and left the content of a RAM location.

```
MYREG EQU $820
SEC                 ;make C = 1
LDAA   #$75         ;A = 75H
STAA   MYREG        ;MYREG = 0111 0101      C = 1
ROR    MYREG        ;MYREG = 1011 1010      C = 1
ROR    MYREG        ;MYREG = 1101 1101      C = 0
ROR    MYREG        ;MYREG = 0110 1110      C = 1
ROR    MYREG        ;MYREG = 1011 0111      C = 0
```

Rotating left through carry

```
ROLA                ;rotate reg A left through carry
```

In ROLA, as bits are shifted from right to left, the carry flag enters the LSB and the MSB exits to the carry flag. In other words, in ROLA the C is moved to the LSB, and the MSB is moved to C. Again the carry flag acts as if it is part of the register, making it a 9-bit register. See the following code:

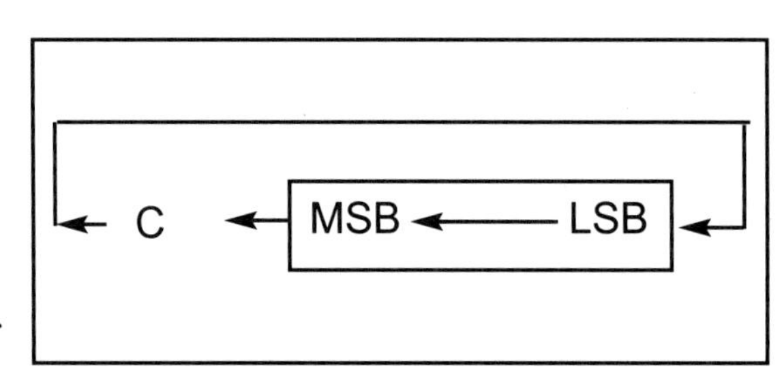


```
SEC                 ;make C = 1
LDAA    #$15        ;A = 0001 0101
ROLA                ;A = 0010 1011    C = 0
ROLA                ;A = 0101 0110    C = 0
ROLA                ;A = 1010 1100    C = 0
ROLA                ;A = 0101 1000    C = 1
```

We can also rotate right and left the content of a RAM location.

```
MREG EQU $820
CLC                  ;make C = 0
LDAA    #$89         ;A = 1000 1001
STAA    MYREG        ;MYREG = 1000 1001
ROL     MYREG        ;MYREG = 0001 0010       C = 1
ROL     MYREG        ;MYREG = 0010 0101       C = 0
ROL     MYREG        ;MYREG = 0100 1010       C = 0
ROL     MYREG        ;MYREG = 1001 0100       C = 0
```

Logical shift vs. arithmetic shift

There are two kinds of shifts: logical and arithmetic. The logical shift is for unsigned operands, and the arithmetic shift is for signed operands. Logical shift will be discussed in this section, and the discussion of arithmetic shift is postponed to the last section in this chapter. Using shift instructions shifts the contents of register A or the memory location right or left.

Shifting the bits right

```
LSRA                 ;logical shift  right reg A
```

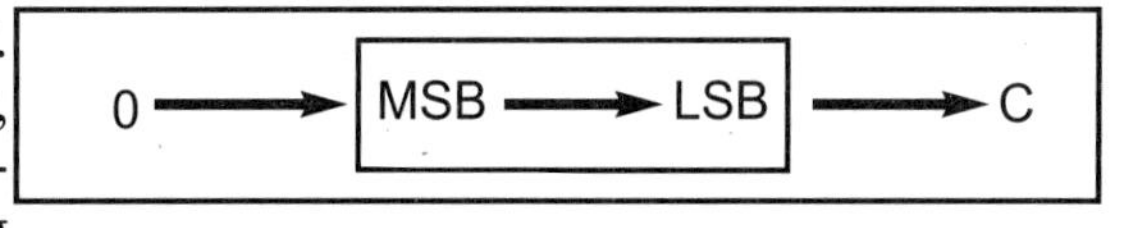

This is the logical shift right. The operand is shifted right bit by bit, and for every shift the LSB (least significant bit) will go to the carry flag (C) and the MSB (most significant bit) is filled with 0. The operand to be shifted can be in a register A or in memory. See the following code:

```
LDAA #$36            ;A = 0011 0110
LSRA                 ;A = 0001 1011       C = 0
LSRA                 ;A = 0000 1101       C = 1
LSRA                 ;A = 0000 0110       C = 1
LSRA                 ;A = 0000 0011       C = 0

MYREG       EQU $820
        LDAA #$75            ;A = 75H
        STAA MYREG           ;MYREG = 0111 0101
        LSR  MYREG           ;MYREG = 0011 1010       C = 1
        LSR  MYREG           ;MYREG = 0001 1101       C = 0
        LSR  MYREG           ;MYREG = 0000 1110       C = 1
        LSR  MYREG           ;MYREG = 0000 0111       C = 0
```

Shifting the bits left

```
LSLA        ;logical shift left reg A
```

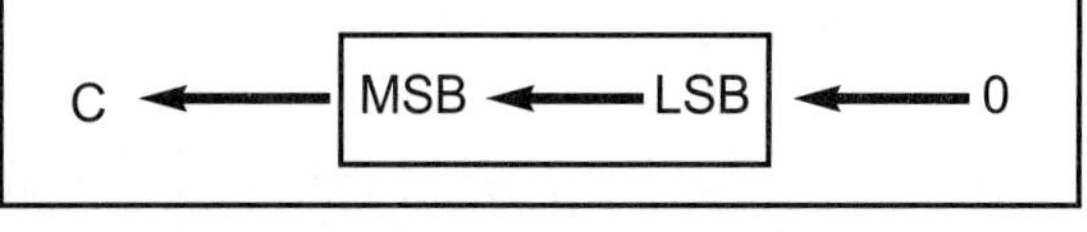

LSLA is also a logical shift. It is the reverse of LSRA. After every shift, the LSB is filled with 0 and the MSB goes to C. All the rules are the same as LSRA. The operand to be shifted can be in a register A or in memory. See the following code:

```
LDAA #$72            ;A = 0111 0010
LSLA                 ;A = 1110 0100     C = 0
LSLA                 ;A = 1100 1000     C = 1
```

```
        LSLA                ;A = 1001 0000    C = 1
        LSLA                ;A = 0010 0000    C = 1
MYREG        EQU $820
        LDAA   #$89         ;A = 1000 1001
        STAA   MYREG        ;MYREG = 1000 1001
        LSL    MYREG        ;MYREG = 0001 0010      C = 1
        LSL    MYREG        ;MYREG = 0010 0100      C = 0
        LSL    MYREG        ;MYREG = 0100 1000      C = 0
        LSL    MYREG        ;MYREG = 1001 0000      C = 0
```

Although the logical shifts affect both the N and Z flags, they are not important in this case. Examine Examples 5-23 through 5-25 for further clarifications.

Example 5-23

Show the result of LSRA in the following:

```
       LDAA   #$9A
       LSRA
       LSRA
       LSRA
```

Solution:

```
    9AH =       10011010
                01001101      C = 0   (shifted right once)
                00100110      C = 1   (shifted right twice)
                00010011      C = 0   (shifted right three times)
```

After three times of shifting right, A = 13H and C = 0.

Example 5-24

Show the results of LSR in the following:

```
       DATA1 EQU    $77
       MYRAM EQU    $800
       LDAA   #DATA ;A=77 hex
       STAA   MYRAM
       LSR    MYRAM
       LSR    MYRAM
       LSR    MYRAM
       LSR    MYRAM
```

Solution:

After the four shifts, the byte at memory location $800 will contain 07. The four LSBs are lost through the carry, one by one, and 0s fill the four MSBs.

Serializing data

Serializing data is a way of sending a byte of data one bit at a time through a single pin of the microcontroller. There are two ways to transfer a byte of data serially:

1. Using the serial port. In using the serial port, programmers have very limited control over the sequence of data transfer. The details of serial port data transfer are discussed in Chapter 10.
2. The second method of serializing data is to transfer data one bit at a time and control the sequence of data and spaces between them. In many new generations of devices such as LCD, ADC, and ROM, the serial versions are becom-

Example 5-25

Show the effects of LSLA in the following:

```
        LDAA   #6
        LSLA
        LSLA
        LSLA
        LSLA
```

Solution:

```
                         00000110
            C=0          00001100     (shifted left once)
            C=0          00011000
            C=0          00110000
            C=0          01100000     (shifted four times)
```

After the four shifts left, the register A has 60H and C = 0.

ing popular because they take less space on a printed circuit board. Next, we discuss how to use rotate instructions in serializing data.

Serializing a byte of data

Serializing data is one of the most widely used applications of the rotate instruction. We can use the shift instruction to transfer a byte of data serially (one bit at a time). Example 5-26 shows how to transfer an entire byte of data serially via any HCS12 pin.

Example 5-26

Write a program to transfer value 41H serially (one bit at a time) via pin PB1. Put one high at the start and end of the data. Send the LSB first.

Solution:

```
RCOUNT EQU  $820                ;loc for counter
       BSET  DDRB,%00000010     ;make RB1 an output bit
       CLC                      ;C = 0
       LDAA  #8                 ;counter
       STAA  RCOUNT             ;load the counter
       LDAA  #$41               ;A = 41, the value to be serialized
       BSET  PORTB,%00000010    ;RB1 = high
AGAIN  LSRA                     ;shift right D0 to carry
       BCC   OVER               ;
       BSET  PORTB,%00000010    ;set the carry bit to PB1
       BRA   NEXT
OVER   BCLR  PORTB,%00000010    ;clear PB1
NEXT   DEC   RCOUNT
       BNE   AGAIN
       BSET  PORTB,%00000010    ;RB1 = high
```

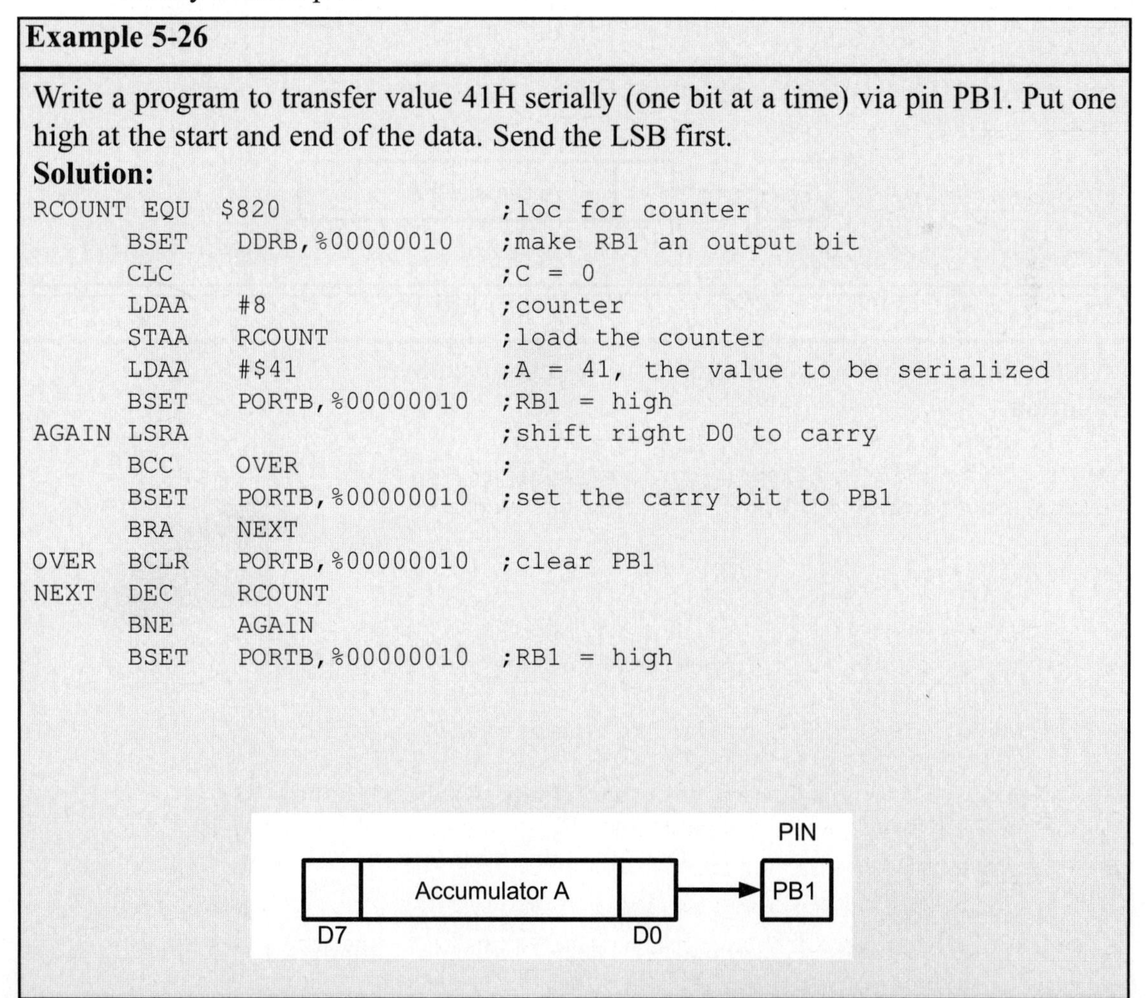

Example 5-27 shows how to bring in a byte of data serially (one bit at a time). We will see how to use these concepts for a serial RTC (real-time clock) chip in Chapter 16. Example 5-28 shows how to scan the bits in a byte.

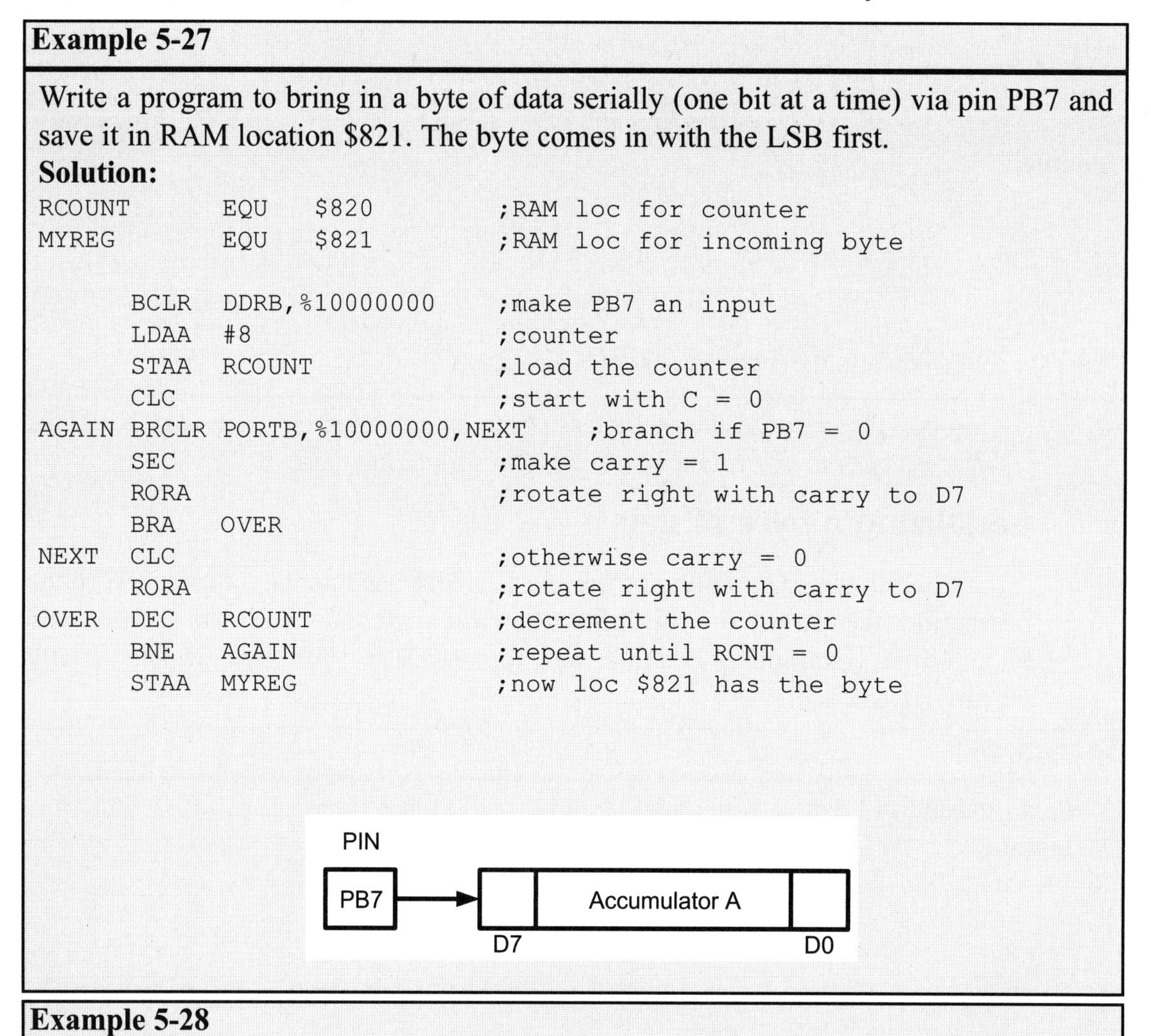

Example 5-27

Write a program to bring in a byte of data serially (one bit at a time) via pin PB7 and save it in RAM location $821. The byte comes in with the LSB first.

Solution:

```
RCOUNT       EQU   $820          ;RAM loc for counter
MYREG        EQU   $821          ;RAM loc for incoming byte

      BCLR  DDRB,%10000000       ;make PB7 an input
      LDAA  #8                   ;counter
      STAA  RCOUNT               ;load the counter
      CLC                        ;start with C = 0
AGAIN BRCLR PORTB,%10000000,NEXT      ;branch if PB7 = 0
      SEC                        ;make carry = 1
      RORA                       ;rotate right with carry to D7
      BRA   OVER
NEXT  CLC                        ;otherwise carry = 0
      RORA                       ;rotate right with carry to D7
OVER  DEC   RCOUNT               ;decrement the counter
      BNE   AGAIN                ;repeat until RCNT = 0
      STAA  MYREG                ;now loc $821 has the byte
```


Example 5-28

Write a program that finds the number of 1s in a given byte.

Solution:

```
R1     EQU  $820  ;fileReg loc for number of 1s
COUNT  EQU  $821  ;fileReg loc for counter
VALREG EQU  $822  ;fileReg loc for the byte

      CLC               ;C = 0
      CLRA
      STAA  R1          ;R1 = 0, keeps the number of 1s
      LDAA  #8          ;counter = 08 to rotate 8 times
      STAA  COUNT
      LDAA  #$97        ;find the number of 1s in 97H
      STAA  VALREG
AGAIN ROLA              ;rotate it through the C once
      BCC   NEXT        ;check for C
      INC   R1          ;if C = 1 then add one to R1 reg
NEXT  DEC   COUNT
      BNE AGAIN         ;go through this 8 times
                        ;now loc $820 has the number of 1s
```

Swap

Another useful application of the shift instruction is for the swap operation. In many applications we need to swap the lower nibble and the higher nibble. In other words, the lower 4 bits are put into the higher 4 bits, and the higher 4 bits are put into the lower 4 bits. See Example 5-29.

Example 5-29

In the absence of a SWAP instruction in HCS12, how would you exchange the nibbles of register A? Write a simple program to show the process.

Solution:

```
MYREG EQU    $820            ;RAM loc
      LDAA #$72              ;A = 0111 0010
      STAA   MYREG           ;MYREG = 0111 0010
                             ;A = 0111 0010
      LSRA                   ;A = 0011 1001
      LSRA                   ;A = 0001 1100
      LSRA                   ;A = 0000 1110
      LSRA                   ;A = 0000 0111
      LSL    MYREG           ;MYREG = 1110 0100
      LSL    MYREG           ;MYREG = 1100 1000
      LSL    MYREG           ;MYREG = 1001 0000
      LSL    MYREG           ;MYREG = 0010 0000
      ORAA   MYREG           ;A = 0010 0111 = $27
```

Review Questions

1. What is the value of MYREG after the following code is executed?

```
MYREG EQU $840
CLRA
SEC                    ;C = 1
ROR     MYREG
SEC                    ;C = 1
ROR     MYREG
```

2. Does "ROL D" give an error in the HCS12?
3. What is in register A after execution of the following code?

```
LDAA  #$85
CLRC
RORA
RORA
RORA
RORA
```

4. Show the step-by-step result if value $A2 were shifted left four times.

```
LDAA  #$A2
LSLA
LSLA
LSLA
LSLA
```

SECTION 5.5: BCD AND ASCII CONVERSION

In this section we provide some real-world examples of how to use arithmetic and logic instructions. We will cover their applications in real-world devices in future chapters. For example, many newer microcontrollers have a real-time clock (RTC), where the time and date are kept even when the power is off. These microcontrollers provide the time and date in BCD. To display them, however, we must convert BCD values to ASCII. Next, we show the application of logic and rotate instructions in the conversion of BCD and ASCII.

ASCII numbers

On ASCII keyboards, when the key "0" is activated, "011 0000" (30H) is provided to the computer. Similarly, 31H (011 0001) is provided for key "1", and so on, as shown in Table 5-8.

It must be noted that BCD numbers are universal although ASCII is standard in the United States (and many other countries). Because the keyboard, printers, and monitors all use ASCII, how is data converted from ASCII to BCD, and vice versa? These are the subjects covered next.

Table 5-8: ASCII and BCD Codes for Digits 0–9

Key	ASCII (hex)	Binary	BCD (unpacked)
0	30	011 0000	0000 0000
1	31	011 0001	0000 0001
2	32	011 0010	0000 0010
3	33	011 0011	0000 0011
4	34	011 0100	0000 0100
5	35	011 0101	0000 0101
6	36	011 0110	0000 0110
7	37	011 0111	0000 0111
8	38	011 1000	0000 1000
9	39	011 1001	0000 1001

Packed BCD to ASCII conversion

In many systems we have what is called a *real-time clock* (RTC). The RTC provides the time of day (hour, minute, second) and the date (year, month, day) continuously, regardless of whether the power is on or off (see Chapter 16). This data, however, is provided in packed BCD. For this data to be displayed on a device such as an LCD, or to be printed by the printer, it must be in ASCII format.

To convert packed BCD to ASCII, you must first convert it to unpacked BCD. Then the unpacked BCD is tagged with 011 0000 (30H). The following demonstrates converting packed BCD to ASCII. See also Example 5-30.

```
Packed BCD   Unpacked BCD      ASCII
29H          $02  , $09        $32  , $39
0010 1001    0000 0010         0011 0010    $32
             0000 1001         0011 1001    $39
```

Example 5-30

Assume that register A has packed BCD. Write a program to convert packed BCD to two ASCII numbers and place them in RAM locations $806 and $807.

Solution:

```
BCD_VAL EQU $29
H_ASC    EQU $806  ;set aside RAM location
L_ASC    EQU $807  ;set aside RAM location

     LDAA #BCD_VAL     ;A = 29H, packed BCD
     ANDA #$0F         ;mask the upper nibble (A = 09)
     ORAA #$30         ;make it an ASCII, A = 39H ('9')
     STAA L_ASC        ;save it (L_ASC = 39H ASCII char)
     LDAA #BCD_VAL     ;A = 29H get BCD data once more
     ANDA #$F0         ;mask the lower nibble (A = 20H)
     LSRA              ;shift right to get upper nibble
     LSRA
     LSRA
     LSRA              ;A = 02H
     ORAA  #$30        ;make it an ASCII, A = 32H ('2')
     STAA H_ASC        ;save it (H_ASC = 32H ASCII char)
```

ASCII to packed BCD conversion

To convert ASCII to packed BCD, you first convert it to unpacked BCD (to get rid of the 3), and then combine it to make packed BCD. For example, for 4 and 7 the keyboard gives 34 and 37, respectively. The goal is to produce $47 or "0100 0111", which is packed BCD. This process is illustrated next.

```
Key   ASCII        Unpacked BCD      Packed BCD
4     34           00000100
7     37           00000111          01000111, which is $47

MYBCD EQU   $820          ;set aside location in RAM
      LDAA  #'4'          ;A = $34, hex for ASCII char 4
      ANDA  #$0F          ;mask upper nibble (A = 04)
      LSLA                ;shift left for lower nibble
      LSLA
      LSLA
      LSLA                ;MYBCD = 40H
      STAA  MYBCD         ;save it in MYBCD loc
      LDAA  #'7'          ;A = $37, hex for ASCII char 7
      ANDA  #$0F          ;mask upper nibble (A = 07)
      ORAA  MYBCD         ;A = $47
      STAA  MYBCD         ;MYBCD = $47, a packed BCD
```

After this conversion, the packed BCD numbers are processed and the result will be in packed BCD format. As we saw earlier in this chapter, a special instruction, "DAA", requires that the data be in packed BCD format.

Review Questions

1. For the following decimal numbers, give the packed BCD and unpacked BCD representations.
   (a) 15 (b) 99
2. Show the binary and hex formats for "76" and its BCD version.
3. Does register A have BCD data after the following instruction is executed?
```
LDAA #'54'
```
4. $67 in BCD when converted to ASCII is ____H and ____H.

SECTION 5.6: SIGNED NUMBER CONCEPTS AND ARITHMETIC OPERATIONS

All data items used so far have been unsigned numbers, meaning that the entire 8-bit operand was used for the magnitude. Many applications require signed data. In this section the concept of signed numbers is discussed along with related instructions. If your applications do not involve signed numbers, you can bypass this section.

Concept of signed numbers in computers

In everyday life, numbers are used that could be positive or negative. For example, a temperature of 5 degrees below zero can be represented as −5, and 20 degrees above zero as +20. Computers must be able to accommodate such numbers. To do that, computer scientists have devised the following arrangement for the representation of signed positive and negative numbers: The most significant bit (MSB) is set aside for the sign (+ or –), while the rest of the bits are used for the magnitude. The sign is represented by 0 for positive (+) numbers and 1 for negative (–) numbers. Signed byte representation is discussed below.

Signed 8-bit operands

In signed byte operands, D7 (MSB) is the sign, and D0 to D6 are set aside for the magnitude of the number. If D7 = 0, the operand is positive, and if D7 = 1, it is negative. The N flag in the status register is a copy of the D7 bit.

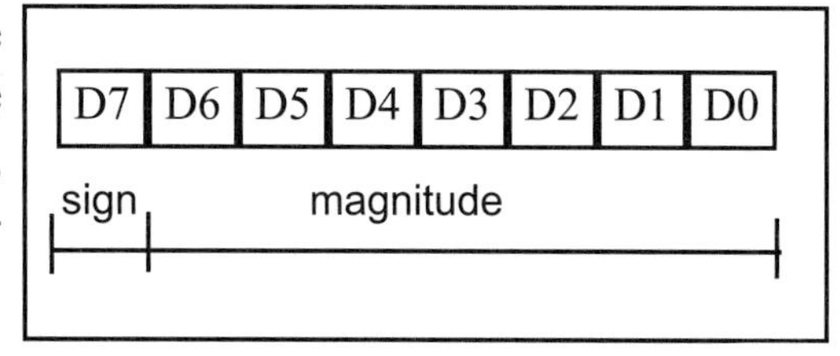

Figure 5-3. 8-Bit Signed Operand

Positive numbers

The range of positive numbers that can be represented by the format shown in Figure 5-3 is 0 to +127. If a positive number is larger than +127, a 16-bit operand must be used.

```
   0     0000 0000
  +1     0000 0001
  ...       ...
  +5     0000 0101
  ...       ...
+127     0111 1111
```

Negative numbers

For negative numbers, D7 is 1; however, the magnitude is represented in its 2's complement. Although the assembler does the conversion, it is still important to understand how the conversion works. To convert to negative number representation (2's complement), follow these steps:

1. Write the magnitude of the number in 8-bit binary (no sign).
2. Invert each bit.
3. Add 1 to it.

Examples 5-31, 5-32, and 5-33 demonstrate these three steps.

Example 5-31

Show how the HCS12 would represent –5.

Solution:
Observe the following steps.

```
1.     0000 0101             5 in 8-bit binary
2.     1111 1010             invert each bit
3      1111 1011             add 1 (which becomes FB in hex)
```

Therefore, –5 = $FB, the signed number representation in 2's complement for –5. D7 = N = 1 indicates that the number is negative.

Example 5-32

Show how the HCS12 would represent –34 in hex.

Solution:
Observe the following steps.

```
1.     0011 0100             $34 given in binary
2.     1100 1011             invert each bit
3      1100 1100             add 1 (which is CC in hex)
```

Therefore, –34 = $CC, the signed number representation in 2's complement for $34. D7 = N = 1 indicates that the number is negative.

Example 5-33

Show how the HCS12 would represent –128.

Solution:
Observe the following steps.

```
1.     1000 0000             128 in 8-bit binary
2.     0111 1111             invert each bit
3      1000 0000             add 1 (which becomes 80 in hex)
```

Therefore, –128 = $80, the signed number representation in 2's complement for –128. D7 = N = 1 indicates that the number is negative. Notice that 128 (binary 10000000) in unsigned representation is the same as signed –128 (binary 10000000).

From the examples above, it is clear that the range of byte-sized negative numbers is –1 to –128. The following lists byte-sized signed number ranges:

```
Decimal     Binary        Hex
-128        1000 0000     80
-127        1000 0001     81
-126        1000 0010     82
...         .........     ..
-2          1111 1110     FE
-1          1111 1111     FF
0           0000 0000     00
+1          0000 0001     01
+2          0000 0010     02
..          .........     ..
+127        0111 1111     7F
```

The above explains the mystery behind the relative address of –128 to +127 in the BNE and other conditional branch instructions discussed in Chapter 3.

Overflow problem in signed number operations

When using signed numbers, a serious problem arises that must be dealt with. This is the overflow problem. The HCS12 indicates the existence of an error by raising the V (overflow) flag, but it is up to the programmer to take care of the erroneous result. The CPU understands only 0s and 1s and ignores the human convention of positive and negative numbers. What is an overflow? If the result of an operation on signed numbers is too large for the register, an overflow has occurred and the programmer must be notified. Look at Example 5-34.

Example 5-34

Examine the following code and analyze the result, including the N and V flags.

```
      LDAA #96            ;A = 0110 0000
      ADDA #70            ;A = (+96) + (+70) = 1010 0110
                          ;A = $A6 = -90 decimal, INVALID!!
```

Solution:

```
     +96    0110 0000
   + +70    0100 0110
   + 166    1010 0110   N = 1 (negative) and OV = 1. Sum = -90
```

According to the CPU, the result is negative (N = 1), which is wrong. The CPU sets V = 1 to indicate the overflow error. Remember that the N flag is the D7 bit. If N = 0, the sum is positive, but if N = 1, the sum is negative.

In Example 5-13, +96 was added to +70 and the result, according to the CPU, was -90. Why? The reason is that the result was larger than what A could contain. Like all other 8-bit registers, A could only contain up to +127. The designers of the CPU created the overflow flag specifically for the purpose of informing the programmer that the result of the signed number operation is erroneous. The N flag is D7 of the result. If N = 0, the sum is positive (+), and if N = 1, the sum is negative.

When is the V (overflow) flag set?

In 8-bit signed number operations, V is set to 1 if either of the following two conditions occurs:

1. There is a carry from D6 to D7 but no carry out of D7 (C = 0).
2. There is a carry from D7 out (C = 1) but no carry from D6 to D7.

In other words, the overflow flag is set to 1 if there is a carry from D6 to D7 or from D7 out, but not both. This means that if there is a carry both from D6 to D7 and from D7 out, V = 0. In Example 5-13, because there is only a carry from D6 to D7 and no carry from D7 out, V = 1. Study Examples 5-35, 5-36, and 5-37 to understand the overflow flag in signed arithmetic.

Example 5-35

Observe the following, noting the role of the V and N flags:

```
        LDAA #-128          ;A = 1000 0000 (A = 80H)
        ADDA #-2            ;A = (-128) + (-2)
                            ;A = 1000000 + 11111110 = 0111 1110,
                            ;N = 0, Acc = $7E = +126, invalid
```

Solution:

```
  -128          1000 0000
 + - 2          1111 1110
 - 130          0111 1110    N = 0 (positive) and V = 1
```

According to the CPU, the result is +126, which is wrong, and V = 1 indicates that.

Example 5-36

Observe the following, noting the V and N flags:

```
        LDAA  #-2           ;A = 1111 1110 (A = $FE)
        ADDA  #-5           ;A = (-2) + (-5) = -7 or $F9
                            ;correct, since V = 0
```

Solution:

```
    -2       1111 1110
  + -5       1111 1011
   - 7       1111 1001   and V = 0 and N = 1. Sum is negative
```

According to the CPU, the result is –7, which is correct, and the V flag indicates that.

Example 5-37

Examine the following, noting the role of the V and N flags:

```
        LDAA #+7            ;A = 0000 0111
        ADDA #+18           ;A = (+7) + (+18)
                            ;A = 00000111 + 00010010 = 0001 1001
                            ;A = (+7) + (+18) = +25, N = 0, positive and
                            ;correct, V = 0
```

Solution:

```
     + 7 0000 0111
   + +18 0001 0010
     +25 0001 1001   N = 0 (positive 25) and V = 0
```

According to the CPU, this is +25, which is correct, and V = 0 indicates that.

From the above examples, we conclude that in any signed number addition, V indicates whether the result is valid or not. If V = 1, the result is erroneous; if V = 0, the result is valid. We can state emphatically that in unsigned number addition, we must monitor the status of C (carry flag), and in signed number addition, the V (overflow) flag must be monitored. In the HCS12, instructions such as BCC and BCS allow the program to branch right after the addition of unsigned numbers, as we saw in Section 5.1. There are also the BVC (branch if V is clear) and the BVS (branch if V is set) instructions for the V flag that allow us to correct the signed number error. We also have two branch instructions for the N flag (negative), BMI (branch if minus) and BPL (branch if plus). See Table 5-9.

Table 5-9: HCS12 Branch Instructions Used for Signed Numbers

Instruction		Action
BMI	Branch if minus	Branch if N = 1
BPL	Branch if plus	Branch if N = 0
BVS	Branch if V set	Branch if V = 1
BVC	Branch if V clear	Branch if V = 0

Word-sized (16-bit) signed numbers

In HCS12 a word is 16 bits in length. Setting aside the MSB (D15) for the sign leaves a total of 15 bits (D14–D0) for the magnitude. See Figure 5-4. This gives a range of −32,768 to +32,767. If a number is larger than this, it must be treated as a multiword operand and be processed chunk by chunk the same way as unsigned numbers.

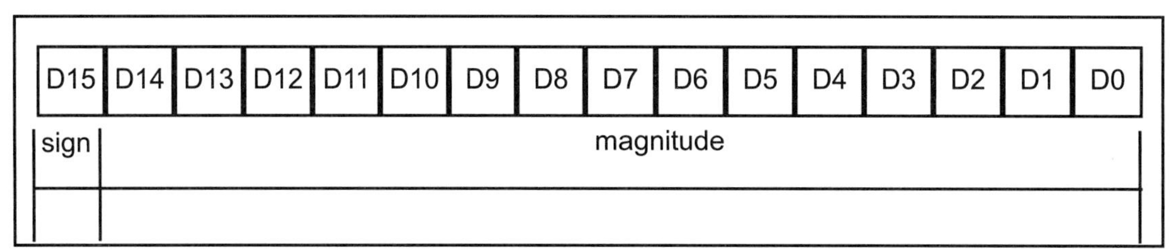

Figure 5-4. 16-Bit Signed Operand

The following shows the range of signed word operands. To convert a negative to its word operand representation, the three steps discussed in negative byte operands are used.

Decimal	Binary	Hex
−32,768	1000 0000 0000 0000	8000
−32,767	1000 0000 0000 0001	8001
−32,766	1000 0000 0000 0010	8002
...	...	...
−2	1111 1111 1111 1110	FFFE
−1	1111 1111 1111 1111	FFFF
0	0000 0000 0000 0000	0000
+1	0000 0000 0000 0001	0001
+2	0000 0000 0000 0010	0002
...	...	...
+32,766	0111 1111 1111 1110	7FFE
+32,767	0111 1111 1111 1111	7FFF

Overflow flag in 16-bit operations

In a 16-bit operation, V is set to 1 in either of two cases:

1. There is a carry from D14 to D15 but no carry out of D15 (C = 0).

2. There is a carry from D15 out (C = 1) but no carry from D14 to D15.

Again the overflow flag is low (not set) if there is a carry from both D14 to D15 and from D15 out. The V is set to 1 only when there is a carry from D14 to D15 or from D15 out, but not from both. Observe the results in the following:

```
            LDD    #$6E2F          ; D = 28,207
            ADDD   #$13D4          ; D = 28,207 + 5,076
                           ;= 33,283 is the expected answer
6E2F              0110 1110 0010 1111
+13D4             0001 0011 1101 0100
8203              1000 0010 0000 0011 = - 32,253 incorrect! V = 1, C = 0, N = 1
```

Also observe the results in the following:

```
            LDD   #$542F           ; D = 21,551
            ADDD #$12E0            ; D = 21,551 + 4,832
                                   ; = 26,383

543F              0101 0100 0010 1111
+12E0             0001 0010 1110 0000
670F              0110 0111 0000 1111 = 26,383 (correct answer) V = 0, C = 0, N = 0
```

Arithmetic shift

As discussed earlier in this chapter, there are two types of shifts: logical and arithmetic. The logical shift, which is used for unsigned numbers, was discussed previously. The arithmetic shift is used for signed numbers. It is basically the same as the logical shift, except that the sign bit is copied to the shifted bits. ASR (arithmetic shift right) and ASL (arithmetic shift left) are two instructions for the arithmetic shift.

ASRA (arithmetic shift right A)

As the bits are shifted to the right into C, the empty bits are filled with the sign bit (MSB). One can use the ASR instruction to divide a signed number by 2, as shown next:

MSB → MSB → LSB → CF

```
            LDAA  #-10          ;A = -10 = F6H = 1111 0110
            ASRA            ;A is shifted right arithmetic once
                            ;A = 1111 1011 = FBH = -5
```

ASLA (arithmetic shift left A) and LSLA (logical shift left A)

These two instructions do exactly the same thing. It is basically the same instruction with two mnemonics. As far as signed numbers are concerned, there is no need for ASL. For a discussion of LSLA, see Section 5.4.

Signed number comparison

Although the CMPA instruction is the same for both signed and unsigned numbers, the Branch condition instruction used to make a decision for the signed numbers is different from the unsigned numbers. While in unsigned number comparisons C and Z are checked for conditions of larger, equal, and smaller, in signed number comparison, V, Z, and N are checked. The following are the flags for CMPA:

```
A > Value   (V Ex-OR N) + Z = 0
A  = Value       Z=1
A  < Value       V Ex-OR N = 1
```

The mnemonics used to detect the conditions are shown in Table 5-10:

Table 5-10: Branch Instructions Used for Signed Numbers

Instruction		Action
BGE	Branch Greater or Equal	Branch if (V Ex-OR N) = 0
BGT	Branch Greater Than	Branch if (V Ex-OR N) + Z = 0
BLE	Branch Less or Equal	Branch if (V Ex-OR N) + Z = 1
BLT	Branch Less Than	Branch if (V Ex-OR N) = 1

Program 5-3 is an example of the application for signed number comparison. It finds the lowest temperature represented by some signed data.

Assume that we have a seven-day list of temperature readings with negative and positive numbers. Find the lowest one and save it in RAM location $810.

```
;Program 5-3: finding the lowest value of signed numbers
            ORG    $850
SIGN_TEMP   DC.B   +13,-10,+19,+14,-18,+12,-19
MYCOUNT     EQU    $800
LOWEST      EQU    $810
 ;-----------------
            LDAA   #7                ;set up loop counter
            STAA   MYCOUNT
            LDX    #SIGN_TEMP        ;X points to TEMP data
            LDAA   #$7F              ;initialize A for the highest
AGAIN:      CMPA   0,X               ;compare next temp to lowest
            BLE    NEXT              ;jump if A still lowest
            LDAA   0,X               ;else A holds new lowest
 NEXT:      INX                      ;point to next temp
            DEC    MYCOUNT
            BNE    AGAIN             ;continue search
            STAA   LOWEST            ;store lowest TEMP
```

Program 5-3. Find the Lowest of Signed Numbers

Multiplication and division of signed numbers

Tables 5-11 and 5-12 show the instructions for signed number multiplication and division, respectively. The concept of signed number arithmetic is so important and widely used that even the RISC processors could not eliminate these instructions in their attempt to streamline the instruction set.

Table 5-11: Signed Number Multiplication Summary

Inst	Multiplication	Operand1	Operand2	Result
EMULS	Word × Word	D	Y	Y:D (Y = high word, D = low word)

Table 5-12: Signed Number Division Summary

Instruc.	Division	Numerator	Denominator	Quotient	Remainder
IDIVS	word / word	D	X	X	D
EDIVS	doubleword/word	Y:D	X	Y	D

(If denominator = 0, then C = 1, indicating an error.)
Word = 16-bit and doubleword = 32-bit

Review Questions

1. In an 8-bit operand, bit _____ is used for the sign bit.
2. Convert –16 (hex) to its 2's complement representation.
3. The range of byte-sized signed operands is – ______ to + _____.
4. Show +9 and –9 in binary.
5. Explain the difference between a carry and an overflow.
6. The instruction for signed multiplication is ________. The instruction for signed division is _______.
7. Explain the difference between the shift logical right and shift arithmetic right instructions.
8. For each of the following instructions, indicate the flag condition necessary for each jump to occur:
   (a) BLE (b) BGE

PROBLEMS

SECTION 5.1: ARITHMETIC INSTRUCTIONS AND PROGRAMS

1. Find the C, Z, and H flags for each of the following:

(a)
```
LDAA    #$3F
ADDA    #$45
```
(b)
```
LDAA    #$99
ADDA    #$58
```
(c)
```
LDAA    #$FF
STAA    MYREG
SEC
LDAA    #0
ADCA   MYREG
```
(d)
```
LDAA    #$FF
ADDA    #$1
```
(e)
```
LDAA    #$FE
STAA    MYREG
SEC
LDAA    0
ADCA    MYREG
```
(f)
```
CLC
LDAA    #$FF
STAA    MYREG
LDAA    #0
ADCA    MYREG
```

2. Write a program to add all the digits of your ID number and save the result in a RAM location. The result must be in BCD.
3. Write a program to add the following numbers and save the result in a RAM location.
   $25, $59, $65
4. Modify Problem 3 to make the result in BCD.
5. Write a program to (a) write the value 25H to RAM locations $1820–$1823, and (b) add all these RAM locations' contents together, and save the result in RAM location $1860.
6. State the steps that the SUBA instruction will go through for each of the fol-

lowing.

(a) $23 – $12 (b) $43 – $53 (c) 99 – 99

7. For Problem 6, write a program to perform each operation.
8. True or false. The "DAA" instruction works only on the A register.
9. Write a program to add $7F9A to $BC48 and save the result in RAM memory locations starting at $1840.
10. Write a program to subtract $7F9A from $BC48, and save the result in RAM memory locations starting at $1840.
11. Write a program to add BCD $7795 to $9548 and save the BCD result in RAM memory locations starting at $1840.
12. Show how to perform 77 × 34 in the HCS12.
13. Show how to perform 77/3 in the HCS12.
14. True or false. The MUL instruction works on any register of the HCS12.
15. The MUL instruction places the result in registers ________ and______.

SECTION 5.2: LOGIC INSTRUCTIONS AND PROGRAMS

16. Assume that A = $F0. Perform the following operations.
    Note: The operations are independent of each other.
    (a) `ANDA #$45` (b) `ORAA #$90`
    (c) `EORA #$76` (d) `ANDA #$90`
    (e) `EORA #$90` (f) `ORAA #$90`
    (g) `ANDA #$FF` (h) `ORAA #$99`
    (i) `EORA #$EE` (j) `EORA #$AA`
17. Find the contents of register A after each of the following instructions:

(a)
```
LDAA #$65
ANDA #$76
```
(b)
```
LDAA #$70
ORAA #$6B
```
(c)
```
LDAA #$95
EORA #$AA
```
(d)
```
LDAA #$5D
ANDA #$78
```
(e)
```
LDAA #$C5
ORAA #$12
```
(f)
```
LDAA #$6A
EORA #$6E
```
(g)
```
LDAA #$37
ORAA #$26
```

SECTION 5.3: COMPARE INSTRUCTION AND PROGRAMS

18. True or false. In using the "CMPA MYRAM" instruction, we must use A as one of the registers.
19. Explain how the BHI instruction works.
20. Does the compare instruction affect the flag bits of the status register?
21. Assume that MYREG = 85H. Indicate if it skips after compare is executed in each of the following cases:

(a)
```
      LDAA   #$90
      CMPA   #$85
      BHI    OVER
      INA
      ADDA   #$2
OVER  ....
```
(b)
```
      LDAA   #$70
      CMPA   #$85
      BLO    OVER
      INA
      ADDA   #$2
OVER  ....
```

4. A2 = 1010 0010
   shift left: 0100 0100 C = 1
   shift again: 1000 1000 C = 0
   shift again: 0001 0000 C = 1
   shift again: 0010 0000 C = 0

   A2 shifted left four times = 20

SECTION 5.5: BCD AND ASCII CONVERSION

1. (a) $15 = 0001 0101 packed BCD, 0000 0001,0000 0101 unpacked BCD
   (b) $99 = 1001 1001 packed BCD, 0000 1001,0000 1001 unpacked BCD
2. $3736 = 00110111 00110110B
   and in BCD we have 76H = 0111 0110B
3. No. We need to write it as $54 (with the $) or 01010100B to make it BCD. The value 54 without the "$" is interpreted as $36 by the assembler.
4. $36, $37

SECTION 5.6: SIGNED NUMBER CONCEPTS AND ARITHMETIC OPERATIONS

1. D7
2. $16 is 00010110 in binary and its 2's complement is 1110 1010 or
   − $16 = EA in hex.
3. −128 to +127
4. +9 = 00001001 and −9 = 11110111 or F7 in hex.
5. An overflow is a carry into the sign bit (D7) but the carry is a carry out of register.
6. EMULS, IDIVS
7. LSR (logical shift right) shifts each bit right one position and fills the MSB with zero.
   ASR (arithmetic shift right) shifts each bit right one position and fills the MSB with the sign bit in each; the LSB is shifted into the carry flag.
8. (a) BLE will branch if (V Ex-OR N) + Z = 1.
   (b) BGE will branch if (V Ex-OR N) = 0.

CHAPTER 6

ADVANCED ADDRESSING MODES, LOOK-UP TABLE, MACROS, AND MODULES

OBJECTIVES

Upon completion of this chapter, you will be able to:

- **>> Code HCS12 Assembly language programs using advanced indexed addressing modes**
- **>> Code HCS12 instructions to manipulate a look-up table**
- **>> Access fixed data residing in the program ROM space**
- **>> Discuss how to create macros and modules**
- **>> Code HCS12 programs for ASCII and BCD data conversion**
- **>> Code HCS12 programs to create and test the checksum byte**
- **>> List the advantages of macros and modules in programming**
- **>> Understand the concepts of GLOBAL and EXTERNAL in modular programming**

SECTION 6.1: ADVANCED INDEXED ADDRESSING MODE

In previous chapters we showed how to use a simple indexed addressing mode. The advanced addressing mode is a very important addressing mode in the HCS12 and will be discussed in this section.

Indexed addressing mode

In the indexed addressing mode, a register is used as a pointer to the data location. In the HCS12, four registers can be used for this purpose: X, Y, SP (stack pointer), and PC (program counter). Since these are 16-bit registers, they allow access to the entire 64K bytes of memory space in the HCS12. In most applications, we use the X, Y, and SP for accessing data, since the PC register is used by the CPU to access the code. One of the advantages of the indexed addressing mode is that it makes accessing data dynamic rather than static, as with the extended addressing mode. Example 6-1 shows three cases of copying 55H into RAM locations $840 to $844. Notice in solution (b) that two instructions are repeated numer-

Example 6-1

Write a program to copy the value 55H into RAM memory locations $840 to $844 using (a) Extended addressing mode. (b) Indexed addressing mode without a loop. (c) A loop.

Solution:

(a)

```
        LDAA    #$55        ;load A with value 55H
        STAA    $840        ;copy A to RAM location $840
        STAA    $841        ;copy A to RAM location $841
        STAA    $842        ;copy A to RAM location $842
        STAA    $843        ;copy A to RAM location $843
        STAA    $844        ;copy A to RAM location $844
```

(b)

```
        LDAA    #$55        ;load A with value $55
        LDX     #$840       ;load the pointer. x = $840
        STAA    0,X         ;copy A to RAM loc X+0 points to
        INX                 ;increment pointer. Now X = 841
        STAA    0,X         ;copy A to RAM loc X+0 points to
        INX                 ;increment pointer. Now X = 842
        STAA    0,X         ;copy A to RAM loc X+0 points to
        INX                 ;increment pointer. Now X = 843
        STAA    0,X         ;copy A to RAM loc X+0 points to
        INX                 ;increment pointer. Now X = 844
```

(c)

```
COUNT   EQU $810    ;location $810 for counter
        LDAA    #$5         ;A = 5
        STAA    COUNT       ;load the counter, Count = 5
        LDX     #$840       ;load pointer. X = $840, RAM address
        LDAA    #$55        ;A = 55h value to be copied
B1      STAA    0,X         ;copy A to RAM loc X+0 points to
        INX                 ;increment pointer
        DEC     COUNT       ;decrement the counter
        BNE     B1          ;loop until counter = zero
```

Use the Codewarrior simulator to examine RAM contents after the above program is run.

840 = (55)   841 = (55)   842 = (55)   843 = (55)   844 = (55)

ous times. We can create a loop with those two instructions as shown in solution (c). Solution (c) is the most efficient and is possible only because of the indexed addressing mode. In Example 6-1, we must use "INX" to increment the pointer. Looping is not possible in extended addressing mode, and that is the main difference between the indexed and extended addressing modes. For example, trying to send a string of data located in consecutive locations of data RAM is much more efficient and dynamic using indexed addressing mode than using extended addressing mode. See Example 6-2.

Example 6-2

Assume that RAM locations $830–834 have a string of ASCII data, as shown below. Write a program to get each character and send it to Port B one byte at a time. Show the program using:
(a) Extended addressing mode.
(b) Indexed addressing mode.

830 = ('H')
831 = ('E')
832 = ('L')
833 = ('L')
834 = ('O')

Solution:

(a) Using extended addressing mode

```
      BSET   DDRB,%11111111      ;make Port B an output
      LDAA   $830           ;get contents of loc $830 to A
      STAA   PORTB          ;send it to PORTB
      LDAA   $831           ;get contents of loc $831 to A
      STAA   PORTB          ;send it to PORTB
      LDAA   $832           ;get contents of loc $832 to A
      STAA   PORTB          ;send it to PORTB
      LDAA   $833           ;get contents of loc $833 to A
      STAA   PORTB          ;send it to PORTB
      LDAA   $834           ;get contents of loc $834 to A
      STAA   PORTB          ;send it to PORTB
```

(b) Using indexed addressing mode

```
COUNT  EQU $810   ;location $810 for counter
       LDAA    #$5         ;A = 5
       STAA    COUNT       ;load the counter, Count = 5
       LDX     #$830       ;load pointer. X = $830, RAM address
B2     LDAA    0,X         ;copy location pointed to by X+0 to A
       STAA    PORTB       ;copy A to PORTB
       INX                 ;increment pointer
       DEC     COUNT       ;decrement the counter
       BNE     B2          ;loop until counter = zero
```

When simulating the above program on the CodeWarrior, make sure that RAM locations $830–834 have the message "HELLO".

Auto-increment option for indexed addressing mode

The HCS12 gives us the auto-increment and auto-decrement options for X, Y, and SP indexed registers. The syntax used for such cases for the CLR instruction is shown in Table 6-1. Examples 6-3 and 6-4 show the use of auto-increment using the X indexed register, while Example 6-5 uses both the X and Y registers. The increment value can be from –8 to –1 or +1 to +8. Example 6-6 shows how to use increment by 2 to move a block of data one word (16-bit) at a time.

Table 6-1: Auto-Increment/Decrement for CLR Instruction

Instruction	Function
CLR 0,X	After clearing loc. pointed to by X, the X stays the same.
CLR 1,X+ (post-inc)	After clearing loc. pointed to by X, the X is incremented by 1.
CLR 1,+X (pre-inc)	The X is incremented by 1, then loc. pointed to by X is cleared.
CLR 1,X- (post-dec)	After clearing loc. pointed to by X, the X is decremented by 1.
CLR 1,-X (pre-dec)	The X is decremented by 1, then loc. pointed to by X is cleared.

Note: This table shows the syntax for the X register. It also works for the Y and SP registers, but PC cannot be used with the auto option. The increment/decrement values can be –8, –7, –6, –5, –4, –3, –2, –1 or +1, +2, +3, +4, +5, +6, +7, and +8.

Example 6-3

Write a program to clear 16 RAM locations starting at RAM address $840. Use the auto-increment.

Solution:

```
COUNT  EQU $810   ;location $810 for counter
       LDAA   #16        ;A = 16
       STAA   COUNT      ;load the counter, Count = 16
       LDX    #$840      ;load pointer. X = $840, RAM address
B1     CLR    1,X+       ;clear RAM loc and increment the pointer by 1
       DEC    COUNT      ;decrement the counter
       BNE    B1         ;loop until counter = zero
```

Example 6-4

Assume that RAM locations 840–843H have the following hex data. Write a program to add them together and place the result in locations 0x806 and 0x807.

840 = (7D)   841 = (EB)   842 = (C5)   843 = (5B)

Solution:

```
H_BYTE EQU $806          ;RAM loc for H_Byte
L_BYTE EQU $807          ;RAM loc for L_Byte
COUNT  EQU $810          ;location $810 for counter
       LDAA   #4         ;A = 4
       STAA   COUNT      ;load the counter, Count = 4
       LDX    #$840      ;load pointer. X = $840, RAM address
       CLRA              ;A = 0, value to be copied
       STAA   H_BYTE     ;clear H_BYTE
B1     ADDA   1,X+       ;ADD A and increment the pointer
       BCC    OVER
       INC    H_BYTE
OVER   DEC    COUNT      ;decrement the counter
       BNE    B1         ;loop until counter = zero
       STAA   L_BYTE
```

Example 6-5

Write a program to copy a block of 5 bytes of data from RAM locations starting at $830 to RAM locations starting at $860.

Solution:

```
COUNT  EQU $810    ;location $810 for counter
       LDAA    #$5         ;A = 5
       STAA    COUNT       ;load the counter, Count = 5
       LDX     #$830       ;load pointer. X = $830, RAM address
       LDY     #$860       ;load pointer. Y = $860, RAM address
B2     LDAA    1,X+        ;get the byte and increment pointer by 1
       STAA    1,Y+        ;copy A and increment pointer by 1
       DEC     COUNT       ;decrement the counter
       BNE     B2          ;loop until counter = zero
```

Before we run the above program.

```
830 = ('H') 831 = ('E') 832 = ('L') 833 = ('L') 834 = ('O')
```

After the program is run, the addresses $860–$864 have the same data as $830–$834.

```
830 = ('H') 831 = ('E') 832 = ('L') 833 = ('L') 834 = ('O')
860 = ('H') 861 = ('E') 862 = ('L') 863 = ('L') 864 = ('O')
```

Example 6-6

Write a program to copy (transfer) a block of 20 bytes of data from RAM locations starting at $830 to RAM locations starting at $860. Transfer a word at a time.

Solution:

```
COUNT  EQU $810    ;location $810 for counter
       LDAA    #10         ;A = 10 word
       STAA    COUNT       ;load the counter, Count = 10 words
       LDX     #$830       ;load pointer. X = $830, RAM address
       LDY     #$860       ;load pointer. Y = $860, RAM address
B3     LDD     2,X+        ;get the word to D and increment pointer by 2
       STD     2,Y+        ;copy D to memory and increment pointer by 2
       DEC     COUNT       ;decrement the counter
       BNE     B3          ;loop until counter = zero
```

Another version of this program was given in Chapter 3.

Effective address in indexed addressing mode

In using the indexed addressing mode, we have the choice of 5-, 9-, or 16-bit signed offset as displacement. The displacement is added to the indexed register to calculate what is called *effective address*. In "LDAA 5,X", X + 5 is the effective address since the fifth byte from the beginning of the location pointed to by X is moved to register A. In this case, the X stays the same and another instruction such as "LDAA 9,X" will load the 9th byte into A. See Example 6-7. This is a widely used concept in accessing the elements of an array, as we will see in the next section. We can also use A, B, and D accumulators for the purpose of holding the offset value. For example, the "STAA B,Y" instruction will store the A register into the RAM location pointed to by the effective address of Y + B. Table 6-2 shows all the options for the indexed addressing mode.

Example 6-7

Assuming X = $1200, find the effective address for the following cases:
a) LDAA 10,X   b) LDAA 14,X   c) LDAA –3,X

Solution:

a) LDAA 10,X   Effective address = $1200 + 10 = $120A
b) LDAA 14,X   Effective address = $1200 + 14 = $120E
c) LDAA –3,X   Effective address = $1200 – 3 = $11FD

Table 6-2: Summary of Indexed Addressing Mode (Courtesy of Freescale.com)

Postbyte Code (xb)	Source Code Syntax	Comments rr; 00 = X, 01 = Y, 10 = SP, 11 = PC
rr0nnnnn	,r n,r -n,r	**5-bit constant offset n = –16 to +15** r can specify X, Y, SP, or PC
111rr0zs	n,r -n,r	**Constant offset (9- or 16-bit signed)** z – 0 = 9-bit with sign in LSB of postbyte(s) –256 <= n <= 255 1 = 16-bit –32,768 <= n <= 65,535 if z = s = 1, 16-bit offset indexed-indirect (see below) r can specify X, Y, SP, or PC
111rr011	[n,r]	**16-bit offset indexed-indirect** rr can specify X, Y, SP, or PC –32,768 <= n <= 65,535
rr1pnnnn	n,-r n,+r n,r- n,r+	**Auto predecrement, preincrement, postdecrement, or postincrement;** p = pre-(0) or post-(1), n = –8 to –1, +1 to +8 r can specify X, Y, or SP (PC not a valid choice) +8 = 0111 ... +1 = 0000 –1 = 1111 ... –8 = 1000
111rr1aa	A,r B,r D,r	**Accumulator offset (unsigned 8-bit or 16-bit)** aa – 00 = A 01 = B 10 = D (16-bit) 11 = see accumulator D offset indexed-indirect r can specify X, Y, SP, or PC
111rr111	[D,r]	**Accumulator D offset indexed-indirect** r can specify X, Y, SP, or PC

Review Questions

1. The instruction "`LDAA 2,X`" uses _________ addressing mode.
2. True or false. We can use the PC register for indexed addressing mode.
3. True or false. We can use the SP register for indexed addressing mode.
4. True or false. We can use the PC register for the auto-increment option in indexed addressing mode.
5. Which registers do we use for indexed addressing mode if the data is in the data RAM area of memory?

SECTION 6.2: ACCESSING LOOK-UP TABLE IN FLASH

So far, we have seen how the HCS12 accesses the data residing in RAM space. In the single-chip mode of the HCS12, we use the Flash ROM for storing the code and RAM for storing data. In single-chip mode, while we normally do not use any of the RAM space for storing code, we can use the Flash code space to store fixed static data. The EEPROM is also used for storing critical data and will be covered in Chapter 14. In this section we discuss how to access fixed data residing in the Flash ROM space of the single-chip HCS12. First we examine how to store fixed data in the program ROM space using the DC.B (define byte) directive.

DC.B (define constant byte) and fixed data in Flash

The DC.B data directive is widely used to allocate ROM program (code) memory in byte-sized chunks. In other words, DC.B is used to define an 8-bit fixed data. When DC.B is used to define fixed data, the numbers can be in decimal, binary, hex, or ASCII formats. The DC.B directive is widely used to define ASCII strings. See Example 6-8. In Example 6-8, notice that we must use single quotes (') for a single character or double quotes (") for a string. CodeWarrior also allows the use of FCB in place of DC.B to define values of 255 ($FF) or less. We can use DC.W (define word) to declare values up to 65,535 ($FFFF). Examples 6-9, 6-10, and 6-11 show how to access the fixed data stored in the Flash ROM.

Example 6-8

Assume that we have burned the following fixed data into program ROM of a HCS12 chip. Give the contents of each ROM location starting at $8500. See Appendix F for the hex values of the ASCII characters.

```
;MY DATA IN ROM
      ORG $8500             ;ROM address
DATA1 DC.B 28               ;DECIMAL(1C in hex)
DATA2 DC.B %00110101        ;BINARY (35 in hex)
DATA3 DC.B $39              ;HEX

      ORG $8510
DATA4 DC.B 'Y'              ;single ASCII char
DATA5 DC.B '2','0','0','5';ASCII numbers

      ORG $8518
DATA6 DC.B "Hello ALI"  ;ASCII string
      END
```

Solution:

```
DATA1          DATA2          DATA3
8500 = (1C), 8501 = (35), 8502 = (39)

DATA4          DATA5
8510 = (59), 8511 = (32), 8512 = (30), 8513 = (30), 8514 = (35)
        Y             2             0             0             5

DATA6
8518 = (48), 8519 = (65), 851A = (6C), 851B = (6C), 851C = (6F)
         H             e             l             l             o
581D = (20), 851E = (41), 851F = (4C), 8520 = (49)
      SPACE             A             L             I
```

Example 6-9

Assuming that program ROM space starting at $8500 contains "USA", write a program to send all the characters to Port B one byte at a time.

Solution:

```
(a) This method uses a counter
COUNT  EQU $810   ;location $810 for counter
       ORG   $8000 ;code is burned into Flash ROM starting at $8000
       BSET   DDRB,%11111111    ; make PORTB an output port
       LDAA   #$3          ;A = 3
       STAA   COUNT        ;load the counter,
       LDX    #MYDATA      ;load pointer. X = $8500 ROM address
B1     LDAA   0,X          ;copy location pointed to by X+0 to A
       STAA   PORTB        ;copy A to PORTB
       INX                 ;increment pointer
       DEC    COUNT        ;decrement the counter
       BNE    B1           ;loop until counter = zero
;data is burned into Flash code space starting at $8500
       ORG   $8500
MYDATA DC.B "USA"
       END                       ;end of program
```

In the above program Flash ROM locations $8500–8502 have the following contents:

```
8500 = ('U')      8501 = ('S')      8502 = ('A')
```

```
b) This method uses null char for end of string
       ORG   $8000        ;code area
       BSET   DDRB,%11111111    ; make PORTB an output port
       LDX    #MYDATA      ;load pointer. X = $8500 ROM address
B2     LDAA   0,X          ;copy location pointed to by X+0 to A
       BEQ    OVER         ;exit if zero is the last character
       STAA   PORTB        ;copy A to PORTB
       INX                 ;increment pointer
       BRA   B2
OVER   BRA    OVER         ;use this or go back to monitor program
       ORG   $8500         ;data area
MYDATA DC.B "USA",0        ;notice null character
       END
```

```
c) This method uses null char for end of string and auto-increment
       ORG   $8000        ;data
       BSET   DDRB,%11111111    ; make PORTB an output port
       LDX    #MYDATA      ;load pointer. X = $8500 ROM address
B3     LDAA   1,X+         ;copy and increment pointer
       BEQ    OVER         ;exit if zero is the last character
       STAA   PORTB        ;copy A to PORTB
       BRA    B3           ;loop until the last char
OVER   BRA    OVER
       ORG   $8500         ;data
MYDATA DC.B "USA",0
       END
```

```
Memory
MYDATA[0]  8500 - 8500                                   Auto
008500 55 53 41 00 -- -- -- --  -- -- -- -- -- -- -- --  USA.------------
008510 -- -- -- -- -- -- -- --  -- -- -- -- -- -- -- --  ----------------
```

Example 6-10

Assume that ROM space starting at $8500 contains the person's information, such as first name, last name, phone number, and so on. Write a program to read the last name and send it to Port B.

Solution:

```
COUNT  EQU $810   ;location $810 for counter
      ORG    $8000        ;Flash ROM starting at $8000
      BSET   DDRB,%11111111   ; make PORTB an output port
      LDAA   #$4          ;A = 4, # of letters in lastname
      STAA   COUNT        ;load the counter
      LDX    #MYDATA      ;load pointer
B2    LDAA   12,X         ;copy location pointed to by X+12 to A
      STAA   PORTB        ;copy A to PORTB
      INX                 ;increment pointer
      DEC    COUNT        ;decrement the counter
      BNE    B2           ;loop until counter = zero
      BRA    $            ;stay here
;---------------------information
      ORG    $8500        ;data burned starting at $8500
MYDATA FCB "John        ";fixed width of 12 characters
       FCB "Dewy        ";fixed width of 12 characters
       FCB "111-999-9999";fixed width of 12 characters
       END
```

Example 6-11

Assume that ROM space starting at $8500 contains the message "The Promise of World Peace". Write a program to bring it into CPU one byte at a time and place the bytes in RAM locations starting at $840.

Solution:

```
      ORG    $8000        ;burn into ROM starting at $8000
      LDX    #MYDATA      ;load pointer. X = $8500, ROM address
      LDY    #$840        ;load pointer. Y = $860, RAM address
B5    LDAA   1,X+  ;get the byte from Flash, increment pointer by 1
      BEQ    EXIT         ;exit if it is the last one
      STAA   1,Y+         ;save it in RAM and increment pointer by 1
      BRA    B5           ;KEEP looping
EXIT  BRA    EXIT

;---------------------message
      ORG    $8500        ;data burned starting at $8500
MYDATA DC.B "The Promise of World Peace",0
      END
```

Examining RAM and ROM using CodeWarrior simulator

The CodeWarrior simulator is a great tool to examine RAM and ROM contents. We encourage its use to examine and verify the results of programs in this chapter.

Look-up table and reading from Flash

The look-up table is a widely used concept in microcontroller programming. It allows access to elements of a frequently used table with minimum operations. As an example, assume that for a certain application we need x^2 values in the range of 0 to 9. We can use a look-up table instead of calculating the values, which takes some time. To get the table element we can add a fixed value to a register such as X to index into the look-up table. For example, the instruction "LDAA A,X" will bring elements of the look-up table pointed to by the effective address formed by the addition of X + A. In this case, A is used as an index into the look-up table. See Examples 6-12, 6-13, and 6-14.

Writing to Flash in HCS12

Because writing to Flash ROM involves manipulating the configuration bits it is discussed in Chapter 14.

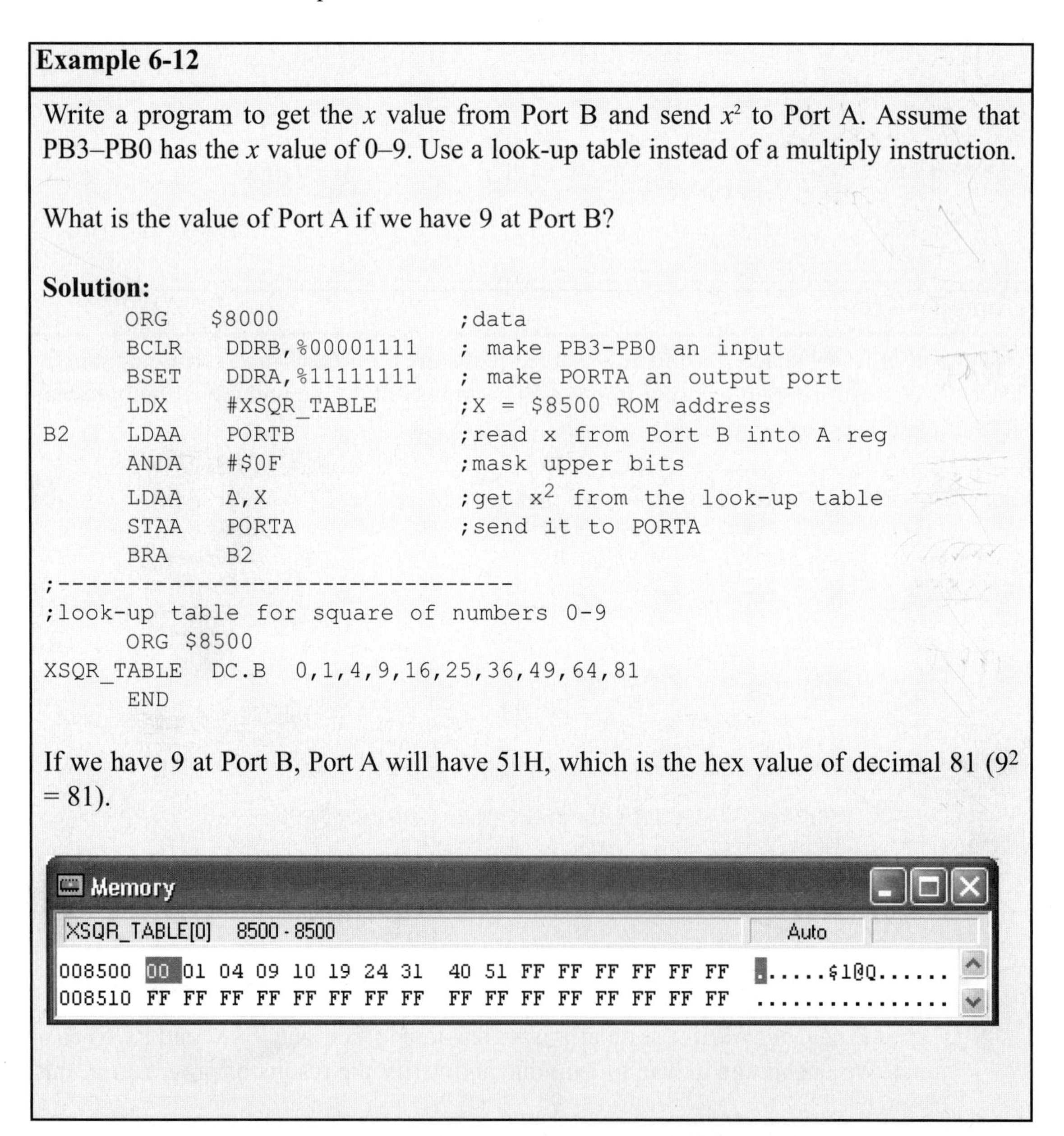

Example 6-12

Write a program to get the x value from Port B and send x^2 to Port A. Assume that PB3–PB0 has the x value of 0–9. Use a look-up table instead of a multiply instruction.

What is the value of Port A if we have 9 at Port B?

Solution:

```
        ORG    $8000                 ;data
        BCLR    DDRB,%00001111       ; make PB3-PB0 an input
        BSET    DDRA,%11111111       ; make PORTA an output port
        LDX     #XSQR_TABLE          ;X = $8500 ROM address
B2      LDAA    PORTB                ;read x from Port B into A reg
        ANDA    #$0F                 ;mask upper bits
        LDAA    A,X                  ;get x2 from the look-up table
        STAA    PORTA                ;send it to PORTA
        BRA     B2
;---------------------------------
;look-up table for square of numbers 0-9
        ORG $8500
XSQR_TABLE  DC.B  0,1,4,9,16,25,36,49,64,81
        END
```

If we have 9 at Port B, Port A will have 51H, which is the hex value of decimal 81 ($9^2 = 81$).

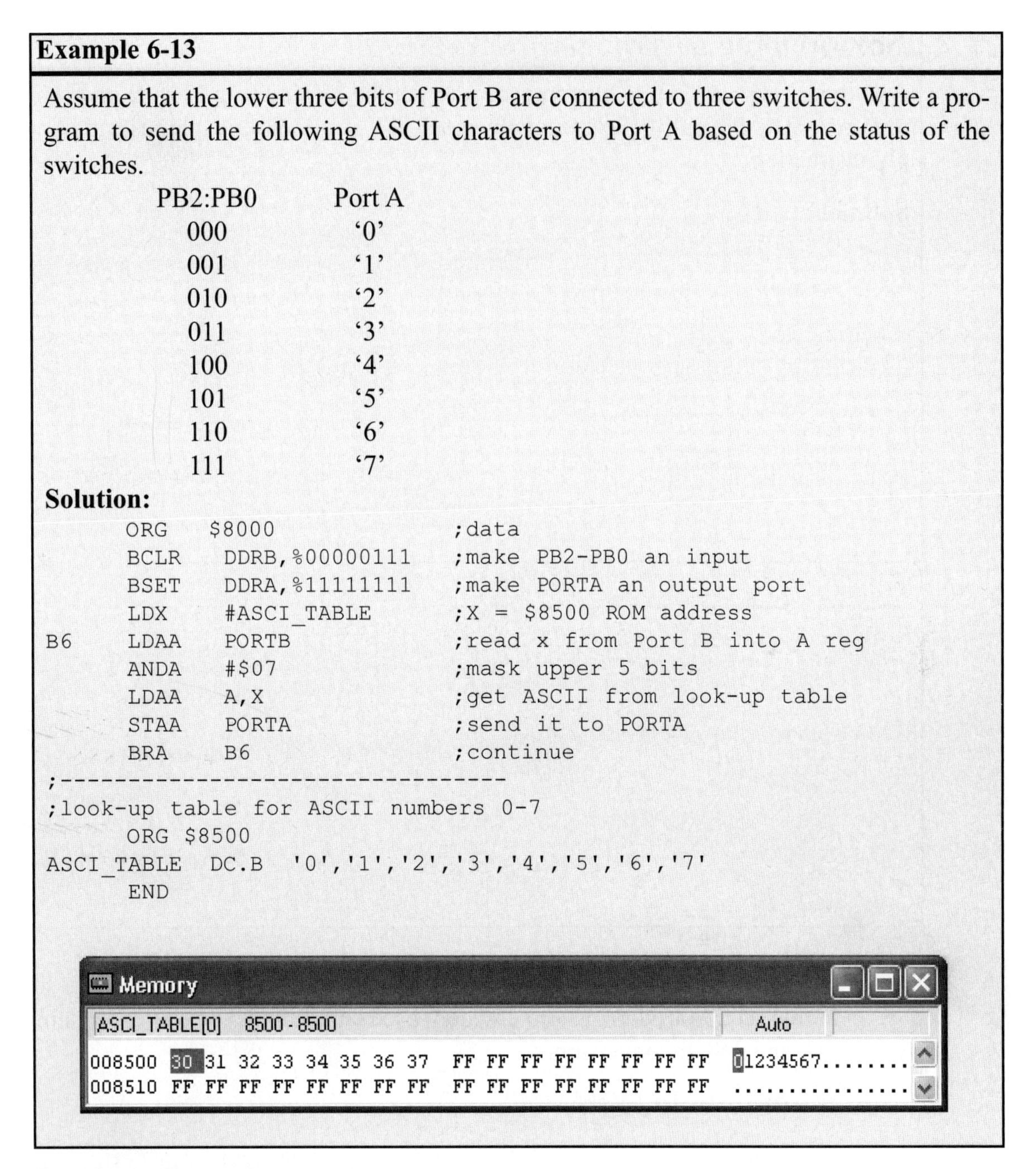

Example 6-13

Assume that the lower three bits of Port B are connected to three switches. Write a program to send the following ASCII characters to Port A based on the status of the switches.

PB2:PB0	Port A
000	'0'
001	'1'
010	'2'
011	'3'
100	'4'
101	'5'
110	'6'
111	'7'

Solution:

```
      ORG    $8000                ;data
      BCLR   DDRB,%00000111       ;make PB2-PB0 an input
      BSET   DDRA,%11111111       ;make PORTA an output port
      LDX    #ASCI_TABLE          ;X = $8500 ROM address
B6    LDAA   PORTB                ;read x from Port B into A reg
      ANDA   #$07                 ;mask upper 5 bits
      LDAA   A,X                  ;get ASCII from look-up table
      STAA   PORTA                ;send it to PORTA
      BRA    B6                   ;continue
;---------------------------------
;look-up table for ASCII numbers 0-7
      ORG $8500
ASCI_TABLE  DC.B  '0','1','2','3','4','5','6','7'
      END
```


Review Questions

1. The instruction "`LDAA 5,X`" uses ______ as address pointer.
2. What register is incremented upon execution of the "LDAA 0,X" instruction?
3. What register holds data once it is read by the "LDAA A,Y" instruction?
4. What is the size of Y? How much ROM space does it cover?
5. What register is incremented upon execution of the "STAA 1,Y+" instruction?

Example 6-14

Assume $y = x^2 + 2x + 3$. Write a program to get the x value from Port B and send the y to Port A. The PB3–PB0 has the x value of 0–9. Use a look-up table instead of a multiply instruction.

Solution:

```
        ORG    $8000                ;data
        BCLR    DDRB,%00001111      ;make PB3-PB0 an input
        BSET    DDRA,%11111111      ;make PORTA an output port
        LDX     #$8500              ;X = $8500 ROM address
B2      LDAA    PORTB               ;read x from Port B into A reg
        ANDA    #$0F                ;mask upper bits
        LDAA    A,X                 ;get y from the look-up table
        STAA    PORTA               ;send it to PORTA
        BRA     B2

        ORG    $8500
Y_TABLE        FCB   3,6,11,18,27,38,51,66,83,102
               ;(0)² + 2(0) + 3 = 3
               ;(1)² + 2(1) + 3 = 6
               ;(2)² + 2(2) + 3 = 11
               ;(3)² + 2(3) + 3 = 18
               ;(4)² + 2(4) + 3 = 27
               ;(5)² + 2(5) + 3 = 38
               ;(6)² + 2(6) + 3 = 51
               ;(7)² + 2(7) + 3 = 66
               ;(8)² + 2(8) + 3 = 83
               ;(9)² + 2(9) + 3 = 102
               END
```

SECTION 6.3: CHECKSUM AND ASCII SUBROUTINES

In this section we look at some widely used subroutines such as checksum byte, BCD, and ASCII conversion.

Checksum byte in ROM

To ensure the integrity of ROM contents, every system must perform a checksum calculation. The checksum will detect any corruption of the contents of ROM. One cause of ROM corruption is current surge, either when the system is turned on, or during operation. To ensure data integrity in ROM, the checksum process uses what is called a *checksum byte*. The checksum byte is an extra byte that is tagged to the end of a series of bytes of data. To calculate the checksum byte of a series of bytes of data, the following steps can be taken:

1. Add the bytes together and drop the carries.
2. Take the 2's complement of the total sum; this is the checksum byte, which becomes the last byte of the series.

To perform a checksum operation, add all the bytes, including the checksum byte. The result must be zero. If it is not zero, one or more bytes of data have been changed (corrupted). To clarify these important concepts, see Example 6-15.

Example 6-15

Assume that we have 4 bytes of hexadecimal data: 25H, 62H, 3FH, and 52H.
(a) Find the checksum byte.
(b) Perform the checksum operation to ensure data integrity.
(c) If the second byte, 62H, has been changed to 22H, show how the checksum method detects the error.

Solution:

(a) Find the checksum byte.

```
      25H
  +   62H
  +   3FH
  +   52H
     118H
```

(Dropping the carry of 1, we have 18H. Its 2's complement is E8H. Therefore the checksum byte is E8H.)

(b) Perform the checksum operation to ensure data integrity.

```
      25H
  +   62H
  +   3FH
  +   52H
  +   E8H
     200H
```

(Dropping the carries, we see 00, indicating that the data is not corrupted.)

(c) If the second byte 62H has been changed to 22H, show how the checksum method detects the error.

```
      25H
  +   22H
  +   3FH
  +   52H
  +   E8H
     1C0H
```

(Dropping the carry, we get C0H, which is not 00. This means that the data is corrupted.)

Checksum program

The checksum generation and testing program is given in the subroutine form below. We have divided Program 6-1 into three subroutines (or subprograms). These three subroutines perform the following operations:

1. Retrieve the data from code ROM.
2. Calculate the checksum byte.
3. Test the checksum byte for any data error.

Each of these subroutines can be used in other applications. Example 6-15 shows how to manually calculate the checksum for a list of values. Also see Program 6-1.

```
;PROG 6-1: CALCULATING AND TESTING CHECKSUM BYTE

MYRAM       EQU $860     ;RAM space to place the bytes
COUNTREG    EQU $820     ;RAM loc for counter
CNTVAL      EQU 4        ;counter value = 4 for adding 4 bytes
CNTVAL1     EQU 5        ;counter value = 5 for adding 5 bytes
                         ;including checksum byte
;------------main program
      LDS   #$900              ;set up stack
      JSR   COPY_DATA
      JSR   CAL_CHKSUM
      JSR   TEST_CHKSUM
      BRA   $

;------copying data from code ROM address 8500H to data RAM loc
COPY_DATA
      LDX    #MYBYTE    ;load pointer. X = $8500, ROM address
      LDY    #MYRAM     ;load pointer. Y = $860, RAM address
      LDAA   #CNTVAL
      STAA   COUNTREG
B5    LDAA   1,X+  ;get the byte from Flash, increment pointer
      STAA   1,Y+       ;save it in RAM and increment pointer
      DEC    COUNTREG
      BNE    B5         ;keep looping
      RTS               ;return

;-----calculating checksum byte
CAL_CHKSUM
      LDAA   #CNTVAL
      STAA   COUNTREG   ;load the counter
      LDX    #MYRAM     ;load pointer. X = $860, RAM address
      CLRA              ;A=0
B2    ADDA   1,X+       ;add and increment pointer
      DEC    COUNTREG   ;decrement the counter
      BNE    B2         ;loop until counter = zero
      NEGA              ;2's comp of sum
      STAA   0,X        ;save the checksum as the last byte
      RTS               ;return

;----------testing checksum byte, place on PORTB
TEST_CHKSUM
      BSET   DDRB,%11111111         ;
      LDAA   #CNTVAL+1  ;adding 5 bytes including checksum byte
      STAA   COUNTREG   ;load the counter
      LDX    #MYRAM     ;load pointer. X = $860, RAM address
      CLRA              ;A=0
B3    ADDA   1,X+       ;add and increment pointer
      DEC    COUNTREG   ;decrement the counter
      BNE    B3         ;loop until counter = zero
      CMPA   #0         ;see if A = zero
      BEQ    G_1        ;is result zero? then good
      LDAA   #'B'       ;then send letter B for Bad to PORTB
      STAA   PORTB            ;if not, data is bad
      RTS
G_1   LDAA   #'G'       ;then send letter G for Good to PORTB
      STAA   PORTB            ;data is not corrupted
      RTS
```

```
;----------my data in Flash ROM
      ORG $8500
MYBYTE FCB $25, $62, $3F, $52, $00
      END
```

BCD to ASCII conversion program

Many RTCs (real-time clocks) provide time and date in BCD format. To display the BCD data on an LCD or a PC screen, we need to convert it to ASCII. Program 6-2 (a) transfers packed BCD data from program ROM to data RAM, (b) converts packed BCD to ASCII, and (c) sends the ASCII to port B for display. We will use a portion of this program in Chapter 16. The displaying of data on an LCD will be shown in Chapter 12. See Chapter 5 for the BCD-to-ASCII conversion algorithm.

```
;PROG 6-2: CONVERTING PACKED BCD TO ASCII

MYRAM        EQU $860          ;RAM space to place the bytes
ASC_RAM      EQU $870
COUNTREG     EQU $820          ;RAM loc for counter
CNTVAL       EQU 4             ;counter value of BCD bytes
CNTVAL1      EQU 8             ;counter value of ASCII bytes

;------------main program
      LDS    #$900             ;set up stack
      JSR    COPY_DATA
      JSR    BCD_ASC_CONV
      JSR    DISPLAY
      BRA    $

;--------copying data from code ROM to data RAM
COPY_DATA
      LDX     #MYBYTE     ;load pointer. X = $8500, ROM address
      LDY     #MYRAM      ;load pointer. Y = $860, RAM address
      LDAA    #CNTVAL
      STAA    COUNTREG
B5    LDAA    1,X+  ;get the byte from Flash, increment pointer
      STAA    1,Y+        ;save it in RAM and increment pointer
      DEC     COUNTREG
      BNE     B5          ;keep looping
      RTS

;-----convert packed BCD to ASCII
BCD_ASC_CONV
      LDX     #MYRAM      ;X = $860, RAM address for BCD
      LDY     #ASC_RAM    ;Y = $870, RAM address for ASCII
      LDAA    #CNTVAL
      STAA    COUNTREG
B6    LDAA    0,X         ;get the BCD from RAM
      ANDA    #$F0        ;mask the lower nibble
      LSRA                ;shift right to get upper nibble
      LSRA                ;into lower 4 bits of A
      LSRA
      LSRA
```

```
        ORAA #$30          ;make it an ASCII
        STAA 1,Y+          ;save it
        LDAA 1,X+         ;get BCD data once more, increment pointer
        ANDA #$0F          ;mask the upper nibble
        ORAA #$30          ;make it an ASCII
        STAA 1,Y+          ;save it
        DEC    COUNTREG
        BNE    B6          ;keep looping
        RTS

;-----send ASCII data to port B
DISPLAY
        BSET  DDRB, %11111111   ;make PORTB output
        LDAA  #CNTVAL1          ;A = 8, send 8 bytes of data
        STAA  COUNTREG          ;load the counter, count = 8
        LDX   #ASC_RAM          ;load pointer
B3      LDAA  1,X+              ;copy RAM to A and inc pointer
        STAA  PORTB             ;copy A to PORTB
        DEC   COUNTREG          ;decrement counter
        BNE   B3                ;loop until counter = zero
        RTS

;----------my BCD data in program ROM
        ORG $8500
MYBYTE FCB $25,$67,$39,$52
        END
```

Binary (hex) to ASCII conversion program

Many ADC (analog-to-digital converter) chips provide output data in binary (hex). To display the data on an LCD or PC screen, we need to convert it to ASCII. The code for the binary-to-ASCII conversion is shown in Program 6-3. Notice that the subroutine gets a byte of 8-bit binary (hex) data from Port B and converts it to decimal digits, and the second subroutine converts the decimal digits to ASCII digits and saves them. We are saving the low digit in the high address location and the high digit in low address location. This is referred to as the big-endian convention (i.e., high-byte to low-location and low-byte to high-location). All HCS12 chips use the big-endian convention. For the binary-to-ASCII conversion algorithm see Chapter 5.

```
;PROG 6-3: CONVERTING BIN(HEX) TO ASCII

RMND_L       EQU   $832        ;the least significant digit loc
RMND_M       EQU   $831        ;the middle significant digit loc
RMND_H       EQU   $830        ;the most significant digit loc
MYDEN        EQU   10          ;value for divide by 10
COUNTREG     EQU   $810        ;RAM loc for counter
CNTVAL       EQU   3           ;counter value
UNPBCD_ADDR EQU   $830
ASCII_RESULT EQU   $840
```

```
;-----------main program
      LDS    #$900
      BCLR   DDRB,%11111111      ;make PORTB input
      LDAB   PORTB               ;get the binary data from PORTB
      JSR    BIN_DEC_CON
      JSR    DEC_ASCII_CON
      BRA    $

;-----converting BIN(HEX) TO DEC (00-FF TO 000-255)
BIN_DEC_CON
      LDX    #MYDEN              ;X = 10, the denominator
      IDIV                       ;D/X = 253/10 = 25, remainder = 3
      STAB RMND_L                ;save the remainder
      XGDX                       ;exchange D and X for new numerator
      LDX   #MYDEN               ;X = 10, the denominator
      IDIV                       ;D/X
      STAB RMND_M                ;save the next digit
      XGDX                       ;exchange D and X
      STAB RMND_H                ;save the final digit
      RTS

;----converting unpacked BCD digits to displayable ASCII digits
DEC_ASCII_CON
      LDAA   #CNTVAL             ;A = 3
      STAA   COUNTREG            ;load the counter, count = 3
      LDX    #UNPBCD_ADDR        ;load pointer X
      LDY    #ASCII_RESULT       ;load pointer Y
B3    LDAA   1,X+  ;copy BCD digit to A, increment X pointer
      ADDA   #$30                ;make it an ASCII
      STAA   1,Y+                ;copy A and increment Y pointer
      DEC    COUNTREG            ;decrement counter
      BNE    B3                  ;loop until counter = zero
      RTS
      END                        ;end of the program
```

Review Questions

1. For the following ASCII numbers, give the ASCII and packed BCD representations.
   (a) '5', '7' (b) '9', '4'
2. Show the hex format for "2005" and its BCD version.
3. Does the A register have BCD data after the following instruction is executed?
   `LDAA #95`
4. 33H in BCD when converted to ASCII is ____ and ____.
5. Find the ASCII value for the binary 11110010 if we want to display it on a computer screen as a 3-digit decimal number.
6. The checksum byte method is used to test data integrity in ____(RAM, ROM).
7. Find the checksum byte for the following hex values: $88, $99, $AA, $BB, $CC, $DD
8. True or false. If we add all the bytes, including the checksum byte, and the result is $FF, there is no error in the data.

SECTION 6.4: MACROS AND MODULES

In this section we explore macros and modules and their use in Assembly language programming. The format and usage of macros are defined, and many examples of their applications are explored. In addition, this section demonstrates modular programming along with rules for writing modules and linking them together. Some very useful modules will be given, along with methods of passing parameters among various modules. Dividing a program into several modules (in C programming these are called *functions*) allows us to use modules in other applications. It is common practice to divide a program into several modules, test each module, and put them into a library.

What is macro and how is it issued?

There are applications in Assembly language programming where a group of instructions performs a task that is used repeatedly. For example, moving data into a RAM location is done repeatedly in the same program. It does not make sense to rewrite this code every time it is needed. Therefore, to reduce the time that it takes to write code and reduce the possibility of errors, the concept of macros was introduced. Macros allow the programmer to write the task (code to perform a specific job) once only, and to invoke it whenever it is needed.

MACRO definition

Every macro definition must have three parts, as follows:

```
name        MACRO
            ......
            ......
            ENDM
```

The MACRO directive indicates the beginning of the macro definition and the ENDM directive signals the end. The code between the MACRO and ENDM directives is called the *body* of the macro. The name must be unique and must follow Assembly language naming conventions. After the macro has been written, it can be invoked (or called) by its name, and appropriate values are substituted for parameters. The parameters in the body of a macro are labeled with a '\' followed by a number between 1 and 9 or a letter. Moving an immediate value into register or data RAM is a widely used service. We can use a macro to do the job as shown in the following code:

```
MOVE  MACRO
      LDAA  #\1
      STAA  \2
      ENDM
```

The above is the macro definition. Note that parameters \1 and \2 are used in the body of macro. Parameter \1 refers to the first argument passed to the macro and \2 refers to the second argument. The labeling can continue from \3 to \Z.

The following are three examples of how to use the above macro:

```
1. MOVE     $55, $820          ;send value 55H to loc 820H

2. VAL_1    EQU $55
   RAM_LOC  EQU $820
   MOVE     VAL_1, RAM_LOC

3. MOVE     $55, PORTB         ;send value 55H to Port B
```

The instruction "MOVE $5, $820" invokes the macro. The assembler expands the macro by providing the following code in the .lst file:

```
12850    36m   000003 8655          +    LDAA   #$55
12851    37m   000005 5A01          +    STAA   PORTB
```

The "m" indicates that the code is from the macro.

Using local labels in a macro

In the discussion of macros so far, the examples chosen do not have a label or name in the body of the macro. To be used within a macro labels must begin with \@.

To clarify these points, look at the following macro for a time delay:

```
DELAY_1 MACRO
        LDAA  #\1
        STAA  \2
\@BACK        NOP
        NOP
        NOP
        NOP
        DEC   \2
        BNE   \@BACK
        ENDM
```

The use of \@ allows the assembler to define the labels separately each time it encounters them. Examining the list file shows that when the macro is expanded for the first time, the list file has "_00002BACK" and for the second time it has "_00004BACK" in place of the "\@BACK" label, indicating that the "\@BACK" label is local. To clarify this concept, look at Program 6-4 and try omitting the \@ in the label to see how the assembler will give an error. The following code is another macro for a time delay with a nested loop:

```
DELAY_2 MACRO
        LDAA  #\2
        STAA  \4
\@AGAIN       LDAA  #\1
        STAA  \3
\@BACK        NOP
        NOP
        DEC   \3
        BNE   \@BACK
        DEC   \4
        BNE   \@AGAIN
        ENDM
```

Now examine Program 6-4 to see how to use a macro in a program.

```
;-------------------------------------------
      ;Program 6-4: toggling Port B using macros

      ;---------------sending data to fileReg macro
      MOVE: MACRO
            LDAA   #\1
            STAA   \2
            ENDM

      ;----------------------------time delay macro
      DELAY_1: MACRO
            LDAA   #\1
            STAA   \2
      \@BACK       NOP
            NOP
            NOP
            NOP
            DEC    \2
            BNE    \@BACK
            ENDM

      ;----------------------------program starts
            BSET     DDRB,%11111111        ;Port B as an output
      OVER  MOVE     $55,PORTB
            DELAY_1 200,$810
            MOVE     $AA,PORTB
            DELAY_1 200,$810
            BRA      OVER
            END
      ;--------------------end of file
```

INCLUDE directive

Assume that several macros are used in every program. Must they be rewritten every time? The answer is no, if the concept of the INCLUDE directive is known. The INCLUDE directive allows a programmer to write macros and save them in a file, and later bring them into any program file. For example, assume that the macros from Program 6-4 were written and then saved under the filename "MYMACRO1.MAC".

Assuming that these macros are saved on a disk under the filename "MYMACRO1.MAC", the INCLUDE directive can be used to bring this file into any ".asm" file and then the program can call upon any of the macros as many times as needed. When a file includes all macros, the macros are listed at the beginning of the ".lst" file and, as they are expanded, will become part of the program. To understand this, see Program 6-5.

```
      ;Program 6-5: toggling Port B using macros
            include 'MYMACRO1.MAC'  ;get macros from macro file

      ;----------------------------program starts
            BSET     DDRB,%11111111        ;Port B as an output
```

```
OVER  MOVE     $55,PORTB
      DELAY_1 200,$10
      MOVE     $AA,PORTB
      DELAY_1 200,$10
      BRA      OVER
      END
;--------------------end of file
```

MLIST directive

When viewing the .lst file with macros, we see them fully displayed. The expansion is set by default, and it shows the macro at every location it is called. This is fine for two or three iterations, but when there are more, it can become cumbersome. Using the `MLIST OFF` directive, we can turn off the display of macros in the list file. Using `MLIST ON` will turn expansion back on. Compare Figures 6-1 and 6-2 to see the difference between these directives.

```
Freescale HC12-Assembler
(c) Copyright Freescale 1987-2006

Rel.  Loc    Obj. code    Source line
----  ------ ---------    -----------
34                        ;Program 6-4: toggling Port B using macros
35
36                        ;---------------sending data to fileReg macro
37                        MOVE: MACRO
38                             LDAA  #\1
39                             STAA  \2
40                             ENDM
41
42                        ;---------------------------time delay macro
43                        DELAY_1: MACRO
44                             LDA\0 #\1
45                             STA\0 \2
46                        \@BACK     NOP
47                             NOP
48                             NOP
49                             NOP
50                             DEC   \2
51                             BNE   \@BACK
52                             ENDM
53
54                        ;---------------------------program starts
55    000000 4C03 FF           BSET DDRB,%11111111     ;Port B as an output
56                        OVER MOVE     $55,PORTB
57                             DELAY_1.A 200,$810
58                             MOVE     $AA,PORTB
59                             DELAY_1.B 200,$810
60    000027 20DA              BRA      OVER
```

Figure 6-1. List File with MLIST OFF Option for Program 6-4

```
Freescale HC12-Assembler
(c) Copyright Freescale 1987-2006

Rel.  Loc    Obj. code  Source line
----  ------ ---------  -----------
33                      ;Program 6-4: toggling Port B using macros
34
35                      ;---------------sending data to fileReg macro
36                        MOVE: MACRO
37                        LDAA #\1
38                        STAA \2
39                        ENDM
40
41                      ;----------------------------time delay macro
42                        DELAY_1: MACRO
43                             LDA\0 #\1
44                             STA\0 \2
45                        \@BACK      NOP
46                             NOP
47                             NOP
48                             NOP
49                             DEC    \2
50                             BNE    \@BACK
51                             ENDM
52
53                      ;----------------------------program starts
54    000000 4C03 FF      BSET DDRB,%11111111;Port B as an output
55                        OVER MOVE   $55,PORTB
37m   000003 8655         +    LDAA   #$55
38m   000005 5A01         +    STAA   PORTB
56                             DELAY_1.A 200,$810
43m   000007 86C8         +    LDAA   #200
44m   000009 7A08 10      +    STAA   $810
45m   00000C A7           +_00002BACK      NOP
46m   00000D A7           +    NOP
47m   00000E A7           +    NOP
48m   00000F A7           +    NOP
49m   000010 7308 10      +    DEC    $810
50m   000013 26F7         +    BNE    _00002BACK
57                             MOVE     $AA,PORTB
37m   000015 86AA         +    LDAA   #$AA
38m   000017 5A01         +    STAA   PORTB
58                             DELAY_1.B 200,$810
43m   000019 C6C8         +    LDAB   #200
44m   00001B 7B08 10      +    STAB   $810
45m   00001E A7           +_00004BACK      NOP
46m   00001F A7           +    NOP
47m   000020 A7           +    NOP
48m   000021 A7           +    NOP
49m   000022 7308 10      +    DEC    $810
50m   000025 26F7         +    BNE    _00004BACK
59    000027 20DA              BRA      OVER
```

Figure 6-2. List File with MLIST ON Option for Program 6-4

Macros vs. subroutines

Macros and subroutines are useful in writing assembly programs, but each has limitations. Macros increase code size every time they are invoked. For example, if you invoke a 10-instruction macro 10 times, the code size is increased by 100 instructions. If you call the same subroutine 10 times, however, the code size is only that of the subroutine instructions. The only problem with subroutines is that they use stack space when called, and this can cause problems when there is a nested call (a subroutine calling another subroutine). The nested call can lead to a stack overflow and cause the program to crash.

Modules

It is common practice in writing software packages to break down the project into small modules and to distribute the task of writing those modules among several programmers. This not only makes the project more manageable but also has other advantages, such as:

1. Each module can be written, debugged, and tested individually.
2. The failure of one module does not stop the entire project.
3. The task of locating and isolating any problem is easier and less time consuming.
4. One can use the modules to link with high-level languages such as C.
5. Parallel development shortens considerably the time required to complete a project.

Next we explain how to write and link modules to create a single executable program.

Writing modules

In the programs given in the last section, a main procedure was written that called many other subroutines. In those examples, if one subroutine did not work properly, the entire program would have to be rewritten and reassembled. A more efficient way to develop software is to treat each subroutine as a separate program (or module) with a separate filename. Then each one can be assembled and tested. After testing each program and making sure that each works, they can all be brought together (linked) to make a single program. To enable these modules to be linked together, certain Assembly language directives must be used. Among these directives, CodeWarrior uses XREF (external reference) and XDEF (external define). The XREF can also be called *EXTERNAL*. The XDEF can also be called *GLOBAL* or *PUBLIC*. Other Assembly languages typically use EXTERN and GLOBAL.

XREF (EXTERNAL) directive

The XREF directive is used to notify the assembler and linker that certain names and variables that are not defined in the present module are defined externally somewhere else. In the absence of the XREF directive, the assembler would show an error because it cannot find where the names are defined. The XREF directive has the following format:

```
XREF name1  ;each name can be in a separate XREF
XREF name2
XREF name3, name4 ;or many can be listed in the same XREF
```

XDEF (GLOBAL) directive

Names or parameters defined as XREF (indicating that they are defined outside the present module) must be defined as XDEF in the module where they are defined. Defining a name as XDEF (GLOBAL) allows the assembler and linker to match it with its XREF counterpart(s). The following is the format for the XDEF directive:

```
XDEF name1        ;each name can be in a separate XDEF
XDEF name2
XDEF name3, name4 ;or many can be listed in the same XDEF
```

Program 6-6 should help to clarify these concepts. It demonstrates that for every XREF definition there is a XDEF directive defined in another module. It is a better version of Program 6-1. Notice the entry and exit points of the program. Modules that are called by the main module have their own END directives. See Program 6-6.

```
;---------------------------------------------------
;PROG 6-6: MAIN.ASM - CALCULATING AND TESTING CHECKSUM BYTE
      XREF CAL_CHKSUM
      XREF TEST_CHKSUM
      XDEF  MYRAM
      XDEF  COUNTREG
      XDEF  CNTVAL
      XDEF  CNTVAL1
MYRAM       EQU $860
COUNTREG    EQU $820            ;RAM loc for counter
CNTVAL      EQU 4               ;counter value
CNTVAL1     EQU 5               ;counter value
;------------main program
      LDS   #$900
      JSR   COPY_DATA           ;this subroutine is in this file
      JSR   CAL_CHKSUM          ;this sub is in external file
      JSR   TEST_CHKSUM         ;this sub is in external file
      BRA   $
;--------copying data from code ROM to data RAM
COPY_DATA
      LDX    #MYBYTE    ;load pointer. X = $8500, ROM address
      LDY    #MYRAM     ;load pointer. Y = $860, RAM address
      LDAA   #CNTVAL
      STAA   COUNTREG
B1    LDAA   1,X+  ;get the byte from Flash, increment pointer
      STAA   1,Y+       ;save it in RAM and increment pointer
      DEC    COUNTREG
      BNE    B1         ;keep looping
      RTS               ;return
```

```
;----------my data in program ROM
      ORG $8500
MYBYTE FCB $25, $62, $3F, $52, $00
      END

;--------------------in a separate file
;PROG 6-6: CALCCSB.ASM - CALCULATING CHECKSUM BYTE
      XREF  MYRAM
      XREF  CNTVAL
      XREF  CNTVAL1
      XREF  COUNTREG
      XDEF  CAL_CHKSUM
CAL_CHKSUM
      LDAA   #CNTVAL
      STAA   COUNTREG   ;load the counter
      LDX    #MYRAM     ;load pointer from RAM address
      CLRA              ;A=0
B2    ADDA   1,X+       ;add and increment pointer
      DEC    COUNTREG   ;decrement the counter
      BNE    B2         ;loop until counter = zero
      NEGA              ;2's comp of sum
      STAA  0,X         ;save the checksum as the last byte
      RTS               ;return
      END

;--------------------in a separate file
;PROG 6-6: TESTCSB.ASM - TESTING CHECKSUM BYTE
     INCLUDE 'mc9s12dp512.inc'
      XREF  MYRAM
      XREF  COUNTREG
      XREF  CNTVAL
      XREF  CNTVAL
      XDEF  TEST_CHKSUM
TEST_CHKSUM
      BSET   DDRB,%11111111
      LDAA   #CNTVAL+1  ;adding 5 bytes including checksum byte
      STAA   COUNTREG   ;load the counter
      LDX    #MYRAM     ;load pointer from RAM address
      CLRA              ;A=0
B3    ADDA   1,X+       ;add and increment pointer
      DEC    COUNTREG   ;decrement the counter
      BNE    B3         ;loop until counter = zero
      CMPA   #0         ;see if A = zero
      BEQ    G_1        ;is result zero? then good
      LDAA  #'B'        ;then send letter B for Bad to PORTB
      STAA  PORTB       ;if not, data is bad
      RTS
G_1   LDAA  #'G'        ;then send letter G for Good to PORTB
      STAA  PORTB       ;data is not corrupted
      RTS
      END
```

Program 6-6 shows how the XREF and XDEF directives can also be applied to data variables. The linker program resolves external references by matching XDEF and XREF names.

Linking modules together in CodeWarrior

Assuming that the program modules in Program 6-6 are saved under the filenames MAIN.ASM, CALCCSB.ASM, and TESTCSB.ASM, we link them together by dropping them in the Sources folder in CodeWarrior in order to generate a single executable file. The linker program will search through the files specified in the Sources folder for the external subroutines. Figure 6-3 shows the project window with the Sources folder.

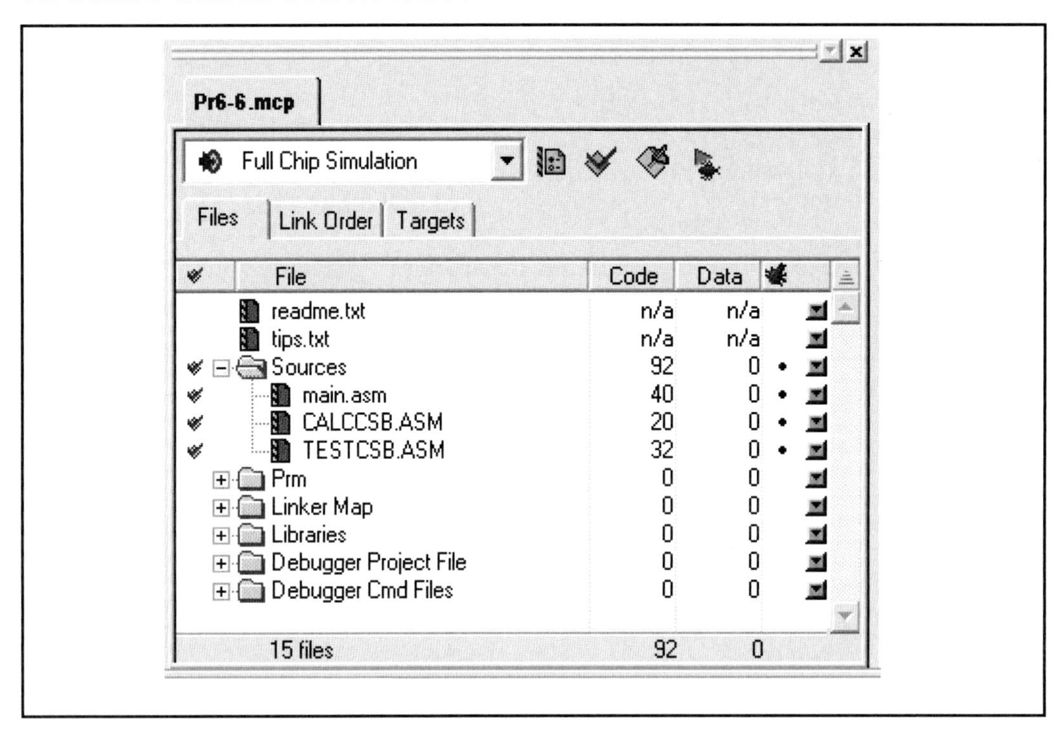

Figure 6-3. Project Window in CodeWarrior

The CodeWarrior IDE handles the compiling and linking in one step. This aids in program development by reducing time and errors in typing the command line call.

Review Questions

1. Discuss the benefits of macro programming.
2. List the three parts of a macro.
3. Explain and contrast the macro definition, invoking the macro, and expanding the macro.
4. True or false. A label defined within a macro is automatically understood by the assembler to be local.
5. The________directive is used within a module to indicate that the named variable or subroutine can be used by another module.
6. The________directive is used within a module to indicate that the named variable or subroutine was defined in another module.

PROBLEMS

SECTION 6.1: ADVANCED INDEXED ADDRESSING MODE

1. Which registers are allowed to be used as a pointer for indexed addressing mode when accessing data RAM? Give their names and show how they are loaded.
2. Write a program to copy $FF into RAM locations 1150H to 116FH.
3. Write a program to copy 10 bytes of data starting at RAM address $840 to RAM locations starting at $870.
4. What is the size of the X register?
5. Give the registers that can be used in indexed addressing mode.
6. Write a program to clear RAM locations $1800 to $187F.
7. Write a program to toggle RAM locations $950 to $95F.
8. Explain the effective address.
9. How much RAM space does the X register cover?

SECTION 6.2: ACCESSING LOOK-UP TABLE IN FLASH

10. Compile and state the contents of each ROM location for the following data:

```
                ORG    $8200
MYDAT_1:        FCB    "Earth"
MYDAT_2:        FCB    "987-65"
MYDAT_3:        FCB    "GABEH 98"
```

11. Compile and state the contents of each ROM location for the following data:

```
                ORG    $8340
DAT_1: FCB $22,$56, %10011001, '32', $F6, %11111011
```

12. Which register is allowed to be used as a pointer for indexed addressing mode when accessing data stored in program ROM? Give the name and show how it is loaded.
13. Explain the auto-increment.
14. What is the size of the X register? How much ROM space does it cover?
15. Give the registers that can be used for auto-increment.
16. Write a program to read the following message from ROM and place it in data RAM starting at $850:

```
                ORG $8600
MYDATA          FCB    "1-800-999-9999",0
```

17. Write a program to find y where $y = x^2 + 2x + 5$, and x is between 0 and 9.
18. Write a program to find y where $y = 2x + 5$, and x is between 0 and 9.
19. Write a program to read the following message from ROM and place it in data RAM starting at $840:

```
                ORG $8700
MYDATA          FCB    "The earth is but one country",0
```

20. True or false. “LDAA 4,X+” is valid instruction.
21. True or false. “LDAA 5,-Y” is a valid instruction.

22. Assume that the lower four bits of PORTB are connected to four switches. Write a program to send the following ASCII characters to PORTA, based on the status of the switches:

0000	‘0’
0001	‘1’
0010	‘2’
0011	‘3’
0100	‘4’
0101	‘5’
0110	‘6’
0111	‘7’
1000	‘8’
1001	‘9’
1010	‘A’
1011	‘B’
1100	‘C’
1101	‘D’
1110	‘E’
1111	‘F’

SECTION 6.3: CHECKSUM AND ASCII SUBROUTINES

23. Find the checksum byte for the following ASCII message: “Hello”
24. True or false. If we add all bytes, including the checksum byte, and the result is 00H, there is no error in the data.
25. Write a program to (a) get the data “Hello, my fellow world citizens” from program ROM, (b) calculate the checksum byte, and (c) test the checksum byte for any data error.
26. To display data on LCD or PC monitors, it must be in ________ (binary, BCD, ASCII).
27. Assume that the lower 4 bits of PortB are connected to switches. Write a program to send the following ASCII characters to Port A based on the status of the switches:

0000	‘0’
0001	‘1’
0010	‘2’
0011	‘3’
0100	‘4’
0101	‘5’
0110	‘6’
0111	‘7’
1000	‘8’
1001	‘9’

28. Write a program to convert a series of packed BCD numbers to ASCII. Assume that the packed BCD is located in ROM locations starting at $8700. Place the ASCII codes in RAM locations starting at $840.

```
                ORG $8700
MYDATA          FCB   $76,$87,$98,$43
```

29. Write a program to get an 8-bit binary number from PORTB, convert it to ASCII, and save the result in some RAM locations. What is the result if PORTB has 1000 1101 binary as input?

SECTION 6.4: MACROS AND MODULES

30. Give two advantages of macros.
31. Which uses more program ROM space: a macro or a module?
32. Give three reasons to write programs with modules.
33. If a label or parameter is not defined in a given module, it must be declared as ________.
34. If a label or parameter is used by other modules, it must be declared as ________ in the present module.

ANSWERS TO REVIEW QUESTIONS

SECTION 6.1: ADVANCED INDEXED ADDRESSING MODE

1. Indexed
2. True
3. True
4. False
5. X and Y

SECTION 6.2: ACCESSING LOOK-UP TABLE IN FLASH

1. X
2. X
3. A
4. 16 bits, 64 KB
5. Y

SECTION 6.3: CHECKSUM AND ASCII SUBROUTINES

1. The $35 and $37 give $57 in BCD. The $39 and $34 give $94.
2. The ASCII data is $32, $30, $30, $35, while 05 and $20 are for BCD.
3. No. To make it BCD, we must have LDAA #$95.
4. $33 and $33
5. 242 or $32, $34, and $32
6. ROM
7. 88 + 99 + AA + BB + CC + DD = 42F. Dropping the carries we have 2F, and its 2's complement is D1.
8. False

SECTION 6.4: MACROS AND MODULES

1. Macro programming can save the programmer time by allowing a set of frequently repeated

instructions to be invoked within the program with a single line. This can also make the code easier to read.

2. The three parts of a macro are the MACRO directive, the body, and the ENDM directive.
3. The macro definition is the list of statements the macro will perform. It begins with the MACRO directive and ends with the ENDM directive. The macro is invoked whenever it is called from within an Assembly language program. The macro is expanded when the Assembly program replaces the line invoking the macro with the Assembly language code in the body of the macro.
4. False. A label that is to be local to a macro must be declared local with the \@ directive.
5. XDEF
6. XREF

CHAPTER 7

HCS12 PROGRAMMING IN C

OBJECTIVES

Upon completion of this chapter, you will be able to:

>> Examine C data types for the HCS12
>> Code C programs for time delay and I/O operations
>> Code C programs for I/O bit manipulation
>> Code C programs for logic and arithmetic operations
>> Code C programs for ASCII and BCD data conversion
>> Code C programs for binary (hex) to decimal conversion
>> Code C programs for data serialization
>> Understand C compiler RAM and ROM allocation

Why program the HCS12 in C?

Compilers produce hex files that we download into the ROM of the microcontroller. The size of the hex file produced by the compiler is one of the main concerns of microcontroller programmers for two reasons:

1. Microcontrollers have limited on-chip ROM.
2. The code space for the HCS12 is limited to 512K.

How does the choice of programming language affect the compiled program size? While Assembly language produces a hex file that is much smaller than C, programming in Assembly language is often tedious and time consuming. On the other hand, C programming is less time consuming and much easier to write, but the hex file size produced is much larger than if we used Assembly language. The following are some of the major reasons for writing programs in C instead of Assembly:

1. It is easier and less time consuming to write in C than in Assembly.
2. C is easier to modify and update.
3. You can use code available in function libraries.
4. C code is portable to other microcontrollers with little or no modification.

Several third-party companies develop C compilers for the HCS12 microcontroller. Our goal is not to recommend one over another, but to provide you with the fundamentals of C programming for the HCS12. You can use the compiler of your choice for the chapter examples and programs. For this book we have chosen Freescale's C compiler to integrate with CodeWarrior IDE. Freescale has a student version of the C compiler available for download from their website. See http://www.MicroDigitalEd.com for tutorials on the C compiler and CodeWarrior simulator.

C programming for the HCS12 is the main topic of this chapter. In Section 7.1, we discuss data types, time delays, and I/O programming. The logic operations AND, OR, XOR, inverter, shift, and I/O bit manipulation are discussed in Section 7.2. Section 7.3 describes ASCII and BCD conversions and checksums. In Section 7.4, data serialization for the HCS12 is shown. In Section 7.5, we show how the C compiler uses program ROM for data storage.

SECTION 7.1: DATA TYPES AND TIME DELAYS IN C

In this section we first discuss C data types for the HCS12 and then provide code for time delay functions.

C data types for the HCS12

One of the goals of C programmers is to create smaller hex files, so it is worthwhile to re-examine C data types. In other words, a good understanding of data types for the C programming language can help programmers to create smaller hex files. In this section we focus on the specific C data types that are the most useful and widely used for the HCS12 microcontroller. Table 7-1 shows data types and sizes.

Table 7-1: Some Data Types Widely Used by C

Data Type	Size in Bits	Data Range/Usage
unsigned char	8-bit	0 to 255
char	8-bit	–128 to +127
unsigned int	16-bit	0 to 65,535
int	16-bit	–32,768 to +32,767
unsigned short	16-bit	0 to 65,535
short	16-bit	–32,768 to +32,767
unsigned short long	24-bit	0 to 16,777,215
short long	24-bit	–8,388,608 to +8,388,607
unsigned long	32-bit	0 to 4,294,967,295
long	32-bit	–2,147,483,648 to +2,147,483,648

Unsigned char

For the HCS12 microcontroller, the character data type is the most natural choice for many applications. The unsigned char is an 8-bit data type that takes a value in the range of 0–255 (00–FFH). It is one of the most widely used data types for the HCS12. In many situations, such as setting a counter value, where there is no need for signed data, we should use the unsigned char instead of the signed char. Remember that C compilers use the signed char as the default unless we put the keyword *unsigned* in front of the char (see Example 7-1). We can also use the unsigned char data type for a string of ASCII characters, including extended ASCII characters. Example 7-2 shows a string of ASCII characters. See Example 7-3 for toggling ports.

In declaring variables, we must pay careful attention to the size of the data and try to use unsigned char instead of int if possible. Because the HCS12 microcontroller has a limited number of data RAM locations, using int in place of char can lead to a larger-size hex file. Such misuse of data types in compilers such as Microsoft Visual C++ for x86 IBM PCs is not a significant issue.

Example 7-1

Write a C program to send values 00–FF to Port B.

Solution:

```
#include <mc9s12dp512.h>        //for DDRB and PORTB declarations
void main(void)
  {
    unsigned char z;
    DDRB = 0xFF;                //make Port B an output
    for(z=0;z<=255;z++)
      PORTB = z;
    while(1);                   //NEEDED IF RUNNING IN HARDWARE
  }
```

Run the above program on your simulator to see how Port B displays values 00–FFH in binary. Notice that "while(1)" is needed if this program is running in hardware.

Example 7-2

Write a C program to send hex values for ASCII characters of 0, 1, 2, 3, 4, 5, A, B, C, and D to Port B.

Solution:

```
#include <mc9s12dp512.h>
void main(void)
  {
    unsigned char mynum[ ] = "012345ABCD";//ASCII string in RAM
    unsigned char z;
    DDRB = 0xFF;                    //make Port B an output
    for(z=0;z<10;z++)
      PORTB = mynum[ z] ;
    while(1);                       //stay here forever
  }
```

Run the above program on your simulator to see how Port B displays values 30H, 31H, 32H, 33H, 34H, 35H, 41H, 42H, 43H, and 44H (the hex values for ASCII 0, 1, 2, etc.). Notice that the last statement "while(1)" is needed only if we run the program in hardware. This is like "GOTO $" or "BRA $" in Assembly language.

Example 7-3

Write a C program to toggle all the bits of Port B continuously.

Solution:

```
// Toggle PB forever
#include <mc9s12dp512.h>
void main(void)
  {
    DDRB = 0xFF;          //make Port B an output
    for(;;)               //repeat forever
      {
        PORTB = 0x55;     //0x indicates the data is in hex (binary)
        PORTB = 0xAA;
      }
  }
```

Run the above program on your simulator to see how Port B toggles continuously.

Signed char

The signed char is an 8-bit data type that uses the most significant bit (D7 of D7–D0) to represent the – or + value. As a result, we have only 7 bits for the magnitude of the signed number, giving us values from –128 to +127. In situations where + and – are needed to represent a given quantity such as temperature, the use of the signed char data type is necessary.

Again, notice that if we do not use the keyword *unsigned,* the default is the signed value. For that reason we should stick with the unsigned char unless the data needs to be represented as signed numbers. See Example 7-4.

Example 7-4

Write a C program to send values of –4 to +4 to Port B.

Solution:

```
//sign numbers
#include <mc9s12dp512.h>
void main(void)
  {
    char mynum[ ] = { +1,-1,+2,-2,+3,-3,+4,-4} ;
    unsigned char z;
    DDRB  = 0xFF;                //make Port B an output
    for(z=0;z<8;z++)
      PORTB = mynum[ z] ;
    while(1);                    //stay here forever
  }
```

Run the above program on your simulator to see how PORTB displays values of 1, FFH, 2, FEH, 3, FDH, 4, and FCH (the hex values for +1, –1, +2, –2, etc.). See Chapter 5 for discussion of signed numbers.

Unsigned int

The unsigned int is a 16-bit data type that takes a value in the range of 0 to 65,535 (0000–FFFFH). In the HCS12, unsigned int is used to define 16-bit variables such as memory addresses. It is also used to set counter values of more than 256. Because the int data type takes two bytes of RAM, we must not use the int data type unless we have to. Because registers and memory accesses are in 8-bit chunks, the misuse of int variables will result in a larger hex file. Such misuse is not a problem in PCs with 512 megabytes of memory, a 32-bit Pentium processor, and a bus speed of 133 MHz. For HCS12 programming, however, do not use signed int in places where unsigned char will do the job. Of course, the compiler will not generate an error for this misuse, but the overhead in hex file size will be noticeable. Also, in situations where there is no need for signed data (such as setting counter values), we should use unsigned int instead of signed int (see Example 7-5). This gives a much wider range for data declaration. Again, remember that the C compiler uses signed int as the default unless we specify the keyword *unsigned.*

Example 7-5

Write a C program to toggle all bits of Port B 50,000 times.

Solution:

```
#include <mc9s12dp512.h>
void main(void)
  {
     unsigned int z;
     DDRB  = 0xFF;              //make Port B an output
     for(z=0;z<=50000;z++)
       {
         PORTB  =  0x55;
         PORTB  =  0xAA;
       }
    while(1);                   //stay here forever
  }
```

Run the above program on your simulator to see how Port B toggles continuously. Notice that the maximum value for unsigned int is 65,535.

Signed int

Signed int is a 16-bit data type that uses the most significant bit (D15 of D15–D0) to represent the – or + value. As a result, we have only 15 bits for the magnitude of the number, or values from –32,768 to +32,767.

Other data types

The unsigned int is limited to values 0–65,535 (0000–FFFFH). The C compiler supports both short long and long data types, if we want values greater than 16-bit. See Table 7-1. The short long value is 24 bits wide, while the long value is 32 bits wide. See Example 7-6 for use of the unsigned long data type.

Example 7-6

Write a C program to toggle all bits of Port B 100,000 times.

Solution:

```
//toggle PB 100,00 times
#include <mc9s12dp512.h>
void main(void)
  {
    unsigned long z;
    DDRB = 0xFF;                //make Port B an output
    for(z=0;z<=100000;z++)
      {
        PORTB = 0x55;
        PORTB = 0xAA;
      }
    while(1);                   //stay here forever
  }
```

Time delay

There are two ways to create a time delay in C:

1. Using a simple for loop
2. Using the HCS12 timers

In either case, when we write a time delay we must use the oscilloscope to measure the duration of our time delay. Next, we use the for loop to create a time delay. The use of the HCS12 timer to create time delays is postponed until Chapter 9.

In creating a time delay using a for loop, we must be mindful of two factors that can affect the accuracy of the delay:

1. The crystal frequency connected to the HCS12 board is the most important factor in the time delay calculation. The duration of the clock period for the instruction cycle is a function of this crystal frequency.
2. The second factor that affects the time delay is the compiler used to compile the C program. When we program in Assembly language, we can control the exact instructions and their sequences used in the delay subroutine. In the case of C programs, it is the C compiler that converts the C statements and functions to Assembly language instructions. As a result, different compilers produce different code. In other words, if we compile a given C program with different compilers, each compiler produces different-size hex code.

For these reasons, when we write time delays for C, we must use the oscilloscope to measure the exact duration. Look at Examples 7-7 through 7-12.

Example 7-7

Write a C program to toggle all the bits of Port B continuously with a 250 ms delay. Bus frequency = 2 MHz.

Solution:

```
#include <mc9s12dp512.h>
void MSDelay(unsigned int);
void main(void)
  {
    DDRB = 0xFF;                  //make Port B an output
    while(1)                      //repeat forever
      {
        PORTB = 0x55;
        MSDelay(250);
        PORTB = 0xAA;
        MSDelay(250);
      }
  }
void MSDelay(unsigned int itime)
  {
    unsigned int i; unsigned int j;
    for(i=0;i<itime;i++)
      for(j=0;j<331;j++);
  }
```

Example 7-8

Write a C program to toggle all the bits of Port A and Port B continuously with a 250 ms delay. Bus frequency = 2 MHz.

Solution:

```
//this program is tested for the MC9S12DP512 with XTAL = 4 MHz
#include <mc9s12dp512.h>
void MSDelay(unsigned int);
void main(void)
  {
    DDRA = 0xFF;
    DDRB = 0xFF;                  //make Ports A and B output
    while(1)                      //another way to do it forever
      {
        PORTA = 0x55;
        PORTB = 0x55;
        MSDelay(250);
        PORTA = 0xAA;
        PORTB = 0xAA;
        MSDelay(250);
      }
  }
void MSDelay(unsigned int itime)
  {
    unsigned int i; unsigned int j;
    for(i=0;i<itime;i++)
      for(j=0;j<331;j++);
  }
```

Example 7-9

LEDs are connected to bits in Port B and Port A. Write a C program that shows the count from 0 to FFH (0000 0000 to 1111 1111 in binary) on the LEDs.

Solution:

```
#include <mc9s12dp512.h>
#define LED PORTA                  //notice how we can define Port A
void MSDelay(unsigned int);
void main(void)
  {
    DDRB = 0xFF;                   //make Port B an output
    DDRA = 0xFF;                   //make Port A an output
    PORTB = 00;                    //clear Port B
    LED = 0;                       //clear Port A
    for(;;)                        //repeat forever
      {
        PORTB++;                   //increment Port B
        LED++;                     //increment Port A
        MSDelay(500);
      }
  }
void MSDelay(unsigned int itime)
  {
    unsigned int i;
    unsigned int j;
    for(i=0;i<itime;i++)
      for(j=0;j<331;j++);
  }
```

Example 7-10

Write a C program to get a byte of data from Port B, wait 1/2 second, and then send it to Port A. Bus frequency = 2 MHz.

Solution:

```
#include <mc9s12dp512.h>
void MSDelay(unsigned int);
void main(void)
  {
    unsigned char mybyte;
    DDRB = 0x00;                   //Port B as input
    DDRA = 0xFF;                   //Port A as output
    while(1)
      {
        mybyte = PORTB;            //get a byte from Port B
        MSDelay(500);
        PORTA = mybyte;            //send it to Port A
      }
  }
void MSDelay(unsigned int itime)
  {
    unsigned int i;
    unsigned int j;
    for(i=0;i<itime;i++)
      for(j=0;j<331;j++);
  }
```

Example 7-11

Write a C program to get a byte of data from Port B. If it is less than 100, send it to Port A; otherwise, send it to Port K.

Solution:

```
#include <mc9s12dp512.h>
void MSDelay(unsigned int); //see Example 7-12 for definition
void main(void)
  {
    unsigned char mybyte;
    DDRB = 0x00;                //make Port B an input
    DDRA = 0xFF;
    DDRK = 0xFF;                //both Ports A and K as output
    while(1)
      {
        mybyte = PORTB;         //get a byte from PORTB
        if(mybyte < 100)
          PORTA = mybyte;       //send it to PORTA if less than 100
        else
          PORTK = mybyte;       //send it to PORTK if more than 100
        MSDelay(250);
      }
  }
```

Example 7-12

Write a C program to toggle all the bits of Port A, B, and K continuously with a 250 ms delay. Bus frequency = 2 MHz.

Solution:

```
#include <mc9s12dp512.h>
void MSDelay(unsigned int);
void main(void)
  {
    DDRA = 0xFF;
    DDRB = 0xFF;
    DDRK = 0xFF;
    while(1)                    //do it forever
      {
        PORTA = 0x55;
        PORTB = 0x55;
        PORTK = 0x55;
        MSDelay(250);           //250 ms delay
        PORTA = 0xAA;
        PORTB = 0xAA;
        PORTK = 0xAA;
        MSDelay(250);
      }
  }
void MSDelay(unsigned int itime)
  {
    unsigned int i;
    unsigned int j;
    for(i=0;i<itime;i++)
      for(j=0;j<331;j++);
  }
```

Review Questions

1. Give the magnitude of the unsigned char and signed char data types.
2. Give the magnitude of the unsigned int and signed int data types.
3. If we are declaring a variable for a person's age, we should use the ___ data type.
4. True or false. Using a for loop to create a time delay is not recommended if you want your code be portable to other HCS12-based systems.
5. Give two factors that can affect the delay size.

SECTION 7.2: LOGIC OPERATIONS IN C

One of the most important and powerful features of the C language is its ability to perform bit manipulation. Many books on C do not cover this feature. But because it is an important topic, it is appropriate to discuss it in this section. This section describes the action of bit-wise logic operators and provides some examples of how they are used.

Bit-wise operators in C

While every C programmer is familiar with the logical operators AND (&&), OR (||), and NOT (!), many C programmers are less familiar with the bit-wise operators AND (&), OR (|), EX-OR (^), inverter (~), shift right (>>), and shift left (<<). These bit-wise operators are widely used in software engineering for embedded systems and control; consequently, their understanding and mastery are critical in microprocessor-based system design and interfacing. See Table 7-2.

The following shows some examples using the C bit-wise operators:

```
1. 0x35 & 0x0F = 0x05           /* ANDing */
2. 0x04 | 0x68 = 0x6C           /* ORing:   */
3. 0x54 ^ 0x78 = 0x2C           /* XORing */
4. ~0x55 = 0xAA                 /* Inverting 55H */
```

Table 7-2: Bit-wise Logic Operators for C

		AND	OR	EX-OR	Inverter
A	**B**	**A&B**	**A|B**	**A^B**	**Y=~B**
0	0	0	0	0	1
0	1	0	1	1	0
1	0	0	1	1	
1	1	1	1	0	

Bit-wise shift operators in C

There are two bit-wise shift operators in C: shift right (>>) and shift left (<<). Their format in C is as follows:

data >> number of bits to be shifted right
data << number of bits to be shifted left

The following shows some examples of shift operators in C:

1. `0x9A >> 3 = 0x13        /* shifting right 3 times */`
2. `0x77 >> 4 = 0x07        /* shifting right 4 times */`
3. `0x6 << 4 = 0x60         /* shifting left 4 times */`

Study Examples 7-13 through 7-16. These show how the bit-wise operators are used in C.

Example 7-13

Run the following program on your simulator and examine the results.

Solution:

```
#include <mc9s12dp512.h>
void main (void)
  {
    DDRA = 0xFF;                //make Ports A, B,
    DDRB = 0xFF;                //and B output ports
    DDRK = 0xFF;
    PORTB = 0x35 & 0x0F;        //ANDing
    PORTA = 0x04 | 0x68;        //ORing
    PORTK = 0x54 ^ 0x78;        //XORing
    PORTA = ~0x55;              //inverting
    PORTB = 0x9A >> 3;          //shifting right 3 times
    PORTK = 0x77 >> 4;          //shifting right 4 times
    PORTB = 0x6 << 4;           //shifting left 4 times
    while(1);                   //stay here forever
  }
```

Example 7-14

Write a C program to toggle all the bits of Port B and Port A continuously with a 250 ms delay. Use the inverting operator.

Solution:

```
#include <mc9s12dp512.h>
void MSDelay(unsigned int);
void main(void)
  {
    DDRB = 0xFF;
    DDRA = 0xFF;                //make Ports B and A output
    PORTB = 0x55;
    PORTA = 0xAA;
    while(1)
      {
        PORTB = ~PORTB;
        PORTA = ~PORTA;
        MSDelay(250);
      }
  }
void MSDelay(unsigned int itime)
  {
    unsigned int i;
    unsigned int j;
    for(i=0;i<itime;i++)
      for(j=0;j<331;j++);
  }
```

Example 7-15

Write a C program to read the PB0 and PB1 bits and issue an ASCII character to Port A according to the following table:

PB1	PB0	
0	0	send '0' to PORTA (notice ASCII '0' is 0x30)
0	1	send '1' to PORTA
1	0	send '2' to PORTA
1	1	send '3' to PORTA

Solution:

```
#include <mc9s12dp512.h>
void main(void)
  {
    unsigned char z;
    DDRB = 0x00;                        //make Port B an input
    DDRA = 0xFF;                        //make Port A an output
    while(1)                            //repeat forever
      {
        z = PORTB;                      //read PORTB
        z = z & 0x3;                    //mask the unused bits
        switch(z)                       //make decision
          {
           case(0):
            {
             PORTA = '0';               //issue ASCII 0
             break;
            }
           case(1):
            {
             PORTA = '1';               //issue ASCII 1
             break;
            }
           case(2):
            {
             PORTA = '2';               //issue ASCII 2
             break;
            }
           case(3):
            {
             PORTA = '3';               //issue ASCII 3
             break;
            }
          }
      }
  }
```

Example 7-16

Write a C program to toggle all the bits of Port A and Port B continuously with a 250 ms delay. Use the EX-OR operator. Bus frequency = 2 MHz.

Solution:

```
#include <mc9s12dp512.h>
void MSDelay(unsigned int);
void main(void)
  {
     DDRB  =  0xFF;
     DDRA  =  0xFF;                    //make Ports A and B output
     PORTB=0x55;
     PORTA=0x55;
     while(1)
       {
         PORTB=PORTB^0xFF;        //XOR will toggle the bits
         PORTA=PORTA^0xFF;
         MSDelay(250);
       }
  }

void MSDelay(unsigned int itime)
  {
     unsigned int i;
     unsigned int j;
     for(i=0;i<itime;i++)
       for(j=0;j<331;j++);
  }
```

Review Questions

1. Find the content of PORTB after the following C code in each case:
   (a) `PORTB=0x37&0xCA;`
   (b) `PORTB=0x37|0xCA;`
   (c) `PORTB=0x37^0xCA;`
2. To mask certain bits we must AND them with ______.
3. To set high certain bits we must OR them with _____.
4. EX-ORing a value with itself results in _______.
5. Find the contents of PORTA after execution of the following code:
   ```
   PORTA = 0;
   PORTA = PORTA | 0x99;
   PORTA = ~PORTA;
   ```
6. Write a short program that toggles all the bits of PORTB.
7. True or false. All the bits of PORTB are bit addressable.
8. True or false. All the bits of DDRB are bit addressable.

SECTION 7.3: DATA CONVERSION PROGRAMS IN C

Recall that BCD numbers were discussed in Chapters 5 and 6. As stated there, many newer microcontrollers have a real-time clock (RTC) where the time and date are kept even when the power is off. Very often the RTC provides the time and date in packed BCD. To display them, however, it must convert them to ASCII. In this section we show the application of logic and rotate instructions in the conversion of BCD and ASCII.

ASCII numbers

On ASCII keyboards, when the "0" key is activated, "011 0000" (30H) is provided to the computer. Similarly, 31H (011 0001) is provided for the "1" key, and so on, as shown in Table 7-3.

Table 7-3: ASCII Code for Digits 0–9

Key	ASCII (hex)	Binary	BCD (unpacked)
0	30	011 0000	0000 0000
1	31	011 0001	0000 0001
2	32	011 0010	0000 0010
3	33	011 0011	0000 0011
4	34	011 0100	0000 0100
5	35	011 0101	0000 0101
6	36	011 0110	0000 0110
7	37	011 0111	0000 0111
8	38	011 1000	0000 1000
9	39	011 1001	0000 1001

Packed BCD to ASCII conversion

The RTC provides the time of day (hour, minute, second) and the date (year, month, day) continuously, regardless of whether the power is on or off. This data is provided in packed BCD, however. To convert packed BCD to ASCII, you must first convert it to unpacked BCD. Then the unpacked BCD is tagged with 011 0000 (30H). The following demonstrates converting from packed BCD to ASCII. See also Example 7-17.

```
Packed BCD    Unpacked BCD         ASCII
0x29           0x02, 0x09          0x32, 0x39
00101001       00000010,00001001   00110010,00111001
```

ASCII to packed BCD conversion

To convert ASCII to packed BCD, you first convert it to unpacked BCD (to get rid of the 3), and then combine to make packed BCD. For example, 4 and 7 on the keyboard give 34H and 37H, respectively. The goal is to produce 47H or "0100 0111", which is packed BCD. See Example 7-18.

```
Key   ASCII       Unpacked BCD      Packed BCD
4     34          00000100
7     37          00000111          01000111 or 47H
```

After this conversion, the packed BCD numbers are processed and the result will be in packed BCD format. Chapter 16 discusses the RTC chip and uses the BCD and ASCII conversion programs shown in Examples 7-17 and 7-18.

Example 7-17

Write a C program to convert packed BCD 0x29 to ASCII and display the bytes on PORTA and PORTB.

Solution:

```
#include <mc9s12dp512.h>
void main(void)
  {
     unsigned char x, y;
     unsigned char mybyte = 0x29;
     DDRA = 0xFF;
     DDRB = 0xFF;                   //make Ports A and B output
     x = mybyte & 0x0F;             //mask upper 4 bits
     PORTB = x | 0x30;              //make it ASCII
     y = mybyte & 0xF0;             //mask lower 4 bits
     y = y >> 4;                    //shift it to lower 4 bits
     PORTA = y | 0x30;              //make it ASCII
  }
```

Example 7-18

Write a C program to convert ASCII digits of '4' and '7' to packed BCD and display the results on PORTB.

Solution:

```
#include <mc9s12dp512.h>
void main(void)
  {
     unsigned char bcdbyte;
     volatile unsigned char w = '4';
     volatile unsigned char z = '7';
     DDRB = 0xFF;               //make Port B an output
     w = w & 0x0F;              //mask 3
     w = w << 4;                //shift left to make upper BCD digit
     z = z & 0x0F;              //mask 3
     bcdbyte = w | z;           //combine to make packed BCD
     PORTB = bcdbyte;
  }
```

Checksum byte in ROM

To ensure the integrity of ROM contents, every system must perform the checksum calculation. The checksum will detect any corruption of the contents of ROM. One of the causes of ROM corruption is current surge, either when the system is turned on or during operation. To ensure data integrity in ROM, the checksum process uses what is called a *checksum byte*. The checksum byte is an extra byte that is tagged to the end of a series of bytes of data.

To calculate the checksum byte of a series of bytes of data, the following steps can be taken:

1. Add the bytes together and drop the carries.
2. Take the 2's complement of the total sum. This is the checksum byte, which becomes the last byte of the series.

To perform the checksum operation, add all the bytes, including the checksum byte. The result must be zero. If it is not zero, one or more bytes of data have been changed (corrupted). To clarify these important concepts, see Examples 7-19 through 7-21.

Example 7-19

Assume that we have 4 bytes of hexadecimal data: $25, $62, $3F, and $52.
(a) Find the checksum byte,
(b) perform the checksum operation to ensure data integrity, and
(c) if the second byte, $62, has been changed to $22, show how checksum detects the error.

Solution:

(a) Find the checksum byte.

```
     25
+    62
+    3F
+    52
    118
```
(dropping carry of 1 and taking 2's complement, we get E8)

(b) Perform the checksum operation to ensure data integrity.

```
     25
+    62
+    3F
+    52
+    E8
    200
```
(dropping the carries we get 00, which means data is not corrupted)

(c) If the second byte, $62, has been changed to $22, show how checksum detects the error.

```
     25
+    22
+    3F
+    52
+    E8
    1C0
```
(dropping the carry, we get $C0, which means data is corrupted)

Example 7-20

Write a C program to calculate the checksum byte for the data given in Example 7-19.

Solution:

```
#include <mc9s12dp512.h>
void main(void)
  {
    unsigned char mydata[] = {0x25,0x62,0x3F,0x52};
    unsigned char sum = 0;
    unsigned char x;
    unsigned char chksumbyte;
    DDRA = 0xFF;
    DDRB = 0xFF;                       //make Ports A and B output
    for(x=0;x<4;x++)
      {
        PORTA = mydata[x];             //issue each byte to PORTA
        sum = sum + mydata[x];         //add them together
        PORTB = sum;                   //issue the sum to PORTB
      }
    chksumbyte = ~sum + 1;             //make 2's complement (invert +1)
    PORTB = chksumbyte;                //show the checksum byte
  }
```

Single-step the above program on the CodeWarrior simulator and examine the contents of PORTA and PORTB. Notice that the bytes are put on PORTA as they are added.

Example 7-21

Write a C program to perform step (b) of Example 7-19. If the data is good, send ASCII character 'G' to PORTB. Otherwise, send 'B' to PORTB.

Solution:

```
#include <mc9s12dp512.h>
void main(void)
  {
    unsigned char mydata[] = {0x25,0x62,0x3F,0x52,0xE8};
    unsigned char chksum = 0;
    unsigned char x;
    DDRB = 0xFF;                             //make Port B an output
    for(x=0;x<5;x++)
      chksum = chksum + mydata[x];           //add them together
    if(chksum == 0)
      PORTB = 'G';
    else
      PORTB = 'B';
  }
```

Change one or two values in the mydata array and simulate the program to see the results.

Binary (hex) to decimal and ASCII conversion in C

The printf function is part of the standard I/O library in C and can do many things including converting data from binary (hex) to decimal, or vice versa. But printf takes a lot of memory space and increases your hex file substantially. For this reason, in systems based on the HCS12 microcontroller, it is better to write your own conversion function instead of using printf.

One of the most widely used conversions is binary to decimal conversion. In devices such as ADCs (Analog-to-Digital Converters), the data is provided to the microcontroller in binary. In some RTCs, the time and date are also provided in binary. In order to display binary data, we need to convert it to decimal and then to ASCII. Because the hexadecimal format is a convenient way of representing binary data, we refer to the binary data as hex. The binary data 00–FFH converted to decimal will give us 000 to 255. One way to do that is to divide it by 10 and keep the remainder, as was shown in Chapters 5 and 6. For example, 11111101 or FDH is 253 in decimal. The following is one version of an algorithm for conversion of hex (binary) to decimal:

Hex	Quotient	Remainder
FD/0A	19	3 (low digit) LSD
19/0A	2	5 (middle digit)
2/0A	0	2 (high digit) (MSD)

Example 7-22 shows the C program for that algorithm.

Example 7-22

Write a C program to convert 11111101 (FD hex) to decimal and display the digits on PORTA, PORTB, and PORTK.

Solution:

```
#include <mc9s12dp512.h>
void main(void)
  {
    unsigned char x, binbyte, d1, d2, d3;
    DDRA = 0xFF;
    DDRB = 0xFF;
    DDRK = 0xFF;                //Ports A, B, and K output
    binbyte = 0xFD;             //binary (hex) byte
    x = binbyte / 10;           //divide by 10
    d1 = binbyte % 10;          //find remainder (LSD)
    d2 = x % 10;                //middle digit
    d3 = x / 10;                //most-significant digit (MSD)
    PORTA = d1;
    PORTB = d2;
    PORTK = d3;
  }
```

Review Questions

1. For the following decimal numbers, give the packed BCD and unpacked BCD representations:
   (a) 15   (b) 99
2. Show the binary and hex formats for “76” and its packed BCD version.

3. 67H in BCD when converted to ASCII is ____H and ____H.
4. Does the following convert unpacked BCD to ASCII?
   `mydata=0x09+0x30;`
5. Why is the use of packed BCD preferable to ASCII?
6. Which takes more memory space: packed BCD or ASCII?
7. In Question 6, which is more universal?
8. Find the checksum byte for the following values: 22H, 76H, 5FH, 8CH, 99H.

SECTION 7.4: I/O BIT MANIPULATION AND DATA SERIALIZATION IN C

I/O bit manipulation in C

As we saw in Chapter 4, the I/O ports of HCS12 are bit-addressable. We can access a single bit without disturbing the rest of the port. We use the AND and OR bit-wise operators to access a single bit of a port. We access the DDRx registers in the same way. Study Examples 7-23 through 7-28 and Table 7-4 to become familiar with the syntax.

Example 7-23

Write a C program to turn bit 5 of Port B on and off 50,000 times.

Solution:

```
#include <mc9s12dp512.h>
#define MYBIT PORTB_BIT5 //defined in the header file
void main(void)
  {
     unsigned int z;
     DDRB_BIT5 = 1;              //make PORTB.5 an output
     for(z=0;z<50000;z++)
       {
         MYBIT = 1;
         MYBIT = 0;
       }
     while(1);                   //stay here forever
  }
```

Example 7-24

Write a C program to toggle only bit PB0 continuously without disturbing the rest of the bits of Port B.

Solution:

```
#include <mc9s12dp512.h>
void main(void)
  {
     DDRB = DDRB | 0x01;              //make PB0 an output
     while(1)
       {
         PORTB = PORTB | 0x01;        //turn on PB0
         PORTB = PORTB & ~0x01;       //turn off PB0
       }
  }
```

Table 7-4: Bit Addressing for HCS12 Ports Using CodeWarrior C Language

Name	Address (hex)	x Range
PORTA	PORTA_BITx	7 >= x >= 0
DDRA	DDRA_BITx	7 >= x >= 0
PORTB	PORTB_BITx	7 >= x >= 0
DDRB	DDRB_BITx	7 >= x >= 0
PORTE	PORTE_BITx	7 >= x >= 0
DDRE	DDRE_BITx	7 >= x >= 2
PTH	PTH_PTHx	7 >= x >= 0
DDRH	DDRH_DDRHx	7 >= x >= 0
PTJ	PTJ_PTJx	x = 7, 6, 1, OR 0
DDRJ	DDRJ_DDRJx	x = 7, 6, 1, OR 0
PORTK	PORTK_BITx	x = 7 OR 5 >= x >= 0
DDRK	DDRK_BITx	x = 7 OR 5 >= x >= 0
PTM	PTM_PTMx	7 >= x >= 0
DDRM	DDRM_DDRMx	7 >= x >= 0
PTP	PTP_PTPx	7 >= x >= 0
DDRP	DDRP_DDRPx	7 >= x >= 0
PTS	PTS_PTSx	7 >= x >= 0
DDRS	DDRS_DDRSx	7 >= x >= 0
PTT	PTT_PTTx	7 >= x >= 0
DDRT	DDRT_DDRTx	7 >= x >= 0

Example 7-25

Write a C program to monitor bit PA7. If it is HIGH, send 55H to Port B; otherwise, send AAH to Port K.

Solution:

```
#include <mc9s12dp512.h>
void main(void)
  {
    unsigned char x;
    DDRA = DDRA & 0x7F;         //PA7 as input
    DDRB = 0xFF;                      //Ports B and K output
    DDRK = 0xFF;
    while(1)
      {
        x = PORTA & 0x80;
        if(x == 0x80)
          PORTB = 0x55;
        else
          PORTK = 0xAA;
      }
  }
```

Example 7-26

A door sensor is connected to the PB1 pin, and a buzzer is connected to PA7. Write a C program to monitor the door sensor and, when it opens, sound the buzzer. You can sound the buzzer by sending a square wave of a few hundred Hz to it.

Solution:

```
#include <mc9s12dp512.h>      /* derivative information */
void MSDelay(unsigned int);//look in previous examples for definition
#define Dsensor PORTB_BIT1
#define buzzer PORTA_BIT7
void main(void)
  {
    DDRB_BIT1 = DDRB & ~0x02;         //PB1 as an input
    DDRA_BIT7 = DDRA | 0x80;          //PA7 as an output
    while(1)
    {
      while(Dsensor == 1)
        {
          buzzer = 0;
          MSDelay(2);
          buzzer = 1;
          MSDelay(2);
        }
    }
  }
```

Example 7-27

The data pins of an LCD are connected to Port B. The information is latched into the LCD whenever its Enable pin goes from HIGH to LOW. Write a C program to send "The Earth is but One Country" to this LCD.

Solution:

```
#include <mc9s12dp512.h>      /* derivative information */
#define LCDData PORTB                 //LCDData declaration
#define En PORTA_BIT2           //the Enable pin
const unsigned char message[] = "The Earth is but One Country";
unsigned char z;
void main(void)
  {
    DDRB = 0xFF;                      //Port B as output
    DDRA = 0b00000100; //PortA.2 as output
    for(z=0;z<28;z++)           //send all the 28 characters
      {
        LCDData = message[z];
        En=1;                         //a HIGH-
        En=0;                   //-to-LOW pulse to latch the LCD data
      }
    while(1);                   //stay here forever
  }
```

Run the above program on your simulator to see how PORTB displays each character of the message. Meanwhile, monitor bit PA2 after each character is issued.

Example 7-28

Write a C program to get the status of bit PORTB.0 and send it to PORTA.7 continuously.

Solution:

```
#include <mc9s12dp512.h>
#define inbit PORTB_BIT0
#define outbit PORTA_BIT7
void main(void)
  {
    DDRB_BIT0 = 0;             //make PB0 an input
    DDRA_BIT7 = 1;             //make PB7 an output
    while(1)
      {
      outbit = inbit;   //get a bit from PB0 and send it to PB7
      }
  }
```

Serializing data in C

Serializing data is a way of sending a byte of data one bit at a time through a single pin of a microcontroller. There are two ways to transfer a byte of data serially:

1. Using the serial port. In using the serial port, the programmer has very limited control over the sequence of data transfer. The details of serial port data transfer are discussed in Chapter 10.
2. The second method of serializing data is to transfer data one bit a time and control the sequence of data and spaces between them. In many new generations of devices such as LCD, ADC, and EEPROM, the serial versions are becoming popular because they take up less space on a printed circuit board. Although we can use standards such as I²C, SPI, and CAN, not all devices support such standards. For this reason we need to be familiar with data serialization using the C language. Examine Examples 7-29 through 7-32 and Figures 7-11 and 7-12.

Example 7-29

Write a C program to send out the value $44 serially one bit at a time via PB0. The LSB should go out first.

Solution:

```
//Serializing data via PB0 (SHIFTING RIGHT)
#include <mc9s12dp512.h>
#define PB0 PORTB_BIT0
void main(void)
  {
    unsigned char conbyte = 0x44;
    unsigned char regALSB;
    unsigned char x;
    regALSB = conbyte;
    DDRB_BIT0 = 1;             //make PB0 an output
    for(x=0;x<8;x++)
      {
        PB0 = regALSB & 0x01; //send out the next bit
        regALSB = regALSB >> 1;//shift the remaining bits right
      }
  }
```

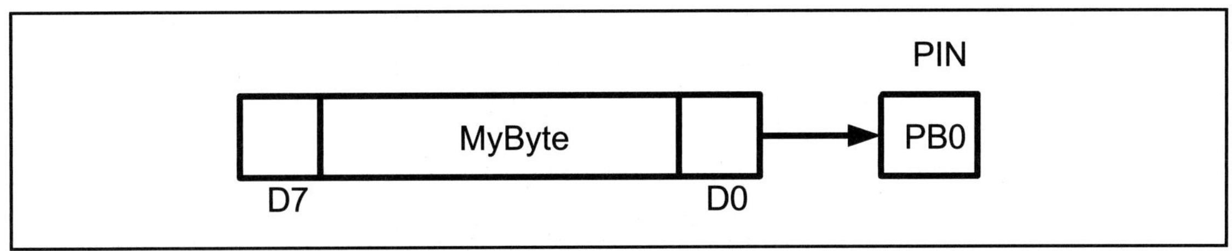

Figure 7-1. Shifting Bits out (LSB Going First)

Example 7-30

Write a C program to send out the value $88 serially one bit at a time via PB0. The MSB should go out first.

Solution:

```
//Serializing data via PB0 (SHIFTING LEFT)
#include <mc9s12dp512.h>
#define PB0 PORTB_BIT0
void main(void)
  {
    unsigned char conbyte = 0x88;
    unsigned char regAMSB;
    unsigned char x;
    regAMSB = conbyte;
    DDRB_BIT0 = 1;              //make PB0 an output
    for(x=0;x<8;x++)
      {
        PB0 = (regAMSB >> 7) & 0x01;//send out next bit
        regAMSB = regAMSB << 1;//shift left remaining bits
      }
  }
```

Example 7-31

Write a C program to bring in a byte of data serially one bit at a time via the PB0 pin. Place the byte on Port A. The LSB should come in first.

Solution:

```
#include <mc9s12dp512.h>
#define PB0 PORTB_BIT0;
void main(void)
  {
    unsigned char x;
    unsigned char temp;
    unsigned char REGA=0;
    DDRB_BIT0 = 0;             //PB0 as input
    DDRA = 0xFF;               //Port A as output
    for(x=0;x<8;x++)
      {
       temp = PB0;             //bring in the bit
       temp = temp & 0x01;     //mask the unused bits
       temp = temp << 7;       //shift left into MSB
       REGA = REGA | temp;     //give it to REGA
       REGA = REGA >> 1;       //shift right to save
      }
    PORTA = REGA;              //display on PORTA
  }
```

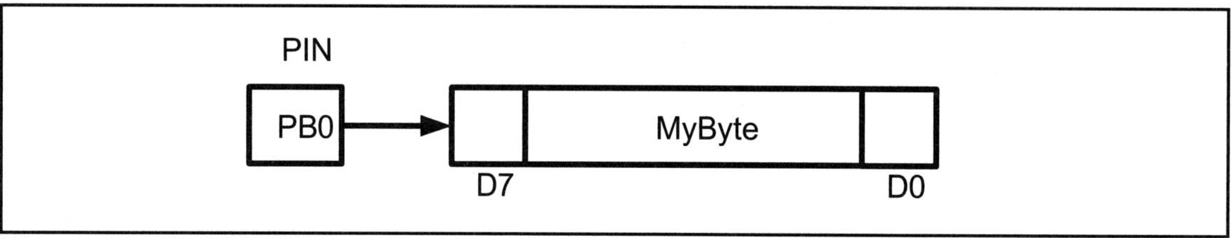

Figure 7-2. Shifting Bits in (Bring in LSB First)

Example 7-32

Write a C program to bring in a byte of data serially one bit at a time via the PB0 pin. The MSB should come in first.

Solution:

```
#include <mc9s12dp512.h>
#define PB0 PORTB_BIT0;
void main(void)
  {
     unsigned char x;
     unsigned char temp;
     unsigned char REGA=0;
     DDRB_BIT0 = 0;              //PB0 as input
     DDRA = 0xFF;                //Port A as output
     for(x=0;x<8;x++)
       {
        temp = PB0;              //bring in the bit
        temp = temp & 0x01;      //mask the unused bits
        REGA = REGA | temp;      //give it to REGA
        REGA = REGA << 1;        //shift left to save
       }
     PORTA = REGA;
  }
```

Review Questions

1. Show the C code to toggle only bit 0 of PORTB.
2. Show the C code to toggle only bit 7 of PORTE.
3. Show the C code to toggle only bit 2 of PORTK.
4. When we bring in a byte serially with LSB first, we must shift (right, left) to get all the bits.
5. When we bring in a byte serially with MSB first, we must shift (right, left) to get all the bits.

SECTION 7.5: PROGRAM ROM ALLOCATION IN C

Using program (code) space for predefined fixed data is a widely used option in the HCS12, as we saw in Chapter 6. In that chapter we saw how to use Assembly language instructions to access the data stored in the program code space. In this chapter, we explore the same concept using the C compiler.

RAM data space vs. code data space

In the HCS12 we have two memory address spaces in which to store data. They are as follows:

1. Several kilobytes of data RAM space. As we have seen in previous chapters, many HCS12 chips have up to 14K bytes for the data RAM. We also have seen how we can read (from) or write (into) this RAM space directly or indirectly.
2. Several hundred kilobytes of code (program) space. This on-chip ROM space is used for storing programs (opcodes) and therefore is directly under control of the program counter (PC). As we have seen in the previous chapters, many HCS12 chips have much more than 64K bytes of on-chip program ROM. We have also seen how to access the program ROM for the purpose of data storage using the indexed addressing mode (see Chapter 6). There is one problem with using this program code space for storage of fixed data: The more code space we use for data, the less is left for our program code. For example, if we have a HCS12 chip with only 32K of on-chip ROM, and we use 8K to store a look-up table, only 24K is left for the program. For some applications, this can be a problem. For this reason Freescale has added EEPROM memory to the HCS12 to be used for data storage. The EEPROM option of HCS12 is discussed in Chapter 14. Next, we will examine how the C compiler uses on-chip ROM space, and discuss how it places data into program ROM.

Allocating program space to data

In all our C examples so far, byte-size variables were stored in the data RAM. As we saw in Chapter 6, it is common practice to use the on-chip program ROM for the purpose of storing fixed data such as strings. To make the C compiler use the program (code) ROM space for the fixed data, we use the keyword *const* as shown in the following lines of C code:

```
const char mynum[ ]  = "Hello"; //use code space for data
```

The following code shows how to use program space for data in C:

```
//Program 7-1
#include <mc9s12dp512.h>
const char mynum[ ]  = "0123456789"; //uses program
//ROM space for fixed (constant) data
void main(void)
  {
    unsigned char z;
    DDRB = 0xFF;          //make Port B an output
    for(z=0;z<10;z++)
      PORTB=mynum[ z] ;
  }
```

Putting data in a specific ROM address

To place the data (containing constants, strings, and look-up tables) at a specific address of program ROM, we use the @ symbol directive. Examine

Program 7-2 to see how the C code assigns the program ROM address of 0xD200 to the string "Hello".

```
//Program 7-2
#include <mc9s12dp512.h>
const char message[ ] @0xD200 = "HELLO"; //uses program
                                //ROM space for fixed (constant) data
void main(void)
  {
    unsigned char z;
    DDRB = 0xFF;              //make Port B an output
    for(z=0;z<5;z++)
      PORTB=message[ z] ;
  }
```

Run the above program on the CodeWarrior simulator. Examine the program code space to see the string "HELLO" located at the ROM address starting at 0x0D200. See Example 7-33.

Review Questions

1. True or false. Using the program ROM space for data means the data is fixed and static.
2. If we have a message string with a size of over 1,000 bytes, then we use ________ (program ROM, data RAM) to store it.
3. Which space would you use to declare the following values for C?
   (a) the number of days in a week
   (b) the number of months in a year
   (c) a character for a video game

See the following website for HCS12 CodeWarrior:

http://www.freescale.com

The following website has a tutorial for CodeWarrior and other C compilers:

http://www.MicroDigitalEd.com

When running a C program on the HCS12 hardware, the following points must be noted:

1. **Place "while(1);" at the end of the program to prevent the program from executing again. This plays the role of "HERE BRA HERE" in Assembly language.**
2. **The method of allocating a fixed address to the string shown in Program 7-2 is strongly discouraged since memory allocation is the job of the compiler.**

Example 7-33

Compare and contrast the following programs and discuss the advantages and disadvantages of each:

(a)

```
#include <mc9s12dp512.h>
void main(void)
  {
     DDRB = 0xFF;                          //make Port B an output
     PORTB = 'H';
     PORTB = 'E';
     PORTB = 'L';
     PORTB = 'L';
     PORTB = 'O';
  }
```

(b)

```
#include <mc9s12dp512.h>
void main(void)
  {
     unsigned char z;
     unsigned char mydata[] = "HELLO";
     DDRB = 0xFF;                          //make Port B an output
     for(z=0;z<5;z++)
        PORTB = mydata[z];
  }
```

(c)

```
#include <mc9s12dp512.h>
const char message[] = "HELLO"; //uses program
                        //ROM space for fixed (constant) data
void main(void)
  {
     unsigned char z;
     DDRB = 0xFF;                          //make Port B an output
     for(z=0;z<5;z++)
        PORTB= message[z];
  }
```

Solution:

All the programs send out "HELLO" to PORTB one character at a time. They do the same thing in different ways. The first way is short and simple, but the individual characters are embedded into the program. If we change the characters, the whole program changes. This method also mixes the code and data together. The second one uses the RAM data space to store array elements; therefore, the size of the array is limited to RAM. The third one uses a separate area of the program code space for data. This allows the size of the array to be as big as you want provided that you have enough on-chip program ROM. The more program code space you use for data, however, the less space is left for your program code. Both the (b) and (c) programs are easily upgradeable if we want to change the string itself or make it longer. That is not the case for program (a).

PROBLEMS

SECTION 7.1: DATA TYPES AND TIME DELAYS IN C

1. Indicate what data type you would use for the following variables:
   (a) the temperature
   (b) the number of days in a week
   (c) the number of days in a year
   (d) the number of months in a year
   (e) the counter to keep the number of people getting on a bus
   (f) the counter to keep the number of people going to a class
   (g) an address of 64K RAM space
   (h) the age of a person
   (i) a string for a message to welcome people to a building
2. Give the hex value that is sent to the port for each of the following C statements:
   (a) PORTB=14; (b) PORTB=0x18; (c) PORTB='A';
   (d) PORTB=7; (e) PORTB=32; (f) PORTB=0x45;
   (g) PORTB=255; (h) PORTB=0x0F;
3. Give two factors that can affect time delay code size in the HCS12 microcontroller.
4. Of the two factors in Problem 3, which can be set by the system designer?
5. Can the programmer set the number of clock cycles used to execute an instruction? Explain your answer.
6. Explain why various C compilers produce different hex file sizes.

SECTION 7.2: LOGIC OPERATIONS IN C

7. Indicate the data on the ports for each of the following:
   Note: The operations are independent of each other.

```
(a) PORTB=0xF0&0x45;        (b)  PORTB=0xF0&0x56;
(c) PORTB=0xF0^0x76;        (d)  PORTA=0xF0&0x90;
(e) PORTA=0xF0^0x90;        (f)  PORTA=0xF0|0x90;
(g) PORTA=0xF0&0xFF;        (h)  PORTA=0xF0|0x99;
(i) PORTA=0xF0^0xEE;        (j)  PORTA=0xF0^0xAA;
```

8. Find the contents of the port after each of the following operations:

```
(a) PORTB=0x65&0x76;        (b)  PORTB=0x70|0x6B;
(c) PORTA=0x95^0xAA;        (d)  PORTA=0x5D&0x78;
(e) PORTA=0xC5|0x12;        (f)  PORTE=0x6A^0x6E;
(g) PORTB=0x37|0x26;
```

9. Find the port value after each of the following is executed:

```
(a) PORTB=0x65>>2;          (b)  PORTA=0x39<<2;
(c) PORTB=0xD4>>3;          (d)  PORTB=0xA7<<2;
```

10. Show the C code to swap 0x95 to make it 0x59.
11. Write a C program that finds the number of zeros in an 8-bit data item.
12. A stepper motor uses the following sequence of binary numbers to move the motor. How would you generate them in C?

```
1100,0110,0011,1001
```

13. Write a C program to toggle all the bits of PORTB every 200 ms.

SECTION 7.3: DATA CONVERSION PROGRAMS IN C

14. Write a program to convert the following series of packed BCD numbers to ASCII. Assume that the packed BCD is located in data RAM.
```
76H,87H,98H,43H
```
15. Write a program to convert the following series of ASCII numbers to packed BCD. Assume that the ASCII data is located in data RAM.
```
"8767"
```
16. Write a program to get an 8-bit binary number from PORTB, convert it to ASCII, and save the result if the input is packed BCD of 00–0x99. Assume that PORTB has 1000 1001 binary as input.

SECTION 7.4: I/O BIT MANIPULATION AND DATA SERIALIZATION IN C

17. Write a C program to toggle bits PB1 and PB7.
18. Write a C program to toggle only bit PB0.
19. Write a C program to count up PORTB from 0–99 continuously.

SECTION 7.5: PROGRAM ROM ALLOCATION IN C

20. Indicate what type of memory (data RAM or code ROM space) you would use for the following variables:
    (a) the temperature
    (b) the number of days in a week
    (c) the number of days in a year
    (d) the name of a country
21. True or false. When using program ROM for data, we should let the compiler set the address location.
22. Why do we use the ROM code space for video game characters and shapes?
23. What is the advantage of using program ROM space for fixed data?
24. What is the drawback of using program ROM space for fixed data?
25. Write a C program to send your first and last names to PORTB. Use the program ROM space for the data.
26. In Problem 25, show how to place the last name at ROM address 0x0D100 and the first name at address 0xD120.

ANSWERS TO REVIEW QUESTIONS

SECTION 7.1: DATA TYPES AND TIME DELAYS IN C

1. 0 to 255 for unsigned char and –128 to +127 for signed char
2. 0 to 65,535 for unsigned int and –32,768 to +32,767 for signed int
3. Unsigned char
4. True
5. (a) Crystal frequency of HCS12 system
   (b) Compiler used for C

SECTION 7.2: LOGIC OPERATIONS IN C

1. (a) 02
   (b) FFH
   (c) FDH
2. Zeros
3. One
4. All zeros
5. 66H
6.

```
#include <mc9s12dp512.h>
void MSDelay(unsigned int);
void main(void)
  {
    DDRB = 0xFF;
    PORTB=0x55;
    while(1)
      {
        PORTB=PORTB^0xFF;
        MSDelay(250);
      }
  }

void MSDelay(unsigned int itime)
  {
    unsigned int i;
    unsigned int j;
    for(i=0;i<itime;i++)
      for(j=0;j<331;j++);
  }
```

7. True
8. True

SECTION 7.3: DATA CONVERSION PROGRAMS IN C

1. (a) 15H = 0001 0101 packed BCD, 0000 0001, 0000 0101 unpacked BCD
   (b) 99H = 1001 1001 packed BCD, 0000 1001, 0000 1001 unpacked BCD
2. 3736H = 00110111 00110110
   and in packed BCD we have 76H = 0111 0110
3. 36, 37
4. Yes, because mydata = 0x39
5. Space savings
6. ASCII
7. BCD
8. E4H

SECTION 7.4: I/O BIT MANIPULATION AND DATA SERIALIZATION IN C

1.

```
#include <mc9s12dp512.h>
#define MYBIT PORTB_BIT0
void main(void)
  {
    DDRB_BIT0 = 1;
    while(1)
      {
        MYBIT = 1;
        MYBIT = 0;
      }
  }
```

2.
```
#include <mc9s12dp512.h>
#define MYBIT PORTE_BIT5
void main(void)
  {
    DDRE_BIT5 = 1;
    while(1)
      {
        MYBIT = 1;
        MYBIT = 0;
      }
  }
```
3.
```
#include <mc9s12dp512.h>
#define MYBIT PORTK_BIT2
void main(void)
  {
    DDRK_BIT2 = 1;
    while(1)
      {
        MYBIT = 1;
        MYBIT = 0;
      }
  }
```
4. Right
5. Left

SECTION 7.5: PROGRAM ROM ALLOCATION IN C

1. True
2. Program ROM
3. a) Data RAM
   b) Data RAM
   c) Code ROM

CHAPTER 8

HCS12 HARDWARE CONNECTION, BDM, AND S19 HEX FILE

OBJECTIVES

Upon completion of this chapter, you will be able to:

>> Explain the function of the reset pin of the HCS12 microcontroller
>> Show the hardware connection of the HCS12 chip
>> Show the use of a crystal oscillator for a clock source
>> Explain the role of the CRG registers in HCS12-based systems
>> Code a test program in Assembly and C for testing the HCS12 systems
>> Code a test program in Assembly and C for setting the PLL clock
>> Explain the S19 hex file characteristics for the HCS12 family
>> Explain the role of the BDM (background debug mode) in the HCS12

SECTION 8.1: HCS12 PIN CONNECTION AND BDM

The HCS12/9S12 family members come in two packages: 80-pin QFP (quad flat package) and 112-pin LQFP (low-profile quad flat package). They all have many pins that are dedicated to various functions such as I/O, ADC, timer, and interrupts. Figures 8-2 and 8-3 show the pins for the HCS12 chips.

Examining Figures 8-2 and 8-3, note that many pins are designated as ports A, B, E, H, and so on, with their alternate functions. The rest of the pins are designated as VDDx, VSSx (Gnd), XTAL, EXTAL, BKGD, and Reset. Next, we describe some of features and functions associated with these pins.

VddR (Vcc) and VssR (Gnd)

These pins are used to provide supply voltage and ground to the voltage regulation section of the chip. The typical voltage source is +5 V. Many of the recent members of microcontroller families have lower voltage for Vdd pins in order to reduce the noise and power dissipation of the system. For chips with 40 pins and more, it is common to have multiple pins for VCC and GND. This will help reduce the noise (ground bounce) in high-frequency systems, as discussed in Appendix C.

XTAL and EXTAL

The HCS12 has two options for the clock source: PLL (phased lock loop) and crystal oscillator. In many trainer boards a quartz crystal oscillator is connected to pins XTAL and EXTAL. The quartz crystal oscillator connected to the XTAL and EXTAL pins also needs three capacitors. The capacitor connection is shown in Figure 8-1a. If you decide to use a frequency source other than a crystal oscillator, such as a CMOS oscillator, it will be connected to EXTAL; XTAL is left unconnected, as shown in Figure 8-1b.

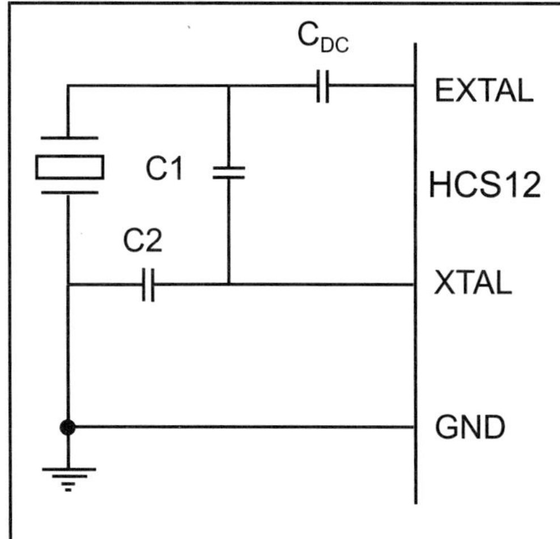

Figure 8-1a. XTAL–EXTAL Connection to Crystal Oscillator

VddPLL and VssPLL

These pins are used to provide supply voltage to the PLL and oscillator part of the chip. The PLL allows us to have another source of working frequency independent of the crystal oscillator frequency. According to the Freescale HCS12 manual, even if PLL is not used, the VddPLL and VssPLL pins must be connected properly.

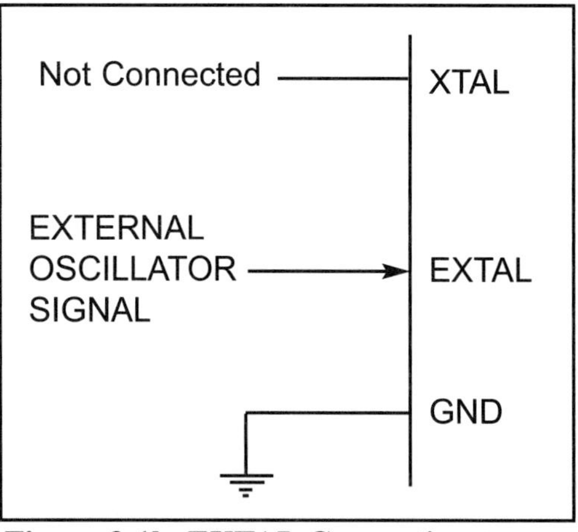

Figure 8-1b. EXTAL Connection to an External Clock Source

We can choose options for the

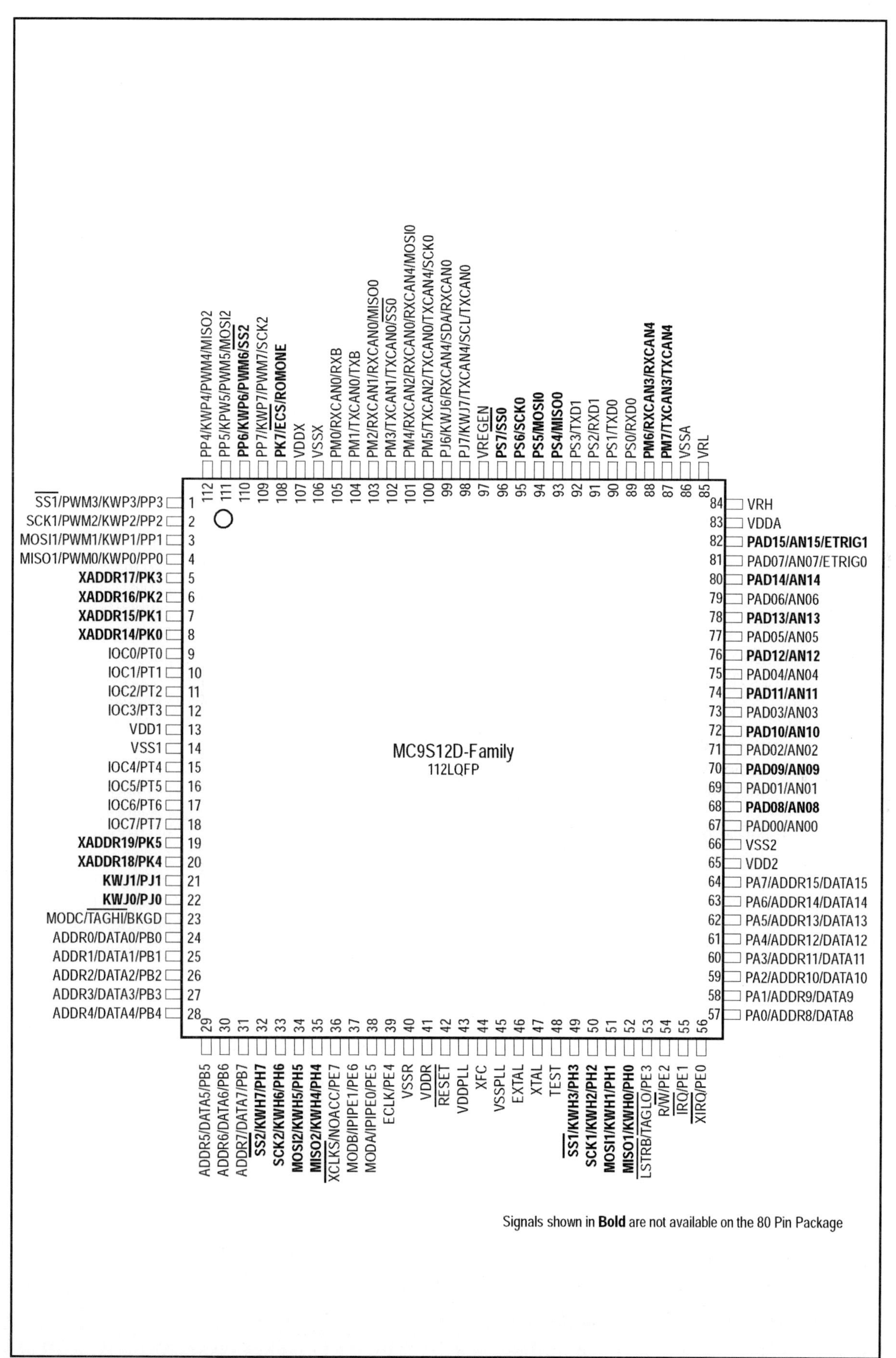

Figure 8-2. MC9S12D 112-Pin Diagram (from Freescale)

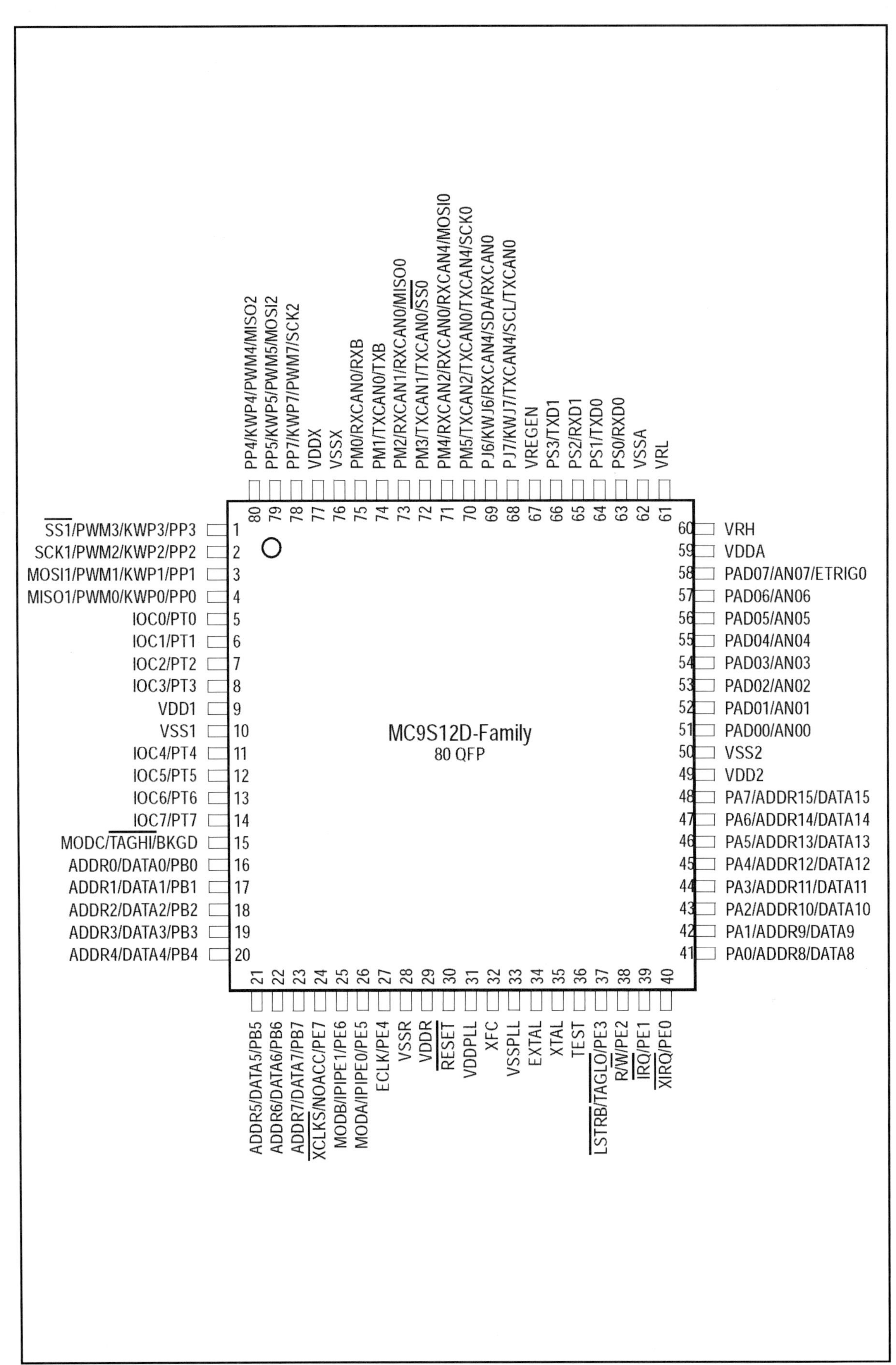

Figure 8-3. MC9S12D 80-Pin Diagram (from Freescale)

clock frequency by setting bits in the CRG (Clock and Reset Generator) register. The CRG register for the oscillator is shown in the HCS12 manual.

Vdd1,2 and Vss1,2

According to the HCS12 manual, these are "core logic supply pins" and are connected to the output of the voltage regulator.

Vregen

According to the HCS12 manual, this is optional and is used "to disable the voltage regulator if the core logic as well as the oscillator are supplied from external" sources.

Reset

Connect the Reset pin to the push button (momentary) switch. The most difficult time for any system is during the power-up. The CPU needs both a stable clock source and a stable voltage level to function properly. The HCS12 chips come with some features that help the reset process. We can choose these features by setting the bits in the Clock and Reset Generator (CRG) registers. In the HCS12 we have four types of reset.

a) Using the Reset pin. The Reset pin is both an input and output. See Figure 8-4. When an external LOW pulse is applied to this pin, the microcontroller terminates all activities and goes to a known state. This is often referred to as the *external reset.* In many HCS12 Trainer boards a push button (momentary) switch is connected to this pin.

b) There is also what is commonly referred to as a *power-on reset (POR).* When a voltage supply is applied to the VddR pin, the HCS12 goes to a known state of reset.

c) Clock monitor reset. The HCS12 monitors the clock source internally, and if the clock falls below a certain frequency or if it stops working, the system is forced into reset state. See Figure 8-5.

d) The COP (computer operating properly) reset. This reset is caused by the COP watchdog timer, which is discussed next.

COP watchdog timer

In recent years, microcontrollers have come with a piece of hardware called a *watchdog timer.* We can use the COP watchdog timer to force the microcontroller into the known state of reset when the system is hung up or out of control due to execution of an incorrect sequence of codes. There are many uses for watchdog timers in embedded systems. One application is to use the watchdog timer to prevent a system from going into an infinite loop due to a software bug. Another application of the watchdog timer can be to catch events that cause the system to hang. These problems can happen due to corruption of the program ROM caused by a power surge, an electrically noisy environment, or inadvertent changes to the program counter. In such situations, the watchdog timer will force the system into a known state of reset, from which the system can recover. In some applications, the system can be put to sleep if there is no activity, thereby saving battery power. In such applications, one can use the watchdog timer to monitor the keyboard and, when there is activity on the keyboard, to awaken the system to process the information.

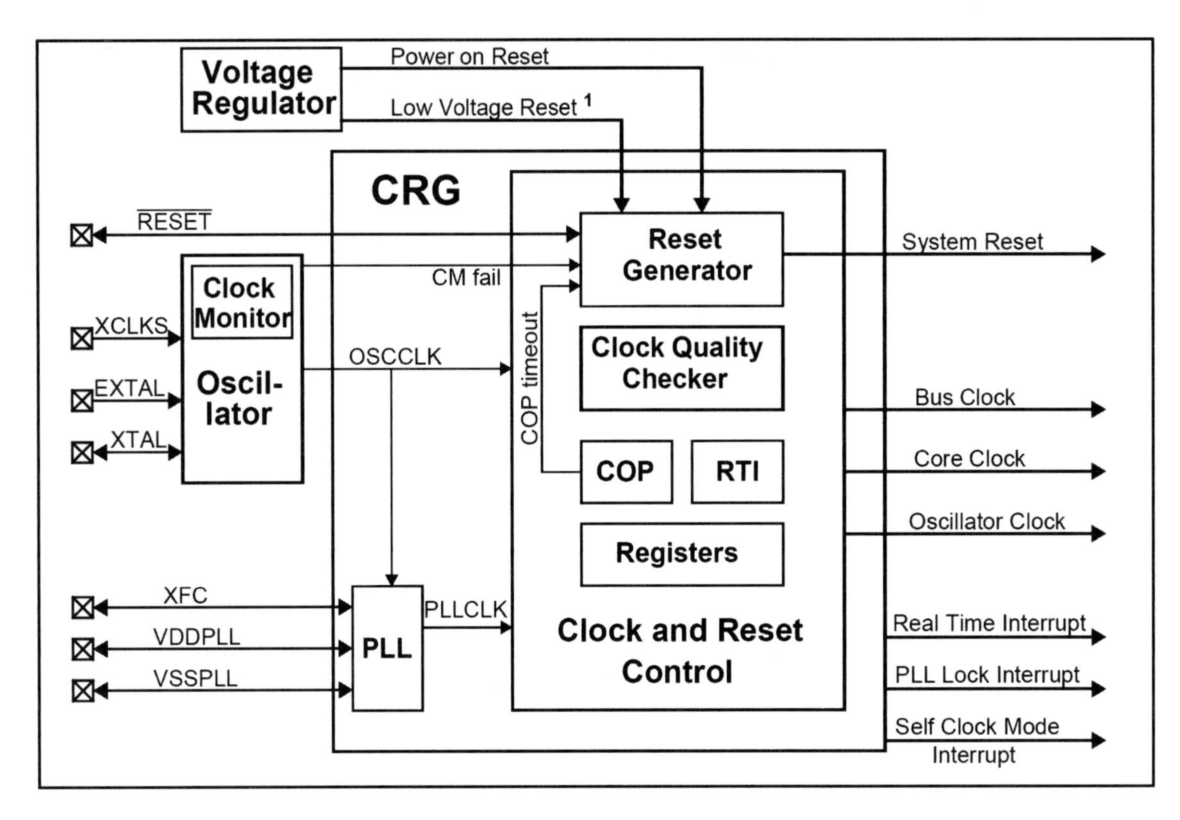

Figure 8-4. CRG Block Diagram in HCS12 (from Freescale)

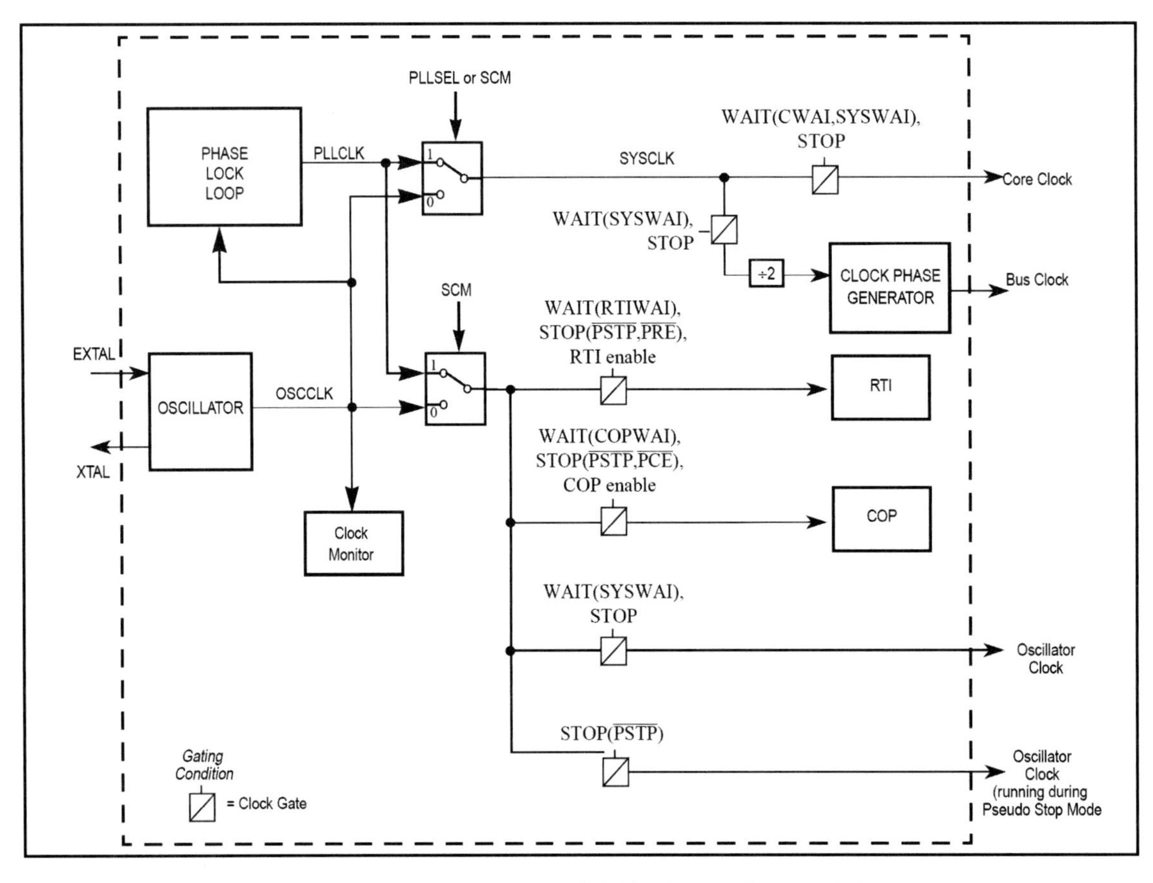

Figure 8-5. System Clock Generator in HCS12 (from Freescale)

BKGD pin and BDM (Background Debugger Mode)

In recent years the Freescale has equipped their CPUs with a powerful feature called BDM (Background Debugger Mode). The BDM is a combination of hardware and software that allows us to trace a program one instruction at a time. Using the BDM we can trace a given program and examine the contents of the CPU registers and memory after the execution of each instruction. This feature is called *background debugger mode* because it tries to use the idle (unused) CPU cycles to perform its task. At the heart of the BDM is a program (firmware) residing in the upper region of the Flash ROM built into the CPU by Freescale. The BDM firmware in Flash uses a single pin called *BKGD* to communicate externally with a host system such as an x86 PC. The IDE softwares such as CodeWarrior send and receive information to the HCS12-based system via this pin serially. Many recent generations of trainer boards come with the BDM and Loader combined ready to be connected to the USB or serial COM ports of the x86 PC. On such a board, we connect the trainer to the USB port of the host PC and use IDE programs such as CodeWarrior to control the processor execution directly, debug code on the processor, program (download) internal or external flash memory, and read/write registers and memory values. The older generations of HCS12 trainer boards come with the BDM 6-pin header, so we have to buy a separate USB type BDM-POD (or serial COM port BDM-POD) to connect these trainers to the host PC. Prior to the advent of BDM (or for the processors that lacked the BDM feature) we had to spend hundreds of dollars for an ICD (in-circuit debugger) or an ICE (in-circuit emulator) to be able to trace the program to find the bug. The BDM is so important in the Freescale products that all their processors come already BDM enabled for the single-chip mode. In other modes such as expanded mode we must enable the BDM option.

Downloading the program into HCS12

Generally there are three ways to download the program into a chip.

a) Using a device programmer (or ROM burner). In this case you must remove the chip from the target system and place it in the device programmer. A ZIF (zero insertion force) socket on both the device programmer and target system makes the removal of the chip quicker and less damaging than with a standard socket. When removing and reinserting, we must observe ESD (electrostatic discharge) procedures. The following are some of the features of a device programmer from https://www.EEtools.com:

1) Auto search device select function supports E(E)PROMs and microcontrollers.

2) Device insertion test identifies improperly inserted device before programming.

3) Checks for incorrect device insertion, backward, incorrect position, and poor pin contact.

4) Device Operations: Read, Blank check, Program, Verify, Checksum, Data compare, Security, Auto(blank check-program-verify), Option Bit program.

5) Displays programming parameters and optional bit information on the screen.

6) Sets device/buffer address ranges before programming devices.

7) Supports binary and all hex files (POF and JEDEC, Intel Hex, Motorola S Records, Tekhex, straight hex, hex-space, Extended Tekhex, and others, automatic file type recognition) with Load, Edit, and Save commands.

b) Using the in-circuit serial programmer (ICSP). This allows the developer to program (load) and debug a microcontroller while it is in the system. This is done by one or two wires communicating with the chip, which has the setup to accept the information.

c) A boot loader is a piece of code burned into the microcontroller's program ROM. Its purpose is to communicate with the host PC to load the program. A boot loader can be written to communicate via a serial port, CAN port, USB port, or even a network connection. In recent years HCS12 trainers have come with a combined BDM and boot loader, which allows users to download and debug programs.

Program counter value upon reset

Activating the Reset pin will cause all values in the registers to be lost. Table 8-1 provides a partial list of HCS12 registers and their values after power-on reset. Note from Table 8-1 that the value of the PC (program counter) is FFFE upon reset, forcing the CPU to fetch the address of the boot program from the ROM memory locations $FFFE and $FFFF. See Chapter 11.

CRG (clock and reset generation)

Inside the HCS12, in addition to the voltage regulator section, we also have the CRG (clock and reset generation). The CRG is responsible for both the reset and the clock. The clock section uses the crystal connection to the XTAL pins and PLL to provide an accurate clock to the core of the CPU for program execution. The most widely used option for clock source is to connect the XTAL and EXTAL pins to a crystal (or ceramic) oscillator, as shown in Figure 8-1. In HCS12 microcontrollers, the clock used for running the CPU is called *Core Clock* and is based on the SYSCLK clock. The SYSCLK comes from the XTAL source, as shown in Figure 8-4. The frequency of the SYSCLK is one half of the frequency of the crystal oscillator connected to the XTAL pins. The CPU (instruction) cycle time is the same as SYSCLK. The CRG also provides a Bus Clock source for connection of the external memory and peripheral devices in expanded mode. The pin for the Bus Clock is called *ECLK*. The frequency of the ECLK (Bus Clock) is one half of the frequency of the crystal oscillator connected to the XTAL pins. This is examined in Example 8-1. We can increase the speed of CPU clock and ECLK using the same crystal oscillator by programming the PLL to provide clocks equal to a multiple of the oscillator clock. Notice that the higher the frequency, the more power is dissipated by the CPU, as discussed in Appendix C. Indeed the PLL not only can multiply the frequency, but also makes sure that clock provided to the CPU and ECLK is stable and clean during the reset phase of the CPU operation. See Section 8.2 for details of how to program the PLL option.

Table 8-1: Reset Values for Program Counter in HCS12

Register	Reset Value (hex)
PC	FFFE
DDRA–DDRT	00

Example 8-1

Find the CPU (instruction) cycle time and ECLK (Bus Clock) for the HCS12 chip with the following crystal oscillator connected to the XTAL and EXTAL pins. Assume we are not using PLL.
(a) 4 MHz (b) 8 MHz

Solution:

The ECLK (the Bus Clock) and the CPU cycle time use the same frequency, which is 1/2 of the crystal oscillator if there is no PLL.

(a) ECLK = CPU cycle time is 1 / (4 MHz / 2) = 0.5 μs
(b) ECLK = CPU cycle time is 1 / (8 MHz / 2) = 0.25 μs

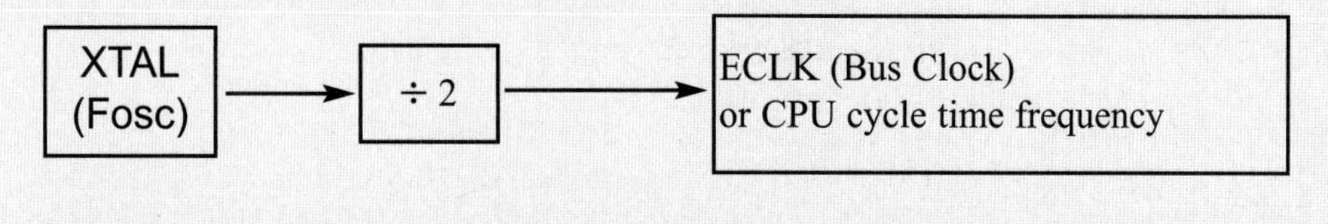

The CRG and CPU operation in critical conditions

The CPU needs both a stable clock source and a stable voltage level to function properly. It is the job of the CRG to make sure that the CPU gets both stable voltage and accurate clock. To do its job, the CRG is equipped with registers that can be programmed by the embedded system designer to ensure proper operation. This proper operation is most critical if we use the HCS12 in the design of medical equipment and automobiles. The CRG supports four modes of operation. They are as follows:

1) Run mode. This is normal run mode and all parts of the CRG and CPU are running in an expected and normal fashion.

2) Wait mode. This mode will stop providing the clock to the CPU and external devices and will go to stand-by mode without losing the register's contents. We can force the CPU into wait mode by using the bits of the CLKREG register or using the WAIT instruction. This mode is widely used for cases where low power consumption can extend battery life.

3) The stop mode has two options: full-stop, and pseudo-stop. They are programmed via the bits of the SSTP register. In the full-stop mode, the oscillator is stopped and as a result both core clock and ECLK are stopped. This will also freeze the COP watchdog timer. In the pseudo-stop mode the oscillator is running, but the clock source to the core clock and ECLK are cut off. Unlike the full-stop, this mode allows the COP watchdog timer to continue functioning.

4) Self-clock mode. In this mode the clock source is internally monitored and if it falls below a certain frequency or if it stops working, the CRG will switch to an internal clock source. This option is used for safety purposes only. This self-clock mode prevents the loss of CPU functionality in case of an emergency in which the system failure can cause life-threatening conditions in medical equipment or automobiles.

Test program for the HCS12 Trainer in Assembly and C

To test your HCS12 trainer, you can run a simple test in which all the bits of PORTB toggle continuously with some delay between the "on" and "off" states. See Programs 8-1 and 8-1C. Notice in these programs that the time delay is based on a 4 MHz crystal with no PLL.

Trainer Test Program in Assembly

```
;Program 8-1
#include <mc9s12dp512.h>
R1 EQU $807
R2 EQU $808
      LDS   #$8FF ;set the stack pointer
      LDAA  #$FF        ;A = FF
      STAA  DDRB        ;make Port B an output port
      LDAA  #$55        ;A = 55H
L2    STAA  PORTB
      JSR   DELAY
      COMA              ;complement reg A
      BRA   L2
;-----------DELAY
DELAY
      PSHA
      LDAA #200         ;start of outer loop
      STAA R1
D1    LDAA #250         ;start of inner loop
      STAA R2
D2    NOP
      NOP
      NOP
      DEC R2
      BNE D2
      DEC R1
      BNE D1
      PULA
      RTS
      END
```

Trainer Test Program in C

```
//Program 8-1C
#include <mc9s12dp512.h>
void MSDelay(unsigned int);
void main(void)
  {
    DDRB = 0xFF;              //make Port B an output
    while(1)                  //repeat forever
      {
        PORTB = 0x55;
        MSDelay(250);
        PORTB = 0xAA;
        MSDelay(250);
      }
  }
void MSDelay(unsigned int itime)
  {
    unsigned int i; unsigned char j;
    for(i=0;i<itime;i++)
      for(j=0;j<165;j++);
  }
```

Some troubleshooting tips

Running the test program on your HCS12-based trainer should toggle all the I/O bits with some delay. If your system does not work, however, follow these steps to find the problem:

1. With the power on, check V_{dd} and GND.
2. Check the Reset pin using an oscilloscope. When the system is powered up, the Reset pin is HIGH. When the momentary switch is pressed, the pin goes LOW. Make sure the momentary switch is connected properly.
3. If all the above steps pass inspection, check the content of the on-chip ROM. It must be the same as the opcodes provided by the lst file. Your assembler produces the lst file, which lists the opcodes and operands on the left side of the assembly instructions. This file's content must match exactly the contents of your on-chip ROM if the proper steps were taken in downloading the program into the on-chip flash ROM.

Review Questions

1. Which pin is used to provide access to the BDM of the HCS12 chip?
2. Upon power-up, the program counter (PC) has a value of ____.
3. Upon power-up, the HCS12 fetches the boot address from ROM address location ____________.
4. Reset is an active-_______ (LOW, HIGH) pin.
5. How many resets does the HCS12 chip have?
6. A given HCS12-based system has a crystal frequency of 6 MHz. What is the CPU (instruction) cycle time for this system?
7. True or false. The higher the clock frequency for the system, the lower the power dissipation.

SECTION 8.2: SETTING THE PLL FREQUENCY IN THE HCS12

Earlier in the chapter we discussed the role of the crystal oscillator in setting the bus speed. In this section, we show how to use the PLL to multiply the XTAL frequency. For the bus clock, we can use either the crystal oscillator (OSCCLK) frequency or the PLL frequency. Upon reset, the HCS12 uses the crystal oscillator frequency. To switch to the PLL frequency, we must set some bits in the CLKSEL and CRGCTL registers. See Figures 8-6 and 8-7. The actual clock frequency of the PLL itself is decided by the SYNR and REFDV registers. Next, we discuss how to set the PLL clock.

Setting the clock for PLL

To set the clock value for the PLL, we must use the following equation:

$$\text{PLLCLOCK} = 2 \times \text{OSCCLK} \times \{[\text{SYNR} + 1]/[\text{REFDV} + 1]\}$$

where the SYNR and REFDV values are set according to the values in the SYNR and REFDV registers. See Figures 8-8 and 8-9. See Examples 8-2 and 8-3.

PLLSEL	PSTP						

PLLSEL D7 PLL select bit
1 = System clock is derived from PLLCLK (Bus Clock = PLLCLK/2)
0 = System clock is derived from OSCCLK (Bus Clock = OSC-CLK/2)

PSTP D6 Pseudo-stop bit
1 = Oscillator continues to run in stop mode
0 = Oscillator is disabled in stop mode

D5–D0 bits are used in wait mode. See the HCS12 manual.
(Upon reset, all the bits in this register are 0s. Notice PLLSEL = 0. That means we must use this register to switch to PLL frequency.)

Figure 8-6. CLKSEL Register

	PLLON	AUTO	ACQ				

PLLON D6 PLL on bit
1 = PLL is turned on.
0 = PLL is turned off (default).

AUTO D5 Automatic bandwidth control
1 = High bandwidth is selected. If PLLON = 1, we must choose this bit.
0 = Use the ACQ bit to set the bandwidth filter.

ACQ D4 Acquisition bit (If AUTO = 1, this bit has no effect. See the HCS12 manual.)

D7, D3–D0 bits are used by other sections of the CRG. See the HCS12 manual.

(Upon reset, all the bits in this register are 0s. Notice PLLON = 0. That means we must turn it on to switch to PLL frequency.)

Figure 8-7. PLLCTL Register

0	0	SYN5	SYN4	SYN3	SYN2	SYN1	SYN0

SYN5–0 D5–D0

SYN5	SYN4	SYN3	SYN2	SYN2	SYN1	SYN0	Decimal value
0	0	0	0	0	0	0	0
0	0	0	0	0	0	1	1
0	0	0	0	0	1	0	2
.	.	.	.	.	.	.	
1	1	1	1	1	1	0	30
1	1	1	1	1	1	1	31

Figure 8-8. SYNR Register

0	0	0	0	REFDV3	REFDV2	REFDV1	REFDV0

REFDV3–0 D3–D0

REFDV3	REFDV2	REFDV2	REFDV1	REFDV0	Decimal
0	0	0	0	0	0
0	0	0	0	1	1
0	0	0	1	0	2
. .	. .	. .	.		
1	1	1	1	0	14
1	1	1	1	1	15

Figure 8-9. REFDV Register

Example 8-2

Assume the XTAL is connected to the 4 MHz crystal oscillator. Show the code for the CRG to get a PLL frequency of 16 MHz.

Solution:

$$PLLCLK = 2 \times OSCCLK \times \{[SYNR + 1]/[REFDV + 1]\}$$
$$PLLCLK = 2 \times 4\ MHz \times \{[1 + 1]/[0 + 1]\} = 16\ MHz$$

```
;in Assembly
      LDAA  #1
      STAA  SYNR              ;SYNR=1
      LDAA  #0
      STAA  REFDV             ;REFDV=0
      LDAA  #$80
      STAA  CLKSEL            ;PLLSEL=1 for the CLKSEL reg.
      LDAA  #$60
      STAA  PLLCTL      ;PLLON=1,high bandwidth in PLLCTL reg.

;in C
      SYNR=1;
      REFDV=0;
      CLKSEL=0x80;      //PLL=1 for the CLKSEL reg.
      PLLCTL=0x60;      //PLLON=1, high bandwidth in PLLCTL reg.
```

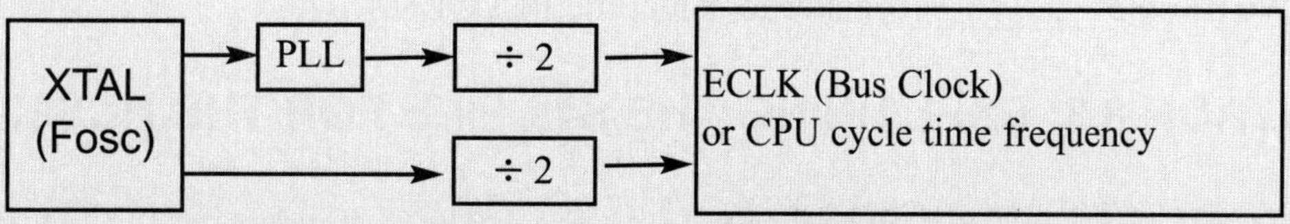

Example 8-3

Assume the XTAL is connected to the 8 MHz crystal oscillator. Show the code for the CRG to get a PLL frequency of 24 MHz.

Solution:

$$PLLCLK = 2 \times OSCCLK \times \{[SYNR + 1]/[REFDV + 1]\}$$
$$PLLCLK = 2 \times 8\text{ MHz} \times \{[2 + 1]/[1 + 1]\} = 24\text{ MHz}$$

```
;in Assembly
        LDAA  #2
        STAA  SYNR              ;SYNR=2
        LDAA  #1
        STAA  REFDV             ;REFDV=1
        LDAA  #$80
        STAA  CLKSEL            ;PLLSEL=1 for the CLKSEL reg.
        LDAA  #$60
        STAA  PLLCTL      ;PLLON=1, high bandwidth in PLLCTL reg.

;in C
        SYNR=2;
        REFDV=1;
        CLKSEL=0x80;      //PLL=1 for the CLKSEL reg.
        PLLCTL=0x60;      //PLLON=1, high bandwidth in PLLCTL reg.
```

Note:
1) In setting the PLL frequency, you must not exceed the maximum frequency rating of the HCS12 chip.
2) In this book, we use 1/2 of XTAL for bus frequency. If you use PLL to multiply the XTAL frequency, then modify all the examples accordingly.

Review Questions

1. True or false. The bus frequency is 1/2 of XTAL frequency.
2. True or false. The bus frequency is 1/2 of PLLCLK if we use PLL.
3. True or false. The PLLCLK is a function of XTAL frequency.
4. True or false. The bus frequency can be either OSCCLK or PLLCLK, but not both.
5. We use the _________register to select the PLLCLK.

SECTION 8.3: EXPLAINING THE S19 FILE FOR THE HCS12

S19 file is a file format designed by Freescale to standardize the loading (transferring) of executable machine code into a ROM chip. Therefore, the loaders that come with a ROM programmers support the S19 file format. In the CodeWarrior IDE environment, the object file is fed into the linker program to produce the S19 file. Notice that the S19 hex file is used by the loader of an EPROM programmer to transfer (load) the file into the ROM chip. The S19 file is commonly referred to as *S-records* in the Freescale literature.

Analyzing the S19 hex file

Since the ROM burner (loader) uses the S19 hex file to download the opcode into ROM, the hex file must provide the following: (1) the number of bytes of information to be loaded, (2) the information itself, and (3) the starting address where the information must be placed. Each line of the S-record hex file consists of five parts as follows:

SxBBAAAAHHHHH.......HHHHCC

The following describes each part:

1. Sx, S-record. Each line starts with a S followed by 1 or 9. If the S-record contains data, then it uses 1. If it is the last record, then it uses 9.
2. BB, the count byte. This tells the loader how many bytes are in the line. BB can range from 00 to FF.
3. AAAA is for the address. This is a 16-bit address for the S19. The loader places the first byte of data into this memory address.
4. HH......H is the real information (data or code). The loader places this information into successive memory locations of ROM.
5. CC is a single byte. This last byte is the checksum byte of everything in that line. Notice that S19 uses 1's complement to calculate the checksum byte rather than 2's complement, which was covered in Chapters 6 and 7. Also notice that the checksum byte at the end of each line represents the checksum byte for everything in that line and not just for the data portion.

Figure 8-10 shows the source file for toggling PORTA. Now, compare the data portion of the S19 file in Figure 8-12 with the information under the OBJ field of the lst file in Figure 8-11. Notice that they are identical, as they should be. The extra

```
        LDAA   #$FF        ;A = FFH
        STAA   DDRA        ;make Port A an output port
L2      LDAA   #$55        ;A = 55H
        STAA   PORTA       ;put 55H on port A pins
        LDAA    #$AA
        STAA PORTA
        BRA    L2
        END
```

Figure 8-10. Source File for Toggling PORTA

```
 Abs. Rel.    Loc     Obj. code    Source line
 ---- ----    ------ ---------    -----------
12826    31    000000 86FF            LDAA   #$FF  ;A = FFH
12827    32    000002 5A02            STAA   DDRA  ;make Port A an output
12828    33    000004 8655         L2 LDAA   #$55  ;A = 55H
12829    34    000006 5A00            STAA   PORTA ;put 55H on port A
12830    35    000008 86AA            LDAA   #$AA
12831    36    00000A 5A00            STAA   PORTA
12832    37    00000C 20F6            BRA    L2
12833    38
```

Figure 8-11. List File for Program in Figure 8-10

0001 0001
1 1H

```
S111C00086FF5A0286555A0086AA5A0020F678
S105FFFEC0003D
S9030000FC

Separating the fields, we get the following:

SS    BB    AAAA  HHHHHHHHHHHHHHHHHHHHHHHHHHHHHHHHHHHHHHHH  CC
S1    11    C000  86FF5A0286555A0086AA5A0020F6              78
S1    05    FFFE  C000                                      3D
S9    03    0000                                            FC
```

Figure 8-12. S19 File for Program in Figure 8-10

Note: S19 uses 1's complement to calculate the checksum byte.

information is added by the S19 file format. You can run the C language version of the test program and verify its operation. Your C compiler will provide you both the lst file and S19 file if you want to explore the S19 file concept. Examine Examples 8-4 through 8-6 to gain insight into the S19 file.

Example 8-4

From Figure 8-12, analyze all the parts of line 1.

Solution:

After the S1, we have 11, which means that 17 bytes (11 hex = 17 decimal) are in this line. C000H is the address at which the data starts. Then the data is as follows: `86FF5A0286555A0086AA5A0020F6`. Finally, the last byte, 78, is the checksum byte.

Example 8-5

Compare the data portion of the S19 file of Figure 8-12 with the opcodes in the list file of the test program given in Figure 8-11. Do they match?

Solution:

In the first line of Figure 8-12, the data portion starts with $86FF. This is the opcode and operand for the instruction "LDAA #$FF", as shown in the list file of Figure 8-11. The last byte of the data is $20F6, which belongs to the "BRA L2" instruction in the list file.

See the following website for the HCS12 Trainers:

http://www.MicroDigitalEd.com

Example 8-6

Verify the checksum byte for line 1 of Figure 8-12. Verify also that the information is not corrupted.

Solution:

`11 + C0 + 00 + 86 + FF + 5A + 02 + 86 + 55 + 5A + 00 + 86 + AA + 5A + 00 + 20 + F6 = 687` in hex. Dropping the carries (6) gives $87, and its 1's complement is $78, which is the last byte of line 1.

If we add all the information in line 1, including the checksum byte, and drop the carries we should get `11 + C0 + 00 + 86 + FF + 5A + 02 + 86 + 55 + 5A + 00 + 86 + AA + 5A + 00 + 20 + F6 + 78 = $FF`.

Review Questions

1. True or false. The S19 file does not use the checksum byte method to ensure data integrity.
2. The first byte of a line in the S19 file represents ____.
3. The last byte of a line in the S19 file represents ____.
4. In the first byte of an S19 file, we have S1. What does it indicate?
5. Find the checksum byte for the following hex values: $22, $76, $5F, $8C, $99.
6. In Question 5, add all the values and the checksum byte. What do you get?

PROBLEMS

SECTION 8.1: HCS12 PIN CONNECTION AND BDM

1. The HCS12 chips come in _______-pin ____ packages.
2. Which pin is assigned to BDM?
3. In the HCS12, how many pins are designated as BDM?
4. The crystal oscillator is connected to pins ____ and ____ .
5. If an HCS12 is rated as 20 MHz, what is the maximum frequency that can be connected to it?
6. Indicate the pin number assigned to the Reset of the HCS12 in both packages.
7. BDM stands for _______ .
8. The Reset pin is normally ______ (LOW, HIGH) and needs a _______ (LOW, HIGH) signal to be activated.
9. What are the contents of the PC (program counter) upon reset of the HCS12?
10. What are the contents of the DDRx register upon reset of the HCS12?
11. What is the wait mode in HCS12?
12. True or false. Upon reset, the ports are configured as input.
13. In HCS12, how many pins are set aside for the V_{dd} of PLL?
14. In HCS12, how many pins are set aside for the V_{ss} of PLL (Gnd)?
15. Indicate the types of reset condition we have in HCS12.
16. Of the XTAL and EXTAL pins, which one can be connected to an external clock source?
17. True or false. In the single-chip mode of HCS12, the BDM is already enabled.

18. True or false. The watchdog timer is a component that we must add to the HCS12 externally.
19. True or false. The BDM is a component that we must add to the HCS12 externally.
20. Find the instruction clock cycle for the following crystal frequencies connected to XTAL and EXTAL. Assume we are not using PLL.
    (a) 4 MHz (b) 8 MHz (c) 6 MHz
21. Find the ECLK cycle for the following crystal frequencies connected to XTAL and EXTAL. Assume we are not using PLL.
    (a) 4 MHz (b) 8 MHz (c) 6 MHz
22. At what program memory location does the HCS12 wake up upon reset? What is the implication of that?
23. Write a program to toggle all the bits of PORTB continuously
    (a) using AAH and 55H (b) using the COM instruction.

SECTION 8.2: SETTING THE PLL FREQUENCY IN THE HCS12

24. What is the role of the CLKSEL register in setting the bus frequency?
25. What is the role of the CRGCTL register in setting the PLL frequency?
26. Explain the role of the SYNR register in setting the PLL frequency.
27. Explain the role of the REFDV register in setting the PLL frequency.
28. Assume the XTAL is connected to the 4 MHz crystal oscillator. Show the code for the CRG to get a PLL frequency of 24 MHz.
29. Assume the XTAL is connected to the 8 MHz crystal oscillator. Show the code for the CRG to get a PLL frequency of 32 MHz.

SECTION 8.3: EXPLAINING THE S19 FILE FOR THE HCS12

30. Analyze the six parts of line 1 of Figure 8-12.
31. Verify the checksum byte for line 1 of Figure 8-12. Verify also that the information is not corrupted.
32. Verify the checksum byte for line 2 of Figure 8-12. Verify also that the information is not corrupted.

ANSWERS TO REVIEW QUESTIONS

SECTION 8.1: HCS12 PIN CONNECTION AND BDM

1. BKGD
2. $FFFE
3. $FFFE
4. LOW
5. 4
6. 6 MHz / 2 = 3 MHz and 1 / 3 MHz = 333 ns
7. False

SECTION 8.2: SETTING THE PLL FREQUENCY IN THE HCS12

1. True

2. True
3. True
4. True
5. CLKSEL

SECTION 8.3: EXPLAINING THE S19 FILE FOR THE HCS12

1. False
2. The number of bytes of data in the line
3. The checksum byte of all the bytes in that line
4. 00 means this is not the last line and that more lines of data follow.
5. $22 + $76 + $5F + $8C + $99 = $21C. Dropping the carries we have $1C and its 1's complement, which is $E3.
6. $22 + $76 + $5F + $8C + $99 + $E3 = $2FF. Dropping the carries we have $FF, which means that the data is not corrupted.

CHAPTER 9

HCS12 TIMER PROGRAMMING IN ASSEMBLY AND C

OBJECTIVES

Upon completion of this chapter, you will be able to:

>> List the timers of the HCS12 and their associated registers
>> Describe the free-running feature of the HCS12 timer
>> Describe the output-compare feature of the HCS12 timer
>> Program the output-compare feature in Assembly and C
>> Describe the input-capture feature of the HCS12 timer
>> Program the input-capture feature in Assembly and C
>> Program the HCS12 timers in Assembly and C to generate time delays
>> Program the HCS12 counters in Assembly and C as event counters

Every timer needs a clock pulse to tick. In the microcontroller, the clock source can be internal or external. If we use the internal clock source, then the frequency of the crystal oscillator on the XTAL and EXTAL pins is fed into the timer. Therefore, it is used for time measurement and time delay generation and for that reason is called a *timer*. By choosing the external clock option, we feed pulses through one of the HCS12's pins; this is called a *counter* since it is being used for event counting. The HCS12 timer can be used either as a timer to measure and generate a time delay or as a counter to count events happening outside the microcontroller.

SECTION 9.1: FREE-RUNNING TIMER AND OUTPUT COMPARE FUNCTION

In this section we discuss the HCS12 timer and Output Compare function and show how to program it in Assembly language.

The free-running timer in HCS12

The HCS12 has a single 16-bit timer (counter) called *free-running timer*. See Figure 9-1a. The 16-bit register for this timer is called *TCNT* (timer count). Upon reset, the TCNT starts with zero and continues to count up with each pulse

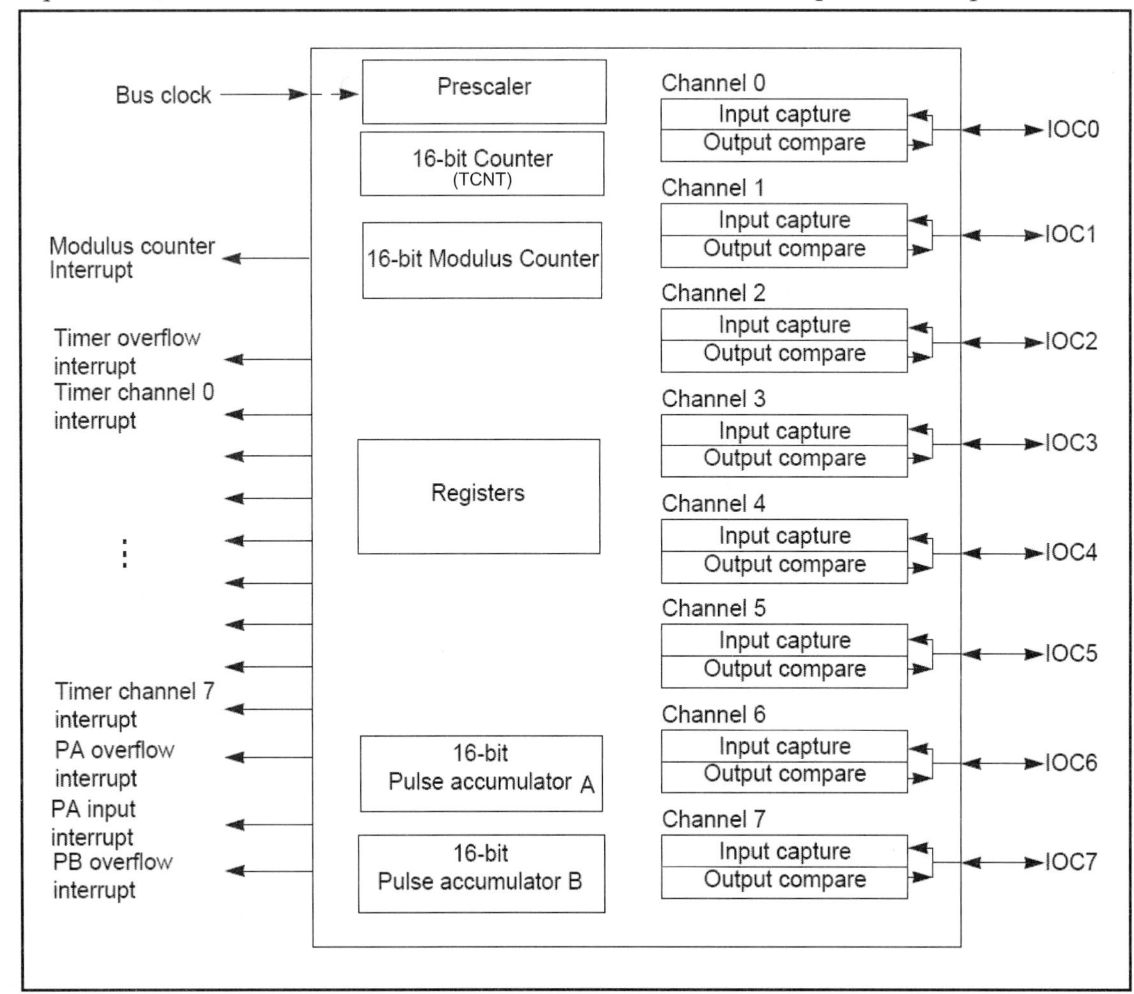

Figure 9-1a. HCS12 Timer Block Diagram (from Freescale)

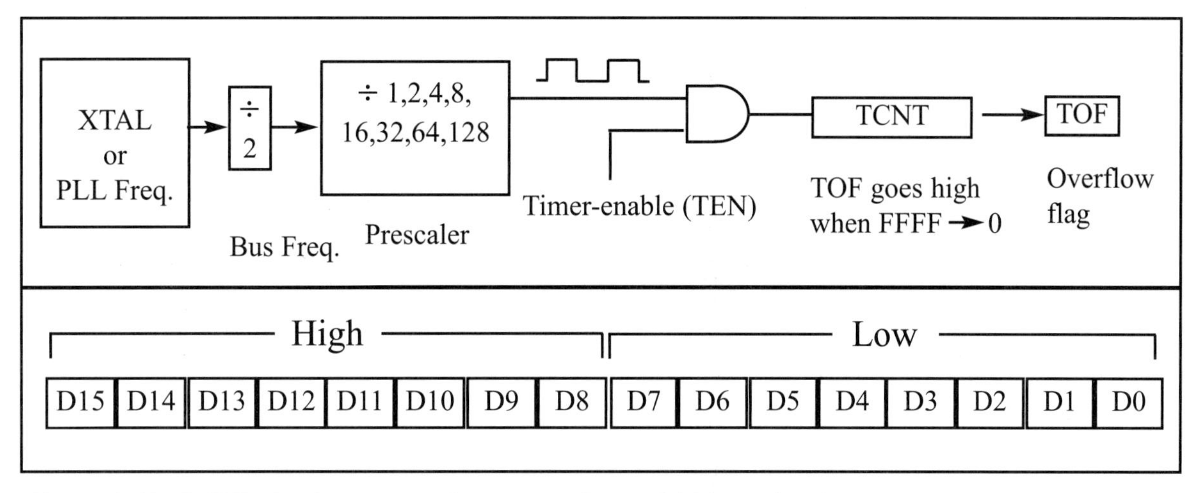

Figure 9-1b. TCNT Register (must be accessed as a 16-bit register)

coming from the oscillator. As we will see, when the timer reaches its maximum value of $FFFF, it rolls over to 0000, and a flag called *timer overflow flag* (TOF) is set to 1. See Figure 9-1b. The free-running timer keeps counting up continuously as long as the HCS12 is powered up, and every time it rolls over from $FFFF to 0 the TOF is set to high. The 16-bit TCNT is read only and must be accessed as a 16-bit register using instructions such as "LDX TCNT" or " LDD TCNT" to avoid inaccurate reading. The TCNT register can be accessed as two separate registers of low byte and high byte, but the reading is not accurate since the timer keeps counting up during the reading of the high byte. That means a separate read for the high byte and the low byte gives us a different result than reading them as a single 16-bit register. Again it must be emphasized that TCNT is a read-only register, so we can read the contents of the TCNT register any time. However, writing to the TCNT is not allowed unless we are in test mode. This is different from other microcontrollers where one can initialize (write to) the timer. We will see how to deal with this issue shortly when we discuss the output-compare feature.

TOF flag bit

The TOF (Timer Overflow Flag) bit is part of the TFLG2 register. See Figure 9-2. As we mentioned before, when the timer reaches its maximum value of $FFFF, it rolls over to 0000, and TOF is set to 1. We can monitor this flag and perform an action such as turning on a port bit when it overflows. See Example 9-1. The strange thing about this flag bit is that in order to clear it for the next round we need to write 1 to it. Indeed this rule applies to all flags of the HCS12 chip. We

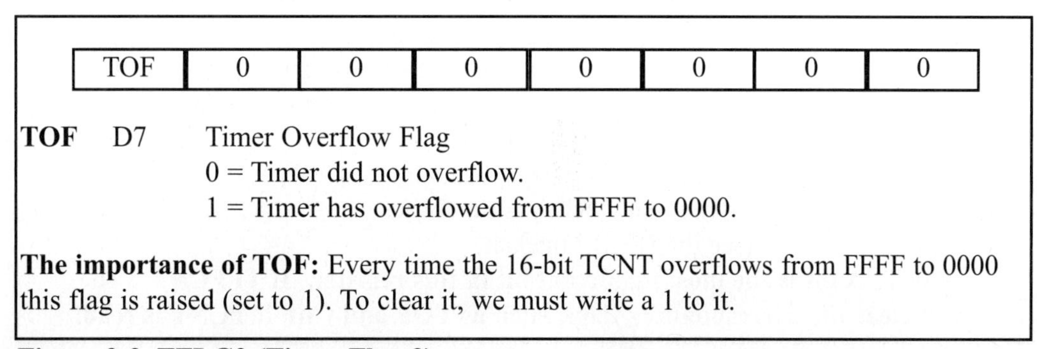

Figure 9-2. TFLG2 (Timer Flag 2)

Example 9-1

Assume TCNT = $FFF2. Explain when the TOF is raised.

Solution:

The timer counts up with the passing of each clock provided by the crystal oscillator. As the timer counts up, it goes through the states of $FFF3, FFF4, FFF5, FFF6, FFF7, FFF8, FFF9, FFFA, FFFB, and so on until it reaches $FFFF. One more clock rolls it to 0, raising the TOF flag (TOF = 1). At that point, the TCNT keeps counting up from 0000 to FFFF; the process continues indefinitely and every time it rolls over the TOF is raised to high. We have no control over the starting point of the free-running timer since it continues to count up upon reset. The only thing we can do is to monitor the TOF and count the number of times it rolls over.

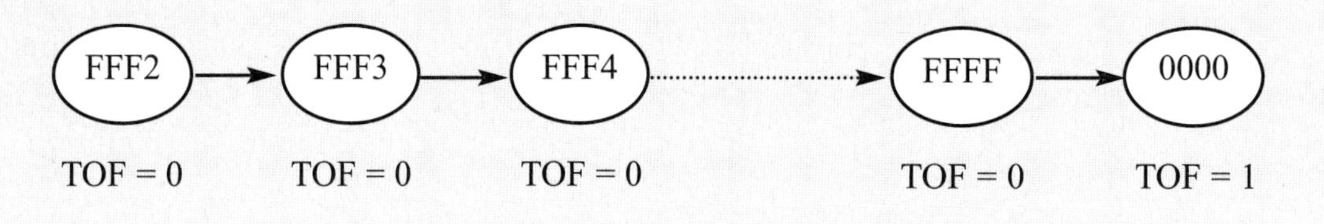

can also use the TOF to cause an interrupt. Chapter 11 shows how to use TOF to cause an interrupt. Next, we describe the prescaler option of the free-running timer.

The system control registers (TSCR1 and TSCR2)

The free-running timer has two control registers called TSCR1 (timer system control reg 1) and TSR2 (timer system control reg 2). The TSCR1 and TSCR2 are 8-bit registers. The bits for TSCR1 and TSCR2 are shown in Figures 9-3 and 9-4, respectively. See Examples 9-2 and 9-3. These registers are widely used for programming the HCS12 timer.

TEN	TSWAI	TSFRZ	TFFCA	0	0	0	0

TEN	D7	Timer enable 0 = Disable timer. Used for reducing the power consumption. 1 = Enable timer. It counts up with each pulse.
TSWAI	D6	Timer stops while in WAIT 0 = Timer continues counting up during the WAIT mode. 1 = Timer stops counting up during the WAIT mode.
TSFRZ	D5	Timer stops during the Freeze mode 0 = Timer continues counting up during the Freeze mode. 1 = Timer stops counting up during the Freeze mode.
TFFCA	D4	Timer Fast Flag Clear All 0 = Clearing the flags manually. 1 = Let the CPU clear the flags. (see the HCS12 manual)

Note: The TEN bit is the most important bit in this register. If TFFCA = 1 the CPU will clear the corresponding flags, such as TOF and CnF if TCNT is read.

Figure 9-3. TSCR1 (Timer System Control Register 1)

TOI	0	0	0	TCRE	PR2	PR1	PR0

TOI D7 — Timer Overflow Interrupt Enable. See Chapter 11.
0 = No hardware interrupt when timer overflows.
1 = Hardware interrupt is requested when timer overflows.

TCRE D6 — Timer Counter Reset Enable. This bit allows the resetting of the Timer/Counter by the Output Compare 7 function.
0 = No resetting of Timer/Counter to zero for Output Compare.
1 = Reset Timer/Counter to zero for Output Compare.

PR20–PR0 D2–D0 — Timer Prescaler Select. These bits are used to set the number of times the bus frequency is divided before it is fed into the timer.

D2	D1	D0	Prescaler factor
0	0	0	1
0	0	1	2
0	1	0	4
0	1	1	8
1	0	0	16
1	0	1	32
1	1	0	64
1	1	1	128

Figure 9-4. TSCR2 (Timer System Control Register 2)

Example 9-2

Find the TSCR1 and TSCR2 values for a) no prescaler, b) count up continuously even during the wait and freeze.

Solution:

From Figures 9-3 and 9-4, we have TSCR1 = 1000000 and TSCR2 = 00000000.

Note: When TSCR1 = 10000000, we must clear the flag bits, such as TOF, by writing 1 to it. When TSCR1 = 10010000, the CPU will clear the flags, such as TOF and CnF, automatically with any reading of the TCNT register. That is the function of TFFCA bit.

Prescaler options of free-running timer

The period of the clock fed to the free-running timer is a function of the crystal frequency connected to the XTAL pins. The crystal frequency (or the E clock, if we use PLL) is divided by 2 before it is fed to the timer. This is the default option when the HCS12 is powered up. We can also use the bits in the TSCR2 register to enable the prescaler option and divide this clock further before it is fed into the free-running timer. The lower 3 bits of the TSCR2 register give the options of the number we can divide by. As shown in Figure 9-4, this number can be 1, 2, 4, 8, 16, 32, 64, and 128. Notice that the lowest factor is 1 and the highest factor is 128. That means the lowest number the XTAL (or the E clock) is divided by is 2 and the highest is 256. See Examples 9-4 and 9-5. Examine Example 9-6 to see how the prescaler options are programmed.

Example 9-3

Write a program to count the number of times the free-running timer overflows (TOF is raised). Display the binary number on PORTB. Show how often the number changes. Assume XTAL = 4 MHz. Use no prescaler.

Solution:

```
       LDAA  #$FF
       STAA  DDRB                  ;make PB an output
       LDAA  #$80
       STAA  TSCR1                 ;enable timer
       LDAA  #$00
       STAA  TSCR2                 ;no interrupt, no prescaler
       LDAA  #0                    ;clear A
OVER   BSET  TFLG2,%10000000       ;TOF=1 to clear it
H1     BRCLR TFLG2,mTFLG2_TOF,H1;stay here until overflow flag is raised
       INCA                        ;increment A every time TCNT overflows
       STAA  PORTB
       BRA   OVER
```

The period for the clock fed to the free-running timer is:

$1/2 \times 4$ MHz $= 2$ MHz and $T = 1/(2$ MHz$) = 0.5$ µs

The free-running timer counts up once every 0.5 µs. It rolls over every 32,768 µs since 65,536 × 0.5 µs = 32,768 µs

To get a more accurate timing, we need to add clock cycles due to the instructions in the loop. **Notice the syntax used to access the bits of HCS12 registers in CodeWarrior.** Examine the content of the header file and you will see the following definition of the bit mask:

```
mTFLG2_TOF: EQU %10000000
```

Example 9-4

Find the timer's clock frequency and its period for various HCS12-based systems, with the following crystal frequencies. Assume that no prescaler is used.

(a) 8 MHz (b) 4 MHz

Solution:

(a) $1/2 \times 8$ MHz $= 4$ MHz for E clock and $T = 1/(4$ MHz$) = 0.25$ µs

(b) $1/2 \times 4$ MHz $= 2$ MHz and $T = 1/(2$ MHz$) = 0.5$ µs

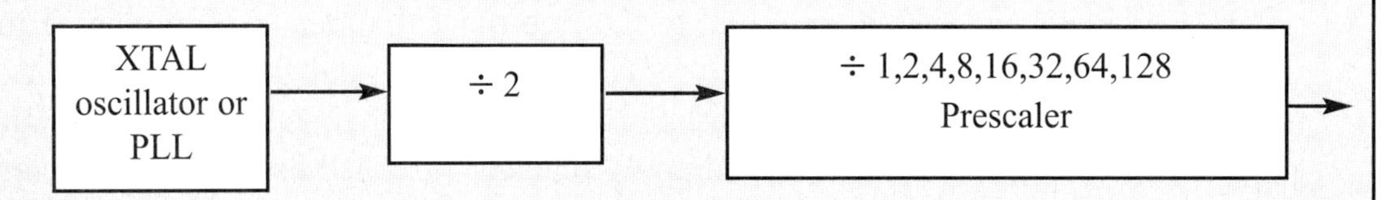

NOTE: HCS12 TIMERS USE 1/2 OF THE CRYSTAL FREQUENCY IN ADDITION TO PRESCALER. IF WE USE PLL, THEN WE MUST USE 1/2 OF THE PLL FREQUENCY IN ADDITION TO THE PRESCALER.

Example 9-5

Find the timer's clock frequency and its period for various HCS12-based systems, with the following crystal frequencies. Assume that a prescaler of 64 is used.
(a) 8 MHz (b) 4 MHz

Solution:

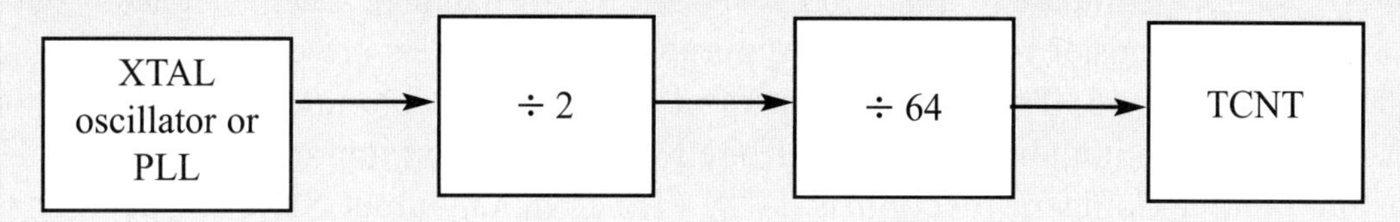

(a) $1/2 \times 8$ MHz = 4 MHz and $1/64 \times 4$ MHz = 62,500 Hz due to 1:64 prescaler and T = 1/62,500 Hz = 16 μs

(b) $1/2 \times 4$ MHz = 2 MHz and $1/64 \times 2$ MHz = 31,250 Hz due to prescaler and T = 1/31,250 Hz = 32 μs

Example 9-6

Repeat Example 9-3 with a prescaler option of 64.

Solution:

```
      LDAA   #$FF
      STAA   DDRB                 ;make PB an output
      LDAA   #$80
      STAA   TSCR1                ;enable timer
      LDAA   #$06
      STAA   TSCR2                ;no interrupt, prescaler of 64
      LDAA   #0                   ;clear A
OVER  BSET   TFLG2,%10000000      ;TOF=1 to clear it
H1    BRCLR  TFLG2,mTFLG2_TOF,H1;stay here until overflow flag is raised
      INCA                        ;increment A every time TCNT overflows
      STAA   PORTB
      BRA    OVER
```

The period for the clock fed to the timer is:
$1/2 \times 4$ MHz = 2 MHz and $1/64 \times 2$ MHz = 31,250 Hz due to prescaler and
T = 1/31,250 Hz = 32 μs. Therefore the counter counts up every $65{,}536 \times 32$ μs = 2,097,152 μs or every 2.10 seconds.

Output Compare function

As we just saw, we have no control over the initialization of the free-running TCNT. In many applications we need to set the initial value and control the exact time. To do that, we must use the Output Compare function of the HCS12 timer. The Output Compare function has its own 16-bit register called TC (timer Compare) and it can be initialized. After the initialization, the TC content is compared with the value in TCNT after each pulse as TCNT counts up. When the values of the TCNT register and TC register are equal, a flag is set high and it will perform some output action such as toggling a bit. The bits that the Output Compare function can manipulate belong to PORTT. That means only bits of PORTT can be used by the Output Compare function.

Output Compare function registers

The Output Compare function in the HCS12 has 8 channels. Each channel has its own 16-bit register for the compare purpose. The registers are called TC0 to TC7. See Figure 9-5. The 16-bit registers of TCs are readable and writable, which means we can initialize them to a desired value. The next important register for the Output Compare function is TIOS (Timer Input/Output Select). See Figure 9-6. Since the timer channel is also used by the Input Capture function, we select the Output Compare option by putting zero for the desired channel. When the Output Compare option is selected, we can monitor the CxF flag to see if the TCNT and TC have matched. The CxF flags are held by the TFLG1 register. See Figure 9-7. We can also use the bits of the TCTL1 and TCTL2 registers to select the kind of action that should be performed on the pin of PORTT. The OL and OM bits decide the action as shown in Figures 9-8 and 9-9.

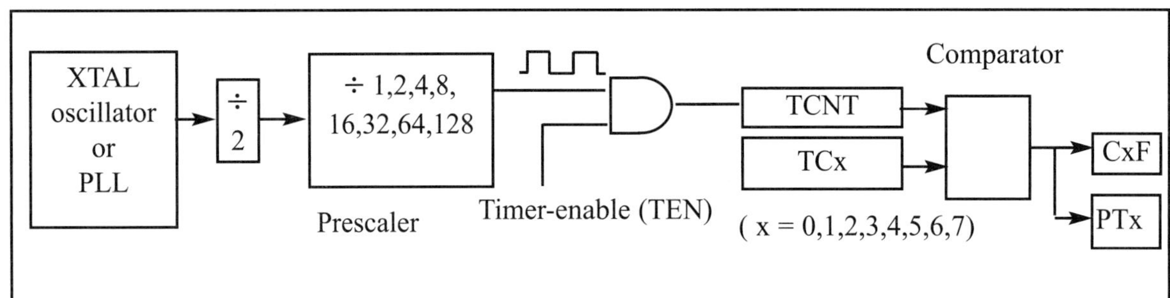

Figure 9-5. Output Compare Function

Programming the Output Compare function

Perform the following steps to program the Output Compare function.

1) Initialize the TSCR1 and TSCR2 registers for the free-running timer.

2) Initialize the TIOS to select the Output Compare option for the desired channel.

3) Initialize the TCTL register bits to select the action you want to perform on the PORTT pin for the desired channel.

4) Read the TCNT of the free-running timer. Remember, you can read but not write to it.

5) Add the desired number of pulses to the value read from TCNT and load it into the TC register of your choice (TC0, TC1, or TC7)

6) Monitor the Cxf flag in TFLG1 to see if the TCNT and TC registers match.

7) Repeat the action by going back to step 4.

IOS7	IOS6	IOS5	IOS4	IOS3	IOS2	IOS1	IOS0

IOS0–IOS7 D7–D0 Input Capture or Output Compare channels 0–7 selection
0 = The corresponding channel acts as an Input Capture.
1 = The corresponding channel acts as Output Compare.

Figure 9-6. TIOS (Timer Input /Output Select) Register

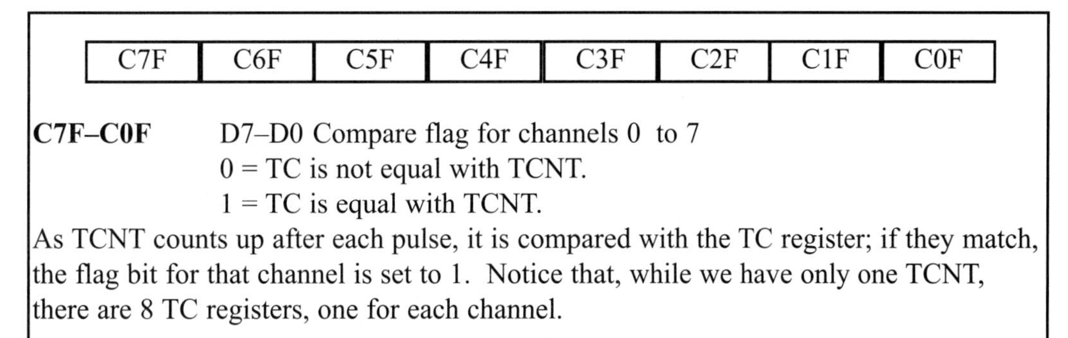

C7F	C6F	C5F	C4F	C3F	C2F	C1F	C0F

C7F–C0F D7–D0 Compare flag for channels 0 to 7
0 = TC is not equal with TCNT.
1 = TC is equal with TCNT.

As TCNT counts up after each pulse, it is compared with the TC register; if they match, the flag bit for that channel is set to 1. Notice that, while we have only one TCNT, there are 8 TC registers, one for each channel.

Figure 9-7. TFLG1 (Timer Flag 1) Register

OM7	OL7	OM6	OL6	OM5	OL5	OM4	OL4

OMx–OLx Selects the action to be performed by channels 4–7 on the PORTT pin

OM	OL	
0	0	Timer disconnected from output pin
0	1	Toggle the pin
1	0	Make output pin 0
1	1	Make output pin 1

When the TCNT matches the TCx register, one of the above actions is performed in addition to making the flag bit high for that channel.

Figure 9-8. TCTL1 (Timer Control 1) Register

OM3	OL3	OM2	OL2	OM1	OL1	OM0	OL0

OMx–OLx Selects the action to be performed by channels 0–3 on the PORTT pin

OM	OL	
0	0	Timer disconnected from output pin
0	1	Toggle the pin
1	0	Make output pin 0
1	1	Make output pin 1

When the TCNT matches the TCx register, one of the above actions is performed in addition to making the flag bit high for that channel.

Figure 9-9. TCTL2 (Timer Control 2) Register

Example 9-7

In the following program we use channel 2 to toggle PT2 once every 500 timer pulses. Find the frequency of the square wave generated by the following program if XTAL = 4 MHz.

```
      LDAA  #$90
      STAA  TSCR1                ;enable timer, and let CPU clear C2F
      LDAA  #$00
      STAA  TSCR2                ;no interrupt, no prescaler
      BSET  TIOS,%00000100       ;select chan 2 for output compare
      LDAA  #$10
      STAA  TCTL2                ;toggle PT2 pin upon match
OVER  LDD   TCNT                 ;read current TCNT value
      ADDD  #500                 ;we need 500 pulses
      STD   TC2                  ;load the TC2 (Timer Compare 2)
HERE  BRCLR TFLG1,mTFLG1_C2F,HERE     ;stay here for chan 2 match
      BRA   OVER
```

Solution:

The period for the clock fed to the timer TCNT is:

1/2 × 4 MHz = 2 MHz and T = 1/(2 MHz) = 0.5 μs

The period for the pulses on PT2 is:

T = 2 × (500 × 0.5 μs) = 500 μs and Freq. = 1/500 μs = 2,000 Hz.

Notice that in the above equation we double the value since the period must include the high and low portions of the pulse. Use an oscilloscope to view and verify the square wave on PT2.

Note: Depending on the XTAL/PLL frequency the LED may appear always on. To slow down the flashing of the LED, set the prescaler to 128 and change the value to 5000.

Example 9-8

Assuming that XTAL = 4 MHz, write a program to generate a square wave with a period of 10 ms on pin PT4.

Solution:

For a square wave with T = 10 ms we must have a time delay of 5 ms. Because XTAL = 4 MHz, the counter counts up every 0.5 μs. This means that we need 5 ms / 0.5 μs = 10,000 clocks.

```
      LDAA  #$80                 ;enable timer and TFFCA = 1 which means
      STAA  TSCR1                ;we must manually clear C4F
      LDAA  #$00
      STAA  TSCR2                ;no interrupt, no prescaler
      BSET  TIOS,%00010000       ;select chan 4 for output compare
      LDAA  #$01
      STAA  TCTL1                ;toggle PT4 pin upon match
OVER  LDD   TCNT                 ;read current TCNT value
      ADDD  #10000               ;we need 10000 pulses
      STD   TC4                  ;load the TC4 (Timer Compare 4)
HERE  BRCLR TFLG1,mTFLG1_C4F,HERE     ;stay here for chan 4 match
      BSET  TFLG1,%00010000      ;clear chan 4 match flag
      BRA   OVER
```

Example 9-9

Modify Example 9-8 to get the largest time delay possible without using a prescaler. Find the delay in ms. In your calculation, exclude the overhead due to the instructions in the loop.

Solution:

To get the largest delay without using a prescaler, it needs to go through a total of 65,536 states. Therefore, we have delay = 65,536 $\times$ 0.5 μs = 32.768 ms. That gives us the smallest frequency of 1 / (2 × 32.768 ms) = 1 / 65.536 ms = 15.25 Hz.

```
      LDAA   #$80
      STAA   TSCR1              ;enable timer
      LDAA   #$00
      STAA   TSCR2              ;no interrupt, no prescaler
      BSET   TIOS,%00010000     ;select chan 4 for output compare
      LDAA   #$01
      STAA   TCTL1              ;toggle PT4 pin upon match
OVER  LDD    TCNT               ;read current TCNT value
      ADDD   #65535             ;we need 65535 pulses
      STD    TC4                ;load the TC4 (Timer Compare 4)
HERE  BRCLR TFLG1,mTFLG1_C4F,HERE   ;stay here for chan 4 match
      BSET   TFLG1,%00010000    ;clear chan 4 match flag
      BRA    OVER
```

Example 9-10

Assuming XTAL = 4 MHz, modify the program in Example 9-8 to toggle PT7 every 1/2 second.

Solution:

0.5 s / 0.5 μs = 1,000,000. That means we must use a prescaler. Using a prescaler of 16 we get 1,000,000/16 = 62,500.

```
      LDAA   #$80
      STAA   TSCR1              ;enable timer
      LDAA   #$04
      STAA   TSCR2              ;no interrupt, prescaler of 16
      BSET   TIOS,%10000000     ;select chan 7 for output compare
      LDAA   #$40
      STAA   TCTL1              ;toggle PT7 pin upon match
OVER  LDD    TCNT               ;read current TCNT valuc
      ADDD   #62500             ;we need 62500 pulses
      STD    TC7                ;load the TC7 (Timer Compare 7)
HERE  BRCLR TFLG1,mTFLG1_C7F,HERE   ;stay here for chan 7 match
      BSET   TFLG1,%10000000    ;clear chan 7 match flag
      BRA    OVER
```

Another way to verify the above value is as follows:
1/2 × 4 MHz = 2 MHz and 1/16 × 2 MHz = 125,000 Hz due to the prescaler and T = 1/125,000 Hz = 8 μs for TCNT. Now, 0.5 s / 8 μs = 62,500

Example 9-11

Assuming XTAL = 4 MHz, write a program to generate a square wave of 50 Hz frequency on pin PT6. Use channel 6 and a prescaler of 128 for the TCNT.

Solution:

Look at the following steps:

(a) $1/2 \times 4$ MHz = 2 MHz and $1/128 \times 2$ MHz = 15,625 Hz and T = 1/15,625 Hz = 64 μs

(b) T = 1 / 50 Hz = 20 ms, the period of the square wave.

(c) 1/2 of it for the high and low portions of the pulse = 10 ms

(d) 10 ms / 64 μs = 156

```
      LDAA  #$80
      STAA  TSCR1                ;enable timer
      LDAA  #$07
      STAA  TSCR2                ;no interrupt, prescaler of 128
      BSET  TIOS,%01000000       ;select chan 6 for output compare
      LDAA  #$10
      STAA  TCTL1                ;toggle PT6 pin upon match
OVER  LDD   TCNT                 ;read current TCNT value
      ADDD  #156                 ;we need 156 pulses
      STD   TC6                  ;load the TC6 (Timer Compare 6)
HERE  BRCLR TFLG1,mTFLG1_C6F,HERE   ;stay here for chan 6 match
      BSET  TFLG1,%01000000      ;clear chan 6 match flag
      BRA OVER
```

Example 9-12

Assuming XTAL = 4 MHz, write a program to toggle all bits of PORTB every second. Use channel 0 and a prescaler of 128 for the TCNT.

Solution:

Look at the following steps:

(a) $1/2 \times 4$ MHz = 2 MHz and $1/128 \times 2$ MHz = 15,625 Hz and T = 1/15,625 Hz = 64 μs

(b) 1 s / 64 μs = 15,625

```
      LDS   #RAMEnd+1            ;stack pointer = last RAM location + 1
      BSET  DDRB,%11111111          ;make PORTB an output
      LDAA  #$90
      STAA  TSCR1                ;enable timer and let CPU clear C0F
      LDAA  #$07
      STAA  TSCR2                ;no interrupt, prescaler of 128
      BSET  TIOS,%00000001       ;select chan 0 for output compare
      LDAA  #$00
      STAA  TCTL2                ;timer disconnected from pin
      LDAA  #$55
OVER  STAA  PORTB
      JSR   DELAY
      COMA                       ;complement reg. A
      BRA   OVER
;-----------------ONE SEC DELAY
DELAY PSHA
   LDD TCNT              ;read current TCNT value
      ADDD  #15625                 ;we need 15625 pulses
      STD   TC0                  ;load the TC0 (Timer Compare 0)
HERE  BRCLR TFLG1,mTFLG1_C0F,HERE   ;stay here for chan 0 match
      PULA
      RTS
```

Example 9-13

Assuming XTAL = 4 MHz, write a program to generate a square wave with a period of 10 ms on pin PT4. Make it 60% duty cycle.

Solution:

For a square wave with T = 10 ms and 60% duty cycle we need 6 ms for the high portion and 4 ms for the low portion of the pulse. Because XTAL = 4 MHz, the counter counts up every 0.5 μs. This means that we need 6 ms / 0.5 μs = 12,000 clocks for high and 4 ms / 0.5 μs = 8,000 clocks for low.

```
      LDS    #RAMEnd+1           ;initialize the stack pointer
      LDAA   #$80
      STAA   TSCR1               ;enable timer
      LDAA   #$00
      STAA   TSCR2               ;no interrupt, no prescaler
      BSET   TIOS,%00010000      ;select chan 4 for output compare
OVER  JSR    HI
      JSR    LO
      BRA    OVER
;-----------------high part of the pulse
HI    LDAA   #$02
      STAA   TCTL1               ;make PT4 pin Low upon match
      LDD    TCNT                ;read current TCNT value
      ADDD   #12000              ;we need 12000 pulses
      STD    TC4                 ;load the TC4 (Timer Compare 4)
HERE1 BRCLR TFLG1,mTFLG1_C4F,HERE1  ;stay here for chan 4 match
      BSET   TFLG1,%00010000     ;clear chan 4 match flag
      RTS
;-----------------low part of the pulse
LO    LDAA   #$03
      STAA   TCTL1               ;make PT4 pin High upon match
      LDD    TCNT                ;read current TCNT value
      ADDD   #8000               ;we need 8000 pulses
      STD    TC4                 ;load the TC4 (Timer Compare 4)
HERE2 BRCLR TFLG1,mTFLG1_C4F,HERE2  ;stay here for chan 4 match
      BSET   TFLG1,%00010000     ;clear chan 4 match flag
      RTS
```

See Examples 9-7 through 9-13 to see how the Output Compare function works.

Notice that in many of the time delay calculations, we have ignored the clocks caused by the overhead instructions in the loop. To get a more accurate time delay, and hence frequency, you need to include them. If you use a digital scope and you don't get exactly the same frequency as we have calculated, it is because of the overhead associated with those instructions.

Review Questions

1. How many free-running timers do we have in the HCS12?
2. True or false. TCNT can be used only as a 16-bit timer.
3. True or false. TCNT can be accessed only as a 16-bit value.

4. True or false. The TFLG2 register is a bit-addressable register.
5. Indicate the selection made if TSCR2 = $80.
6. In TCNT, the counter rolls over when the counter goes from ____ to ____.
7. The TOF is raised when TCNT rolls over from ____ to ____.
8. In the selection for TSCR2 = $06, find the value for the prescaler.
9. To get a 100 μs delay, what number should be loaded into the TC4 register using no prescaler? Assume that XTAL = 4 MHz.

SECTION 9.2: INPUT CAPTURE PROGRAMMING

In the last section, we used the timers of the HCS12 to generate time delays. The HCS12 timer can also be used to count, detect, and measure the time of events happening outside the HCS12 chip. The use of the timer as an event detector and counter is covered in this section.

Programming the Input Capture function

The Input Capture function is widely used for many applications. Among them are a) recording the arrival-time of an event, b) pulse width measurement, and c) period measurement. In the last section we saw how the Output Control function is used to control the pins of PORTT. Many of the Output Compare function registers are also used by the Input Capture function to detect and measure the events happening outside the chip. Since the Input Capture and Output Compare functions use many of the same registers to do their job, we must use the TIOS register to select the function. As shown in Figure 9-6, we use 0 or 1 in the TIOS register to select Input Capture or Output Compare. There are two more important registers dedicated to the Input Capture function and they are called TCTL3 and TCTL4. As shown in Figures 9-10 and 9-11, we use these registers to select the channel and the type of edge detection. See Example 9-14. Just like Output Compare, Input Capture also uses the free-running timer (TCNT) to measure the time. Upon detection of an edge, the TCNT value is loaded into the TCx register to record the arrival of a pulse edge. Just as with Output Compare, we use the PORTT pins for Input Capture. The only difference is that we must make the

EDG7B	EDG7A	EDG6B	EDG6A	EDG5B	EDG5A	EDG4B	EDG4A

EDGnB–EDGnA Selects the edge detection for the Input Capture function

EDGnB	EDGnA	
0	0	Capture disabled
0	1	Capture on rising edge only
1	0	Capture on falling edge only
1	1	Capture on any edge (rising or falling)

When an edge is detected, the TCNT is loaded into the TCx register. It also raises the CxF (Capture flag) for that channel. Notice that the Compare flag is also used for the Capture flag.

Figure 9-10. TCTL3 (Timer Control 3) Register

EDG3B	EDG3A	EDG2B	EDG2A	EDG1B	EDG1A	EDG0B	EDG0A

EDGnB – EDGnA Selects the edge detection for the Input Capture function

EDGnB	EDGnA	
0	0	Capture disabled
0	1	Capture on rising edge only
1	0	Capture on falling edge only
1	1	Capture on any edge (rising or falling)

When an edge is detected, the TCNT is loaded into the TCx register. It also raises the CxF (Capture flag) for that channel. Notice that the Compare flag is also used for the Capture flag.

Figure 9-11. TCTL4 (Timer Control 4) Register

Example 9-14

Find a) TCTL3 for channel 6 on rising edge only, b) TCTL4 for channel 1 on falling edge only.

Solution:

From Figures 9-10 and 9-11, we have TCTL3 = 00010000 and TCTL4 = 00001000.

PORTT pin an input pin (using TIOS and TCTL3 or 4) since we are feeding it the pulse whose period is being measured.

Steps to program the Input Capture function

Perform the following steps to measure the edge arrival time for the Input Capture function.

1) Initialize the TSCR1 and TSCR2 registers for the free-running timer.

2) Initialize the TIOS to select the Input Capture option for the desired channel.

3) Initialize the TCTL3 or 4 register bits to select the edge we want to measure the arrival time for.

4) Monitor the the Cxf flag in TFLG1 to see if the edge has arrived. Upon the arrival of the edge, the TCNT is loaded into the TCx register automatically by the HCS12.

5) Save the TCx.

6) Monitor the Cxf flag in TFLG1 to see if the second edge has arrived. Upon the arrival of the edge, the TCNT is loaded into the TCx register automatically by the HCS12.

7) Save the TCx for the second edge. Subtract the second edge value from the first edge value to get the time.

See Example 9-15 to see how the Input Capture function works. The Input

Example 9-15

Assuming that clock pulses are fed into pin PT6, write a program for channel 6 to read the TCNT value on every rising edge. Place the result on PORTA and PORTB. Assume XTAL = 4 MHz.

Solution:

```
       LDAA   #$FF
       STAA   DDRA                ;PORTA as output
       STAA   DDRB                ;PORTB as output
       LDAA   #$90
       STAA   TSCR1               ;enable timer
       LDAA   #$0
       STAA   TSCR2               ;no interrupt, no prescaler
       BCLR   TIOS,%01000000      ;select chan 6 for input capture
       LDAA   #%00010000
       STAA   TCTL3         ;select channel 6 for capture on rising edge
HERE1  BRCLR  TFLG1,mTFLG1_C6F,HERE1  ;stay here for chan 6 rising edge
       LDD    TC6   ;get TCNT value which was loaded into TC6
       STAB   PORTB ;display the low byte on PORTB
       STAA   PORTA ;display the high byte on PORTA
       BRA    HERE1
```

HCS12
PORTA
PORTB
PT6
Pulses
to LEDs

Note: Upon the detection of each rising edge, the TCNT value is loaded into the TC6. Also notice that by reading the TC6 value we clear the C6F flag bit since TSCR1 =.$90.

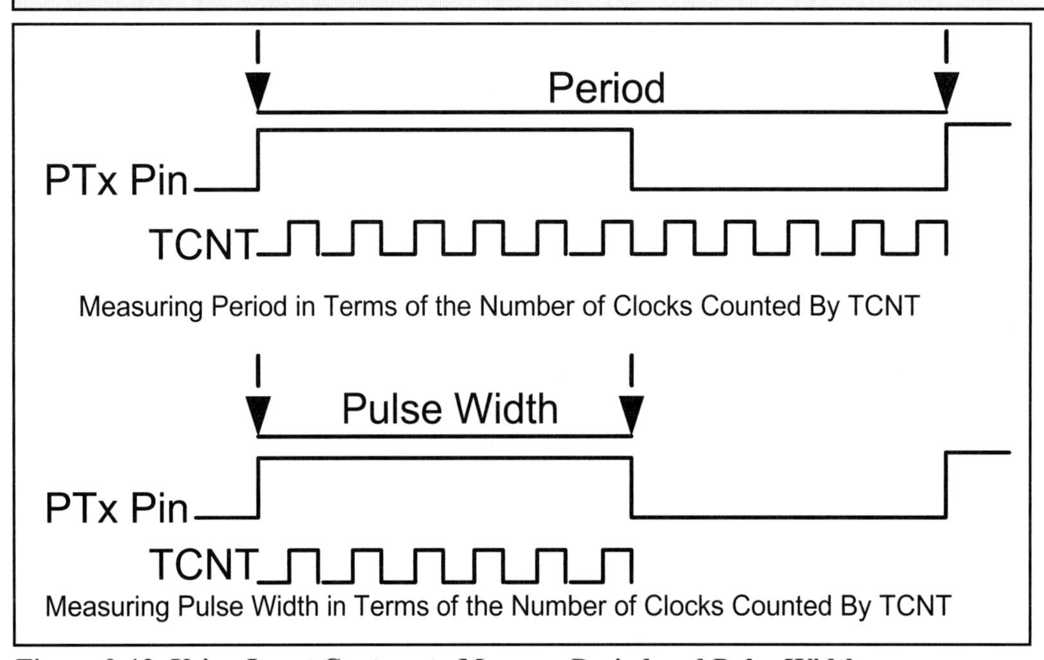

Figure 9-12. Using Input Capture to Measure Period and Pulse Width

Example 9-16

Assuming that clock pulses are fed into pin PT6, write a program for channel 6 to measure the period of the pulses. Place the binary result on PORTA and PORTB. Assume XTAL = 4 MHz.

Solution:

```
       ORG    $1000
R_edge_1        DS.W 1
       LDAA   #$FF
       STAA   DDRA                 ;PORTA as output
       STAA   DDRB                 ;PORTB as output
       LDAA   #$80
       STAA   TSCR1                ;enable timer
       LDAA   #$0
       STAA   TSCR2                ;no interrupt, no prescaler
       BCLR   TIOS,%01000000       ;select chan 6 for input capture
       LDAA   #%00010000
       STAA   TCTL3          ;select channel 6 for capture on rising edge
       BSET   TFLG1,%01000000      ;clear chan 6 input capture flag
HERE1  BRCLR  TFLG1,mTFLG1_C6F,HERE1   ;stay here for chan 6 rising edge
       LDD    TC6    ;get TCNT value which was loaded into TC6
       STD    R_edge_1
       BSET   TFLG1,%01000000      ;clear chan 6 input capture flag
HERE2  BRCLR  TFLG1,mTFLG1_C6F,HERE2   ;stay here for the second edge
       LDD    TC6    ;read TCNT value for second rising edge
       SUBD   R_edge_1       ;D = D-(R_edge_1)
       STAA   PORTA
       STAB   PORTB
       BRA $
```


Capture function is widely used to measure the period or the pulse width of an incoming signal. See Figure 9-12 and Examples 9-16 through 9-19 to see how it is done.

Review Questions

1. How many channels of Input Capture do we have in the HCS12?
2. True or false. TCNT is also used by the Input Capture function.
3. True or false. The Output Compare function can also be used for the Input Capture function.
4. Indicate the registers used by the Input Capture function.
5. True or false. The Input Capture function can capture the timing of an incoming pulse on the rising edge only.

Example 9-17

Assume that a 75 Hz pulse is connected to the input for channel 6 (pin PT6). Write a program to measure its period and verify the frequency. Use the prescaler value that gives the result in a single byte. Display the result on PORTB. Assume XTAL = 4 MHz.

Solution:

The frequency of 75 Hz gives us the period of 1/75 Hz = 13.3 ms. Now, $1/2 \times 4$ MHz = 2 MHz and $1/128 \times 2$ MHz = 15,625 Hz due to prescaler and T = 1/15,625 Hz = 64 μs for TCNT. That means we get the value of 208 (1101 0000 binary) on PORTB since 13.3 ms / 64 μs = 208

```
      ORG   $1000
R_edge_1       DS.W 1
      LDAA  #$FF
      STAA  DDRB                ;PORTB as output
      LDAA  #$80
      STAA  TSCR1               ;enable timer
      LDAA  #$07
      STAA  TSCR2               ;no interrupt, prescaler 128
      BCLR  TIOS,%01000000      ;select chan 6 for input capture
      LDAA  #%00100000
      STAA  TCTL3         ;select channel 6 for capture on rising edge
      BSET  TFLG1,%01000000    ;clear chan 6 input capture flag
HERE1 BRCLR TFLG1,mTFLG1_C6F,HERE1  ;stay here for chan 6 rising edge
      LDD   TC6   ;get TCNT value which was loaded into TC6
      STD   R_edge_1
      BSET  TFLG1,%01000000    ;clear chan 6 input capture flag
HERE2 BRCLR TFLG1,mTFLG1_C6F,HERE2  ;stay here for the second edge
      LDD   TC6   ;read TCNT value for second rising edge
      SUBD  R_edge_1     ;D = D-(R_edge_1)
      STAB  PORTB
      BRA $
```

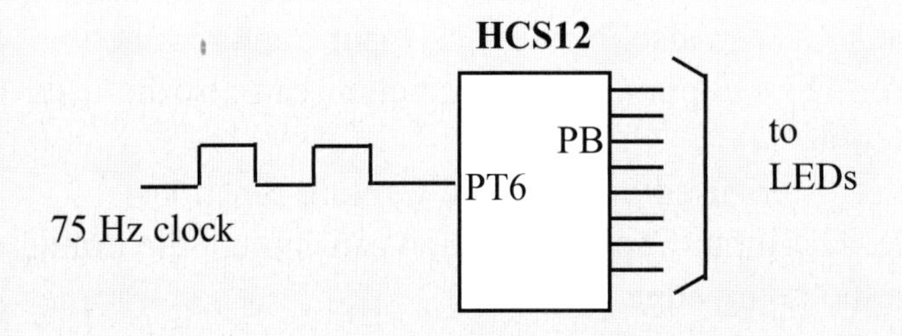

Example 9-18

Assume that a 75 Hz frequency pulse is connected to input for channel 6 (pin PT6). Write a program to measure its pulse width. Use the prescaler value that gives the result in a single byte. Display the result on PORTB. Assume XTAL = 4 MHz.

Solution:

The frequency of 75 Hz gives us the period of 1/75 Hz = 13.3 ms. Now, $1/2 \times 4$ MHz = 2 MHz and $1/128 \times 2$ MHz = 15,625 Hz due to the prescaler and T = 1/15,625 Hz = 64 μs for TCNT. That means we get the value of 208 (1101 0000 binary) for the period since 13.3 ms / 64 μs = 208. Now the pulse width can be anywhere from 1 to 207 clock ticks.

```
      ORG   $1000
R_edge_1       DS.W 1
      LDAA  #$FF
      STAA  DDRB                ;PORTB as output
      LDAA  #$80
      STAA  TSCR1               ;enable timer
      LDAA  #$07
      STAA  TSCR2               ;no interrupt, prescaler 128
      BCLR  TIOS,%01000000      ;select chan 6 for input capture
      LDAA  #%00010000
      STAA  TCTL3         ;select channel 6 for capture on rising edge
      BSET  TFLG1,%01000000     ;clear chan 6 input capture flag
HERE1 BRCLR TFLG1,mTFLG1_C6F,HERE1  ;stay here for chan 6 rising edge
      LDD   TC6    ;get TCNT value which was loaded into TC6
      STD   R_edge_1
      LDAA  #%00100000
      STAA  TCTL3         ;select channel 6 for capture on falling edge
      BSET  TFLG1,%01000000     ;clear chan 6 input capture flag
HERE2 BRCLR TFLG1,mTFLG1_C6F,HERE2  ;stay here for the second edge
      LDD   TC6    ;read TCNT value for second rising edge
      SUBD  R_edge_1      ;D = D-(R_edge_1)
      STAB  PORTB
      BRA $
```

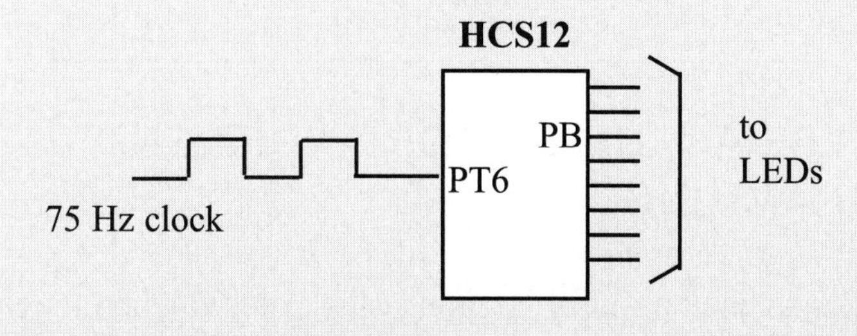

Example 9-19

Assume that a temperature sensor is connected to pin PT6. The temperature provided by the sensor is proportional to pulse width and is in the range of 1 μs to 250 μs. Write a program to measure the temperature if 1 μs is equal to 1 degree. Use the prescaler value that gives the result in a single byte. Display the result on PORTB. Assume XTAL = 4 MHz.

Solution:

1/2 × 4 MHz = 2 MHz and 1/2 × 2 MHz = 1,000,000 Hz due to the prescaler and T = 1/1,000,000 Hz = 1 μs for TCNT. That means we get values from 1 to 65,536 μs for the TCNT, but since the pulse width never goes beyond 250 μs we should be able to display the temperature value on PORTB.

```
      ORG    $1000
R_edge_1       DS.W 1
      LDAA   #$FF
      STAA   DDRB                ;PORTB as output
      LDAA   #$80
      STAA   TSCR1               ;enable timer
      LDAA   #$07
      STAA   TSCR2               ;no interrupt, prescaler 128
      BCLR   TIOS,%01000000      ;select chan 6 for input capture
      LDAA   #%00010000
      STAA   TCTL3          ;select channel 6 for capture on rising edge
      BSET   TFLG1,%01000000     ;clear chan 6 input capture flag
HERE1 BRCLR  TFLG1,mTFLG1_C6F,HERE1  ;stay here for chan 6 rising edge
      LDD    TC6    ;get TCNT value which was loaded into TC6
      STD    R_edge_1
      LDAA   #%00100000
      STAA   TCTL3          ;select channel 6 for capture on falling edge
      BSET   TFLG1,%01000000     ;clear chan 6 input capture flag
HERE2 BRCLR  TFLG1,mTFLG1_C6F,HERE2  ;stay here for the second edge
      LDD    TC6    ;read TCNT value for second rising edge
      SUBD   R_edge_1       ;D = D-(R_edge_1)
      STAB   PORTB
      BRA $
```

HCS12

PB

PT6

to LEDs

variable pulse width

SECTION 9.3: PULSE ACCUMULATOR AND EVENT COUNTER PROGRAMMING

The HCS12 timer has the option of counting clock pulses when they come from a source outside the HCS12 chip. These clock pulses could represent the number of people passing through an entrance, or the number of wheel rotations, or any other event that can be converted to pulses. For this reason it is commonly referred to as an *event counter* in microcontroller literature. The HCS12 calls this function *Pulse*

Accumulator. There are four 8-bit registers for Pulse Accumulator in the HCS12 and they are designated as Pulse Accumulator numbers 0–3 (PACN0–PACN3). See Figure 9-13. There is also a pin associated with each of the PACN0–PACN3 registers where pulses can be fed into the register. The pins that can be used for feeding pulses into HCS12 Pulse Accumulators are PT0–PT3, where PACN0 uses the PT0 pin, PACN1 uses PT1, and so on. We use the ICPAR register to enable the 8-bit pulse accumulator. See Figure 9-14. Notice that the pin belongs to PORTT, just like the Output Compare/Input Capture functions. Pulses can be counted on a rising edge or a falling edge. The edge selection of the 8-bit Pulse Accumulator is done using the TCTL4 register. Example 9-20 shows how to use the 8-bit Pulse Accumulator PAC3 to count pulses being fed to the HCS12. We can make a 16-bit pulse counter by combining two 8-bit registers together, as we will see next.

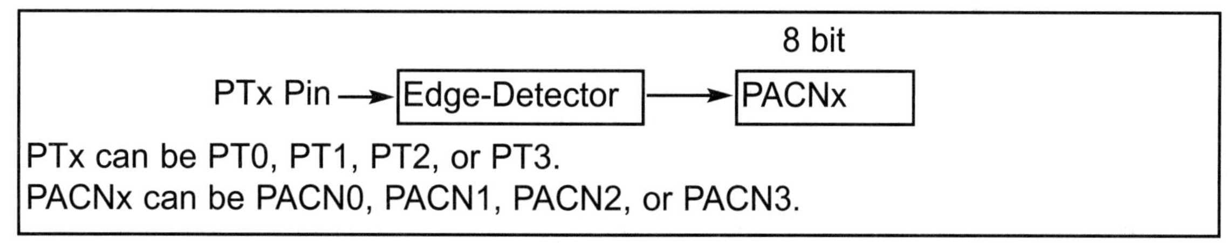

Figure 9-13. 8-bit Pulse Accumulator (Counter)

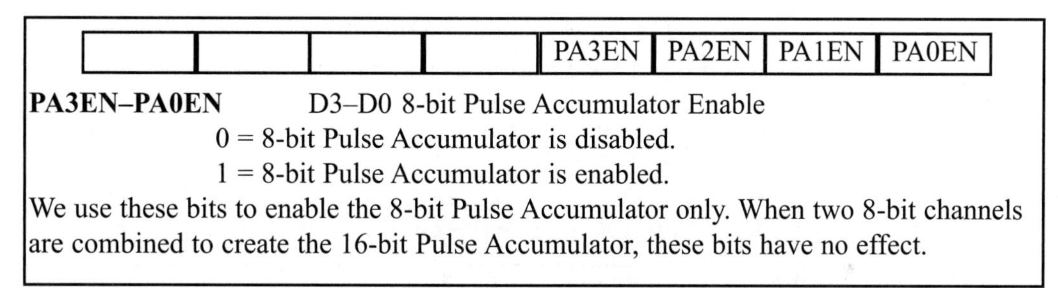

Figure 9-14. ICPAR (Input Control Pulse Accumulator) Register

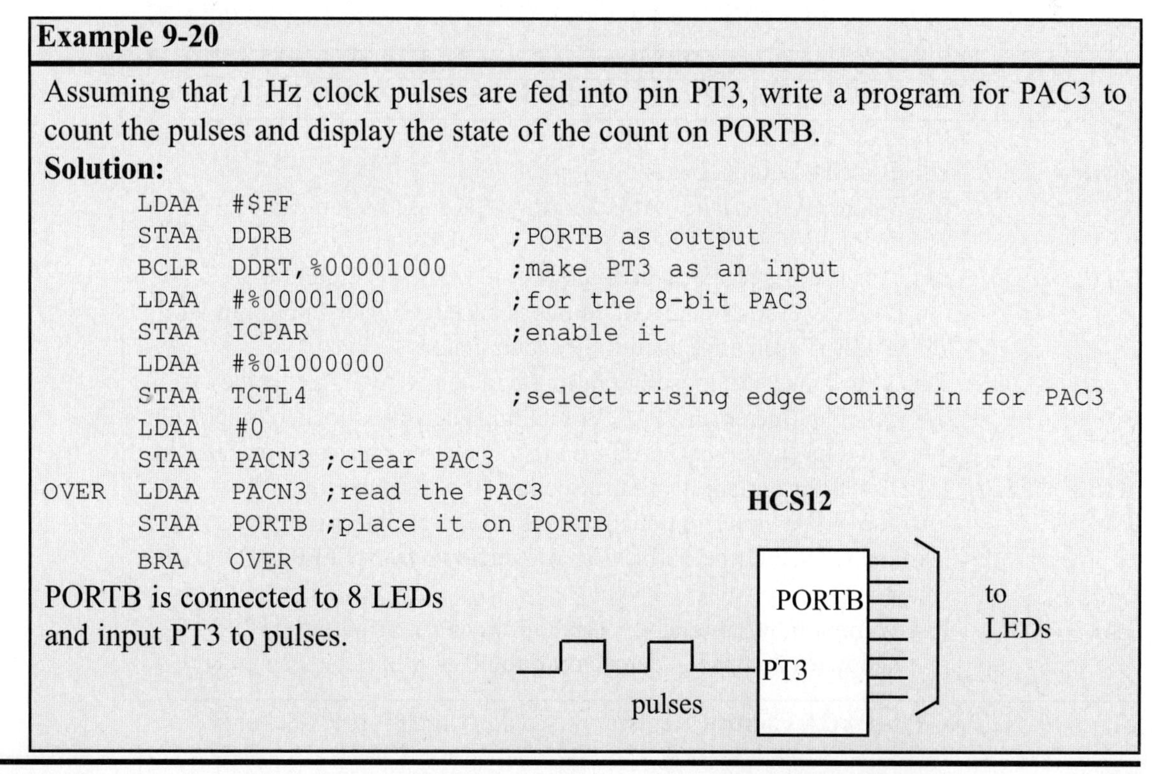

Example 9-20

Assuming that 1 Hz clock pulses are fed into pin PT3, write a program for PAC3 to count the pulses and display the state of the count on PORTB.

Solution:

```
       LDAA  #$FF
       STAA  DDRB              ;PORTB as output
       BCLR  DDRT,%00001000    ;make PT3 as an input
       LDAA  #%00001000        ;for the 8-bit PAC3
       STAA  ICPAR             ;enable it
       LDAA  #%01000000
       STAA  TCTL4             ;select rising edge coming in for PAC3
       LDAA  #0
       STAA  PACN3 ;clear PAC3
OVER   LDAA  PACN3 ;read the PAC3
       STAA  PORTB ;place it on PORTB
       BRA   OVER
```

PORTB is connected to 8 LEDs and input PT3 to pulses.

Programming 16-bit Pulse Accumulator A

In the HCS12 we can have two 16-bit pulse accumulators. They are designated as PACA (Pulse Accumulator A) and PACB (Pulse Accumulator B). We first describe the PACA, and the PACB will be examined shortly. The Pulse Accumulator A (PACA) is formed by cascading the 8-bit PACN2 and PACN3, where the PACN3 is the upper 8-bit block and PACN2 is the lower 8-bit block. See Figure 9-15. We can access the entire 16-bit PACA using an instruction such as "LDD PACA". The PACA has its own control register called PACTL. See Figure 9-16. The PACA counts the pulses fed to it via pin PT7. As it can bc seen from the PEDGE bit of PACTL in Figure 9-16, we can choose the count-up to be done on the rising edge or the falling edge of the pulses fed to pin PT7. See Example 9-21. The PACA can also be used as a gated time accumulator instead of an event counter. See the HCS12 manual.

We monitor the PACA overflow flag using the PAFLG register. See Figure 9-17. In Example 9-22, we use PACA as an event counter that counts up as clock pulses are fed into pin PT7. Examples 9-23 through 9-26 show how timers are used as counters.

As another example of the application of the event counter, we can feed an external square wave of 60 Hz frequency into the timer. The program will generate the second, the minute, and the hour out of this input frequency and display the result on an LCD. This will be a nice digital clock, but not a very accurate one.

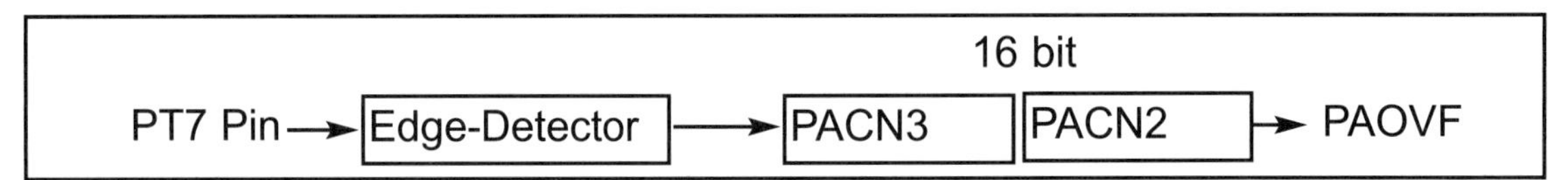

Figure 9-15. 16-bit Pulse Accumulator A

0	PAEN	PAMOD	PEDGE	CLK1	CLK0	PAOVI	PAI

PAEN D6 Pulse Accumulator A enable
0 = Disable PACA. The PAC3 and PAC2 are used as an 8-bit PAC.
1 = Enable PACA. It allows PACN3:PACN2 to form a 16-bit PACA.

PAMOD D5 PACA is used as an event counter or gated time accumulator.
0 = PACA is used as an event counter counting pulses fed to pin PT7.
1 = PACA is used as a gated time accumulator.

PEDGE D4 This allows one to choose the falling edge or rising edge for the PACA.
0 = Count up (increment) PACA on falling edge of pulses fed to PT7 pin.
1 = Count up (increment) PACA on rising edge of pulses fed to PT7 pin.

CLK1,CLK0 D4:D3 00 for our applications. See the HCS12 manual.

PAOVI D1 PACA overflow interrupt enable. This allows an interrupt to be invoked whenever the 16-bit PACA overflows from FFFF to 0000. See Chapter 11.

PAI D1 PACA Input interrupt enable. This allows an interrupt to be invoked when an edge is detected on the PT7 pin. See Chapter 11.

Figure 9-16. PACTL (PACA Control Register) used for the 16-bit PACA only

Example 9-21

Find the PACTL (PACA Control) register value for rising edge count-up, with the PACA enabled for 16-bit event counter and no interrupts to be invoked.

Solution:

From Figure 9-16, we have PACTL = 01010000.

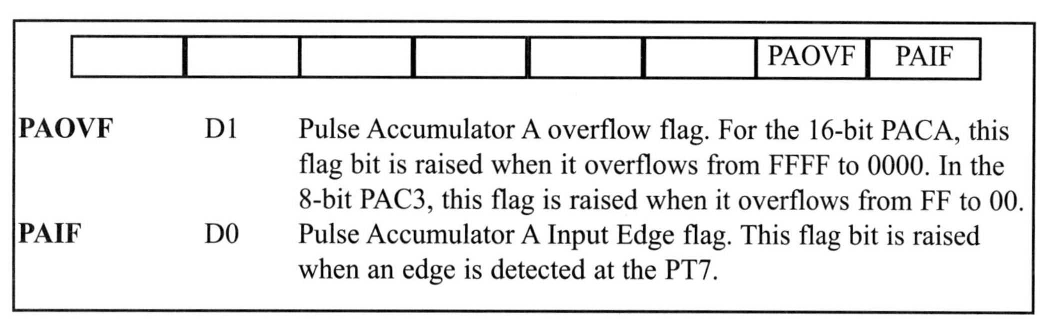

PAOVF	D1	Pulse Accumulator A overflow flag. For the 16-bit PACA, this flag bit is raised when it overflows from FFFF to 0000. In the 8-bit PAC3, this flag is raised when it overflows from FF to 00.
PAIF	D0	Pulse Accumulator A Input Edge flag. This flag bit is raised when an edge is detected at the PT7.

Figure 9-17. PAFLG (Pulse Accumulator Flag) Register

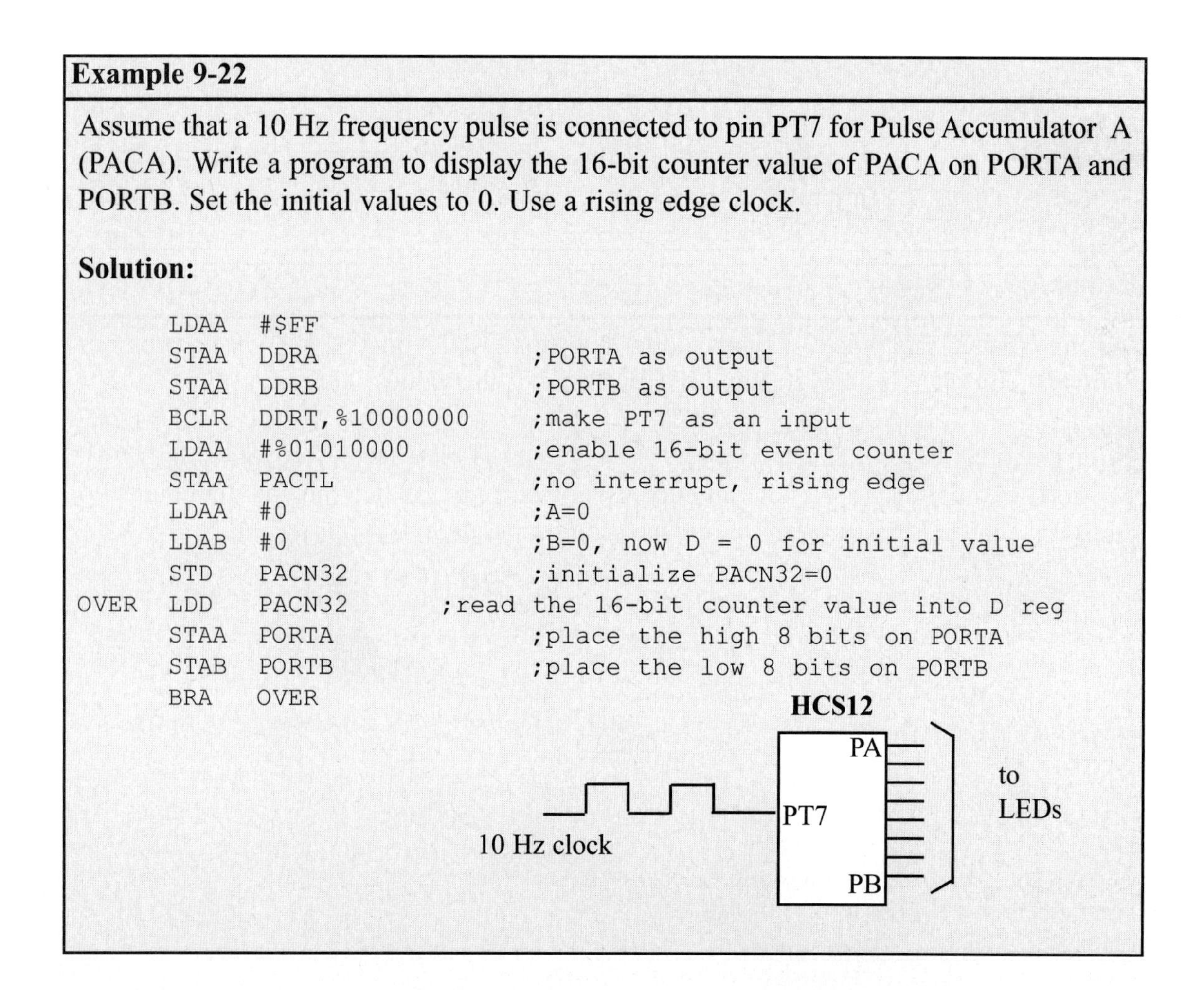

Example 9-22

Assume that a 10 Hz frequency pulse is connected to pin PT7 for Pulse Accumulator A (PACA). Write a program to display the 16-bit counter value of PACA on PORTA and PORTB. Set the initial values to 0. Use a rising edge clock.

Solution:

```
        LDAA   #$FF
        STAA   DDRA               ;PORTA as output
        STAA   DDRB               ;PORTB as output
        BCLR   DDRT,%10000000     ;make PT7 as an input
        LDAA   #%01010000         ;enable 16-bit event counter
        STAA   PACTL              ;no interrupt, rising edge
        LDAA   #0                 ;A=0
        LDAB   #0                 ;B=0, now D = 0 for initial value
        STD    PACN32             ;initialize PACN32=0
OVER    LDD    PACN32        ;read the 16-bit counter value into D reg
        STAA   PORTA              ;place the high 8 bits on PORTA
        STAB   PORTB              ;place the low 8 bits on PORTB
        BRA    OVER
```

Example 9-23

Assuming that clock pulses are fed into pin PT7 for PACA and a buzzer is connected to pin PORTB.1, write a program to sound the buzzer after 1,000 pulses.

Solution:

To sound the buzzer every 1,000 pulses, we set the initial event counter value to –1,000 (FC18 in hex), then the counter counts up until it reaches FFFF. Upon overflow, we can sound the buzzer by toggling the PORTB.1 pin.

```
      BCLR  DDRT,%10000000     ;make PT7 as an input
      BSET  DDRB,%00000010     ;make PB1 as an output
      LDAA  #%01010000         ;enable 16-bit event counter
      STAA  PACTL              ;no interrupt, rising edge
      LDD   #$FC18             ;D = -1000 (FC18 in hex) since we need
                               ;1000 pulses
      STD   PACN32             ;initialize PACN32=-1000
HERE  BRCLR PAFLG,mPAFLG_PAOVF,HERE ;stay here until PACA overflows
      BSET  PAFLG,mPAFLG_PAOVF
OVER  BSET  PORTB,%00000010    ;keep toggling
      BCLR  PORTB,%00000010    ;the buzzer
      BRA   OVER
```

Bit 1 of PORTB is connected to a buzzer and input PT7 to pulses.

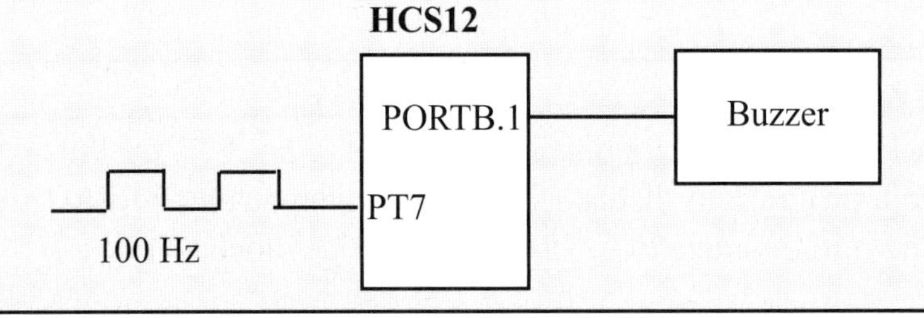

Example 9-24

Assume that a 1 Hz pulse is connected to input for PACA (pin PT7). Write a program to turn on pin PORTB.4 every time PACA reaches the value 50 (decimal).

Solution:

We set the initial event counter value to –50 (FFCE in hex), then the counter counts up until it reaches FFFF. Upon overflow, we turn on the PORTB.4 pin.

```
      BCLR  DDRT,%10000000     ;make PT7 as an input
      BSET  DDRB,%00010000     ;make PB4 as an output
      LDAA  #%01010000         ;enable 16-bit event counter
      STAA  PACTL              ;no interrupt, rising edge
OVER  LDD   #$FFCE     ;D = -50 (FFCE in hex) since we need 50 pulses
      STD   PACN32             ;initialize PACA=-50
      BCLR  PORTB,%00010000    ;PORTB.4 = 0
HERE  BRCLR PAFLG,mPAFLG_PAOVF,HERE ;stay here until PACA overflows
      BSET  PAFLG,mPAFLG_PAOVF;clear the flag for the next round
      BSET  PORTB,%00010000    ;PORTB.4 = 1
      BRA   OVER
```

Example 9-25

Assume that a 1 Hz pulse is connected to input for PACA (pin PT7). The pulses represent the number of bottles being packaged together. Write a program to turn on PORTB.4 every time 12 bottles go by. The program should display the binary count on PORTA.

Solution:

We set the initial event counter value to –12 (FFF4 in hex), then the counter counts up until it reaches FFFF. Upon overflow, we turn on the PORTB.4 pin.

```
Counter      DS.B 1
       LDAA  #$FF
       STAA  DDRA                ;PORTA as output
       BCLR  DDRT,%10000000      ;make PT7 as an input
       CLR   Counter
       LDAA  #%01010000          ;enable 16-bit event counter
       STAA  PACTL               ;no interrupt, rising edge
OVER   LDD   #$FFF4     ;D = -12 (FFF4 in hex) since we need 12 pulses
       STD   PACN32              ;initialize PACA=-12
       BCLR  PORTB,%00010000     ;PORTB.4 = 0 (turn off)
HERE   BRCLR PAFLG,mPAFLG_PAOVF,HERE ;stay here until PACA overflows
       BSET  PORTB,%00010000     ;turn on PORTB.4 for every 12 pulses
       BSET  PAFLG,mPAFLG_PAOVF;clear the flag for the next round
       INC   Counter
       LDAA  Counter
       STAA  PORTA        ;display the count on PORTA
       BRA   OVER
```


Programming 16-bit Pulse Accumulator B

The Pulse Accumulator B is another 16-bit event counter in HCS12. The Pulse Accumulator B (PACB) is formed by cascading the 8-bit PACN0 and PACN1, where the PACN1 is the upper 8-bit block and PACN0 is the lower 8-bit block. We can access the entire 16-bit PACB using an instruction such as "LDD PACB". The PACB counts the pulses fed to it via the PT0 pin. The PACB has its own control register called PBCTL. See Figure 9-18. But unlike PACA, we cannot choose the rising edge or the falling edge for pulses. We monitor the overflow of the 16-bit PACB from FFFF to 0000 using the PBOVF flag bit in the PBFLG register. See Figure 9-19. Example 9-27 shows the programming of the 16-bit PACB.

Before we finish this section, we need to state an important point. You might think monitoring the PAVOF and PBOVF flags are a waste of the microcontroller's time. You are right. There is a solution to this: the use of interrupts. Using interrupts enables us to do other things with the microcontroller. When a counter interrupt flag such as PAOVF is raised it will inform us. This important and powerful feature of the HCS12 is discussed in Chapter 11.

Example 9-26

Modify the Pulse Accumulator A in Example 9-25 to display the data in ASCII on PORTA and PORTH.

Solution:

We need to convert the binary data to decimal and then to ASCII before we display it.

```
Counter     DS.B 1
RMND_L      DS.B 1
RMND_M      DS.B 1

      LDS   #$4000              ;initialize the stack pointer
      LDAA  #$FF
      STAA  DDRA                ;PORTA as output
      STAA  DDRH                ;PORTH as output
      CLR   Counter
      BCLR  DDRT,%10000000      ;make PT7 as an input
      LDAA  #%01010000          ;enable 16-bit event counter
      STAA  PACTL         ;no interrupt, rising edge
OVER  LDD   #$FFF4     ;D = -12 (FFF4 in hex) since we need 12 pulses
      STD   PACN32              ;initialize PACA=-12
      BCLR  PORTB,%00010000     ;PORTB.4 = 0 (turn off)
HERE  BRCLR PAFLG,mPAFLG_PAOVF,HERE ;stay here until PACA overflows
      BSET  PAFLG,mPAFLG_PAOVF;clear the flag for the next round
      BSET  PORTB,%00010000     ;turn on PORTB.4 for every 12 pulses
      INC   Counter
      LDAB  Counter             ;B holds the count
      JSR   BIN_DEC_CON         ;convert to decimal value in reg A
      JSR   DEC_ASCII_CON       ;convert to ASCII
      LDAA  RMND_L
      STAA  PORTA
      LDAA  RMND_M
      STAA  PTH
      BRA   OVER
;-----converting BIN(HEX) TO DEC (0 TO 99)
BIN_DEC_CON
      CLRA                ;clear high byte of D
      LDX  #10            ;X = 10, the denominator
      IDIV                ;example D/X = 53/10 = 5, remainder = 3
      STAB RMND_L         ;save the remainder
      XGDX                ;exchange D and X for new numerator
      LDX  #10            ;X = 10, the denominator
      IDIV                ;D/X
      STAB RMND_M         ;save the next digit
      RTS
;----converting unpacked BCD digits to displayable ASCII digits
DEC_ASCII_CON
      LDAA RMND_L         ;load the low byte
      ADDA #$30           ;convert to ASCII
      STAA RMND_L         ;put it back
      LDAA RMND_M         ;load the middle byte
      ADDA #$30           ;convert to ASCII
      STAA RMND_M         ;put it back
      RTS
```

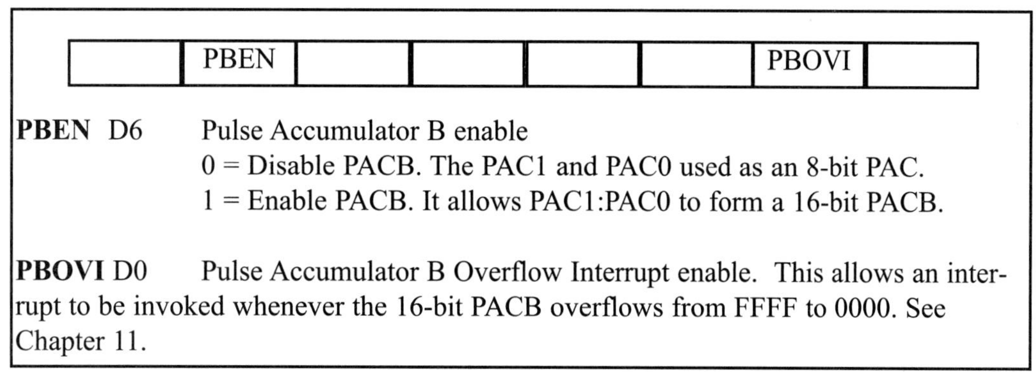

PBEN D6 Pulse Accumulator B enable
0 = Disable PACB. The PAC1 and PAC0 used as an 8-bit PAC.
1 = Enable PACB. It allows PAC1:PAC0 to form a 16-bit PACB.

PBOVI D0 Pulse Accumulator B Overflow Interrupt enable. This allows an interrupt to be invoked whenever the 16-bit PACB overflows from FFFF to 0000. See Chapter 11.

Figure 9-18. PBCTL (Pulse Accumulator B Control) Register

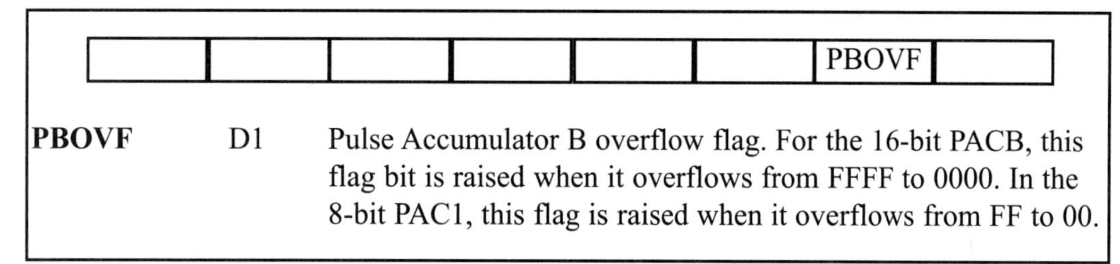

PBOVF D1 Pulse Accumulator B overflow flag. For the 16-bit PACB, this flag bit is raised when it overflows from FFFF to 0000. In the 8-bit PAC1, this flag is raised when it overflows from FF to 00.

Figure 9-19. PBFLG (Pulse Accumulator B Flag) Register

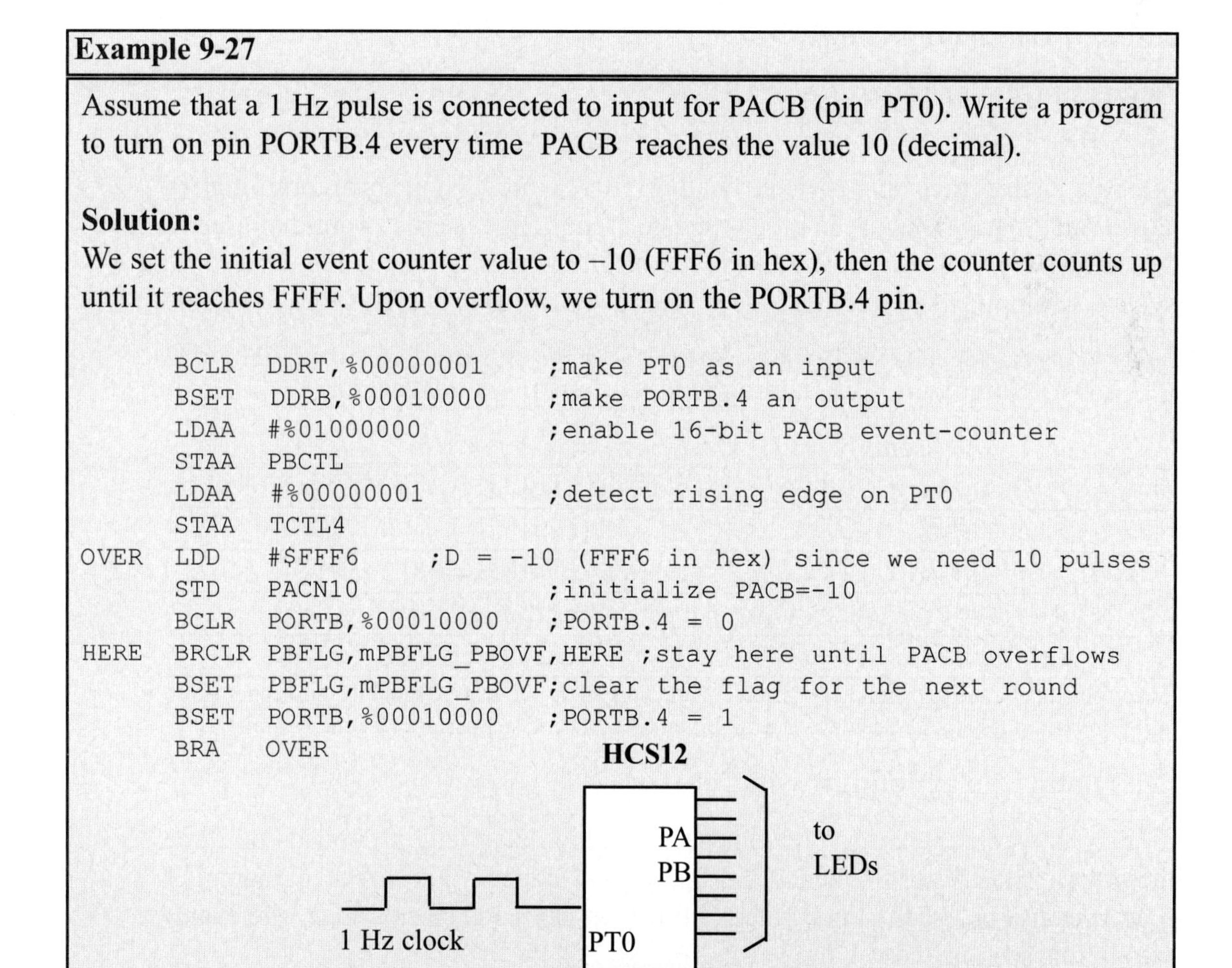

Example 9-27

Assume that a 1 Hz pulse is connected to input for PACB (pin PT0). Write a program to turn on pin PORTB.4 every time PACB reaches the value 10 (decimal).

Solution:

We set the initial event counter value to –10 (FFF6 in hex), then the counter counts up until it reaches FFFF. Upon overflow, we turn on the PORTB.4 pin.

```
      BCLR  DDRT,%00000001      ;make PT0 as an input
      BSET  DDRB,%00010000      ;make PORTB.4 an output
      LDAA  #%01000000          ;enable 16-bit PACB event-counter
      STAA  PBCTL
      LDAA  #%00000001          ;detect rising edge on PT0
      STAA  TCTL4
OVER  LDD   #$FFF6     ;D = -10 (FFF6 in hex) since we need 10 pulses
      STD   PACN10              ;initialize PACB=-10
      BCLR  PORTB,%00010000     ;PORTB.4 = 0
HERE  BRCLR PBFLG,mPBFLG_PBOVF,HERE ;stay here until PACB overflows
      BSET  PBFLG,mPBFLG_PBOVF;clear the flag for the next round
      BSET  PORTB,%00010000     ;PORTB.4 = 1
      BRA   OVER
```

Review Questions

1. What pin provides the clock pulses to the HCS12 event counter if ICPAR = $08?
2. What pin provides the clock pulses to the HCS12 event-counter if ICPAR = $01?
3. Does the discussion about the free-running timer in Section 9.1 apply to PACA if it used as an event counter?
4. To allow PT7 to be used as an input for the PACA clock, what must be done, and why?
5. Do we have a choice of counting up on the positive or negative edge of the clock?

SECTION 9.4: HCS12 TIMER PROGRAMMING IN C

In Chapter 7 we showed some examples of C programming for the HCS12. In this section we show C programming for the HCS12 Output Compare, Input Capture, and Pulse Accumulator features. As we saw in the examples in Chapter 7, the general-purpose registers and memory of the HCS12 are under the control of the C compiler and are not accessed directly by C statements. All of the special function registers for peripherals, however, are accessible directly using C statements. As an example of accessing these special registers directly, we saw how to access ports PORTB–PORTE in Chapter 7. Next, we discuss how to access the HCS12 timers directly using the C compiler.

Accessing timer registers in C

In C we can access timer registers such as TCNT, TSCR1, and so on, directly using the appropriate header file. Note when programming in CodeWarrior using C that the bit naming convention is different from that used in Assembly. We must make sure to use the bit mask version of the registers for C syntax. Examples 9-28 through 9-46 repeat the previous examples in C. See Table 9-1.

Table 9-1: Comparing Bit Naming Conventions in CodeWarrior

Assembly	C
mTFLG1_C0F	TFLG1_C0F_MASK
mTFLG1_C1F	TFLG1_C1F_MASK
mTFLG1_CxF	TFLG1_CxF_MASK
mPAFLG_PAOVF	PAFLG_PAOVF_MASK
mPBFLG_PBOVF	PBFLG_PBOVF_MASK

Note: This may not apply in other assemblers and compilers. Check the include and header files provided with these tools. See www.MicroDigitalEd.com for other assemblers and compilers.

Calculating delay length using timers

As we have seen in previous sections, the delay length depends on three factors: (a) the crystal frequency, (b) the prescaler factor, and (c) the C compiler (because various C compilers generate different hex code sizes). Study the examples in this section and verify them.

Example 9-28 (C version of Examples 9-3 and 9-6)

Write a C program to count the number of times the free-running timer overflows (TOF is raised). Display the binary number on PORTB. Show how often the number changes. Assume XTAL = 4 MHz. Use a) no prescaler and, b) a prescaler of 64. This is the C version of Examples 9-3 and 9-6.

Solution:

a)

```
#include <mc9s12dp512.h>
#ifndef TFLG1_TOF_MASK
  #define TFLG1_TOF_MASK 0x80
#endif
unsigned char overflow = 0;
void main(void)
  {
    DDRB = 0xFF;    //make PB an output
    TSCR1 = 0x80; //enable timer
    TSCR2 = 0x00; //no interrupt, no prescaler

    while(1)
     {
        TFLG2 = 0x80;    //clear TOF
        //wait for overflow flag  to be raised
        while (!(TFLG2 & TFLG2_TOF_MASK));
        overflow = overflow + 1;
        PORTB = overflow;
     }
  }
```

The period for the clock fed to the free-running timer is:

$1/2 \times 4$ MHz = 2 MHz and T = 1/(2 MHz) = 0.5 μs

The free-running timer counts up once every 0.5 μs. It rolls over every 32,768 μs since 65,536 × 0.5 μs = 32,768 μs

To get a more accurate timing, we need to add clock cycles due to the instructions in the loop.

b) In the above program make TSCR2 = 0x06 and run it again to test it.

The period for the clock fed to the timer is:

$1/2 \times 4$ MHz = 2 MHz and $1/64 \times 2$ MHz = 31,250 Hz due to prescaler and T = 1/31,250 Hz = 32 μs. Therefore the counter counts up every 65,536 × 32 μs = 2,097,152 μs or every 2.10 seconds.

Example 9-29 (C version of Example 9-7)

Assuming XTAL = 4 MHz, write a C program for channel 2 to toggle PT2 every 500 pulses. This is the C version of Example 9-7.

Solution:

```
#include <mc9s12dp512.h>
#ifndef TFLG1_C2F_MASK                    //we use this for C compilers
  #define TFLG1_C2F_MASK 0x04             // other than CodeWarrior
#endif
unsigned int Tcount;
void main(void)
  {
    TSCR1 = 0x90; //enable timer and let CPU clear C2Fs
    TSCR2 = 0x00; //no interrupt, no prescaler
    TIOS = 0x04;  //select chan 2 for output compare
    TCTL2 = 0x10;  //toggle PT2 pin upon match

    while(1)
     {
       Tcount = TCNT;
       Tcount = Tcount + 500;
       TC2 = Tcount;
       while (!(TFLG1 & TFLG1_C2F_MASK));//wait for chan 2 match
     }
  }
```

The period for the clock fed to the timer TCNT is:

$1/2 \times 4$ MHz $= 2$ MHz and $T = 1/(2 \text{ MHz}) = 0.5\ \mu s$

The period for the pulses on PT2 is:

$T = 2 \times (500 \times 0.5\ \mu s) = 500\ \mu s$ and Freq. $= 1/500\ \mu s = 2{,}000$ Hz

To get a more accurate timing, we need to add clock cycles due to the instructions in the loop. Use an oscilloscope to view and verify the square wave on PT2.

Note: Depending on the XTAL/PLL frequency the LED may appear always on. To slow down the flashing of the LED, set the prescaler to 128 and change the value to 5,000.

Note: When TSCR1 = 10000000, we must clear the flag bits, such as TOF, by writing 1 to it. When TSCR1 = 10010000, the CPU will clear the flags, such as TOF and CnF, automatically with any reading of the TCNT register. That is the function of TFFCA bit.

Example 9-30 (C version of Example 9-8)

Assuming XTAL = 4 MHz, write a program to generate a square wave with a period of 10 ms on pin PT4. This is the C version of Example 9-8.

Solution:

For a square wave with T = 10 ms we must have a time delay of 5 ms. Because XTAL = 4 MHz, the counter counts up every 0.5 μs. This means that we need 5 ms / 0.5 μs = 10,000 clocks.

```
#include <hidef.h>        /* common defines and macros */
#include <mc9s12dp512.h>      /* derivative information */
#pragma LINK_INFO DERIVATIVE "mc9s12dp512"

unsigned int Tcount;
void main(void)
  {
    DDRT = DDRT | 0x10; //make PT4 an output
    TSCR1 = 0x80; //Enable Timer
    TSCR2 = 0x00; //No Interrupt, no prescaler
    TIOS = 0x10;  //Select chan 4 for output compare
    TCTL1 = 0x01;  //Toggle PT4 pin upon match

while(1)
    {
       Tcount = TCNT;
       Tcount = Tcount + 10000;
        TC4 = Tcount;
       while (!(TFLG1 & TFLG1_C4F_MASK));//wait for chan 4 match
        TFLG1 = TFLG1 | TFLG1_C4F_MASK;   //clear C4F
    }
  }
```

Example 9-31 (C version of Example 9-10)

Assuming XTAL = 4 MHz, modify the program in Example 9-30 to toggle PT7 every 1/2 second. This is the C version of Example 9-10.

Solution:

0.5 s / 0.5 μs = 1,000,000. That means we must use prescaler. Using a prescaler of 16 we get 1,000,000/16 = 62,500.

```
#include <mc9s12dp512.h>
unsigned int Tcount;
#ifndef TFLG1_C7F_MASK
  #define TFLG1_C7F_MASK 0x80
#endif
void main(void)
  {
   TSCR1 = 0x80; //enable timer
   TSCR2 = 0x04; //no interrupt, prescaler 16
   TIOS = 0x80;  //select chan 7 for output compare
   TCTL1 = 0x40;  //toggle PT7 pin upon match

   while(1)
    {
      Tcount = TCNT;
      Tcount = Tcount + 62500;
      TC7 = Tcount;
      while (!(TFLG1 & TFLG1_C7F_MASK));//wait for chan 7 match
      TFLG1 = TFLG1 | TFLG1_C7F_MASK;   //clear C7F
    }
  }
```

Another way to verify the above values is as follows:
1/2 × 4 MHz = 2 MHz and 1/16 × 2 MHz = 125,000 Hz due to the prescaler and T = 1/125,000 Hz = 8 μs for TCNT. Now, 0.5 s / 8 μs = 62,500.

Example 9-32 (C version of Example 9-11)

Assuming XTAL = 4 MHz, write a program to generate a square wave of 50 Hz frequency on pin PT6. Use channel 6 and a prescaler of 128 for the TCNT. This is the C version of Example 9-11.

Solution:

Look at the following steps:

(a) $1/2 \times 4$ MHz = 2 MHz and $1/128 \times 2$ MHz = 15,625 Hz and T = 1/15,625 Hz = 64 μs

(b) T = 1 / 50 Hz = 20 ms, the period of the square wave.

(c) 1/2 of it for the high and low portions of the pulse = 10 ms

(d) 10 ms / 64 μs = 156

```
#include <mc9s12dp512.h>
#ifndef TFLG1_C6F_MASK
  #define TFLG1_C6F_MASK 0x40
#endif
unsigned int Tcount;
void main(void)
  {
   TSCR1 = 0x80; //enable timer
   TSCR2 = 0x07; //no interrupt, 128 prescaler
   TIOS = 0x40;  //select chan 6 for output compare
   TCTL1 = 0x10;  //toggle PT6 pin upon match

   while(1)
    {
      Tcount = TCNT;
      Tcount = Tcount + 156;
      TC6 = Tcount;
      while (!(TFLG1 & TFLG1_C6F_MASK));//wait for chan 6 match
      TFLG1 = TFLG1 = TFLG1_C6F_MASK;   //clear C6F
    }
  }
```

Example 9-33 (C version of Example 9-12)

Assuming XTAL = 4 MHz, write a program to toggle all bits of PORTB every second. Use channel 0 and a prescaler of 128 for the TCNT. This is the C version of Example 9-12.

Solution:

Look at the following steps:

(1) 1/2 × 4 MHz = 2 MHz and 1/128 × 2 MHz = 15,625 Hz and T = 1/15,625 Hz = 64 μs

(2) 1s / 64 μs = 15,625

```
#include <mc9s12dp512.h>
#ifndef TFLG1_C0F_MASK
  #define TFLG1_C0F_MASK 0x01
#endif
unsigned int Tcount;
void main(void)
  {
   DDRB = 0xFF;   //make PB an output
   TSCR1 = 0x90; //enable timer and let CPU clear C0F
   TSCR2 = 0x07; //no interrupt, 128 prescaler
   TIOS = 0x01;  //select chan 0 for output compare
   TCTL2 = 0x00;  //timer disconnected from pin
   PORTB = 0x55;
   while(1)
    {
      Tcount = TCNT;
      Tcount = Tcount + 15625;
      TC0 = Tcount;
      while (!(TFLG1 & TFLG1_C0F_MASK));//wait for chan 0 match
      PORTB = ~PORTB;
    }
  }
```

Example 9-34 (C version of Example 9-13)

Assuming XTAL = 4 MHz, write a program to generate a square wave with a period of 10 ms on pin PT4. Make it 60% duty cycle. This is the C version of Example 9-13.

Solution:

For a square wave with T = 10 ms and 60% duty cycle we need 6 ms for the high portion and 4 ms for the low portion of the pulse. Because XTAL = 4 MHz, the counter counts up every 0.5 μs. This means that we need 12,000 clocks (6 ms / 0.5 μs = 12,000) for high and 8,000 clocks (4 ms / 0.5 μs = 8,000) for low.

```
#include <mc9s12dp512.h>
#ifndef TFLG1_C4F_MASK
  #define TFLG1_C4F_MASK 0x10
#endif
unsigned int Tcount;
void main(void)
  {
   TSCR1 = 0x80; //enable timer
   TSCR2 = 0x00; //no interrupt, no prescaler
   TIOS = 0x10;  //select chan 4 for output compare
   while(1)
    {
      TCTL1 = 0x02;  //make PT4 pin low upon match
      Tcount = TCNT;
      Tcount = Tcount + 12000;
      TC4 = Tcount;
      while (!(TFLG1 & TFLG1_C4F_MASK));//wait for chan 4 match
      TFLG1 = TFLG1 | TFLG1_C4F_MASK;   //clear C4F
      TCTL1 = 0x03;  //make PT4 pin high upon match
      Tcount = TCNT;
      Tcount = Tcount + 8000;
      TC4 = Tcount;
      while (!(TFLG1 & TFLG1_C4F_MASK));//wait for chan 4 match
      TFLG1 = TFLG1 | TFLG1_C4F_MASK;   //clear C4F
    }
  }
```

Example 9-35 (C version of Example 9-15)

Assuming that clock pulses are fed into pin PT6, write a program for channel 6 to read the TCNT value on every rising edge. Place the result on PORTA and PORTB. Assume XTAL = 4 MHz. This is the C version of Example 9-15.

Solution:

```
#include <mc9s12dp512.h>
#ifndef TFLG1_C6F_MASK
  #define TFLG1_C6F_MASK 0x40
#endif
unsigned int Tcount;
void main(void)
  {
    DDRA = 0xFF;   //make PA an output
    DDRB = 0xFF;   //make PB an output
    TSCR1 = 0x90; //enable timer and let the CPU clear the flag
    TSCR2 = 0x00; //no interrupt, no prescaler
    TIOS = TIOS & ~0x40;  //select chan 6 for Input Capture
    TCTL3 = 0x10;  //select channel 6 for capture on rising edge
 while(1)
     {
      while (!(TFLG1 & TFLG1_C6F_MASK));//wait for chan 6 rising edge
      TFLG1 = TFLG1 | TFLG1_C6F_MASK;
      Tcount = TC6; //get the TCNT value
      PORTB = Tcount & 0x00FF;  //display the low byte
      PORTA = ((Tcount & 0xFF00) >> 8);  //display the upper byte

     }
  }
```

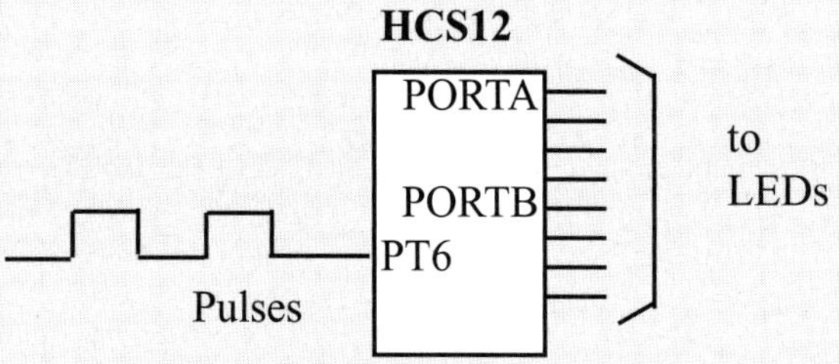

Note: Upon the detection of each rising edge, the TCNT value is loaded into TC6. Also notice that by reading the TC6 value we clear the C6F flag bit since TSCR1 = $90.

Example 9-36 (C version of Example 9-16)

Assuming that clock pulses are fed into pin PT6, write a program for channel 6 to measure the period of the pulses. Place the binary result on PORTA and PORTB. Assume XTAL = 4 MHz. This is the C version of Example 9-16.

Solution:

```
#include <mc9s12dp512.h>
#ifndef TFLG1_C6F_MASK
  #define TFLG1_C6F_MASK 0x40
#endif
unsigned int Tcount, T1, T2;
void main(void)
  {
   DDRA = 0xFF;   //make PA an output
   DDRB = 0xFF;   //make PB an output
   TSCR1 = 0x80; //enable timer
   TSCR2 = 0x00; //no interrupt, no prescaler
   TIOS = TIOS & ~0x40;  //select chan 6 for Input Capture
   TCTL3 = 0x10;  //select channel 6 for capture on rising edge
   TFLG1 = TFLG1 | TFLG1_C6F_MASK;  //clear C6F
   while (!(TFLG1 & TFLG1_C6F_MASK));//chan 6 first rising edge
   T1 = TC6; //get the TC6 value
   TFLG1 = TFLG1 | TFLG1_C6F_MASK;  //clear C6F
   while (!(TFLG1 & TFLG1_C6F_MASK));//chan 6 2nd rising edge
   T2 = TC6; //get the TCNT value
   Tcount = T2-T1;
   PORTB = Tcount & 0x00FF;  //display the low byte
   PORTA = ((Tcount & 0xFF00) >> 8);  //display the upper byte
   for(;;);
  }
```

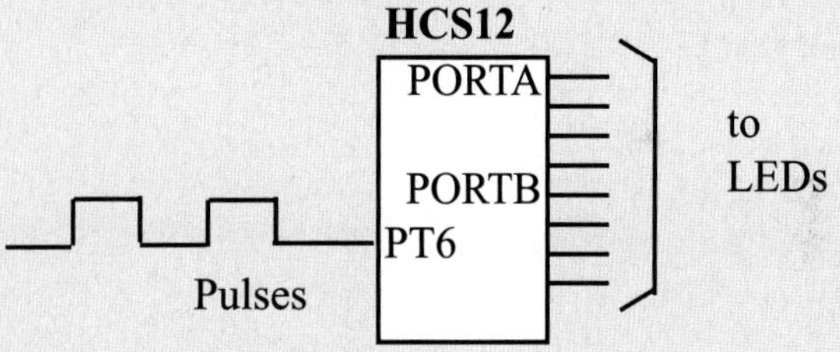

Example 9-37 (C version of Example 9-17)

Assume that a 75 Hz pulse is connected to input for channel 6 (pin PT6). Write a program to measure its period and verify the frequency. Display the result on PORTA and PORTB. Assume XTAL = 4 MHz. This is the C version of Example 9-17.

Solution:

The frequency of 75 Hz gives us the period of 1/75 Hz = 13.3 ms. Now, $1/2 \times 4$ MHz = 2 MHz and $1/128 \times 2$ MHz = 15,625 Hz due to the prescaler and T = 1/15,625 Hz = 64 μs for TCNT. That means we get the value of 208 (1101 0000 binary) on PORTB since 13.3 ms / 64 μs = 208

```
#include <mc9s12dp512.h>
#ifndef TFLG1_C6F_MASK
  #define TFLG1_C6F_MASK 0x40
#endif
unsigned int Tcount, T1, T2;
void main(void)
  {
   DDRB = 0xFF;   //make PB an output
   TSCR1 = 0x80; //enable timer
   TSCR2 = 0x07; //no interrupt, prescaler 128
   TIOS = TIOS & ~0x40;  //select chan 6 for Input Capture
   TCTL3 = 0x10;  //select channel 6 for capture on rising edge
 while(1)
    {
     TFLG1 = TFLG1 | TFLG1_C6F_MASK;   //clear C6F
     while (!(TFLG1 & TFLG1_C6F_MASK));//chan 6 first rising edge
     T1 = TC6; //get the TCNT value
     TFLG1 = TFLG1 | TFLG1_C6F_MASK;   //clear C6F
     while (!(TFLG1 & TFLG1_C6F_MASK));//chan 6 2nd rising edge
     T2 = TC6;
     Tcount = T2 - T1;
     PORTAB = Tcount;   //display the result
    }
  }
```

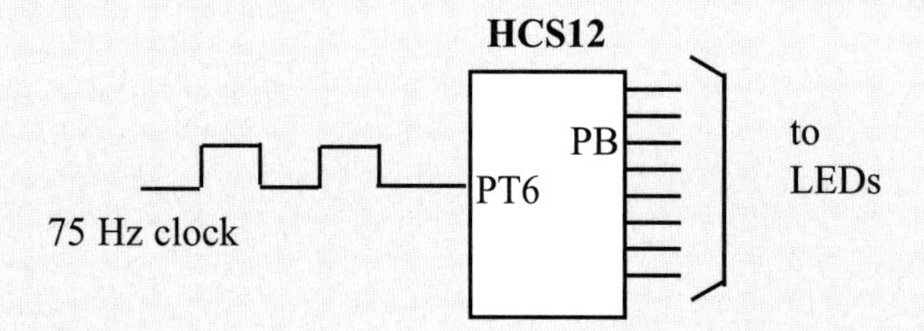

Example 9-38 (C version of Example 9-18)

Assume that a 75 Hz pulse is connected to input for channel 6 (pin PT6). Write a program to measure its pulse width. Use the prescaler value that gives the result in a single byte. Display the result on PORTA and PORTB. Assume XTAL = 4 MHz. This is the C version of Example 9-18.

Solution:

The frequency of 75 Hz gives us the period of 1/75 Hz = 13.3 ms. Now, $1/2 \times 4$ MHz = 2 MHz and $1/128 \times 2$ MHz = 15,625 Hz due to the prescaler and T = 1/15,625 Hz = 64 μs for TCNT. That means we get the value of 208 (1101 0000 binary) for the period since 13.3 ms / 64 μs = 208. Now the pulse width can be anywhere from 1 to 207 clock ticks.

```
#include <mc9s12dp512.h>
#ifndef TFLG1_C6F_MASK
  #define TFLG1_C6F_MASK 0x40
#endif
unsigned int Tcount, T1, T2;
void main(void)
  {
    DDRB = 0xFF;   //make PB an output
    TSCR1 = 0x80; //enable timer
    TSCR2 = 0x07; //no interrupt, prescaler 128
    TIOS = TIOS & ~0x40;  //select chan 6 for Input Capture
  while(1)
     {
      TCTL3 = 0x10;  //select channel 6 for capture on rising edge
      while (!(TFLG1 & TFLG1_C6F_MASK));//chan 6 first rising edge
      TFLG1 = TFLG1 | TFLG1_C6F_MASK;  //clear C6F
      T1 = TC6; //get the TCNT value
      TCTL3 = 0x20;  //select channel 6 for capture on falling edge
      while (!(TFLG1 & TFLG1_C6F_MASK));//wait for chan 6 falling
edge
      TFLG1 = TFLG1 | TFLG1_C6F_MASK;  //clear C6F
      T2 = TC6; //get the TCNT value
      Tcount = T2 - T1;
      PORTAB = Tcount;  //display the result
     }
  }
```

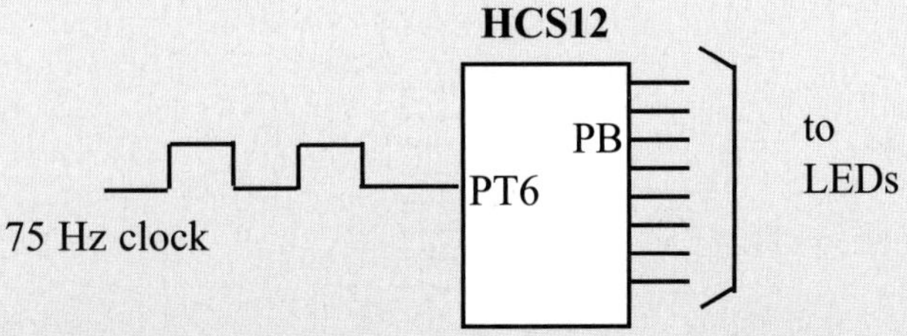

HCS12

PB

PT6

to LEDs

75 Hz clock

Example 9-39 (C version of Example 9-19)

Assume that a temperature sensor is connected to pin PT6. The temperature provided by the sensor is proportional to pulse width and is in the range of 1 μs to 250 μs. Write a program to measure the temperature if 1 μs is equal to 1 degree. Use the prescaler value that gives the result in a single byte. Display the result on PORTB. Assume XTAL = 4 MHz. This is the C version of Example 9-19.

Solution:

$1/2 \times 4$ MHz = 2 MHz and $1/2 \times 2$ MHz = 1,000,000 Hz due to prescaler and T = 1/1,000,000 Hz = 1 μs for TCNT. That means we get values from 1 to 65,536 μs for the TCNT, but since the pulse width never goes beyond 250 μs we should be able to display the temperature value on PORTB.

```
#include <mc9s12dp512.h>
#ifndef TFLG1_C6F_MASK
  #define TFLG1_C6F_MASK 0x40
#endif
unsigned int Tcount, T1, T2;
void main(void)
  {
   DDRT = DDRT & 0xBF; //make PT6 an output
   TSCR1 = 0x80; //enable timer
   TSCR2 = 0x01; //no interrupt, prescaler 2
   TIOS = 0x40;  //select chan 6 for Input Capture
 while(1)
    {
     TCTL3 = 0x10;  //select channel 6 for capture on rising edge
     TFLG1 = TFLG1 | TFLG1_C6F_MASK;  //clear C6F
     while (!(TFLG1 & TFLG1_C6F_MASK));//chan 6 first rising edge
     T1 = TC6; //get the TCNT value
     TCTL3 = 0x20;  //select channel 6 for capture on falling edge
     TFLG1 = TFLG1 | TFLG1_C6F_MASK;  //clear C6F
     while (!(TFLG1 & TFLG1_C6F_MASK));//chan 6 falling edge
     T2 = TC6; //get the TCNT value
     Tcount = T2 - T1;
     PORTAB = Tcount;  //display the result
    }
  }
```

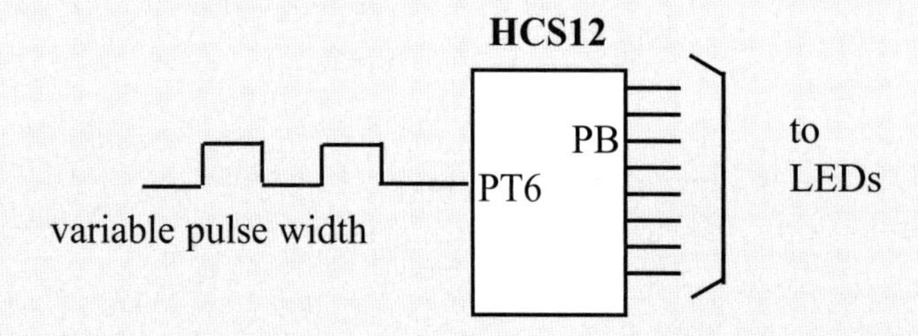

Example 9-40 (C version of Example 9-20)

Assuming that 1 Hz clock pulses are fed into pin PT3, write a program for PAC3 to count the pulses and display the state of the count on PORTB. This is the C version of Example 9-20.

Solution:

```
#include <mc9s12dp512.h>
void main(void)
  {
    DDRB = 0xFF;    //make PB an output
    DDRT = DDRT & 0xF7; //make PT3 an input
    ICPAR = 0x08; //enable PAC3
    TCTL4 = 0x40;   //select rising edge for PAC3
    PACN3 = 0x00; //clear PAC3
 while(1)
     {
      PORTB = PACN3;    //display the low byte
     }
  }
```

PORTB is connected to 8 LEDs and input PT3 to pulses.

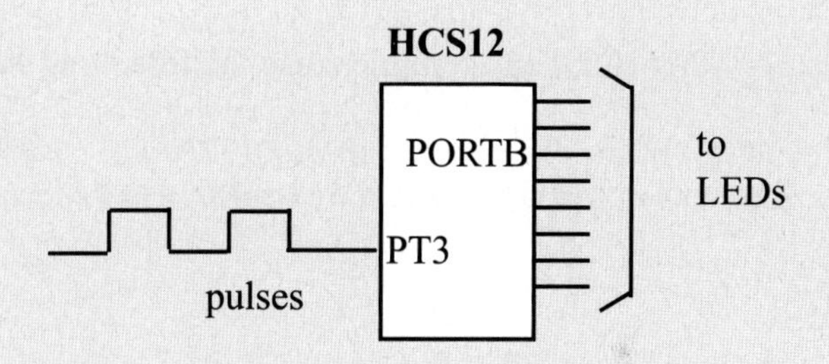

Example 9-41 (C version of Example 9-22)

Assume that a 10 Hz frequency pulse is connected to pin PT7 for Pulse Accumulator A (PACA). Write a program to display the 16-bit counter value of PACA on PORTA and PORTB. Set the initial values to 0. Use rising edge clock. This is the C version of Example 9-22.

Solution:

```
#include <mc9s12dp512.h>
unsigned int Mycount;
void main(void)
  {
    DDRA = 0xFF;    //make PA an output
    DDRB = 0xFF;    //make PB an output
    DDRT = DDRT & 0x7F; //make PT7 an input
    PACTL = 0x50; //enable 16-bit PACA event counter,
                    //no interrupt, count pulses on rising edge
    PACN32 = 0;     //clear PACA 16-bit register
  while(1)
     {
     Mycount = PACN32;    //read 16-bit pulse counter value
     PORTAB = Mycount;//display on PORTA and PORTB (CodeWarrior)
      }
  }
```

Note: For other HCS12 C compilers use the following in place of the last line:

```
     PORTB = (byte)(Mycount&0x00FF);    //display the low byte
     PORTA = (byte)((Mycount&0xFF00)>>8);//display the high byte
```

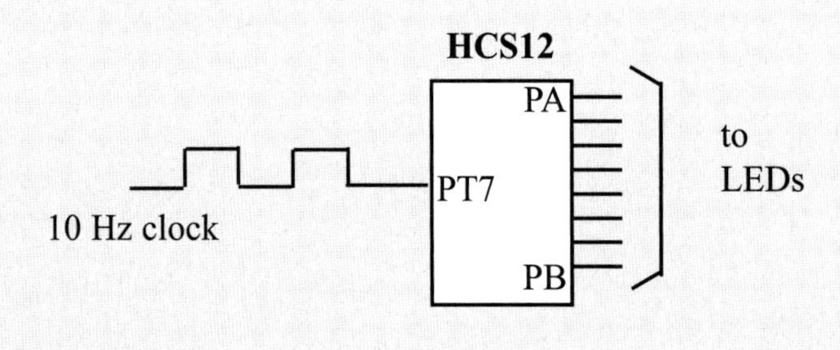

Example 9-42 (C version of Example 9-23)

Assuming that clock pulses are fed into pin PT7 for PACA and a buzzer is connected to pin PORTB.1, write a program to sound the buzzer after 1,000 pulses. This is the C version of Example 9-23.

Solution:

To sound the buzzer every 1,000 pulses, we set the initial event counter value to –1,000 (FC18 in hex), then the counter counts up until it reaches $FFFF. Upon overflow, we can sound the buzzer by toggling the PORTB.1 pin.

```
#include <mc9s12dp512.h>
unsigned int Mycount;
void main(void)
  {
   DDRB = DDRB | 0x02; //make PORTB.1 an output
   DDRT = DDRT & ~0x80; //make PT7 an input
   PACTL = 0x50;        //enable 16-bit PACA event counter,
                        //no interrupt, count pulses on rising edge
   PACN32 = 0xFC18;     //set PACA 16-bit register to -1000
 while (!(PAFLG & PAFLG_PAOVF_MASK));//wait for PACA to overflow
 while(1)               //keep sounding the buzzer
    {
      PORTB = PORTB | 0x02;  //make PB1 high
      PORTB = PORTB & 0xFD;  //make PB1 low
    }
  }
```

Bit 1 of PORTB is connected to a buzzer and input PT7 to pulses.

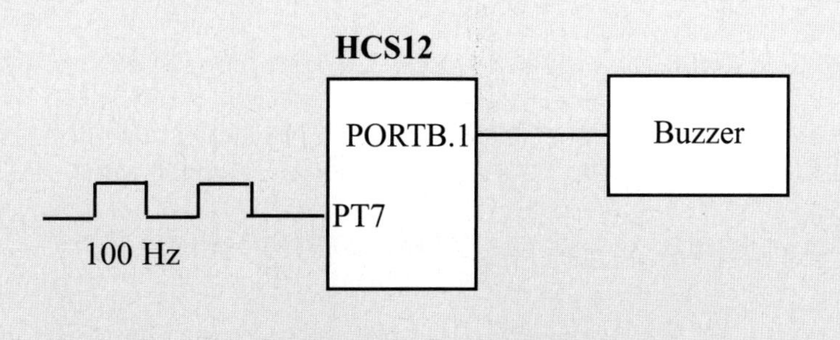

Example 9-43 (C version of Example 9-24)

Assume that a 1 Hz frequency pulse is connected to input for PACA (pin PT7). Write a program to turn on pin PORTB.4 every time PACA reaches the value 50 (decimal). This is the C version of Example 9-24.

Solution:

We set the initial event counter value to –50 (FFCE in hex), then the counter counts up until it reaches FFFF. Upon overflow, we turn on the PORTB.4 pin.

```
#include <mc9s12dp512.h>
void main(void)
  {
    DDRB = DDRB | 0x10; //make PORTB.4 an output
    DDRT = DDRT & ~0x80; //make PT7 an input
    PACTL = 0x50; //enable 16-bit PACA event counter,
                  //no interrupt, count pulses on rising edge
  while(1)        //keep checking counter and make PB4=1 if overflow
    {
      PACN32 = 0xFFCE; //set PACA 16-bit register to -50
      PORTB = PORTB & 0xEF;  //make PB4 low
      while (!(PAFLG & PAFLG_PAOVF_MASK));
      //wait for PACA to overflow
      PAFLG = PAFLG | PAFLG_PAOVF_MASK;//clear PAOVF flag
      PORTB = PORTB | 0x10;  //make PB4 high
    }
  }
```

Example 9-44 (C version of Example 9-25)

Assume that a 1 Hz frequency pulse is connected to input for PACA (pin PT7). The pulses represent the number of bottles being packaged together. Write a program to turn on PORTB.4 every time 12 bottles go by. The program should display the binary count on PORTA. This is the C version of Example 9-25.

Solution:

We set the initial event counter value to –12 (FFF4 in hex), then the counter counts up until it reaches FFFF. Upon overflow, we turn on the PORTB.4 pin.

```
#include <mc9s12dp512.h>
unsigned int MyCount;
void main(void)
  {
    DDRA = 0xFF;    //make PORTA output
    PACTL = 0x50; //enable 16-bit PACA counter, no inter, rising edge
    while(1)      //keep checking counter and make PB4=1 if overflow
     {
      PACN32 = 0xFFF4;   //set PACA 16-bit register to -12
      PORTB = PORTB & ~0x10;   //make PB4 low
      while (!(PAFLG & PAFLG_PAOVF_MASK));//wait for PACA to overflow
      PAFLG = PAFLG | PAFLG_PAOVF_MASK;//clear PAOVF flag
      PORTA = ++MyCount;
      PORTB = PORTB | 0x10;    //make PB4 high
     }
  }
```

Example 9-45 (C version of Example 9-26)

Modify Example 9-44 to display the data in ASCII on PORTA and PORTH. This is the C version of Example 9-26.

Solution:

We need to convert the binary data to decimal and then to ASCII before we display it.

```
#include <mc9s12dp512.h>
unsigned int Mycount, temp;
unsigned char y,z;
void main(void)
  {
    DDRA = 0x0FF;  //make PORTB output
    DDRH = 0x0FF;  //make PORTH output
    DDRB = DDRB | 0x02; //make PORTB.1 an output
    PACTL = 0x50; //enable 16-bit PACA counter, no inter, rising edge
    while(1) //keep checking counter and make PB1=1 if overflow
      {
      PACN32 = 0xFFF4;    //set PACA 16-bit register to -12
      PORTB = PORTB & ~0x10;   //make PB4 low
      while (!(PAFLG & PAFLG_PAOVF_MASK));//wait for PACA to overflow
      PAFLG = PAFLG | PAFLG_PAOVF_MASK; //clear PAOVF flag
      PORTB = PORTB | 0x10;    //make PB4 high
      ++Mycount;
      temp = Mycount;
      z = temp % 10;
      temp = temp/10;
      y = temp % 10;
      PORTA = z | 0x30;
      PTH = y | 0x30;
      }
  }
```

Example 9-46 (C version of Example 9-27)

Assume that a 1 Hz pulse is connected to input for PACB (pin PT0). Write a program to turn on pin PORTB.4 every time PACB reaches the value 10 (decimal). This is the C version of Example 9-27.

Solution:

We set the initial event counter value to –10 (FFF6 in hex), then the counter counts up until it reaches FFFF. Upon overflow, we turn on the PORTB.4 pin.

```
#include <mc9s12dp512.h>
void main(void)
  {
   unsigned int Mycount;
   DDRT = DDRT & ~0x01;
   DDRB = DDRB | 0x10; //make PORTB.4 an output
   PBCTL = 0x40; //enable 16-bit PACB event counter
   TCTL4 = 0x01; //detect rising edge on PT0
   while(1) //keep checking counter and make PB1=1 if overflow
     {
     PACN10 = 0xFFF6;   //set PACB 16-bit register to -10
     PORTB = PORTB & 0xEF;    //make PB4 low
     while (!(PBFLG & PBFLG_PBOVF_MASK));//wait for PACB to overflow
     PBFLG = PBFLG | PBFLG_PBOVF_MASK;    //clear PAOVF flag
     PORTB = PORTB | 0x10;    //make PB4 high
     }
  }
```

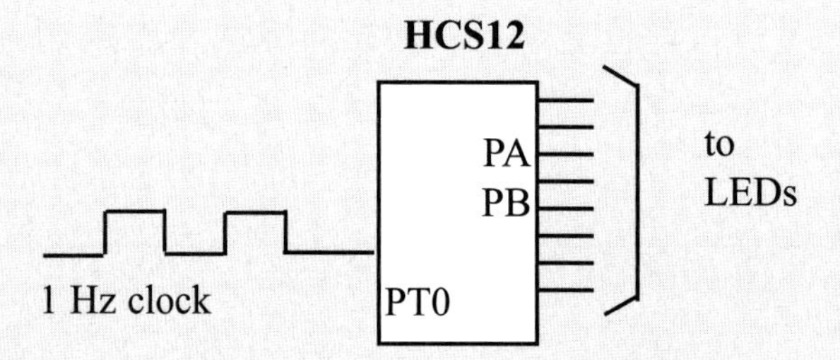

PROBLEMS

SECTION 9.1: FREE-RUNNING TIMER AND OUTPUT COMPARE FUNCTION

1. How many free-running timers do we have in the HCS12?
2. TCNT of the HCS12 is ____-bit.
3. For accurate reading, the TCNT is accessed as (8-bit, 16-bit).
4. The free-running timer supports a highest prescaler value of ______.
5. The free-running timer supports a lowest prescaler value of ______.
6. Which register holds the TOF?
7. How do we enable the free-running timer?
8. True or false. The TEN bit is part of the TSCR1 register.
9. Find the TSCR2 value for no prescaler, no interrupt, count up during the freeze, and wait.
10. Find the frequency and period used by the free-running timer if the crystal attached to the HCS12 has the following values:
    (a) XTAL = 4 MHz (b) XTAL = 8 MHz
    (c) XTAL = 10 MHz
    No prescaler is used.
11. How many Output Compare channels do we have in HCS12? Give the name of the TCx register for each channel and the output pin associated with it.
12. Indicate the rollover value (in hex and decimal) for the 16-bit free-running timer.
13. Explain the role of the TCx register in the Output Compare function and how the CxF flag is raised.
14. True or false. TCNT has its own overflow flag.
15. True or false. Each Output Compare channel has its own match flag.
16. The selection of Output Compare/Input Capture function is done by manipulating the bits of the ______ registers.
17. Explain the role of TCTL1 and TCTL2 in the Output Compare function.
18. Assume that XTAL = 8 MHz. Find the value needed to generate a time delay of 5 ms. Use the largest prescaler possible.
19. Assume that XTAL = 8 MHz. Find the value needed to generate a time delay of 5 ms. Use prescaler of 16.
20. Assume that XTAL = 4 MHz. Find the value needed to generate a time delay of 5 ms. Use the largest prescaler possible.
21. Assuming that XTAL = 8 MHz, and we are generating a square wave on pin PB7, find the lowest square wave frequency that we can generate.
22. Assuming that XTAL = 8 MHz, and we are generating a square wave on pin PB2, find the highest square wave frequency that we can generate.
23. Assuming that XTAL = 8 MHz, find the period of clock pulse fed to the TCNT if we use a prescaler of 4.
24. Assuming that XTAL = 8 MHz, find the period of clock pulse fed to the TCNT if we use a prescaler of 64.
25. Program channel 2 for Output Compare to generate a square wave of 1 kHz with 50% duty cycle. Assume that XTAL = 4 MHz.

26. Program channel 4 for Output Compare to generate a square wave of 1 kHz with 60% duty cycle. Assume that XTAL = 4 MHz. Use the largest prescaler possible.
27. Assuming that TCNT = FFF1H, indicate which states TCNT goes through until the overflow flag is raised. How many states is that?
28. Assuming that XTAL = 8 MHz, find the period of clock pulse fed to the TCNT for the following prescalers:

(a)	2	(b)	128
(c)	8	(d)	16
(e)	32	(f)	64

SECTION 9.2: INPUT CAPTURE PROGRAMMING

29. Explain the role of the TIOS register for the Input Capture function.
30. True or false. The Input Capture function can detect only the falling edge.
31. How many Input Capture channels do we have in HCS12? Give the name of the TCx register for each channel and the input pin associated with it.
32. Explain the role of TCTL3 and TCLT4 in Input Capture function programming.
33. Explain the role of the TCx register in the Input Capture function and how the CxF flag is raised.
34. True or false. We can use a given TCx channel for both the Output Compare and the Input Capture function at the same time.
35. True or false. Each Input Capture channel has its own edge detect flag.
36. The selection of Output Compare/Input Capture function is done by manipulating the bits of the ______ register. Show how we select the Input Capture function for channel 3.

SECTION 9.3: PULSE ACCUMULATOR AND EVENT COUNTER PROGRAMMING

37. How many 8-bit Pulse Accumulator channels do we have in the HCS12? Give their names and the input pins associated with them.
38. Explain the role of the PACTL in Pulse Accumulator function programming.
39. Explain the role of the ICPAR register in Pulse Accumulator programming.
40. True or false. We can use a given PACx channel for both the 8-bit and 16-bit pulse accumulation at the same time.
41. True or false. Each Pulse Accumulator channel has its own edge detect flag.
42. The selection of 16-bit and 8-bit Pulse Accumulator function is done by manipulating the bits of the ______ register. Show how we select the 16-bit Pulse Accumulator A.
43. How many 16-bit Pulse Accumulators can we have in HCS12?
44. Explain how the PACA is formed.
45. For the PACA counter, which pin is used for the input pulse?
46. For the PAC1 counter, which pin is used for the input pulse?
47. True or false. The Pulse Accumulator can detect only the falling edge.
48. Program PACA to be an event counter. Use 16-bit mode, and display the bina-

ry count on PORTB and PORTA continuously. Set the initial count to 20,000.

49. Program the PAC2 counter to be an event counter. Display the binary count on PORTB continuously. Set the initial count to 20.
50. Explain the role of the PAOVF bit in the PAFLG register for the Pulse Accumulator function.

SECTION 9.4: HCS12 TIMER PROGRAMMING IN C

51. Write the C version of Problem 25.
52. Write the C version of Problem 26.
53. Write the C version of Problem 48.
54. Write the C version of Problem 49.

ANSWERS TO REVIEW QUESTIONS

SECTION 9.1: FREE-RUNNING TIMER AND OUTPUT COMPARE FUNCTION

1. 1
2. False
3. False
4. True
5. 10,000,000 indicates timer enable, timer continues counting during the WAIT, and freeze.
6. $FFFF, 0000
7. $FFFF, 0000
8. 64
9. 100 μs / 0.5 μs = 200; therefore TC0 = TCNT + 200.

SECTION 9.2: INPUT CAPTURE PROGRAMMING

1. 8
2. True
3. True
4. TIOS
5. False

SECTION 9.3: PULSE ACCUMULATOR AND EVENT COUNTER PROGRAMMING

1. PT3
2. The clock source for the timer comes from pin PT0.
3. Yes
4. We must configure the pin as input to allow the clocks to come in from an external source.
5. Yes

CHAPTER 10

HCS12 SERIAL PORT PROGRAMMING IN ASSEMBLY AND C

OBJECTIVES

Upon completion of this chapter, you will be able to:

>> Contrast and compare serial versus parallel communication
>> List the advantages of serial communication over parallel
>> Explain serial communication protocol
>> Contrast synchronous versus asynchronous communication
>> Contrast half- versus full-duplex transmission
>> Explain the process of data framing
>> Describe data transfer rate and bps rate
>> Define the RS232 standard
>> Explain the use of the MAX232 and MAX233 chips
>> Interface the HCS12 with an RS232 connector
>> Discuss the baud rate of the HCS12
>> Describe the serial communication features of the HCS12
>> Describe the main registers used for serial communication on the HCS12
>> Program the HCS12 serial port in Assembly and C

Computers transfer data in two ways: parallel and serial. In parallel data transfers, often eight or more lines (wire conductors) are used to transfer data to a device that is only a few feet away. Devices that use parallel transfers include printers and hard disks; each uses cables with many wire strips. Although a lot of data can be transferred in a short amount of time by using many wires in parallel, the distance cannot be great. To transfer to a device located many meters away, the serial method is used. In serial communication, the data is sent one bit at a time, in contrast to parallel communication, in which the data is sent a byte or more at a time. Serial communication of the HCS12 is the topic of this chapter. The HCS12 has serial communication capability built into it, thereby making possible fast data transfer using only a few wires. In this chapter we discuss the basics of serial communication for HCS12 and show how to program it in Assembly and C.

SECTION 10.1: BASICS OF SERIAL COMMUNICATION

When a microprocessor communicates with the outside world, it provides the data in byte-sized chunks. In some cases, such as printers, the information is simply grabbed from the 8-bit data bus and presented to the 8-bit data bus of the printer. This can work only if the cable is not too long, because long cables diminish and even distort signals. Furthermore, an 8-bit data path is expensive. For these reasons, serial communication is used for transferring data between two systems located at distances of hundreds of feet to millions of miles apart. Figure 10-1 diagrams serial versus parallel data transfers.

The fact that a single data line is used in serial communication instead of the 8-bit data line of parallel communication makes serial transfer not only much cheaper but also enables two computers located in two different cities to communicate over a telephone line.

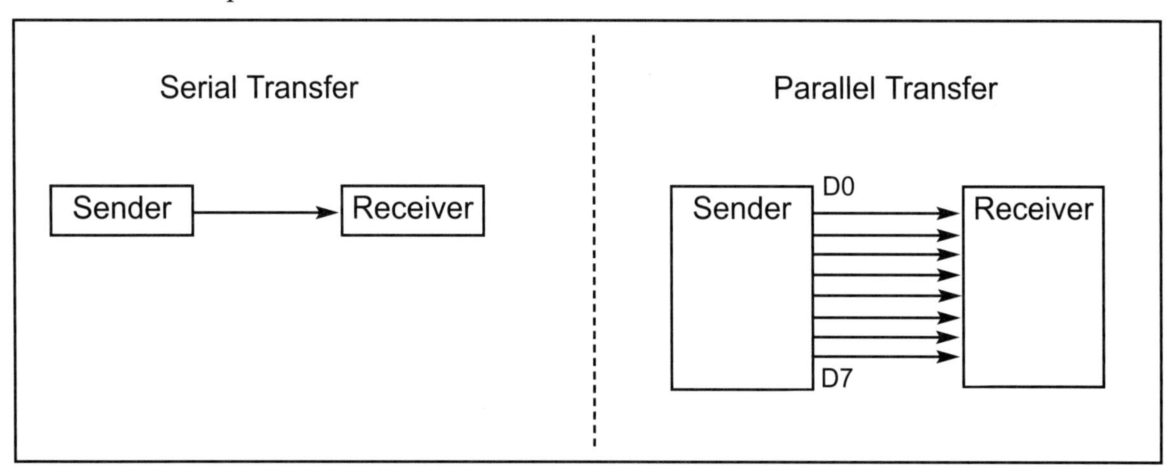

Figure 10-1. Serial versus Parallel Data Transfer

For serial data communication to work, the byte of data must be converted to serial bits using a parallel-in-serial-out shift register; then it can be transmitted over a single data line. This also means that at the receiving end there must be a serial-in-parallel-out shift register to receive the serial data bits and pack them into a byte. Of course, if data is to be transferred on the telephone line, it must be converted from 0s and 1s to audio tones, which are sinusoidal signals. This conversion is performed by a peripheral device called a *modem*, which stands for "modulator/demodulator."

When the distance is short, the digital signal can be transferred as it is on a simple wire and requires no modulation. This is how IBM PC keyboards transfer data to the motherboard. For long-distance data transfers using communication lines such as a telephone, however, serial data communication requires a modem to *modulate* (convert from 0s and 1s to audio tones) and *demodulate* (convert from audio tones to 0s and 1s).

Serial data communication uses two methods, asynchronous and synchronous. The *synchronous* method transfers a block of data (characters) at a time whereas the *asynchronous* method transfers a single byte at a time. It is possible to write software to use either of these methods, but the programs can be tedious and long. For this reason, special IC chips are made by many manufacturers for serial data communications. These chips are commonly referred to as UART (universal asynchronous receiver-transmitter) and USART (universal synchronous-asynchronous receiver-transmitter). The HCS12 chip has a built-in USART, which is discussed in detail in Section 10.3.

Half- and full-duplex transmission

A data transmission in which the data can be both transmitted and received is a *duplex* transmission. This is in contrast to *simplex* transmissions such as with printers, in which the computer only sends data. Duplex transmissions can be half or full duplex, depending on whether or not the data transfer can be simultaneous. If data is transmitted one way at a time, it is referred to as *half duplex*. If the data can go both ways at the same time, it is *full duplex*. Of course, full duplex requires two wire conductors for the data lines (in addition to the signal ground), one for transmission and one for reception, in order to transfer and receive data simultaneously. See Figure 10-2.

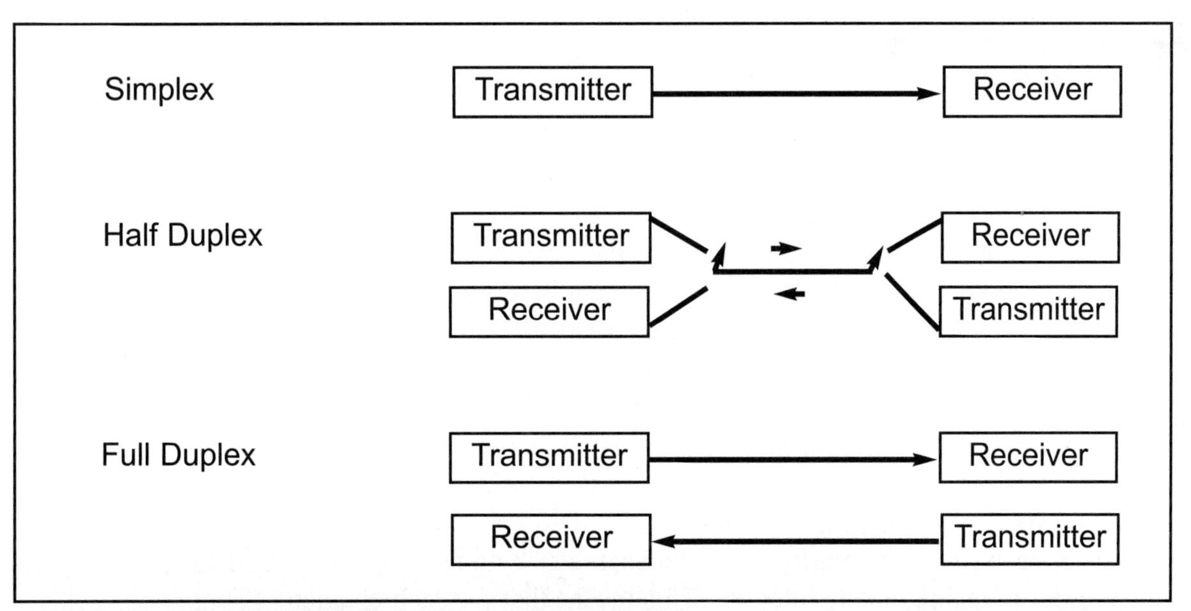

Figure 10-2. Simplex, Half-, and Full-Duplex Transfers

Asynchronous serial communication and data framing

The data coming in at the receiving end of the data line in a serial data transfer is all 0s and 1s. It is difficult to make sense of the data unless the sender and receiver agree on a set of rules, a *protocol*, on how the data is packed, how many bits constitute a character, and when the data begins and ends.

Start and stop bits

Asynchronous serial data communication is widely used for character-oriented transmissions, while block-oriented data transfers use the synchronous method. In the asynchronous method, each character is placed between start and stop bits. This is called *framing*. In data framing for asynchronous communications, the data, such as ASCII characters, are packed between a start bit and a stop bit. The start bit is always one bit, but the stop bit can be one or two bits. The start bit is always a 0 (low) and the stop bit(s) is 1 (high). For example, look at Figure 10-3 in which the ASCII character "A" (8-bit binary 0100 0001) is framed between the start bit and a single stop bit. Notice that the LSB is sent out first.

Notice in Figure 10-3 that when there is no transfer, the signal is 1 (high), which is referred to as *mark*. The 0 (low) is referred to as *space*. Notice that the transmission begins with a start bit followed by D0, the LSB, then the rest of the bits until the MSB (D7), and finally, the one stop bit indicating the end of the character "A".

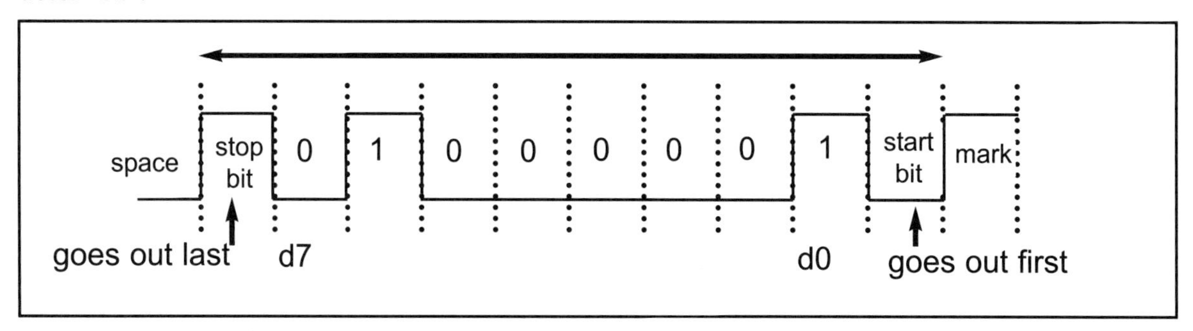

Figure 10-3. Framing ASCII 'A' (41H)

In asynchronous serial communications, peripheral chips and modems can be programmed for data that is 7 or 8 bits wide. This is in addition to the number of stop bits, 1 or 2. While in older systems ASCII characters were 7-bit, in recent years, 8-bit data has become common due to the extended ASCII characters. In some older systems, due to the slowness of the receiving mechanical device, two stop bits were used to give the device sufficient time to organize itself before transmission of the next byte. In modern PCs, however, the use of one stop bit is standard. Assuming that we are transferring a text file of ASCII characters using 1 stop bit, we have a total of 10 bits for each character: 8 bits for the ASCII code, and 1 bit each for the start and stop bits. Therefore, each 8-bit character has an extra 2 bits, which gives 25% overhead.

In some systems, the parity bit of the character byte is included in the data frame in order to maintain data integrity. This means that for each character (7- or 8-bit, depending on the system) we have a single parity bit in addition to start and stop bits. The parity bit is either odd or even. In the case of an odd parity bit the number of 1s in the data bits, including the parity bit, is odd. Similarly, in an even parity bit system the total number of bits, including the parity bit, is even. For example, the ASCII character "A", binary 0100 0001, has 0 for the even parity bit. UART chips allow programming of the parity bit for odd-, even-, and no-parity options.

Data transfer rate

The rate of data transfer in serial data communication is stated in *bps* (bits per second). Another widely used terminology for bps is *baud rate*. However, the baud and bps rates are not necessarily equal. This is because baud rate is the modem terminology and is defined as the number of signal changes per second. In modems, sometimes a single change of signal transfers several bits of data. As far as the conductor wire is concerned, the baud rate and bps are the same, and for this reason we use the terms bps and baud interchangeably in this book.

The data transfer rate of a given computer system depends on the communication ports incorporated into that system. For example, the early IBM PC/XT could transfer data at the rate of 100 to 9,600 bps. In recent years, however, Pentium-based PCs have transferred data at rates as high as 56K. Note that in asynchronous serial data communication, the baud rate is generally limited to 100,000 bps.

RS232 standards

To allow compatibility among data communication equipment made by various manufacturers, an interfacing standard called RS232 was set by the Electronics Industries Association (EIA) in 1960. In 1963 it was modified and called RS232A. RS232B and RS232C were issued in 1965 and 1969, respectively. In this book we refer to it simply as RS232. Today, RS232 is the most widely used serial I/O interfacing standard. This standard is used in PCs and numerous types of equipment. Because the standard was set long before the advent of the TTL logic family, however, its input and output voltage levels are not TTL compatible. In RS232, a 1 is represented by –3 to –25 V, while a 0 bit is +3 to +25 V, making -3 to +3 undefined. For this reason, to connect any RS232 to a microcontroller system we must use voltage converters such as MAX232 to convert the TTL logic levels to the RS232 voltage level, and vice versa. MAX232 IC chips are commonly referred to as *line drivers*. RS232 connection to MAX232 is discussed in Section 10.2.

RS232 pins

Table 10-1 provides the pins and their labels for the RS232 cable, commonly referred to as the DB-25 connector. In labeling, DB-25P refers to the plug connector (male) and DB-25S is for the socket connector (female). See Figure 10-4.

Because not all the pins are used in PC cables, IBM introduced the DB-9 version of the serial I/O standard, which uses only 9 pins, as shown in Table 10-2. The DB-9 pins are shown in Figure 10-5.

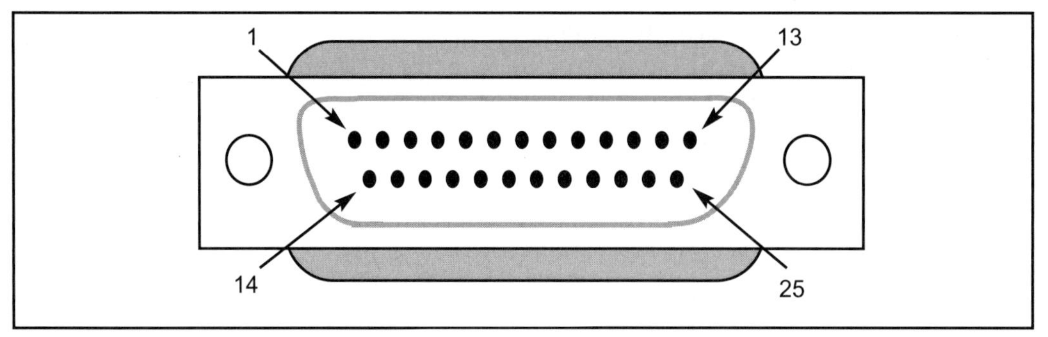

Figure 10-4. RS232 Connector DB-25

Data communication classification

Current terminology classifies data communication equipment as DTE (data terminal equipment) or DCE (data communication equipment). DTE refers to terminals and computers that send and receive data, while DCE refers to communication equipment, such as modems, that are responsible for transferring the data. Notice that all the RS232 pin function definitions of Tables 10-1 and 10-2 are from the DTE point of view.

The simplest connection between a PC and a microcontroller requires a minimum of three pins, TX, RX, and ground, as shown in Figure 10-6. Notice in that figure that the RX and TX pins are interchanged.

Table 10-1: RS232 Pins (DB-25)

Pin	Description
1	Protective ground
2	Transmitted data (TxD)
3	Received data (RxD)
4	Request to send ($\overline{\text{RTS}}$)
5	Clear to send ($\overline{\text{CTS}}$)
6	Data set ready ($\overline{\text{DSR}}$)
7	Signal ground (GND)
8	Data carrier detect ($\overline{\text{DCD}}$)
9/10	Reserved for data testing
11	Unassigned
12	Secondary data carrier detect
13	Secondary clear to send
14	Secondary transmitted data
15	Transmit signal element timing
16	Secondary received data
17	Receive signal element timing
18	Unassigned
19	Secondary request to send
20	Data terminal ready ($\overline{\text{DTR}}$)
21	Signal quality detector
22	Ring indicator
23	Data signal rate select
24	Transmit signal element timing
25	Unassigned

Examining RS232 handshaking signals

To ensure fast and reliable data transmission between two devices, the data transfer must be coordinated. Just as in the case of the printer, because the receiving device may have no room for the data in serial data communication, there must be a way to inform the sender to stop sending data. Many of the pins of the RS-232 connector are used for handshaking signals. Their descriptions are provided below only as a reference, and they can be bypassed because they are not supported by the HCS12 UART chip.

1. DTR (data terminal ready). When the terminal (or a PC COM port) is turned on, after going through a self-test, it sends out signal DTR to indicate that it is ready for communication. If there is something wrong with the COM port, this sig-

nal will not be activated. This is an active-LOW signal and can be used to inform the modem that the computer is alive and kicking. This is an output pin from DTE (PC COM port) and an input to the modem.

2. DSR (data set ready). When the DCE (modem) is turned on and has gone through the self-test, it asserts DSR to indicate that it is ready to communicate. Thus, it is an output from the modem (DCE) and an input to the PC (DTE). This is an active-LOW signal. If for any reason the modem cannot make a connection to the telephone, this signal remains inactive, indicating to the PC (or terminal) that it cannot accept or send data.
3. RTS (request to send). When the DTE device (such as a PC) has a byte to transmit, it asserts RTS to signal the modem that it has a byte of data to transmit. RTS is an active-LOW output from the DTE and an input to the modem.
4. CTS (clear to send). In response to RTS, when the modem has room to store the data it is to receive, it sends out signal CTS to the DTE (PC) to indicate that it can receive the data now. This input signal to the DTE is used by the DTE to start transmission.

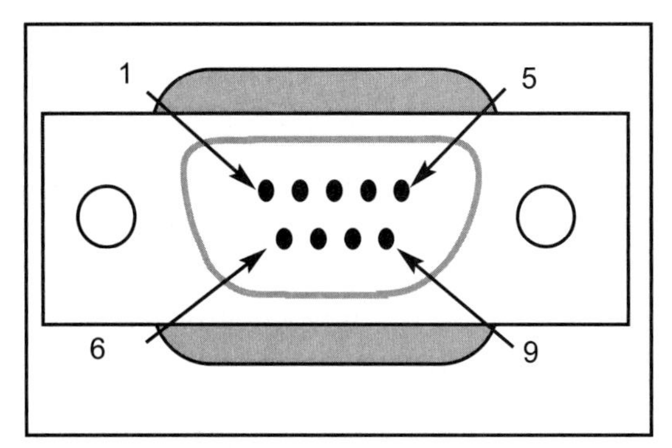

Figure 10-5. DB-9 9-Pin Connector

Table 10-2: IBM PC DB-9 Signals

Pin	Description
1	Data carrier detect ($\overline{\text{DCD}}$)
2	Received data (RxD)
3	Transmitted data (TxD)
4	Data terminal ready (DTR)
5	Signal ground (GND)
6	Data set ready ($\overline{\text{DSR}}$)
7	Request to send ($\overline{\text{RTS}}$)
8	Clear to send ($\overline{\text{CTS}}$)
9	Ring indicator (RI)

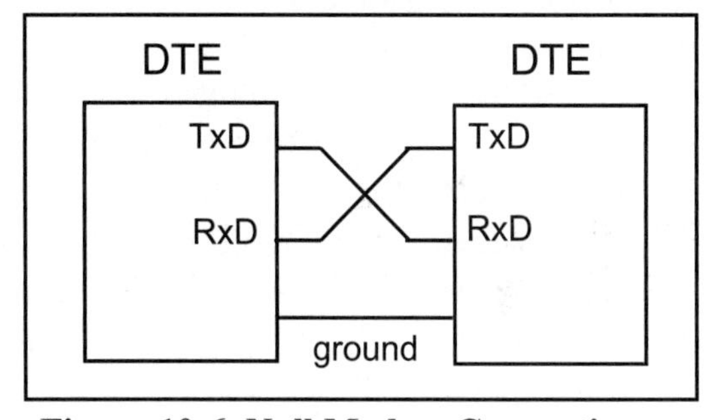

Figure 10-6. Null Modem Connection

5. DCD (data carrier detect). The modem asserts signal DCD to inform the DTE (PC) that a valid carrier has been detected and that contact between it and the other modem is established. Therefore, DCD is an output from the modem and an input to the PC (DTE).
6. RI (ring indicator). An output from the modem (DCE) and an input to a PC (DTE) that indicates that the telephone is ringing. RI goes on and off in synchronization with the ringing sound. Of the six handshake signals, this is the least often used because modems take care of answering the phone. If in a given system the PC is in charge of answering the phone, however, this signal can be used.

From the above description, PC and modem communication can be summarized as follows: While signals DTR and DSR are used by the PC and modem, respectively, to indicate that they are alive and well, RTS and CTS are the signals that actually control the flow of data. When the PC wants to send data it asserts RTS, and in response, the modem, if it is ready (has room) to accept the data, sends back CTS. If, for lack of room, the modem does not activate CTS, the PC will desert DTR and try again. RTS and CTS are also referred to as *hardware flow control signals*.

This concludes the description of the most important pins of the RS232 handshake signals plus TxD, RxD, and ground. Ground is also referred to as SG (signal ground).

IBM PC/compatible COM ports

IBM PC/compatible computers based on x86 (8086, 286, 386, 486, and all Pentium) microprocessors used to have two COM ports. Both COM ports were RS232-type connectors. Many PCs used one each of the DB-25 and DB-9 RS232 connectors. The COM ports were designated as COM 1 and COM 2. In recent years, one of these has been replaced with the USB port, and COM 1 is the only serial port available, if any. We can connect the HCS12 serial port to the COM 1 port of a PC for serial communication experiments. In the absence of a COM port, we can use a COM-to-USB converter module.

With this background in serial communication, we are ready to look at the HCS12. In the next section we discuss the physical connection of the HCS12 and RS232 connector, and in Section 10.3 we see how to program the HCS12 serial communication port.

Review Questions

1. The transfer of data using parallel lines is ___________ (faster, slower) but ________________ (more expensive, less expensive).
2. True or false. Sending data to a printer is duplex.
3. True or false. In full duplex we must have two data lines, one for transfer and one for receive.
4. The start and stop bits are used in the ___________ (synchronous, asynchronous) method.
5. Assuming that we are transmitting the ASCII letter "E" (0100 0101 in binary) with no parity bit and one stop bit, show the sequence of bits transferred serially.
6. In Question 5, find the overhead due to framing.
7. Calculate the time it takes to transfer 10,000 characters as in Question 5 if we use 9,600 bps. What percentage of time is wasted due to overhead?
8. True or false. RS232 is not TTL compatible.
9. What voltage levels are used for binary 0 in RS232?
10. True or false. The HCS12 has a built-in UART.
11. On the back of x86 PCs, we normally have ____ COM port connectors.
12. The PC COM ports are designated by DOS and Windows as ______ and ______.

SECTION 10.2: HCS12 CONNECTION TO RS232

In this section, the details of the physical connections of the HCS12 to RS232 connectors are given. As stated in Section 10.1, the RS232 standard is not TTL compatible; therefore, a line driver such as the MAX232 chip is required to convert RS232 voltage levels to TTL levels, and vice versa. The interfacing of the HCS12 with RS232 connectors via the MAX232 chip is the main topic of this section.

RXD and TXD pins in the HCS12

The HCS12 has two SCI (serial communication interface) ports and they are designated as SCI0 and SCI1. Each of the SCI ports has two pins that are used specifically for transferring and receiving data serially. These two pins are called TXD and RXD and are part of the port S group (PS0–PS3) of the QFP package. For SCI0 we have pins 63 (PS0/RXD0) and 64 (PS1/TXD0). In the case of the second serial port of SCI1, pin 65 (PS2/RXD1) is assigned to RXD1 and pin 66 (PS3/TXD1) is designated as TXD1. These four pins are not TTL compatible; therefore, they require a line driver to make them RS232 compatible. One such line driver is the MAX232 chip. This is discussed next.

MAX232

Because the RS232 is not compatible with today's microprocessors and microcontrollers, we need a line driver (voltage converter) to convert the RS232's signals to TTL voltage levels that will be acceptable to the HCS12's TXD and RXD pins. One example of such a converter is MAX232 from Maxim Corp. (www.maxim-ic.com). The MAX232 converts from RS232 voltage levels to TTL voltage levels, and vice versa. One advantage of the MAX232 chip is that it uses a +5 V power source, which is the same as the source voltage for the HCS12. In other words, with a single +5 V power supply we can power both the HCS12 and MAX232, with no need for the dual power supplies that are common in many older systems.

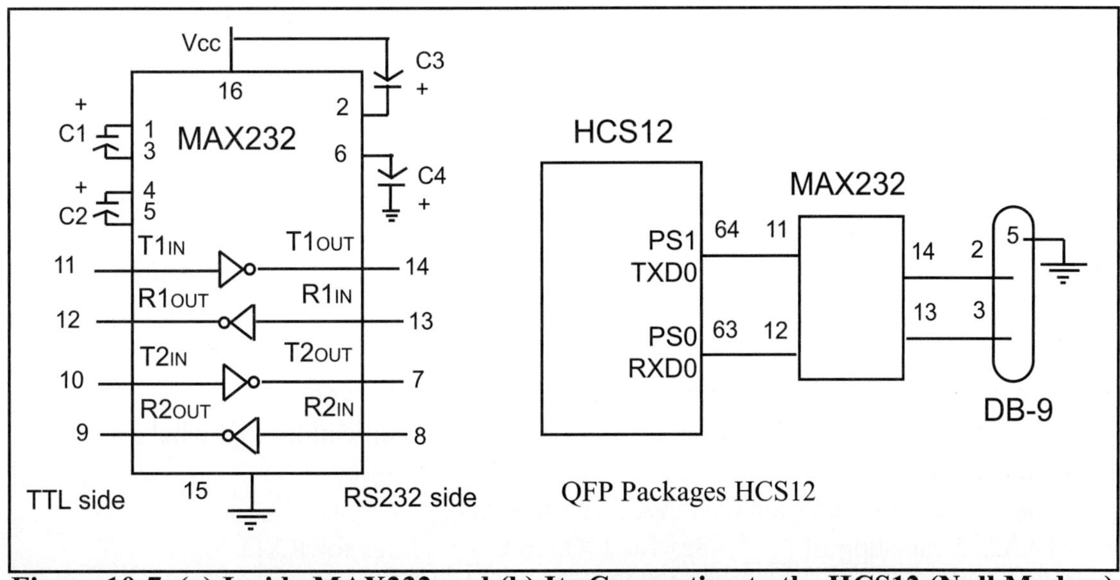

Figure 10-7. (a) Inside MAX232 and (b) Its Connection to the HCS12 (Null Modem)

The MAX232 has two sets of line drivers for transferring and receiving data, as shown in Figure 10-7. The line drivers used for TXD are called T1 and T2, while the line drivers for RXD are designated as R1 and R2. In many applications only one is used. For example, T1 and R1 are used together for TXD0 and RXD0 of the HCS12, and the second set is either left unused or used for RXD1 and TXD1. Notice in MAX232 that the T1 line driver has a designation of T1in and T1out on pin numbers 11 and 14, respectively. The T1in pin is the TTL side and is connected to TX of the microcontroller, while T1out is the RS232 side that is connected to the RX pin of the RS232 DB connector. The R1 line driver has a designation of R1in and R1out on pin numbers 13 and 12, respectively. The R1in (pin 13) is the RS232 side that is connected to the TX pin of the RS232 DB connector, and R1out (pin 12) is the TTL side that is connected to the RXD pin of the microcontroller. See Figure 10-7. Notice the null modem connection where RXD for one is TXD for the other.

MAX232 requires four capacitors ranging from 1 to 22 μF. The most widely used value for these capacitors is 22 μF.

MAX233

To save board space, some designers use the MAX233 chip from Maxim. The MAX233 performs the same job as the MAX232 but eliminates the need for capacitors. However, the MAX233 chip is much more expensive than the MAX232. Notice that MAX233 and MAX232 are not pin compatible. You cannot take a MAX232 out of a board and replace it with a MAX233. See Figure 10-8 for MAX233 with no capacitor used.

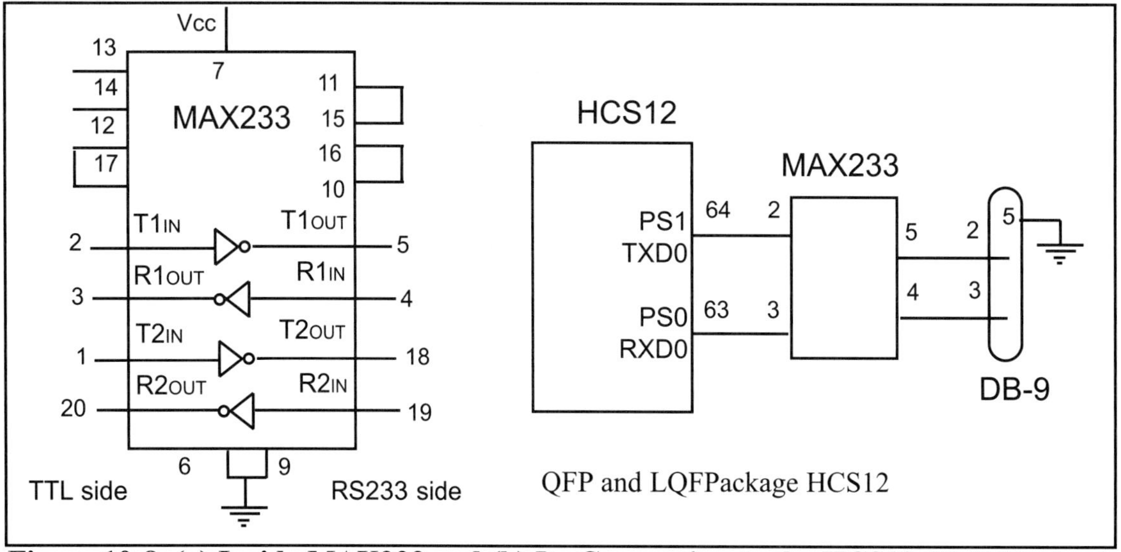

Figure 10-8. (a) Inside MAX233 and (b) Its Connection to the HCS12 (Null Modem)

Review Questions

1. True or false. The PC COM port connector is the RS232 type.
2. Which pins of the HCS12 are set aside for serial communication, and what are their functions?
3. What are line drivers such as MAX 232 used for?
4. MAX232 can support ____ lines for TXD and ____ lines for RXD.
5. What is the advantage of the MAX233 over the MAX232 chip?

SECTION 10.3: HCS12 SERIAL PORT PROGRAMMING IN ASSEMBLY

In this section we discuss the serial communication registers of the HCS12 and show how to program them to transfer and receive data using the asynchronous mode. The UART (universal asynchronous receiver) in the HCS12 is called SCI (serial communication interface). This is in contrast to the SPI (serial peripheral interface), which is the synchronous type and is covered in Chapter 16. The asynchronous type is the one we use to connect the HCS12-based system to the IBM PC serial COM port for the purpose of full-duplex serial data transfer. In the HCS12 microcontroller there are some major registers associated with the UART that we deal with in this chapter. They are (a) SCIDR (SCI data register), (b) SCICR (SCI control register), (c) SCISR (SCI status register), and (d) SCIBD (SCI Baud register). We examine each of them and show how they are used in full-duplex serial data communication.

SCI (Serial Com Interface) baud rate register in the HCS12

Because IBM PC/compatible computers are so widely used to communicate with HCS12-based systems, we will emphasize serial communications of the HCS12 with the COM port of the PC. Some of the baud rates supported by PC HyperTerminal are listed in Table 10-3. You can examine these baud rates by going to the Microsoft Windows HyperTerminal program and clicking on the Communication Settings option. The HCS12 transfers and receives data serially at many different baud rates. The baud rate in the HCS12 is programmable. This is done with the help of two 8-bit registers called SCIBDH (SCI baud rate high) and SCIBDL (SCI baud rate low). For a given crystal frequency, the value loaded into the SCIBDH:SCIBDL registers decides the baud rate. The relation between the value loaded into SCIBDH:SCIBDL and Fosc (frequency of the crystal connected to the XTAL and EXTAL pins) is dictated by the following formula:

Table 10-3: Some PC Baud Rates in HyperTerminal

HyperTerminal
1,200
2,400
4,800
9,600
19,200
38,400
57,600
115,200

Desired Baud Rate = Fosc/2(16 × X)

where X is the value we load into the SCIBDH:SCIBDL registers. The two registers of SCIBDH and SCIBDL give us 16 bits, but only 13 bits are used. That means they can take values from 1 to 8191 (0 to $1FFF). Assuming that Fosc = 8 MHz (ECLK = 4 MHz), we have the following:

Desired Baud Rate = 8 MHz/ 2(16 × X) = 250 kHz /X

To get the X value for different baud rates we can solve the equation as follows:

X = 250 kHz/Desired Baud Rate

Table 10-4 shows the X values for the different baud rates if ECLK = 4 MHz

SBR7	SBR6	SBR5	SBR4	SBR3	SBR2	SBR1	SBR0

SBR7–SBR0 These are the lower 8 bits of the 13 used for setting the baud rate.

0	0	0	SBR12	SBR11	SBR10	SBR9	SBR8

SBR12–SBR8 These are the upper 5 bits of the 13 used for setting the baud rate.

Note: We must load (write) both of these registers even if the upper 5 bits are all zeros.

Figure 10-9. SCI Baud Rate Registers (SCIBDL and SCIBDH)

(XTAL frequency = 8 MHz with no PLL). Another way to understand the SCIBD values in Table 10-4 is to look at them from the perspective of the bus cycle time. As we discussed in Chapter 8, the HCS12 divides the crystal frequency (Fosc) by 2 to get the ECLK (or bus) frequency. In the case of XTAL = 8 MHz, the ECLK cycle frequency is 4 MHz. The HCS12's UART circuitry divides the ECLK frequency by 16 once more before it is used by an internal timer to set the baud rate. Therefore, 4 MHz divided by 16 gives 250 kHz. This is the number we use to find the SCIBD value shown in Table 10-4. Example 10-1 shows how to verify the data in Table 10-4. Table 10-5 shows the SCIBD values with the crystal frequency of 4 MHz (ECLK = 2 MHz).

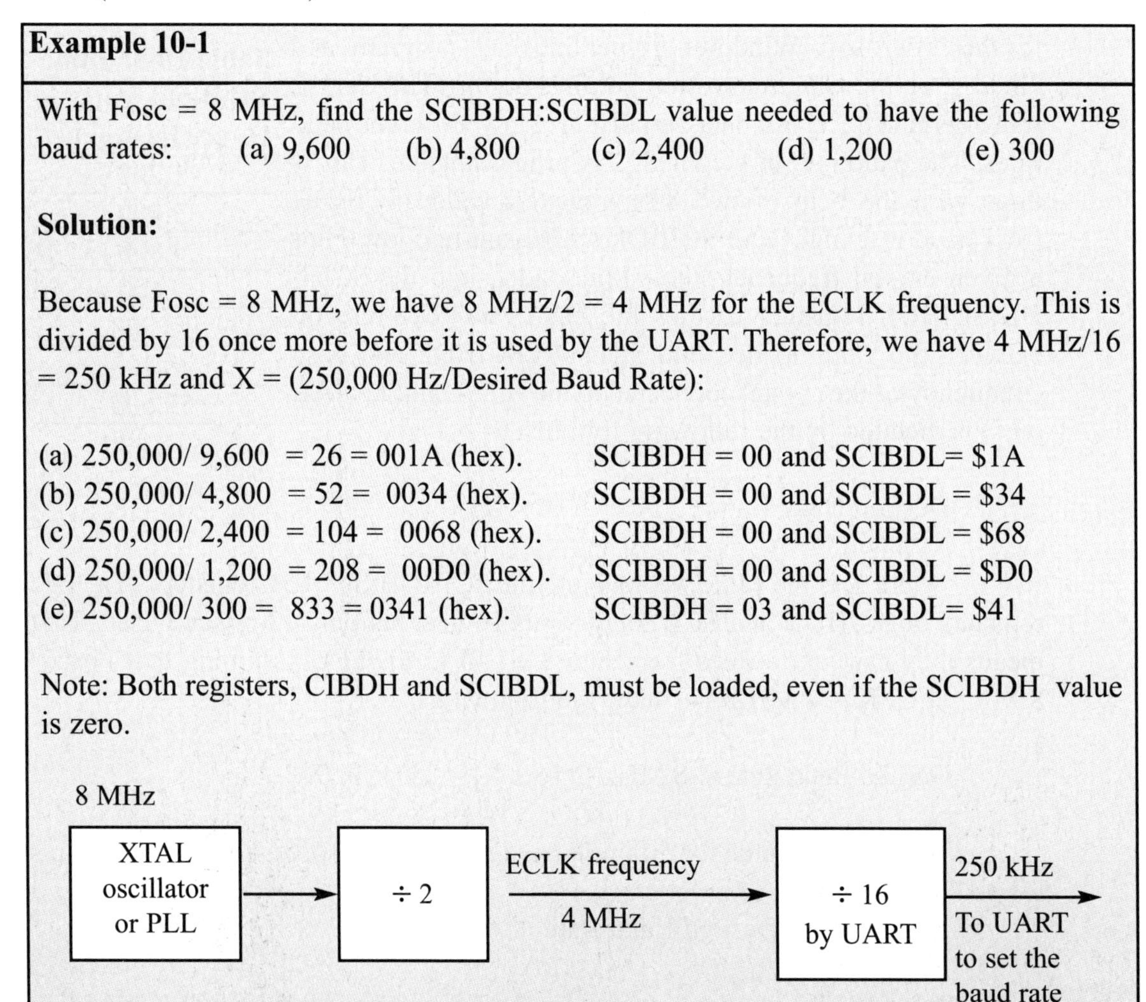

Example 10-1

With Fosc = 8 MHz, find the SCIBDH:SCIBDL value needed to have the following baud rates: (a) 9,600 (b) 4,800 (c) 2,400 (d) 1,200 (e) 300

Solution:

Because Fosc = 8 MHz, we have 8 MHz/2 = 4 MHz for the ECLK frequency. This is divided by 16 once more before it is used by the UART. Therefore, we have 4 MHz/16 = 250 kHz and X = (250,000 Hz/Desired Baud Rate):

(a) 250,000/ 9,600 = 26 = 001A (hex). SCIBDH = 00 and SCIBDL= $1A
(b) 250,000/ 4,800 = 52 = 0034 (hex). SCIBDH = 00 and SCIBDL = $34
(c) 250,000/ 2,400 = 104 = 0068 (hex). SCIBDH = 00 and SCIBDL = $68
(d) 250,000/ 1,200 = 208 = 00D0 (hex). SCIBDH = 00 and SCIBDL = $D0
(e) 250,000/ 300 = 833 = 0341 (hex). SCIBDH = 03 and SCIBDL= $41

Note: Both registers, CIBDH and SCIBDL, must be loaded, even if the SCIBDH value is zero.

Table 10-4: SCIBD Values for Various Baud Rates (Fosc = 8 MHz, ECLK = 4 MHz)

Baud Rate	SCIBDH:SCIBDL (Decimal)	SCIBDH:SCIBDL (Hex)
38400	6	6
19200	13	$0D
9600	26	$1A
4800	52	$34
2400	104	$68
1200	208	$D0
300	833	$341

Note: For Fosc = 8 MHz, SCIBDH:SCIBDL = 8 MHz/ 2(16 × BaudRate) = 250,000/(Baud Rate)

Table 10-5: SCIBD Values for Various Baud Rates (Fosc = 4 MHz, ECLK = 2 MHz)

Baud Rate	SCIBDH:SCIBDL (Decimal)	SCIBDH:SCIBDL (Hex)
28800	4	4
9600	13	$0D
4800	26	$1A
2400	52	$34
1200	104	$68

Note: For Fosc = 4 MHz, SCIBDH:SCIBDL = 4/2(16 × BaudRate) = 125,000/ (Baud Rate)

SCI Data Register for transmit

SCI Data Register (SCIDR) is another register used for serial communication in the HCS12. For a byte of data to be transferred via the TXD pin, it must be placed in the SCIDR register. SCIDR has a low byte (SCIDRL) and a high byte (SCIDRH). For 8-bit data format, we use only the SCIDRL. The SCIDRH is used for the 9-bit data format. See the HCS12 manual since 9-bit data is not widely used.

SCI Data Register for receive

Similarly, when the bits are received serially via the RXD pin, the HCS12 deframes them by eliminating the stop and start bits, making a byte out of the data received, and then placing it in the SCIDRL register. Notice that HCS12 uses the same SCI Data Register, for both receive and transmit.

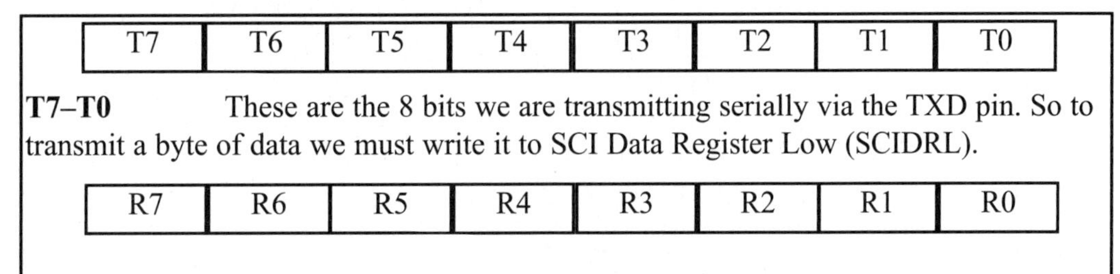

T7–T0 These are the 8 bits we are transmitting serially via the TXD pin. So to transmit a byte of data we must write it to SCI Data Register Low (SCIDRL).

R7–R0 R7–R0 holds the 8-bit data received serially via RXD. That means we use the **same register,** SCI Data Register Low (SCIDRL), for both send and receive.

Note: If we use a 9-bit data frame, then we use the second SCI data register SCI Data Register High (SCIDRH). See the HCS12 manual.

Figure 10-10. SCI Data Register Low (SCIDRL)

SCI Control Registers 1 and 2

We have two SCI Control Registers: SCICR1 and SCICR2. They are 8-bit registers. The SCICR1 is used to select the data framing size among other things. See Figure 10-11. Notice that the M bit (D4) of the SCICR1 register determines the framing of data by specifying the number of bits per character. In this textbook we use the no parity option with a data size of 8 bits. The SCICR2 register is used to enable the serial port to send and receive data, among other things. Figure 10-12 describes various bits of the SCICR2 register. Several of the SCICR2 register bits are widely used by the UART. The TE (transmit enable) and RE (receive enable) are the most important bits in this register. The rest of the bits are used for interrupt-driven serial communication. In Chapter 11 we will see how these flags are used with interrupts instead of polling.

LOOPS	SCISWAI	RSRC	M	WAKE	ILT	PE	PT

LOOPS	D7	0 = Normal operation 1 = LOOP operation enabled. See the HCS12 manual.
SCISWAI	D6	Using this we can disable or enable SCI in wait mode 0 = SCI enabled in Wait mode 1 = SCI disabled in Wait mode
RSRC	D5	Receiver source bit. Used only when LOOPS=1 (see HCS12 manual) 0 = For internally connected loop 1 = For externally connected loop
M	D4	Data format mode bit. We must use this to select 8-bit data frame size. 0 = Select 8-bit data frame, one stop bit and one start bit 1 = Select 9-bit data frame, one stop bit and one start bit
WAKE	D3	Wake-up condition bit. See the HCS12 manual. 0 = Idle line wake up 1 = Address mark wake up
ILT	D2	Idle line type bit. See the HCS12 manual. 0 = Idle character bit count begins after start bit. 1 = Idle character bit count begins after stop bit.
PE	D1	Parity Enable bit. This will allow us to insert a parity bit right after the 8th (MSB) bit. 0 = No parity bit 1 = Parity bit
PT	D0	Parity bit type (used only if PE is one) 0 = Even parity bit 1 = Odd parity bit

Note: The most important bit in this register is the M bit. The vast majority of the applications use M = 0 for 8-bit data size. The rest of the bits are for testing purposes and we do not use them unless we are writing SCI diagnostic test software. For that reason, we make them all zeros and we use SCICR1 = $00.

Figure 10-11. SCI Control Register 1 (SCICR1)

TIE	TCIE	RIE	ILIE	TE	RE	RWU	SBK

TIE	D7	Transmit Interrupt Enable bit. Used for interrupt-driven SCI. See Chapter 11. 0 = TDRE Interrupt Request is disabled. 1 = TDRE Interrupt Request is enabled.
TCIE	D6	Transmission Complete Interrupt Enable bit. Used for interrupt-driven SCI. See Chapter 11. 0 = TC Interrupt Request is disabled. 1 = TC Interrupt Request is enabled.
RIE	D5	Receiver Full Interrupt Enable bit. Used for interrupt-driven SCI. See Chapter 1. 0 = RDRF Interrupt Request is disabled. 1 = RDRF Interrupt Request is enabled.
ILIE	D4	Idle Line Interrupt Enable bit. Used for interrupt-driven SCI. 0 = IDLE Interrupt Request is disabled. 1 = IDLE Interrupt Request is enabled.
TE	D3	Transmitter Enable bit. **We must enable this bit to transmit data.** 0 = Transmitter is disabled. 1 = Transmitter is enabled.
RE	D2	Receiver Enable bit. **We must enable this bit to receive data.** 0 = Receiver is disabled. 1 = Receiver is enabled.
RWU	D1	Used for wake-up condition in stand-by mode. See the HCS12 manual. 0 = Normal operation 1 = RWU is enabled.
SBK	D0	Used for break bit. See the HCS12 manual. 0 = No break character 1 = Transmit break character

Note: The most important bits in this register are the TE and RE bits. In applications using the polling method we make the interrupt request bits all zeros. For the polling method, we use SCICR2 = 0C in hex. The rest of the bits are for testing purposes. To use interrupt-driven SCI, see Chapter 11.

Figure 10-12. SCI Control Register 2 (SCICR2)

SCI Status Registers 1 and 2

We have two SCI Status Registers: SCISR1 and SCISR2. They are 8-bit registers. The most important SCI status register is the SCISR1, which is used to monitor the arrival of data among other things. Figure 10-13 describes various bits of the SCISR1. Several of the SCISR1 register bits are widely used by the UART. We monitor (poll) the TDRE flag bit to make sure that all the bits of the last byte are transmitted before we write another byte into the SCI Data Register. By the same logic, we monitor (poll) the RDRF flag to see if a byte of data has come in yet. In Chapter 11 we will see how these flags are used with interrupts instead of polling. The SCI Status Register 2 is used for single-wire operation and is discussed in the HCS12 manual. We do not use it in this textbook.

TDRE	TC	RDRF	IDLE	OR	NF	FE	PF

TDRE D7 Transmit Data Register Empty
0 = No byte is given to transmit shift register from SCI Data Register.
1 = A byte is given to transmit shift register and SCI Data Register is empty and ready for the next byte.

TC D6 Transmit complete flag bit
0 = Transmission is in progress.
1 = No transmission in progress

RDRF D5 Receive Data Register Full flag. This indicates a byte has been received and is sitting in SCI Data Register and ready to be picked up.
0 = No data is available in SCI Data Register.
1 = Data is available in SCI Data Register and ready to be picked up.

IDLE D4 Idle line flag. See the HCS12 manual.

OR D3 Overrun error
0 = No overrun
1 = Overrun error

NF D2 Noise Flag error bit
0 = No noise
1 = Noise error

FE D1 Framing Error bit
0 = No framing error
1 = Framing error

PF D0 Parity flag error bit
0 = No parity error
1 = Parity error

The importance of the TDRE. To transmit a byte of data serially via the TXD pin, we must write it into the SCI Data Register. The transmit shift register is an internal register whose job is to get the data from the SCI Data Register, frame it with the start and stop bits, and send it out one bit at a time via the TXD pin. Notice that the transmit shift register is a parallel-in-serial-out shifter and is not accessible to the programmer. We can only write to the SCI Data Register. Whenever the shifter is empty, it gets its new data from the SCI Data Register and clears the SCI Data Register immediately, so it does not send out the same data twice. When the shifter fetches the data from the SCI Data Register, it clears the TDRE flag to indicate it is empty and the SCI Data Register is ready for the next character. We must check the TDRF flag before we write another byte to the SCI Data Register.

The importance of the RDRF. The internal serial-in-parallel-out receive register receives data via the RXD pin. It gets rid of the start and stop bits and writes the received byte to the SCI Data Register and makes RDRF high. We must check the RDRF flag to see if we need to pick up the received byte.

Figure 10-13. SCI Status Register 1 (SCISR1)

SCI registers for Serial #0 and Serial #1

As mentioned earlier, the HCS12 has two serial communication interface (SCI) ports designated as SCI0 and SCI1. The registers associated with the SCI ports and their addresses are shown in Table 10-6. These registers are commonly referred to as *special function registers* (SFRs) and can be accessed like any other

Table 10-6A: SCI0 (Number 0) Registers and their Addresses

SCI #0 Register Name	Function	Addresses
SCI0BDH	SCI Baud High byte	$00C8
SCI0BDL	SCI Baud Low byte	$00C9
SCI0CR1	SCI Control Reg 1	$00CA
SCI0CR2	SCI Control Reg 2	$00CB
SCI0SR1	SCI Status Reg 1	$00CC
SCI0SR2	SCI Status Reg 2	$00CD
SCI0DRH	SCI Data Reg High byte	$00CE
SCI0DRL	SCI Data Reg Low byte	$00CF

Table 10-6B: SCI1 (Number 1) Registers and their Addresses

SCI #1 Register Name	Function	Addresses
SCI1BDH	SCI Baud High byte	$00D0
SCI1BDL	SCI Baud Low byte	$00D1
SCI1CR1	SCI Control Reg 1	$00D2
SCI1CR2	SCI Control Reg 2	$00D3
SCI1SR1	SCI Status Reg 1	$00D4
SCI1SR2	SCI Status Reg 2	$00D5
SCI1DRH	SCI Data Reg High byte	$00D6
SCI1DRL	SCI Data Reg Low byte	$00D7

register in the HCS12. Look at the following example of how SCI#0 Data Register is accessed:

```
LDAA #$45       ;Acc=$45, ASCII for letter 'E'
STAA SCI0DR     ;copy 'E" into SCI0DR. Note SCI#0
```

The moment a byte is written into SCI0DR, it is fetched into the transmit shifter register. The shifter frames the 8-bit data with the start and stop bits and the 10-bit data is transferred serially via the TXD pin. Similarly, the following code will dump the received byte sitting in SCI#1 Data Register onto PORTB:

```
LDAA SCI1DR
STAA PORTB ;copy SCI1 to PORTB
```

Programming serial port #0 to transfer data serially

In programming the HCS12 to transfer character bytes serially, the following steps must be taken:

1. The SCI0BDH:SCI0BDL registers are loaded with one of the values in Table 10-4 (or Table 10-5 if ECLK = 2 MHz) to set the baud rate for serial data transfer.
2. The SCI0CR1 register is loaded with the value $00, indicating 8-bit data frame, no parity bit, and no loop.
3. The SCI0CR2 register is loaded with the value $0C to enable the serial port for

both send and receive. This also disables the interrupt.

4. Monitor the TDRE bit of the SCI0SR register to make sure UART is ready for a byte before writing another byte to SCI0DRL. If TDRE = 1, then go to the next step.
5. The character byte to be transmitted serially is written into the SCI0DRL register.
6. To transfer the next character, go to Step 4.

Example 10-2 shows the program to transfer data serially at 9,600 baud. Example 10-3 shows how to transfer "YES" continuously.

Example 10-2

Write a program for the HCS12 to transfer the letter 'G' serially at 9,600 baud, continuously. Use SCI # 0 and assume XTAL = 4 MHz.

Solution:

```
        LDAA   #$0           ;set the baud rate
        STAA   SCI0BDH       ;we must write to high byte reg even if zero
        LDAA   #$0D          ;set the baud rate to 9600
        STAA   SCI0BDL       ;write to low byte
        LDAA   #0            ;8-bit data, no parity and no interrupt
        STAA   SCI0CR1       ;write to control reg 1
        LDAA   #$0C          ;enable both transmit and receive
        STAA   SCI0CR2       ;write to control reg 2
OVER    LDAA   #'G'          ;ASCII letter 'G' to be transferred
HERE    BRCLR SCI0SR1, mSCI0SR1_TDRE, HERE  ;check for Data reg empty
        STAA   SCI0DRL       ;load it into SCI data reg
        BRA OVER             ;keep sending letter 'G'
```

Importance of the TDRE (Transmit Data Register Empty) flag

To understand the importance of the role of TDRE, look at the following sequence of steps that the HCS12 goes through in transmitting a character via TXD:

1. The byte character to be transmitted is written into the SCI Data Register.
2. The TDRE flag is set to 0 internally to indicate that the SCI Data Register has a byte and will not accept another byte until this one is transmitted.
3. The transmit shift register reads the byte from the SCI Data Register and begins to transfer the byte starting with the start bit. See Figure 10-14.
4. The TDRE flag is set to 1 to indicate that the last byte is being transmitted and the SCI Data Register is ready to accept another byte.
5. The 8-bit character is transferred one bit at a time.
6. By monitoring the TDRE flag, we make sure that we are not overloading the SCI Data Register. If we write another byte into the SCI Data Register before the shifter has fetched the last one, the old byte could be lost before it is transmitted.

Example 10-3

Write a program to transmit the message "YES" serially at 9,600 baud, 8-bit data, and 1 stop bit. Do this forever. Assume XTAL = 4 MHz.

Solution:

```
      LDS    #$4000
      LDAA   #$0          ;set the baud rate
      STAA   SCI0BDH      ;we must write to high byte reg even if zero
      LDAA   #$0D         ;set the baud rate to 9600
      STAA   SCI0BDL      ;write to low byte
      LDAA   #0           ;8-bit data, no parity and no interrupt
      STAA   SCI0CR1      ;write to control reg 1
      LDAA   #$0C         ;enable both transmit and receive
      STAA   SCI0CR2      ;write to control reg 2
OVER  LDAA   #'Y'         ;ASCII letter 'Y' to be transferred
      JSR    TRANS
      LDAA   #'E'         ;ASCII letter 'E' to be transferred
      JSR    TRANS
      LDAA   #'S'         ;ASCII letter 'S' to be transferred
      JSR    TRANS
      BRA    OVER         ;keep doing it

;----serial data transfer subroutine
TRANS BRCLR SCI0SR1, mSCI0SR1_TDRE, HERE  ;check for Data reg empty
      STAA   SCI0DRL      ;load it into SCI data reg
      RTS                 ;return to caller
```

From the above discussion we conclude that by checking the TDRE flag bit, we know whether or not the HCS12 is ready to transfer another byte. The TDRE flag bit can be checked by the instruction BRCLR (branch clear) such as "`BRCLR SCI0SR,TDRE,HERE`" or we can use an interrupt, as we will see in Chapter 11. In Chapter 11 we will show how to use interrupts to transfer data serially, and avoid tying down the microcontroller with instructions such as "`BRCLR SCI0SR,TDRE,HERE`". Notice that we can also check the TC (transmit complete) flag before loading the SCI data register with a new byte.

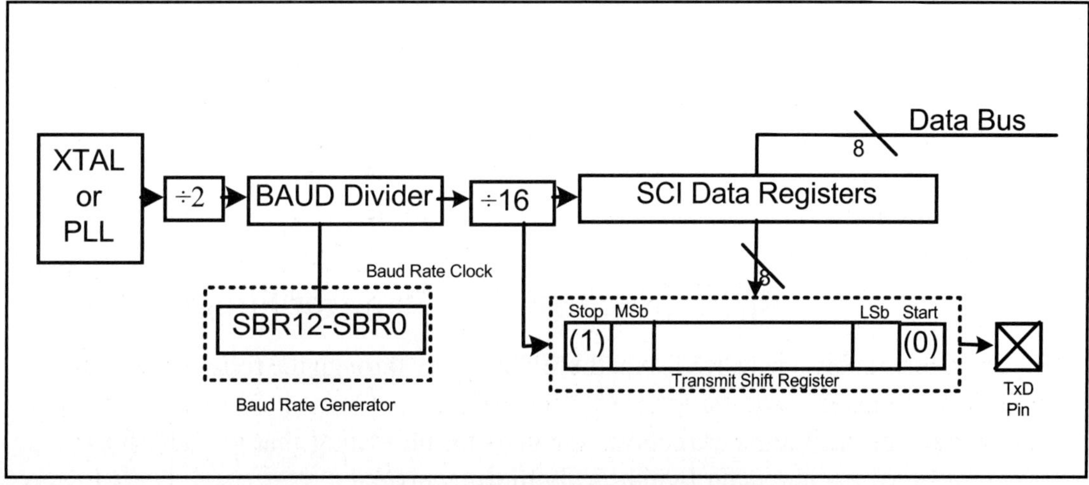

Figure 10-14. Simplified USART Transmit Block Diagram

Programming the HCS12 to receive data serially

In programming the HCS12 to receive character bytes serially, the following steps must be taken:

1. The SCI0BDH:SCI0BDL registers are loaded with one of the values in Table 10-4 (or Table 10-5 if Fosc = 4 MHz) to set the baud rate for serial data transfer.
2. The SCI0CR1 register is loaded with the value $00, indicating 8-bit data frame, no parity bit, and no loop.
3. The SCI0CR2 register is loaded with the value $0C to enable the serial port for both transmit and receive. This also disables the interrupt.
5. The RDRE (receive data reg full) flag bit of the SCI Status Register 2 is monitored for a HIGH to see if an entire character has been received yet. If RDRE = 1, then go to the next step.
6. When the RDRE flag is raised, the SCI Data Register has the byte. Its contents are moved into a safe place.
7. To receive the next character, go to Step 5.

Example 10-4 shows the coding of the above steps.

Example 10-4

Program the HCS12 to receive bytes of data serially and put them on PORTB. Set the baud rate at 9,600, 8-bit data, and 1 stop bit. Use serial port # 0.

Solution:

```
        LDAA   #$0          ;set the baud rate
        STAA   SCI0BDH      ;we must write to high byte reg even if zero
        LDAA   #$0D         ;set the baud rate to 9600
        STAA   SCI0BDL      ;write to low byte
        LDAA   #$0          ;8-bit data, no parity and no interrupt
        STAA   SCI0CR1      ;write to control reg 1
        LDAA   #$0C         ;enable both transmit and receive
        STAA   SCI0CR2      ;write to control reg 2
        LDAA   #$FF         ;make PORTB output
        STAA   DDRB

;get a byte from serial port and place it on PORTB
HERE    BRCLR SCI0SR1, mSCI0SR1_RDRF ,HERE ;check for data
        LDAA   SCI0DRL
        STAA   PORTB        ;save value into PORTB
        BRA    HERE         ;keep doing that
```

Importance of the RDRF (Receive Data Register Full) flag bit

In receiving bits via its RXD pin, the HCS12 goes through the following steps:

1. The receiver's shift register receives the start bit indicating that the next bit is the first bit of the character byte it is about to receive.

2. The 8-bit character is received one bit at time. When the last bit is received, a byte is formed and placed in SCI Data Register and the HCS12 makes RDRF = 1, indicating that an entire character byte has been received and must be picked up before it gets overwritten by another incoming character.
3. By checking the RDRF flag bit when it is raised, we know that a character has been received and is sitting in the SCI Data Register. We copy the SCIDR contents to a safe place in some other register or memory before it is lost. Notice that for receiving data we use the same register as when sending data. See Figure 10-15.

From the above discussion we conclude that by checking the RDRF flag bit we know whether or not the HCS12 has received a character byte. If we fail to copy SCI Data Register into a safe place, we risk the loss of the received byte. The RDRF flag bit can be checked by the BRCLR (branch if clear) instruction or by using an interrupt, as we will see in Chapter 11.

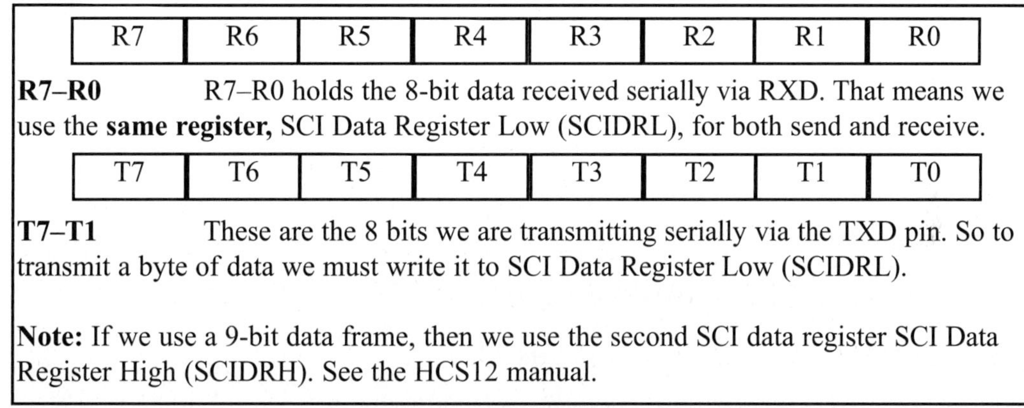

R7 R6 R5 R4 R3 R2 R1 R0
R7–R0 R7–R0 holds the 8-bit data received serially via RXD. That means we use the same register, SCI Data Register Low (SCIDRL), for both send and receive.
T7 T6 T5 T4 T3 T2 T1 T0
T7–T1 These are the 8 bits we are transmitting serially via the TXD pin. So to transmit a byte of data we must write it to SCI Data Register Low (SCIDRL).
Note: If we use a 9-bit data frame, then we use the second SCI data register SCI Data Register High (SCIDRH). See the HCS12 manual.

Figure 10-15. SCI Data Register Low (SCIDRL) Is Also Used for Receive

Programming the second serial port in HCS12

The programming of the second serial port in HCS12 is identical to the first one. The only difference is that we use SCI1 instead of SCI0. Example 10-5 shows the syntax.

Example 10-5

Write a program for the HCS12 to transfer the letter ‘G’ serially at 9,600 baud, continuously. Assume XTAL = 8 MHz. Use the second serial (SCI1) port.

Solution:

```
          LDAA   #$0           ;set the baud rate
          STAA   SCI1BDH       ;we must write to high byte reg even if zero
          LDAA   #$1A          ;set the baud rate to 9600
          STAA   SCI1BDL       ;write to low byte
          LDAA   #0            ;8-bit data, no parity and no interrupt
          STAA   SCI1CR1       ;write to control reg 1
          LDAA   #$0C          ;enable both transmit and receive
          STAA   SCI1CR2       ;write to control reg 2
OVER      LDAA   #'G'          ;ASCII letter 'G' to be transferred
HERE      BRCLR SCI1SR1,mSCI0SR1_TDRE, HERE   ;check for Data reg empty
          STAA   SCI1DRL       ;load it into SCI data reg
          BRA    OVER          ;keep sending letter 'G'
```

Baud rate error calculation

In calculating the baud rate, we have used the integer number for the SCIBD register values because HCS12 microcontrollers can only use integer values. By dropping the decimal portion of the calculated values we run the risk of introducing error into the baud rate. One way to calculate this error is to use the following formula.

Error = (Calculated value for the SCIBD – Integer part)/Integer part

For example, with XTAL = 4 MHz we have the following for the 9,600 baud rate:

SCIBD Value = (XTAL/(2 × 16))/(desired baud rate)
SCIBD value = (125,000/9,600) = 26.04 = 26
and the error is (26.04 – 26)/26 = 0.1%. See Example 10-6.

Example 10-6

Assuming XTAL = 8 MHz, calculate the baud rate error for the following:
(a) 2,400 (b) 1,200 (c) 57,600

Solution:

SCIBD Value = (XTAL/(2 × 16))/(desired baud rate)

(a) SCIBD Value = (250,000/2,400) = 104.16

Error = (104.16 – 104)/104 = 0.16%

(b) SCIBD Value (250,000/1,200) = 208.33

Error = (208.33 – 208)/208 = 0.15%

(c) SCIBD Value (250,000/57,600) = 4.34

Error = (4.34 – 4)/4 = 8.5%

Transmit and receive

Assume that the HCS12 serial port is connected to the COM port of the IBM PC, and we are using the HyperTerminal program on the PC to send and receive data serially. The ports PORTB and PORTA of the HCS12 are connected to LEDs and switches, respectively. Program 10-1 shows a HCS12 program with the following parts: (a) sends the message "YES" once to the PC screen, (b) gets data on switches and transmits it via the serial port to the PC's screen, and (c) receives any key press sent by HyperTerminal and puts it on LEDs. The program performs part (a) once, but parts (b) and (c) continuously. It uses the 9,600 baud rate for XTAL = 4 MHz. Example 10-7 shows how to transfer a string of bytes located in FLASH ROM.

```
;Program 10-1 Transmit and Receive
       LDAA   #$0           ;set the baud rate
       STAA   SCI0BDH ;we must write to high byte reg even if zero
       LDAA   #$0D          ;set the baud rate to 9600
       STAA   SCI0BDL       ;write to low byte
       LDAA   #0            ;8-bit data, no parity and no interrupt
```

```
      STAA   SCI0CR1      ;write to control reg 1
      LDAA   #$0C         ;enable both transmit and receive
      STAA   SCI0CR2      ;write to control reg 2
      LDAA   #$FF
      STAA   DDRB         ;make PORTB an output port for LEDs
      LDAA   #$0
      STAA   DDRA         ;make PORTA an input port for SW
;send the message "YES"
      LDAA   #'Y'         ;ASCII letter 'Y' to be transferred
      JSR    TRANS
      LDAA   #'E'         ;ASCII letter 'E' to be transferred
      JSR    TRANS
      LDAA   #'S'         ;ASCII letter 'S' to be transferred
      JSR    TRANS
;get a byte from switches and transmit data to PC screen
OVER  LDAA   PORTA        ;get a byte from SW
      JSR    TRANS        ;transmit it via serial port
;as keys are pressed on PC receive data and put it on LEDs
      JSR    RECV         ;receive the byte from serial port
      STAA   PORTB        ;display it on LEDS of PORTB
      BRA    OVER         ;keep doing it
;--serial transfer (Reg A needs the byte to be transmitted)
TRANS
S1    BRCLR SCI0SR1, mSCI1SR1_TDRE,S1 ;wait until Data Reg empty
      STAA   SCI0DRL      ;load it into SCI data reg
      RTS                 ;return to caller
;----serial data receive subroutine (Reg A = received byte)
RECV
R1    BRCLR SCI0SR1, mSCI1SR1_RDRF,R1 ;check for data
      LDAA   SCI0DRL      ;save value in Reg A
      RTS                 ;RETURN
```

Interrupt-based data transfer

By now you might have noticed that it is a waste of the microcontroller's time to poll the TDRE and RDRF flags. In order to avoid wasting the microcontroller's time we use interrupts instead of polling. In Chapter 11, we will show how to use interrupts to program the HCS12's serial communication port.

Idle and break characters

In the HCS12 manual we see the mention of some terminology such as idle and break. The idle is when the SCI sends out all 1s with no start and stop bits. That is 10 ones when the data frame size is 8-bit. The break character is when the SCI sends out all 0s with no start and stop bits. That is 10 zeros when the data frame size is 8-bit. The idle and break characters are used for the testing purpose and writing diagnostic software.

Review Questions

1. Which register of the HCS12 is used to set the baud rate?
2. If XTAL = 8 MHz, what frequency is used by the UART to set the baud rate?
3. Which bit of the SCICR2 register is used to enable the serial transmit?
4. With XTAL = 8 MHz, what value should be loaded into SCIBD to have a 9,600 baud rate? Give the answer in both decimal and hex.
5. To transmit a byte of data serially, it must be placed in register _______.

Example 10-7

Assume a switch is connected to pin PA7. Write a program to monitor its status and send two messages to the serial port continuously as follows:
SW = 0 send "HELLO"
SW = 1 send "GOOD BYE"
Assume XTAL = 4 MHz, and set the baud rate to 9,600 and 8-bit data.

Solution:

```
              ORG    $4000
       BCLR   DDRA,%10000000     ;PORTA.7 as in input for SW
       LDAA   #$0           ;set the baud rate
       STAA   SCI0BDH       ;we must write to high byte reg even if zero
       LDAA   #$0D          ;set the baud rate to 9600
       STAA   SCI0BDL       ;write to low byte
       LDAA   #0            ;8-bit data, no parity and no interrupt
       STAA   SCI0CR1       ;write to control reg 1
       LDAA   #$0C          ;enable both transmit and receive
       STAA   SCI0CR2       ;write to control reg 2
AGAIN  BRSET PORTA, %10000000,OVER
       LDX    #$8500        ;load the pointer for MESS1
S1     LDAA   1,X+          ;get the byte and increment pointer
       BEQ    AGAIN         ;start over if it is the last char
       JSR    TRANS         ;send it out
       BRA    S1            ;keep doing it
OVER   LDX    #$8510        ;load the pointer for MESS2
S2     LDAA   1,X+
       BEQ    AGAIN
       JSR    TRANS
       BRA    S2            ;keep doing it

;----serial data transfer subroutine
TRANS BRCLR SCI0SR1,mSCI0SR1_TDRE,TRANS ;wait until Data Reg. empty
       STAA   SCI0DRL       ;load it into SCI data reg
       RTS                  ;return to caller

;------------------
              ORG    $8500
MESS1 DC.B   "HELLO",0
              ORG    $8510
MESS2 DC.B   "GOOD BYE",0
```

We can also use the following for the AGAIN portion:

```
AGAIN BRSET PORTA, %10000000,OVER
       LDX    #$8500              ;load the pointer for MESS1
       JSR    SEND
OVER   LDX    #$8510              ;load the pointer for MESS2
SEND   LDAA   1,X+
       BEQ    AGAIN
       JSR    TRANS
       BRA    SEND                ;keep doing it
;Note: In CodeWarrior, use mSCI0SR1_TDRE to monitor TDRE.
```

6. SCI0CR1 stands for __________ and it is a(n) ____-bit register.
7. Which register is used to set the data frame size?
8. True or false. SCISR1 is a bit-addressable register.
9. When is TDRE raised? When is it cleared?
10. How many serial ports does the HCS12 have?

SECTION 10.4: HCS12 SERIAL PORT PROGRAMMING IN C

This section shows C programming of the serial ports for the HCS12 chip.

Transmitting and receiving data in C

As we saw in Chapter 7, all the special function registers (SFRs) of the HCS12 are accessible directly in C compilers by using the appropriate header file. Examples 10-8 through 10-11 show how to program the serial port in C. Connect your HCS12 Trainer to the PC's COM port and use HyperTerminal to test the operation of these examples. Notice that these examples are C versions of the Assembly programs in the last section.

Example 10-8 (C version Example 10-2)

Write a C program for the HCS12 to transfer the letter 'G' serially at 9,600 baud, continuously. Use 8-bit data and 1 stop bit for serial port #0. Assume XTAL = 4 MHz.

Solution:

```
#include <mc9s12dp512.h>        /* derivative information */
void main(void)
  {
    SCI0BDH=0x0;                        //choose
    SCI0BDL=0x0D;                       //9600 baud rate
    SCI0CR1=0x0;                        //8-bit data, no interrupt
    SCI0CR2=0X0C;   //enable transmit and receive and no parity
    while(1)
      {
        SCI0DRL='G';                        //place value in buffer
        while(!(SCI0SR1 & SCI0SR1_TDRE_MASK));//is data empty?
      }
  }
```

Review Questions

1. True or false. All the SFR registers of HCS12 are accessible in the C compiler.
2. True or false. C compilers support the bit-addressable registers of the HCS12.
3. True or false. All members of the HCS12 come with two serial ports.
4. Which register is used to set the baud rate?
5. To which register does the RDRF bit belong, and what is its role?

Example 10-9 (C version Example 10-3)

Write a HCS12 C program to transfer the message "YES" serially at 9,600 baud, 8-bit data, and 1 stop bit. Do this continuously. Assume XTAL = 4 MHz.

Solution:

```
#include <mc9s12dp512.h>      /* derivative information */
void SerTx(unsigned char);
void main(void)
  {
    SCI0BDH=0x0;                          //choose
    SCI0BDL=0x0D;                         //9600 baud rate
    SCI0CR1=0x0;                          //8-bit data, no interrupt
    SCI0CR2=0X0C;  //enable transmit and receive and no parity
    while(1)
      {
        SerTx('Y');
        SerTx('E');
        SerTx('S');
      }
  }
void SerTx(unsigned char c)
  {
    while(!(SCI0SR1 & SCI0SR1_TDRE_MASK));   //is data empty?
    SCI0DRL=c;                            //place value in buffer
  }
```

Example 10-10

Write a C program to send two different strings to the serial port. Assuming that SW is connected to pin PORTB.5, monitor its status and make a decision as follows:
SW = 0: send your first name
SW = 1: send your last name
Assume XTAL = 4 MHz, a baud rate of 9,600, and 8-bit data.

Solution:

```
#include <mc9s12dp512.h>        /* derivative information */
#define MYSW PORTB_BIT5              //INPUT SWITCH

void main(void)
{
  unsigned char z;
  unsigned char fname[]="ALI";
  unsigned char lname[]="SMITH";
  DDRB_BIT5 = 0;                     //an input
  SCI0BDH=0x0;                  //choose
  SCI0BDL=0x0D;                 //9600 baud rate
  SCI0CR1=0x0;                  //8-bit data, no interrupt
  SCI0CR2=0x0C;  //enable transmit and receive and no parity
  if(MYSW==0)                        //check switch
  {
    for(z=0;z<3;z++)            //write name
    {
      while(!(SCI0SR1 & SCI0SR1_TDRE_MASK)); //is data empty?
      SCI0DRL=fname[z];         //place value in buffer
    }
  }
  else
  {
    for(z=0;z<5;z++)                 //write name
    {
      while(!(SCI0SR1 & SCI0SR1_TDRE_MASK));//is data empty?
      SCI0DRL=lname[z];         //place value in buffer
    }
  }
  while(1);
}
```

Example 10-11

Program the HCS12 in C to receive bytes of data serially and put them on PORTB. Set the baud rate at 9,600, 8-bit data, and 1 stop bit.

Solution:

```
#include <mc9s12dp512.h>              /* derivative information */
void main (void)
  {
    DDRB=0xFF; //PORTB output
    SCI0BDH=0x0;                       //choose
    SCI0BDL=0x0D;                      //9600 baud rate
    SCI0CR1=0x0;                       //8-bit data, no interrupt
    SCI0CR2=0x0C;  //enable transmit and receive and no parity
    while(1)                           //repeat forever
      {
        while(!(SCI0SR1 & SCI0SR1_RDRF_MASK)); //wait to receive
        PORTB=SCI0DRL;                 //save value
      }
  }
```

Example 10-12

Repeat Example 10-9 using the second serial port of HCS12.

Solution:

```
#include <mc9s12dp512.h>    /* derivative information */
void SerTx(unsigned char);
void main(void)
  {
    SCI1BDH=0x0;                       //choose
    SCI1BDL=0x0D;                      //9600 baud rate
    SCI1CR1=0x0;                       //8-bit data, no interrupt
    SCI1CR2=0X0C;  //enable transmit and receive and no parity
    while(1)
      {
        SerTx('Y');
        SerTx('E');
        SerTx('S');
      }
  }
void SerTx(unsigned char c)
  {
    while(!(SCI1SR1 & SCI1SR1_TDRE_MASK));//wait until all gone
    SCI1DRL=c;                         //place value in buffer
  }
```

PROBLEMS

SECTION 10.1: BASICS OF SERIAL COMMUNICATION

1. Which is more expensive, parallel or serial data transfer?
2. True or false. 0- and 5-V digital pulses can be transferred on the telephone without being converted (modulated).
3. Show the framing of the letter ASCII 'Z' (0101 1010), no parity, 1 stop bit.
4. If there is no data transfer and the line is high, it is called ___________ (mark, space).
5. True or false. The stop bit can be 1, 2, or none at all.
6. Calculate the overhead percentage if the data size is 7, 1 stop bit, no parity.
7. True or false. RS232 voltage specification is TTL compatible.
8. What is the function of the MAX 232 chip?
9. True or false. DB-25 and DB-9 are pin compatible for the first 9 pins.
10. How many pins of the RS232 are used by the IBM serial cable, and why?
11. True or false. The longer the cable, the higher the data transfer baud rate.
12. State the absolute minimum number of signals needed to transfer data between two PCs connected serially. Give their names.
13. If two PCs are connected through the RS232 without the modem, both are configured as a ________ (DTE, DCE) -to- _______ (DTE, DCE) connection.
14. State the nine most important signals of the RS232.
15. Calculate the total number of bits transferred if 200 pages of ASCII data are sent using asynchronous serial data transfer. Assume a data size of 8 bits, 1 stop bit, and no parity. Assume each page has 80x25 of text characters.
16. In Problem 15, how long will the data transfer take if the baud rate is 9,600?

SECTION 10.2: HCS12 CONNECTION TO RS232

17. The MAX232 DIP package has _____ pins.
18. For the MAX232, indicate the V_{CC} and GND pins.
19. The MAX233 DIP package has ____ pins.
20. For the MAX233, indicate the V_{CC} and GND pins.
21. Is the MAX232 pin compatible with the MAX233?
22. State the advantages and disadvantages of the MAX232 and MAX233.
23. MAX232/233 has ____ line driver(s) for the RX wire.
24. MAX232/233 has ____ line driver(s) for the TX wire.
25. Show the connection of pins TXD and RXD of the HCS12 to a DB-9 RS232 connector via the second set of line drivers of MAX232.
26. Show the connection of the TXD and RXD pins of the HCS12 to a DB-9 RS232 connector via the second set of line drivers of MAX233.
27. What is the advantage of the MAX233 over the MAX232 chip?
28. Which pins of the HCS12 are set aside for serial communication, and what are their functions?

SECTION 10.3: HCS12 SERIAL PORT PROGRAMMING IN ASSEMBLY

29. Which of the following baud rates are supported by the HyperTerminal program in PCs?
    (a) 4,800 (b) 3,600 (c) 9,600
    (d) 1,800 (e) 1,200 (f) 19,200
30. Which timer of the HCS12 is used for baud rate programming?
31. Which bits of the SCIBDH are used for baud rate speed?
32. What is the role of the SCIDR register in serial data transfer?
33. SCIDR is a(n) ____-bit register.
34. What is the role of the SCICR1 register in serial data transfer?
35. SCICR1 is a(n) ____-bit register.
36. For XTAL = 8 MHz, find the SCIBD value (in both decimal and hex) for each of the following baud rates.
    (a) 9,600 (b) 4,800 (c) 1,200
37. What is the baud rate if we use SCIBDH:SCIBD = 4 to program the baud rate? Assume XTAL = 4 MHz.
38. Write a HCS12 program to transfer serially the letter 'Z' continuously at 1,200 baud rate. Assume XTAL = 4 MHz.
39. Write a HCS12 program to transfer serially either of these messages "The earth is but one country and mankind its citizens" or "The promise of world peace" continuously at 1,200 baud rate. Use a switch to choose the message. Assume XTAL = 4 MHz.
40. When is the TDRE flag bit raised or cleared?
41. When is the RDRF flag bit raised or cleared?
42. To which register do TDRE and RDRF belong? Is that register bit-addressable?
43. What is the role of the TE bit in the SCICR2 register?
44. In a given situation we cannot accept reception of any serial data. How do you block such a reception with a single instruction?
45. Find the SCIBD for the following baud rates if XTAL = 4 MHz. Assume no PLL is being used.
    (a) `9600` (b) `19200`
    (c) `38400` (d) `57600`
46. Find the SCIBD for the following baud rates if XTAL = 16 MHz. Assume no PLL is being used.
    (a) `9600` (b) `19200`
    (c) `38400` (d) `57600`
47. Find the SCIBD for the following baud rates if PLL = 20 MHz.
    (a) `9600` (b) `19200`
    (c) `38400` (d) `57600`
48. Find the SCIBD for the following baud rates if PLL = 24 MHz.
    (a) `9600` (b) `19200`
    (c) `38400` (d) `57600`
49. Find the baud rate error for Problem 47.
50. Find the baud rate error for Problem 48.

ANSWERS TO REVIEW QUESTIONS

SECTION 10.1: BASICS OF SERIAL COMMUNICATION

1. Faster, more expensive
2. False; it is simplex.
3. True
4. Asynchronous
5. With 0100 0101 binary the bits are transmitted in the sequence:
   (a) 0 (start bit) (b) 1 (c) 0 (d) 1 (e) 0 (f) 0 (g) 0 (h) 1 (i) 0 (j) 1 (stop bit)
6. 2 bits (one for the start bit and one for the stop bit). Therefore, for each 8-bit character, a total of 10 bits is transferred.
7. 10,000 × 10 = 100,000 total bits transmitted. 100,000 / 9,600 = 10.4 seconds; 2 / 10 = 20%.
8. True
9. +3 to +25 V
10. True
11. One
12. COM 1, COM 2

SECTION 10.2: HCS12 CONNECTION TO RS232

1. True
2. Pins PS0 and PS1. Pin PS1 is for TXD and pin PS0 for RXD.
3. They are used for converting from RS232 voltage levels to TTL voltage levels and vice versa.
4. Two, two
5. It does not need the four capacitors that the MAX232 must have.

SECTION 10.3: HCS12 SERIAL PORT PROGRAMMING IN ASSEMBLY

1. SCIBDH:SCIBDL
2. 8MHz/2 × 16= 250,000
3. TE
4. 26 in decimal (or 1A in hex) because 250,000 / 9,600 = 26
5. SCIDR
6. SCI Control Register, 8
7. SCI Control Register
8. True
9. It is raised when the byte in the SCI Data Register is transferred to the transmit shift register. It is cleared when we write a byte to SCIDR to be transmitted.
10. 2

SECTION 10.4: HCS12 SERIAL PORT PROGRAMMING IN C

1. True
2. True
3. True
4. SCIBD
5. SCI Status Register. It lets us know that a byte has been received and we need to pick it up from the SCI Data Register.

CHAPTER 11

INTERRUPT PROGRAMMING IN ASSEMBLY AND C

OBJECTIVES

Upon completion of this chapter, you will be able to:

>> Contrast and compare interrupts versus polling
>> Explain the purpose of the ISR (interrupt service routine)
>> List the major interrupts of the HCS12
>> Explain the purpose of the interrupt vector table
>> Enable or disable HCS12 interrupts
>> Program the HCS12 timers using interrupts
>> Describe the external hardware interrupts of the HCS12
>> Program the HCS12 for interrupt-based serial communication
>> Define the interrupt priority of the HCS12
>> Program HCS12 interrupts in C
>> Program HCS12 RTI interrupt in Assembly and C

In this chapter we explore the concept of the interrupt. The HCS12 interrupt programming in Assembly and C are also discussed.

SECTION 11.1: HCS12 INTERRUPTS

In this section, first we examine the difference between polling and interrupts and then describe the various interrupts of the HCS12.

Interrupts vs. polling

A single microcontroller can serve several devices. There are two ways to do that: interrupts or polling. In the *interrupt* method, whenever any device needs service, the device notifies the microcontroller by sending it an interrupt signal. Upon receiving an interrupt signal, the microcontroller interrupts whatever it is doing and serves the device. The program associated with the interrupt is called the *interrupt service routine* (ISR) or *interrupt handler*. In *polling*, the microcontroller continuously monitors the status of a given device; when the status condition is met, it performs the service. After that, it moves on to monitor the next device until each one is serviced. Although polling can monitor the status of several devices and serve each of them as certain conditions are met, it is not an efficient use of the microcontroller. The advantage of interrupts is that the microcontroller can serve many devices (not all at the same time, of course); each device can get the attention of the microcontroller based on the priority assigned to it. The polling method cannot assign priority since it checks all devices in a round-robin fashion. More importantly, in the interrupt method the microcontroller can also ignore (mask) a device request for service. This is again not possible with the polling method. The most important reason that the interrupt method is preferable is that the polling method wastes much of the microcontroller's time by polling devices that do not need service. So in order to avoid tying down the microcontroller, interrupts are used. For example, in discussing timers in Chapter 9 we used the instruction "`BRCLR  TFLG2,TOF,target`", and waited until the timer rolled over, and while we were waiting we could not do anything else. That is a waste of the microcontroller's time that could have been used to perform some useful tasks. In the case of the timer, if we use the interrupt method, the microcontroller can go about doing other tasks, and when the TOF flag is raised the timer will interrupt the microcontroller in whatever it is doing.

Interrupt service routine

For every interrupt, there must be an interrupt service routine (ISR), or interrupt handler. When an interrupt is invoked, the microcontroller runs the interrupt service routine. For every interrupt, there is a fixed location in memory that holds the starting address of its ISR. The group of memory locations set aside to hold the addresses of ISRs is called the *interrupt vector table*.

Interrupts and the flag register

Among bits D0 to D7 of the CCR (condition code register) register, there are two bits that are associated with the interrupt: I and X. See Figure 11-1. The I flag is used to mask (ignore) globally the interrupt requests that may come in

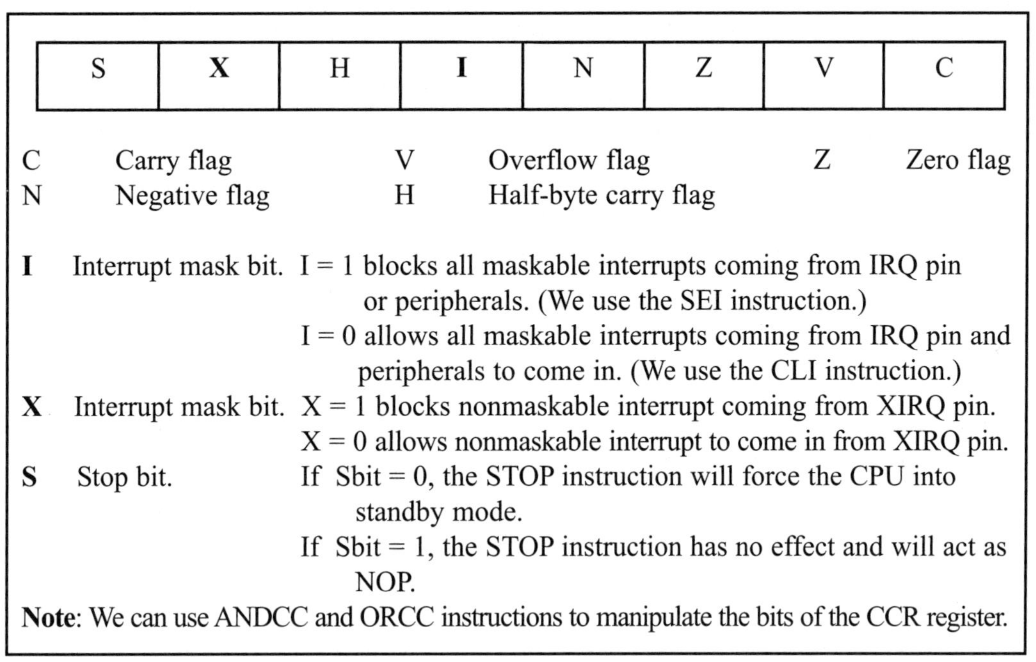

Figure 11-1. The I and X Bits in the CCR Register

from the IRQ (Interrupt Request) pin or other maskable interrupts belonging to peripherals. To allow interrupt requests through the IRQ pin and peripherals, this flag must be cleared (I = 0). The CLI (clear interrupt flag) instruction will make I = 0. If I = 1, all interrupts from the IRQ pin and peripherals are ignored. According to the HCS12 manual, the CLI is really "ANDCC #%11101111", which is ANDing the CCR register with an immediate value of 11101111 (binary). The SEI (set interrupt flag) instruction can be used to set the I bit to HIGH to block these interrupts. The SEI is really "ORCC #%00010000", which is ORing the CCR register with an immediate value of 00010000 (binary). The SEI instruction has no effect on the interrupt coming from the XIRQ pin. The XIRQ is called a *non-maskable interrupt* and is controlled by the X bit in the CCR register. We will have more discussions on the IRQ and XIRQ interrupts in Section 11.3.

Steps in executing an interrupt

When the HCS12 processes an interrupt (software or hardware), it goes through the following steps:

1. It finishes the instruction it is currently executing.
2. The current PC register is pushed onto the stack and SP is decremented by 2.
3. The current Y register is pushed onto the stack and SP is decremented by 2.
4. The current X register is pushed onto the stack and SP is decremented by 2.
5. The current A:B registers are pushed onto the stack and SP is decremented by 2.
6. The current CCR is pushed onto the stack and SP is decremented by 1.
7. The CPU fetches the address of the interrupt service routine (ISR) from the interrupt vector table. The CPU places this address in its Program Counter (PC).

Stack Pointer	Stack Content	
SP-9	CCR	Condition Code Register
SP-8	A	Register A
SP-7	B	Register B
SP-6	XH	Register X High Byte
SP-5	XL	Register Y Low Byte
SP-4	YH	Register X High Byte
SP-3	YL	Register Y Low Byte
SP-2	PCH	Program Counter High Byte
SP-1	PCL	Program Counter Low Byte

Figure 11-2. Stack after Interrupt Execution

8. The I bit in CCR is set to high to make sure no other interrupt can interrupt the CPU while serving the current one. Depending on the nature of the interrupt procedure, a programmer can unmask the interrupts by using the CLI instruction. Doing this means that we are allowing an interrupt inside of an interrupt.
9. With the new PC, the CPU starts to fetch and execute instructions belonging to the ISR program.
10. The last instruction of the ISR must be RTI. Upon execution of RTI, the original Y, X, A:B, CCR, and PC are pulled out of the stack and placed into the CPU, and the CPU continues to run the code from where it left off before the interrupt.

Notice from Step 2 the critical role of the stack. See Figure 11-2. For this reason, we must be careful in manipulating the stack contents from within the ISR. Specifically, in the ISR, just as in any CALL subroutine, the number of pushes and pulls must be equal.

Interrupt vector table in the HCS12

Table 11-1 shows the interrupt vector table for the HCS12. Some of the most widely used interrupts in the HCS12 are allocated as follows:

1. Reset. Locations $FFFE and $FFFF are set aside for the address of the hardware reset. This is the power-up reset discussed in Chapter 8.
2. Several interrupts are set aside for the timers: 8 for the channel 0 through channel 7 Output Compare/Input Capture functions. An interrupt is assigned to each of the Pulse Accumulators A and B. Memory locations for these interrupts are shown in Figure 11-1.
3. Two interrupts are set aside for external hardware interrupts. Pin numbers 56 (PE0) and 55 (PE1) in port E are for the external hardware interrupts XIRQ and IRQ, respectively. Memory locations $FFF2 and $FFF4 in the interrupt vector table are assigned to XIRQ and IRQ, respectively.
4. Serial Communication #0 (SCI0) has a single interrupt that belongs to both receive and transmit.
5. Two interrupts are set aside for the Analog-to-Digital Converters ATD0 and ATD1. See Table 11-1.

Table 11-1: Partial Listing of Interrupt Vector Table for the HCS12

Interrupt	ROM Location (Hex)	Global Enable (CCR Mask)	Local Enable
Reset	FFFE–FFFF	None	None
Clock Monitor Fail reset	FFFC–FFFD	None	PLLCTL
COP Failure reset	FFFA–FFFB	None	COP rate sele.
Unimplemented instruc. opcode trap	FFF8–FFF9	None	None
SWI instruction	FFF6–FFF7	None	None
XIRQ External hardware interrupt	FFF4–FFF5	X bit	None
IRQ External hardware interrupt	FFF2–FFF3	I bit	IRQCR (IRQEN)
Real Time interrupt	FFF0–FFF1	I bit	CRGINT (RTIE)
Capture Timer channel 0	FFEE–FFEF	I bit	TIE (C0I)
Capture Timer channel 1	FFEC–FFED	I bit	TIE (C1I)
Capture Timer channel 2	FFEA–FFEB	I bit	TIE (C2I)
Capture Timer channel 3	FFE8–FFE9	I bit	TIE (C3I)
Capture Timer channel 4	FFE6–FFE7	I bit	TIE (C4I)
Capture Timer channel 5	FFE4–FFE5	I bit	TIE (C5I)
Capture Timer channel 6	FFE2–FFE3	I bit	TIE (C6I)
Capture Timer channel 7	FFE0–FFE1	I bit	TIE (C7I)
TCNT timer overflow	FFDE–FFDF	I bit	TSRC2(TOI)
Pulse Accumulator A overflow	FFDC–FFDD	I bit	PACTL (PAOVI)
Pulse Accumulator input edge	FFDA–FFDB	I bit	PACTL (PAI)
SPI	FFD8–FFD9	I bit	SPICR1 (SPIE, ...)
SCI0 (Serial COM # 0)	FFD6–FFD7	I bit	SCICR2 (TIE, RIE)
SCI1 (Serial COM #1)	FFD4–FFD5	I bit	SCIR2 (TIE, RIE)
ATD0 (Analog-To-Digital #0)	FFD2–FFD3	I bit	ATDCTL2 (ASCIE)
ATD1 (Analog-To-Digital #1)	FFD0–FFD1	I bit	ATDCTL2 (ASCIE)
PORTJ	FFCE–FFCF	I bit	PIEJ (PIEJ7,6,1,0)
PORTH	FFCC–FFCD	I bit	PIEH (PIEH0–7)
....	....	...	....
PORTP	FF8E–FF8F	I bit	PIEP (PIEP0–7)

For the complete list see the HCS12 manual.

From Table 11-1, notice that only two bytes of ROM space are assigned to the reset pin. They are ROM address locations $FFFE and $FFFF. Also notice in Table 11-1 that 2 bytes of address location are set aside for each interrupt. Because these address locations belong to the Flash ROM in the HCS12, we have two choices: a) place the 2-byte address of the ISR in the vector table to point to the ISR itself, or b) place the 2-byte address of a RAM to redirect the processor away from the interrupt vector table in Flash to some user-accessible RAM locations. The HCS12 trainers employ the second method to allow access to the HCS12 interrupt vector table. In the next section we will see how this works in the context of some examples.

Categories of interrupts

The following are some of the major categories of interrupts in HCS12:

Maskable and nonmaskable hardware pin-based interrupts

There are two pins in the HCS12 that are associated with hardware interrupts. They are IRQ (interrupt request) and XIRQ. The XIRQ is also referred to as a *nonmaskable interrupt* (NMI). See Section 11-3 for a detailed discussion of IRQ, XIRQ, and other hardware interrupts in HCS12.

Software interrupt (SWI)

An ISR can be called upon as a result of the execution of an SWI (software interrupt) instruction. This is referred to as a *software interrupt* since it was invoked from software, not from external hardware or any peripheral. Whenever the SWI instruction is executed, the CPU will go to memory locations $FFF6 and FFF7 to get the address of the ISR associated with the SWI. See Figure 11-3 for an example of the SWI software interrupt.

Invalid instruction exception (trap)

This interrupt belongs to the category of interrupts referred to as *exception interrupts*. Exception interrupts are invoked internally by the CPU whenever there are conditions (exceptions) that the CPU is unable to handle. One such situation is an attempt to execute an opcode that does not exist. Since the result is undefined, and the CPU has no way of handling it, it automatically invokes the invalid instruction exception interrupt. This is commonly referred to as a *trap* in the HCS12 literature. Whenever an invalid instruction is executed, the CPU will go to memory locations $FFF8 and $FFF9 to get the address of the ISR to handle the situation. Notice from Table 11-1 that the invalid instruction trap cannot be masked.

Peripheral interrupts

An ISR can be called upon as the result of a condition set by a peripheral device such as a timer or analog-to-digital converter. The largest number of the interrupts in the HCS12 belong to this category. For these interrupts to take effect, we must have the I bit in CCR cleared (I = 0) as well as the interrupt flag for the desired peripheral enabled.

```
Main      LDS  #$4000     ;initialize stack pointer
          LDAA #$FF
          STAA DDRB       ;make PORTB an output port
          LDAA #$00
          STAA DDRA       ;make PORTA an input port
BACK      SWI             ;execute the SWI instruction
          BRA BACK
SWI_ISR   LDAA PORTA
          STAA PORTB
          RTI
;**************Interrupt Vectors*****************;
          ORG   $FFF6     ;vector location for SWI
          DC.W  SWI_ISR   ; goto SWI_ISR
          ORG   $FFFE     ;vector location for RESET
          DC.W  Main      ; goto MAIN ISR
```

Figure 11-3. Redirecting the HCS12 from the Interrupt Vector Table at Power-up

Enabling and disabling the interrupts globally

Upon reset, all interrupts are disabled (masked), meaning that none will be responded to by the microcontroller if they are activated. The interrupts must be enabled by software (instruction CLI) in order for the microcontroller to respond to them. The vast majority of the interrupts are globally enabled (unmasked) by the I bit in the CCR register. The I bit is responsible for enabling (unmasking) and disabling (masking) the IRQ and many of the peripheral interrupts. To enable the interrupt belonging to the I bit, we take the following steps:

1. The I bit of the CCR register must be made low to allow the local interrupt flag to take effect. We do that with the CLI instruction.
2. If I = 0, interrupts are unmasked and will be responded to if their corresponding bits in the local interrupt flag are also enabled. If I = 1, no interrupt will be responded to, even if the associated bit in the local interrupt flag register is enabled.

D-BUG12 vector table address map

D-BUG12 is a popular monitor program from Freescale. Many of the Trainers come with the D-BUG12 monitor program already installed. Table 11-2 shows the interrupt vector locations for D-BUG12. See Appendix B for the full list.

Table 11-2: Partial Listing of Interrupt Vector Table from the D-BUG12 Manual

Interrupt	RAM Location (Hex)	
Reset	3E7E	Not Available
Clock Monitor Fail reset	3E7C	Not Available
COP Failure reset	3E7A	Not Available
Unimplemented instruc. opcode trap	3E78	
SWI instruction	3E76	
XIRQ External hardware interrupt	3E74	
IRQ External hardware interrupt	3E72	
Real Time interrupt	3E70	
Capture Timer channel 0	3E6E	
Capture Timer channel 1	3E6C	
Capture Timer channel 2	3E6A	
....	.....	... .
Capture Timer channel 6	3E62	
Capture Timer channel 7	3E60	
TCNT timer overflow	3E5E	
Pulse Accumulator A overflow	3E5C	
Pulse Accumulator input edge	3E5A	
SPI0	3E58	
SCI0 (Serial COM #0)	3E56	
SCI1 (Serial COM #1)	3E54	
ATD0 (Analog-To-Digital #0)	3E52	
ATD1 (Analog-To-Digital #1)	3E50	
PORTJ	3E4E	
PORTH	3E4C	
....	.....	... .
PORTP	3E0E	

Review Questions

1. Of the interrupt and polling methods, which one avoids tying down the microcontroller?
2. Besides reset, how many hardware pin-based interrupts do we have in the HCS12?
3. In the HCS12, what memory location in the interrupt vector table is assigned to the Reset pin? Can the programmer change the memory space assigned to the table?
4. What instruction is used to change the contents of register CCR?
5. Show the instruction used to enable the IRQ interrupt.
6. Which pin of the HCS12 is assigned to the external hardware interrupt of IRQ?
7. True or false. The I bit is used to block IRQ and peripheral interrupts.

SECTION 11.2: PROGRAMMING TIMER INTERRUPTS

In Chapter 9 we discussed how to use HCS12 timers with the polling method. In this section we use interrupts to program the HCS12 timers. Please review Chapter 9 before you study this section.

Using interrupts for roll-over timer flag

In Chapter 9 we stated that the TCNT timer overflow flag (TOF) is raised when the timer rolls over. In that chapter, we also showed how to monitor TOF with the instruction "`BRCLR TFLG2,TOF,target`". In polling TOF, we have to wait until the TOF is raised. The problem with this method is that the microcontroller is tied down while waiting for TOF to be raised, and cannot do anything else. Using interrupts solves this problem and avoids tying down the controller. If the timer interrupt flag of TOI in the TSCR2 register is enabled, whenever the

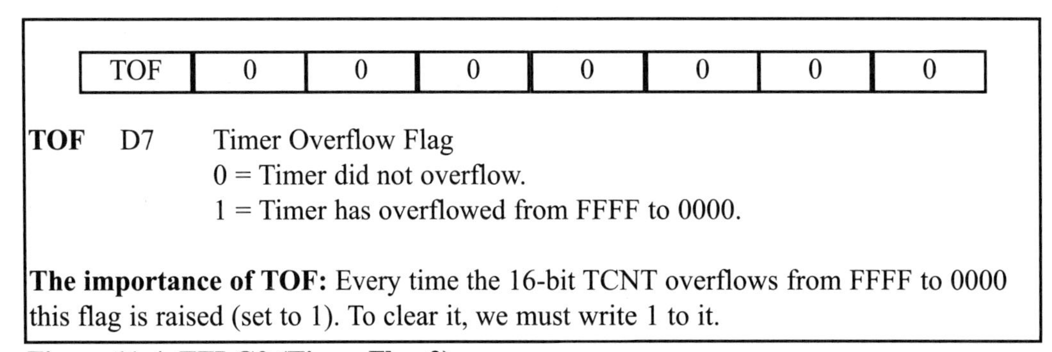

Figure 11-4. TFLG2 (Timer Flag 2)

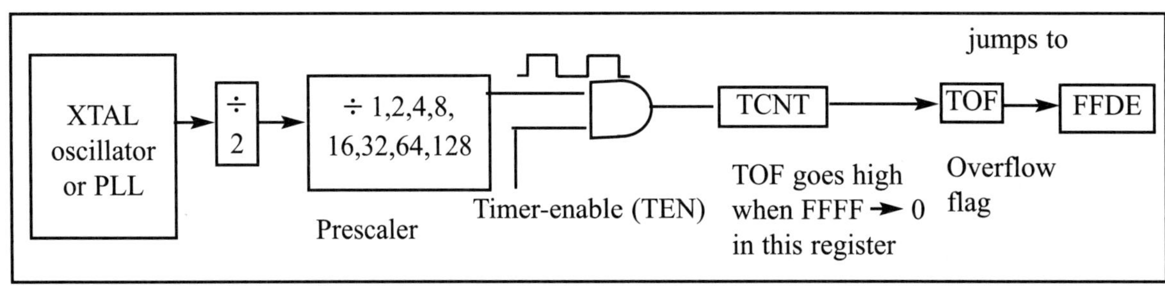

Figure 11-5. TCNT Overflow Interrupt

TOI	0	0	0	TCRE	PR2	PR1	PR0

TOI D7 Timer Overflow Interrupt Enable
0 = No hardware interrupt when timer overflows
1 = Hardware interrupt is requested when timer overflows.

For the TOF (TCNT overflow flag) to interrupt the CPU, we must set TOI = 1 in addition to I = 0 in CCR.

TCRE D3 Timer Counter Reset Enable. This bit allows the resetting of the Timer/Counter by the Output Compare 7 function.
0 = No resetting of Timer/Counter to zero for Output Compare
1 = Reset Timer/Counter to zero for Output Compare

PR2–PR0 D2–D0 Timer Prescaler Select. These bits are used to set the number of times the Fosc/2 frequency is divided by 2 before it is fed into the timer.

D2	D1	D0	Prescaler factor
0	0	0	1
0	0	1	2
0	1	0	4
0	1	1	8
1	0	0	16
1	0	1	32
1	1	0	64
1	1	1	128

Figure 11-6. TSCR2 (Timer System Control Register 2)

timer rolls over, TOF is raised, and the microcontroller is interrupted in whatever it is doing and jumps to the address $FFDE in the interrupt vector table to service the ISR. In this way, the microcontroller can do other things until it is notified that the timer has rolled over. Figure 11-4 shows the TOI bit in the TSCR2 register.

Notice that the TCNT interrupt is referred to as Enhanced Capture Timer Overflow in the interrupt vector table. See Figure 11-5 and Example 11-1.

Notice the following points about the program in Example 11-1.

1. In the MAIN section, we enabled the TCNT timer interrupt located in the TSCR2 register in addition to unmasking the I bit in the CCR register.
2. The ISR for TCNT is located starting at memory location $8200.
3. While the PORTH data is brought in and issued to PORTA continuously, whenever TCNT is rolled over, the TOF flag is raised, and the microcontroller gets out of the "H1" loop and goes to $FFDE to get the address of the ISR associated with TCNT. From there it goes to the vector table in RAM. At the RAM vector table we redirect it to the ISR.
4. In the ISR for TCNT, notice that we must make TOF = 1 to clear it for the next round. This is because the HCS12 does not clear the TOF flag internally upon jumping to the interrupt vector table.

In Example 11-1, we direct the CPU to the interrupt service routine by placing its address in the RAM location set aside by the Trainer. Check the vector address for your Trainer.

Example 11-1

Using interrupts, write a program to count the number of times the TCNT free-running timer overflows. Display the binary number on PORTB. The program should get 8-bit data from PORTH and send it to PORTA while simultaneously counting up the overflow. Assume that XTAL = 4 MHz. This is similar to Example 9-6 except that it uses an interrupt count-up TCNT timer.

Solution:

```
;--The main program for initialization
      ORG   RAMStart
COUNT DC.B  1
      ORG   ROMStart
Main: LDS   #RAMEnd+1            ;initialize the stack pointer
      LDAA  #$FF
      STAA  DDRB                 ;make PORTB an output
      STAA  DDRA                 ;make PORTA an output
      LDAA  #$00
      STAA  DDRH                 ;make PTH an input
      STAA  COUNT                ;COUNT = 0
      LDAA  #$80
      STAA  TSCR1                ;enable timer
      LDAA  #$86
      STAA  TSCR2                ;interrupt enabled, prescaler of 64
      LDAA  #0                   ;clear A
      BSET  TFLG2,%10000000      ;TOF=1 to clear it
      CLI                        ;unmask the interrupts in CCR
H1    LDAB  PTH                  ;get data from PH
      STAB  PORTA                ;send it to PA
      BRA   H1                   ;loop unless interrupted by TFO
;--ISR for TCNT timer to count up
;control came here because overflow flag was raised
TOF_ISR INC COUNT                ;increment A every time TCNT overflows
      LDAA  COUNT
      STAA  PORTB
      BSET  TFLG2,%10000000      ;TOF=1 to clear it for next round
      RTI                        ;return from interrupt
;--------------------
      ORG   $FFDE
      DC.W  TOF_ISR
      ORG   $FFFE
      DC.W  Main            ; reset vector
```

The period for the clock fed to the timer is:

$1/2 \times 4$ MHz = 2 MHz and $1/64 \times 2$ MHz = 31,250 Hz due to prescaler and T = 1/31,250 Hz = 32 μs. Therefore the counter counts up every 65,536 × 32 μs = 2,097,152 μs or every 2.10 seconds. This is similar to Example 9-6 except that it uses an interrupt.

TIE (Timer Interrupt Enable) register

In Chapter 9 we used the polling method for the Output Compare and Input Capture features of the timer channels 0–7. To implement the interrupt version of these features, we must enable the bits in TIE (Timer Interrupt Enable) register. Figure 11-7 shows the bits of the TIE register. Examples 11-2 through 11-6 show the interrupt versions of some of the programs in Chapter 9.

C7I	C6I	C5I	C4I	C3I	C2I	C1I	C0I

C7I–C0I D7–D0 Enable hardware interrupt for channels 0 to 7 in Output Compare
0 = Disable hardware interrupt for Input Capture/output Compare pin overflow
1 = Enable hardware interrupt for Input Capture/output Compare pin overflow
These are the hardware versions of the CxF in the TFLG1 register.

Figure 11-7. TIE (Timer Interrupt Enable) Register

Example 11-2

Write a program to toggle pin PT7 every 1/2 second. This is similar to Example 9-10 except that it uses an interrupt for Output Compare of channel 7. Assume that XTAL = 4 MHz.

Solution:

0.5 s / 0.5 μs = 1,000,000. That means we must use a prescaler. Using a prescaler of 16 we get 1,000,000/16 = 62,500.

```
Main: LDS    #RAMEnd+1          ;initialize the stack pointer
      LDAA   #$80
      STAA   TSCR1              ;enable timer
      LDAA   #$04
      STAA   TSCR2              ;prescaler of 16
      BSET   TIOS,%10000000     ;select Channel 7 for Output Compare
      LDAA   #$40
      STAA   TCTL1              ;toggle PT7 pin upon match
      CLI           ;unmask the interrupts in CCR
      LDAA   #$80
      STAA   TIE           ;enable interrupt for Channel 7
OVER  BRA    OVER          ;keep busy and wait for interrupt
;--ISR for Output Compare Channel 7
;it came here because of Output Compare channel 7 match.
TC7_ISR LDD TCNT           ;read current TCNT value
      ADDD   #62500             ;we need 62500 pulses
      STD    TC7                ;load the TC7 (Timer Compare 7)
      BSET   TFLG1,%10000000    ;clear Channel 7 match flag
      RTI                       ;return from interrupt
;Interrupt Vector Table
      ORG $FFE0
      DC.W TC7_ISR              ;Timer Channel 7 vector location
      ORG    $FFFE
      DC.W   Main               ;reset vector
```

Example 11-3

Assuming that clock pulses are fed into pin PT7, using interrupts, write a program for channel 7 to read the TCNT value on every rising edge. Place the result on PORTA and PORTB. This is the interrupt version of Example 9-15, except it uses channel 7.

Solution:

```
Entry:LDS    #RAMEnd+1          ;initialize the stack pointer
      LDAA   #$FF
      STAA   DDRB
      STAA   DDRA
      LDAA   #$80
      STAA   TSCR1              ;enable timer
      LDAA   #$00
      STAA   TSCR2              ;no interrupt, no prescaler
      BSET   TIOS,%10000000     ;select Channel 7 for input capture
      LDAA   #$40
      STAA   TCTL3              ;select Channel 7 for input capture
      CLI                       ;unmask the interrupts in CCR
      LDAA   #$80
      STAA   TIE                ;enable interrupt for Channel 7
OVER  BRA    OVER               ;keep busy and wait for interrupt
;--ISR for Input Capture Channel 7
;it came here because of Input Capture channel 7

TC7_ISR LDD TCNT                ;get the TC7
      STAB   PORTB              ;put it on PORTB
      STAA   PORTA              ;  and PORTA
      BSET   TFLG1,%10000000    ;clear Channel 7 match flag
      RTI                       ;return from interrupt
;----------------------
      ORG    $FFE0
      DC.W   TC7_ISR
      ORG    $FFFE
      DC.W   Entry              ;reset vector
```

Note: Upon the detection of each rising edge, the TCNT value is loaded into the TC7. Also notice that by reading the TC7 value we clear the C7F flag bit.

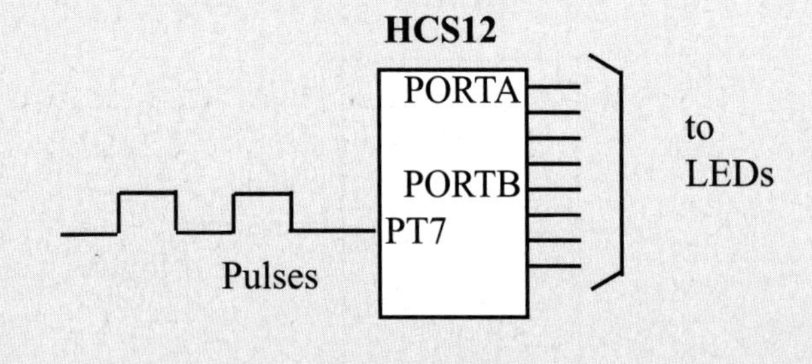

Example 11-4

What is the difference between the RTS and RTI instructions?

Solution:

Both perform the same actions of pulling off the top two bytes of the stack into the program counter, and making the HCS12 return to where it left off. However, RTI also performs the task of restoring the registers of X, Y, A:B, and CCR to their original values before the interrupt was invoked.

Example 11-5

Assume that a 1-Hz pulse is connected to the input for PACA (pin PT7). The pulses represent the number of bottles being packaged together. Using interrupts, write a program to turn on PORTB.4 every time 12 bottles go by. The program should display the binary count on PORTA. This is the interrupt version of Example 9-25.

Solution:

We set the initial event counter value to –12 (FFF4 in hex), then the counter counts up until it reaches FFFF. Upon overflow, we turn on the PORTB.4 pin.

```
Counter      DS.B 1
Entry:LDS    #RAMEnd+1              ;initialize the stack pointer
      LDAA   #$FF
      STAA   DDRA                   ;PORTA as output
      BSET   DDRB,%00010000         ;PORTB.4 output
      BCLR   DDRT,%10000000         ;make PT7 as input
      CLR    Counter
      LDAA   #%01010010             ;enable 16-bit event counter
      STAA   PACTL                  ;with interrupt PAOVI, rising edge
      CLI
      LDD    #$FFF4 ;D=-12 D = FFF4 (in hex) since we need 12 pulses
      STD    PACN32                 ;initialize PACA=-12
      BCLR   PORTB,%00010000        ;PORTB.4 = 0, turn off PB4
      BRA $                         ;stay here for interrupt
;---------------------
PACA_ISR
      BSET   PORTB,%00010000        ;turn on PORTB.4 for every 12 pulses
      BSET   PAFLG,mPAFLG_PAOVF;clear the flag for the next round
      INC    Counter
      LDAA   Counter
      STAA   PORTA          ;display the count on PORTA
      LDD    #$FFF4 ;D=-12 D = FFF4 (in hex) since we need 12 pulses
      STD    PACN32                 ;initialize PACA=-12
      BCLR   PORTB,%00010000        ;PORTB.4 = 0, turn off PB4
      RTI
;---------------------
      ORG $FFDC
      DC.W PACA_ISR         ;PACA overflow vector location
      ORG    $FFFE
      DC.W   Entry
```

HCS12

PA

PT7

PB4

1 Hz clock

Example 11-6

Assume that a 10-Hz pulse is connected to pin PT7 for Pulse Accumulator A (PACA). Using interrupts, write a program to display the 16-bit counter value of PACA on PORTA and PORTB. Set the initial values to 0. Use a rising edge clock. This is the interrupt version of Example 9-22.

Solution:

```
Entry:LDS    #RAMEnd+1          ;initialize the stack pointer
      LDAA   #$FF
      STAA   DDRA               ;PORTA as output
      STAA   DDRB               ;PORTB as output
      BCLR   DDRT,%10000000     ;make PT7 as an input
      LDAA   #%01010001         ;enable 16-bit event counter
      STAA   PACTL              ;with interrupt PAI, rising edge
      LDAA   #0                 ;A=0
      LDAB   #0                 ;B=0, now D = 0 for initial value
      STD    PACN32             ;initialize PACN32=0
      CLI
      BRA $
PACA_EDGE
      LDD    PACN32       ;read the 16-bit counter value into D reg
      STAB   PORTA              ;place the lower 8 bits on PORTA
      STAA   PORTB              ;place the upper 8 bits on PORTB
      BSET   PAFLG,mPAFLG_PAIF;clear the flag
      RTI

      ORG    $FFDA
      DC.W   PACA_EDGE          ;PACA input edge detect vector location
      ORG    $FFFE
      DC.W   Entry              ;reset vector
```


Review Questions

1. True or false. There is only a single interrupt in the interrupt vector table assigned to TCNT timer.
2. What address in the interrupt vector table is assigned to channel 7 of the Output Compare/Input Capture function?
3. Which bit of CCR can be used to block the timer interrupts globally? Show how it is done.
4. To allow the channels 0–7 hardware interrupt we enable the bits in register_______.
5. True or false. The last instruction of the ISR is RTS.

SECTION 11.3: PROGRAMMING EXTERNAL HARDWARE INTERRUPTS

The HCS12 has two hardware interrupts: IRQ and XIRQ. They can be used as external hardware interrupts. The IRQ is an input signal into the CPU that can be masked (ignored) and unmasked through the use of CLI and SEI instructions. However, XIRQ, which is also an input signal into the CPU, cannot be masked and unmasked using CLI and SEI instructions, and for this reason it is called a *non-maskable interrupt* (NMI). In this section we study these two as well as other external hardware interrupts of the HCS12 .

External interrupt IRQ

The HCS12 pin PE1 belongs to the IRQ external hardware interrupt. The interrupt vector table location $FFF2 and $FFF3 belongs to the IRQ. The IRQ is part of the group of the maskable interrupts that are enabled and disabled globally by the I bit of CCR register. In addition to the I bit of CCR register, we must also enable the IRQ locally using the IRQE bit of the INTCR (INT Control Register) register. The INTCR register is shown in Figure 11-8. How is it activated? There are two types of activation for the IRQ: (1) level triggered, and (2) edge triggered. We use the IRQEN bit of the INTCR register to make it edge triggered or level triggered. Let's look at each one. First, we see how the level-triggered interrupt works.

IRQE	IRQEN						

IRQE D7 IRQ Edge
0 = Level triggered. IRQ pin is activated by a low-level pulse (default).
1 = Edge triggered. IRQ pin is activated by falling edge pulse.

IRQEN D6 IRQ enable
0 = IRQ pin interrupt disabled (default)
1 = IRQ pin interrupt enabled

Figure 11-8. INTCR (INT Control Register)

Level-triggered IRQ

In the level-triggered mode, the IRQ pin is normally high, and if a low-level signal is applied to them, it triggers the interrupt. Then the microcontroller stops whatever it is doing and jumps to the interrupt vector table to service that interrupt. This is called a *level-triggered* or *level-activated* interrupt. The low-level signal at the IRQ pin must be removed before the execution of the last instruction of the interrupt service routine, RTI; otherwise, another interrupt will be generated. In other words, if the low-level interrupt signal is not removed before the ISR is finished, it is interpreted as another interrupt and the HCS12 jumps to the vector table to execute the ISR again. Look at Example 11-7.

In this program, the microcontroller is looping continuously in the HERE loop. Whenever the switch on the IRQ pin is activated, the microcontroller gets out of the loop and jumps to vector location FFF2. The ISR for IRQ toggles the LED. If the IRQ pin is still low by the time the microcontroller executes the RTI instruction, the microcontroller initiates the interrupt again. Therefore, to end this problem, the IRQ pin must be brought back to high by the time RTI is executed. Using the level-triggered interrupt allows us to tie together interrupts from several sources and feed them to the IRQ pin. See Figure 11-9. This is not the case with the edge-triggered option, as we see next.

Example 11-7

Assume that the IRQ pin is connected to a switch that is normally high. Write a program to perform the following: Whenever the SW is activated (goes LOW), it should toggle an LED. The LED is connected to PB.4 and is normally off.

Solution:

```
TEMP  DC.B  1
Entry:LDS    #RAMEnd+1         ;initialize the stack pointer
      BSET  INTCR,#%01000000 ;enable the IRQ, level trigger
      CLR   TEMP               ;temporary storage
      CLI                      ;unmask the interrupts in CCR
      BSET  DDRB,%00010000     ;make PORTB.4 an output
      BCLR  PORTB,%00010000    ;PORTB.4 = 0
HERE  BRA   HERE               ;loop unless interrupted by IRQ
;--ISR for IRQ to turn on PORTB.4
;it came here because IRQ was activated
IRQ_ISR
      LDAA  TEMP              ;get the last value
      EORA  #%00010000         ;toggle bit 4
      STAA  TEMP               ;store the new value
      STAA  PORTB              ;display on PORTB
      RTI                      ;return from interrupt
      ORG $FFF2
      DC.W IRQ_ISR             ;IRQ vector location
      ORG    $FFFE
      DC.W   Entry             ;reset vector
```

Pressing the switch will blink the LED. If the switch is kept activated, the LED keeps blinking.

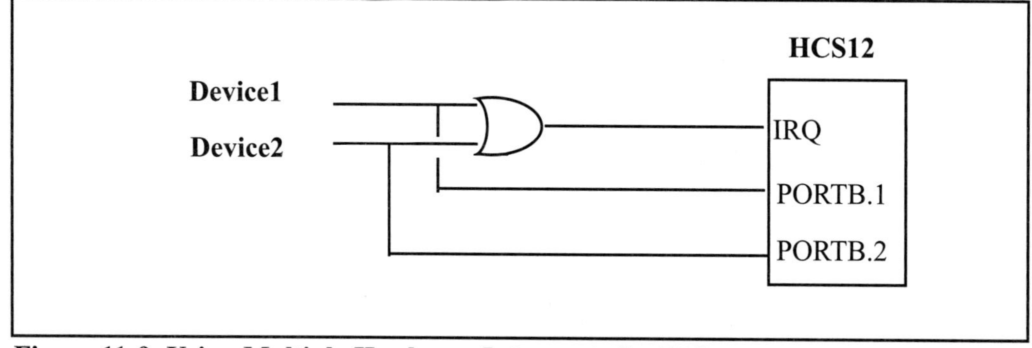

Figure 11-9. Using Multiple Hardware Interrupts for IRQ

Edge-triggered IRQ

As stated before, the IRQE bit in the INTCR register determines level- or edge-triggered mode of the IRQ hardware interrupt. For example, the instruction "`BSET INTCR, #%10000000`" makes IRQ what is called an *edge-triggered interrupt*, which a high-to-low signal applied to pin IRQ will interrupt the controller and force it to jump to location $FFF2 in the vector table to service the ISR (assuming that the I bit is enabled in the CCR register).

Compare Example 11-7 with Example 11-8. Notice that the only difference between them is in the first line of MAIN where the "`BSET INTCR,%11000000`" instruction makes IRQ an edge-triggered interrupt. When the falling edge of the signal is applied to pin IRQ, the LED will toggle. To toggle the LED again, another high-to-low pulse must be applied to pin IRQ. This is the opposite of Example 11-7. In Example 11-7, due to the level-triggered nature of the interrupt, as long as IRQ is kept at a low level, the LED keeps blinking.

Example 11-8

Assume that the IRQ pin is connected to a switch that is normally high. Write a program to perform the following: Whenever the SW goes LOW, it should toggle an LED. The LED is connected to PB.4. Use an edge trigger for IRQ.

Solution:

```
TEMP  DC.B  1
Entry:LDS   #RAMEnd+1         ;initialize the stack pointer
      BSET  INTCR,%11000000   ;enable the IRQ, edge trigger
      CLR   TEMP
      CLI                     ;unmask the interrupts in CCR
      BSET  DDRB,%00010000    ;make PORTB.4 an output
      BCLR  PORTB,%00010000   ;PORTB.4 = 0
HERE  BRA   HERE              ;loop unless interrupted by IRQ
;--ISR for IRQ to turn on PORTB.4
;it came here because IRQ was activated
IRQ_EDGE
      LDAA  TEMP              ;get the last value
      EORA  #%00010000        ;toggle bit 4
      STAA  TEMP              ;store the new value
      STAA  PORTB             ;display on PORTB
      RTI                     ;return from interrupt

      ORG   $FFF2
      DC.W  IRQ_EDGE          ;IRQ vector location
      ORG   $FFFE
      DC.W  Entry
```

Pressing the switch will toggle the LED. The LED will not toggle again until the next falling edge (press of the switch).

External interrupt XIRQ

The HCS12 pin PE0 belongs to the XIRQ external hardware nonmaskable interrupt. The interrupt vector table locations $FFF4 and $FFF5 belong to XIRQ. The XIRQ is a nonmaskable interrupt and is disabled upon reset. We must enable it by clearing the X bit of the CCR register. Upon reset, the X bit is HIGH, which makes it unavailable. After reset, when the CPU is stablized we enable it only once, and from then on anytime the XIRQ pin is activated (goes LOW) the CPU will go to the vector table to fetch the address of the ISR. After enabling the XIRQ (X bit = 0) we cannot disable it to block XIRQ and that is the reason it is called the nonmaskable interrupt. Contrast this with the IRQ pin, which we can use the I bit to mask or unmask. Also notice that unlike the IRQ, there is no need for enabling it locally. It must be emphasized that the enabling of XIRQ is done once and only through the X bit of the CCR register. The XIRQ is level triggered only; there is no edge-triggered option available for this interrupt. See Example 11-9.

Example 11-9

Assume that the XIRQ pin is connected to a switch that is normally high. Write a program to perform the following: Whenever the SW goes LOW, it should toggle an LED. The LED is connected to PB.4 and is normally off.

Solution:

```
TEMP    DS.B 1
Entry: LDS      #RAMEnd+1 ;initialize the stack pointer
     ANDCC      #%10111111;enable XIRQ in CCR
     BSET DDRB,%00010000 ;make PORTB.4 an output
     BCLR PORTB,%00010000;PORTB.4 = 0
HERE BRA   HERE           ;loop unless interrupted by XRQ
;--ISR for XIRQ to turn on PORTB.4
;control came here because XIRQ was activated
XIRQ_ISR
     LDAA TEMP            ;get the last value
     EORA #%00010000      ;toggle bit 4
     STAA TEMP            ;store the new value
     STAA PORTB           ;display on PORTB
     RTI                  ;return from interrupt
;----------------
     ORG    $FFF4
     DC.W   XIRQ_ISR ;XIRQ Vector
     ORG    $FFFE
     DC.W   Entry            ;reset vector
```

Pressing the switch will turn the LED on. If the switch is kept activated, the LED stays on.

PORTH, PORTJ, and PORTP external hardware interrupts

One might complain that there are not enough external hardware interrupts in the HCS12. Beside the IRQ and XIRQ, we have three more ports that are equipped with external hardware interrupts. They are PORTH, PORTJ, and

PORTP. Port J has only four pins: PJ7, PJ6, PJ1, and PJ0. Ports H and P each have 8 pins. See Table 11-1. Notice that there is only one interrupt associated with the entire group of pins for any of these ports. We must enable the interrupt feature of these ports using the PIEx registers as well as the I bit of CCR. See Figures 11-10 through 11-12. We can also choose the falling edge or rising edge for interrupt activation using the PPSx register. Figure 11-13 shows the PPSH for PORTH. Notice that the default is falling edge upon reset. Figure 11-14 shows the PIFH for PORTH. The HCS12 manual shows the PPSx and PIFx registers associated with PORTJ and PORTP.

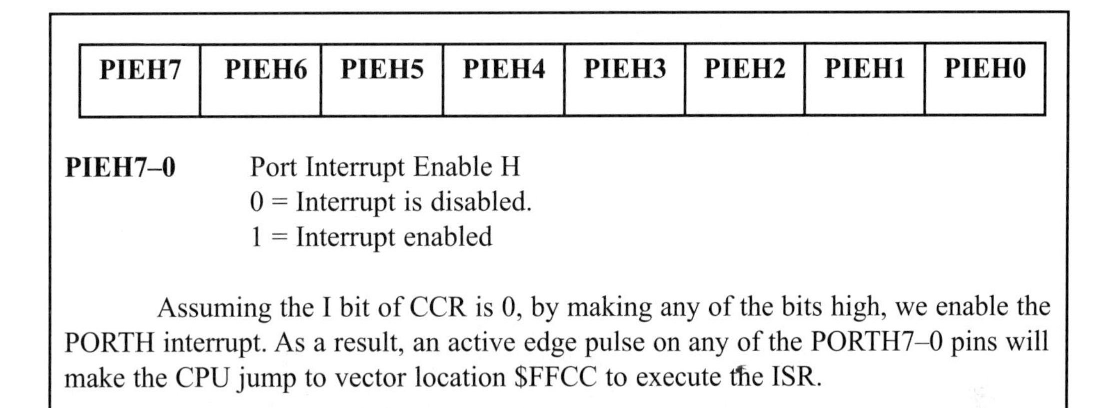
PIEH7
PIEH6
PIEH5
PIEH4
PIEH3
PIEH2
PIEH1
PIEH0
PIEH7–0
Port Interrupt Enable H
0 = Interrupt is disabled.
1 = Interrupt enabled
Assuming the I bit of CCR is 0, by making any of the bits high, we enable the PORTH interrupt. As a result, an active edge pulse on any of the PORTH7–0 pins will make the CPU jump to vector location $FFCC to execute the ISR.

Figure 11-10. PIEH (Port Interrupt Enable H) Register

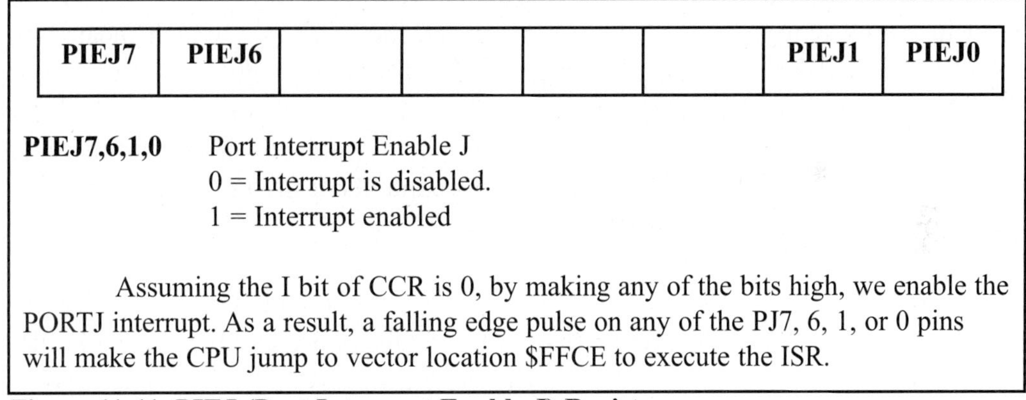
PIEJ7
PIEJ6
PIEJ1
PIEJ0
PIEJ7,6,1,0
Port Interrupt Enable J
0 = Interrupt is disabled.
1 = Interrupt enabled
Assuming the I bit of CCR is 0, by making any of the bits high, we enable the PORTJ interrupt. As a result, a falling edge pulse on any of the PJ7, 6, 1, or 0 pins will make the CPU jump to vector location $FFCE to execute the ISR.

Figure 11-11. PIEJ (Port Interrupt Enable J) Register

PIEP7	PIEP6	PIEP5	PIEP4	PIEP3	PIEP2	PIEP1	PIEP0

PIEP7–0 Port Interrupt Enable P
0 = Interrupt is disabled.
1 = Interrupt enabled

Assuming the I bit of CCR is 0, by making any of the bits high, we enable the PORTP interrupt. As a result, a falling edge pulse on any of the PORTP7–0 pins will make the CPU jump to vector location $FF8E to execute the ISR.

Figure 11-12. PIEP (Port Interrupt Enable P) Register

PPSH7	PPSH6	PPSH5	PPSH4	PPSH3	PPSH2	PPSH1	PPSH0

PPSH7–0 Port Polarity Select register for Port H
0 = Falling edge pulse will activate the interrupt (default upon reset).
1 = Rising edge pulse will activate the interrupt.

Figure 11-13. PPSH (Port Polarity Interrupt Select H) Register

PIFH7	PIFH6	PIFH5	PIFH4	PIFH3	PIFH2	PIFH1	PIFH0

PIFH7–0 Port Interrupt Flag for H
0 = No active edge pending
1 = Active edge on associated pin has occurred.

An active edge is captured by the flag bit. Assuming the I bit of CCR is 0 and the PIEHx bit is enabled, the CPU jumps to vector location $FFCC to execute the ISR.

Figure 11-14. PIFH (Port Interrupt Flag H) Register

STOP and WAIT instructions and wake-up key

We can reduce the power consumption of the HCS12 by using the STOP and WAIT instructions. The execution of the STOP instruction in the HCS12 will stop (freeze) the oscillator and cut off the frequency to the CPU, therefore reducing the power dissipation. The STOP instruction will not take effect unless the S bit in the CCR register is cleared. Upon reset, the S bit = 1, which means the HCS12 will treat the STOP instruction as a NOP. If S bit = 0, upon the execution of the STOP instruction the HCS12 will stop (freeze) the oscillator and put the CPU in standby mode. The standby mode reduces the system power consumption to absolute minimum without losing the register's contents or the I/O states. Any activation of the interrupts such as IRQ or XIRQ will force the CPU out of the standby mode. Another instruction that freezes clock to the CPU is the WAI (wait) instruction. The WAI instruction forces the CPU into the wait state. According to the HCS12 manual "during the wait state CPU clocks are stopped, just like the standby mode, but other MCU clocks can continue to run." Just like the STOP instruction, any activation of the interrupts such as IRQ or XIRQ will force the CPU out of the wait state. In many embedded systems if there is no activity, the system will go into standby mode in order to reduce power consumption. In such systems any keyboard activity will cause the system to wake up and exit the standby mode. This is often referred to as wake-up key operation. One can implement the wake-up key concept using the interrupts associated with PORTJ, PORTH, or PORTP.

Review Questions

1. True or false. There is a single interrupt in the interrupt vector table assigned to both external hardware interrupts, IRQ and XIRQ.
2. What address in the interrupt vector table is assigned to IRQ and XIRQ?
3. Which bit of CCR belongs to the external hardware interrupt IRQ?
4. Assume that the CCR bit for the external hardware interrupt IRQ is enabled

and is active-LOW. Explain how this interrupt works when it is activated.
5. True or false. Upon reset, the IRQ is disabled.
6. In Question 4, how do we make sure that a single interrupt is not recognized as multiple interrupts?
7. True or false. The last instruction of the ISR for IRQ is `RET`.
8. Explain the role that each of the two bits IRQE and IRQTN play in the execution of external IRQ.
9. True or false. There is a single interrupt in the interrupt vector table assigned to all the hardware interrupts PORTJ, PORTH, and PORTP.
10. True or false. There is a single interrupt in the interrupt vector table assigned to all pins of PORTH.

SECTION 11.4: PROGRAMMING THE SERIAL COMMUNICATION INTERRUPT

In Chapter 10 we studied the serial communication of the HCS12. All examples in that chapter used the polling method. In this section we explore interrupt-based serial communication, which allows the HCS12 to do many things, in addition to sending and receiving data from the serial communication port.

TDRE and RDRF flags and interrupts

As you may recall from Chapter 10, TDRE (transmitter data register empty) is raised indicating that the SCIDR register got the byte from the transmitter register and it is ready for the next byte. RDRF (received data register full) is raised when the entire frame of data, including the stop bit, is received. In other words, when the SCIDR register has a byte, RDRF is raised to indicate that the received byte needs to be picked up before it is lost (overrun) by new incoming serial data. As far as serial communication is concerned, all of the above concepts apply equally when using either polling or an interrupt. The only difference is in how the serial communication needs are served. In the polling method, we wait for the flag (TDRE or RDRF) to be raised; while we wait we cannot do anything else. In the interrupt method, we are notified when the HCS12 has received a byte, or is ready to send the next byte; we can do other things while the serial communication needs are served.

In the HCS12, only one interrupt is set aside for each of the SCI0 and SCI1 serial communications. That means a single interrupt is used for both send and receive data. To use interrupts instead of polling we must enable the I bit in addition to the TIE and RIE bits of the SCICR2 register. See Example 11-10. If these bits are enabled, when TDRE or RDRF is raised the HCS12 gets interrupted and jumps to memory address location $FFD6 to execute the ISR for SCI0. In that ISR we must examine the TDRE and RDRF flags to see which one caused the inter-

Example 11-10

What is the value for the SCICR2 register if we want to use interrupt-driven serial communication?

Solution:

Assuming the I bit of CCR is enabled, we need SCICR2 = 10101100.

rupt and respond accordingly. See Example 11-11. Figure 11-15 shows the TIE and RIE bits of the SCICR2 register. Figure 11-16 shows how RIE and TIE use a single interrupt to get the attention of the HCS12 CPU.

TIE	TCIE	RIE	ILIE	TE	RE	RWU	SBK

TIE D7 Transmit Interrupt enable bit. Used for interrupt-driven SCI.
0 = TDRE Interrupt Request is disabled.
1 = TDRE Interrupt Request is enabled.

TCIE D6 Transmission Complete Interrupt Enable bit. Used for interrupt-driven SCI.
0 = TC Interrupt Request is disabled.
1 = TC Interrupt Request is enabled.

RIE D5 Receiver Full Interrupt Enable bit. Used for interrupt-driven SCI.
0 = RDRF Interrupt Request is disabled.
1 = RDRF Interrupt Request is enabled.

ILIE D4 Idle Line Interrupt Enable bit. Used for interrupt-driven SCI.
0 = IDLE Interrupt Request is disabled.
1 = IDLE Interrupt Request is enabled.

TE D3 Transmitter Enable bit. **We must enable this bit to transmit data.**
0 = Transmitter is disabled.
1 = Transmitter is enabled.

RE D2 Receiver Enable bit. **We must enable this bit to receive data.**
0 = Receiver is disabled.
1 = Receiver is enabled.

RWU D1 Used for wake-up condition in stand-by mode. See the HCS12 manual.
0 = Normal operation
1 = RWU is enabled.

SBK D0 Used for break bit. See the HCS12 manual.
0 = No break character
1 = Transmit break character

Note: The most important bits in this register are the TE and RE bits. In applications using the interrupt method we make the interrupt request bits all 1s. For the interrupt method, we use SCICR2 = AC in hex. The rest of the bits are for testing purposes. To use polling-driven SCI, see Chapter 10.

Figure 11-15. SCI Control Register 2 (SCICR2)

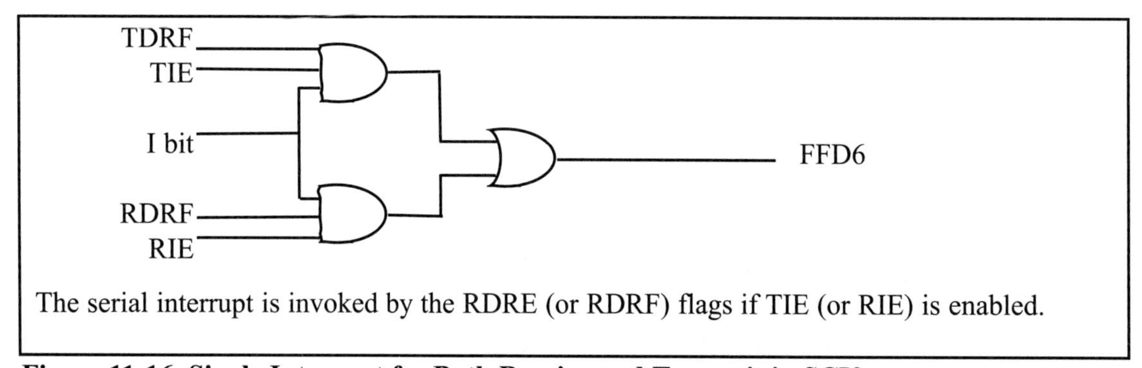

The serial interrupt is invoked by the RDRE (or RDRF) flags if TIE (or RIE) is enabled.

Figure 11-16. Single Interrupt for Both Receive and Transmit in SCI0

Example 11-11

Write a program in which the HCS12 reads data from PORTA and writes it to PORTB continuously while giving a copy of it to the serial COM port to be transferred serially. Assume that XTAL = 4 MHz. Set the baud rate at 9,600.

Solution:

```
;--The main program for initialization
      LDS  #RAMEnd+1 ;initialize the stack pointer
      LDAA #$FF
      STAA DDRB      ;make PB an output
      LDAA #$00
      STAA DDRA      ;make PA an input
      LDAA #$0       ;set the baud rate
      STAA SCI0BDH   ;we must write to high byte reg
      LDAA #$0D      ;set the baud rate to 9600
      STAA SCI0BDL   ;write to low byte
      LDAA #0        ;8-bit data, no parity
      STAA SCI0CR1   ;write to control reg 1
      LDAA #$AC      ;enable both transmit and receive
      STAA SCI0CR2   ;write to control reg 2
      LDAA PORTA     ;get data from PA
      STAA SCI0DRL   ;load SW status into SCI data reg
      CLI            ;unmask the interrupts in CCR
H1    STAA PORTB     ;send it to PB
      LDAA PORTA     ;get data from PA
      BRA  H1        ;loop unless interrupted by TIE

;--ISR for serial transmit
;it came here because TDRE or RDRF flag was raised
SCI0_ISR BRCLR SCI0SR1,mSCI0SR1_TDRE,EXIT ;jump if receive
      LDAA PORTA
      STAA SCI0DRL
EXIT RTI
;-----------------------------------
      ORG   $FFD6
      DC.W  SCI0_ISR
      ORG   $FFFE
      DC.W  Entry             ;reset vector
```

In the above program notice the role of TDRE. The moment a byte is written into the serial transmitter register TDRE is raised, which causes the serial interrupt to be invoked since the corresponding bit in the TIE register is high. In the serial ISR, we check for both TDRE and RDRF since both could have invoked the interrupt. In other words, there is only one interrupt for both transmit and receive.

Use of serial COM interrupt in the HCS12

In the vast majority of applications, the serial interrupt is used mainly for receiving data and is rarely used for sending data serially. This is like receiving a telephone call, where we need a ring to be notified. If we need to make a phone call there are other ways to remind ourselves and so no need for ringing. In receiving the phone call, however, we must respond immediately no matter what we are doing or we will miss the call. Similarly, we use the serial interrupt to receive incoming data so that it is not lost. Look at Example 11-12.

Example 11-12

Write a program in which the HCS12 gets data from PORTA and sends it to PORTB continuously while incoming data from the serial port is sent to PTH. Assume that XTAL = 4 MHz. Set the baud rate at 9,600.

Solution:

```
      ;--The main program for initialization
      LDS    #RAMEnd+1   ;initialize the stack pointer
      LDAA   #$FF
      STAA   DDRB         ;make PB an output
      STAA   DDRH
      LDAA   #$00
      STAA   DDRA         ;make PA an input
      LDAA   #$0          ;set the baud rate
      STAA   SCI0BDH;we must write to high byte reg even if zero
      LDAA   #$0D         ;set the baud rate to 9600
      STAA   SCI0BDL      ;write to low byte
      LDAA   #0           ;8-bit data, no parity
      STAA   SCI0CR1      ;write to control reg 1
      LDAA   #$AC  ;enable both transmit and receive interrupt
      STAA   SCI0CR2      ;write to control reg 2
      CLI                 ;unmask the interrupts in CCR
H1    LDAA   PORTA        ;get data from PA
      STAA   PORTB        ;send it to PB
      BRA    H1           ;loop unless interrupted by RIE
;--ISR for serial transmit
;it came here because TDRE or RDRF flag was raised
SCI0_ISR BRCLR   SCI0SR1,mSCI0SR1_RDRF,EXIT    ;jump if not receive
      LDAA SCI0DRL
      STAA PTH
EXIT  RTI
;--------------------------
      ORG    $FFD6
      DC.W   SCI0_ISR
      ORG    $FFFE
      DC.W   Entry              ;reset vector
```

Clearing TDRE and RDRF before the RTI instruction

Notice that in the previous examples TDRE and RDRF are checked to find out which caused the SCI0_ISR to be executed. Neither case contains an instruction to explicitly clear the flag. This is because these flags can only be cleared by reading SCI0SR1 followed by the appropriate action on SCI0DRL. This was accomplished by first checking which flag was set and then writing to SCI0DRL when transmitting (or reading from SCI0DRL when receiving). One might wonder how to clear TDRE once all data has been sent. A simple solution is to disable the interrupts by clearing TIE as is the case with any interrupt we wish to ignore.

Review Questions

1. True or false. There is a single interrupt in the interrupt vector table assigned to both the transmit and receive of the SCI.
2. What address in the interrupt vector table is assigned to the SCI0 interrupt?
3. Which bit of the SCICR2 register belongs to the transmit interrupt?
4. Assume that the TIE bit for the serial transmit interrupt is enabled. Explain how this interrupt gets activated and also explain its actions upon activation.
5. True or false. Upon reset, the serial interrupt is active and ready to go.
6. True or false. The last two instructions of the ISR for the receive interrupt are:

```
BSET SCI0SR1,TDRE
RTI
```

SECTION 11.5: INTERRUPT PRIORITY IN THE HCS12

The next topic that we must deal with is, What happens if two interrupts are activated at the same time? Which of these two interrupts is responded to first? Interrupt priority is the main topic of discussion in this section.

Interrupt priority upon reset

When the HCS12 is powered up, the priorities are assigned according to Table 11-3. Table 11-3 shows the hardware Reset pin has the highest priority. From Table 11-3 we also see, for example, that if external hardware interrupts XIRQ and IRQ are activated at the same time, the XIRQ is responded to first. Only after XIRQ has been serviced is the IRQ serviced since it has the lower priority. Indeed, all interrupts associated with the I bit have lower priority than XIRQ. See Figure 11-17 and Table 11-4. In reality, the priority scheme in the table is nothing but an internal polling sequence in which the HCS12 polls the interrupts in the sequence listed in Table 11-4, and responds accordingly.

Table 11-3: HCS12 Interrupt Priority Upon Reset

Highest to Lowest Priority
Reset
Clock Monitor Fail Reset
COP Failure Reset
Invalid Instruction Opcode Trap
SWI
XIRQ
IRQ
Other peripherals have lower priority. Their rankings are shown in Table 11-4. We can raise the priority by loading a value into the HPRIO register.

Setting interrupt priority with the HPRIO register

We cannot alter the priority assigned to the first six interrupts shown in

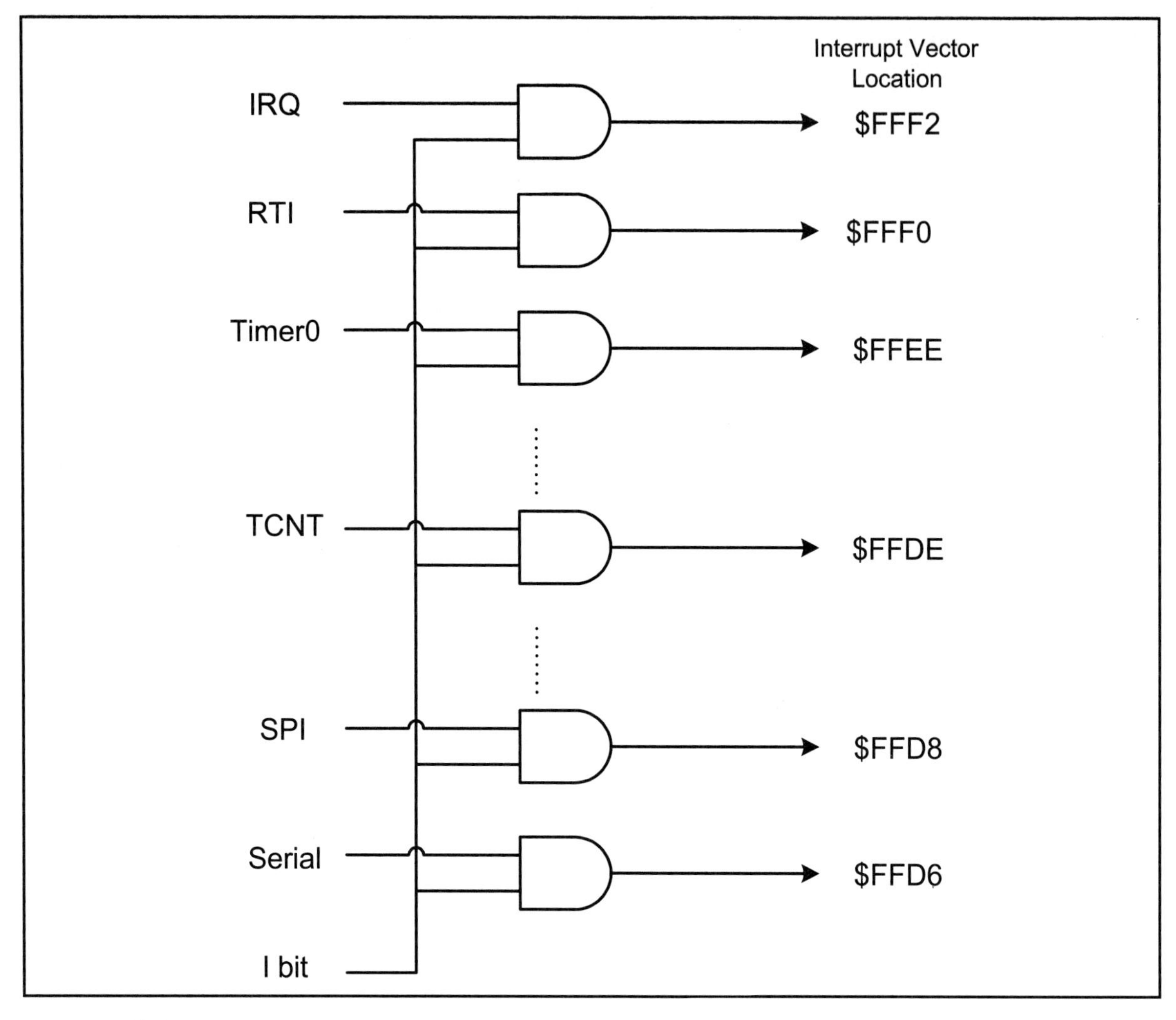

Figure 11-17. Interrupt Priority For I Bit Group of Interrupts

Table 11-4. That means XIRQ has lower priority than the Reset at all times. However, we can change the priority assigned to IRQ and peripherals. This is done by programming a register called HPRIO (highest interrupt priority I/O). Figure 11-18 shows the bits of the HPRIO register. Upon power-up reset, the HPRIO register contains value $F2, making the IRQ the highest priority among the peripherals. To give a higher priority to any of the peripherals, we load the value shown in Table 11-4 into the HPRIO register. Look at Examples 11-13 and 11-14.

Example 11-13

Discuss what happens if interrupts Timer Channel 0, XIRQ, and SCI0 (serial 0) are activated at the same time. Assume priority levels were set by the power-up reset.

Solution:

If these three interrupts are activated at the same time, they are latched and kept internally. Then the HCS12 checks all three interrupts according to the sequence listed in Table 11-3. Therefore, when the above three interrupts are activated, XIRQ (external interrupt) is serviced first, then Timer Channel 0 (C0I), and finally SCI0 (Serial Com #0).

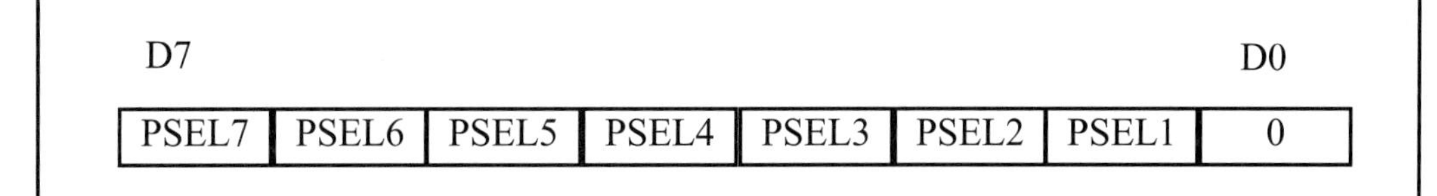

Upon reset it has value $F2, making the IRQ the highest priority among the I/O peripherals. Use the values in Table 11-4 to raise the priority.

Figure 11-18. HPRIO (High Priority Register for I/O Peripherals)

Table 11-4: Interrupt Priority List for the HCS12

Interrupt	ROM Location (Hex)	Global Enable (CCR Mask)	HPRIO Value to Elevate (Hex)
Reset	FFFE–FFFF	None	
Clock Monitor Fail reset	FFFC–FFFD	None	
COP Failure reset	FFFA–FFFB	None	
Unimplemented instruction trap	FFF8–FFF9	None	
SWI instruction	FFF6–FFF7	None	
XIRQ External hardware interrupt	FFF4–FFF5	X bit	
IRQ External hardware interrupt	FFF2–FFF3	I bit	F2
Real Time interrupt	FFF0–FFF1	I bit	F0
Capture Timer channel 0	FFEE–FFEF	I bit	EE
Capture Timer channel 1	FFEC–FFED	I bit	EC
Capture Timer channel 2	FFEA–FFEB	I bit	EA
Capture Timer channel 3	FFE8–FFE9	I bit	E8
Capture Timer channel 4	FFE6–FFE7	I bit	E6
Capture Timer channel 5	FFE4–FFE5	I bit	E4
Capture Timer channel 6	FFE2–FFE3	I bit	E2
Capture Timer channel 7	FFE0–FFE1	I bit	E0
TCNT timer overflow	FFDE–FFDF	I bit	DE
Pulse Accumulator A overflow	FFDC–FFDD	I bit	DC
Pulse Accumulator input edge	FFDA–FFDB	I bit	DA
SPI	FFD8–FFD9	I bit	D8
SCI0 (Serial COM # 0)	FFD6–FFD7	I bit	D6
SCI1 (Serial COM #1)	FFD4–FFD5	I bit	D4
ATD0 (Analog-To-Digital #0)	FFD2–FFD3	I bit	D2
ATD1 (Analog-To-Digital #1)	FFD0–FFD1	I bit	D0
PORTJ	FFCE–FFCF	I bit	CE
PORTH	FFCC–FFCD	I bit	CC
PORTP	FF8E–FF8F	I bit	8E

See the HCS12 manual for the complete list.

Fast context saving in task switching

In many applications, such as multitasking real-time operating systems (RTOS), the CPU brings in one task (job or process) at a time and executes it

Example 11-14

(a) Show the HPRIO register value if we want to assign the highest priority to SCI0 (Serial Com 0). Then (b) discuss what happens if SCI0, IRQ, and Timer Channel 7 are activated at the same time.

Solution:

(a) According to Table 11-4 we must have `HPRIO = $D6`

(b) Step (a) assigned a higher priority to SCI0 than the others; therefore, when the SCI0, IRQ, and Timer Channel 7 interrupts are activated at the same time, the HCS12 services SCI0 first, then it services IRQ, then Timer Channel 7. This is due to the fact that SCI0 has a higher priority than the other two because of Step (a). As a result, the sequence in Table 11-4 is followed, which gives a higher priority to IRQ over Timer Channel 7.

before it moves to the next one. In executing each task, which is often organized as an interrupt service routine, access to all the resources of the CPU is critical in performing the task in a timely manner. It is the job of the programmers to make sure the entire contents of the CPU are saved on the stack before execution of the new task. This saving of the CPU contents before switching to a new task is called *context saving* (or *context switching*). The use of the stack as a place to save the CPU's contents is tedious, time consuming, and slow. For this reason some CPUs such as x86 microprocessors have instructions such as PUSHA (Push All) and POPA (Pop All), which will push and pop all the main registers onto the stack with a single instruction. In the HCS12, each task generally needs the key registers of A, B, X, Y, CCR, and so on. For that reason the HCS12 automatically saves these registers on stack when an interrupt is activated. This way, these key registers of the main task are saved internally. To restore the original contents of these key registers, we use the instruction "RTI" instead of "RET" at the end of the ISR.

Interrupt latency

The time from the moment an interrupt is activated to the moment the CPU starts to execute the ISR code is called the *interrupt latency*. This latency can be as high as ten clock cycles depending on whether the source of the interrupt is an internal (e.g., timers) or an external hardware (e.g., hardware IRQ and XIRQ) interrupt. The duration of an interrupt latency can also be affected by the type of the instruction in which the CPU was executing when the interrupt comes in. It takes slightly longer in cases where the instruction being executed lasts for many cycles compared to the instructions that last for only one cycle time. In the HCS12, we also have extra clocks added to the latency due to the fact that it saves the main registers (A, B, X, and so on) on stack. See the HCS12 manual for the timing data sheet.

Interrupt inside an interrupt

What happens if the HCS12 is executing an ISR belonging to an interrupt and another interrupt is activated? In such cases, a high-priority interrupt can interrupt a low-priority interrupt. This is an interrupt inside an interrupt. Generally a low-priority interrupt can be interrupted by a higher-priority interrupt, but not by another lower-priority interrupt. In HCS12 systems, it is up to the software engineer to implement such a policy since it does not support nested interrupts. To do that, we must make I bit = 0 in the lower priority ISR since the CPU makes I bit = 1 upon activation of any maskable interrupt. We can design our software in a way that no lower-priority interrupt can get the immediate attention of the CPU until the HCS12 has finished servicing all the higher-priority interrupts. Program 11-1 shows multiple interrupts in the same program.

```
;Program 11-1: Multiple Interrupt Sources
CHAR   DC.B 1

     ORG    ROMStart
Entry:
     LDS  #RAMEnd+1 ; initialize the stack pointer
     LDAA #$FF
     STAA DDRB      ;make PB an output
     LDAA #$00
     STAA DDRA      ;make PA an input
     JSR  SCI0_INIT ;initial SCI0
     JSR  TC7_INIT  ;initial TC7
     LDAA #$D6      ;the low byte SCI0 vector
     STAA HPRIO
     CLI            ;unmask the interrupts in CCR
H1   LDAA PORTA     ;get data from PA
     STAA PORTB     ;send it to PB
     BRA  H1        ;loop unless interrupted by TIE
;---------------------
SCI0_INIT
     LDAA #$0       ;set the baud rate
     STAA SCI0BDH   ;we must write to high byte
     LDAA #$0D      ;set the baud rate to 9600
     STAA SCI0BDL   ;write to low byte
     LDAA #0        ;8-bit data, no parity
     STAA SCI0CR1   ;write to control reg 1
     LDAA #$AC ;enable transmit and receive with intrp
     STAA SCI0CR2   ;write to control reg 2
     RTS
;---------------------
TC7_INIT
     LDAA #$80
     STAA TSCR1     ;enable timer
     LDAA #$04
     STAA TSCR2     ;interrupt enabled, prescaler 16
     BSET TIOS,%10000000 ;select Channel 7
     LDAA #$40
     STAA TCTL1     ;toggle PT7 pin upon match
     LDAA #$80
     STAA TIE       ;enable interrupt for Channel 7
```

```
      RTS
;---------------------
SCI0_ISR
      BRCLR SCI0SR1,mSCI0SR1_TDRE,RCV ;jump not trans
      LDAA CHAR       ;get the character
      STAA SCI0DRL
RCV   BRCLR SCI0SR1,mSCI0SR1_RDRF,EXIT ;jump not recv
      LDAA SCI0DRL
      STAA CHAR       ;save the character
EXIT RTI
;---------------------
TC7_ISR
      LDD  TCNT       ;read current TCNT value
      ADDD #62500     ;we need 62500 pulses
      STD  TC7        ;load the TC7 (Timer Compare 7)
      BSET TFLG1,%10000000     ;clear Channel 7 match
flag
      RTI             ;return from interrupt
;---------------------
      ORG   $FFD6
      DC.W  SCI0_ISR ;Serial 0 vector location
      ORG   $FFE0
      DC.W  TC7_ISR  ;Timer Channel 7 vector location
      ORG   $FFFE
      DC.W  Entry    ;reset vector
```

Review Questions

1. True or false. Upon reset, all interrupts have the same priority.
2. What register is used to assign interrupt priority in the HCS12?
3. Which byte belongs to the SCI0 (Serial Com 0)? Show how to assign it the highest priority.
4. Assume that the HPRIO register contains $E0. Explain what happens if both IRQ and Timer Channel 7 are activated at the same time.
5. Explain what happens if a higher-priority interrupt is activated while the HCS12 is serving a lower-priority interrupt (that is, executing a lower-priority ISR). Assume I bit = 0 at the beginning of the lower-priority ISR.

SECTION 11.6: INTERRUPT PROGRAMMING IN C

So far, all the programs in this chapter have been written in Assembly. In this section, we provide the C versions of the programs for HCS12 interrupts. In reading this section, it is assumed that you already know the material in the first five sections of this chapter.

Interrupt numbering in CodeWarrior

Notice in Examples 11-1C through 11-12C the way the interrupt service routine (ISR) is defined. CodeWarrior allows the keyword "interrupt" and the equation that follows to begin the definition of the ISR. The "interrupt" tells the compiler to end the following code block with opcode for return from interrupt (RTI). The equation calculates the interrupt number. See Example 11-15 for more details about the equation. The function itself should be defined as "void functionName(void)" as the interrupt does not receive any data nor return any data. In CodeWarrior that is all that is needed to define the ISR in the code file. CodeWarrior also allows defining the address in the vector table in the PRM file. Other compilers may have different standards for defining interrupts. Check their documentation for explanations on how to define ISRs.

Example 11-15

Calculate the interrupt number for XIRQ used by CodeWarrior.

Solution:

The interrupt vector can be found in the data sheet and in the CodeWarrior header file. The interrupt vector for XIRQ is $FFF4. Interrupt numbers start at 0, which is the Reset vector $FFFE. To find the interrupt number for XIRQ, count the number of vectors to $FFF4, like this:

Vector	Number
$FFFE	0
$FFFC	1
$FFFA	2
$FFF8	3
$FFF6	4
$FFF4	5

This means the interrupt number for XIRQ is 5. In the programming examples, we use a calculation instead, so that the code is compatible with other chips programmed with CodeWarrior. The equation subtracts the vector location from the top of the vector table, in this case $10000. Then the value returned is divided by 2, since the vector locations are two bytes. Finally, we subtract 1 from the value, since the vector numbers begin at 0. For XIRQ, the equation looks like this:

0x10000 – 0xFFF4 = 12 (decimal)
12 / 2 = 6
6 – 1 = 5

In the examples we will be using the vector name provided by the header file. For XIRQ this is "Vxirq". Use of the vector name in the equation allows the header file to define the vector number for the HCS12 chip. The equation for XIRQ is:

(((0x10000 – Vxirq) / 2) – 1) (in C syntax)

Example 11-1C (C version of Example 11-1)

Using interrupts, write a C program to count the number of times the TCNT free-running timer overflows. Display the binary number on PORTB. The program should get 8-bit data from PORTH and send it to PORTA while simultaneously counting up the overflow. Assume that XTAL = 4 MHz. This is the C version of Example 11-1.

Solution:

```
#include <mc9s12dp512.h>          /* derivative information */

unsigned char COUNT;

void main(void)
{
   /* put your own code here */
   DDRB = 0xFF; //make PORTB an output
   DDRA = 0xFF; //make PORTA an output
   DDRH = 0x00; //make PTH an input
   COUNT = 0;    //initialize the count
   TSCR1 = 0x80;//enable the timer
   TSCR2 = 0x86;//enable interrupt, prescaler=64
   TFLG2 = TFLG2 | TFLG2_TOF_MASK; //clear flag
   __asm CLI; //enable interrupts globally
   for(;;)
   {
     PORTA = PTH;
   } /* do forever */
   /* please make sure that you never leave this function */
}

interrupt (((0x10000-Vtimovf)/2)-1) void TOF_ISR(void)
{
   COUNT++; //increment COUNT
   PORTB = COUNT; //display on PORTB
   TFLG2 = TFLG2 | TFLG2_TOF_MASK; //clear flag
}
```

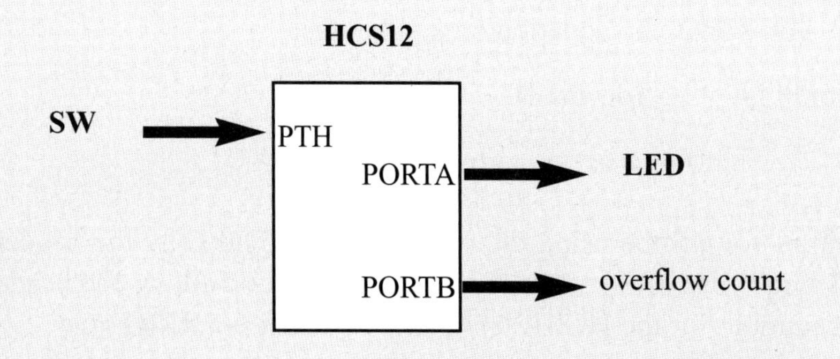

Example 11-2C (C version of Example 11-2)

Write a C program to toggle pin PT7 every 1/2 second. This is similar to Example 9-10 except that it uses an interrupt for Output Compare of channel 7. Assume that XTAL = 4 MHz.

Solution:

0.5 s / 0.5 μs = 1,000,000. That means we must use a prescaler. Using a prescaler of 16 we get 1,000,000/16 = 62,500.

```
#include <mc9s12dp512.h>      /* derivative information */

void main(void)
{
   /* put your own code here */
   TSCR1 = 0x80;    //enable the timer
   TSCR2 = 0x04;    //prescaler=16
   TIOS = TIOS | TIOS_IOS7_MASK; //select channel 7
   TCTL1 = 0x40; //toggle PT7 upon match
   TIE = 0x80; //enable interrupt for Channel 7
   __asm CLI; //enable interrupts globally
   for(;;)
   {
   } /* wait forever */
   /* please make sure that you never leave this function */
}

interrupt (((0x10000-Vtimch7)/2)-1) void TC7_ISR(void)
{
   TC7 = TCNT + (word)62500; //we need 62500 pulses
   TFLG1 = TFLG1 | TFLG1_C7F_MASK; //clear the flag
}
```

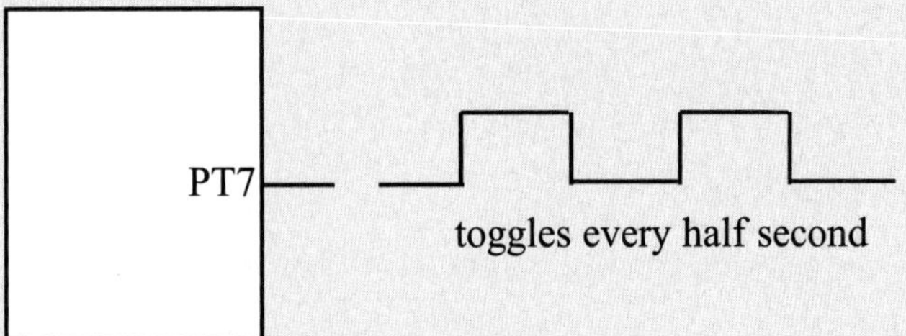

Example 11-3C (C version of Example 11-3)

Assuming that pin PT7 is connected to a pulse generator, write a C program in which the rising edge of the pulse will send a high to PB.4, which is connected to an LED (or buzzer). In other words, the LED is turned on and off at the same rate as the pulses are applied to the PT7 pin.

Solution:

```
void main(void)
{
  /* put your own code here */
  DDRT = DDRT & ~DDRT_DDRT7_MASK;   //make PT7 an input
  DDRA = DDRA | DDRA_BIT4_MASK; //make PA4 an output
  DDRB = DDRB | DDRB_BIT4_MASK; //make PB4 an output
  TSCR1 = 0x80;    //enable timer
  TSCR2 = 0x00;    //no overflow interrupt, no prescaler
  TIOS = TIOS | TIOS_IOS7_MASK; //select channel 7
  TCTL3 = 0x40;    //channel 7 input capture
  TIE = 0x80; //enable channel 7 interrupt
  __asm CLI;  //enable interrupts globally
  for(;;)
  {
  } /* wait forever */
}

interrupt (((0x10000-Vtimch7)/2)-1) void TC7_ISR(void)
{
  PORTB = PORTB ^ PORTB_BIT4_MASK; //toggle PB4
  TFLG1 = TFLG1 | TFLG1_C7F_MASK; //clear the flag
}
```

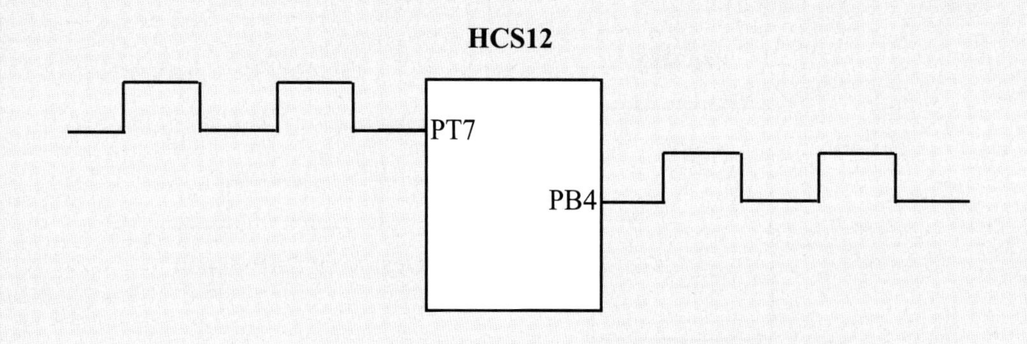

Example 11-5C (C version of Example 11-5)

Assume that a 1 Hz pulse is connected to input for PACA (pin PT7). The pulses represent the number of bottles being packaged together. Using interrupts, write a program to turn on PORTB.4 every time 12 bottles go by. The program should display the binary count on PORTA. This is the interrupt version of Example 11-5.

Solution:

```
unsigned char Counter;

void main(void)
{
  /* put your own code here */
  DDRA = 0xFF;   //PORTA output
  DDRB = DDRB | DDRB_BIT4_MASK; //PB4 output
  DDRT = DDRT & ~DDRT_DDRT7_MASK; //PT7 as input
  Counter = 0;   //clear the counter
  PACTL = 0b01010010; //enable 16-bit event counter
  PACN32 = 0xFFF4;   //(D=-12 or 0xFFF4) we need 12 pulses
  PORTB = PORTB & ~PORTB_BIT4_MASK; //PB4 = 0
  __asm CLI;//enable interrupts globally
  for(;;)
  {
  } /* wait forever */
}

interrupt (((0x10000-Vtimpaaovf)/2)-1) void PACA_ISR(void)
{
  PORTB = PORTB | PORTB_BIT4_MASK; //PB4 = 1
  PAFLG = PAFLG | PAFLG_PAOVF_MASK; //clear the flag
  Counter++; //increment the counter
  PORTA = Counter; //display the count on PORTA
  PACN32 = 0xFFF4; //reload for 12 pulses
  PORTB = PORTB & ~PORTB_BIT4_MASK; //PB4 = 0
}
```

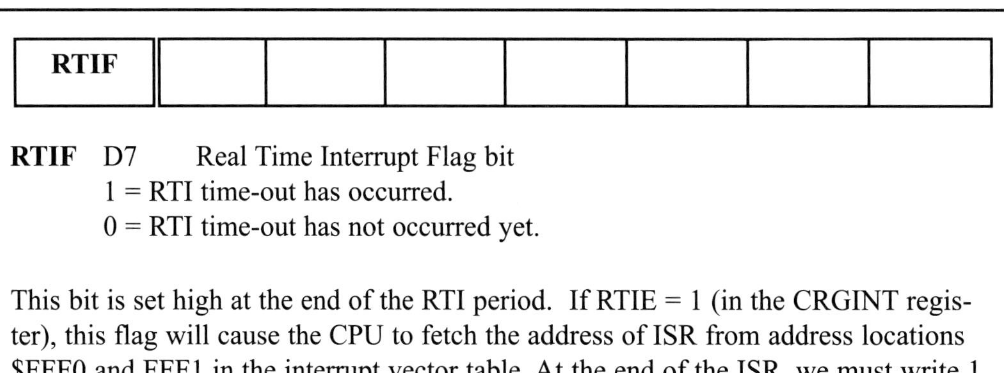

RTIF D7 Real Time Interrupt Flag bit
1 = RTI time-out has occurred.
0 = RTI time-out has not occurred yet.

This bit is set high at the end of the RTI period. If RTIE = 1 (in the CRGINT register), this flag will cause the CPU to fetch the address of ISR from address locations $FFF0 and FFF1 in the interrupt vector table. At the end of the ISR, we must write 1 to this flag to clear it.

Figure 11-20. CRGFLG Register

Example 11-18

Assume that XTAL = 4 MHz. Using RTI write a program to toggle an LED approximately every 2 seconds. For the RTICTL value use option B in Example 11-16. The LED is connected to PB.4.

Solution:

```
TEMP  DC.B  1
COUNT DC.B  1
Entry:LDS   #RAMEnd+1          ;initialize the stack pointer
      BSET  CRGINT,%10000000   ;enable RTI interrupt
      LDAA  #%01111111         ;every 0.262144 s
      STAA  RTICTL             ;load RTICTL register
      CLR   TEMP               ;clear the TEMP register
      CLR   COUNT              ;clear COUNT
      BSET  DDRB,%00010000     ;make PORTB.4 an output
      CLI                      ;unmask the interrupts in CCR
HERE  BRA   HERE               ;loop unless interrupted by RTI
;--ISR for RTI to turn on PORTB.4
;it came here because RTI was activated
RTI_ISR
      INC   COUNT              ;increment the COUNT
      LDAA  COUNT              ;A = COUNT
      CMPA  #8                 ;is it 2 s? (8 × 0.262144 s)
      BNE   OVER               ;if not clear RTIF and return
      LDAA  TEMP               ;get the last value
      EORA  #%00010000         ;toggle bit 4
      STAA  TEMP               ;store the new value
      STAA  PORTB              ;display on PORTB
      CLR   COUNT
OVER  BSET  CRGFLG,%10000000   ;clear RTIF
      RTI                      ;return from interrupt
;--------------------------------
      ORG   $FFF0
      DC.W  RTI_ISR            ;RTI vector location
      ORG   $FFFE
      DC.W  Entry              ;reset vector
```

RTIE							

RTIE D7 Real Time Interrupt Enable bit
1 = Interrupt will be requested whenever RTIF is raised.
0 = Interrupt request from RTI is denied (disabled).

Figure 11-21. CRGINT Register

Example 11-18C (C version of Example 11-18)

Assume that XTAL = 4 MHz. Using RTI write a C program to toggle an LED approximately every 2 seconds. For the RTICTL value use option B in Example 11-16. The LED is connected to PB.4.

Solution:

```
unsigned int COUNT;
void main(void)
{
  /* put your own code here */
  CRGINT = CRGINT | CRGINT_RTIE_MASK;      //enable RTI
  RTICTL = 0b01111111;                  //every 0.262144 s
  COUNT = 0;                            //clear the COUNT
  DDRB = DDRB | DDRB_BIT4_MASK;            //PB4 output
  __asm CLI;                            //enable interrupts globally.
  for(;;) {}  /* wait forever */
}

interrupt (((0x10000-Vrti)/2)-1) void RTI_ISR(void)
{
  COUNT++;                              //increment the count
  if(COUNT==8)                //is it 2 s? (8 × 0.262144 s)
  {
    PORTB = PORTB ^ PORTB_BIT4_MASK; //toggle bit 4
    COUNT = 0;                          //clear the COUNT
  }
  CRGFLG = CRGFLG | CRGFLG_RTIF_MASK;      //clear RTI flag
}
```

Review Questions

1. True or false. RTI uses bus frequency.
2. True or false. We only use 7 bits of the RTICTL register.
3. True or false. To make RTI work, we must set the RTIE bit HIGH.
4. RTIE is part of the ________________ register.
5. The last instruction for the ISR of RTI is ____________________.

PROBLEMS

SECTION 11.1: HCS12 INTERRUPTS

1. Which technique, interrupt or polling, avoids tying down the microcontroller?
2. Including hardware reset, how many interrupts does the HCS12 have for reset?
3. In the HCS12, what memory location in the interrupt vector table is assigned to Reset?
4. True or false. The HCS12 programmer cannot change the memory space assigned to the interrupt vector table.
5. What memory address in the interrupt vector table is assigned to invalid opcode?
6. What memory address in the interrupt vector table is assigned to IRQ?
7. What memory address in the interrupt vector table is assigned to XIRQ?
8. What memory address in the interrupt vector table is assigned to Timer Channel. 1?
9. What memory address in the interrupt vector table is assigned to the second serial COM?
10. How many bytes of address space in the interrupt vector table are assigned to each interrupt?
11. How many bytes of address space in the interrupt vector table are assigned to the Timer Channel 0 and Timer Channel 1 interrupts?
12. With a single instruction, show how to disable all the peripheral interrupts.
13. With a single instruction, show how to enable the XIRQ interrupt.
14. True or false. Upon reset, all interrupts are enabled by the HCS12.
15. True or false. Each interrupt is given two bytes of memory space in the interrupt vector table.

SECTION 11.2: PROGRAMMING TIMER INTERRUPTS

16. True or false. Each Channel of 0–7 has an interrupt assigned to it in the interrupt vector table.
17. What address in the interrupt vector table is assigned to Channel 1?
18. Which bit of TIE belongs to the Channel 0 interrupt? Show how it is enabled.
19. Which bit of TIE belongs to the Channel 7 interrupt? Show how it is enabled.
20. Assume that Timer Channel 0 is enabled. Explain how the interrupt for the timer works if we are using the Output Compare function.
21. True or false. The last instruction of the ISR for timer is RTI.

SECTION 11.3: PROGRAMMING EXTERNAL HARDWARE INTERRUPTS

22. True or false. A single interrupt is assigned to both IRQ and XIRQ.
23. What address in the interrupt vector table is assigned to IRQ and XIRQ?
24. Which bit of CCR belongs to the IRQ interrupt? Show how it is enabled.
25. Which bit of CCR belongs to the XIRQ interrupt? Show how it is enabled.
26. Show how to enable both external hardware interrupts.
27. Assume that the I bit for external hardware interrupt IRQ is enabled and is low-

level triggered. Explain how this interrupt works when it is activated. How can we make sure that a single interrupt is not interpreted as multiple interrupts?

28. True or false. Upon reset, the external hardware interrupt is edge triggered.
29. In level-triggered interrupts, how do we make sure that a single interrupt is not recognized as multiple interrupts?
30. Which bits of INTCR belong to IRQ?
31. Which register is used to decide edge activation of the PORTH interrupt?
32. True or false. The PORTH hardware interrupt has its own location in the vector table.
33. Explain how we can prevent the STOP instruction from taking effect.
34. Explain the role of the IRQE and IRQEN bits of INTCR in the execution IRQ interrupt.
35. Assume that the I bit for external hardware interrupt IRQ is enabled and is edge triggered. Explain how this interrupt works when it is activated. How can we make sure that a single interrupt is not interpreted as multiple interrupts?
36. Explain the difference between the low-level and edge-triggered interrupts.
37. How do we make the IRQ interrupt edge triggered?
38. True or false. Both the STOP and WAIT instructions save the main registers of HCS12 on stack before execution.
39. How do we make the IRQ interrupt level triggered?
40. Besides IRQ and XIRQ, how many hardware interrupts do we have?

SECTION 11.4: PROGRAMMING THE SERIAL COMMUNICATION INTERRUPT

41. True or false. There are two interrupts assigned to interrupts for send and receive.
42. What address in the interrupt vector table is assigned to the SCI0 serial interrupt? How many bytes are assigned to it?
43. Which bit of the CCR register is used to enable the serial interrupt? Show how it is enabled.
44. In Problem 43, what other flags need to be set?
45. True or false. Upon reset, the serial interrupt is blocked.
46. True or false. The last instruction of the ISR for the transmit interrupt is RTI.
47. Answer Problem 46 for the receive interrupt.
48. Assuming that the I bit in the CCR register is enabled, when TDRE is raised, what happens subsequently?
49. Assuming that the I bit in the CCR register is enabled, when RDFE is raised, what happens subsequently?

SECTION 11.5: INTERRUPT PRIORITY IN THE HCS12

50. True or false. Upon reset, IRQ has the highest priority among the peripherals.
51. What register is used to assign priority in the HCS12?
52. What byte is used to assign higher priority to Timer Channel 7?
53. Give the five interrupts with the highest priority.
54. What byte is used to assign higher priority to SCI0?
55. Explain what happens if both IRQ and XIRQ are activated at the same time.
56. Explain what happens if both Timer Channel 0 and XIRQ are activated at the same time.

ANSWERS TO REVIEW QUESTIONS

SECTION 11.1: HCS12 INTERRUPTS

1. Interrupts
2. Two. IRQ and XIRQ.
3. Address locations $FFFE and $FFFF. No. They are set when the processor is designed. However, we can redirect it to other locations.
4. ORCC and ANDCC
5. CLI
6. Pin PE1
7. True

SECTION 11.2: PROGRAMMING TIMER INTERRUPTS

1. True
2. $FFE0 and $FFE1
3. The I bit and "SEI" will enable all the peripheral interrupts, including the timers.
4. TIE
5. False, RTI

SECTION 11.3: PROGRAMMING EXTERNAL HARDWARE INTERRUPTS

1. False. There is an interrupt for each of the external hardware interrupts IRQ and XIRQ.
2. $FFF2–FFFF3 and $FFF4–FFF5
3. X bit
4. Upon application of a low pulse to the IRQ pin, the HCS12 is interrupted in whatever it is doing and jumps to address location $FFF2 to fetch the address of the ISR.
5. True
6. Make sure that the low pulse applied to the IRQ pin is brought back to HIGH by the time the HCS12 executes the RTI instruction in the ISR.
7. False
8. We use these bits to enable the IRQ and decide whether it is edge triggered or level triggered.
9. False
10. True

SECTION 11.4: PROGRAMMING THE SERIAL COMMUNICATION INTERRUPT

1. True. There is only one interrupt for both the transfer and receive.
2. $FFD6–FFD7
3. TIE bit D7
4. The RDRF (received reg full) flag is raised when the entire frame of data, including the stop bit, is received. As a result the received byte is delivered to the SCIDRL register and the HCS12 jumps to the memory location in the vector table to fetch the address of the ISR belonging to this interrupt.
5. False
6. True

SECTION 11.5: INTERRUPT PRIORITY IN THE HCS12

1. False. They are assigned priority according to Table 11-4.
2. HPRIO
3. $D6,
   LDAA #$D6
   STAA HPRIO
4. If both are activated at the same time, IRQ is serviced first since it has a higher priority. After

IRQ is serviced, Channel 7 is serviced.

5. We have an interrupt inside an interrupt, meaning that the lower-priority interrupt is put on hold and the higher one is serviced. After servicing this higher-priority interrupt, the HCS12 resumes servicing the lower-priority ISR.

SECTION 11.7: PROGRAMMING THE REAL TIME INTERRUPT

1. False. They are assigned priority according to Table 11-3.
2. True
3. True
4. CGRINT
5. BSET CRGFLG,%10000000 ;clear RTIF

CHAPTER 12

LCD AND KEYBOARD INTERFACING

OBJECTIVES

Upon completion of this chapter, you will be able to:

- >> List reasons that LCDs have gained widespread use, replacing LEDs
- >> Describe the functions of the pins of a typical LCD
- >> List instruction command codes for programming an LCD
- >> Interface an LCD to the HCS12
- >> Program an LCD in Assembly and C
- >> Explain the basic operation of a keyboard
- >> Describe the key press and detection mechanisms
- >> Interface a 4×4 keypad to the HCS12 using C and Assembly

This chapter explores some real-world applications of the HCS12. We explain how to interface the HCS12 to devices such as an LCD and a keyboard. We use C and Assembly for both.

SECTION 12.1: LCD INTERFACING

This section describes the operation modes of LCDs, then describes how to program and interface an LCD to an HCS12 using Assembly and C.

LCD operation

In recent years the LCD has found widespread use replacing LEDs (seven-segment LEDs or other multisegment LEDs). This is due to the following reasons:

1. The declining prices of LCDs.
2. The ability to display numbers, characters, and graphics. This is in contrast to LEDs, which are limited to numbers and a few characters.
3. Incorporation of a refreshing controller into the LCD, thereby relieving the CPU of the task of refreshing the LCD. In contrast, the LED must be refreshed by the CPU (or in some other way) to keep displaying the data.
4. Ease of programming for characters and graphics.

LCD pin descriptions

The LCD discussed in this section has 14 pins. The function of each pin is given in Table 12-1. Figure 12-1 shows the pin positions for various LCDs.

V_{CC}, V_{SS}, and V_{EE}

While V_{CC} and V_{SS} provide +5 V and ground, respectively, V_{EE} is used for controlling LCD contrast.

RS, register select

There are two very important registers inside the LCD. The RS pin is used for their selection as follows. If RS = 0, the instruction command code register is selected, allowing the user to send a command such as clear display, cursor at home, etc. If RS = 1, the data register is selected, allowing the user to send data to be displayed on the LCD.

R/W, read/write

R/W input allows the user to write information to the LCD or read information from it. R/W = 1 when reading; R/W = 0 when writing.

E, enable

The enable pin is used by the LCD to latch information presented to its data pins. When data is supplied to the data pins, a

Table 12-1: Pin Descriptions for LCD

Pin	Symbol	I/O	Description
1	V_{SS}	--	Ground
2	V_{CC}	--	+5 V power supply
3	V_{EE}	--	Power supply to control contrast
4	RS	I	RS = 0 to select command register, RS = 1 to select data register
5	R/W	I	R/W = 0 for write, R/W = 1 for read
6	E	I/O	Enable
7	DB0	I/O	The 8-bit data bus
8	DB1	I/O	The 8-bit data bus
9	DB2	I/O	The 8-bit data bus
10	DB3	I/O	The 8-bit data bus
11	DB4	I/O	The 8-bit data bus
12	DB5	I/O	The 8-bit data bus
13	DB6	I/O	The 8-bit data bus
14	DB7	I/O	The 8-bit data bus

high-to-low pulse must be applied to this pin in order for the LCD to latch in the data present at the data pins. This pulse must be a minimum of 450 ns wide.

D0–D7

The 8-bit data pins, D0–D7, are used to send information to the LCD. To display letters and numbers, we send ASCII codes for the letters A–Z, a–z, and numbers 0–9 to these pins while making RS = 1.

There are also instruction command codes that can be sent to the LCD to clear the display, force the cursor to the home position, or blink the cursor. Table 12-2 lists the instruction command codes. Before issuing a command or data character to the LCD, we need a 2 ms delay to allow the LCD to recover, or we can use the busy flag to see if the LCD is ready to receive information. The busy flag is accessed via D7 and can be read when R/W = 1 and RS = 0. When D7 = 1 (busy flag = 1), the LCD is busy taking care of internal operations and will not accept any new information. When D7 = 0, the LCD is ready to receive new information.

Table 12-2: LCD Command Codes

Code (Hex)	Command to LCD Instruction Register
1	Clear display screen
2	Return home
4	Decrement cursor (shift cursor to left)
6	Increment cursor (shift cursor to right)
5	Shift display right
7	Shift display left
8	Display off, cursor off
A	Display off, cursor on
C	Display on, cursor off
E	Display on, cursor blinking
F	Display on, cursor blinking
10	Shift cursor position to left
14	Shift cursor position to right
18	Shift the entire display to the left
1C	Shift the entire display to the right
80	Force cursor to beginning of 1st line
C0	Force cursor to beginning of 2nd line
38	2 lines and 5x7 matrix (D0–D7, 8-bit)
28	2 lines and 5x7 matrix (D4–D7, 4-bit)

Note: This table is extracted from Table 12-4.

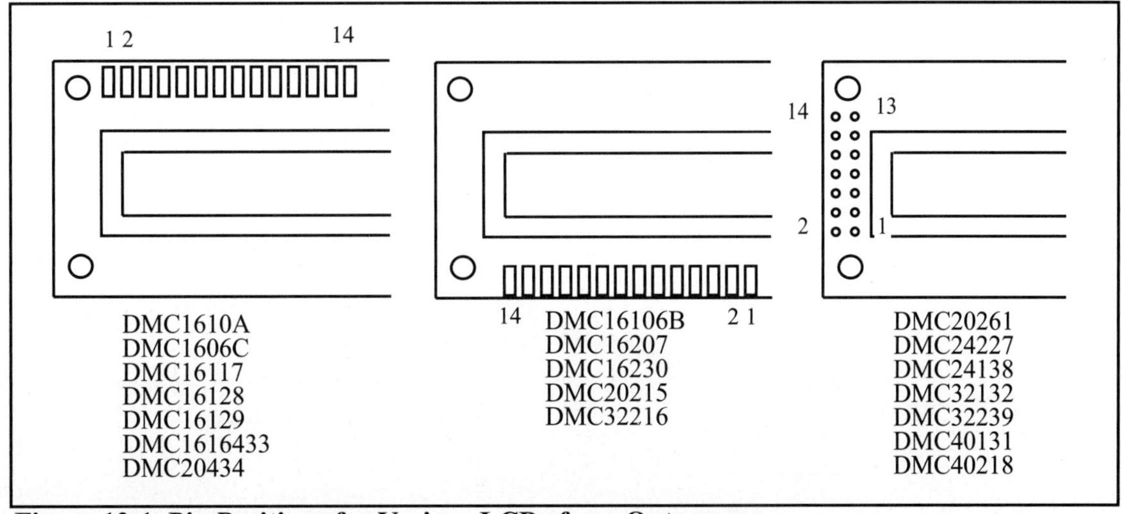

Figure 12-1. Pin Positions for Various LCDs from Optrex

Sending commands and data to LCDs with a time delay

To send any of the commands from Table 12-2 to the LCD, make pin RS = 0. For data, make RS = 1. Then send a HIGH-to-LOW pulse to the E pin to enable the internal latch of the LCD. This is shown in Program 12-1. See Figure 12-2a for LCD connections.

```
;Program 12-1: Communicating with LCD using a fixed delay
;calls a time delay before sending next data/command
D250mU   EQU 5
D250mH   EQU 56      ;250ms delay high byte of value
D250mL   EQU 254          ;     low byte of value
D15mH    EQU 17      ;15ms delay high byte of value
D15mL    EQU 250          ;     low byte of value
LCD_DATA EQU PORTT        ;LCD data pins D0-D7
LCD_CTRL EQU PORTB        ;LCD control pins
RS       EQU mPORTB_BIT5 ;RS pin of LCD
RW       EQU mPORTB_BIT6 ;R/W pin of LCD
EN       EQU mPORTB_BIT7 ;E pin of LCD
DR250mU  DS.B 1
DR250mH  DS.B 1           ;registers for 250ms delay
DR250mL  DS.B 1
DR15mH   DS.B 1           ;registers for 15ms delay
DR15mL   DS.B 1
         LDS    #RAMEnd+1 ;initialize the stack pointer
         LDAA   #$FF
         STAA   DDRT      ;PORTT=Output
         STAA   DDRB      ;PORTB=Output
         JSR    LDELAY    ;wait for LCD power up
         LDAA   #$38      ;init. LCD 2 lines,5x7 matrix
         JSR    COMNWRT   ;call command subroutine
         JSR    LDELAY    ;initialization hold
         LDAA   #$0E      ;display on, cursor on
         JSR    COMNWRT   ;call command subroutine
         JSR    DELAY     ;give LCD some time
         LDAA   #$01      ;clear LCD
         JSR    COMNWRT   ;call command subroutine
         JSR    DELAY     ;give LCD some time
         LDAA   #$06      ;shift cursor right
         JSR    COMNWRT   ;call command subroutine
         JSR    DELAY     ;give LCD some time
         LDAA   #$84      ;cursor at line 1,pos. 4
         JSR    COMNWRT   ;call command subroutine
         JSR    DELAY     ;give LCD some time
         LDAA   #'N'      ;display letter N
         JSR    DATAWRT   ;call display subroutine
         JSR    DELAY     ;give LCD some time
         LDAA   #'O'      ;display letter O
         JSR    DATAWRT   ;call display subroutine
AGAIN:
         BRA    AGAIN     ;stay here
```

```
COMNWRT:                               ;send command to LCD
         STAA   LCD_DATA               ;copy to LCD DATA pin
         BCLR   LCD_CTRL,RS            ;RS=0 for command
         BCLR   LCD_CTRL,RW            ;R/W=0 for write
         BSET   LCD_CTRL,EN            ;E=1 for high pulse
         NOP                           ;make a wide En pulse
         BCLR   LCD_CTRL,EN            ;E=0 for H-to-L pulse
         RTS
DATAWRT:                               ;write data to LCD
         STAA   LCD_DATA               ;copy to LCD DATA pin
         BSET   LCD_CTRL,RS            ;RS=1 for data
         BCLR   LCD_CTRL,RW            ;R/W=0 for write
         BSET   LCD_CTRL,EN            ;E=1 for high pulse
         NOP                           ;make a wide En pulse
         BCLR   LCD_CTRL,EN            ;E=0 for H-to-L pulse
         RTS
;--------------------
DELAY:   LDAA   #D15mH                 ;high byte of delay
         STAA   DR15mH                 ;store in register
D2:      LDAA   #D15mL                 ;low byte of delay
         STAA   DR15mL                 ;store in register
D1:      DEC    DR15mL            ;stay until DR15mL becomes 0
         BNE    D1
         DEC    DR15mH                 ;loop until 0x0000
         BNE    D2
         RTS
;--------------------
LDELAY:  LDAA   #D250mU           ;upper most byte of delay
         STAA   DR250mU                ;store in register
DL3:     LDAA   #D250mH                ;high byte of delay
         STAA   DR250mH                ;store in register
DL2:     LDAA   #D250mL                ;low byte of delay
         STAA   DR250mL                ;store in register
DL1:     DEC    DR250mL           ;stay until DR250mL becomes 0
         BNE    DL1
         DEC    DR250mH
         BNE    DL2
         DEC DR250mU
         BNE DL3
         RTS
```

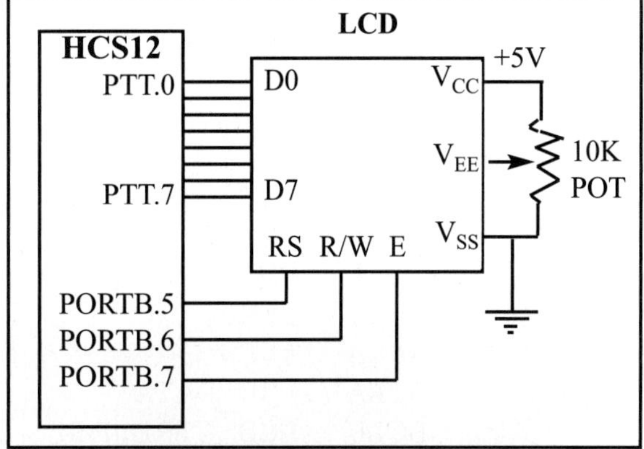

Figure 12-2a. LCD Connections (8-bit)

Sending code or data to the LCD 4 bits at a time

The above code showed how to send commands to the LCD with 8 bits. In most cases it is preferred to use 4-bit data to conserve pins. The LCD may be forced into the 4-bit mode even with an 8-bit data port as shown in Program 12-2. Notice that initialization differs from the 8-bit mode and that data is sent out on the high nibble, high nibble first.

```
;Program 12-2: Communicating with LCD using 4 bits
;Include EQUs and DS.Bs from Program 12-1 and the delays
TEMP        DS.B 1
            LDS     #RAMEnd+1;initialize the stack pointer
            LDAA    #$FF
            STAA    DDRT            ;PORTT=Output
            STAA    DDRB            ;PORTB=Output
            LDAA    #$33
            JSR     COMWRT4         ;call command subroutine
            JSR     LDELAY
            LDAA    #$32
            JSR     COMWRT4         ;call command subroutine
            JSR     DELAY
            LDAA    #$28        ;init. LCD 2 lines,5x7 matrix
            JSR     COMWRT4         ;call command subroutine
            JSR     DELAY           ;initialization hold
            LDAA    #$0E            ;display on, cursor on
            JSR     COMWRT4         ;call command subroutine
            JSR     DELAY
            LDAA    #$01            ;clear LCD
            JSR     COMWRT4         ;call command subroutine
            JSR     DELAY
            LDAA    #$06            ;shift cursor right
            JSR     COMWRT4         ;call command subroutine
            JSR     DELAY
            LDAA    #$84            ;cursor at line 1,pos. 4
            JSR     COMWRT4         ;call command subroutine
            JSR     DELAY
            LDAA    #'N'            ;display letter N
            JSR     DATWRT4         ;call display subroutine
            JSR     DELAY
            LDAA    #'O'            ;display letter O
            JSR     DATWRT4         ;call display subroutine
AGAIN:      BRA     AGAIN           ;stay here

COMWRT4:                            ;send command to LCD
            STAA    TEMP            ;save the byte
            LDAA    #BITS
            STAA    CNT
            LDAA    TEMP
            ANDA    #$F0
            STAA    LCD_DATA
            BCLR    LCD_CTRL,RS     ;RS=0 for command
            BCLR    LCD_CTRL,RW     ;R/W=0 for write
```

```
        BSET   LCD_CTRL,EN     ;E=1 for high pulse
        NOP                    ;make a wide En pulse
        BCLR   LCD_CTRL,EN     ;E=0 for H-to-L pulse
        LDAA   #BITS
        STAA   CNT
        LDAA   TEMP
        ANDA   #$0F
CLO:    LSLA
        DEC    CNT
        BNE    CLO
        STAA   LCD_DATA
        BSET   LCD_CTRL,EN     ;E=1 for high pulse
        NOP                    ;make a wide En pulse
        BCLR   LCD_CTRL,EN     ;E=0 for H-to-L pulse
        RTS
;--------------------------
DATWRT4:                       ;write data to LCD
        STAA    TEMP           ;save the byte
        LDAA    #BITS
        STAA    CNT
        LDAA    TEMP
        ANDA    #$F0
        STAA    LCD_DATA
        BSET    LCD_CTRL,RS    ;RS=1 for data
        BCLR    LCD_CTRL,RW    ;R/W=0 for write
        BSET    LCD_CTRL,EN    ;E=1 for high pulse
        NOP                    ;make a wide En pulse
        BCLR    LCD_CTRL,EN    ;E=0 for H-to-L pulse
        LDAA    #BITS
        STAA    CNT
        LDAA    TEMP
        ANDA    #$0F
DLO:    LSLA
        DEC     CNT
        BNE     DLO
        STAA    LCD_DATA
        BSET    LCD_CTRL,EN    ;E=1 for high pulse
        NOP                    ;make a wide En pulse
        BCLR    LCD_CTRL,EN    ;E=0 for H-to-L pulse
        RTS
```

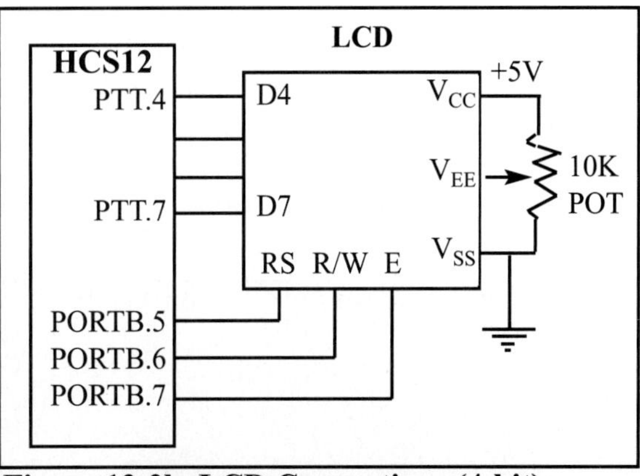

Figure 12-2b. LCD Connections (4-bit)

In the above program we initialized the LCD with the series 33, 32, and 28 in hex. This represents nibbles 3, 3, 3, and 2, which tells the LCD to go into 4-bit mode. The value $28 initializes the display mode in 4-bit as required by LCD datasheet. The remainder of the program is the same as Program 12-1. The write routines sends the high nibble first, then shifts the low nibble to the high nibble before it is sent to data pins D4-D7. Notice that the write routines do not require a delay between the high and low nibbles. This is because the LCD does not process the command or data until both nibbles have been received. Contrast the Write timing for the 8-bit and 4-bit modes in Figures 12-3 and 12-4.

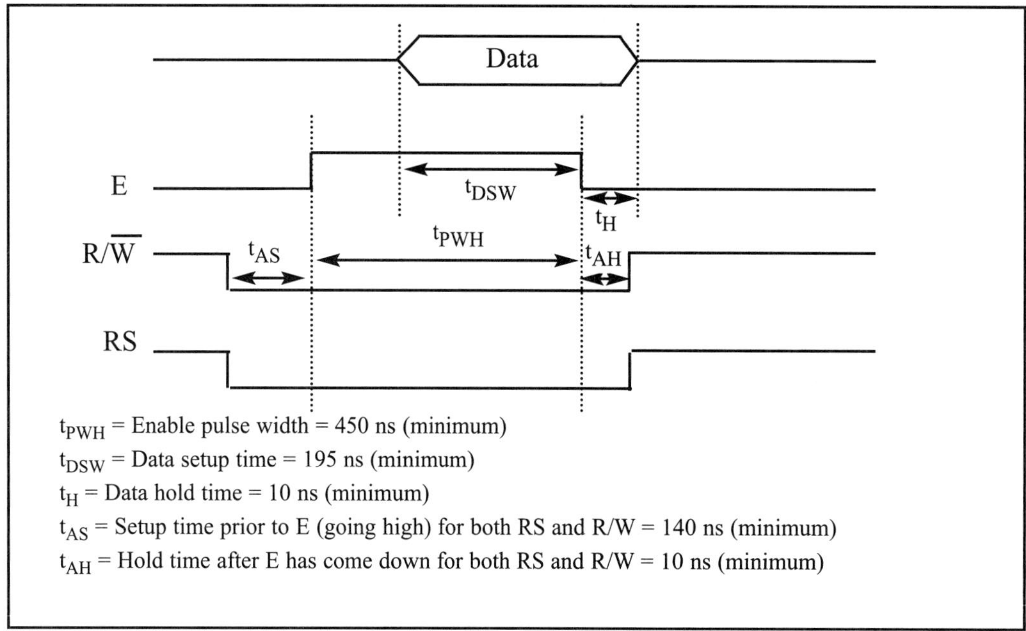

Figure 12-3. LCD Timing for 8-bit Write

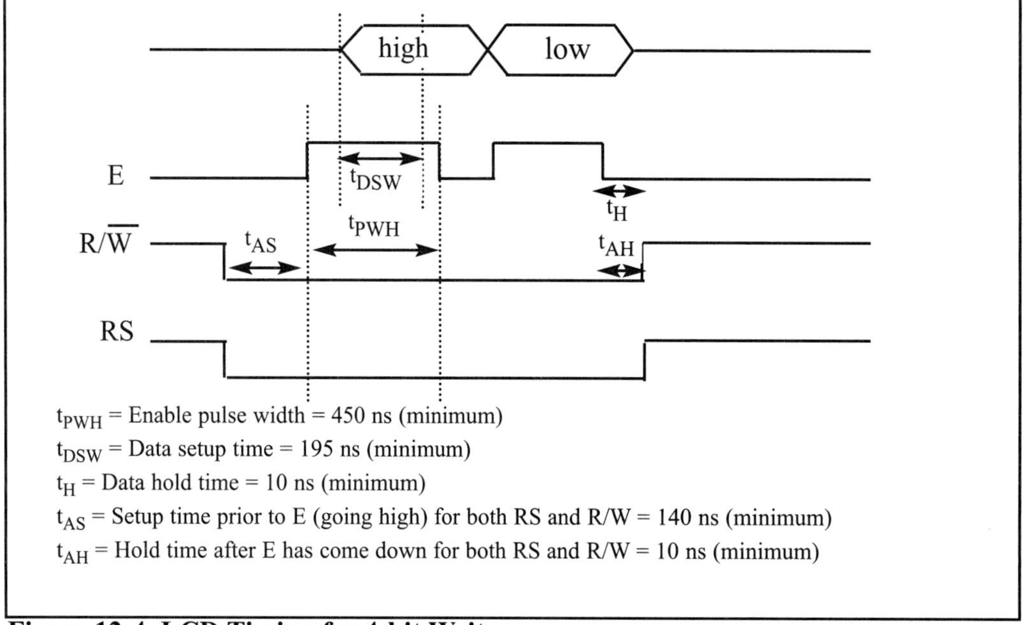

Figure 12-4. LCD Timing for 4-bit Write

LCD data sheet

In the LCD, one can put data at any location on the display. The following shows address locations and how they are accessed.

RS	R/W	DB7	DB6	DB5	DB4	DB3	DB2	DB1	DB0
0	0	1	A	A	A	A	A	A	A

where AAAAAAA = 0000000 to 0100111 for line 1 and AAAAAAA = 1000000 to 1100111 for line 2. See Table 12-3.

Table 12-3: LCD Addressing

	DB7	DB6	DB5	DB4	DB3	DB2	DB1	DB0	HEX
Line 1 (min)	1	0	0	0	0	0	0	0	$80
Line 1 (max)	1	0	1	0	0	1	1	1	$A7
Line 2 (min)	1	1	0	0	0	0	0	0	$C0
Line 2 (max)	1	1	1	0	0	1	1	1	$E7

The upper address range can go as high as 0100111 for the 40-character-wide LCD, while for the 20-character-wide LCD it goes up to 010011 (19 decimal = 10011 binary). Notice that the upper range 0100111 (binary) = 39 decimal, which corresponds to locations 0 to 39 for the LCDs of 40×2 size.

From the above discussion we can get the addresses of cursor positions for various sizes of LCDs. See Figure 12-5 for the cursor addresses for common types of LCDs. Note that all the addresses are in hex. Table 12-4 provides a detailed list of LCD commands and instructions. Table 12-2 is extracted from this table.

16 x 2 LCD	80	81	82	83	84	85	86 through 8F
	C0	C1	C2	C3	C4	C5	C6 through CF
20 x 1 LCD	80	81	82	83	through 93		
20 x 2 LCD	80	81	82	83	through 93		
	C0	C1	C2	C3	through D3		
20 x 4 LCD	80	81	82	83	through 93		
	C0	C1	C2	C3	through D3		
	94	95	96	97	through A7		
	D4	D5	D6	D7	through E7		
40 x 2 LCD	80	81	82	83	through A7		
	C0	C1	C2	C3	through E7		

Note: All data is in hex.

Figure 12-5. Cursor Addresses for Some LCDs

Table 12-4: List of LCD Instructions

Instruction	RS	R/W	DB7	DB6	DB5	DB4	DB3	DB2	DB1	DB0	Description	Execution Time (Max)
Clear Display	0	0	0	0	0	0	0	0	0	1	Clears entire display and sets DD RAM address 0 in address counter.	1.64 ms
Return Home	0	0	0	0	0	0	0	0	1	-	Sets DD RAM address 0 as address counter. Also returns display being shifted to original position. DD RAM contents remain unchanged.	1.64 ms
Entry Mode Set	0	0	0	0	0	0	0	1	1/D	S	Sets cursor move direction and specifies shift of display. These operations are performed during data write and read.	40 μs
Display On/ Off Control	0	0	0	0	0	0	1	D	C	B	Sets On/Off of entire display (D), cursor On/Off (C), and blink of cursor position character (B).	40 μs
Cursor or Display Shift	0	0	0	0	0	1	S/C	R/L	-	-	Moves cursor and shifts display without changing DD RAM contents.	40 μs
Function Set	0	0	0	0	1	DL	N	F	-	-	Sets interface data length (DL), number of display lines (L), and character font (F).	40 μs
Set CG RAM Address	0	0	0	1	AGC						Sets CG RAM address. CG RAM data is sent and received after this setting.	40 μs
Set DD RAM Address	0	0	1	ADD							Sets DD RAM address. DD RAM data is sent and received after this setting.	40 μs
Read Busy Flag & Address	0	1	BF	AC							Reads Busy flag (BF) indicating internal operation is being performed and reads address counter contents.	40 μs
Write Data CG or DD RAM	1	0	Write Data								Writes data into DD or CG RAM.	40 μs
Read Data CG or DD RAM	1	1	Read Data								Reads data from DD or CG RAM.	40 μs

Notes:

1. Execution times are maximum times when Fcp or Fosc is 250 kHz.
2. Execution time changes when frequency changes. Ex: When Fcp or Fosc is 270 kHz: 40 μs × 250 / 270 = 37 μs.
3. Abbreviations:

DD RAM	Display data RAM		
CG RAM	Character generator RAM		
ACC	CG RAM address		
ADD	DD RAM address, corresponds to cursor address		
AC	Address counter used for both DD and CG RAM addresses.		
1/D = 1	Increment	1/D = 0	Decrement
S = 1	Accompanies display shift		
S/C = 1	Display shift;	S/C = 0	Cursor move
R/L = 1	Shift to the right;	R/L = 0	Shift to the left
DL = 1	8 bits, DL = 0: 4 bits		
N = 1	1 line, N = 0: 2 lines		
F = 1	5 × 10 dots, F = 0: 5 × 7 dots		
BF = 1	Internal operation;	BF = 0	Can accept instruction

Optrex is one of the largest manufacturers of LCDs. You can obtain data sheets from their website, www.optrex.com.

The LCDs can be purchased from the following websites:

www.digikey.com
www.jameco.com
www.elexp.com

Sending information to LCD from look-up table

Program 12-3 shows how to use the index registers X and Y to send data and commands to an LCD. For a C version of LCD programming see Programs 12-1C through 12-3C.

```
;Program 12-3: Writing to LCD from look-up table
;Include EQUs and DS.Bs from Program 12-2 and the routines
MYCOM     DC.B $33,$32,$28,$0E,$01,$06,$84,0 ;commands
MYDATA    DC.B "HELLO",0                      ;data

     LDS  #RAMEnd+1 ;initialize the stack pointer
     LDAA #$FF
     STAA DDRT      ;PORTT=Output
     STAA DDRB      ;PORTB=Output
     LDX  #MYCOM    ;load commands
     LDY  #MYDATA
SEND_COMM:
     LDAA 1,X+      ;get command
     BEQ  SEND_DATA ;branch if command = 0
     JSR  COMWRT4   ;call command subroutine
     JSR  LDELAY    ;give LCD some time
     BRA  SEND_COMM
```

```
SEND_DATA:
     LDAA 1,Y+      ;get the data
     BEQ  HOLD
     JSR  DATWRT4    ;call display subroutine
     JSR  DELAY      ;give LCD some time
     BRA  SEND_DATA
HOLD: BRA HOLD      ;stay here
```

```
//Program 12-1C: C Version of Program 12-1
#define LCD_DATA PTT //LCD data pins D0-D7
#define LCD_CTRL PORTB    //LCD control pins
#define RS PORTB_BIT5_MASK     //RS pin of LCD
#define RW PORTB_BIT6_MASK     //R/W pin of LCD
#define EN PORTB_BIT7_MASK     //E pin of LCD
void COMNWRT(unsigned char);
void DATAWRT(unsigned char);
void MSDelay(int);

void main(void) {
     DDRT =0xFF; //PORTT=Output
     DDRB =0xFF; //PORTB=Output
     MSDelay(250); //wait for LCD power up
     COMNWRT(0x38);//init. LCD 2 lines,5x7 matrix
     MSDelay(250); //initialization hold
     COMNWRT(0x0E);  //display on, cursor on
     MSDelay(15);  //give LCD some time
     COMNWRT(0x01);  //clear LCD
     MSDelay(15);  //give LCD some time
     COMNWRT(0x06);  //shift cursor right
     MSDelay(15);  //give LCD some time
     COMNWRT(0x84);  //cursor at line 1, pos. 4
     MSDelay(15);  //give LCD some time
     DATAWRT('N'); //display letter N
     MSDelay(15);  //give LCD some time
     DATAWRT('O'); //display letter O
  while(1);
}

void COMNWRT(unsigned char command) //send com. to LCD
  {
     LCD_DATA = command; //copy to LCD DATA pin
     LCD_CTRL = LCD_CTRL & ~RS;//RS=0 for command
     LCD_CTRL = LCD_CTRL & ~RW;//R/W=0 for write
     LCD_CTRL = LCD_CTRL | EN; //E=1 for high pulse
     LCD_CTRL = LCD_CTRL & ~EN;//E=0 for H-to-L pulse
  }
```

```
void DATAWRT(unsigned char data) //write data to LCD
  {
     LCD_DATA = data;  //copy to LCD DATA pin
     LCD_CTRL = LCD_CTRL | RS; //RS=1 for data
     LCD_CTRL = LCD_CTRL & ~RW;//R/W=0 for write
     LCD_CTRL = LCD_CTRL | EN; //E=1 for high pulse
     LCD_CTRL = LCD_CTRL & ~EN;//E=0 for H-to-L pulse
  }

void MSDelay(int ms)
  {
  int i,j;
  for(i=0;i < ms; i++)
    for(j=0;j < 329; j++);
  }
```

```
//Program 12-2C : C Version of Program 12-2
//Include MSDelay from Program 12-1C
#define LCD_DATA PTT //LCD data pins D4-D7
#define LCD_CTRL PORTB   //LCD control pins
#define RS PORTB_BIT5_MASK     //RS pin of LCD
#define RW PORTB_BIT6_MASK     //R/W pin of LCD
#define EN PORTB_BIT7_MASK     //E pin of LCD
void COMWRT4(unsigned char); //replaces COMNWRT()
void DATWRT4(unsigned char); //replaces DATAWRT()
void MSDelay(int);

void main(void)
  {
     DDRT =0xFF; //PORTT=Output
     DDRB =0xFF; //PORTB=Output
     COMWRT4(0x33);//init. LCD 2 lines,5x7 matrix
     MSDelay(250); //initialization hold
     COMWRT4(0x32);//init. LCD 2 lines,5x7 matrix
     MSDelay(250); //initialization hold
     COMWRT4(0x28);//init. LCD 2 lines,5x7 matrix
     MSDelay(15); //initialization hold
     COMWRT4(0x0E);   //display on, cursor on
     MSDelay(15);  //give LCD some time
     COMWRT4(0x01);   //clear LCD
     MSDelay(15);  //give LCD some time
     COMWRT4(0x06);   //shift cursor right
     MSDelay(15);  //give LCD some time
     COMWRT4(0x84);   //cursor at line 1, pos. 4
     MSDelay(15);  //give LCD some time
     DATWRT4('N'); //display letter N
     MSDelay(15);  //give LCD some time
```

```
    DATWRT4('O'); //display letter O
    while(1);
  }

void COMWRT4(unsigned char command)//send com. to LCD
  {
    LCD_DATA = command & 0xF0; //high nibble to LCD
    LCD_CTRL = LCD_CTRL & ~RS; //RS=0 for command
    LCD_CTRL = LCD_CTRL & ~RW; //R/W=0 for write
    LCD_CTRL = LCD_CTRL | EN; //E=1 for high pulse
    LCD_CTRL = LCD_CTRL & ~EN;//E=0 for H-to-L pulse
    LCD_DATA = (command & 0x0F)<<4;//low nib. to LCD
    LCD_CTRL = LCD_CTRL | EN; //E=1 for high pulse
    LCD_CTRL = LCD_CTRL & ~EN;//E=0 for H-to-L pulse
  }

void DATWRT4(unsigned char data) //write data to LCD
  {
    LCD_DATA = data & 0xF0;    //high nibble to LCD
    LCD_CTRL = LCD_CTRL | RS; //RS=1 for data
    LCD_CTRL = LCD_CTRL & ~RW;//R/W=0 for write
    LCD_CTRL = LCD_CTRL | EN; //E=1 for high pulse
    LCD_CTRL = LCD_CTRL & ~EN;//E=0 for H-to-L pulse
    LCD_DATA = (data & 0x0F) << 4;//low nib. to LCD
    LCD_CTRL = LCD_CTRL | EN; //E=1 for high pulse
    LCD_CTRL = LCD_CTRL & ~EN;//E=0 for H-to-L pulse
  }
```

```
//Program 12-3C: C Version of Program 12-3
//Include #defines and routines from Program 12-2C
const unsigned char MYCOM[] =
    {0x33,0x32,0x28,0x0E,0x01,0x06,0x84,0};
const unsigned char MYDATA[]={'H','E','L','L','O',0};
int i;

void main(void) {
    DDRT = 0xFF; //PORTT=Output
    DDRB = 0xFF; //PORTB=Output
    i=0;
    while(MYCOM[i]!=0)
      {
      COMWRT4(MYCOM[i++]);//send the command
      MSDelay(250);   //give LCD some time
      }
    i=0;
    while(MYDATA[i]!=0)
      {
```

```
        DATWRT4(MYDATA[ i++] );//send the data
        MSDelay(15);   //give LCD some time
        }
  while(1);
}
```

Review Questions

1. The RS pin is an _______ (input, output) pin for the LCD.
2. The E pin is an ________ (input, output) pin for the LCD.
3. The E pin requires an _______ (H-to-L, L-to-H) pulse to latch in information at the data pins of the LCD.
4. For the LCD to recognize information at the data pins as data, RS must be set to _____ (high, low).
5. Give the command codes for line 1, first character, and line 2, first character.

SECTION 12.2: KEYBOARD INTERFACING

Keyboards and LCDs are the most widely used input/output devices of the HCS12, and a basic understanding of them is essential. In this section, we first discuss keyboard fundamentals, along with key press and key detection mechanisms. Then we show how a keyboard is interfaced to an HCS12.

Interfacing the keyboard to the HCS12

At the lowest level, keyboards are organized in a matrix of rows and columns. The CPU accesses both rows and columns through ports; therefore, with two 8-bit ports, an 8×8 matrix of keys can be connected to a microprocessor. When a key is pressed, a row and a column make a contact; otherwise, there is no connection between rows and columns. In IBM PC keyboards, a single microcontroller takes care of hardware and software interfacing of the keyboard. In such systems, it is the function of programs stored in the EPROM of the microcontroller to scan the keys continuously, identify which one has been activated, and present it to the motherboard. In this section we look at the mechanism by which the HCS12 scans and identifies the key.

Scanning and identifying the key

Figure 12-6 shows a 4×4 matrix connected to two ports. The rows are connected to an output port and the columns are connected to an input port. If no key has been pressed, reading the input port will yield 1s for all columns since they are all connected to high (V_{CC}). If all the rows are grounded and a key is pressed, one of the columns will have 0 since the key pressed provides the path to ground. It is the function of the microcontroller to scan the keyboard continuously to detect and identify the key pressed. How it is done is explained next.

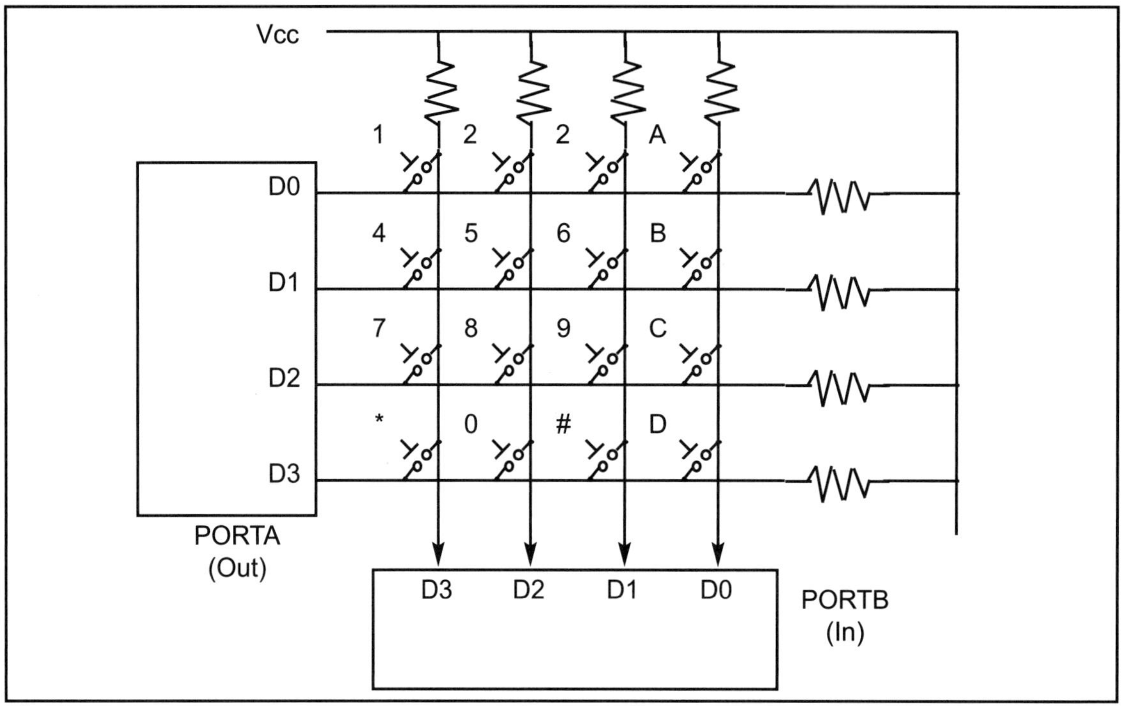

Figure 12-6. Matrix Keyboard Connection to Ports

Grounding rows and reading the columns

To detect a pressed key, the microcontroller grounds all rows by providing 0 to the output latch, then it reads the columns. If the data read from the columns is D3–D0 = 1111, no key has been pressed and the process continues until a key press is detected. However, if one of the column bits has a zero, this means that a key press has occurred. For example, if D3–D0 = 1101, this means that a key in the D1 column has been pressed. After a key press is detected, the microcontroller will go through the process of identifying the key. Starting with the top row, the microcontroller grounds it by providing a low to row D0 only; then it reads the columns. If the data read is all 1s, no key in that row is activated and the process is moved to the next row. It grounds the next row, reads the columns, and checks for any zero. This process continues until the row is identified. After identification of the row in which the key has been pressed, the next task is to find out which column the pressed key belongs to. This should be easy since the microcontroller knows at any time which row and column are being accessed. Look at Example 12-3.

Example 12-3

From Figure 12-6, identify the row and column of the pressed key for each of the following.
(a) D3–D0 = 1110 for the row, D3–D0 = 1011 for the column
(b) D3–D0 = 1101 for the row, D3–D0 = 0111 for the column

Solution:

From Figure 12-6 the row and column can be used to identify the key.
(a) The row belongs to D0 and the column belongs to D2; therefore, key number 2 was pressed.
(b) The row belongs to D1 and the column belongs to D3; therefore, key number 7 was pressed.

Program 12-4 is the HCS12 Assembly language program for detection and identification of key activation. In this program, it is assumed that PORTA and PORTB are initialized as output and input, respectively. Figure 12-7 provides the flowchart for Program 12-4 for scanning and identifying the pressed key. Program 12-4 goes through the following four major stages:

1. To make sure that the preceding key has been released, 0s are output to all rows at once, and the columns are read and checked repeatedly until all the columns are high. When all columns are found to be high, the program waits for a short amount of time before it goes to the next stage of waiting for a key to be pressed.
2. To see if any key is pressed, the columns are scanned over and over in an infinite loop until one of them has a 0 on it. Remember that the output latches connected to rows still have their initial zeros (provided in stage 1), making them grounded. After the key press detection, the microcontroller waits 20 ms for the bounce and then scans the columns again. This serves two functions: (a) it ensures that the first key press detection was not an erroneous one due to a spike noise, and (b) the 20-ms delay prevents the same key press from being interpreted as a multiple key press. If after the 20-ms delay the key is still pressed, it goes to the next stage to detect which row it belongs to; otherwise, it goes back into the loop to detect a real key press.
3. To detect which row the key press belongs to, the microcontroller grounds one row at a time, reading the columns each time. If it finds that all columns are high, this means that the key press cannot belong to that row; therefore, it grounds the next row and continues until it finds the row the key press belongs to. Upon finding the row that the key press belongs to, it sets up the starting address for the look-up table holding the scan codes (or the ASCII value) for that row and goes to the next stage to identify the key.
4. To identify the key press, the microcontroller rotates the column bits, one bit at a time, into the carry flag and checks to see if it is low. Upon finding the zero, it pulls out the ASCII code for that key from the look-up table; otherwise, it increments the pointer to point to the next element of the look-up table.

While the key press detection is standard for all keyboards, the process for determining which key is pressed varies. The look-up table method shown in Program 12-4 can be modified to work with any matrix up to 8 × 8.

There are IC chips such as National Semiconductor's MM74C923 that incorporate keyboard scanning and decoding all in one chip. Such chips use combinations of counters and logic gates (no microcontroller) to implement the underlying concepts presented in Program 12-4. Program 12-4C shows keypad programming in HCS12 C.

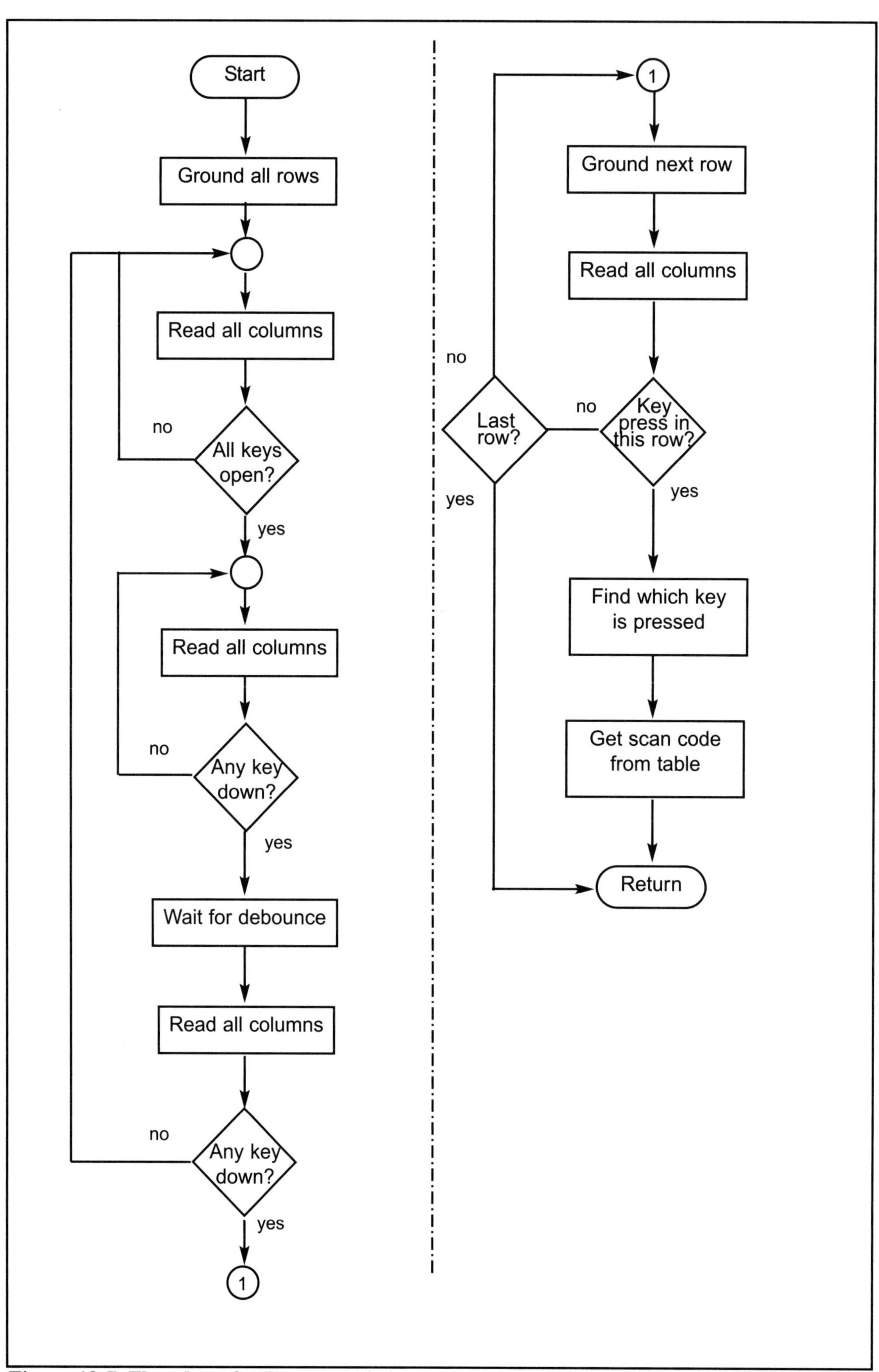

Figure 12-7. Flowchart for Program 12-4

```
;Program 12-4:  4x4 keypad programming using polling

D15mH EQU 17   ;15ms delay high byte of value
D15mL EQU 250  ;low byte of value
COLS  EQU PORTB;keypad columns
ROWS  EQU PORTA;keypad rows
ROW0  EQU %11111110;mask for row0
ROW1  EQU %11111101;mask for row1
ROW2  EQU %11111011;mask for row2
ROW3  EQU %11110111;mask for row3
COLM  EQU %00001111;column mask

DR15mH  DS.B 1 ;registers for 15ms delay
DR15mL  DS.B 1
PDELAY  DS.B 1
KCODE0  DC.B '123A' ;ROW 0
KCODE1  DC.B '456B' ;ROW 1
KCODE2  DC.B '789C' ;ROW 2
KCODE3  DC.B '*0#D' ;ROW 3

; code section
             ORG   ROMStart
Entry:
     LDS  #RAMEnd+1 ;initialize the stack pointer
     LDAA #$FF
     STAA DDRA      ;PORTA Output
     STAA DDRT      ;PORTT=Output
     LDAA #$00
     STAA DDRB      ;PORTB Input
K1:  CLR ROWS       ;ground all rows at once
     LDAA COLS ;read all col. ensure keys are open
     ANDA #COLM     ;mask unused bits
     CMPA #COLM
     BNE K1         ;wait until key release
K2:  CLR ROWS       ;ground all rows at once
     JSR  DELAY     ;call 15 ms delay
     LDAA COLS      ;check for a key press
     ANDA #COLM     ;mask unused bits
     CMPA #COLM
     BNE OVER       ;if not equal, a key is pressed
     BRA K2         ;keep checking
OVER: JSR DELAY     ;wait 15 ms debounce time
     LDAA COLS      ;check for a key press
     ANDA #COLM     ;mask unused bits
     CMPA #COLM
     BNE OVER1      ;key pressed, find row
     BRA K2         ;if none, keep polling
```

```
OVER1: LDAA #ROW0
     STAA ROWS      ;ground ROW 0
     MOVB #$08,PDELAY
P1:  DEC PDELAY     ;allow time for stability
     BNE P1
     LDAA COLS      ;check for a key press
     CMPA #COLM
     BNE R0         ;if not equal, key in ROW 0
     LDAA #ROW1
     STAA ROWS      ;ground ROW 1
     MOVB #$08,PDELAY
P2   DEC PDELAY     ;allow time for stability
     BNE P2
     LDAA COLS      ;check for a key press
     CMPA #COLM
     BNE R1         ;if not equal, key in ROW 1
     LDAA #ROW2
     STAA ROWS      ;ground ROW 2
     MOVB #$08,PDELAY
P3  DEC PDELAY      ;allow time for stability
     BNE P3
     LDAA COLS      ;check for a key press
     CMPA #COLM
     BNE R2         ;if not equal, key in ROW 2
     LDAA #ROW3
     STAA ROWS      ;ground ROW 3
     MOVB #$08,PDELAY
P4   DEC PDELAY     ;allow time for stability
     BNE P4
     LDAA COLS      ;check for a key press
     CMPA #COLM
     BNE R3         ;if not equal, key in ROW 3
     BRA K2         ;else false input
R0:  LDX #KCODE0    ;load pointer for row 0
     BRA FIND       ;find the column
R1:  LDX #KCODE1    ;load pointer for row 1
     BRA FIND       ;find the column
R2:  LDX #KCODE2    ;load pointer for row 2
     BRA FIND       ;find the column
R3:  LDX #KCODE3    ;load pointer for row 3
     BRA FIND       ;find the column
FIND: ANDA #COLM    ;mask the high nibble
SHIFT: LSRA         ;shift to find the low
     BCC MATCH      ;if 0 get the ASCII code
     INX            ;point to next character
     BRA SHIFT      ;keep searching
MATCH: LDAA 0,X     ;get the character
```

```
     STAA PTT       ;display on PTT
     LBRA K1        ;keep monitoring for key press
;--------------------
DELAY: LDAA #D15mH  ;high byte of delay
     STAA DR15mH    ;store in register
D2:  LDAA #D15mL    ;low byte of delay
     STAA DR15mL    ;store in register
D1:  DEC DR15mL     ;stay until DR15mL becomes 0
     NOP
     NOP
     BNE  D1
     DEC DR15mH     ;loop until all DR15m = 0x0000
     BNE  D2
     RTS
```

```
//Program 12-4C: C version of Program 12-4 with
//display on serial port. Assume XTAL = 4 MHz.
#include <mc9s12dp512.h>/* derivative information */

#define ROWS PORTA
#define COLS PORTB
void MSDelay(int);
void SerTX(unsigned char);

const unsigned char keypad[4][4] =
  {
    '1','2','3','A',
    '4','5','6','B',
    '7','8','9','C',
    '*','0','#','D'
  };
unsigned char column,row;

void main(void) {
  SCI0BDH=0x0;      //choose
  SCI0BDL=0x0D;     //9600 baud rate
  SCI0CR1=0x0;      //8-bit data, no Int
  SCI0CR2=0x0C;//enable trans. and recv., no parity
  DDRA =0xFF;       //PORTA=Output
  DDRT =0xFF;       //PORTT=Output
  DDRB =0;          //PORTB=Input
  while(1)
  {
    do
    {
      ROWS = 0;     //ground all rows at once
      column = COLS;     //read the columns
```

```
    column = column & 0x0F;      //mask unused bits
  }while(column != 0x0F);//wait until key release
  do
  {
    do
    {
      MSDelay(15);           //wait a little
      column = COLS;         //read the columns
      column = column & 0x0F; //mask unused bits
    }while(column == 0x0F);//check for key press
    MSDelay(15);             //wait for debounce
    column = COLS;           //read the columns
    column = column & 0x0F;//mask unused bits
  }while(column == 0x0F);   //false key press
  while(1)
  {
    ROWS = 0xFE;        //ground ROW 0
    column = COLS;           //read the columns
    column = column & 0x0F;//mask unused bits
    if(column != 0x0F)       //key is in ROW 0
    {
      row = 0;
      break;
    }
    ROWS = 0xFD;        //ground ROW 1
    column = COLS;           //read the columns
    column = column & 0x0F;//mask unused bits
    if(column != 0x0F)       //key is in ROW 1
    {
      row = 1;
      break;
    }
    ROWS = 0xFB;        //ground ROW 2
    column = COLS;           //read the columns
    column = column & 0x0F;//mask unused bits
    if(column != 0x0F)       //key is in ROW 2
    {
      row = 2;
      break;
    }
    ROWS = 0xF7;        //ground ROW 3
    column = COLS;           //read the columns
    column = column & 0x0F;//mask unused bits
    if(column != 0x0F)       //key is in ROW 3
    {
      row = 3;
      break;
```

```
        }
        column = 0;              //key not found
        break;
      }
      if(column == 0x0E)         //first column
      {
        PTT = keypad[row][0];    //display on PTT
        SerTX(keypad[row][0]);   //send to COM port
      }
      else if(column == 0x0D)    //second column
      {
        PTT = keypad[row][1];    //display on PTT
        SerTX(keypad[row][1]);   //send to COM port
      }
      else if(column == 0x0B)    //third column
      {
        PTT = keypad[row][2];    //display on PTT
        SerTX(keypad[row][2]);   //send to COM port
      }
      else if(column == 0x07)    //fourth column
      {
        PTT = keypad[row][3];    //display on PTT
        SerTX(keypad[row][3]);   //send to COM port
      }
    }
}
void SerTX(unsigned char x)
{
  while(!(SCI0SR1 & SCI0SR1_TDRE_MASK)); //send all
  SCI0DRL=x;                     //place value in buffer
}
void MSDelay(int ms)
{
  int i,j;
  for(i=0;i < ms; i++)
    for(j=0;j < 329; j++);
}
```

For keypad programs on how to:

A) Use a single port for keypad,
B) Use an interrupt instead of polling,

see http://www.MicroDigitalEd.com

Review Questions

1. True or false. To see if any key is pressed, all rows are grounded.
2. If D3–D0 = 0111 is the data read from the columns, which column does the pressed key belong to?
3. True or false. Key press detection and key identification require two different processes.
4. In Figure 12-6, if the rows are D3–D0 = 1110 and the columns are D3–D0 = 1110, which key is pressed?
5. True or false. To identify the pressed key, one row at a time is grounded.

PROBLEMS

SECTION 12.1: LCD INTERFACING

1. The LCD discussed in this section has _______ (4, 8) data pins.
2. Describe the function of pins E, R/W, and RS in the LCD.
3. What is the difference between the V_{CC} and V_{EE} pins on the LCD?
4. "Clear LCD" is a __________ (command code, data item) and its value is ___ hex.
5. What is the hex value of the command code for "display on, cursor on"?
6. Give the state of RS, E, and R/W when sending a command code to the LCD.
7. Give the state of RS, E, and R/W when sending data character 'Z' to the LCD.
8. Which of the following is needed on the E pin in order for a command code (or data) to be latched in by the LCD?
   (a) H-to-L pulse (b) L-to-H pulse
9. True or false. For the above to work, the value of the command code (data) must already be at the D0–D7 pins.
10. There are two methods of sending streams of characters to the LCD: (1) checking the busy flag, or (2) putting some time delay between sending each character without checking the busy flag. Explain the difference and the advantages and disadvantages of each method. Also explain how we monitor the busy flag.
11. For a 16×2 LCD, the location of the last character of line 1 is 8FH (its command code). Show how this value was calculated.
12. For a 16×2 LCD, the location of the fifth character of line 2 is C4H (its command code). Show how this value was calculated.
13. For a 20×2 LCD, the location of the last character of line 1 is 93H (its command code). Show how this value was calculated.
14. For a 20×2 LCD, the location of the third character of line 2 is C2H (its command code). Show how this value was calculated.
15. For a 40×2 LCD, the location of the last character of line 1 is A7H (its command code). Show how this value was calculated.
16. For a 40×2 LCD, the location of the last character of line 2 is E7H (its command code). Show how this value was calculated.
17. Show the value (in hex) for the command code for the 10th location, line 1 on a 20×2 LCD. Show how you got your value.
18. Show the value (in hex) for the command code for the 20th location, line 2 on

a 40×2 LCD. Show how you got your value.

19. True or false. An 8-bit LCD configuration uses #38 for initialization command.
20. True or false. A 4-bit LCD configuration uses #28 for initialization command in addition to values 3, 3, 3, and 2.

SECTION 12.2: KEYBOARD INTERFACING

21. In reading the columns of a keyboard matrix, if no key is pressed we should get all ______ (1s, 0s).
22. In Figure 12-6, to detect the key press, which of the following is grounded?
    (a) all rows (b) one row at time (c) both (a) and (b)
23. In Figure 12-6, to identify the key pressed, which of the following is grounded?
    (a) all rows (b) one row at time (c) both (a) and (b)
24. Indicate the steps to detect the key press.
25. Indicate the steps to identify the key pressed.

ANSWERS TO REVIEW QUESTIONS

SECTION 12.1: LCD INTERFACING

1. Input
2. Input
3. H-to-L
4. High
5. 80H and C0H

SECTION 12.2: KEYBOARD INTERFACING

1. True
2. Column 3
3. True
4. 0
5. True

CHAPTER 13

ADC, DAC, AND SENSOR INTERFACING

OBJECTIVES

Upon completion of this chapter, you will be able to:

- >> Discuss the ADC (analog-to-digital converter) section of the HCS12 chip
- >> Interface temperature sensors to the HCS12
- >> Explain the process of data acquisition using ADC
- >> Describe factors to consider in selecting an ADC chip
- >> Program the HCS12's ADC in C and Assembly
- >> Explain the function of precision IC temperature sensors
- >> Describe signal conditioning and its role in data acquisition
- >> Describe the basic operation of a DAC (digital-to-analog converter) chip
- >> Interface a DAC chip to the HCS12
- >> Program a DAC chip to produce a sine wave on an oscilloscope
- >> Program DAC chips in HCS12 C and Assembly

This chapter explores real-world devices such as ADCs (analog-to-digital converters), DACs (digital-to-analog converters), and sensors. In this chapter, we will explain how the ADC section of the HCS12 works. We will also show the interfacing of sensors and discuss the issue of signal conditioning.

SECTION 13.1: ADC CHARACTERISTICS

This section will explore the ADC characteristics of analog-to-digital converter chips. First, we describe some general aspects of the ADC itself, then examine the serial and parallel ADC chips.

ADC devices

Analog-to-digital converters are among the most widely used devices for data acquisition. Digital computers use binary (discrete) values, but in the physical world everything is analog (continuous). Temperature, pressure (wind or liquid), humidity, and velocity are a few examples of physical quantities that we deal with every day. A physical quantity is converted to electrical (voltage, current) signals using a device called a *transducer*. Transducers are also referred to as *sensors*. Sensors for temperature, velocity, pressure, light, and many other natural quantities produce an output that is voltage (or current). Therefore, we need an analog-to-digital converter to translate the analog signals to digital numbers so that the microcontroller can read and process them. See Figures 13-1 and 13-2.

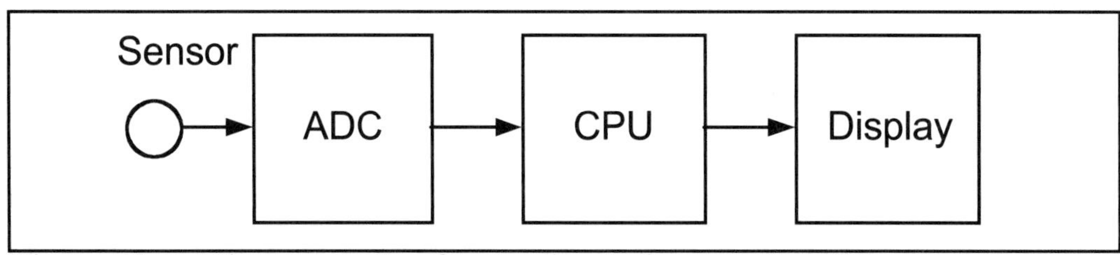

Figure 13-1. Microcontroller Connection to Sensor via ADC

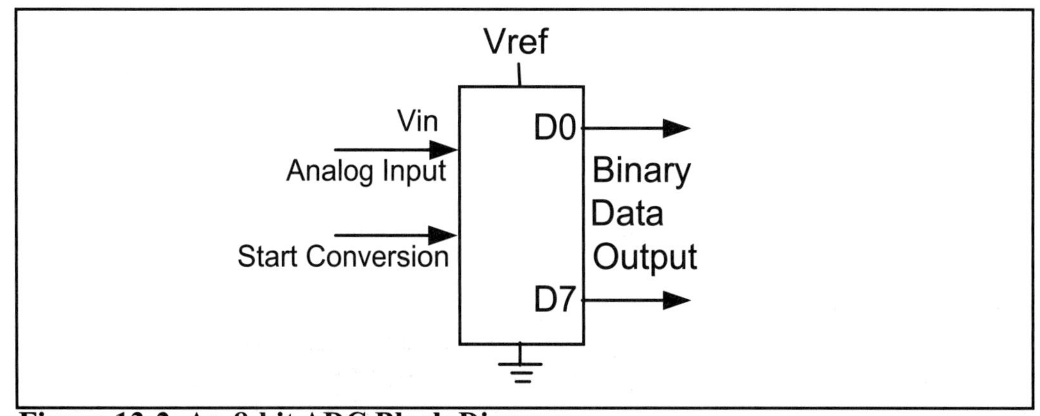

Figure 13-2. An 8-bit ADC Block Diagram

Table 13-1: Resolution versus Step Size for ADC (V_{ref} = 5 V)

n-bit	Number of steps	Step size (mV)
8	256	5/256 = 19.53
10	1,024	5/1,024 = 4.88
12	4,096	5/4,096 = 1.2
16	65,536	5/65,536 = 0.076

Notes: V_{CC} = 5 V

Step size (resolution) is the smallest change that can be discerned by an ADC.

Some of the major characteristics of the ADC are as follows:

Resolution

ADCs have *n*-bit resolution, where *n* can be 8, 10, 12, 16, or even 24 bits. The higher-resolution ADC provides a smaller step size, where *step size* is the smallest change that can be discerned by an ADC. Some widely used resolutions for ADCs are shown in Table 13-1. Although the resolution of an ADC chip is decided at the time of its design and cannot be changed, we can control the step size with the help of what is called V_{ref}. This is discussed below.

Conversion time

In addition to resolution, conversion time is another major factor in ADC performance. *Conversion time* is defined as the time it takes the ADC to convert the analog input to a digital (binary) number. The conversion time is dictated by the clock source connected to the ADC as well as the method used for data conversion and technology used in the fabrication of the ADC chip such as MOS or TTL technology.

V_{ref}

V_{ref} is an input voltage used for the reference voltage. The voltage connected to this pin, along with the resolution of the ADC chip, dictate the step size. For an 8-bit ADC, the step size is V_{ref}/256 because it is an 8-bit ADC, and 2 to the power of 8 gives us 256 steps. See Table 13-1. For example, if the analog input range needs to be 0 to 4 volts, V_{ref} is connected to 4 volts. That gives 4 V/256 = 15.62 mV for the step size of an 8-bit ADC. In another case, if we need a step size of 10 mV for an 8-bit ADC, then V_{ref} = 2.56 V, because 2.56 V/256 = 10 mV. For the 10-bit ADC, if V_{ref} = 5 V, then the step size is 4.88 mV as shown in Table 13-

Table 13-2: V_{ref} Relation to V_{in} Range for an 8-bit ADC

V_{ref} (V)	V_{in} Range (V)	Step Size (mV)
5.00	0 to 5	5/256 = 19.53
4.0	0 to 4	4/256 = 15.62
3.0	0 to 3	3/256 = 11.71
2.56	0 to 2.56	2.56/256 = 10
2.0	0 to 2	2/256 = 7.81
1.28	0 to 1.28	1.28/256 = 5
1	0 to 1	1/256 = 3.90

Table 13-3: V_{ref} Relation to V_{in} Range for an 10-bit ADC

V_{ref} (V)	V_{in} Range (V)	Step Size (mV)
5.00	0 to 5	5/1,024 = 4.88
4.096	0 to 4.096	4.096/1,024 = 4
3.0	0 to 3	3/1,024 = 2.93
2.56	0 to 2.56	2.56/1,024 = 2.5
2.048	0 to 2.048	2.048/1,024 = 2
1.28	0 to 1.28	1/1,024 = 1.25
1.024	0 to 1.024	1.024/1,024 = 1

1. Tables 13-2 and 13-3 show the relationship between V_{ref} and step size for the 8- and 10-bit ADCs, respectively. In some input applications, we need the differential reference voltage where $V_{ref} = V_{ref}(+) - V_{ref}(-)$. Often the $V_{ref}(-)$ pin is connected to ground and the Vref (+) pin is used as the Vref.

Digital data output

In an 8-bit ADC we have an 8-bit digital data output of D0–D7 while in the 10-bit ADC the data output is D0–D9. To calculate the output voltage, we use the following formula:

$$D_{out} = \frac{V_{in}}{step\ size}$$

where D_{out} = digital data output (in decimal), V_{in} = analog input voltage, and step size (resolution) is the smallest change, which is $V_{ref}/256$ for an 8-bit ADC. See Example 13-1. This data is brought out of the ADC chip either one bit at a time (serially), or in one chunk, using a parallel line of outputs. This is discussed next.

Example 13-1

For an 8-bit ADC, we have V_{ref} = 2.56 V. Calculate the D0–D7 output if the analog input is: (a) 1.7 V, and (b) 2.1 V.

Solution:

Because the step size is 2.56/256 = 10 mV, we have the following:

(a) D_{out} = 1.7 V/10 mV = 170 in decimal, which gives us 10101011 in binary for D7–D0.

(b) D_{out} = 2.1 V/10 mV = 210 in decimal, which gives us 11010010 in binary for D7–D0.

Parallel versus serial ADC

The ADC chips are either parallel or serial. In a parallel ADCs, we have 8 or more pins dedicated to bringing out the binary data, but in a serial ADC we have only one pin for data out. That means that inside the serial ADC, there is a paral-

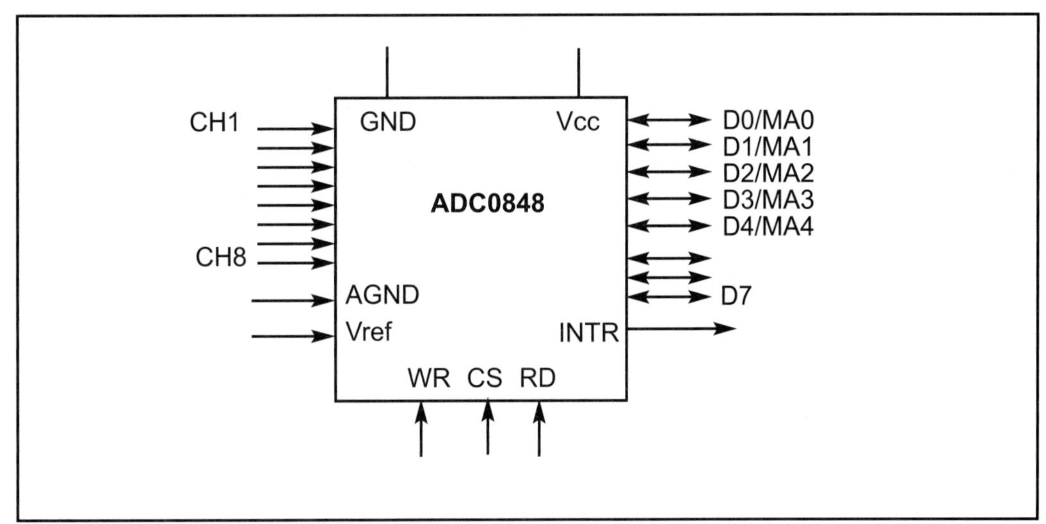

Figure 13-3. ADC0848 Parallel ADC Block Diagram

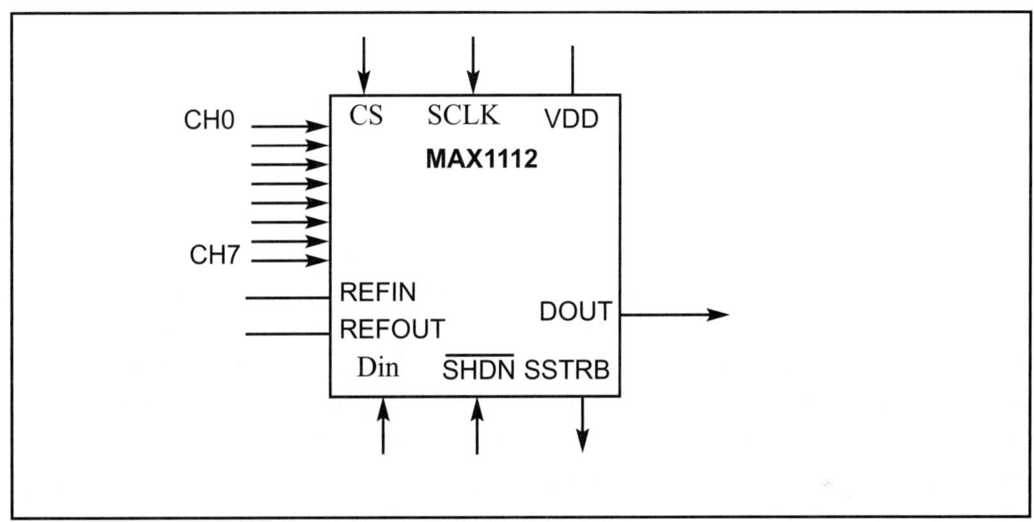

Figure 13-4. MAX1112 Serial ADC Block Diagram

lel-in-serial-out shift register responsible for sending out the binary data one bit at a time. The D0–D7 data pins of the 8-bit ADC provide an 8-bit parallel data path between the ADC chip and the CPU. In the case of the 16-bit parallel ADC chip, we need 16 pins for the data path. In order to save pins, many 12- and 16-bit ADCs use pins D0–D7 to send out the upper and lower bytes of the binary data. In recent years, for many applications where space is a critical issue, using such a large number of pins for data is not feasible. For this reason, serial devices such as the serial ADC are becoming widely used. While the serial ADCs use fewer pins and their smaller packages take much less space on the printed circuit board, more CPU time is needed to get the converted data from the ADC because the CPU must get data one bit at a time, instead of in one single read operation as with the parallel ADC. ADC848 is an example of a parallel ADC with 8 pins for the data output, while the MAX1112 is an example of a serial ADC with a single pin for D_{out}. Figures 13-3 and 13-4 show the block diagram for ADC848 and MAX1112, respectively.

Analog input channels

Many data acquisition applications need more than one ADC. For this reason, we see ADC chips with 2, 4, 8, or even 16 channels on a single chip. Multiplexing of analog inputs is widely used as shown in the ADC848 and MAX1112. In these chips, we have 8 channels of analog inputs, allowing us to monitor multiple quantities such as temperature, pressure, heat, and so on.

Start conversion and end-of-conversion signals

The fact that we have an analog input channel and a single digital output register makes start conversion (SC) and end-of-conversion (EOC) signals necessary. When SC is activated, the ADC starts converting the analog input value of V_{in} to an n-bit digital number. The amount of time it takes to convert varies depending on the conversion method as was explained earlier. When the data conversion is complete, the end-of-conversion signal notifies the CPU that the converted data is ready to be picked up.

Successive Approximation ADC

Successive Approximation is a widely used method of converting an analog input to digital output. It has three main components: a) successive approximation register (SAR), b) comparator, and c) control unit. See Figure 13-5. Assuming a step size of 10 mv, the 8-bit successive approximation ADC will go through the following steps to convert an input of 1 volt:

(1) It starts with binary 10000000. Since 128 × 10 mv = 1.28 V is greater than the 1 V input bit 7 is cleared (dropped). (2) 01000000 gives us 64 × 10 mv = 640 mv and bit 6 is kept since it is smaller than the 1 V input. (3) 01100000 gives us 96 × 10 mv = 960 mv and bit 5 is kept since it is smaller than the 1 V input. (4) 01110000 gives us 112 × 10 mv = 1,120 mv and bit 4 is dropped since it is greater than the 1 V input. (5) 01101000 gives us 108 × 10 mv = 1,080 mv and bit 3 is dropped since it is greater than the 1 V input. (6) 01100100 gives us 100 × 10 mv = 1,000 mv = 1 V and bit 2 is kept since it is equal to input. Even though the answer is found it does not stop. (7) 011000110 gives us 102 × 10 mv = 1,020 mv and bit 1 is dropped since it is greater than the 1 V input. (8) 01100101 gives us 101 × 10 mv = 1,010 mv and bit 0 is dropped since it is greater than the 1 V input.

Notice that the Successive Approximation method goes through all the steps even if the answer is found in one of the earlier steps. The advantage of the Successive Approximation method is that the conversion time is fixed since it has to go through all the steps.

Review Questions

1. Give two factors that affect the step size calculation.
2. The ADC0848 is a(n) _______-bit converter.
3. True or false. While the ADC0848 has 8 pins for D_{OUT}, the MAX1112 has only one D_{OUT} pin.
4. Indicate the number of analog input channels for each of the following ADC chips.
   (a) ADC0848 (b) MAX1112
5. Find the step size for an 8-bit ADC, if V_{ref} = 1.28 V
6. For question 5, calculate the D0–D7 output if the analog input is: (a) 0.7 V, and (b) 1 V.

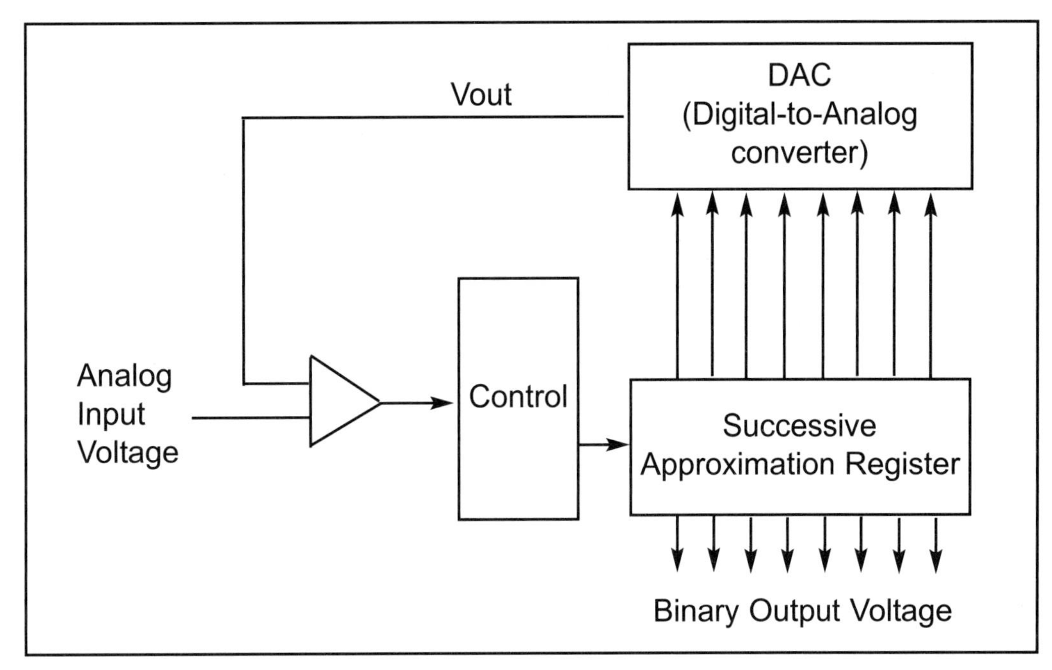

Figure 13-5. Successive Approximation ADC

SECTION 13.2: ATD PROGRAMMING IN THE HCS12

Because the ADC is widely used in data acquisition, in recent years an increasing number of microcontrollers have an on-chip ADC peripheral, just like timers and UARTs. An on-chip ADC eliminates the need for an external ADC connection, which leaves more pins for other I/O activities. The majority of the HCS12 chips come with 8 channels of ADC, and some HCS12s have as many as 16 channels of ADC. The Freescale HCS12 literature uses the ATD (analog-to-digital) designation instead of ADC (analog-to-digital converter). The HCS12 chips have two sets of ADC called ATD0 and ATD1. ATD0 and ATD1 each have 8 channels of ADC. That means we can have up to 16 channels of ADC in the HCS12. In this section we discuss the ATD feature of the HCS12 and show how it is programmed in both Assembly and C.

HCS12 ATD features

The ADC peripheral of the HCS12 has the following characteristics:

(a) It is a 10-bit ADC. It can be configured as 8-bit, too.

(b) It has two sets of 8 channels of analog input: ATD0 and ATD1.

(c) The converted output binary data is held by two special function registers called ATDDRL (ATD Result Low) and ATDDRH (ATD Result High).

(d) Because the ATDDRH:ATDDRL registers give us 16 bits and the ATD data out is only 10 bits wide, 6 bits of the 16 are unused. We have the option of making either the upper 6 bits or the lower 6 bits unused. We also have the option of using 8-bit resolution. In that case we need only the 8-bit register of ATDDRL.

(e) It allows the implementation of the differential V_{ref} voltage using the V_{rh} and V_{rl} pins, where $V_{ref} = V_{rh} - V_{rl}$. See Figure 13-6.

(f) The conversion time is dictated by the bus frequency. While the bus fre-

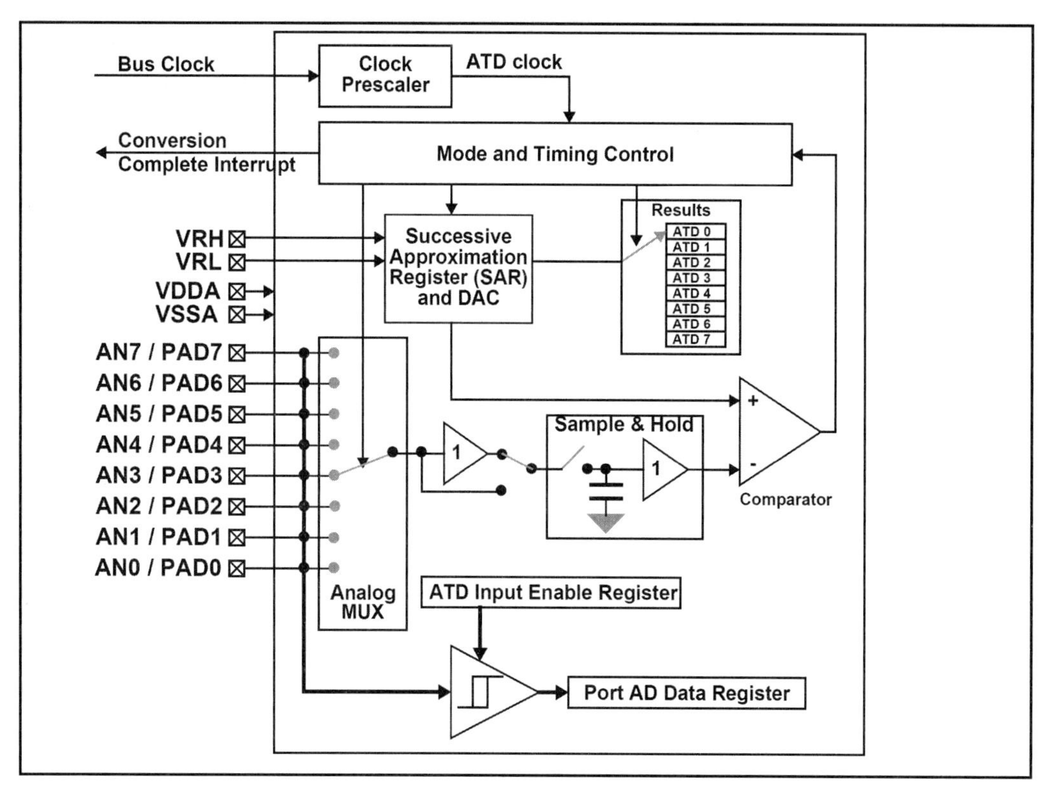

Figure 13-6. HCS12 ATD Blocking Diagram

quency for HCS12 can be as high as 40 MHz, the conversion speed cannot be faster than 2 MHz. The minimum speed for the conversion frequency is 500 kHz. See Figure 13-7.

Many of the above features can be programmed using the control registers of ATD. These are ATDCTL2, ATDCTL3, ATDCTL4, and ATDCTL5. We start with the ATDCTL4, which is responsible for the A/D conversion speed.

ATDCTL4 register and A/D conversion speed

The ATDCTL4 register is used to select the A/D resolution and conversion speed. It is shown in Figure 13-8. We can choose 8-bit or 10-bit resolution using the SRES8 bit. The most important function of ATDCTL4 is to set the conversion time. Next, we discuss this important topic.

A/D conversion frequency

The ATD conversion speed depends on the bus frequency (Fosc/2 if we are not using the PLL). The bus frequency goes through the prescaler and is divided by 2 once more before it is fed to the ATD unit, as shown in Figure 13-7. The prescaler value can be anywhere from 0 to 32, as indicated by the bits in the ATDCTL4 registers. The following equation is used to calculate ATD conversion frequency:

ATD Clock = Bus Freq / 2(N Prescaler value + 1) (Equation 13-1)

It must be noted that the ATD conversion frequency cannot be less than 500 kHz. It also cannot be greater than 2 MHz regardless of the bus frequency. See Examples 13-2 and 13-3.

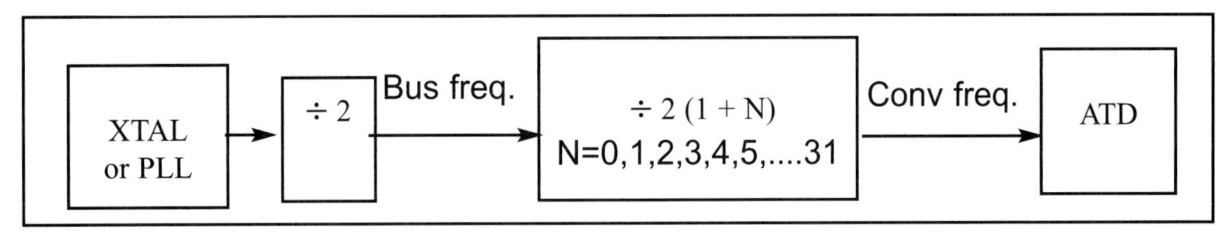

Figure 13-7. HCS12 Conversion Frequency

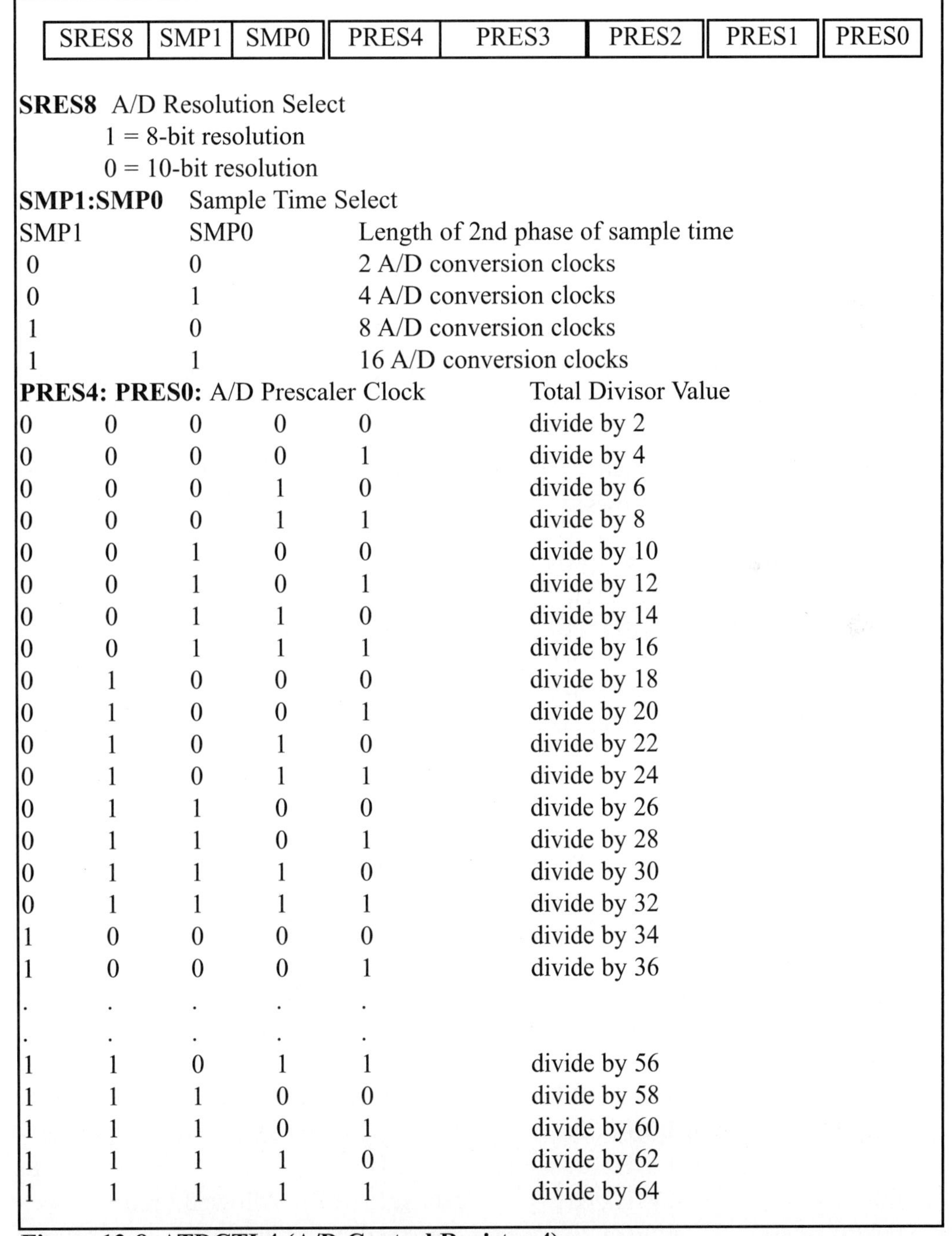

SRES8	SMP1	SMP0	PRES4	PRES3	PRES2	PRES1	PRES0

SRES8 A/D Resolution Select
1 = 8-bit resolution
0 = 10-bit resolution

SMP1:SMP0 Sample Time Select

SMP1	SMP0	Length of 2nd phase of sample time
0	0	2 A/D conversion clocks
0	1	4 A/D conversion clocks
1	0	8 A/D conversion clocks
1	1	16 A/D conversion clocks

PRES4: PRES0: A/D Prescaler Clock

					Total Divisor Value
0	0	0	0	0	divide by 2
0	0	0	0	1	divide by 4
0	0	0	1	0	divide by 6
0	0	0	1	1	divide by 8
0	0	1	0	0	divide by 10
0	0	1	0	1	divide by 12
0	0	1	1	0	divide by 14
0	0	1	1	1	divide by 16
0	1	0	0	0	divide by 18
0	1	0	0	1	divide by 20
0	1	0	1	0	divide by 22
0	1	0	1	1	divide by 24
0	1	1	0	0	divide by 26
0	1	1	0	1	divide by 28
0	1	1	1	0	divide by 30
0	1	1	1	1	divide by 32
1	0	0	0	0	divide by 34
1	0	0	0	1	divide by 36
.	.	.	.	.	
.	.	.	.	.	
1	1	0	1	1	divide by 56
1	1	1	0	0	divide by 58
1	1	1	0	1	divide by 60
1	1	1	1	0	divide by 62
1	1	1	1	1	divide by 64

Figure 13-8. ATDCTL4 (A/D Control Register 4)

Calculating conversion time

The conversion time for the ATD has three parts. They are as follows:

1) In the first phase, a sample amplifier of unity gain samples the analog input for a total of 2 clock cycles. This buffering of the analog input charges the sample capacitor and up to the input potential.

2) In the second phase, the sample buffer is disconnected and connected to the storage node for a certain number of clock cycles. The number of clock cycles can be 2, 4, 8, or 16. We program this number via the ATDCTL4 register. For the fast input analog signals where dv/dt is large (fast slew rate), we should choose the 2-clock option. For the slow analog inputs we should choose the 16-clock option to ensure the accurate reading.

3) In the third phase, the analog input is converted to binary numbers using the successive approximation method. In this phase, the number of clocks used depends on how many bits are in the binary output. For each bit we need one clock. That means we need 8 clocks for the 8-bit output and 10 clocks for the 10-bit output. We choose the *n*-bit option using the ATDCTL4 register. Look at Examples 13-2 through 13-7 for clarification.

Example 13-2

A HCS12 is connected to the 4 MHz crystal oscillator. Calculate the conversion clock frequency and period of ATD for all possible prescalers in order to stay within the 500 kHz–2 MHz limit.

Solution:

The Bus Freq = Fosc/2 = 4 MHz / 2 = 2 MHz. Now, the options for the conversion clock source available are as follows:

Prescaler (N Bin Value)	ADT Conversion Freq. ATD Freq = Bus Freq / 2(N + 1)	ATD Clock period
a) 00000	2 MHz/ 2(0 + 1) = 1 MHz	1 μs
b) 00001	2 MHz/ 2(1 + 1) = 500 kHz (Min. Allowed Freq)	2 μs
c) 00010	2 MHz/ 2(2 + 1) = 333.3 kHz (Not Allowed)	

Example 13-3

A HCS12 is connected to the 8 MHz crystal oscillator. Calculate the conversion clock frequency and period of ATD for all possible prescalers.

Solution:

The Bus Freq = Fosc/2 = 8 MHz / 2 = 4 MHz. Now, the options for the conversion clock source available are as follows:

Prescaler (N Bin Value)	ADT Conversion Freq. ATD Freq = Bus Freq / 2(N +1)	ATD Clock period
a) 00000	4 MHz/ 2(0 + 1) = 2 MHz (Max. Allowed Freq)	0.5 μs
b) 00001	4 MHz/ 2(1 + 1) = 1 MHz	1 μs
c) 00010	4 MHz/ 2(2 + 1) = 666 kHz	1.5 μs
d) 00011	4 MHz/ 2(3 + 1) = 500 kHz (Min. Allowed Freq)	2 μs
e) 00100	4 MHz/ 2(4 + 1) = 400 kHz (Not Allowed)	

Example 13-4

A HCS12 has a PLL frequency of 16 MHz. Calculate the conversion clock frequency and period of ATD for all possible prescalers.

Solution:

The Bus Freq = PLL/2 = 16 MHz / 2 = 8 MHz.

Prescaler (N Bin Value)	ADT Conversion Freq. ATD Freq = Bus Freq/2 (N + 1)	ATD Clock period
a) 00000	8 MHz/ 2(0 + 1) = 4 MHz (Not Allowed)	
b) 00001	8 MHz/ 2(1 + 1) = 2 MHz (Max. Allowed Freq.)	0.5 μs
c) 00010	8 MHz/ 2(2 + 1) = 1.333 MHz	0.75 μs
d) 00011	8 MHz/ 2(3 + 1) = 1 MHz	1 μs
e) 00100	8 MHz/ 2(4 + 1) = 800 kHz	1.25 μs
f) 00101	8 MHz/ 2(5 + 1) = 666 kHz	1.5 μs
g) 00110	8 MHz/ 2(6 + 1) = 571 kHz	1.75 μs
h) 00111	8 MHz/ 2(7 + 1) = 500 kHz (Min. Allowed Freq,)	2 μs
i) 01000	8 MHz/ 2(8 + 1) = 444 kHz (Not Allowed)	

Example 13-5

A HCS12 is connected to the 4 MHz crystal oscillator. Calculate the conversion time for all possible prescaler options shown in Example 13-2. Use the 4 clocks for the second phase of the conversion. Assume an 8-bit binary output.

Solution:

From Example 13-2a, we have T = 1/1 MHz = 1 μs for conversion clock if we use the option of 00000 for the prescaler. Now, the conversion time = 2 clocks + 4 clocks + 8 clocks = 2 × 1 μs + 4 × 1 μs + 8 × 1 μs = 14 μs.

b) From Example 13-2b, we have T = 1/500 kHz = 2 μs for conversion clock if we use the option of 00001 for the prescaler. Now, the conversion time = 2 clocks + 4 clocks + 8 clocks = 2 × 2 μs + 4 × 2 μs + 8 × 2 μs = 28 μs.

Example 13-6

Assume the ATD conversion clock is 500 kHz. Calculate the fastest and slowest conversion times possible for the 8-bit resolution.

Solution:

The difference between the fastest and slowest conversion times is the clocks for the second phase of conversion. The fastest is 16 clocks and slowest is 2 clocks.

From Example 13-5, we have T = 1/500 kHz = 2 μs. Now, the fastest conversion time = 2 clocks + 2 clocks + 8 clocks = 2 × 2 μs + 2 × 2 μs + 8 × 2 μs = 24 μs. The slowest conversion time = 2 clocks + 16 clocks + 8 clocks = 2 × 2 μs + 16 × 2 μs + 8 × 2 μs = 52 μs.

Example 13-7

Repeat Example 13-6 for 10-bit resolution.

Solution:

Now, the fastest conversion time = 2 clocks + 2 clocks + 10 clocks = 2 × 2 μs + 2 × 2 μs + 10 × 2 μs = 28 μs. The slowest conversion time = 2 clocks + 16 clocks + 10 clocks = 2 × 2 μs + 16 × 2 μs + 10 × 2 μs = 56 μs.

ATDCTL2 register

The ATDCTL2 register is used to turn on the ATD among other things. Figure 13-9 shows the ATDCTL2 register. The most important bit of ATDCTL2 is the ADPU bit. In order to reduce the power consumption of the HCS12, the A/D section is powered down upon reset. We must power up (turn on) the ADC section of the HCS12 with the ADPU bit of the ATDCTL2 register. We also have the option of allowing the conversion to continue when the CPU goes into the wait state. Using the options available in the ATDCTL2 register, the ATD conversion can be triggered from an external source. Instead of polling, we can also use an interrupt to get the data out of the ATD when conversion is done.

ATDCTL3 register

Another major register of the HCS12's ADC feature is ATDCTL3. The ATDCTL3 register is used to select the number of conversions per sequence among other things. It is shown in Figure 13-10. We have the option of stopping conversion or continuing to convert during the breakpoint. The FRZ1:FRZ0 bits are used for this purpose. The FIFO mode allows us to store the consecutive conversions in consecutive registers and wrap around at the end.

ADPU	AFFC	AWAI	ETRIGLE	ETRIGP	ETRIGE	ASCIE	ASCIF

ADPU ATD Power Up

1 = A/D feature is powered up (turned on).

0 = A/D part of the HCS12 is powered down. This is the default and we should leave it powered down for applications in which A/D is not used.

AFFC ATD Fast Flag Clear. See the bits of the ATDSTAT1 register.

1 = Fast flag clearing

0 = Normal flag clearing (default)

AWAI ATD Power down in Wait mode

1 = Halt conversion and power down the A/D during Wait mode.

0 = ATD continues to run in Wait mode.

ETRIGLE:ETRIGP External Trigger Level/Edge Control/Polarity

ETRIGLE	ETRIGP	External Trigger Sensitivity
0	0	Falling edge
0	1	Rising edge
1	0	Low level
1	1	High level

ETRIGE External Trigger mode enable

1 = Enable External Trigger

0 = Disable External Trigger

ASCIE ATD Sequence Complete Interrupt Enable

1 = ATD Interrupt will be requested whenever ASCIF = 1 is set.

0 = ATD Sequence Complete Interrupt disable

ASCIF ATD Sequence Complete Interrupt Flag

1 = ATD Sequence Complete Interrupt pending

0 = No ATD interrupt occurred.

Figure 13-9. ATDCTL2 (A/D Control Register 2)

0	S8C	S4C	S2C	S1C	FIFO	FRZ1	FRZ0

S8C	S4C	S2C	S1C	Number of Conversions per Sequence
0	0	0	1	1
0	0	1	0	2
0	0	1	1	3
0	1	0	0	4
0	1	0	1	5
0	1	1	0	6
0	1	1	1	7
1	x	x	x	8
0	0	0	0	8

FRZ1	FRZ0	Behavior in Freeze Mode (Break Point)
0	0	Continue conversion
0	1	Reserved
1	0	Finish current conversion, then freeze
1	1	Freeze immediately

FIFO Result register FIFO mode.
1 = Conversion results are placed in consecutive result registers (wrap around at the end).
0 = Conversion results are placed in the corresponding result register up to the selected sequence length.

Figure 13-10. ATDCL3 (A/D Control Register 3)

ATDCTL5 register

Another major register of the HCS12's ADC feature is ATDCTL5. The ATDCTL5 register is used to select the input analog channel among other things. It is shown in Figure 13-11. The A/D channel selection is handled by the CC:CB:CA bits. After the A/D conversion is complete, the result sits in registers ATDDRL (A/D Result Low Byte) and ATDDH (A/D Result High Byte). The DSGN allows us to have the result in signed or unsigned numbers. In unsigned, the 8-bit value can go from 00 to $FF. For the signed data the result can go from $7F to $80 (+127 to –128). For the 10-bit, the unsigned value can go from 000 to $3FF. For signed 10-bit data, the value can range from –1,024 to +1,023. See Chapter 5 for a discussion of signed numbers. The SCAN bit allows us to have one conversion sequence or to have continuous conversion sequences. The MULT bit allows us to have conversion across multiple channels instead of a single channel. It must be noted that in the absence of an externally triggered start-conversion (see the option in ATDCTL2), writing to the ATDCTL5 register will start the conversion of analog input. In that case, if we choose the option of SCAN=0, then we must write the ATDCTL5 every time we want to convert the analog input. The DJM bit of the ATDCTL5 is used for making it right-justified or left-justified because we need only 10 bits of the 16. See Figure 13-12.

DJM	DSGN	SCAN	MULT	0	CC	CB	CA

DJM A/D Result Register Data Justified (format select bit)

1 = Right-justified: The 10-bit result is in the ATDDRL register and the upper 2 bits in ATDDRH. That means the 6 most significant bits of the ATDDRH register are all 0s. For the 8-bit resolution, only ATDDRL is used.

0 = Left-justified: The 10-bit result is in the ATDDRH register and the lower 2 bits in ATDDRL. That means the 6 least significant bits of the ATDDRL register are all 0s. For the 8-bit resolution, only ATDDRH is used.

DSGN A/D result Data Signed or Unsigned. The result is in signed or unsigned format

1 = Signed data representation

–128 to +127 ($80 to $7F) for 8-bit and –512 to +511 ($8000 to $7FFF) for 10-bit

0 = Unsigned data representation

SCAN A/D Continuous Conversion Sequence mode

1 = Scan mode. Continuous conversion sequence.

0 = Single Conversion sequence. We must write to the ATDCTL5 register every time we need to convert the analog input. The CC:CB:CA bits indicate the channel we are converting.

MULT A/D Multichannel sample mode

1 = Sample across several channels

0 = Sample only one channel

CC	CB	CA	CHANNEL SELECTION
0	0	0	CHAN0 (AN0)
0	0	1	CHAN1 (AN1)
0	1	0	CHAN2 (AN2)
0	1	1	CHAN3 (AN3)
1	0	0	CHAN4 (AN4)
1	0	1	CHAN5 (AN5)
1	1	0	CHAN6 (AN6)
1	1	1	CHAN7 (AN7)

Note: To start conversion, we write the byte to the ATDCTL5 register, unless the external trigger option in the ATDCTL2 register is chosen.

Figure 13-11. ATDCTL5 (A/D Control Register 5)

ATD result registers

In the HCS12, for each channel we have an ATD conversion result register called ATDDRHx:ATDDRLx where x is 0, 1, 2, 3 and so on. As we mentioned earlier, not all bits of the 16-bit ATDDRH:ATDDRL register are used. In the case of 10-bit resolution 6 bits are unused. For the 8-bit resolution only one of the ATDDRHigh or ATDDRLow registers is used. These are called ATDDRH0:ATDDRL0, ATDDRH1:ATDDRL1, ATDDRH2:ATDDRL2, and so on. As we saw earlier, the SRES8 bit in ATDCTL4 and the DJM bit in ATDCTL5 decide the resolution and the format, respectively. See Figure 13-12.

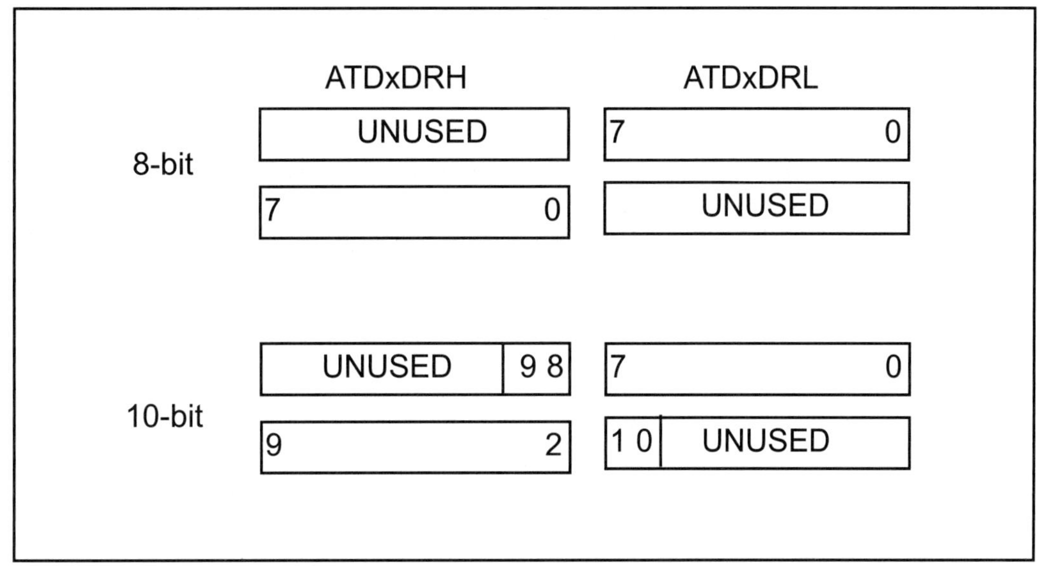

Figure 13-12. Right- and Left-Justified Data for 8-bit and 10-bit Data

ATD status registers

The most important status register in ATD is ATDSTAT0. See Figure 13-13. After the ATD converts the analog input, it places the result in the ATDDRH:ATDDRL registers and raises the flag SCF (sequence conversion flag). The SCF is the D7 bit of ATDSTAT0 register. To read the binary result out of the ATD chip, we must monitor (poll) the SCF bit. The vast majority of the applications use the SCF flag for end-of-conversion to get the final result out of the ATD. Now, to know when to read the data for a specific sequence, we must use the bits of the ATDSTAT1 register. See Figure 13-14. It has a total of 8 flags of CCF7–CCF0. Each flag belongs to one of the sequences. After the conversion for each sequence, the flag for that sequence is raised. To see at what sequence the ATD is operating in at a given time, we can use the lower 3 bits of the ATDSTAT0 register.

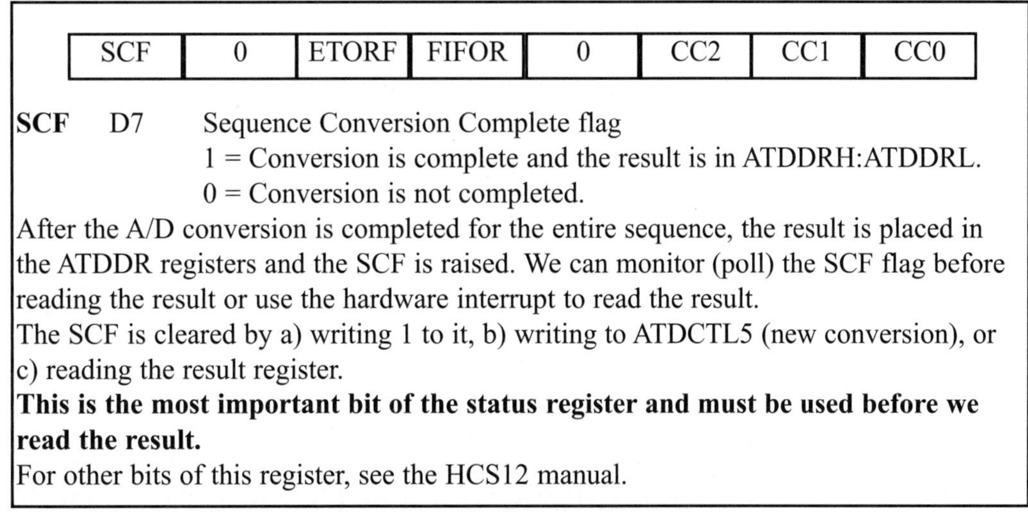

Figure 13-13. ATDSTAT0 (A/D Status Register 0) Register

CC7F	CC6F	CC5F	CC4F	CC3F	CC2F	CC1F	CC0F

CC7F–CC0F D7–D0 Conversion Complete Flag for sequence 0 to 7
1 = Conversion for sequence x is complete and the result is in ATDDR.
0 = Conversion for sequence x is not complete.

After the completion of A/D conversion sequence x, the result is placed in the ATDDRx register and the CCFx is raised. We can use the option of the fast flag clear in ATDCTL2.

Figure 13-14. ATDSTAT1 (A/D Status register 1) Register

First and second 8-channel ATD

Many of the members of the HCS12 family have two 8-channel A/D converters. They are designated as ATD0 and ATD1. Tables 13-4 and 13-5 show the partial listing of the registers and their addresses. For a complete list of the registers, see the HCS12 manual.

Voltage connection for ATD

We use the VRH and VRL for the V_{ref} of the ATD. In addition to the VRH and VRL, the ATD of the HCS12 has its own power and ground, which must be connected. See Figure 13-15 and Example 13-8.

Example 13-8

Find the step size values for the following if V_{rh} = 5 V and V_{rl} = V_{ss}.
(a) 8-bit resolution, and (b) 10-bit resolution.

Solution:

We have $V_{ref} = V_{rh} - V_{rl}$ = 5 V – 0 = 5 V.
(a) 5 V / 256 = 19.53 mV
(b) 5 V/1,024 = 4.82 mV

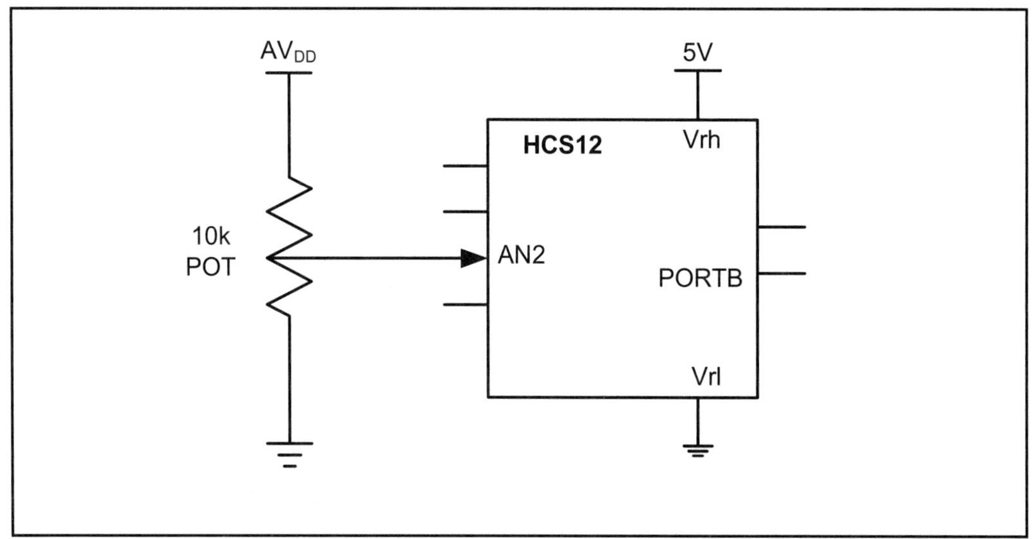

Figure 13-15. A/D Connection for Program 13-1

Table 13-4: Partial Listing of the ATD0 Registers and Their Addresses

ATD0 Register Name	Function	Addresses
ATD0CTL0	ATD0 Control Reg0 (reserved)	$0080
ATD0CTL1	ATD0 Control Reg1 (reserved)	$0081
ATD0CTL2	ATD0 Control Reg2	$0082
ATD0CTL3	ATD0 Control Reg3	$0083
ATD0CTL4	ATD0 Control Reg4	$0084
ATD0CTL5	ATD0 Control Reg5	$0085
ATD0STAT0	ATD0 Status Reg0	$0086
ATD0STAT1	ATD0 Status Reg1	$008B
ATD0DR0H	ATD0 Data Result Reg High byte	$0090
ATD0DR0L	ATD0 Data Result Reg Low byte	$0091
ATD0DR1H	ATD0 Data Result Reg High byte	$0092
ATD0DR1L	ATD0 Data Result Reg Low byte	$0093
ATD0DR2H	ATD0 Data Result Reg High byte	$0094
ATD0DR2L	ATD0 Data Result Reg Low byte	$0095
ATD0DR3H	ATD0 Data Result Reg High byte	$0096
ATD0DR3L	ATD0 Data Result Reg Low byte	$0097
ATD0DR4H	ATD0 Data Result Reg High byte	$0098
ATD0DR4L	ATD0 Data Result Reg Low byte	$0099
ATD0DR5H	ATD0 Data Result Reg High byte	$009A
ATD0DR5L	ATD0 Data Result Reg Low byte	$009B
ATD0DR6H	ATD0 Data Result Reg High byte	$009C
ATD0DR6L	ATD0 Data Result Reg Low byte	$009D
ATD0DR7H	ATD0 Data Result Reg High byte	$009E
ATD0DR7L	ATD0 Data Result Reg Low byte	$009F

Table 13-5: Partial Listing of the ATD1 Registers and Their Addresses

ATD1 Register Name	Function	Addresses
ATD1CTL0	ATD1 Control Reg0 (reserved)	$0120
ATD1CTL1	ATD1 Control Reg1 (reserved)	$0121
ATD1CTL2	ATD1 Control Reg2	$0122
ATD1CTL3	ATD1 Control Reg3	$0123
ATD1CTL4	ATD1 Control Reg4	$0124
ATD1CTL5	ATD1 Control Reg5	$0125
ATD1STAT0	ATD1 Status Reg0	$0126
ATD1STAT1	ATD1 Status Reg1	$012B
ATD1DR0H	ATD1 Data Result Reg High byte	$0130
ATD1DR0L	ATD1 Data Result Reg Low byte	$0131
ATD1DR1H	ATD1 Data Result Reg High byte	$0132
ATD1DR1L	ATD1 Data Result Reg Low byte	$0133
ATD1DR2H	ATD1 Data Result Reg High byte	$0134
ATD1DR2L	ATD1 Data Result Reg Low byte	$0135
...	....	....
...	....	...
ATD0DR7H	ATD1 Data Result Reg High byte	$013E
ATD0DR7L	ATD1 Data Result Reg Low byte	$013F

Steps in programming the A/D converter using polling

To program the ATD converter of the HCS12, the following steps must be taken:

1. Turn on the ATD section of the HCS12. (It is disabled upon power-on reset to save power.) To do that, we use the bits of ATDCTL2.
2. Wait for 20 μs for the ATD to get ready.
3. Use the ATDCTL3 register to select the number of conversion sequences.
4. Use ATDCTL4 to select the conversion speed and resolution.
5. Select the channel and start conversion by writing to ATDCTL5.
6. Wait for the conversion to be completed by polling the SCF bit in ATDSTAT0. The SCF indicates the end-of-conversion.
7. After the SCF bit has gone HIGH, read the ATDDRL and ATDDRH registers to get the digital data output.
8. Go back to step 5.

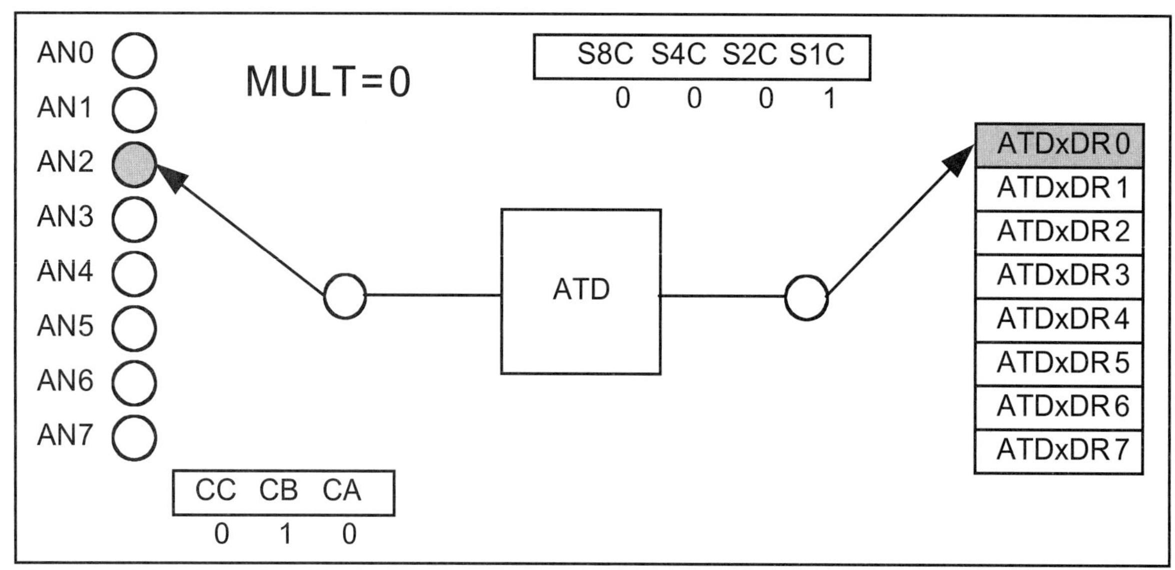

Figure 13-16. Selecting Channel 2 For a Sequence Length of 1. (For Program 13-1.)

Programming HCS12 ATD0 in Assembly

Figure 13-16 shows the channel selection. The Assembly language Program 13-1 illustrates the steps for ADC conversion shown above. The C version of the program is shown in Program 13-1C.

```
;Program 13-1: This program gets data from channel 2 of
;ATD0 and displays the result on PORTB. Assume XTAL = 8 MHz.
;Use the following values for control registers
;ATDCTL2 = %10000000. ATD on, Normal flag clear, run
;during Wait, no External Trig. No Interrupt
;ATDCTL3 = % 00001000. one conversion, run during break-
;point, no FIFO
;ATDCTL4 = %11100011. 8-bit resolution, 16-clock for 2nd
;phase, prescaler 00011 for conversion freq. of 500 kHz.
```

```
;ATDCTL5 = % 10000010, right justified, unsigned result,
;single conversion, one channel only, convert channel 2

;Program 13-1
Entry:LDS    #RAMEnd+1          ;initialize the stack pointer
      BSET  DDRB,%11111111 ;make PORTB an output
      MOVB  #$80,ATD0CTL2   ;turn on ATD0
      JSR   DELAY20uS  ;20 microsec for ATD0 to get ready
      MOVB  #$08,ATD0CTL3
      MOVB  #$E3,ATD0CTL4
H1    MOVB  #$82,ATD0CTL5    ;convert Channel 2 once (SCAN=0)
H2    BRCLR ATD0STAT0,mATD0STAT0_SCF,H2;wait for conversion
      LDAA  ATD0DR0L          ;read the binary result
      STAA  PORTB             ;display it on PORTB
      JSR    TDELAY           ;wait so we can see the output
      BRA    H1               ;keep doing that
;------------
DELAY20uS:
      LDAB #40
D1    DECB
      BNE D1
      RTS
;----------------
TDELAY:
      LDD #$0000       ;gives some time to view the result
D2    DECB
      BNE D2
      DECA
      BNE D2
      RTS
```

Note: If we make SCAN = 1, we select the channel once. See Program 13-1B.

```
;Program 13-1B : Using SCAN = 1
Entry:
      LDS   #RAMEnd+1         ;initialize the stack pointer
      BSET  DDRB,%11111111   ;make PORTB an output
      MOVB  #$80,ATD0CTL2    ;turn on ATD0
      JSR   DELAY20uS  ;20 microsec for ATD0 to get ready
      MOVB  #$08,ATD0CTL3
      MOVB  #$E3,ATD0CTL4
      MOVB  #$A2,ATD0CTL5    ;SCAN=1, keep converting Chan 2
H2    BRCLR ATD0STAT0,mATD0STAT0_SCF,H2 ;wait for conversion
      LDAA  ATD0DR0L         ;read the binary result
      STAA  PORTB            ;display it on PORTB
      JSR   TDELAY           ;wait so we can see the output
      BRA   H2               ;keep doing that
```

Programming HCS12 ATD in C

Program 13-1C is the C version of the ATD conversion for Program 13-1.

```
Program 13-1C: This program gets data from channel 2 (AN2)
of ADC and displays the result on PORTB. This is the C
version of Program 13-1.
```

```
//Program 13-1C
#include <mc9s12dp512.h>      /* derivative information */
#pragma LINK_INFO DERIVATIVE "mc9s12dp512"
void DELAY20uS(void);
void TDELAY(void);
void main(void)
{
  DDRB = 0xFF;                //make PORTB an output
  ATD0CTL2 = 0x80;            //turn on ATD0
  DELAY20uS();          //20 microsec for ATD0 to get ready
  ATD0CTL3 = 0x08;
  ATD0CTL4 = 0xE3;
  for(;;)               //do it forever
    {
    ATD0CTL5 = 0x82;        //convert Channel 2 once (SCAN=0)
     //wait for conversion
    while(!(ATD0STAT0 & ATD0STAT0_SCF_MASK));
    PORTB = ATD0DR0L;         //display result on PORTB
    TDELAY();                 //wait so we can see the output
    }
}
void DELAY20uS(void)
  {
  int x = 40;
  while(x--);
  }
void TDELAY(void)
 {
  int x = 0;
  while(--x);
  }
```

Programming ATD converter using interrupts

In Chapter 11, we showed how to use interrupts instead of polling to avoid tying down the microcontroller. To program the A/D using the interrupt method, we need to set HIGH the ASCIE flag in the ATDCTL2 register. If ASCIE = 1, then upon completion of the conversion, the ASCIF flag bit in ATDCTL2 becomes HIGH, which will force the CPU to jump to location $FFD2 in the vector table, if the I bit is enabled. In this case the ASCIF bit of ATDCTL2 is the same as the SCF bit of the ATDSTAT register. To best use the interrupt, we also need to make SCAN = 1 in the ATDCTL5 register. If SCAN = 0, it converts the analog input only once and it stops. That means we must write to the ATDCTL5 register every time we want to convert the analog input and that defies the whole notion of using an interrupt. However, when SCAN = 1, the analog input is converted continuously and there is no need to write to ATDCTL5 every time. This process lends itself nicely to the interrupt method since it converts continuously and whenever it finishes each conversion it will generate an interrupt. Program 13-2 is an interrupt version of 13-1.

```
//Program 13-2(Interrupt version of Program 13-1)
Entry: LDS #RAMEnd+1          ;initialize the stack pointer
      BSET DDRB,%11111111   ;make PORTB an output
      MOVB #$82,ATD0CTL2    ;turn on ATD0 with interrupt
      JSR   DELAY20uS ;20 microsec for ATD0 to get ready
      MOVB #$08,ATD0CTL3
      MOVB #$E3,ATD0CTL4
      MOVB #$82,ATD0CTL5    ;convert Channel 2 once
      CLI                   ;enable interrupts globally
      BRA  $                ;wait for interrupt
;-------------------------
AD_ISR:
      LDAA ATD0DR0L         ;read the binary result
      STAA PORTB            ;display it on PORTB
      JSR  TDELAY           ;wait so we can see the output
      BSET ATD0CTL2,mATD0CTL2_ASCIF; clear the flag
      MOVB #$82,ATD0CTL5    ;start converting Channel 2
      RTI                   ;keep doing that
;-------------------------
      ORG $FFD2
      DC.W AD_ISR
      ORG $FFFE
      DC.W Entry
;for delays see Program 13-1
```

```
//Program 13-2C (This is the C version of Program 13-2)
#include <mc9s12dp512.h>      /* derivative information */
void DELAY20uS(void);
void TDELAY(void);
void main(void)
{
  DDRB = 0xFF;           //make PORTB an output
  ATD0CTL2 = 0x82;       //turn on ATD0 with interrupt
  DELAY20uS();           //20 microsec for ATD0 to get ready
  ATD0CTL3 = 0x08;
  ATD0CTL4 = 0xE3;
  ATD0CTL5 = 0x82;       //start converting Channel 2
  __asm CLI;             //enable interrupts globally
  for(;;) {}             //wait for interrupt
}
interrupt (((0x10000-Vatd0)/2)-1) void AD_ISR(void)
  {
  PORTB = ATD0DR0L;      //display result on PORTB
  TDELAY();              //wait so we can see the output
  ATD0CTL5 = 0x82;       //start converting Channel 2
  }
```

```
void DELAY20uS(void) {
  int x = 40;
  while(x--);
  }
void TDELAY(void)
  {
  int x = 0;
  while(--x);
  }
```

Multiple conversion for a single channel

In Program 13-1, we showed how to convert a single analog input channel only one time. In Program 13-2, we used the interrupt method to show how to convert a single channel continuously. Now, how about converting a single channel several times? The number of conversions per sequence is set by the bits in ATD-CTL3. For example, if we set the sequence to 2, the ATD will convert the input and place the results in the ATDxDR registers. The question is, Which register is used for storing the two results? In the single sequence conversion of a single channel, the ATD placed the result in registers starting with ATDxDR0. In the two-conversion sequence of a single channel the ATD places the result in ATDxDR0 for the first round and in ATDxDR1 for the second round. This is done if we set FIFO = 0 in the ATDCTL3 register. For FIFO = 1, the ATD uses ATDxDR0 and ATDxDR1 for the first round, ATDxDR2 and ATDxDR3 for the second round, ATDxDR4 and ATDxDR5 for third round, and so on. In converting a single channel several times, one can calculate the average of results to get a more accurate reading. See Figure 13-17. In Program 13-3, we use the 4-sequence option in ATD0CTL3, take the average of the four results, and display the result. Program 13-3C is the C version of the same program.

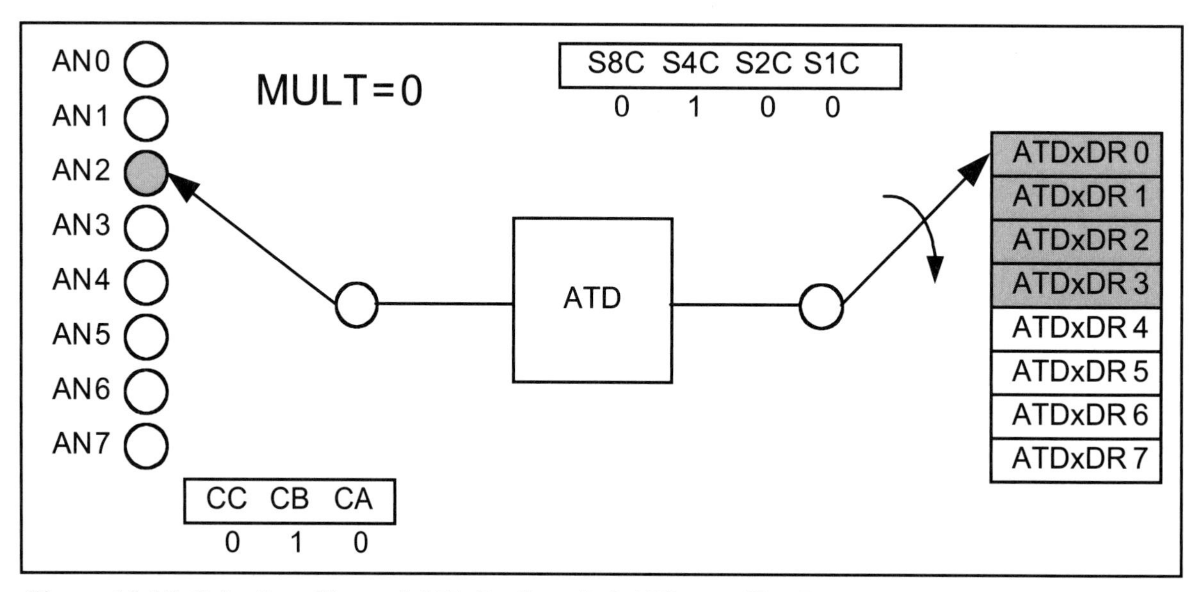

Figure 13-17. Selecting Channel 2 To Be Sampled 4 Times. (For Program 13-3.)

;Program 13-3

```
Entry:
      LDS   #RAMEnd+1          ;initialize the stack pointer
      BSET  DDRB,%11111111     ;make PORTB an output
      MOVB  #$80,ATD0CTL2      ;turn on ATD0
      JSR   DELAY20uS          ;20 microsec for ATD0 to get ready
      MOVB  #$20,ATD0CTL3      ;4-sequence
      MOVB  #$E3,ATD0CTL4
H1    MOVB  #$82,ATD0CTL5      ;start converting Channel 2
H2    BRCLR ATD0STAT0,mATD0STAT0_SCF,H2 ;wait for 4 conversions
      LDD   ATD0DR0            ;read the binary result
      ADDD  ATD0DR1            ;read the second result
      ADDD  ATD0DR2            ;read the third result
      ADDD  ATD0DR3            ;read the fourth result
      LDX   #4                 ;we need to average 4 results
      IDIV                     ;divide by the number of results
      STX   PORTAB             ;display it on PORTA and PORTB
      JSR   TDELAY             ;wait so we can see the output
      BRA   H1                 ;keep doing that
;-----------------------
DELAY20uS:
      LDAB #40
D1    DECB
      BNE D1
      RTS
;-------------------------
TDELAY:
      LDD #$0000
D2    DECB
      BNE D2
      DECA
      BNE D2
      RTS
```

//Program 13-3C

```
#include <hidef.h>              /* common defines and macros */
#include <mc9s12dp512.h>       /* derivative information */
#pragma LINK_INFO DERIVATIVE "mc9s12dp512"
void DELAY20uS(void);
void TDELAY(void);
int result;
void main(void)
{
  DDRB = 0xFF;                       //make PORTB an output
  ATD0CTL2 = 0x80;                   //turn on ATD0
  DELAY20uS();                 //20 microsec for ATD0 to get ready
  ATD0CTL3 = 0x20;                   //4-sequence
  ATD0CTL4 = 0xE3;
  for(;;)
   {
    ATD0CTL5 = 0x82;                 //start converting Channel 2
    //wait for conversion
    while(!(ATD0STAT0 & ATD0STAT0_SCF_MASK));
    result = ATD0DR0;                //get the binary result
    result = result + ATD0DR1;  //get second result and add it
    result = result + ATD0DR2;  //get third result and add it
```

```
      result = result + ATD0DR3;  //get fourth result and add it
      result = result / 4; //divide by the number of results
      PORTAB = result; //display on PORTB
      TDELAY();
    } /* do it forever */
}
void DELAY20uS(void)
 {
   int x = 40;
   while(x--);
 }
void TDELAY(void)
 {
   int x = 0;
   while(--x);
   }
```

Programming ATD for multiple input channels (MULT = 1)

In Program 13-1, we showed how to convert a single analog input channel. To program the ATD using multiple input channels, we need to understand the MULT bit option in the ATDxCTL5. If MULT = 0, then only one channel is converted and the channel number is given by the CC:CB:CA bits of the ATDxCTL5. Now, if MULT = 1, we can convert two or more channels. In that case the address of the first channel is given by the CC:CB:CA bits and the ATD automatically increments that number to get the address of the next channel. Now the sequence options in ATDxCTL3 will tell it how far to go. Program 13-4 shows the reading for channels 2, 3, and 4. Notice that the ATDxCTL3 the option is set for sequence length = 3. Figure 13-19 shows where the result are stored.

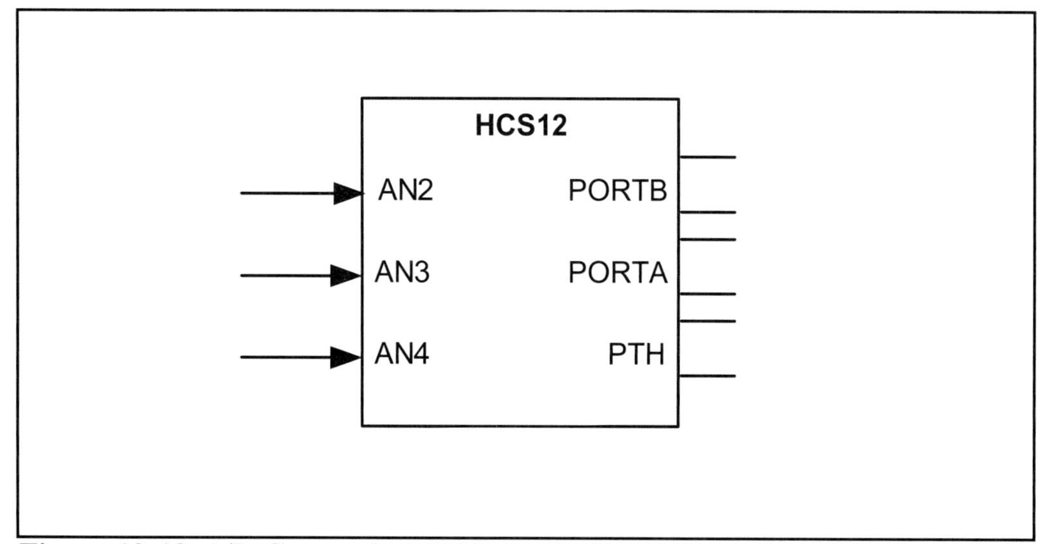

Figure 13-18. A/D Connection for Program 13-4

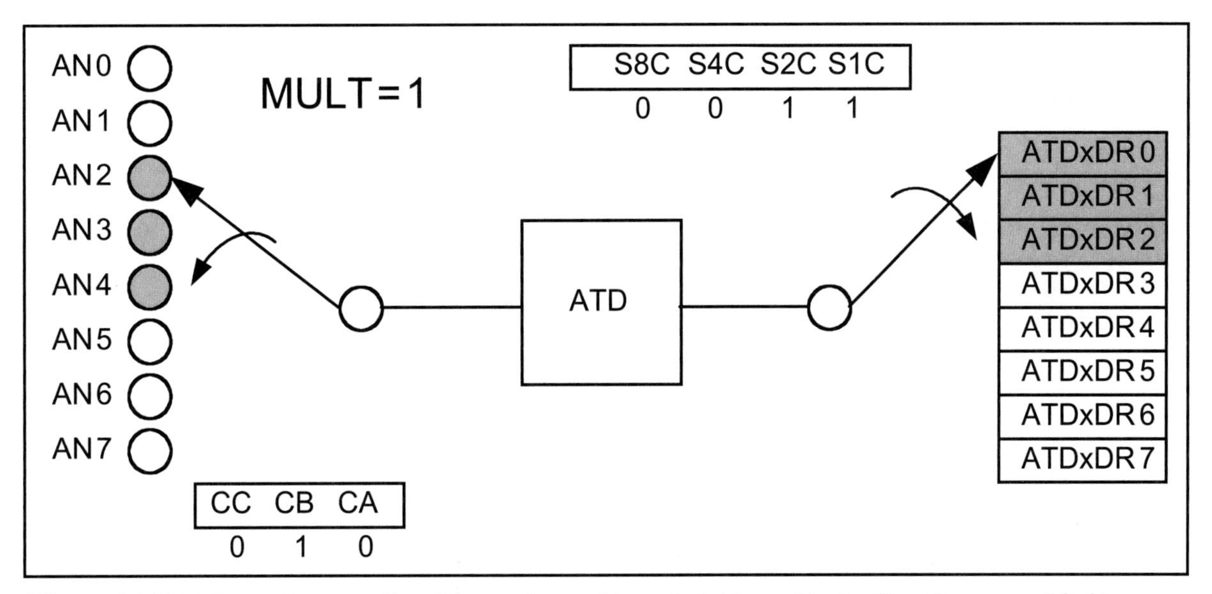

Figure 13-19. Three Consecutive Channels are Sampled, Once Each. (For Program 13-4.)

```
;Program 13-4 : Converting channels 2, 3, and 4
Entry: LDS  #RAMEnd+1          ;initialize the stack pointer
       BSET DDRB,%11111111     ;make PORTB an output
       BSET DDRA,%11111111     ;make PORTA an output
       BSET DDRH,%11111111     ;make PTH an output
       MOVB #$80,ATD0CTL2      ;turn on ATD0
       JSR  DELAY20uS          ;20 microsec for ATD0 to get ready
       MOVB #$18,ATD0CTL3      ;convert 3 channels
       MOVB #$E3,ATD0CTL4
H1     MOVB #$92,ATD0CTL5      ;start converting from Channel 2
H2     BRCLR ATD0STAT0,mATD0STAT0_SCF,H2 ;wait for conversion
       LDAA ATD0DR0L           ;read the binary result
       STAA PORTB              ;display it on PORTB
       LDAA ATD0DR1L           ;read the next result
       STAA PORTA              ;display it on PORTA
       LDAA ATD0DR2L           ;read the next result
       STAA PTH                ;display on PTH
       JSR  TDELAY             ;wait so we can see the output
       BRA  H1                 ;keep doing that
```

```
//Program 13-4C: Converting channels 2, 3, and 4 in C
#include <hidef.h>      /* common defines and macros */
#include <mc9s12dp512.h>     /* derivative information */
#pragma LINK_INFO DERIVATIVE "mc9s12dp512"
void DELAY20uS(void);
void TDELAY(void);
void main(void)
{
  DDRB = 0xFF;//make PORTB an output
  DDRA = 0xFF;//make PORTA an output
  DDRH = 0xFF;//make PTH an output
  ATD0CTL2 = 0x80;//turn on ATD0
  DELAY20uS();//20 microsec for ATD0 to get ready
  ATD0CTL3 = 0x18;//convert 3 channels
  ATD0CTL4 = 0xE3;

  for(;;)
```

```
    {
       ATD0CTL5 = 0x92;//start converting from channel 2
       //wait for conversion
       while(!(ATD0STAT0&ATD0STAT0_SCF_MASK));
       PORTB = ATD0DR0L; //read the binary result
       PORTA = ATD0DR1L; //get the next result
       PTH = ATD0DR2L; //get the next result
       TDELAY(); //wait so we can see the output
    } /* wait forever */
    /* please make sure that you never leave this function */
}
```

Review Questions

1. Give the main factors affecting the step size of ATD in HCS12.
2. The ATD of the HCS12 is a(n) _______-bit converter.
3. True or false. The ATD of HCS12 has a minimum conversion frequency.
4. True or false. ATD in the HCS12 has a maximum conversion frequency of 10 MHz.
5. Find the step size for an HCS12 ATD if V_{ref} = 1.024 V.
6. For Question 5, calculate the D0–D9 output if the analog input is: (a) 0.7 V, and (b) 1 V.
7. True or false. Upon reset the ATD in the HCS12 is off.
8. True or false. In HCS12, the ATDCTL5 is used for channel selection and start of conversion.
9. The minimum conversion clock frequency allowed is ________ .
10. Which bit is used to poll for the end of conversion?

SECTION 13.3: SENSOR INTERFACING AND SIGNAL CONDITIONING

This section will show how to interface sensors to the microcontroller. We examine some popular temperature sensors and then discuss the issue of signal conditioning. Although we concentrate on temperature sensors, the principles discussed in this section are the same for other types of sensors such as light and pressure sensors.

Temperature sensors

Transducers convert physical data such as temperature, light intensity, flow, and speed to electrical signals. Depending on the transducer, the output produced is in the form of voltage, current, resistance, or capacitance. For example, temperature is converted to electrical signals using a transducer called a *thermistor*. A thermistor responds to temperature change by changing its resistance, but its response is not linear, as seen in Table 13-6.

The complexity associated with writing software for such nonlinear devices has led many manufacturers to market a linear temperature sensor. Simple and widely used linear temperature sensors include the LM34 and LM35 series from National Semiconductor Corp. They are discussed next.

LM34 and LM35 temperature sensors

The sensors of the LM34 series are precision integrated-circuit temperature sensors whose output voltage is linearly proportional to the Fahrenheit temperature. See Table 13-7. The LM34 requires no external calibration because it is internally calibrated. It outputs 10 mV for each degree of Fahrenheit temperature. Table 13-7 is a selection guide for the LM34.

Table 13-6: Thermistor Resistance vs. Temperature

Temperature (C)	Tf (kilohms)
0	29.490
25	10.000
50	3.893
75	1.700
100	0.817

From William Kleitz, *Digital Electronics*

The LM35 series sensors are precision integrated-circuit temperature sensors whose output voltage is linearly proportional to the Celsius (centigrade) temperature. The LM35 requires no external calibration because it is internally calibrated. It outputs 10 mV for each degree of centigrade temperature. Table 13-8 is the selection guide for the LM35. (For further information see http://www.national.com.)

Signal conditioning and interfacing the LM35 to the HCS12

Signal conditioning is widely used in the world of data acquisition. The most common transducers produce an output in the form of voltage, current, charge, capacitance, and resistance. We need to convert and scale these signals to voltage, however, in order to send input to an A-to-D converter. This conversion and scaling is commonly called *signal conditioning*. See Figure 13-20. Signal conditioning can be a current-to-voltage conversion or a signal amplification. For example, the thermistor changes resistance with temperature. The change of resistance must be translated into voltages of 0–5 V in order to be of any use to an ADC.

Table 13-7: LM34 Temperature Sensor Series Selection Guide

Part	Temperature Range	Accuracy	Output Scale
LM34A	–50 F to +300 F	+2.0 F	10 mV/F
LM34	–50 F to +300 F	+3.0 F	10 mV/F
LM34CA	–40 F to +230 F	+2.0 F	10 mV/F
LM34C	–40 F to +230 F	+3.0 F	10 mV/F
LM34D	–32 F to +212 F	+4.0 F	10 mV/F

Note: Temperature range is in degrees Fahrenheit.

Table 13-8: LM35 Temperature Sensor Series Selection Guide

Part	Temperature Range	Accuracy	Output Scale
LM35A	–55 C to +150 C	+1.0 C	10 mV/C
LM35	–55 C to +150 C	+1.5 C	10 mV/C
LM35CA	–40 C to +110 C	+1.0 C	10 mV/C
LM35C	–40 C to +110 C	+1.5 C	10 mV/C
LM35D	0 C to +100 C	+2.0 C	10 mV/C

Note: Temperature range is in degrees Celsius.

Look at the case of connecting an LM34 to an ATD of the HCS12. Since the HCS12 has an option for 8-bit resolution with a maximum of 256 (2^8) steps and the LM34 (or LM35) produces 10 mV for every degree of temperature change, we can condition V_{in} of the ADC in the HCS12 to produce a V_{out} of 2,560 mV (2.56 V) for full-scale output. Therefore, in order to produce the full-scale V_{out} of 2.56 V for the HCS12, we need to set $V_{rh} = 2.56$ and $V_{rl} = 0$. This makes V_{out} of the ATD correspond directly to the temperature as monitored by the LM34. See Table 13-9.

Figure 13-21 shows the connection of a temperature sensor to the HCS12. Notice that we use the LM336-2.5 zener diode to fix the voltage across the 10 kilohms pot at 2.5 volts. The use of the LM336-2.5 should overcome any fluctuations in the power supply.

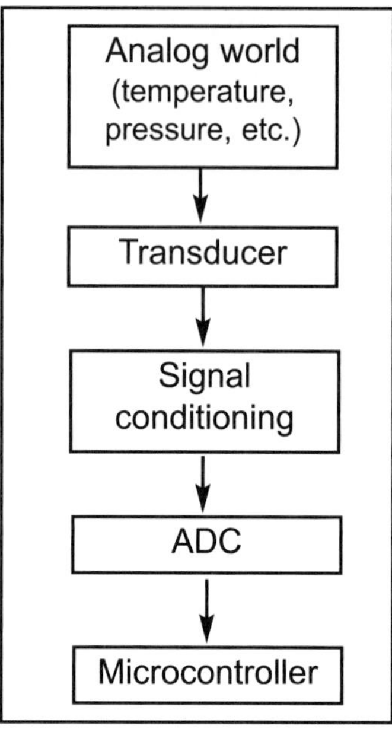

Figure 13-20. Getting Data from the Analog World

Table 13-9: Temperature vs. V_{out} for HCS12 with $V_{rh} = 2.56$ V, $V_{rl} = 0$ V. (8-bit ATD option, Step Size = 10 mV using LM34)

Temp. (F)	V_{in} (mV)	# of Steps	Binary V_{out} (D7–D0)	Temp. in Binary
0	0	0	00000000	00000000
1	10	1	00000001	00000100
2	20	2	00000010	00000010
3	30	3	00000011	00000011
10	100	10	00001010	00001010
20	200	20	00010100	00010100
30	300	30	00011110	00011110
40	400	40	00101000	00101000
55	550	55	00110111	00110111
85	850	85	01010101	01010101
90	900	90	01011010	01011010
255	2550	255	11111111	11111111

Example 13-9

In Table 13-9, verify the HCS12 output for a temperature of 70 degrees. Find the values in the HCS12 ATD registers of ATDDR.

Solution:

The step size is 2.56/256 = 10 mV because $V_{rh} = 2.56$ V.

For the 70 degrees temperature we have 700 mV output because the LM34 provides 10 mV output for every degree. Now, the number of steps are 700 mV/10 mV = 70 in decimal. Now 70 = 01000110 in binary and the HCS12 ATD output registers have ATDxDRL = 01000110.

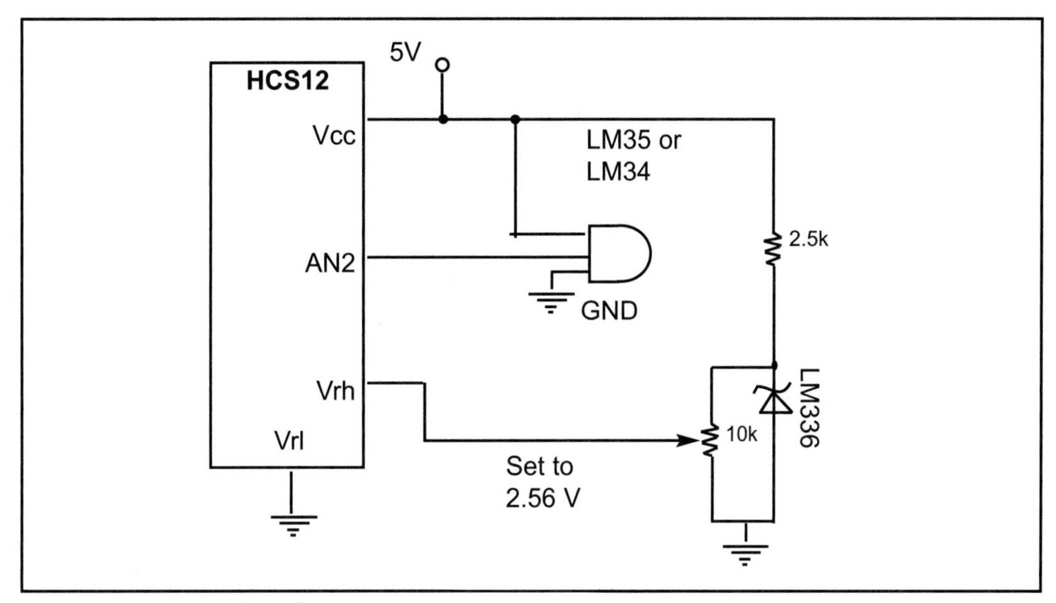

Figure 13-21. HCS12 Connection to Temperature Sensor

Reading and converting temperature

Programs 13-5 and 13-5C show code for reading and converting temperature in Assembly and C, respectively.

The programs correspond to Figure 13-21. Regarding these two programs, the following points must be noted:

(1) The LM34 (or LM35) is connected to channel 2 (ATD0 pin).
(2) The V_{rh} is connected to the V_{ref} of 2.56 V.
(3) The 8-bit option of the ATD is used.

```
;Program 13-5
;this program reads temperature and converts to ASCII
Entry:LDS    #RAMEnd+1       ; initialize the stack pointer
      MOVB  #$80,ATD0CTL2    ;turn on ATD0
      JSR   DELAY20uS  ;20 microsec for ATD0 to get ready
      MOVB  #$08,ATD0CTL3    ;sequence length = 1
      MOVB  #$E3,ATD0CTL4
H1    MOVB  #$82,ATD0CTL5    ;start converting Channel 2
H2    BRCLR ATD0STAT0,mATD0STAT0_SCF,H2;wait for conversion
      LDAA  ATD0DR0L          ;read the binary result
      JSR   BIN_DEC_CON      ;see Program 6-3
      JSR   DEC_ASCII_CON    ;see Program 6-3
      JSR    TDELAY           ;wait so we can see the output
      BRA    H1               ;keep doing that
```

The following C program reads temperature and converts to ASCII.

```
//Program 13-5C
#include <hidef.h>              /* common defines and macros */
#include <mc9s12dp512.h>        /* derivative information */
#pragma LINK_INFO DERIVATIVE "mc9s12dp512"
void DELAY20uS(void);
```

```
void TDELAY(void);
unsigned char binbyte,x,d1,d2,d3;
void main(void) {
  ATD0CTL2 = 0x80;          //turn on ATD0
  DELAY20uS();          //20 microsec for ATD0 to get ready
  ATD0CTL3 = 0x08;
  ATD0CTL4 = 0xE3;
  for(;;)
  {
    ATD0CTL5 = 0x82;          //start converting Channel 2
    //wait for conversion
    while(!(ATD0STAT0 & ATD0STAT0_SCF_MASK));
    //display result on PORTB
    binbyte = ATD0DR0L;       //binary (hex) byte
    x = binbyte / 10;         //divide by 10
    d1 = binbyte % 10;        //find remainder (LSD)
    d1 = d1 | 0x30;           //convert to ASCII
    d2 = x % 10;              //middle digit
    d2 = d2 | 0x30;           //convert to ASCII
    d3 = x / 10;              //most-significant digit (MSD)
    d3 = d3 | 0x30;           //convert to ASCII
   TDELAY();                  //wait so we can see the output
  }
}
```

Review Questions

1. True or false. The transducer must be connected to signal conditioning circuitry before its output is sent to the ADC.
2. The LM35 provides ______ mV for each degree of ______ (Fahrenheit, Celsius) temperature.
3. The LM34 provides ____ mV for each degree of _____ (Fahrenheit, Celsius) temperature.
4. Why do we set the V_{ref} of the PIC to 2.56 V if the analog input is connected to the LM35?
5. In Question 4, what is the temperature if the ADC output is 0011 1001?

SECTION 13.4: DAC INTERFACING

This section will show how to interface a DAC (digital-to-analog converter) to the HCS12. Then we demonstrate how to generate a sine wave on the scope using the DAC.

Digital-to-analog converter (DAC)

The digital-to-analog converter (DAC) is a device widely used to convert digital pulses to analog signals. See Figure 13-22. In this section we discuss the basics of interfacing a DAC to the HCS12.

Recall from your digital electronics course the two methods of creating a DAC: binary-weighted and R/2R ladder. The vast majority of integrated circuit DACs, including the MC1408 (DAC0808) used in this section, use the R/2R method because it can achieve a much higher degree of precision. The first crite-

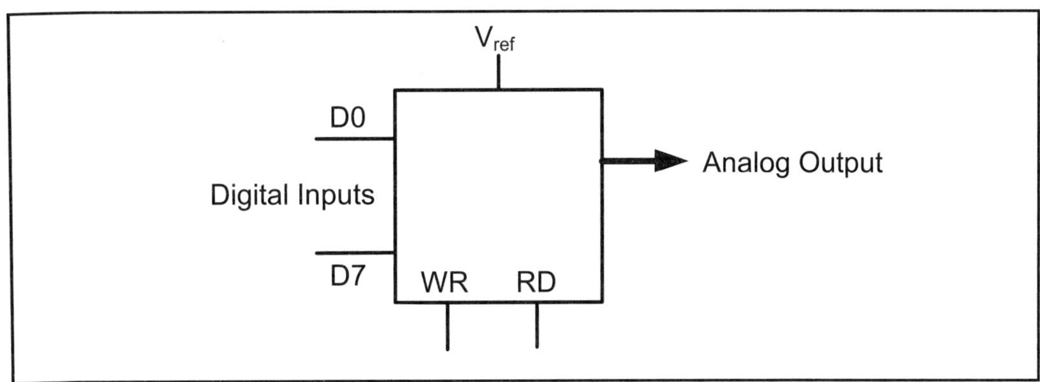

Figure 13-22. DAC Block Diagram

rion for judging a DAC is its resolution, which is a function of the number of binary inputs. The common ones are 8, 10, and 12 bits. The number of data bit inputs decides the resolution of the DAC because the number of analog output levels is equal to 2^n, where n is the number of data bit inputs. Therefore, an 8-input DAC such as the DAC0808 provides 256 discrete voltage (or current) levels of output. See Figure 13-23. Similarly, the 12-bit DAC provides 4,096 discrete voltage levels. There are also 16-bit DACs, but they are more expensive.

MC1408 DAC (or DAC0808)

In the MC1408 (DAC0808), the digital inputs are converted to current (I_{out}), and by connecting a resistor to the I_{out} pin, we can convert the result to a voltage. The total current provided by the I_{out} pin is a function of the binary numbers at the D0–D7 inputs of the DAC0808 and the reference current (I_{ref}), and is as follows:

$$I_{out} = I_{ref}\left(\frac{D7}{2}+\frac{D6}{4}+\frac{D5}{8}+\frac{D4}{16}+\frac{D3}{32}+\frac{D2}{64}+\frac{D1}{128}+\frac{D0}{256}\right)$$

where D0 is the LSB, D7 is the MSB for the inputs, and I_{ref} is the input current that must be applied to pin 14 ($V_{ref}(+)$). The I_{ref} current is generally set to 2.0 mA. Figure 13-23 shows the generation of current reference (setting I_{ref} = 2 mA) by using the standard 5 V power supply. Now assuming that I_{ref} = 2 mA, if all the inputs to the DAC are high, the maximum output current is 1.992 mA (verify this for yourself).

Example 13-10

Assuming that R_f = 5 kΩ and I_{ref} = 2 mA, calculate V_{out} for the following binary inputs:

(a) 10011001 binary (99H) (b) 11001000 (C8H)

Solution:

(a) I_{out} = 2 mA (153/256) = 1.195 mA and V_{out} = 1.195 mA × 5 kΩ = 5.975 V

(b) I_{out} = 2 mA (200/256) = 1.562 mA and V_{out} = 1.562 mA × 5 kΩ = 7.8125 V

SECTION 13.2: ATD PROGRAMMING IN THE HCS12

17. True or false. The HCS12 has an on-chip ATD converter.
18. True or false. The ATD of the HCS12 is an 8-bit ADC.
19. True or false. The ATD0 of the HCS12 has 8 channels of analog input.
20. True or false. The ATD section of HCS12 does not need an external source of power since it uses the VDD of the HCS12.
21. True or false. The ATD conversion speed in the HCS12 depends on the bus frequency.
22. True or false. Upon power-on reset, the ATD section of the HCS12 is turned on and ready to go.
23. True or false. The ATD module of the HCS12 has an external pin for the start-conversion signal.
24. True or false. The ATD module of the HCS12 can convert only one channel at a time.
25. True or false. The ATD section of the HCS12 has its own VDD and VSS pins.
26. True or false. The ATD module of the HCS12 has an internal VDD for V_{rh}.
27. In the ATD of HCS12 what happens to the converted analog data? How do we know that the ATD is ready to provide us the data?
28. In the ATD of HCS12 what happens to the old data if we start conversion again before we pick up the last data?
29. Assume V_{rl} = Gnd and 10-bit resolution. For the ATD of HCS12, find the step size for each of the following V_{rh}:
    (a) V_{rh} = 1.024 V (b) V_{rh} = 2.048 V (c) V_{rh} = 2.56 V
30. In the HCS12, what should be the V_{rh} value if we want a step size of 2 mV and 10-bit resolution? Assume V_{rl} = Gnd.
31. In the HCS12, what should be the V_{rh} value if we want a step size of 3 mV and 10-bit resolution? Assume V_{rl} = Gnd.
32. With a step size of 1 mV, what is the analog input voltage if all outputs are 1? Assume 10-bit resolution.
33. With V_{rh} = 1.024 V, find the V_{in} for the following outputs:
    (a) D9–D0 = 0011111111 (b) D9–D0 = 0010011000 (c) D9–D0 = 0011010000
34. In the ATD of HCS12, what should be the V_{rh} value if we want a step size of 4 mV and 10-bit resolution? Assume V_{rl} = Gnd.
35. With V_{rh} = 2.56 V and V_{rl} = Gnd, find the V_{in} for the following outputs:.
    (a) D9–D0 = 1111111111 (b) D9–D0 = 1000000001 (c) D9–D0 = 1100110000
36. Find the ATD frequency and period for all possible cases of prescaler values if PLL frequency = 12 MHz.
37. In Problem 36, find the conversion time for the following cases of the second phase of conversion:
    (a) 2-clock (b) 4-clock (c) 8-clock (d) 16-clock
    Use a prescaler value of 5 and 8-bit resolution.
38. How do we start conversion in the HCS12?
39. How do we recognize the end of conversion in the HCS12?
40. The HCS12 can have a maximum of _______ channels of analog input.

41. In the HCS12, how many clocks are used for the successive approximation phase of analog conversion?
42. Which register of the HCS12 is used to turn on and off the ATD?
43. Which register of the HCS12 is used to select the A/D's conversion speed?
44. Which register of the HCS12 is used to select the analog channel to be converted?
45. Find the value for the ATDCTL4 register if we want Fbus/8, 16 clocks for the second phase of conversion, and 8-bit resolution.
46. Find the value for the ATDCTL5 register if we want channel 4, unsigned, SCAN = 0, single channel conversion, and right-justified output.
47. Find the value for the ATDCTL4 register if we want Fbus/2, 4 clocks for the second phase of conversion, and 10-bit resolution.
48. Find the value for the ATDCTL5 register if we want channel 5, unsigned, SCAN = 0, single channel conversion, and left-justified output.
49. Give the name of the end-of-conversion flag for the ATD0 of the HCS12. State which register it belongs to.
50. Upon power-on reset, the ATD of the HCS12 is (on, off).

SECTION 13.3: SENSOR INTERFACING AND SIGNAL CONDITIONING

51. What does it mean when a given sensor is said to have a linear output?
52. The LM34 sensor produces _______ mV for each degree of temperature.
53. What is signal conditioning?
54. What is the purpose of the LM336 Zener diode around the pot setting the V_{ref} in Figure 13-21?

SECTION 13.4: DAC INTERFACING

55. True or false. DAC0808 is the same as DAC1408.
56. Find the number of discrete voltages provided by the n-bit DAC for the following:
    (a) $n = 8$ (b) $n = 10$ (c) $n = 12$
57. For DAC1408, if $I_{ref} = 2$ mA, show how to get an I_{out} of 1.99 when all inputs are HIGH.
58. Find the I_{out} for the following inputs. Assume $I_{ref} = 2$ mA for DAC0808.
    (a) 10011001 (b) 11001100 (c) 11101110
    (d) 00100010 (e) 00001001 (f) 10001000
59. To get a smaller step, we need a DAC with ______ (more, fewer) digital inputs.
60. To get full-scale output, what should be the inputs for DAC?

ANSWERS TO REVIEW QUESTIONS

SECTION 13.1: ADC CHARACTERISTICS

1. Number of steps and V_{ref} voltage
2. 8
3. True

4. (a) 8 (b) 8
5. 1.28 V/256 = 5 mV
6. (a) 0.7 V/ 5 mV = 140 in decimal and D7–D0 = 10001100 in binary.
   (a) 1 V/ 5 mV = 200 in decimal and D7–D0 = 11001000 in binary.

SECTION 13.2: ATD PROGRAMMING IN THE HCS12

1. V_{rh}, V_{rl}, and the resolution
2. 10
3. True
4. True
5. 1 mV
6. (a) 700 mV (1010111100), (b) 1,000 mV (1111101000)
7 True
8. True
9. 500 kHz
10. SCF bit of the ATDSTAT0 register

SECTION 13.3: SENSOR INTERFACING AND SIGNAL CONDITIONING

1. True
2. 10, Celsius
3. 10, Fahrenheit
4. Using the 8-bit part of the 10-bit ADC, it gives us 256 steps, and 2.56 V/256 = 10 mV. The LM35 produces 10 mV for each degree of temperature, which matches the ADC's step size.
5. 00111001 = 57, which indicates it is 57 degrees.

SECTION 13.4: DAC INTERFACING

1. Digital, analog
2. Analog, digital
3. 8
4. (a) current (b) true

CHAPTER 14

ACCESSING FLASH AND EEPROM, AND PAGE SWITCHING

OBJECTIVES

Upon completion of this chapter, you will be able to:

- >> Explain how to access the entire Flash memory of the HCS12 using page switching
- >> Code HCS12 Assembly and C programs for writing data into HCS12 Flash memory space
- >> Code HCS12 Assembly and C programs for erasing the Flash memory in HCS12
- >> Explain how to write data to the EEPROM memory of the HCS12
- >> Explain how to read data from the EEPROM memory of the HCS12

In this chapter we discuss how to access data stored in both Flash and EEPROM memories of the HCS12. We will discuss how to access Flash memory space beyond 64 KB using page switching. The process of writing to EEPROM and Flash memories in the HCS12 is also explored.

SECTION 14.1: PAGE SWITCHING OF FLASH MEMORY IN HCS12

In Chapter 0 we discussed various types of semiconductor memories and their characteristics such as capacity, organization, and access time. In this section we discuss how to access the HCS12 Flash memory space beyond the 64 KB linear address space.

Flash memory paging in HCS12

The HCS12 physical memory space is limited to 64K bytes. The 64 KB space is divided into four 16K byte spaces. Generally, the lower 16K bytes from \$0000 to \$3FFF are used by I/O ports, special function registers of peripherals, RAM, and EEPROM. The 16 KB address space of \$4000–\$7FFF is used by Flash. The upper 32K bytes from \$8000 to \$FFFF are used exclusively by the Flash memory. The upper 32 KB is divided into two equal sections. The upper 16K bytes from \$C000 to \$FFFF is used for Flash memory, and some portion of this space is used by the CPU to store many important functions such as the BDM program, reset subroutine, and interrupt vector table information. The 16 KB address space of \$8000–BFFF is also used for Flash memory. We also use this space as a window into a memory space larger than 64 KB. This is called *paging* in the HCS12 literature. The page size in the HCS12 is defined as 16K bytes. See Figure 14-1. In order to use the paging concept to access memory beyond the 64 KB space, we must use the PPAGE (program page) register. Therefore, the term unpaged refers tothe memory space that is directly accessible by the CPU without using the PPAGE register. Although PPAGE is an 8-bit register, only 6 bits of it are used by the HCS12 for the purpose of paging. See Figure 14-2. The 6 bits of the PPAGE register give us a total of $2^6 = 64$ pages. Since each page is 16K bytes, we can have a maximum of 1M byte (64×16 KB = 1 MB) of paged memory. Not all chips have that much on-chip memory. See Table 14-1. We can use window paging to access the Flash memory space beyond 64 KB only via the memory address space of \$8000–\$BFFF. That means the rest of the 64 KB space (0000–\$7FFF and \$C000–\$FFFF) is not paged. For this reason, the HCS12 manual recommends that we keep the stack and I/O register addresses in the unpaged memory space of 0000–\$3FFF to make them accessible from any page. It is also recommended that all reset and interrupt vectors point to locations in the unpaged memory space of \$C000–FFFF. The memory spaces of \$4000–7FFF and \$C000–FFFF are called *fixed Flash memory* since we do not use the PPAGE register to access them. The address space of \$8000–BFFF is called *windowed page,* since it is used as a window to memory larger than 64 KB. In assigning the page numbers, the HCS12 assigned the last 4 page numbers, \$3F, \$3E, \$3D, and \$3C, to 64 KB space memory even though we never use the 0000–\$7FFF and \$C000–FFFF Flash memory as windowed pages. See Figure 14-3.

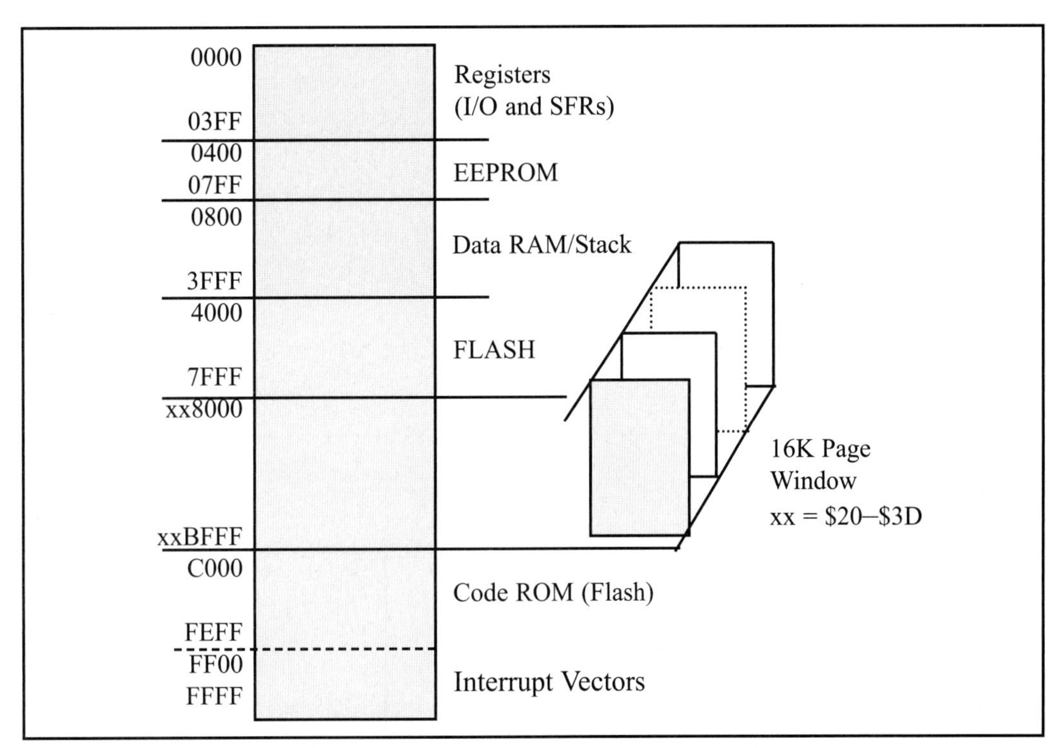

Figure 14-1. MC9S12Dx512 Memory Allocation (Using Page Window)

0	0	PIX5	PIX4	PIX3	PIX2	PIX1	PIX0

PIX5	PIX4	PIX3	PIX2	PIX1	PIX0	Hex	Page Selected
0	0	0	0	0	0	00	16 KB Page 0
0	0	0	0	0	1	01	16 KB Page 1
0	0	0	0	1	0	02	16 KB Page 2
0	0	0	0	1	1	03	16 KB Page 3
0	0	0	1	0	0	04	16 KB Page 4
.	.	.	.	.	.		
.	.	.	.	.	.		
1	1	1	0	1	0	3A	16 KB Page 58
1	1	1	0	1	1	3B	16 KB Page 59
1	1	1	1	0	0	3C	16 KB Page 60
1	1	1	1	0	1	3D	16 KB Page 61
1	1	1	1	1	0	3E	16 KB Page 62
1	1	1	1	1	1	3F	16 KB Page 63

Figure 14-2. PPAGE (Page Program) Register

Table 14-1: On-chip Memories for HCS12 Chips

Part No.	Flash	EEPROM	RAM
MC9S12Dx32	32 KB	1 KB	2 KB
MC9S12Dx64	64 KB	1 KB	4 KB
MC9S12Dx128	128 KB	2 KB	8 KB
MC9S12Dx256	256 KB	4 KB	12 KB
MC9S12Dx512	512 KB	4 KB	14 KB

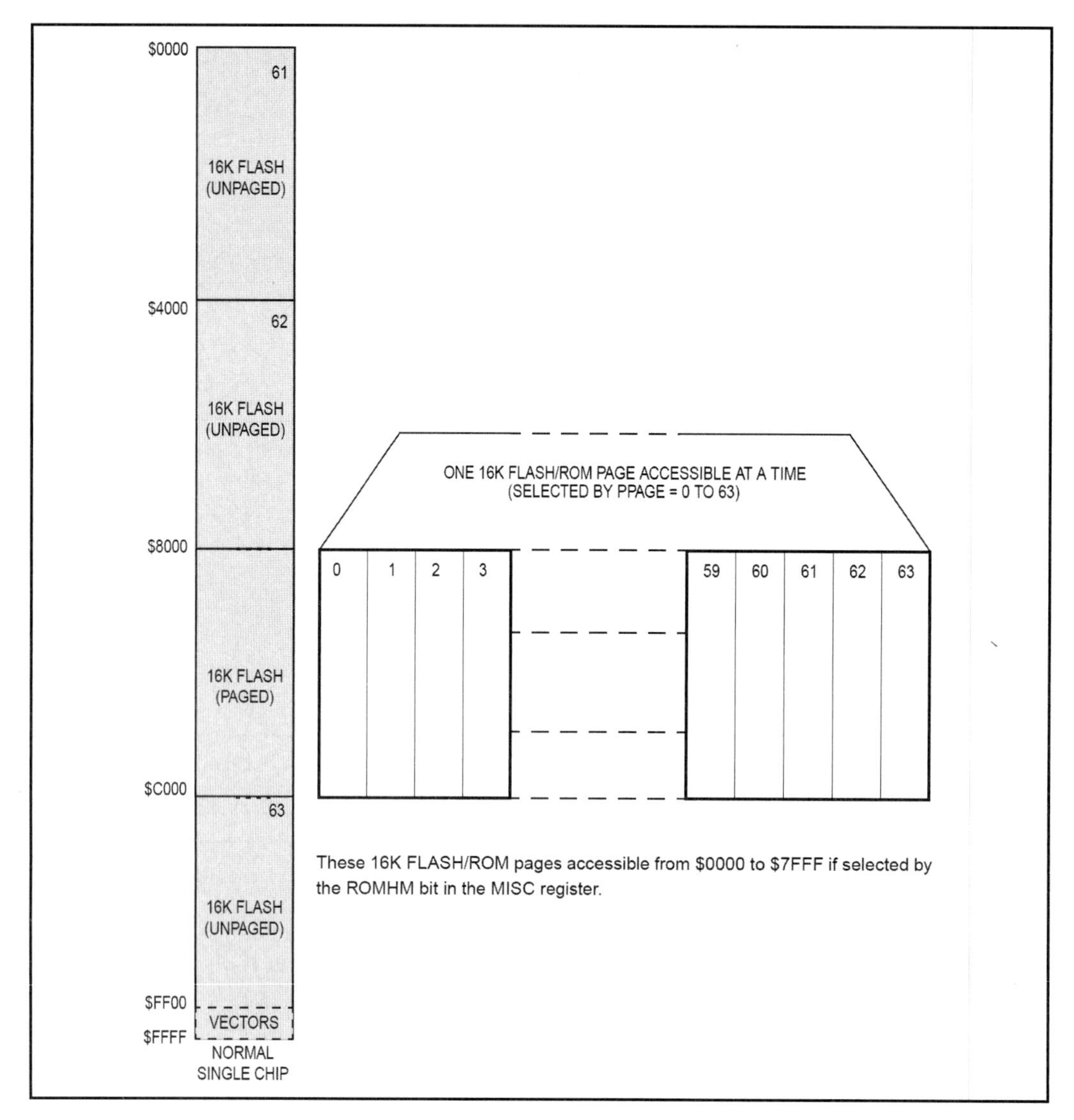

Figure 14-3. Paging in HCS12

CALL instruction and paging

In Chapter 3 we mentioned that the JSR is used for calling the subroutines located in the unpaged memory space of the 64 KB physical space. To access the flash ROM space beyond the 64 KB, we must use page switching with the help of the PPAGE register. The CALL instruction is a 4-byte instruction in contrast to JSR, which is 3 bytes. See Table 14-2. Because CALL is a 4-byte instruction, the target address of the subroutine can be within 1 MB of expanded memory since 1 byte of the instruction is used for the PPAGE value. There is also a difference

Table 14-2: Call vs. JSR in HCS12

Instruction	Action	Instruction size
JSR	branch (call) to an address within 64 KB	3-byte
RTS	return from subroutine (used with JSR)	1-byte
CALL	call an address beyond 64 KB	4-byte
RTC	return from subroutine (used with CALL)	2-byte

Table 14-3: CodeWarrior Flash Memory Paging for MC9S12Dx512

Page Name		Address Location
EEPROMStart:	equ	$00000400
EEPROMEnd:	equ	$000007FF
RAMStart:	equ	$00000800
RAMEnd:	equ	$00003FFF
ROM_4000Start:	equ	$00004000
ROM_4000End:	equ	$00007FFF
ROM_C000Start:	equ	$0000C000
ROM_C000End:	equ	$0000FEFF
PAGE_20Start:	equ	$00208000
PAGE_20End:	equ	$0020BFFF
PAGE_21Start:	equ	$00218000
PAGE_21End:	equ	$0021BFFF
PAGE_22Start:	equ	$00228000
PAGE_22End:	equ	$0022BFFF
PAGE_23Start:	equ	$00238000
PAGE_23End:	equ	$0023BFFF
....	...	....
PAGE_3CStart:	equ	$003C8000
PAGE_3CEnd:	equ	$003CBFFF
PAGE_3DStart:	equ	$003D8000
PAGE_3DEnd:	equ	$003DBFFF

between JSR and CALL in terms of saving the program counter on the stack or the function of the return instruction. The difference is that JSR uses 2 bytes of stack, while CALL uses 3 bytes of stack since it must also save the PPAGE register. We also must use the RTC (return from call) instruction instead of RTS (return from subroutine). RTC pulls the PPAGE value from stack in addition to the program counter. See Program 14-1 and its listing file. Program 14-1 is the same as Program 6-1 (Chapter 6) except subroutines and data are placed in seperate pages. Table 14-3 shows the memory map of 9S12DP512 as shown by CodeWarrior.

```
;Program 14-1: Using CALL and PPAGE register
           ORG    $C000
Entry:
RAM_ADDR   EQU $840   ;RAM space to place the bytes
COUNTREG   EQU $820   ;RAM loc for counter
CNTVAL     EQU 4      ;counter value = 4 for 4 bytes
CNTVAL1    EQU 5      ;5 bytes including checksum byte
;------------main program
  LDS  #RAMEnd+1      ;initialize the stack pointer
  LDAA #$20
  STAA PPAGE   ;place following subroutine in Page $20
  CALL COPY_DATA
  LDAA #$21
  STAA PPAGE   ;place following subroutine in Page $21
  CALL CAL_CHKSUM
  LDAA #$22
  STAA PPAGE   ;place following subroutine in Page $22
  CALL TEST_CHKSUM
  BRA  $
```

```
;copying data from code ROM to data RAM locations
  ORG PAGE_20Start
COPY_DATA
  LDX  #$8500 ;load pointer. X = $8500, ROM address
  LDY  #$860  ;load pointer. Y = $860, RAM address
  LDAA #CNTVAL
  STAA COUNTREG
B5 LDAA 1,X+ ;get the byte from Flash, incr. pointer
  STAA 1,Y+  ;save it in RAM and increment pointer
  DEC  COUNTREG
  BNE  B5     ;keep looping
  RTC         ;return
;calculating checksum byte
     ORG PAGE_21Start
CAL_CHKSUM
  LDAA #CNTVAL
  STAA COUNTREG;load the counter
  LDX  #$860  ;load pointer. X = $860, RAM address
  CLRA        ;A=0
B2 ADDA 1,X+  ;add and increment pointer
  DEC  COUNTREG;decrement the counter
  BNE  B2     ;loop until counter = zero
  NEGA        ;2's comp of sum
  STAA 0,X    ;save the checksum as the last byte
  RTC         ;return
;testing checksum byte
     ORG PAGE_22Start
TEST_CHKSUM
  BSET DDRB,%11111111
  LDAA #CNTVAL+1;adding 5 bytes and checksum byte
  STAA COUNTREG;load the counter
  LDX  #$860 ;load pointer. X = $860, RAM address
  CLRA        ;A=0
B3 ADDA 1,X+  ;add and increment pointer
  DEC  COUNTREG;decrement the counter
  BNE  B3     ;loop until counter = zero
  CMPA #0     ;see if A = zero
  BEQ  G_1    ;is result zero? then good
  LDAA #'B'   ;then send letter B for Bad to PORTB
  STAA PORTB  ;if not, data is bad
  RTC
G_1 LDAA #'G' ;then send letter G for Good to PORTB
  STAA PORTB  ;data is not corrupted
  RTC
;--my data in FLASH ROM
     ORG $8500
MYBYTE FCB $25, $62, $3F, $52, $00
```

The list file below shows the memory usage for various subroutines of Program 14-1.

```
;List file for Program 14-1 as shown by CodeWarrior
; (modified for clarity)
12822    27                                  ;------------main program
12823    28   a00C000 CF40 00                  LDS    #RAMEnd+1
12824    29   a00C003 8620                     LDAA   #$20
12825    30   a00C005 5A30                     STAA   PPAGE
12826    31   a00C007 4A80 0020                CALL   COPY_DATA
12827    32   a00C00B 8621                     LDAA   #$21
12828    33   a00C00D 5A30                     STAA   PPAGE
12829    34   a00C00F 4A80 0021                CALL   CAL_CHKSUM
12830    35   a00C013 8622                     LDAA   #$22
12831    36   a00C015 5A30                     STAA   PPAGE
12832    37   a00C017 4A80 0022                CALL   TEST_CHKSUM
12833    38   a00C01B 20FE                     BRA    $
12834    39                                  ;------copying data from ...
12835    40                                           ORG PAGE_20Start
12836    41                                  COPY_DATA
12837    42   a208000 CE85 00                  LDX     #$8500
12838    43   a208003 CD08 60                  LDY     #$860
12839    44   a208006 8604                     LDAA    #CNTVAL
...      ..   ...     ...           ...                ...
12845    50   a208014 0A            RTC
12846    51                                  ;----calculating checksum byte
12847    52                                           ORG PAGE_21Start
12848    53                                  CAL_CHKSUM
12849    54   a218000 8604                     LDAA    #CNTVAL
12850    55   a218002 7A08 20                  STAA    COUNTREG
12851    56   a218005 CE08 60                  LDX     #$860
...      ..   ...     ...           ...                ...
12858    63   a218013 0A                       RTC
12859    64                                  ;------testing checksum byte
12860    65                                           ORG PAGE_22Start
12861    66                                  TEST_CHKSUM
12862    67   a228000 4C03 FF                  BSET    DDRB,%11111111
12863    68   a228003 8605                     LDAA    #CNTVAL+1
12864    69   a228005 7A08 20                  STAA    COUNTREG
...      ..   ...     ...           ...                ...
12877    82   a228020 0A                       RTC
12878    83                                  ;-------my data in FLASH ROM
12879    84                                    ORG $8500
12880    85   a008500 2562 3F52 MYBYTE FCB $25, $62, $3F, $52, $00
                008504 00
```

Remapping of I/O registers, RAM, and EEPROM addresses

In HCS12 chips we can remap the addresses assigned to the I/O registers, RAM, and EEPROM to different address locations as long as they are on the right boundaries. For example, the I/O registers in the 1 KB space of 000–$FFF can be assigned to any 2 KB space boundary. To do that, we use the INITRG register. See Figure 14-4. We can do the same thing for on-chip RAM and EEPROM. Figures 14-5 and 14-6 show the registers used for this purpose. This remapping scheme allows us to solve the problem of overlapping memory in some HCS12 chips. For example, in the MC9S12Dx512 chip we have 1 KB of I/O registers, 14 KB of RAM, and 4 KB of EEPROM. That adds up to 19 KB of memory space, which means some of them will overlap since we need to fit all of them into the lower 16 KB memory space. The good news is that in the HCS12 trainers, these addresses are already mapped and there is no need to remap them. But, if you are designing the HCS12-based system from scratch, you need to consult the HCS12 manual.

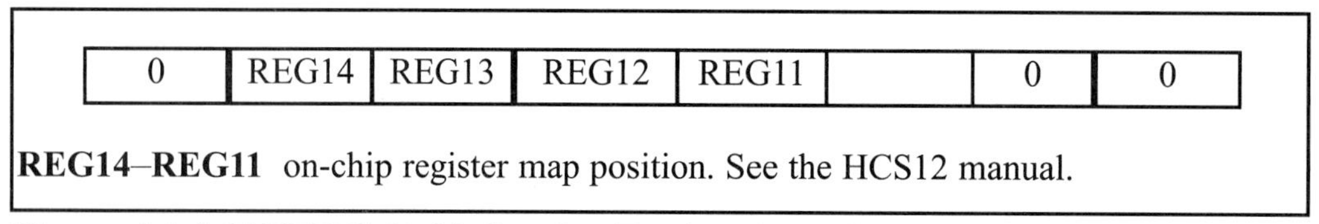

Figure 14-4. INITRG (Initialization Register) Register

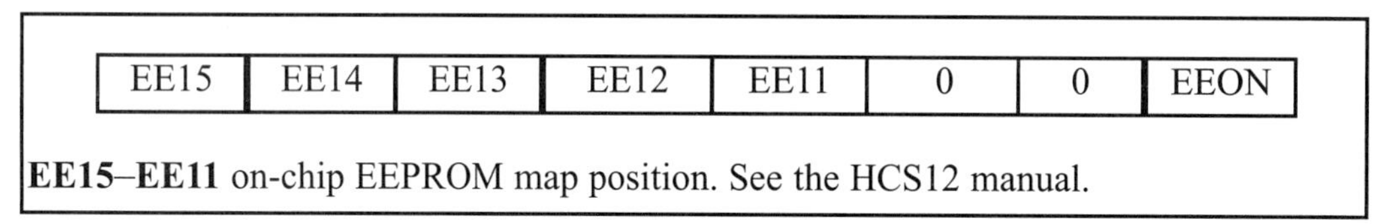

RAM15
RAM14
RAM13
RAM12
RAM11
0
0
RAMHAL
RAM15–RAM11 on-chip RAM map position. See the HCS12 manual.

Figure 14-5. INITRM (Initialization RAM) Register

EE15	EE14	EE13	EE12	EE11	0	0	EEON

EE15–EE11 on-chip EEPROM map position. See the HCS12 manual.

Figure 14-6. INITEE (Initialization EEPROM) Register

Review Questions

1. What is the size of Program Counter in the HCS12?
2. In the HCS12, how much memory can we access without using paging?
3. With the help of paging, we can access up to ______ of memory.
4. True or false. To access the paged memory we must use the PPAGE register.
5. True or false. The CALL instruction uses the PPAGE register.

SECTION 14.2: ERASING AND WRITING TO FLASH

The HCS12 comes with three types of memory (a) SRAM, (b) Flash, and (c) EEPROM. The SRAM is for general-purpose usage including function registers of peripherals, as we have seen throughout the book. The EEPROM is used mainly for storing data. While the Flash memory is used primarily to store program (code), we can also use it for storing fixed data such as look-up tables, as we have seen throughout the book. In Flash memory, the write or erase process is done on a block of data, while in EEPROM we can write or erase one byte at a time, which means it is byte-accessible memory. Read access for both Flash and EEPROM memories are in byte sizes. The block size for Flash memory varies among the Flash memories depending on their size and intended application. In Chapter 6 we discussed how to read the fixed data stored in program Flash. In this section, we discuss how to write to Flash memory. In the next section, we discuss how to access the EEPROM memory in the HCS12.

There are two ways to store (write) information (code or data) to the Flash memory or erase its content: (a) using an external Flash programmer (burner), and (b) using HCS12 instructions. In this section, we show how to use HCS12 instructions to write to Flash memory. We will also show how to erase the contents of Flash memory. Due to similarities between the read and write instructions, it is very helpful to understand the material in Section 6.2 of Chapter 6, where we showed how to read data stored in the Flash ROM.

Flash memory registers

The timing of programming (writing to) the Flash or erasing its contents is a function of the bus clock and crystal oscillator. Regardless of the oscillator clock speed or the bus clock speed, this timing must be in the range of 150 kHz–200 kHz. In the HCS12, we use the FCLKDIV register to make sure that write (program)/erase timing is within the 150 kHz–200 kHz range. See Figure 14-7. In many Trainers, this timing is already set by the monitor program. Section 14.4 in this chapter shows how to do it if you are designing an HCS12 from the ground up. The next important register associated with the Flash memory is the FCMD register, shown in Figure 14-8.

FDIVLD	PRDIV8	FDIV5	FDIV4	FDIV3	FDIV2	FDIV1	FDIV0

FDIVLD Flash clock divider loaded
1 = Register has been written to since the last reset.
(Now, Flash can be programmed (written to) and erased.)
0 = Register has not been written to.

PRDIV8 Enable the divide-by-8 prescaler
1 = Enables the divide-by-8 prescaler. The bus clock is divided by 8 before it is divided again by the FDIVx bits.
0 = The bus clock is divided by the FDIVx bits. See below.

FDVI5–FDIV0 Flash clock divider bits

Notes:
1) We cannot program (write) / erase the Flash until we write to this register. The FDIVLD lets us know if this register has been written to.
2) For the purpose of on-chip Flash timing, the bus clock cannot be less than 1 MHz and the Flash clock must be in the range of 150 kHz–200 kHz. Otherwise, the program (write) / erase of Flash will not work and it can even damage the Flash.
3) A combination of PRDIV8 and FDIVx bits are used to set the timing of Flash/EEPROM to the range of 150 kHz–200 kHz.

Figure 14-7. FCLKDIV (Flash Clock Divide) Register

Using FCMD we can perform the following actions on the Flash:

a) Erase Verify. This ensures that there is no data in a given Flash location before we write new data.

0	CMDB6	CMDB5	0	0	CMDB2	0	CMDB0

CMDB CMDB bits are used to perform the following actions:

- $05 Erase Verify
- $20 Word Program (Writes 2 bytes to an even address location.)
- $40 Sector Erase (Sector size varies from chip to chip.)
- $41 Mass Erase (Block Erase. Block size varies from chip to chip.)

Note:
In HCS12Dx256 the size of a block is 64 KB and the sector size is 512 B.
In HCS12Dx512 the size of a block is 128 KB and the sector size is 1,024 B.

Figure 14-8. FCMD (Flash Command) Register

b) To write (program) 2 bytes of data to a given address location. The address location must be an even address. This is the option used to program (write) new data to Flash after it has been erased. Notice, we can write only 2 bytes at a time.

c) Erase a sector of Flash. The sector size varies among family members, is shown in Table 14-4. For example, in the MC9S12Dx512 the sector is 1,024 bytes.

d) Erase a block of Flash. In the HCS12, several pages grouped together are referred to as a *block*. For example, in the MC9S12Dx512, a block is equal to 8 pages of 16KB, which gives us a block size of 128 KB. See Figure 14-9. That means, in the MC9S1Dx512, there are four blocks. They are numbered as blocks 0, 1, 2, and 3. To erase a block of Flash memory, we need to know the block number. The FCNFG (Flash Configuration) register is used to identify the block number we want to erase. For details of the FNCFG register see the MC9S12Dx512 manual. Block erasing is ideal for using Flash in applications such as mass storage devices like memory stick drives or MP3 players. The only problem with the HCS12 Flash is the data writing to Flash, which is limited to 2 bytes at a time.

Table 14-4: Flash Size for Some HCS12 Chips

Part No.	On-chip Flash	Block size	Sector size	Write Size
MC9S12Dx128	128 KB	32 KB	256 B	Word (2 bytes)
MC9S12Dx256	256 KB	64 KB	512 B	Word (2 bytes)
MC9S12Dx512	512 KB	128 KB	1,024 B	Word (2 bytes)

Note: For erase, we use block size or sector size. The write size is for programming (writing to) the on-chip Flash memory of the HCS12.

The next important register we use for programming the flash is the FSTAT register, shown in Figure 14-10. The CBEIFbit is the most important bit of this register. After writing to Flash using the write option of the FCMD registers, we must check the CBEIFbit to make sure that the Flash is ready for the next write. The HCS12 uses the other bits of the FSTAT register to let us know if there is a problem in programming the Flash. It must be noted that the HCS12 Flash memory can be secured and protected against accidental programming or erasing. See the HCS12 manual.

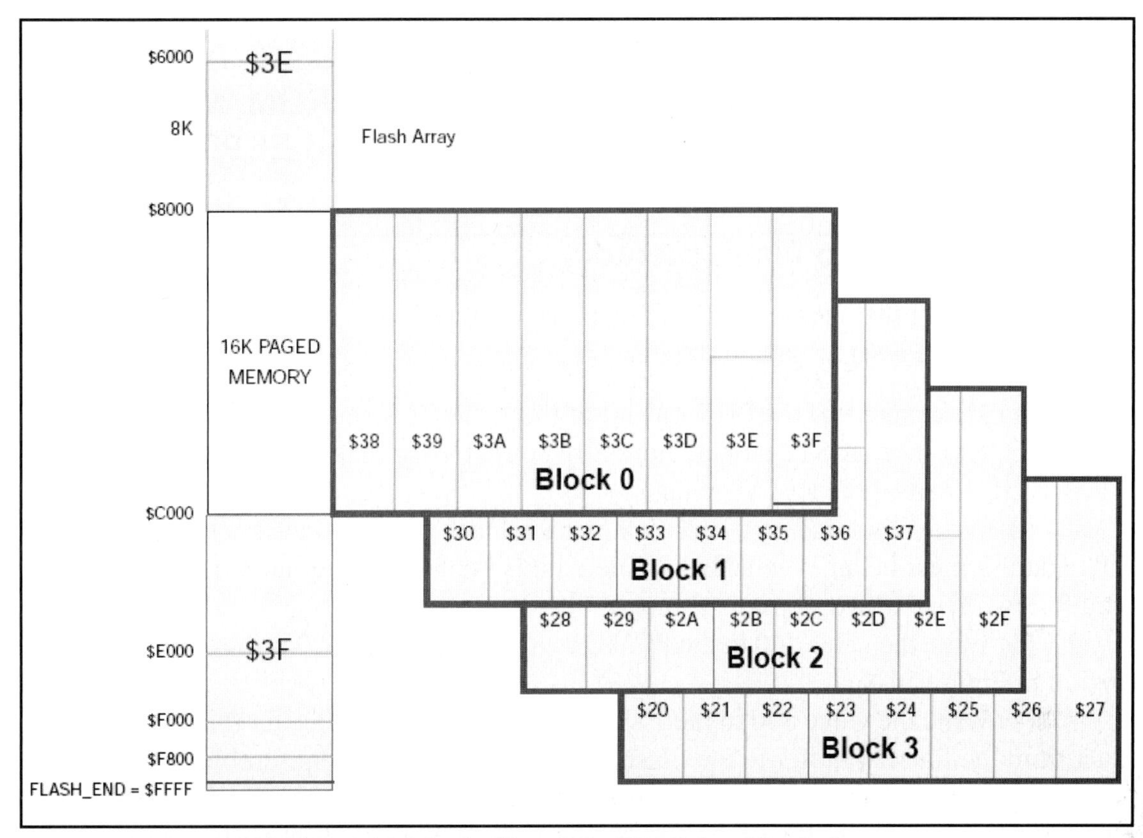

Figure 14-9. MC9S12Dx512 Block Size (Modified for Clarity)

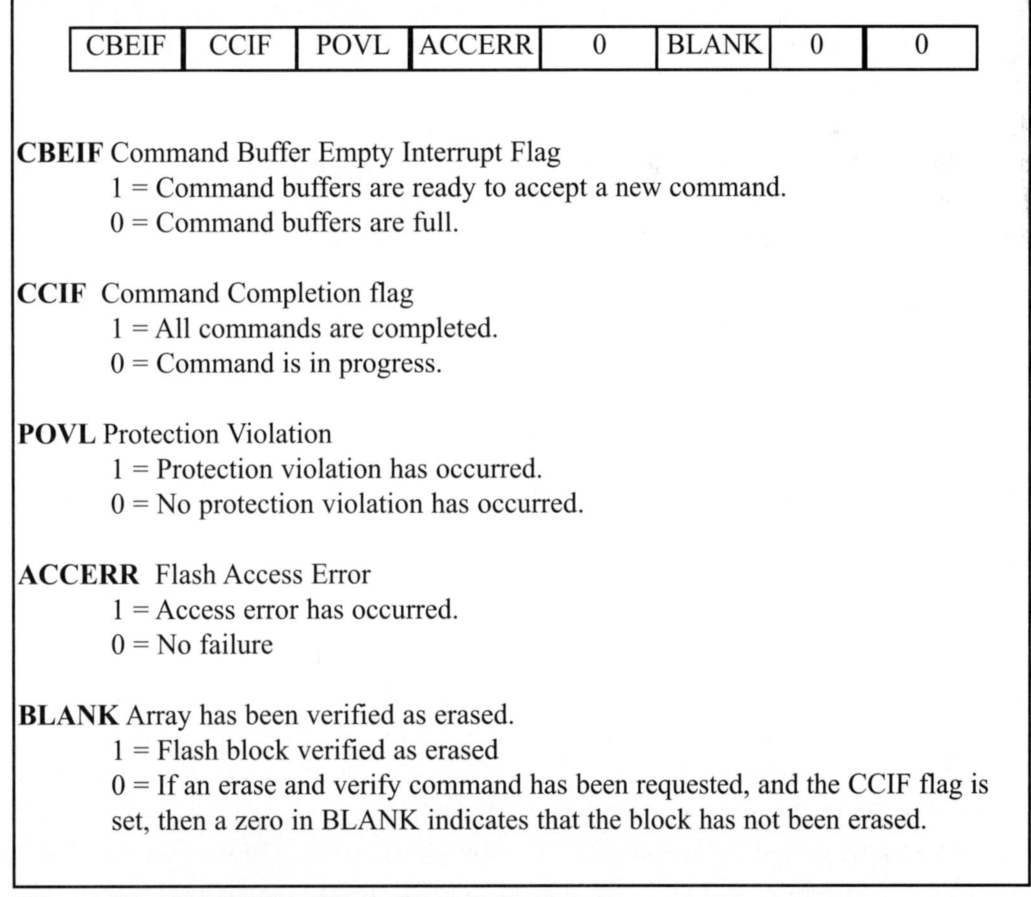

CBEIF	CCIF	POVL	ACCERR	0	BLANK	0	0

CBEIF Command Buffer Empty Interrupt Flag
1 = Command buffers are ready to accept a new command.
0 = Command buffers are full.

CCIF Command Completion flag
1 = All commands are completed.
0 = Command is in progress.

POVL Protection Violation
1 = Protection violation has occurred.
0 = No protection violation has occurred.

ACCERR Flash Access Error
1 = Access error has occurred.
0 = No failure

BLANK Array has been verified as erased.
1 = Flash block verified as erased
0 = If an erase and verify command has been requested, and the CCIF flag is set, then a zero in BLANK indicates that the block has not been erased.

Figure 14-10. FSTAT (Flash Status) Register

Caution: Check the documentation for your HCS12 Trainer before erasing or writing to the Flash or EEPROM. Many HSC12 Trainers use a portion of the Flash and EEPROM for the bootloader.

Steps in writing to Flash memory

Assuming that an area of Flash memory is erased, we can use the following steps to write a set of data to the Flash memory:

(1) Make sure that the FDIVLD bit of the FDIVCLK register is set to high (FDIVLD = 1). The HCS12 trainers normally do that. Also make sure CBEIF= 1. The CBEIFis part of the FSTAT register.

(2) Write a word (2 bytes) of data and its Flash memory address location. The address must be an even address. The data is kept in an internal buffer until you follow the next two steps.

(3) Write the value $20 to the FCMD register. The value $20 is the program option in Figure 14-8.

(4) Write the value $80 to the FSTAT register. This will transfer (write) the data from the internal buffer to the Flash unless there is a problem. Problems can occur a) if we try to write to a secure area of Flash (indicated by the PVIOL flag) or b) Flash is not accessible (indicated by the ACCERR flag). That is the reason we need to check these two flags in the next steps.

(5) Check the PVIOL and ACCERR flag bits to ensure that there was no error.

(6) Check the CBEIF flag bit to ensure the process is complete and Flash is ready for the next write. Go back to step 2 to write the next word. See Figure 14-11.

Notice from step 2 that the internal buffer cannot be accessed by the programmer. It is internal and used exclusively for the purpose of writing/erasing the Flash memory. Program 14-2 shows how to write 8 bytes of data to Flash locations starting at address $4400. After writing the bytes, we read and display them on PORTB one byte at a time to verify the write operation. The C language version of Program 14-2 is given at the end of this section.

Program 14-2 (a) writes the message "GOOD BYE" to Flash memory starting at location $4400, and (b) reads the data from Flash and places it in PORTB one byte at a time.

```
;Program 14-2
R1    DS.B 1
      ORG    ROMStart
Entry:
   LDS #RAMEnd+1       ;initialize the stack pointer
   BSET DDRB,#$FF
   LDAA #'G'
   LDAB #'O'
   STD  $4400
```

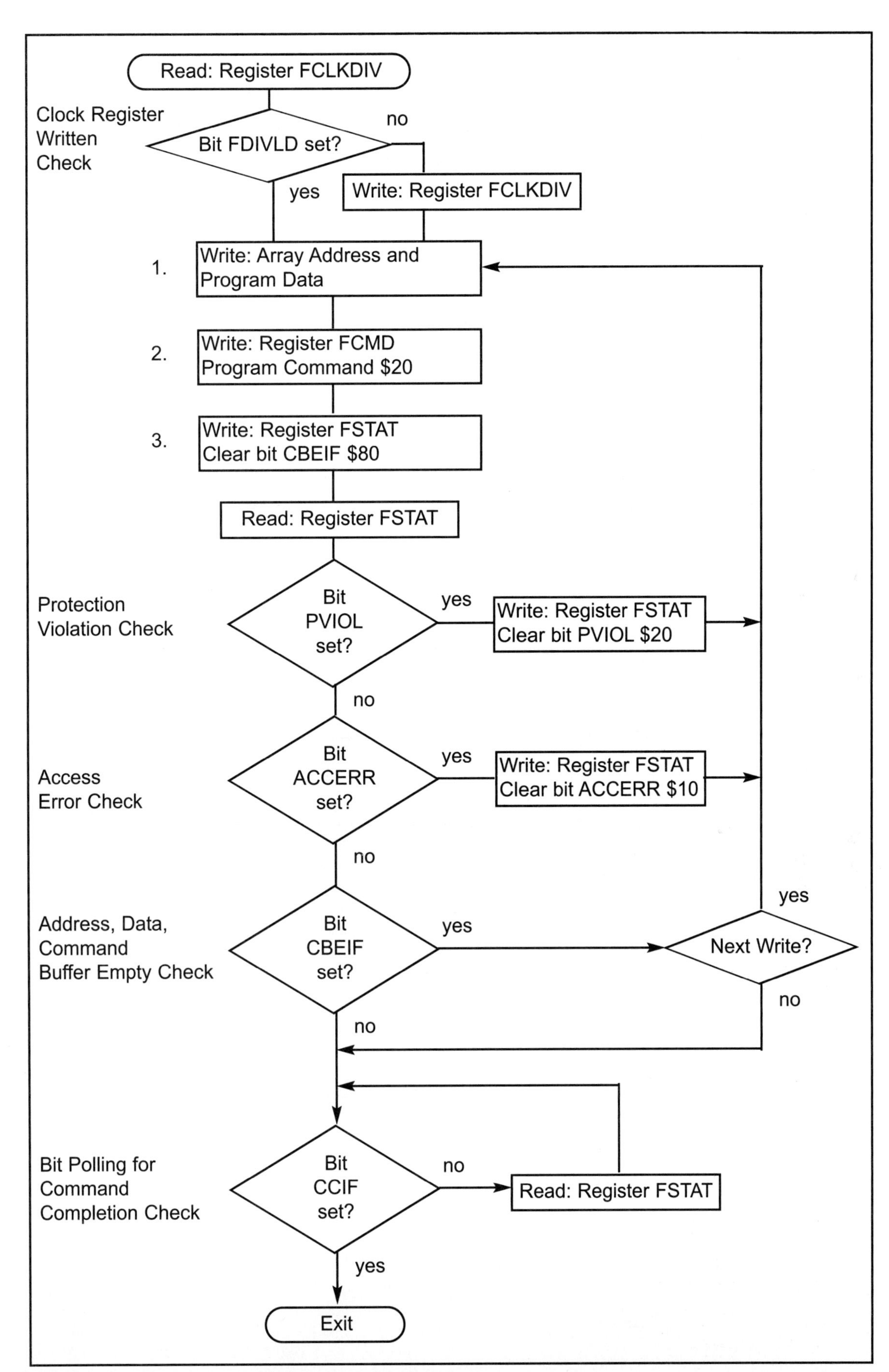

Figure 14-11. Steps in Writing to Flash Memory

```
   JSR   F_WRT
   LDAA #'O'
   LDAB #'D'
   STD  $4402
   JSR   F_WRT
   LDAA #' '
   LDAB #'B'
   STD  $4404
   JSR   F_WRT
   LDAA #'Y'
   LDAB #'E'
   STD  $4406
   JSR   F_WRT
;read them back one byte at a time and examine the
;bytes on PORTB
   LDX   #$4400
   MOVB #8,R1        ;load counter
H1 LDAA 1,X+
   STAA PORTB
   JSR   DELAY
   DEC   R1
   BNE   H1
HERE BRA  HERE

;-----write to Flash. A=0 if no error, A=FF if error
F_WRT
   LDAA #$20
   STAA FCMD         ;write (program) to Flash
   LDAA #$80
   STAA FSTAT   ;make sure it is OK
   LDAA #$FF                  ;A=FF if error
   BRSET FSTAT,mFSTAT_PVIOL,OVER
   BRSET FSTAT,mFSTAT_ACCERR,OVER
   CLRA                       ;A=00 if no error
OVER RTS
```

If the number of bytes is not an even number, then the unused byte will be untouched. We can use a loop to write an array of data to flash. See Program 14-3.

Program 14-3 (a) writes the message "HELLO" to Flash memory starting at location $4450, and (b) reads the data from Flash and places it in PORTB one byte at a time.

```
;Program 14-3
R1   DS.B   1
R2   DS.B  1
MYDATA FCB      "HELLO!"
; code section
     ORG    ROMStart
```

```
Entry:
  LDS   #RAMEnd+1           ;initialize the stack pointer
  BSET DDRB,#$FF            ;make PORTB output
  LDX   #MYDATA             ;set the pointer X
  LDY   #$4500              ;set the pointer Y
  MOVB #3,R1                ;set the counter
OVER MOVW 0,X,0,Y    ;move a word from source to dest.
  JSR   F_WRT
  INX                       ;point to next word
  INX
  INY
  INY
  DEC   R1                  ;decrement counter
  BNE   OVER                ;repeat
;read them back one byte at a time and examine the
;bytes on PORTB
  LDX   #$4500
  MOVB #6,R2                ;load counter
H1 LDAA 1,X+
  STAA PORTB
  JSR   DELAY
  DEC   R2
  BNE   H1
HERE BRA HERE

;-----write to Flash. A=0 if no error, A=FF if error
F_WRT
  LDAA #$20                 ;code for write
  STAA FCMD
  LDAA #$80
  STAA FSTAT
  LDAA #$FF                 ;A=FF if error
  BRSET FSTAT,mFSTAT_PVIOL,OVER2
  BRSET FSTAT,mFSTAT_ACCERR,OVER2
  CLRA                      ;A=00 if no error
OVER2     RTS
```

Program 14-4 (a) transfers a block of data from the code space (FLASH) of the HCS12 chip into RAM, (b) then reads the data from new RAM locations and sends it to the serial port of the HCS12 one byte at a time.

```
;Program 14-4
R1    DS.B 1
R2    DS.B 1
MYDATA DS.B 6
; code section
     ORG    ROMStart
```

```
Entry:
   LDS   #RAMEnd+1 ;initialize the stack pointer
   BSET  DDRB,#$FF ;make PORTB output
   LDX   #MYDATA ;set the pointer X
   LDY   #DATA   ;set the pointer Y
   MOVB  #3,R1   ;set the counter
OVER MOVW 0,Y,0,X ;move a word from src. to dest.
   INX           ;point to next word
   INX
   INY
   INY
   DEC   R1      ;decrement counter
   BNE   OVER    ;repeat
;read them back one byte at a time and place the
;bytes on PORTB and send them to the COM port
   LDX   #MYDATA
   MOVB  #6,R2         ;load counter
H1 LDAA 1,X+
   STAA PORTB
   JSR   SERIAL_COM    ;see Chapter 10
   DEC   R2
   BNE   H1
HERE BRA HERE

     ORG $4500
DATA FCB   "HELLO!"
```

Steps in erasing Flash memory

Although we can use external Flash programmers to erase the Flash memory contents, the HCS12 allows us to write a program to erase the Flash memory. The erasure process works on both block-size and sector-size data. The block size varies for different family members of HCS12, as was shown in Table 14-4. The sector size also varies among HCS12 chips. For example, the sector size for the MC9S12Dx512 is 1,024 bytes. That means the lowest 10 bits of addresses are all zeros, making them 1,024-byte sector boundaries. We can use the following steps to erase a single sector of Flash memory:

(1) Make sure that the FDIVLD bit of the FDIVCLK register is set to high (FDIVLD = 1). The HCS12 trainers normally does that. Also make sure the CBEIF = 1. The CBEIFbit is part of the FSTAT register.

(2) Write an array of words of data and their Flash memory address locations. The data is a don't care value, and the address is the starting address of a sector.

(3) Write value $40 to the FCMD register. The value $40 is the erase option for the sectors in Figure 14-8.

(4) Write value $80 to the FSTAT register. This will erase the data in the Flash unless there is a problem. Problems can occur a) if we try to erase a secure (protected) area of Flash (indicated by the PVIOL flag) or b) Flash is not accessi-

ble (indicated by the ACCERR flag). That is the reason we need to check these two flags in the next steps.

(5) Check the PVIOL and ACCERR flag bits to ensure that there was no error.

(6) Check the CBEIF flag bit to ensure the erase process is complete.

We can use option 05 in the FMCD register to verify the erasure of the Flash. Program 14-5 shows how to erase a sector.

```
;Program 14-5: This program erases the Flash
;memory starting at location 0x4500.
;read the erased bytes one byte at a time and
;send it to PORTB
R1   DS.B   1
R2   DS.B  1
     ORG    ROMStart
Entry: LDS #RAMEnd+1        ;initialize the stack pointer
  BSET DDRB,#$FF            ;make PORTB output
  LDX   #DATA
  MOVB #8,R1                ;load counter
  LDAA #$00     ;for erase, the value does not matter
H1 STAA 1,X+
  JSR   F_ERASE
  DEC   R1
  BNE   H1
;read the bytes one byte at a time and place them
;on PORTB
  LDX   #DATA
  MOVB #8,R2                ;load counter
H2 LDAA 1,X+
  STAA PORTB
  JSR   DELAY
  DEC   R2
  BNE   H2
HERE BRA HERE

F_ERASE
  LDAA #$40                 ;code for erase
  STAA FCMD                 ;erase the word
  LDAA #$80
  STAA FSTAT
  LDAA #$FF                 ;A=FF if error
  BRSET FSTAT,mFSTAT_PVIOL,OVER
  BRSET FSTAT,mFSTAT_ACCERR,OVER
  CLRA                      ;A=00 if no error
OVER RTS
     ORG $4500         ;on-chip FLASH data
DATA FCB "GOOD BYE"
```

Erasing and writing to Flash memory in C

Programs 14-6C through 14-8C are the C versions of earlier programs.

```
/*Program 14-6C: This C program (a) writes the mes-
sage "GOOD BYE" to Flash memory starting at location
$4400, (b) reads the data from Flash and places it in
PORTB one byte at a time. */
#include <mc9s12dp512.h>/* derivative information */

unsigned char F_WRT(void);
unsigned char data[8] @0x4400;
unsigned char R1;
unsigned char ret;
void main(void) {
  DDRB = 0xFF;
  R1 = 0;
  data[R1++] = 'G';
  data[R1++] = 'O';
  ret =F_WRT();
  data[R1++] = 'O';
  data[R1++] = 'D';
  ret = F_WRT();
  data[R1++] = ' ';
  data[R1++] = 'B';
  ret = F_WRT();
  data[R1++] = 'Y';
  data[R1++] = 'E';
  ret = F_WRT();
  for(R1=0;R1 < 8; R1++)
    {
      PORTB = data[R1];
    }
  for(;;) {} /* wait forever */
}

unsigned char F_WRT(void)
  {

     FCMD = 0x20; //write (program) to Flash
     FSTAT = 0x80; //make sure it is OK
     ret = 0xFF; //ret=ff if  error
     if(!FSTAT&&FSTAT_PVIOL_MASK)
       if(!FSTAT&&FSTAT_ACCERR_MASK)
         ret = 0x00; //ret=00 if no error
  return ret;
  }
```

```
//Program 14-7C: This C program writes and erases the
//Flash memory starting at location 0x4400.
#include <mc9s12dp512.h>/* derivative information */

unsigned char F_WRT(void);
unsigned char F_ERASE(void);
unsigned char data[4] @0x4400;
unsigned char R1;
unsigned char ret;
void main(void) {
  DDRB = 0xFF;
  R1 = 0;
  data[R1++] = 'Y';       //write to Flash
  data[R1++] = 'E';
  ret =F_WRT();
  data[R1++] = 'S';
  data[R1++] = ' ';
  ret = F_WRT();
  for(R1=0;R1 < 4; R1++) //erase Flash
    { data[R1] = 0;
      ret = F_ERASE();
    }

  //reads out erased Flash locations
  for(R1=0;R1 < 4; R1++)
    { PORTB = data[R1]; }
  for(;;) {} /* wait forever */
}
unsigned char F_ERASE(void) {
  FCMD    = 0x40; //erase Flash
  FSTAT   = 0x80; //make sure it is OK
  ret = 0xFF; //ret=ff if  error
  if(!FSTAT&&FSTAT_PVIOL_MASK)
    if(!FSTAT&&FSTAT_ACCERR_MASK)
      ret = 0x00; //ret=00 if no error
  return ret;
  }
unsigned char F_WRT(void) {
  FCMD    = 0x20; //write (program) to Flash
  FSTAT   = 0x80; //make sure it is OK
  ret = 0xFF; //ret=ff if  error
  if(!FSTAT&&FSTAT_PVIOL_MASK)
    if(!FSTAT&&FSTAT_ACCERR_MASK)
      ret = 0x00; //ret=00 if no error
  return ret;
  }
```

Review Questions

1. True or false. The HCS12 Flash memory can be used for both program code and data.
2. True or false. In the Flash memory programming, use of the FCMD register is optional.
3. True or false. In the HCS12, writing (programming) to Flash in byte-size data is not allowed.
4. True or false. Data read from Flash memory is in byte size, while data written to it is in word size.
5. True or false. During the write or erase, the Flash cannot be read.
6. What is the size of the block for erasing the Flash memory in the MC9S12Dx512?
7. What is the size of the sector for erasing the Flash memory in the MC9S12Dx512?

SECTION 14.3: WRITING TO EEPROM IN THE HCS12

The vast majority of the members of the HCS12 family come with some EEPROM memory. The amount varies from 1,024 bytes to a few K depending on the family member. For example, the MC9S12Dx128 has 2K bytes of EEPROM, while the MC9S12Dx512 has only 4K bytes. Table 14-5 shows some of the family members and their EEPOM size. While the Flash memory in the HCS12 can be used for storing both code and data, the EEPROM space is used generally for storing data. Of the three memory spaces that the HCS12 has, the SRAM and EEPROM are used for data while the Flash is used mainly for programs and sometimes for storage of fixed data such as look-up tables and string messages.

Table 14-5: EEPROM Size for Some HCS12 Chips

Part No.	EEPROM	Write Size	Sector Erase Size	Mass Erase
MC9S12Dx64	1 KB	2 bytes	4 bytes	1 KB
MC9S12Dx128	2 KB	2 bytes	4 bytes	2 KB
MC9S12Dx256	4 KB	2 bytes	4 bytes	4 KB
MC9S12Dx512	4 KB	2 bytes	4 bytes	4 KB

EEPROM registers

Many of the registers associated with the EEPROM are similar to Flash registers. For example, we have the FDIVCLK register in Flash, and the EEPROM version of that register is called EDIVCLK. See Figures 14-12 through 14-15. Due to the many similarities between the bits of these registers we will not go into a detailed discussion of the EEPROM registers. From the ECMD register shown in Figure 14-14, notice the write (program) size is still 2 bytes (word), just like the Flash. The sector size for erasing is 4 bytes. We also have the option of erasing the entire EEPROM. To do that, however, we must enable some bits in the EPROT (EEPROM Protection) register. The EPROT register allows us to secure a section of the EEPROM to protect it from accidental program (write) or erasure. See Figure 14-15.

EDIVLD	PRDIV8	EDIV5	EDIV4	EDIV3	EDIV2	EDIV1	EDIV0

EDIVLD EEPROM Clock divider loaded
1 = Register has been written to since the last reset.
(Now, Flash can be programmed (written to) and erased.)
0 = Register has not been written to.

PRDIV8 Enable the divide-by-8 prescaler
1 = Enables divide-by-8 prescaler. The bus clock is divided by 8 before it is divided again by the EDIVx bits.
0 = The bus clock is divided by the EDIVx bits. See below.

EDVI5–EDIV0 EEPROM clock divider bits

Notes:
1) Upon reset, we cannot program (write) or erase the EEPROM until we write to this register. The EDIVLD lets us know if this register has been written to.
2) For the purpose of on-chip EEPROM timing, the bus clock cannot be less than 1 MHz and the EEPROM clock must be in the range of 150 kHz–200 kHz. Otherwise, the program (write) / erase of the EEPROM will not work and it can even damage the EEPROM.
3) A combination of PRDIV8 and EDIVx bits are used to set the timing of EEPROM to the range of 150 kHz–200 kHz.

Figure 14-12. ECLKDIV (EEPROM Clock Divide) Register

CBEIF	CCIF	PVIOL	ACCERR	0	BLANK	0	0

CBEIF Command Buffer Empty Interrupt Flag
1 = Command buffers are ready to accept a new command.
0 = Command buffers are full.

CCIF Command Completion Flag
1 = All commands are completed.
0 = Command is in progress.

PVIOL Protection Violation
1 = Protection violation has occurred.
0 = No protection violation has occurred.

ACCERR EEPROM Access Error
1 = Access error has occurred.
0 = No failure

BLANK Array has been verified as erased
1 = EEPROM verified as erased
0 = If an erase and verify command has been requested, and the CCIF flag is set, then a zero in BLANK indicates that the EEPROM has not been erased.

Figure 14-13. ESTAT (EEPROM Status) Register

0	CMDB6	CMDB5	0	0	CMDB2	0	CMDB0

CMDB CMDB bits are used to perform the following actions:

$05 Erase Verify. Verify that all memory bytes of EEPROM are erased. Upon verification the BLANK bit of the ESTAT register is set to HIGH

$20 Word Program. Write a word (2 bytes) to EEPROM. The starting address location must be an even number.

$40 Sector Erase. Erase 2 words (4 bytes) of EEPROM.

$41 Mass Erase. Erase all of the EEPROM. This can happen only when the EPDIS and EPOPEN bits of the EPROT are set to HIGH.

$61 Sector Modify. Erase 2 words of EEPROM, reprogram one word.

Figure 14-14. ECMD (EEPROM Command) Register

EPOPEN	NV6	NV5	NV4	EPDIS	EP2	EP1	EP0

EPOPEN EEPROM Open for program or erase

1 = The EEPROM sectors not protected are enabled for program or erase.

0 = The whole EEPROM array is protected.

NV[6:4] Non-Volatile flag bits

These three bits are available to the user as nonvolatile flags.

EPDIS EEPROM Protection address range Disable

1 = Protection disabled

0 = Protection enabled

EP[2:0] EEPROM Protection address size

The EP[2:0] bits determine the size of the protected sector. See the table below:

000	$FC0–$FFF	64 bytes
001	$F80–$FFF	128 bytes
010	$F40–$FFF	192 bytes
011	$F00–$FFF	256 bytes
100	$EC0–$FFF	320 bytes
101	$E80–$FFF	384 bytes
110	$E40–$FFF	448 bytes
111	$E00–$FFF	512 bytes

Figure 14-15. EPROT (EEPROM Protection) Register

Steps in writing to EEPROM memory

Figure 14-16 shows the steps in writing to EEPROM. Assuming that an area of EEPROM memory is erased, we can use the following steps to write a set of data to the EEPROM memory:

(1) Make sure that the EDIVLD bit of the EDIVCLK register is set to high (EDIVLD = 1). The HCS12 trainers normally do that. Also make sure the CBEIF = 1. The CBEIFbit is part of the ESTAT register.

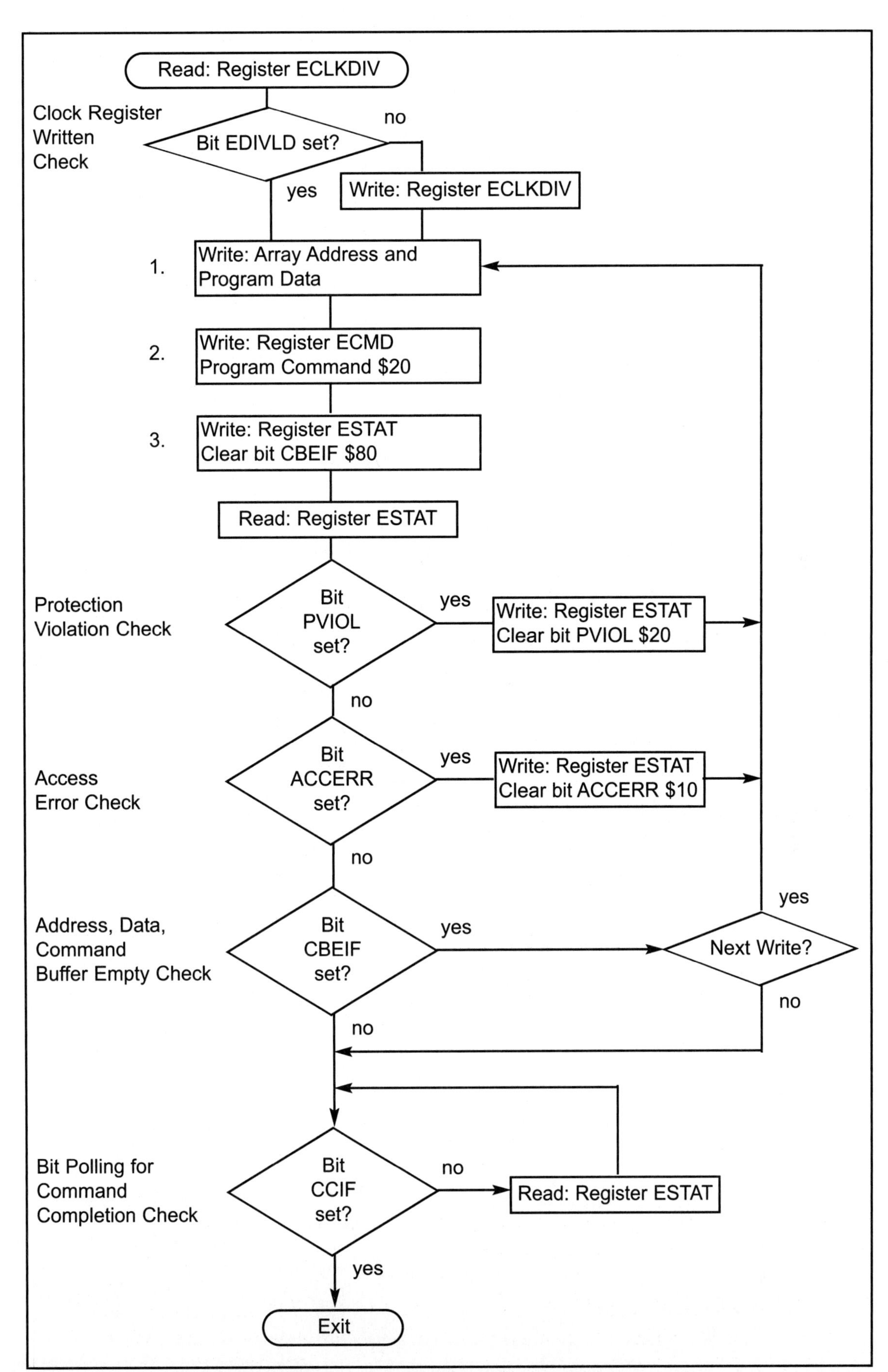

Figure 14-16. Steps in Writing to EEPROM Memory

(2) Write a word (2 bytes) of data and its EEPROM memory address location. The address must be an even address. The data is kept in an internal buffer until you follow the next two steps.

(3) Write value $20 to the ECMD register.

(4) Write value $80 to the ESTAT register. This will transfer (write) the data from the internal buffer to the EEPROM unless there is a problem. Problems can occur a) if we try to write to a secure area of the EEPROM (indicated by the PVIOL flag) or b) the EEPROM is not accessible (indicated by the ACCERR flag). That is the reason we need to check these two flags in the next steps.

(5) Check the PVIOL and ACCERR flag bits to ensure that there was no error.

(6) Check the CBEIF flag bit to ensure the process is complete and the EEPROM is ready for the next write. Go back to step 2 to write the next word. See Figure 14-16.

Notice from step 2 that the internal buffer cannot be accessed by the programmer. It is internal and used exclusively for the purpose of writing/erasing the EEPROM memory. Program 14-8 (a) writes the message "HELLO" to EEPROM memory starting at location $400, and (b) reads the data back from EEPROM and places it in PORTB one byte at a time.

```
;Program 14-8: Writing to EEPROM
   LDS #RAMEnd+1        ;initialize the stack pointer
   BSET DDRB,#$FF
   LDAA #'G'
   LDAB #'O'
   STD $400
   JSR EE_WRT
   LDAA #'O'
   LDAB #'D'
   STD $402
   JSR EE_WRT
   LDAA #' '
   LDAB #'B'
   STD $404
   JSR EE_WRT
   LDAA #'Y'
   LDAB #'E'
   STD $406
   JSR EE_WRT
;read them back one byte at a time and examine the
;bytes on PORTB
   LDX #$400
   MOVB #8,R1            ;load counter
H1 LDAA 1,X+
   STAA PORTB
   JSR DELAY
```

```
   DEC R1
   BNE H1
HERE BRA  HERE
;-----write to Flash. A=0 if no error, A=FF if error
EE_WRT
   LDAA #$20
   STAA ECMD                ;write (program) to Flash
   LDAA #$80
   STAA ESTAT               ;make sure it is OK
   LDAA #$FF                ;A=FF if error
   BRSET ESTAT,mESTAT_PVIOL,OVER
   BRSET ESTAT,mESTAT_ACCERR,OVER
   CLRA                     ;A=00 if no error
OVER RTS
```

Program 14-9 (a) reads a block of data from the code space (Flash) and writes it to EEPROM, and (b) then reads the same data from EEPROM and sends it to PORTB of the HCS12, one byte at a time.

```
;Program 14-9
R1   DS.B 1
R2   DS.B 1
MYDATA FCB     "HELLO!"  ;data in Flash
     ORG   ROMStart
Entry: LDS #RAMEnd+1     ;initialize the stack pointer
   BSET DDRB,#$FF           ;make PORTB output
   LDX #MYDATA  ;set the pointer X (Flash address)
   LDY #$400    ;set the pointer Y (EEPROM address)
   MOVB #3,R1               ;set the counter
OVER MOVW 0,X,0,Y ;move a word from source to dest.
   JSR EE_WRT
   INX                      ;point to next word
   INX
   INY
   INY
   DEC R1                   ;decrement counter
   BNE OVER                 ;repeat
;read them back one byte at a time and examine the
;bytes on PORTB
   LDX #$400
   MOVB #6,R2               ;load counter
H1 LDAA 1,X+
   STAA PORTB
   JSR DELAY
   DEC R2
   BNE H1
HERE BRA HERE
```

```
;-----write to Flash. A=0 if no error, A=FF if error
EE_WRT
  LDAA #$20
  STAA FCMD
  LDAA #$80
  STAA FSTAT
  LDAA #$FF                    ;A=FF if error
  BRSET ESTAT,mESTAT_PVIOL,OVER2
  BRSET ESTAT,mESTAT_ACCERR,OVER2
  CLRA                         ;A=00 if no error
OVER2     RTS
```

Accessing the EEPROM in C

Program 14-10C shows how to write and read the EEPROM memory in the C language. This is the C version of an earlier program. Program 14-10C (a) writes the message "YES" to EEPROM memory, and (b) then reads the same data from EEPROM and sends it to PORTB one byte at a time.

```
//Program 14-10C
#include <mc9s12dp512.h>/* derivative information */
unsigned char E_WRT(void);
unsigned char data[8] @0x500; //EEPROM space
unsigned char R1;
unsigned char ret;
void main(void) {
  DDRB = 0xFF;
  R1 = 0;
  data[R1++] = 'G';
  data[R1++] = 'O';
  ret =E_WRT();
  data[R1++] = 'O';
  data[R1++] = 'D';
  ret = E_WRT();
  data[R1++] = ' ';
  data[R1++] = 'B';
  ret = E_WRT();
  data[R1++] = 'Y';
  data[R1++] = 'E';
  ret = E_WRT();
  for(R1=0;R1 < 8; R1++)
    {
      PORTB = data[R1];
    }
  for(;;) {} /* wait forever */
}
unsigned char E_WRT(void)
  {
     ECMD = 0x20; //write (program) to EEPROM
     ESTAT= 0x80; //make sure it is OK
     ret = 0xFF; //ret=ff if error
```

```
    if(!ESTAT&&ESTAT_PVIOL_MASK)
      if(!ESTAT&&ESTAT_ACCERR_MASK)
        ret = 0x00; //ret=00 if no error
  return ret;
  }
```

Program 14-11C (a) transfers a block of data from Flash to RAM, (b) writes the block to EEPROM memory, and (c) then reads the same data from EEPROM and sends it to PORTB one byte at a time.

```
//Program 14-11C
#include <mc9s12dp512.h>/* derivative information */
unsigned char E_WRT(void);
const unsigned char fdata[8] @0x4400 =
  {
  'G','O','O','D',' ','B','Y','E' //data in Flash
  };
unsigned char rdata[8]; //RAM space
unsigned char edata[8] @0x500; //EEPROM space
unsigned char R1;
unsigned char ret;
void main(void) {
  DDRB = 0xFF;
  R1 = 0;
  for(R1=0;R1 < 8; R1++)      //read from Flash
    {
      rdata[R1] = fdata[R1]; //write to RAM
    }
  for(R1=0;R1 < 8; R1++)      //read from RAM
    {
      edata[R1] = rdata[R1]; //write to EEPROM
      ret = E_WRT();
    }
  for(R1=0;R1 < 8; R1++)      //read from EEPROM
    {
      PORTB = edata[R1];      //send to PORTB
    }
  for(;;) {}                  /* wait forever */
}
unsigned char E_WRT(void)
  {
     ECMD = 0x20;         //write (program) to EEPROM
     ESTAT = 0x80;        //make sure it is OK
     ret = 0xFF;                  //ret=ff if  error
     if(!ESTAT&&ESTAT_PVIOL_MASK)
       if(!ESTAT&&ESTAT_ACCERR_MASK)
         ret = 0x00;              //ret=00 if no error
  return ret;
  }
```

Review Questions

1. True or false. The HCS12 EEPROM memory is used for both program code and data.
2. True or false. The MC9S12Dx512 has 4K bytes of EEPROM memory.
3. True or false. In the HCS12, EEPROM contents are lost when power is cut off to the chip.
4. True or false. In EEPROM memory programming, the use of the ECMD register is optional.
5. True or false. Every HCS12 chip comes with at least 1 KB of EEPROM.

SECTION 14.4: CLOCK SPEED FOR FLASH AND EEPROM

In programming or erasing the Flash and EEPROM, we must make sure that the clock speed is set in the range of 150 kHz to 200 kHz. In this section, we show how to do that. As we said before, the monitor program in HCS12 trainers sets this value, so you do not need to set it again. Use this section only if you are designing an HCS12-based system from scratch.

Setting the Flash clock (FCLK)

According to the HCS12 manual, "prior to issuing any program or erase command, it is first necessary to write the FCLKDIV register to divide the oscillator down to within the 150 kHz to 200 kHz range." The FCLK (Flash timing Clock) is a function of Fosc and Fbus. We use the PRDIV8 and FDIV5–FDIV0 bits of the FCLKDIV register to set FCLK within the range of 150 kHz–200 kHz. See Figure 14-17. The algorithm used for the calculation of FCLK is shown in Figure 14-18. From Figure 14-17, notice that the Fbus cannot be less than 1 MHz. In Figure 14-18, PRDCLK = Fosc if Fosc is less than 12.8 MHz. If Fosc is greater than 12.8 MHz, then we need to divide it by the prescaler value of 8 to get the PRDCLK value. See Examples 14-1 and 14-2.

Setting the EEPROM clock

Setting the EEPROM clock is very similar to setting the Flash clock. For the EEPROM, we have ECLK instead of FCLK. Figure 14-19 shows the ECLKDIV register. The algorithm used for the calculation of ECLK is shown in Figure 14-20. See Examples 14-3 and 14-4.

Review Questions

1. True or false. The clock speed for EEPROM is 150 kHz and 200 kHz.
2. True or false. The clock speed for Flash is 150 kHz and 200 kHz.
3. True or false. In setting the clock for Flash, the use of the FCLKDIV register is optional.
4. True or false. In setting the clock for EEPROM, the use of the ECLKDIV register is optional.
5. True or false. Upon reset, we cannot write or erase Flash until we write FCLKDIV.

FDIVLD	PRDIV8	FDIV5	FDIV4	FDIV3	FDIV2	FDIV1	FDIV0

FDIVLD Flash clock divider loaded
1 = Register has been written to since the last reset.
(Now, Flash can be programmed (written to) and erased.)
0 = Register has not been written to.
PRDIV8 Enable the divide-by-8 prescaler
1 = Enables divide-by-8 prescaler. The bus clock is divided by 8 before it is divided again by the FDIVx bits.
0 = The bus clock is divided by the FDIVx bits. See below.
FDVI5–FDIV0 Flash clock divider bits

FDIV5	FDIV4	FDIV3	FDIV2	FDIV1	FDIV0	
0	0	0	0	0	0	divide by 1 (FDIVx + 1)
0	0	0	0	0	1	divide by 2
0	0	0	0	1	0	divide by 3
.	.	.	.	.	.	
1	1	1	1	1	1	divide by 32 (FDIVx + 1)

Notes:
1) We cannot program (write) / erase the Flash until we write to this register. The FDVLD lets us know if this register has been written to.
2) For the purpose of on-chip Flash timing, the bus clock cannot be less than 1 MHz and the Flash clock must be in the range of 150 kHz–200 kHz. Otherwise, the program (write) / erase of Flash will not work and it can even damage the Flash.
3) A combination of PRDIV8 and FDIVx bits are used to set the timing of Flash to the range of 150 kHz –200 kHz.

Figure 14-17. FCLKDIV (Flash Clock Divide) Register

Example 14-1

For an HCS12-based system, we have Fosc = 4 MHz and Fbus = 2 MHz. Find the values for the FCLKDIV register if we want the FCLK to be in the 150 kHz–200 kHz range.

Solution:

From Figure 14-18, we have the following:
1) Fbus = 2 MHz, which is greater than 1 MHz. In other words Tbus = 1/2 MHz = 0.5 μs, which is less than 1 μs.
2) Fosc = 4 MHz. It is less than 12.8 MHz, therefore we make PRDIV8 = 0.
That means PRDCLK = 4 MHz.
3) PRDCLK × [5 + 0.5 μs] = 4 MHz × 5.5 μs = 22 is an integer.
Therefore FDIV5–FDIV0 values are set at 4 MHz × (5 + 0.5 μs) – 1 = 21 or 010101 in binary.
5) FCLK = 4 MHz / (21 + 1) = 4 MHz /22 = 181 kHz
6) FCLK = 181 kHz is > 150 kHz. Also 1/(181 kHz) + 0.5 μs = 5.5 μs + 0.5 μs = 6 μs, which is greater than 5 μs.
Therefore we set the FCLKDIV register to 00010101 in binary or 15 in hex.

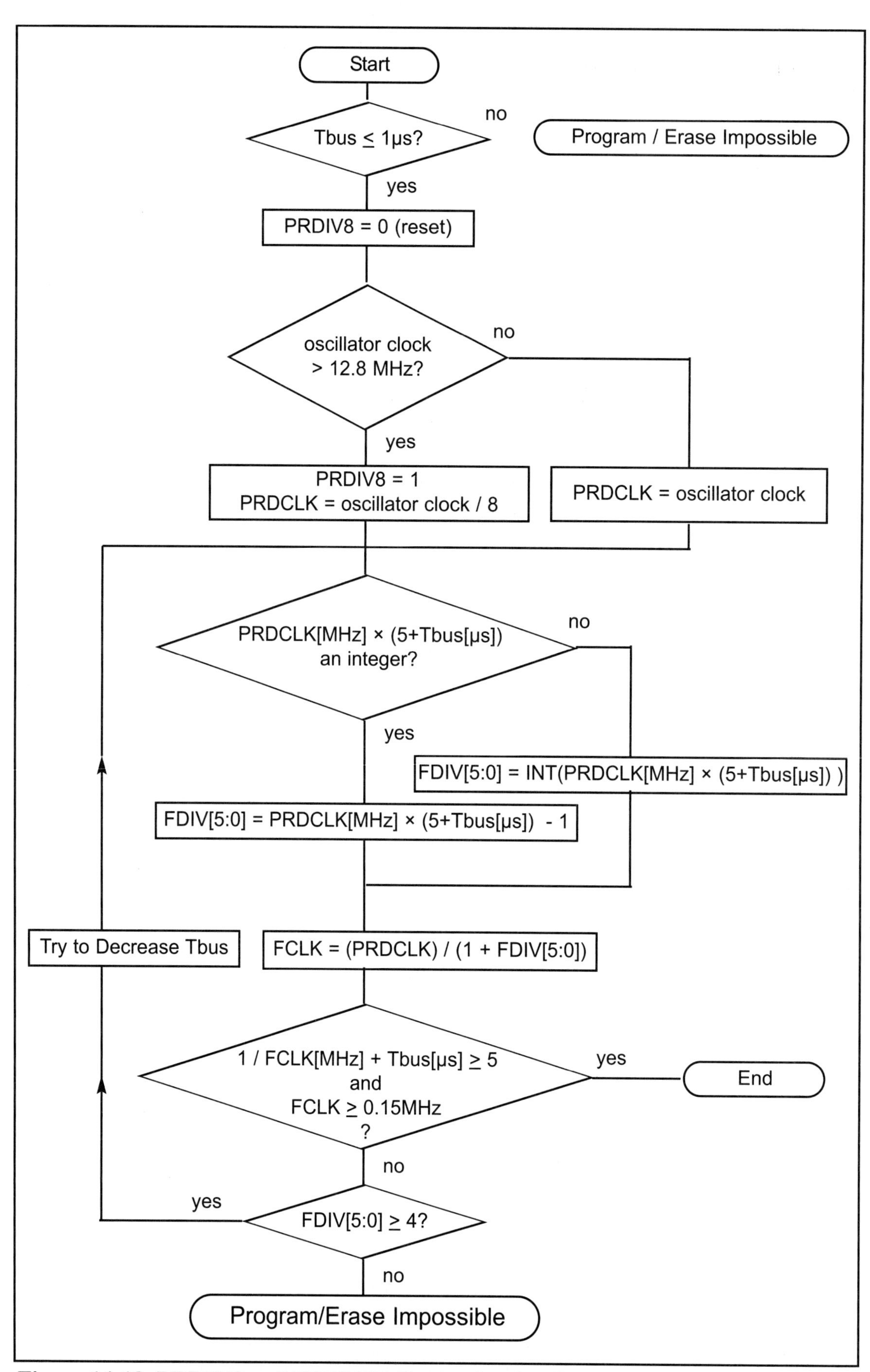

Figure 14-18. PRDIV8 and FDIV Bits Determination Procedure

Example 14-2

For an HCS12-based system, we have Fosc = 16 MHz and Fbus = 20 MHz using PLL. Find the values for the FCLKDIV register if we want FCLK to be in the 150 kHz–200 kHz range.

Solution:

From Figure 14-18, we have the following:
1) Fbus = 20 MHz, which is greater than 1 MHz. In other words Tbus = 1/20 MHz = 0.05 μs, which is less than 1 μs.
2) Fosc = 20 MHz. It is more than 12.8 MHz, therefore we make PRDIV8 = 1 and PRDCLK = 20 MHz/8 = 2.5 MHz.
3) PRDCLK × [5 + 0.05 μs] = 2.5 MHz × 5.05 μs = 12.625 is not an integer. Therefore FDIV5–FDIV0 values are set at 12 decimal or 001100 in binary.
5) FCLK = 2.5 MHz / (12 + 1) = 2 MHz /13 = 192 kHz
6) FCLK= 192 kHz is > 150 kHz. Also 1/(192 kHz) + 0.05 μs = 5.2 μs + 0.05 μs = 5.25 μs, which is greater than 5 μs.
Therefore we set the FCLKDIV register to 01001100 in binary or 4C in hex.

EDIVLD	PRDIV8	EDIV5	EDIV4	EDIV3	EDIV2	EDIV1	EDIV0

EDIVLD EEPROM lock divider loaded
1 = Register has been written to since the last reset. (Now, EEPROM can be programmed (written to) and erased.)
0 = Register has not been written to.

PRDIV8 Enable the divide-by-8 prescaler
1 = Enables divide-by-8 prescaler. The Fosc clock is divided by 8 before it is divided again by the EDIVx bits.
0 = The bus clock is divided by the EDIVx bits. See below.

EDVI5–EDIV0 Flash clock divider bits

EDIV5	EDIV4	EDIV3	EDIV2	EDIV1	EDIV0	
0	0	0	0	0	0	divide by 1 (EDIVx + 1)
0	0	0	0	0	1	divide by 2
0	0	0	0	1	0	divide by 3
.	.	.	.	.	.	
1	1	1	1	1	1	divide by 32 (EDIVx + 1)

Notes:
1) Upon reset, we cannot program (write) or erase the EEPROM until we write to this register. The EDVLD lets us know if this register has been written to.
2) For the purpose of on-chip Flash timing, the bus clock cannot be less than 1 MHz and the Flash clock must be in the range of 150 kHz–200 kHz. Otherwise, the program (write) / erase of Flash will not work and it can even damage the Flash.
3) A combination of PRDIV8 and EDIVx bits are used to set the timing of EEPROM to the range of 150 kHz–200 kHz.

Figure 14-19. ECLKDIV (EEPROM Clock Divide) Register

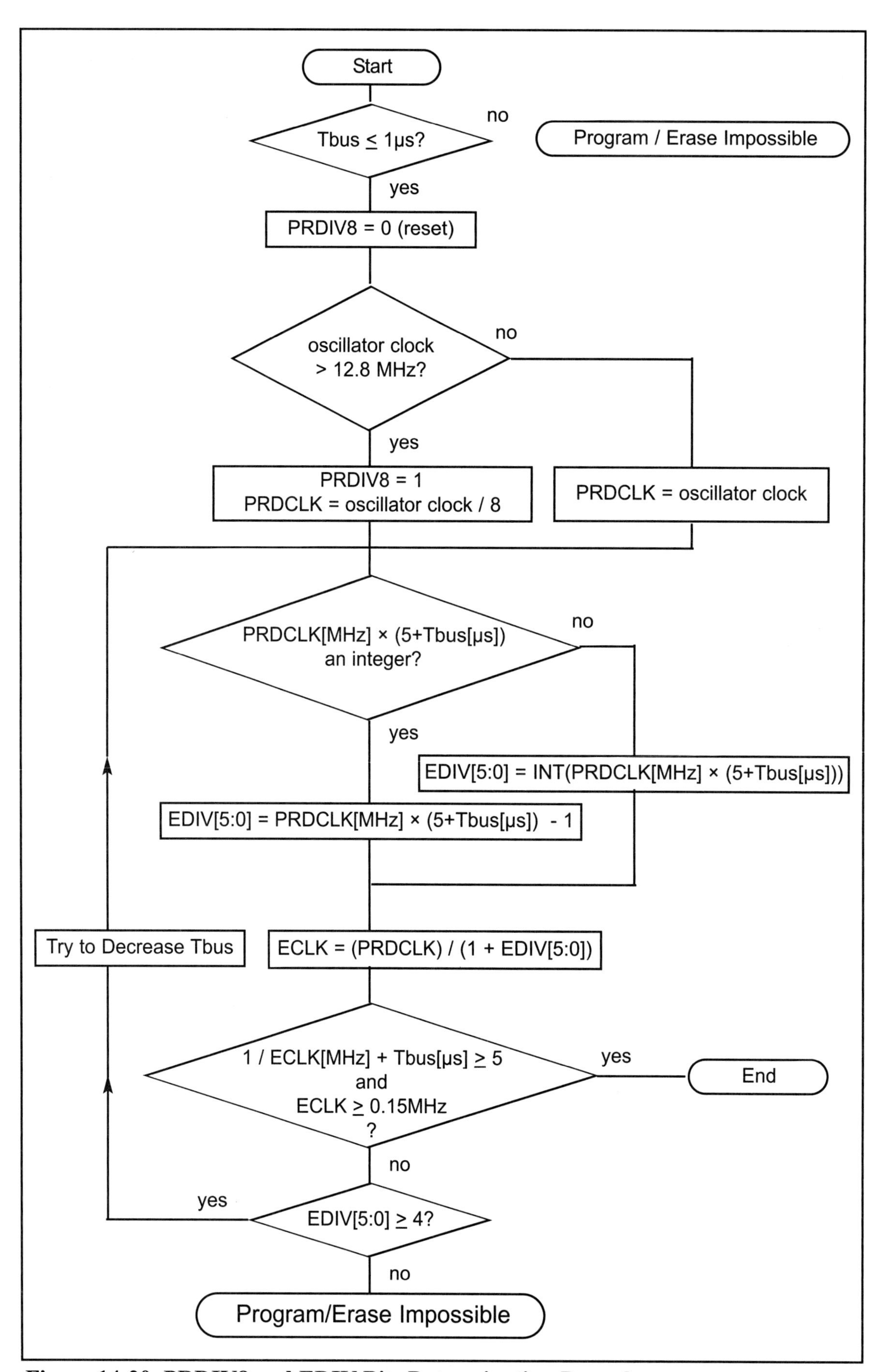

Figure 14-20. PRDIV8 and EDIV Bits Determination Procedure

Example 14-3

For an HCS12-based system, we have Fosc = 8 MHz and Fbus = 4 MHz. Find the values for the ECLKDIV register if we want ECLK to be in the 150 kHz–200 kHz range.

Solution:

From Figure 14-20, we have the following:
1) Fbus = 4 MHz, which is greater than 1 MHz. In other words Tbus = 1/4 MHz = 0.25 μs, which is less than 1 μs.
2) Fosc = 8 MHz. It is less than 12.8 MHz, therefore we make PRDIV8 = 0.
That means PRDCLK = 8 MHz.
3) PRDCLK × [5 + 0.25 μs] = 8 MHz × 5.25 μs = 42 is an integer.
Therefore EDIV5–EDIV0 values are set at 8 MHz × (5 + 0.25 μs) – 1 = 41 or 101001 in binary.
5) ECLK = 8 MHz / (41 + 1) = 8 MHz /42 = 190 kHz
6) ECLK = 190 kHz is > 150 kHz. Also 1/(190 kHz) + 0.25 μs = 5.25 μs + 0.25 μs = 5.5 μs, which is greater than 5 μs.

Therefore we set the ECLKDIV register to 00101001 in binary or 29 in hex.

Example 14-4

For an HCS12-based system, we have Fosc = 16 MHz and Fbus = 20 MHz using PLL. Find the values for the ECLKDIV register if we want ECLK to be in the 150 kHz–200 kHz range.

Solution:

From Figure 14-20, we have the following:
1) Fbus = 20 MHz, which is greater than 1 MHz. In other words Tbus = 1/20 MHz = 0.05 μs, which is less than 1 μs.
2) Fosc = 20 MHz. It is more than 12.8 MHz, therefore we make PRDIV8 = 1 and PRDCLK = 20 MHz/8 = 2.5 MHz.
3) PRDCLK × [5 + 0.05 μs] = 2.5 MHz × 5.05 μs = 12.625 is not an integer. Therefore EDIV5–EDIV0 values are set at 12 decimal or 001100 in binary.
5) ECLK = 2.5 MHz / (12 + 1) = 2 MHz /13 = 192 kHz
6) ECLK = 192 kHz is > 150 kHz. Also 1/192 kHz + 0.05 μs = 5.2 μs + 0.05 μs = 5.25 μs, which is greater than 5 μs.

Therefore we set the ECLKDIV register to 01001100 in binary or 4C in hex.

PROBLEMS

SECTION 14.1: PAGE SWITCHING OF FLASH MEMORY IN HCS12

1. What is the difference between the 64 KB memory space and 512 KB of Flash memory in the MC9S12Dx512 chip?
2. True or false. In the HCS12, any memory space beyond 64 KB is off-chip.
3. True or false. To access the memory beyond 64 KB, we must use page switching.
4. True or false. PPAGE is an 8-bit register.
5. True or false. Only 6 bits of PPAGE are used by the HCS12.
6. The memory space of 64 KB is accessed using the ___________ register.
7. True or false. All the HCS12 chips have the same 64 KB directly accessible memory space.
8. What is the difference between RTC and RTS?
9 How many kilobytes is the size of the page in the HCS12?
10. Which of the following, EEPROM, Flash, or SRAM, can be remapped?
11. How many kilo bytes is the size of the Flash for MC9S12Dx512?
12. Which of the following, RAM, EEPROM, or Flash, is volatile memory?
13. In accessing memory space beyond 64 KB we must use the _________register.
14. Using paging in the HCS12, we can have a maximum memory of_____ bytes.
15. Find the number of pages for the Flash memory MC9S12Dx256.
16. Find the number of pages for the Flash memory MC9S12Dx512.

SECTION 14.2: ERASING AND WRITING TO FLASH

17. True or false. The Flash memory in HCS12 is used primarily for the program code.
18. True or false. The Flash memory in HCS12 can also be used for storing fixed data.
19. True or false. The maximum memory space accessible using paging is 2M bytes.
20. True or false. Reading data from Flash memory can be done one byte at a time.
21. True or false. Writing data to Flash memory can be done one byte at a time.
22. True or false. Writing data to Flash memory must be done in words.
23. True or false. Erasing of Flash memory can be done one byte at a time.
24. True or false. The use of the FSTAT register in writing/erasing Flash memory is optional.
25. True or false. In reading the contents of Flash memory, we must use the FSTAT register.
26. What option in the FCMD register is used in writing (programming) data to Flash memory?
27. What option in the FCMD register is used in erasing the Flash memory?
28. What is the difference between the sector and the block in the Flash memory erase?
29. What bits of the FSTAT register are used to make sure writing to Flash was successful?

30. What register is used to set the Flash clock?
31. Explain the difference between the protected and unprotected Flash in the HCS12.
32. True or false. During the process of writing to Flash, reading the Flash by the CPU is suspended?
33. What is the size of the block of data for erasing of Flash memory in MC9S12Dx256?
34. What is the size of the sector of data for erasing the Flash memory in MC9S12Dx256?
35. Indicate all the addresses that have a word (2-byte) boundary:
    (a) $511 (b) $516 (c) $6001 (d) $510
36. Indicate all the addresses that have a 1,024-byte sector boundary:
    (a) $C400 (b) $1C00 (c) $C500
37. True or false. All HCS12 chips have the same block size for the Flash memory.
38. True or false. All HCS12 chips have the same sector size for the Flash memory.

SECTION 14.3: WRITING TO EEPROM IN THE HCS12

39. True or false. The EEPROM memory in the HCS12 is used primarily for the program code.
40. True or false. The EEPROM memory in the HCS12 is used for data only.
41. True or false. Every MC9S12Dxx member has at least 1,024 bytes of EEPROM memory.
42. True or false. Reading data from EEPROM memory can be done one byte at a time.
43. True or false. Writing data to EEPROM memory can be done one byte at a time.
44. True or false. Writing data to EEPROM memory must be done in blocks of 2 bytes.
45. True or false. Erasing of data in EEPROM memory can be done one byte at a time.
46. True or false. The use of the ECMD register in reading and writing of EEPROM memory is optional.
47. True or false. The ECMD register is used by both the Flash and EEPROM memory write operations.
48. Give the major differences between Flash and EEPROM in the HCS12.
49. What is the size of the sector for erasing the EEPROM memory in the HCS12?
50. Which option of the ECMD is used for erasing a sector of the EEPROM?
51. What is the size of the EEPROM in MC9S12Dx256?
52. What is the size of the EEPROM in MC9S12D512?

ANSWERS TO REVIEW QUESTIONS

SECTION 14.1: PAGE SWITCHING OF FLASH MEMORY IN HCS12

1. 16 bits
2. 64 KB
3. 1 MB
4. True
5. True

SECTION 14.2: ERASING AND WRITING TO FLASH

1. True
2. False
3. True
4. True
5. True
6. 128 KB
7. 1,024 B

SECTION 14.3: WRITING TO EEPROM IN THE HCS12

1. False
2. True
3. False
4. False
5. True

SECTION 14.4: CLOCK SPEED FOR FLASH AND EEPROM

1. True
2. True
3. True
4. True
5. True

CHAPTER 15

RELAY, OPTOISOLATOR, AND STEPPER MOTOR INTERFACING WITH HCS12

OBJECTIVES

Upon completion of this chapter, you will be able to:

>> Describe the basic operation of a relay
>> Interface the HCS12 with a relay
>> Describe the basic operation of an optoisolator
>> Interface the HCS12 with an optoisolator
>> Describe the basic operation of a stepper motor
>> Interface the HCS12 with a stepper motor
>> Code HCS12 programs to control and operate a stepper motor
>> Define stepper motor operation in terms of step angle, steps per revolution, tooth pitch, rotation speed, and RPM

Microcontrollers are widely used in motor control. In motor control we also use relays and optoisolators. This chapter discusses motor control and shows HCS12 interfacing with relays, optoisolators, and stepper motors. We use both Assembly and C in our programming examples.

SECTION 15.1: RELAYS AND OPTOISOLATORS

This section begins with an overview of the basic operation of electromechanical relays, solid-state relays, reed switches, and optoisolators. Then we describe how to interface them to the HCS12. We use both Assembly and C language programs to demonstrate their control.

Electromechanical relays

A *relay* is an electrically controllable switch widely used in industrial controls, automobiles, and appliances. It allows the isolation of two separate sections of a system with two different voltage sources. For example, a +5 V system can be isolated from a 120 V system by placing a relay between them. One such relay is called an *electromechanical* (or *electromagnetic*) *relay* (EMR) as shown in Figure 15-1. The EMRs have three components: the coil, spring, and contacts. In Figure 15-1, a digital +5 V on the left side can control a 12 V motor on the right side without any physical contact between them. When current flows through the coil, a magnetic field is created around the coil (the coil is energized), which causes the armature to be attracted to the coil. The armature's contact acts like a switch and closes or opens the circuit. When the coil is not energized, a spring pulls the armature to its normal state of open or closed. In the block diagram for electro mechanical relays (EMR) we do not show the spring, but it does exist internally. There are all types of relays for all kinds of applications. In choosing a relay the following characteristics need to be considered:

1. The contacts can be normally open (NO) or normally closed (NC). In the NC type, the contacts are closed when the coil is not energized. In the NO type, the contacts are open when the coil is unenergized.
2. There can be one or more contacts. For example, we can have SPST (single pole, single throw), SPDT (single pole, double throw), and DPDT (double pole, double throw) relays.
3. The voltage and current needed to energize the coil. The voltage can vary from a few volts to 50 volts, while the current can be from a few mA to 20 mA. The relay has a minimum voltage, below which the coil will not be energized. This minimum voltage is called the "pull-in" voltage. In the data sheet for relays we might not see current, but rather coil resistance. The V/R will give you the pull-in current. For example, if the coil voltage is 5 V, and the coil resistance is 500 ohms, we need a minimum of 10 mA (5 V/500 ohms = 10 mA) pull-in current.
4. The maximum DC/AC voltage and current that can be handled by the contacts. This is in the range of a few volts to hundreds of volts, while the current can be from a few amps to 40 A or more, depending on the relay. Notice the difference between this voltage/current specification and the voltage/current needed for energizing the coil. The fact that one can use such a small amount of volt-

age/current on one side to handle a large amount of voltage/current on the other side is what makes relays so widely used in industrial controls. Examine Table 15-1 for some relay characteristics.

Table 15-1: Selected DIP Relay Characteristics (http://www.Jameco.com)

Part No.	Contact Form	Coil Volts	Coil Ohms	Contact Volts-Current
106462CP	SPST-NO	5 VDC	500	100 VDC-0.5 A
138430CP	SPST-NO	5 VDC	500	100 VDC-0.5 A
106471CP	SPST-NO	12 VDC	1,000	100 VDC-0.5 A
138448CP	SPST-NO	12 VDC	1,000	100 VDC-0.5 A
129875CP	DPDT	5 VDC	62.5	30 VDC-1 A

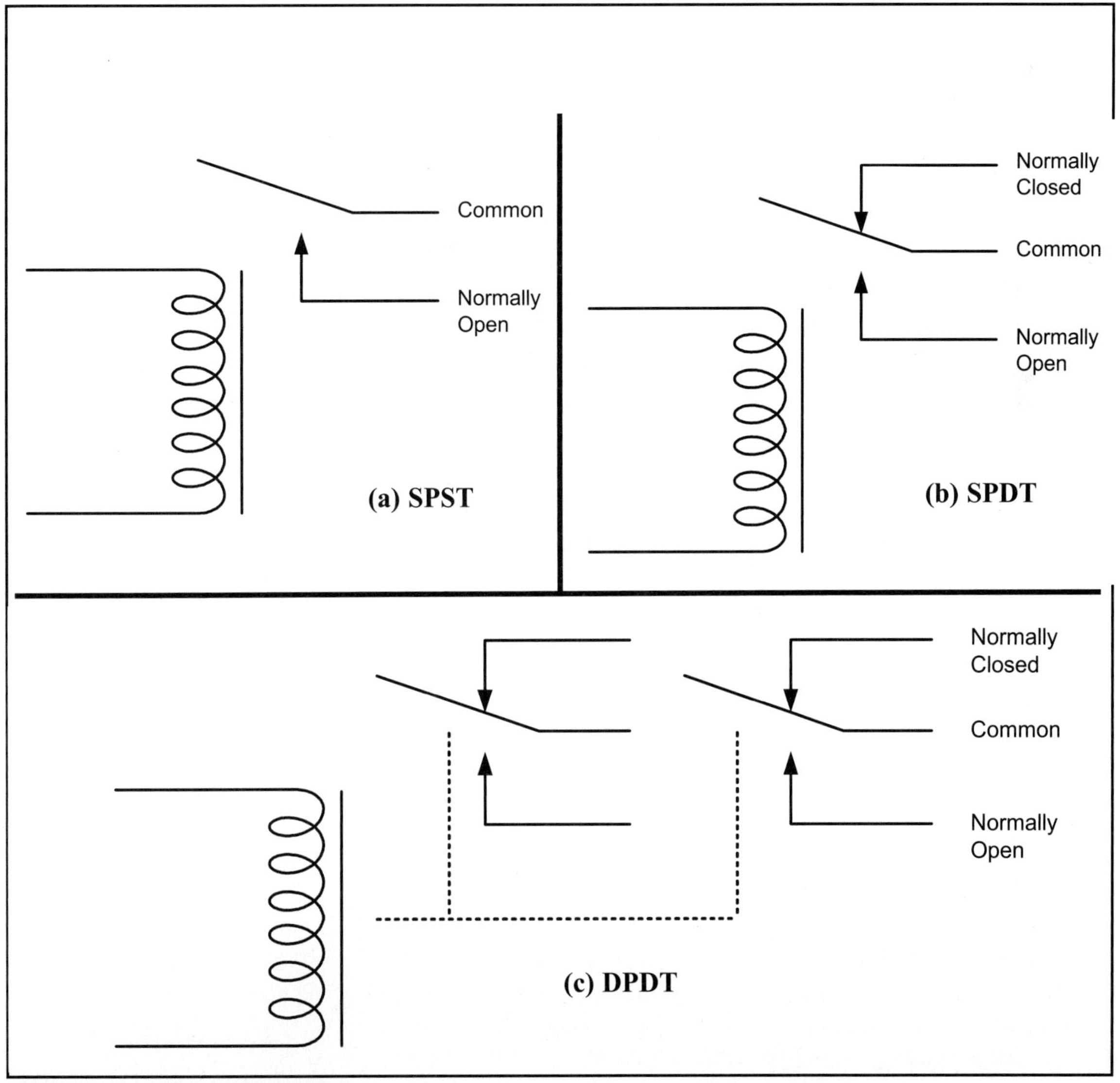

Figure 15-1. Relay Diagrams

Driving a relay

Digital systems and microcontroller pins lack sufficient current to drive the relay. While a relay's coil needs around 10 mA to be energized, a microcontroller's pin can provide a maximum of 1–2 mA current. For this reason, we place a driver, such as the ULN2803, or a power transistor between the microcontroller and the relay as shown in Figure 15-2.

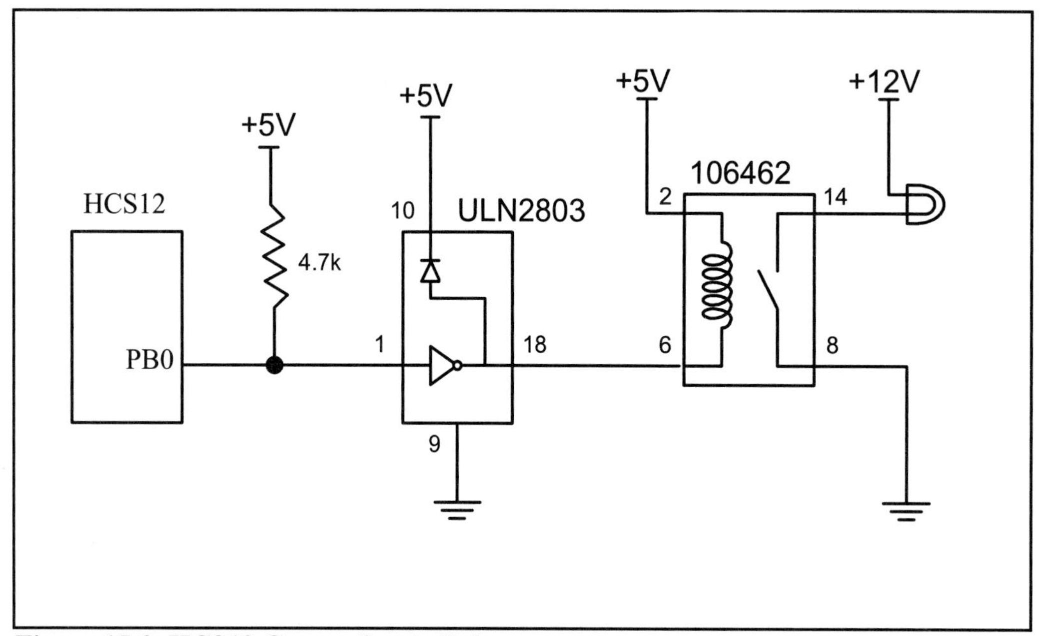

Figure 15-2. HCS12 Connection to Relay

Program 15-1 turns the lamp shown in Figure 15-2 on and off by energizing and de-energizing the relay every few ms.

```
;Program 15-1
R3      DS.B 1
R4      DS.B 1
     ORG $C000
     BSET DDRB,$01    ;PB0 as output
OVER BSET PORTB,$01   ;turn on the lamp
     JSR  DELAY
     BCLR PORTB,$01   ;turn off the lamp
     JSR  DELAY
     BRA  OVER
DELAY LDAA #$FF
     STAA R4
D1   LDAA #$FF
     STAA R3
D2   NOP
     NOP
     DEC  R3
     BNE  D2
     DEC  R4
     BNE  D1
     RTS
```

Solid-state relay

Another widely used relay is the solid-state relay. See Table 15-2. In this relay, there is no coil, spring, or mechanical contact switch. The entire relay is made out of semiconductor materials. Because no mechanical parts are involved in solid-state relays, their switching response time is much faster than that of electromechanical relays. Another advantage of the solid-state relay is its greater life expectancy. The life cycle for the electromechanical relay can vary from a few hundred thousand to a few million operations. Wear and tear on the contact points can cause a relay to malfunction after a while. Solid-state relays, however, have no such limitations. Extremely low input current and small packaging make solid-state relays ideal for microprocessor and logic control switching. They are widely used in controlling pumps, solenoids, alarms, and other power applications. Some solid-state relays have a phase control option, which is ideal for motor-speed control and light-dimming applications. Figure 15-3 shows the control of a fan using a solid-state relay (SSR).

Table 15-2: Selected Solid-State Relay Characteristics (http://www.Jameco.com)

Part No.	Contact Style	Control Volts	Contact Volts	Contact Current
143058CP	SPST	4–32 VDC	240 VAC	3 A
139053CP	SPST	3–32 VDC	240 VAC	25 A
162341CP	SPST	3–32 VDC	240 VAC	10 A
172591CP	SPST	3–32 VDC	60 VDC	2 A
175222CP	SPST	3–32 VDC	60 VDC	4 A
176647CP	SPST	3–32 VDC	120 VDC	5 A

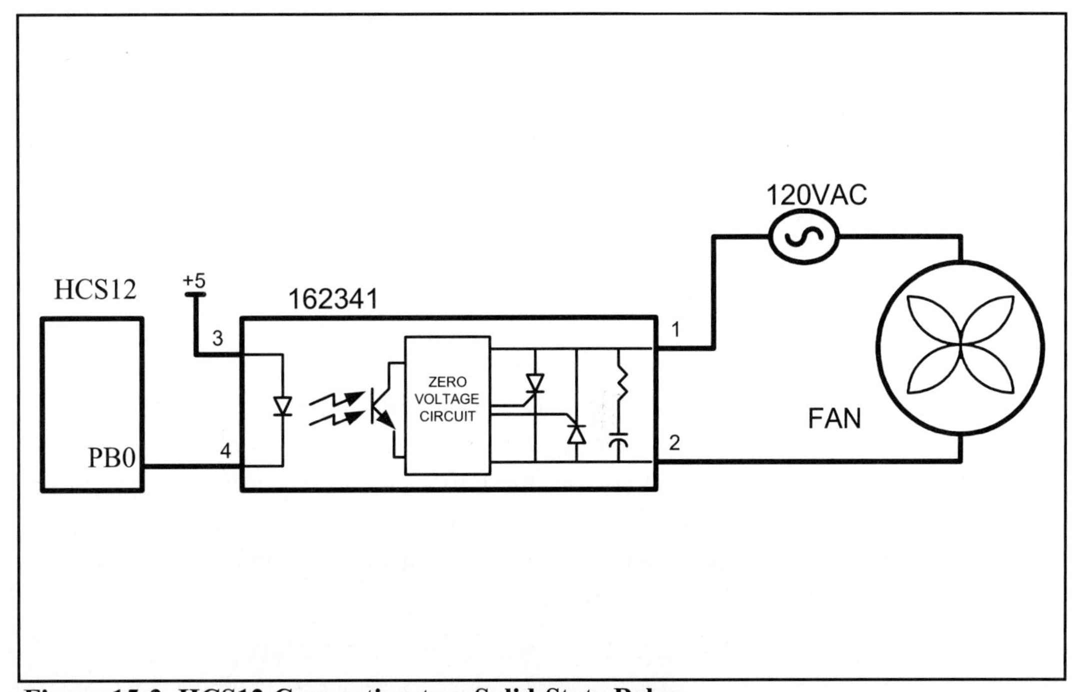

Figure 15-3. HCS12 Connection to a Solid-State Relay

Reed switch

Another popular switch is the reed switch. When the reed switch is placed in a magnetic field, the contact is closed. When the magnetic field is removed, the contact is forced open by its own spring. See Figure 15-4. The reed switch is ideal for moist and marine environments where it can be submerged in fuel or water. Reed switches are also widely used in dirty and dusty atmospheres because they are tightly sealed.

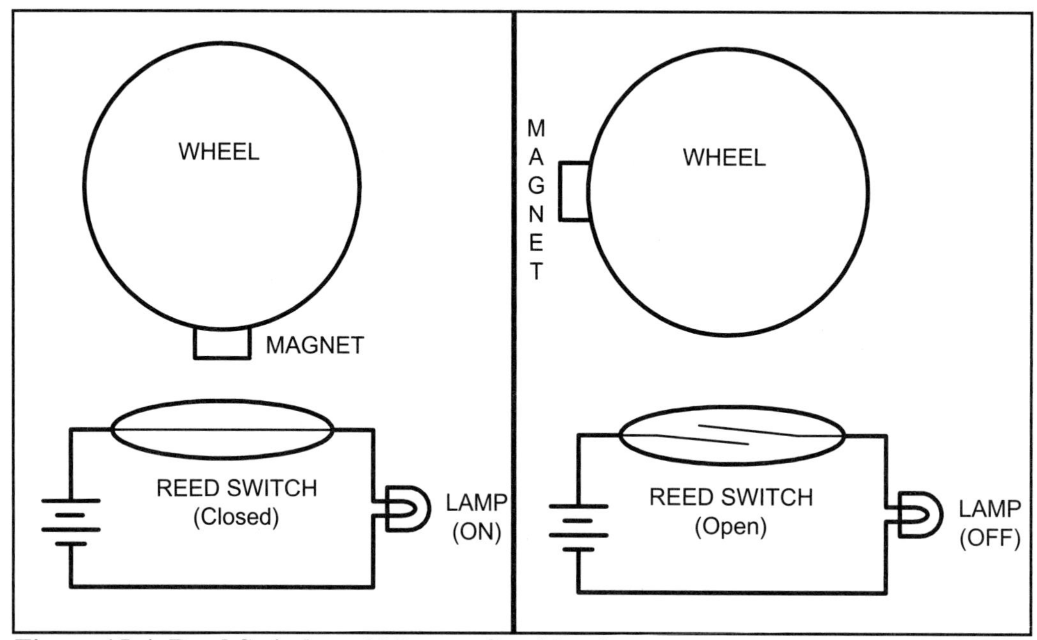

Figure 15-4. Reed Switch and Magnet Combination

Optoisolator

In some applications we use an optoisolator (also called an *optocoupler*) to isolate two parts of a system. An example is driving a motor. Motors can produce what is called *back EMF*, a high-voltage spike produced by a sudden change of current as indicated in the V = Ldi/dt formula. In situations such as printed circuit board design, we can reduce the effect of this unwanted voltage spike (called *ground bounce*) by using decoupling capacitors (see Appendix C). In systems that have inductors (coil windings), such as motors, a decoupling capacitor or a diode will not do the job. In such cases we use optoisolators. An optoisolator has an LED (light-emitting diode) transmitter and a photosensor receiver, separated from each other by a gap. When current flows through the diode, it transmits a light signal across the gap and the receiver produces the same signal with the same phase but a different current and amplitude. See Figure 15-5. Optoisolators are also widely used in communication equipment such as modems. This device allows a computer to be connected to a telephone line without risk of damage from power surges. The gap between the transmitter and receiver of optoisolators prevents the electrical current surge from reaching the system.

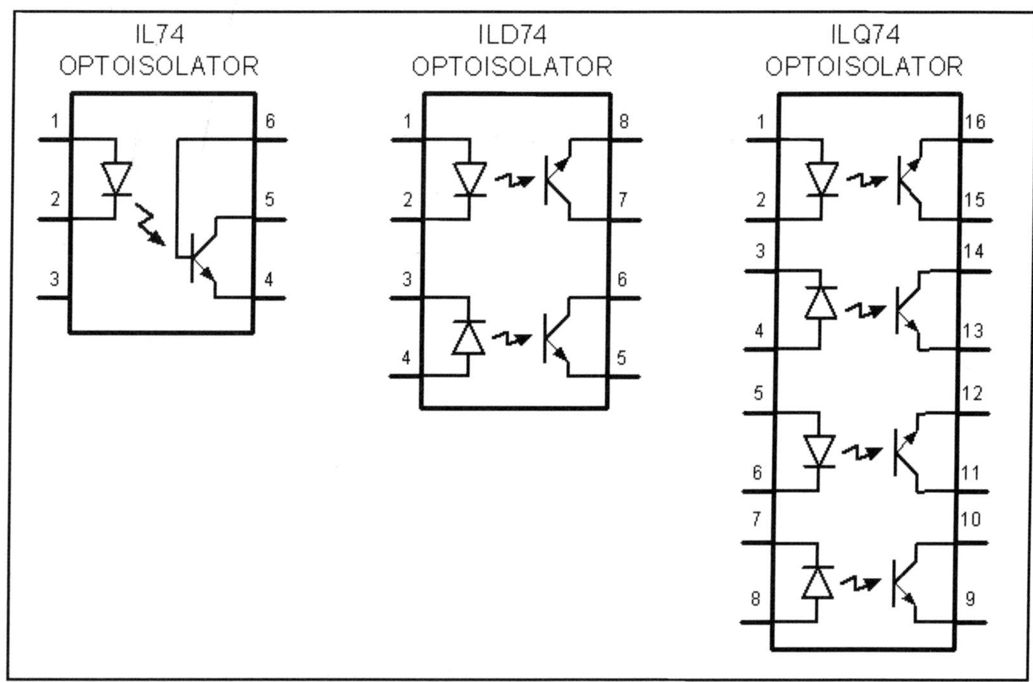

Figure 15-5. Optoisolator Package Examples

Interfacing an optoisolator

The optoisolator comes in a small IC package with four or more pins. There are also packages that contain more than one optoisolator. When placing an optoisolator between two circuits, we must use two separate voltage sources, one for each side, as shown in Figure 15-6. Unlike relays, no drivers need to be placed between the microcontroller/digital output and the optoisolators.

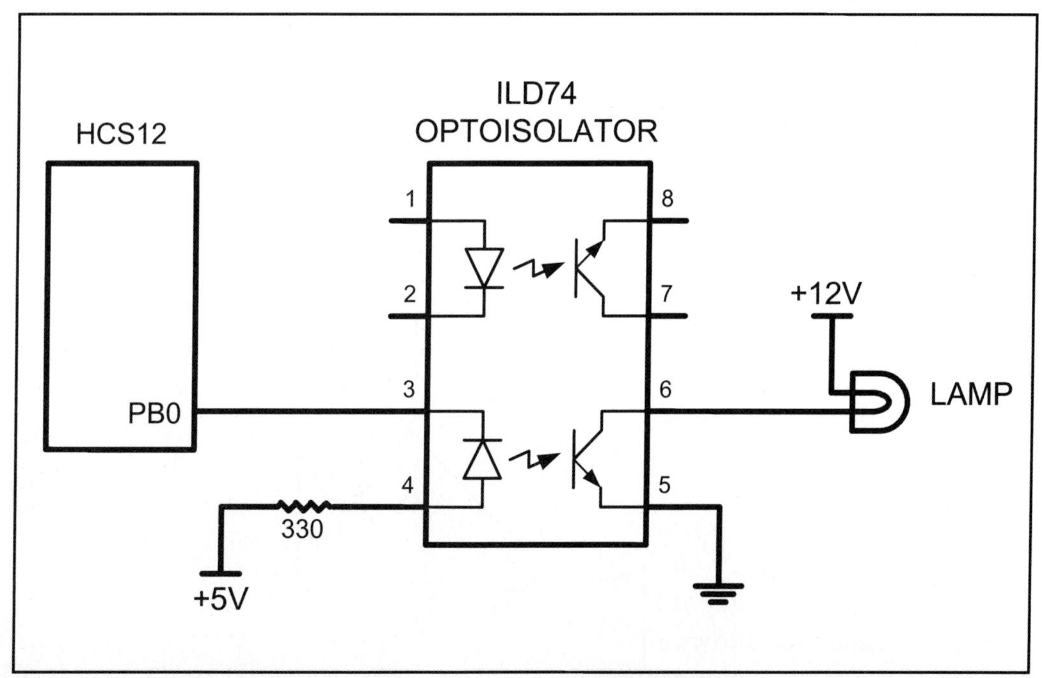

Figure 15-6. Controlling a Lamp via an Optoisolator

Review Questions

1. Give one application where would you use a relay.
2. Why do we place a driver between the microcontroller and the relay?
3. What is an NC relay?
4. Why are relays that use coils called *electromechanical* relays?
5. What is the advantage of a solid-state relay over an EMR?
6. What is the advantage of an optoisolator over an EMR?

SECTION 15.2: STEPPER MOTOR INTERFACING

This section begins with an overview of the basic operation of stepper motors. Then we describe how to interface a stepper motor to the HCS12. Finally, we use Assembly language programs to demonstrate control of the angle and direction of stepper motor rotation.

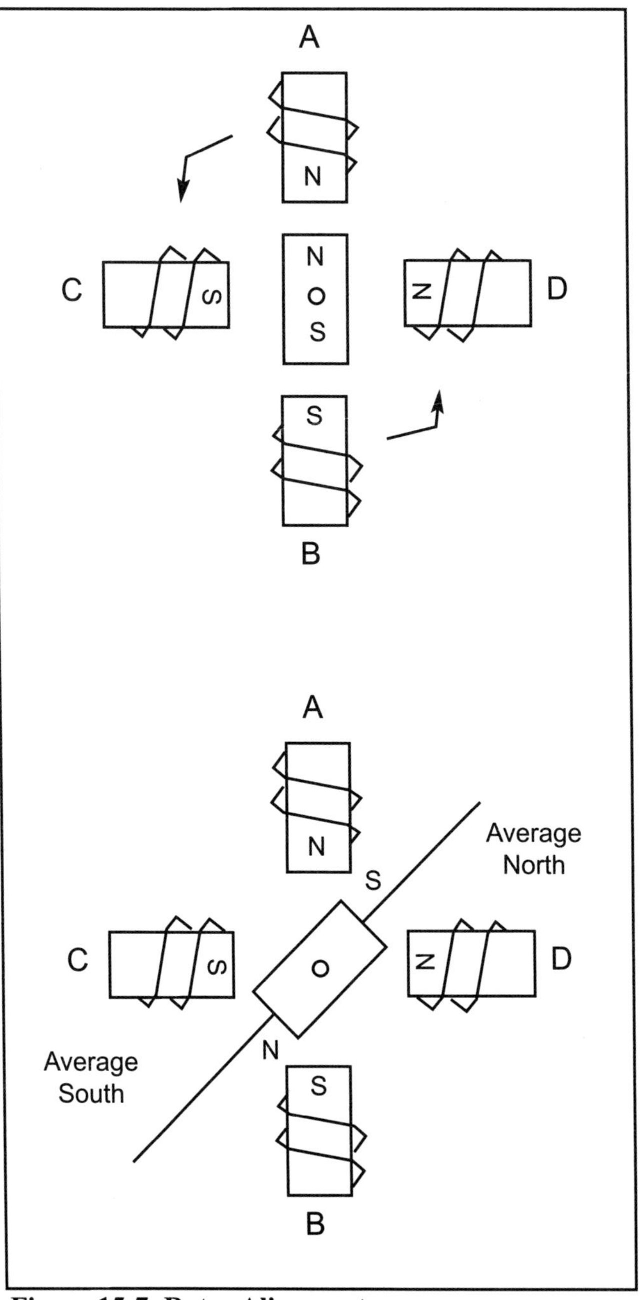

Figure 15-7. Rotor Alignment

Stepper motors

A *stepper motor* is a widely used device that translates electrical pulses into mechanical movement. In applications such as disk drives, dot matrix printers, and robotics, the stepper motor is used for position control. Stepper motors commonly have a permanent magnet *rotor* (also called the *shaft*) surrounded by a *stator* (see Figure 15-7). There are also steppers called *variable-reluctance stepper motors* that do not have a permanent magnet rotor. The most common stepper motors have four stator windings that are paired with a center-tapped common as shown in Figure 15-8. This type of stepper motor is commonly referred to as a *four-phase* or *unipolar* stepper motor. The center tap allows a change of current direction in each of two coils when a winding is grounded, thereby resulting in a polarity change of the stator. Notice that while a conventional motor shaft runs freely, the stepper motor shaft moves in a fixed repeatable increment, which allows one to move it to a precise position. This repeatable fixed movement is possible as a result of basic magnetic theory where poles of the same polarity repel and opposite poles attract. The direction of the rotation is dictated by the stator poles. The stator poles are determined by the current sent through the wire coils. As the direction of the current is changed, the polarity is also changed causing the reverse motion of the rotor. The stepper motor discussed here has a total of six leads: four leads representing the four stator windings and two commons for the center-tapped leads. As the sequence of power is applied to each stator winding, the rotor will rotate. There are several widely used sequences, each of which has a different degree of precision. Table 15-3 shows a two-phase, four-step stepping sequence.

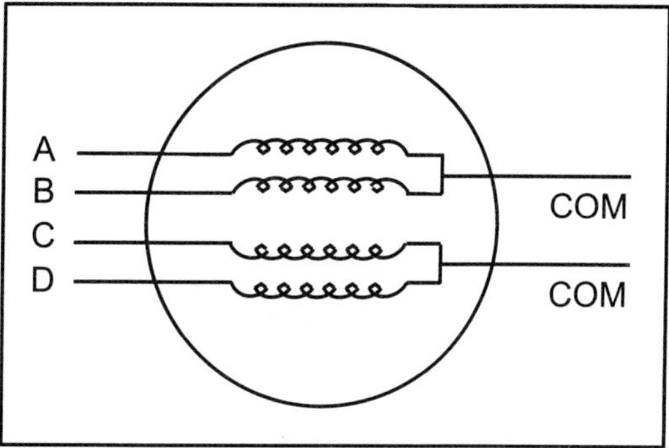

Figure 15-8. Stator Winding Configuration

Note that although we can start with any of the sequences in Table 15-3, once we start we must continue in the proper order. For example, if we start with step 3 (0110), we must continue in the sequence of steps 4, 1, 2, etc.

Table 15-3: Normal Four-Step Sequence

Clockwise ↓ (steps 1 to 4); Counter-clockwise ↑ (steps 4 to 1)

Step #	Winding A	Winding B	Winding C	Winding D
1	1	0	0	1
2	1	1	0	0
3	0	1	1	0
4	0	0	1	1

Step angle

How much movement is associated with a single step? This depends on the internal construction of the motor, in particular the number of teeth on the rotor, and on the stator. The *step angle* is the minimum degree of rotation associated with a single step. Various motors have different step angles. Table 15-4 shows some step angles for various motors. In Table 15-4, notice the term *steps per revolution*. This is the total number of steps needed to rotate one complete rotation or 360 degrees (e.g., 180 steps × 2 degrees = 360).

Table 15-4: Stepper Motor Step Angles

Step Angle	Steps per Revolution
0.72 degrees	500
1.8	200
2.0	180
2.5	144
5.0	72
7.5	48
15	24

It must be noted that perhaps contrary to one's initial impression, a stepper motor does not need more terminal leads for the stator to achieve smaller steps. All the stepper motors discussed in this section have four leads for the stator winding and two COM wires for the center tap. Although some manufacturers set aside only one lead for the common signal instead of two, they always have four leads for the stators. See Example 15-1. Next we discuss some associated terminology in order to understand the stepper motor further.

Example 15-1

Describe the HCS12 connection to the stepper motor of Figure 15-9 and code a program to rotate it continuously.

Solution:

The following steps show the HCS12 connection to the stepper motor and its programming:

1. Use an ohmmeter to measure the resistance of the leads. This should identify which COM leads are connected to which winding leads.
2. The common wire(s) are connected to the positive side of the motor's power supply. In many motors, +5 V is sufficient.
3. The four leads of the stator winding are controlled by four bits of the HCS12 port (PB0–PB3). Because the HCS12 lacks sufficient current to drive the stepper motor windings, we must use a driver such as the ULN2003 (or ULN2008) to energize the stator. Instead of the ULN2003, we could have used transistors as drivers, as shown in Figure 15-11. However, notice that if transistors are used as drivers, we must also use diodes to take care of inductive current generated as the coil is turned off. One reason that using the ULN2003 is preferable to the use of transistors as drivers is that the ULN2003 has an internal diode to take care of back EMF.

```
R2          DS.B 1              ;loc 820H for R2 Reg
            MOVB #$FF,DDRB      ;Port B as output
            CLC
            LDAA #$66           ;load step sequence
BACK        STAA PORTB          ;issue sequence to motor
            LSRA                ;rotate right clockwise
            BCC OVER
            ORAA #$80
OVER        JSR  DELAY          ;wait
            BRA  BACK           ;keep going
DELAY       MOVB R2,$FF
D1          NOP
            NOP
            NOP
            NOP
            NOP
            NOP
            DEC R2
            BNEQ D1
            RTS
```

Change the value of DELAY to set the speed of rotation.

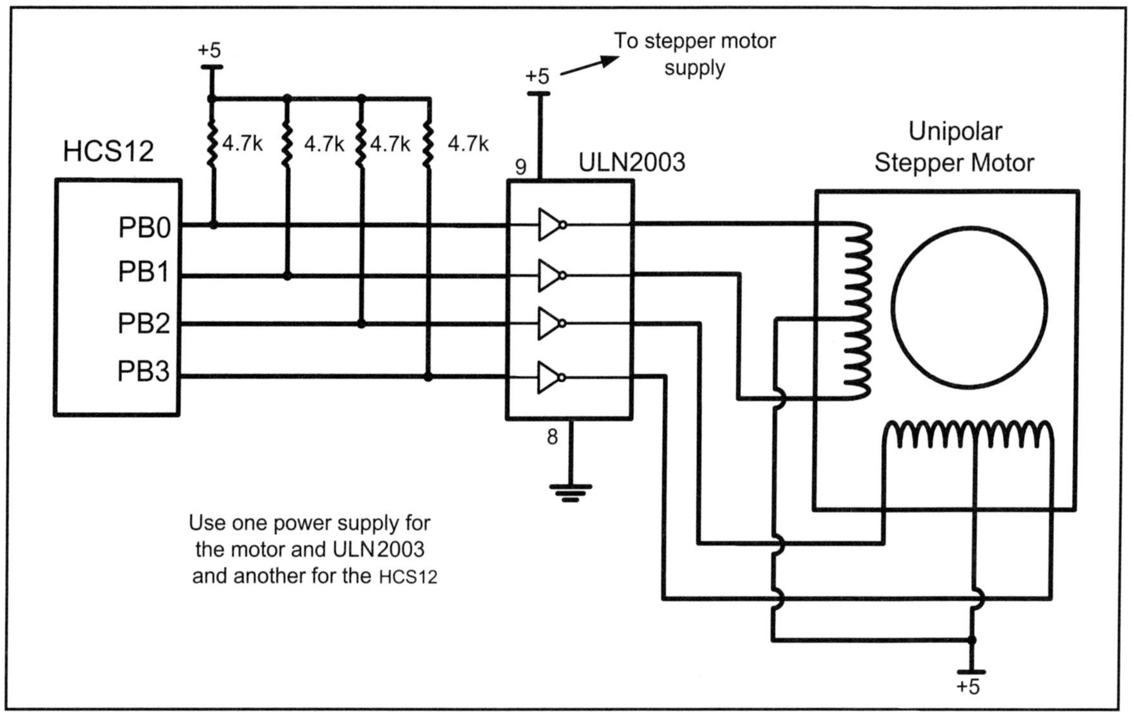

Figure 15-9. HCS12 Connection to Stepper Motor

Steps per second and rpm relation

The relation between rpm (revolutions per minute), steps per revolution, and steps per second is as follows.

$$\text{Steps per second} = \frac{\text{rpm} \times \text{Steps per revolution}}{60}$$

The 4-step sequence and number of teeth on rotor

The switching sequence shown earlier in Table 15-3 is called the *4-step* switching sequence because after four steps the same two windings will be "ON". How much movement is associated with these four steps? After completing every four steps, the rotor moves only one tooth pitch. Therefore, in a stepper motor with 200 steps per revolution, the rotor has 50 teeth because $4 \times 50 = 200$ steps are needed to complete one revolution. This leads to the conclusion that the minimum step angle is always a function of the number of teeth on the rotor. In other words, the smaller the step angle, the more teeth the rotor passes. See Example 15-2.

Example 15-2

Give the number of times the four-step sequence in Table 15-3 must be applied to a stepper motor to make an 80-degree move if the motor has a 2-degree step angle.

Solution:

A motor with a 2-degree step angle has the following characteristics:

Step angle:	2 degrees	Steps per revolution:	180
Number of rotor teeth:	45	Movement per 4-step sequence:	8 degrees

To move the rotor 80 degrees, we need to send 10 consecutive 4-step sequences, because 10×4 steps $\times$ 2 degrees = 80 degrees.

Looking at Example 15-2, one might wonder what happens if we want to move 45 degrees, because the steps are 2 degrees each. To allow for finer resolutions, all stepper motors allow what is called an *8-step* switching sequence. The 8-step sequence is also called *half-stepping,* because in the 8-step sequence each step is half of the normal step angle. For example, a motor with a 2-degree step angle can be used as a 1-degree step angle if the sequence of Table 15-5 is applied.

Table 15-5: Half-Step 8-Step Sequence

Clockwise ↓ / Counter-clockwise ↑

Step #	Winding A	Winding B	Winding C	Winding D
1	1	0	0	1
2	1	0	0	0
3	1	1	0	0
4	0	1	0	0
5	0	1	1	0
6	0	0	1	0
7	0	0	1	1
8	0	0	0	1

Motor speed

The motor speed, measured in steps per second (steps/s), is a function of the switching rate. Notice in Example 15-1 that by changing the length of the time delay loop, we can achieve various rotation speeds.

Holding torque

The following is a definition of holding torque: "With the motor shaft at standstill or zero rpm condition, the amount of torque, from an external source, required to break away the shaft from its holding position. This is measured with rated voltage and current applied to the motor." The unit of torque is ounce-inch (or kg-cm).

Wave drive 4-step sequence

In addition to the 8-step and the 4-step sequences discussed earlier, there is another sequence called the *wave drive 4-step sequence*. It is shown in Table 15-6. Notice that the 8-step sequence of Table 15-5 is simply the combination of the wave drive 4-step and normal 4-step normal sequences shown in Tables 15-6 and 15-3, respectively. Experimenting with the wave drive 4-step sequence is left to the reader.

Table 15-6: Wave Drive 4-Step Sequence

Clockwise ↓ / Counter-clockwise ↑

Step #	Winding A	Winding B	Winding C	Winding D
1	1	0	0	0
2	0	1	0	0
3	0	0	1	0
4	0	0	0	1

Table 15-7: Selected Stepper Motor Characteristics (http://www.Jameco.com)

Part No.	Step Angle	Drive System	Volts	Phase Resistance	Current
151861CP	7.5	unipolar	5 V	9 ohms	550 mA
171601CP	3.6	unipolar	7 V	20 ohms	350 mA
164056CP	7.5	bipolar	5 V	6 ohms	800 mA

Unipolar versus bipolar stepper motor interface

There are three common types of stepper motor interfacing: universal, unipolar, and bipolar. They can be identified by the number of connections to the motor. A universal stepper motor has eight, while the unipolar has six and the bipolar has four. The universal stepper motor can be configured for all three modes, while the unipolar can be either unipolar or bipolar. Obviously the bipolar cannot be configured for the universal or the unipolar mode. Table 15-7 shows selected stepper motor characteristics. Figure 15-10 shows the basic internal connections of all three type of configurations.

Unipolar stepper motors can be controlled using the basic interfacing shown in Figure 15-11, whereas the bipolar stepper requires H-Bridge circuitry. Bipolar stepper motors require a higher operational current than the unipolar; the advantage of this is a higher holding torque.

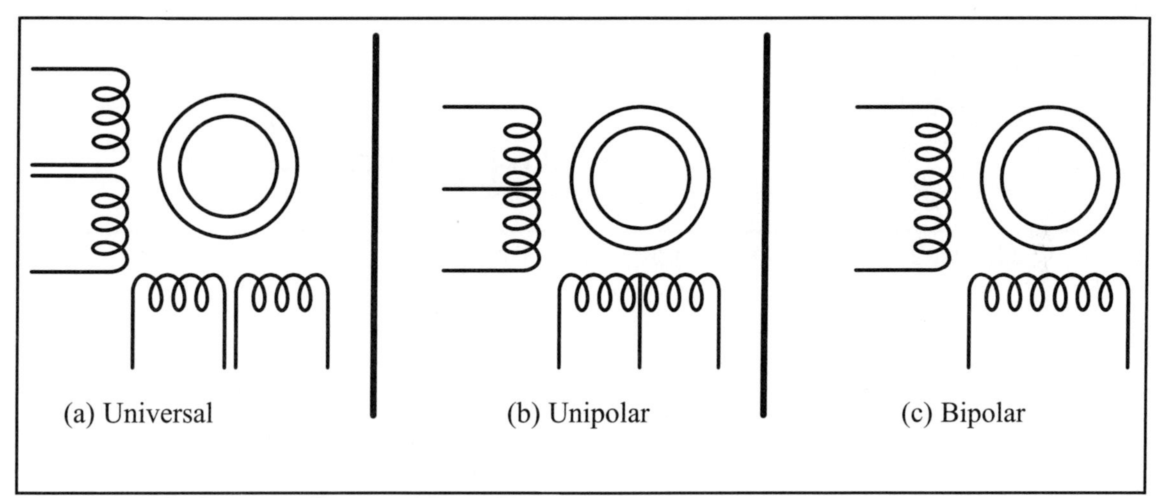

Figure 15-10. Common Stepper Motor Types

Using transistors as drivers

Figure 15-11 shows an interface to a unipolar stepper motor using transistors. Diodes are used to reduce the back EMF spike created when the coils are energized and de-energized, similar to the electromechanical relays discussed earlier. TIP transistors can be used to supply higher current to the motor. Table 15-8 lists the common industrial Darlington transistors. These transistors can accommodate higher voltages and currents.

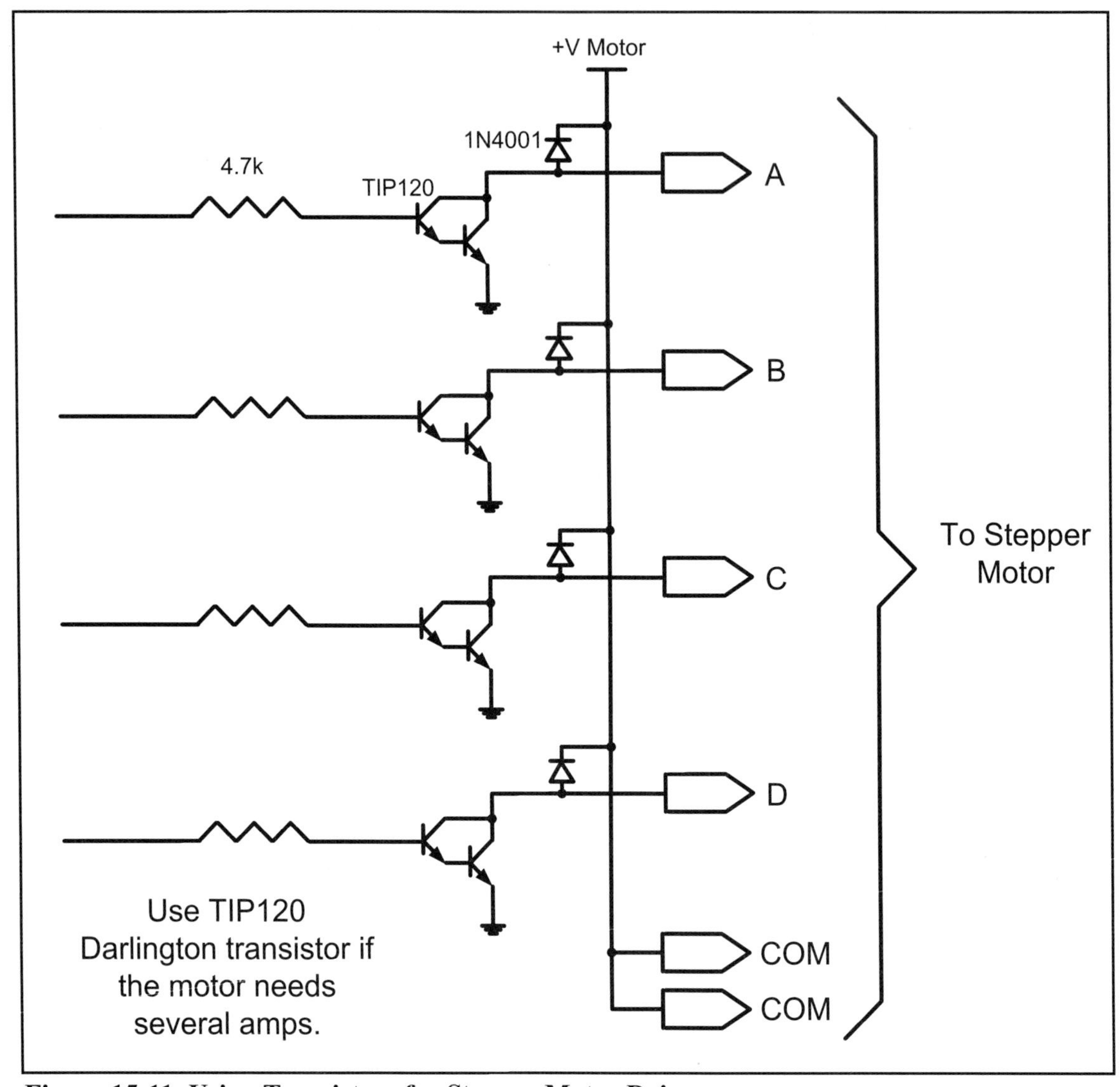

Figure 15-11. Using Transistors for Stepper Motor Driver

Table 15-8: Darlington Transistor Listing

NPN	PNP	Vceo (volts)	Ic (amps)	hfe (common)
TIP110	TIP115	60	2	1,000
TIP111	TIP116	80	2	1,000
TIP112	TIP117	100	2	1,000
TIP120	TIP125	60	5	1,000
TIP121	TIP126	80	5	1,000
TIP122	TIP127	100	5	1,000
TIP140	TIP145	60	10	1,000
TIP141	TIP146	80	10	1,000
TIP142	TIP147	100	10	1,000

Controlling stepper motor via optoisolator

In the first section of this chapter we examined the optoisolator and its use. Optoisolators are widely used to isolate the stepper motor's EMF voltage and keep it from damaging the digital/microcontroller system. This is shown in Figure 15-12. See Examples 15-3 and 15-4.

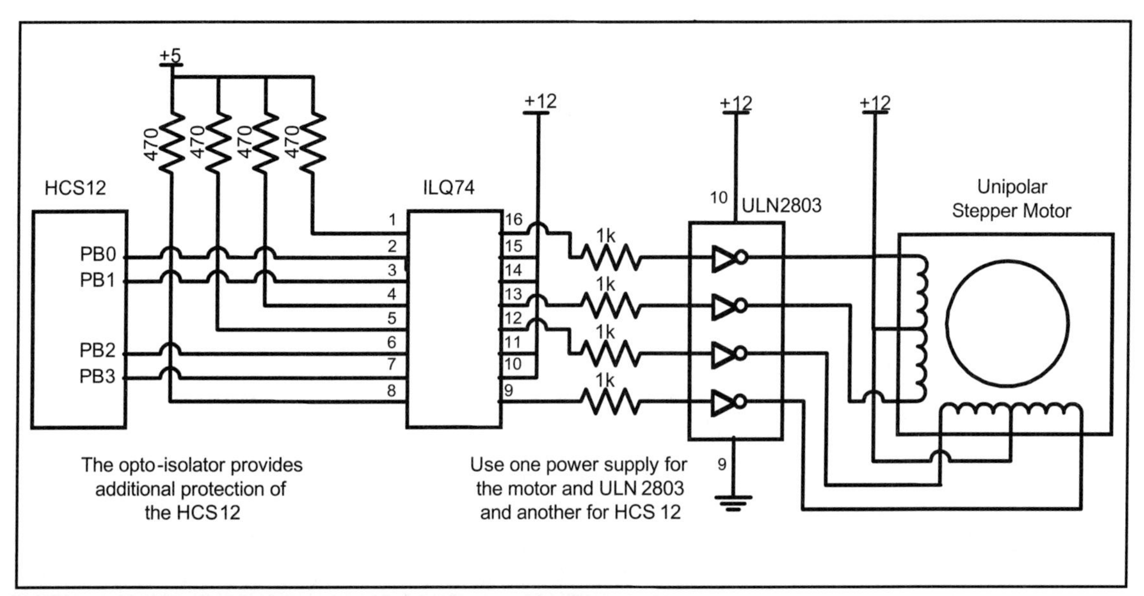

Figure 15-12. Controlling Stepper Motor via Optoisolator

Example 15-3

A switch is connected to pin PA7 (PORTA.7). Write a program to monitor the status of SW and perform the following:
(a) If SW = 0, the stepper motor moves clockwise.
(b) If SW = 1, the stepper motor moves counterclockwise.

Solution:

```
      MOVB  #$FF,DDRB       ;Port B as output
      MOVB  #$7F,DDRA       ;make PA7 an input
      CLC
      LDAA  #$66            ;starting phase value
      STAA  PORTB           ;send value to port
TURN  BRCLR PORTA,$80,CW    ;check switch result
      LSRA                  ;rotate right
      BCC   OVER
      ORAA  #$80
OVER  JSR   DELAY           ;call delay
      STAA  PORTB           ;write value to port
      BRA   TURN            ;repeat
CW    LSLA                  ;rotate left
      BCC   OV1
      ORAA  #$01
OV1   JSR   DELAY           ;call delay
      STAA  PORTB           ;write value to port
      BRA   TURN            ;repeat
```

Stepper motor control with HCS12 C

The HCS12 C version of the stepper motor control is given below. In this program we could have used << (shift left) and >> (shift right) as was shown in Chapter 7.

```
//Program 15-1
#include <hcs12.h>
void main()
  {
     DDRB=0xFF;          //PORTB as output
     while(1)
       {
          PORTB = 0x66;
          MSDelay(100);
          PORTB = 0xCC;
          MSDelay(100);
          PORTB = 0x99;
          MSDelay(100);
          PORTB = 0x33;
          MSDelay(100);
       }
  }
```

Example 15-4

A switch is connected to pin PA7. Write a C program to monitor the status of SW and perform the following:
(a) If SW = 0, the stepper motor moves clockwise.
(b) If SW = 1, the stepper motor moves counterclockwise.

Solution:

```
#include <hcs12.h>
void MSDelay(unsigned int ms);
void main()
  {
     unsigned char x;
     DDRA=0x7F;          //PA7 as input pin
     DDRB=0xFF;          //PORTB as output
     while(1)
      {
         x = PORTA;
         x = x & 0x80;
         if(x == 0)
           {
           PORTB = 0x66;
           MSDelay(100);
           PORTB = 0xCC;
           MSDelay(100);
           PORTB = 0x99;
           MSDelay(100);
           PORTB = 0x33;
           MSDelay(100);
           }
```

Example 15-4 Cont.

```
            else
              {
              PORTB = 0x66;
              MSDelay(100);
              PORTB = 0x33;
              MSDelay(100);
              PORTB = 0x99;
              MSDelay(100);
              PORTB = 0xCC;
              MSDelay(100);
              }
          }
    }

void MSDelay(unsigned int value)
    {
      unsigned int x, y;
      for(x=0;x<1275;x++)
        for(y=0;y<value;y++);
    }
```

Review Questions

1. Give the 4-step sequence of a stepper motor if we start with 0110.
2. A stepper motor with a step angle of 5 degrees has ____ steps per revolution.
3. Why do we put a driver between the microcontroller and the stepper motor?

PROBLEMS

SECTION 15.1: RELAYS AND OPTOISOLATORS

1. True or false. The minimum voltage needed to energize a relay is the same for all relays.
2. True or false. The minimum current needed to energize a relay depends on the coil resistance.
3. Give the advantages of a solid-state relay over an EMR.
4. True or false. In relays, the energizing voltage is the same as the contact voltage.
5. Find the current needed to energize a relay if the coil resistance is 1,200 ohms and the coil voltage is 5 V.
6. Give two applications for an optoisolator.
7. Give the advantages of an optoisolator over an EMR.
8. Of the EMR and solid-state relay, which has the problem of back EMF?
9. True or false. The greater the coil resistance, the worse the back EMF voltage.
10. True or false. We should use the same voltage sources for both the coil voltage and contact voltage.

SECTION 15.2: STEPPER MOTOR INTERFACING

11. If a motor takes 90 steps to make one complete revolution, what is the step

angle for this motor?

12. Calculate the number of steps per revolution for a step angle of 7.5 degrees.
13. Finish the normal four-step sequence clockwise if the first step is 0011 (binary).
14. Finish the normal four-step sequence clockwise if the first step is 1100 (binary).
15. Finish the normal four-step sequence counterclockwise if the first step is 1001 (binary).
16. Finish the normal four-step sequence counterclockwise if the first step is 0110 (binary).
17. What is the purpose of the ULN2003 placed between the HCS12 and the stepper motor? Can we use that for 3A motors?
18. Which of the following cannot be a sequence in the normal four-step sequence for a stepper motor?
    (a) CCH (b) DDH (c) 99H (d) 33H
19. What is the effect of a time delay between issuing each step?
20. In Problem 19, how can we make a stepper motor go faster?

ANSWERS TO REVIEW QUESTIONS

SECTION 15.1: RELAYS AND OPTOISOLATORS

1. With a relay we can use a 5 V digital system to control 12 V–120 V devices such as horns and appliances.
2. Because microcontroller/digital outputs lack sufficient current to energize the relay, we need a driver.
3. When the coil is not energized, the contact is closed.
4. When current flows through the coil, a magnetic field is created around the coil, which causes the armature to be attracted to the coil.
5. It is faster and needs less current to get energized.
6. It is smaller and can be connected to the microcontroller directly without a driver.

SECTION 15.2: STEPPER MOTOR INTERFACING

1. 0110, 0011, 1001, 1100 for clockwise; and 0110, 1100, 1001, 0011 for counterclockwise
2. 72
3. Because the microcontroller pins do not provide sufficient current to drive the stepper motor

CHAPTER 16

SPI PROTOCOL AND RTC INTERFACING WITH HCS12

OBJECTIVES

Upon completion of this chapter, you will be able to:

- >> Understand the Serial Peripheral Interfacing (SPI) protocol
- >> Explain how the SPI read and write operations work
- >> Examine the SPI pins SDO, SDI, CE, and SCLK
- >> Code programs in Assembly and C for SPI
- >> Explain how the real-time clock (RTC) chip works
- >> Explain the function of the DS1306 RTC pins
- >> Explain the function of the DS1306 RTC registers
- >> Understand the interfacing of the DS1306 RTC to the HCS12
- >> Code programs to display time and date in Assembly and C
- >> Explore and program the alarm and interrupt features of the RTC

This chapter discusses the SPI bus and shows the interfacing and programming of the DS1306 real-time clock (RTC), an SPI chip. We will describe the DS1306 RTC's pin functions and show its interfacing and programming with the HCS12. Will use both Assembly and C for programs.

SECTION 16.1: SPI BUS PROTOCOL

The SPI (serial peripheral interface) is a bus interface connection incorporated into many devices such as ADC, DAC, and EEPROM. In this section we examine the pins of the SPI bus and show how the read and write operations in the SPI work.

SPI bus

The SPI bus was originally started by Motorola Corp. (now Freescale), but in recent years has become a widely used standard adapted by many semiconductor chip companies. SPI devices use only 2 pins for data transfer, called SDI (Din) and SDO (Dout), instead of the 8 or more pins used in traditional buses. This reduction of data pins reduces the package size and power consumption drastically, making them ideal for many applications in which space is a major concern. The SPI bus has the SCLK (shift clock) pin to synchronize the data transfer between two chips. The last pin of the SPI bus is CE (chip enable), which is used to initiate and terminate the data transfer. These four pins, SDI, SDO, SCLK, and CE, make the SPI a 4-wire interface. See Figure 16-1. There is also a widely used standard called a *3-wire interface bus*. In a 3-wire interface bus, we have SCLK and CE, and only a single pin for data transfer. The SPI 4-wire bus can become a 3-wire interface when the SDI and SDO data pins are tied together. However, there are some major differences between the SPI and 3-wire devices in the data transfer protocol. For that reason, a device must support the 3-wire protocol internally in order to be used as a 3-wire device. Many devices such as the DS1306 RTC (real-time clock) support both SPI and 3-wire protocols.

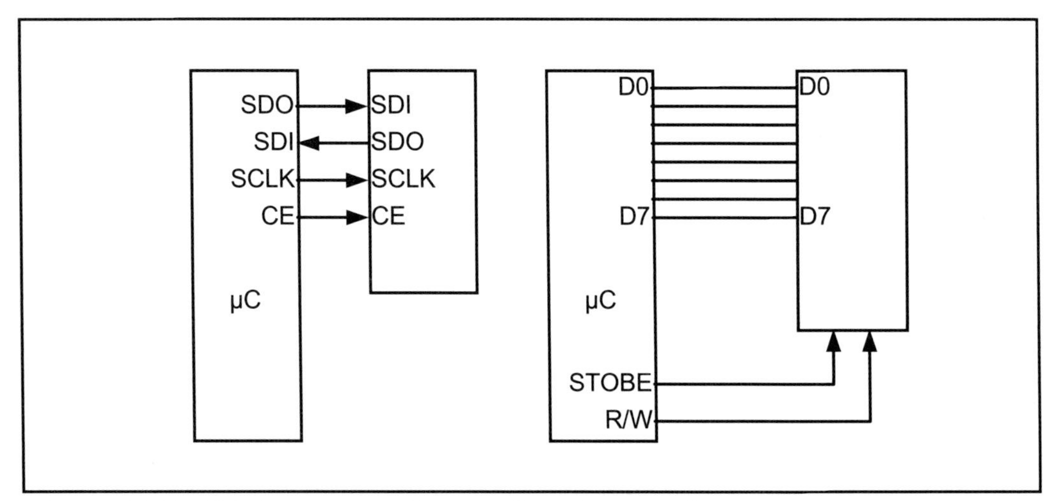

Figure 16-1. SPI Bus vs. Traditional Parallel Bus Connection to Microcontroller

SPI read and write protocol

In connecting a device with an SPI bus to a microcontroller, we use the microcontroller as the master while the SPI device acts as a slave. This means that the microcontroller generates the SCLK, which is fed to the SCLK pin of the SPI device. The SPI protocol uses SCLK to synchronize the transfer of information one bit at a time, where the most-significant bit (MSB) goes in first. During the transfer, the CE must stay HIGH. The information (address and data) is transferred between the microcontroller and the SPI device in groups of 8 bits, where the address byte is followed immediately by the data byte. To distinguish between the read and the write, the D7 bit of the address byte is always 1 for write, while for the read, the D7 bit is LOW, as we will see next.

Steps for writing data to an SPI device

In accessing SPI devices, we have two modes of operation: single-byte and multibyte. We will explain each one separately.

Single-byte write

The following steps are used to send (write) data in single-byte mode for SPI devices, as shown in Figure 16-2:

1. Make CE = 1 to begin writing.
2. The 8-bit address is shifted in one bit at a time, with each edge of SCLK. Notice that A7 = 1 for the write operation, and the A7 bit goes in first.
3. After all 8 bits of the address are sent in, the SPI device expects to receive the data belonging to that address location immediately.
4. The 8-bit data is shifted in one bit at a time, with each edge of the SCLK.
5. Make CE = 0 to indicate the end of the write cycle.

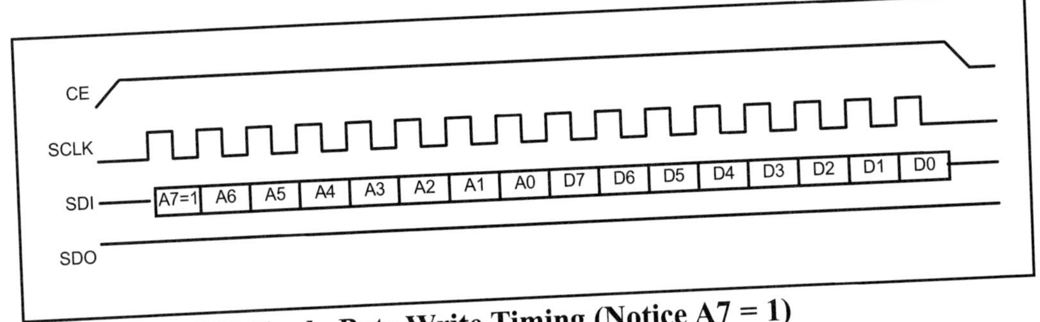

Figure 16-2. SPI Single-Byte Write Timing (Notice A7 = 1)

Multibyte burst write

Burst mode writing is an effective means of loading consecutive locations. In burst mode, we provide the address of the first location, followed by the data for that location. From then on, while CE = 1, consecutive bytes are written to consecutive memory locations. In this mode, the SPI device internally increments the

SPI baud rate selection

We use the SPIBR (SPI Baud Rate) register to set the SCLK rate for data transfer and receive. See Figure 16-6 and Example 16-1. Notice from Figure 16-6, we have multiple options for the same baud rate.

0	SPPR2	SPPR1	SPPR0	0	SPR2	SPR1	SPR0

SPPR2–SPPR0 D6–D4 SPI Baud Rate Preselection bits

SPR2–SPR0 D2–D0 SPI Baud Rate Selection bits

These bits specify the SPI baud rates as shown in the following two equations:

$$\text{BaudRateDivisor} = (\text{SPPR} + 1) \times 2^{(\text{SPR} + 1)}$$

$$\text{Baud Rate} = \text{BusClock} / \text{BaudRateDivisor}$$

SPPR2	SPPR1	SPPR0	SPR2	SPR1	SPR0	Hex	BaudRateDivisor
0	0	0	0	0	0	00	2
0	0	0	0	0	1	01	4
0	0	0	0	1	0	02	8
0	0	0	0	1	1	03	16
0	0	0	1	0	0	04	32
0	0	0	1	0	1	05	64
.	.	.	.	.	.	.	.
1	1	0	1	1	1	67	1792
1	1	1	0	0	0	70	16
1	1	1	0	0	1	71	32
1	1	1	0	1	0	72	64
1	1	1	0	1	1	73	128
1	1	1	1	0	0	74	256
1	1	1	1	0	1	75	512
1	1	1	1	1	0	76	1024
1	1	1	1	1	1	77	2048

Note: The highest Baud Rate = BusFreq/2 and the lowest Baud Rate is BusFreq/2048.

Figure 16-6. SPIBR–SPI Baud Rate Register

Example 16-1

Using BusFreq = 2 MHz, find the value for the SPI Baud Rate (SPIBR) register for the following baud rates: a) 1 MHz, b) 500 kHz c) 7,812.5 Hz

Solution:

(a) For 2 MHz/1 MHz = 2, we have SPIBR = 00000000 = 00 hex.
(b) For 2 MHz/500 kHz = 4, we have SPIBR = 0000001= 01 hex.
(c) For 2 MHz/7,812.5 Hz = 256, we have SPIBR = 01110100 = 74 hex.

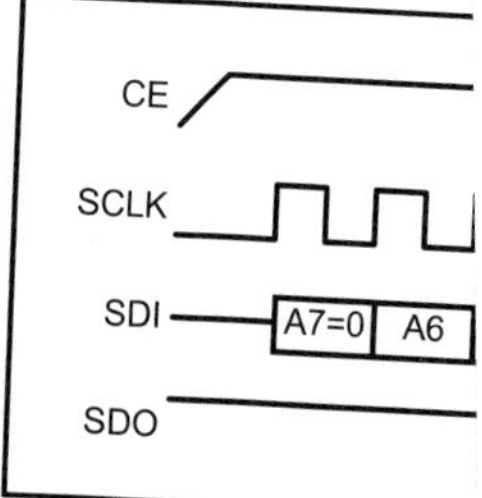

Multibyte burst read

Burst mode reading is an effective means of bringing out the contents of consecutive locations. In burst mode, we provide the address of the first location only. From then on, while CE = 1, consecutive bytes are brought out from consecutive memory locations. In this mode, the SPI device internally increments the address location as long as CE is HIGH. The following steps are used to get (read) multiple bytes of data in burst mode for SPI devices, as shown in Figure 16-5:

1. Make CE = 1 to begin reading.
2. The 8-bit address of the first location is provided and shifted in one bit at a time, with each edge of SCLK. Notice that A7 = 0 for the read operation, and the A7 bit goes in first.
3. The 8-bit data for the first location is shifted out one bit at a time, with each edge of the SCLK. From then on, we simply keep getting consecutive bytes of data belonging to consecutive memory locations. In the process, CE must stay HIGH to indicate that this is a burst mode multibyte read operation.
4. Make CE = 0 to end reading.

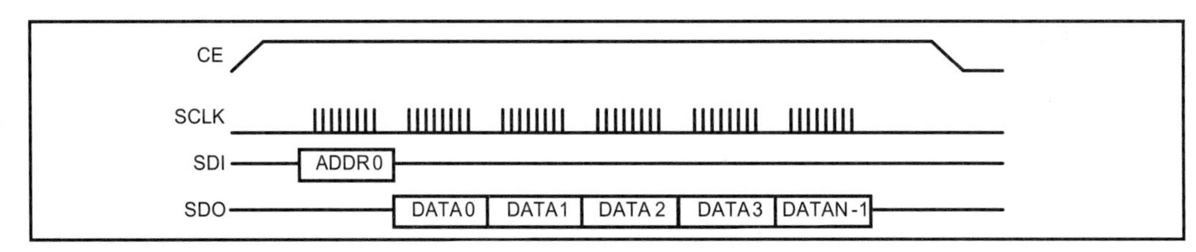

Figure 16-5. SPI Burst (Multibyte) Mode Reading

Review Questions

1. True or false. The SPI protocol writes and reads information in 8-bit chunks.
2. True or false. In SPI, the address is immediately followed by the data.
3. True or false. In an SPI write cycle, bit A7 of the address is LOW.
4. True or false. In an SPI write, the LSB goes in first.
5. State the difference between the single-byte and burst modes in terms of the CE signal.

SECTION 16.2: SPI MODULES IN THE HCS12

The SPI module inside the HCS12 supports SPI bus protocol. Five registers are associated with SPI of the HCS12. They are SPIDR (SPI Data Register), SPICR1 (SPI Control Register 1), SPICR2 (SPI Control Register 2), SPIBR (SPI Baud Rate register), and SPISR (SPI Status Register). Next, we describe each one.

SPI baud rate selection

We use the SPIBR (SPI Baud Rate) register to set the SCLK rate for data transfer and receive. See Figure 16-6 and Example 16-1. Notice from Figure 16-6, we have multiple options for the same baud rate.

0	SPPR2	SPPR1	SPPR0	0	SPR2	SPR1	SPR0

SPPR2–SPPR0 D6–D4 SPI Baud Rate Preselection bits

SPR2–SPR0 D2–D0 SPI Baud Rate Selection bits

These bits specify the SPI baud rates as shown in the following two equations:

BaudRateDivisor = (SPPR + 1) × $2^{(SPR + 1)}$

Baud Rate = BusClock / BaudRateDivisor

SPPR2	SPPR1	SPPR0	SPR2	SPR1	SPR0	Hex	BaudRateDivisor
0	0	0	0	0	0	00	2
0	0	0	0	0	1	01	4
0	0	0	0	1	0	02	8
0	0	0	0	1	1	03	16
0	0	0	1	0	0	04	32
0	0	0	1	0	1	05	64
.	.	.	.	.	.	.	.
.	.	.	.	.	.	.	.
1	1	0	1	1	1	67	1792
1	1	1	0	0	0	70	16
1	1	1	0	0	1	71	32
1	1	1	0	1	0	72	64
1	1	1	0	1	1	73	128
1	1	1	1	0	0	74	256
1	1	1	1	0	1	75	512
1	1	1	1	1	0	76	1024
1	1	1	1	1	1	77	2048

Note: The highest Baud Rate = BusFreq/2 and the lowest Baud Rate is BusFreq/2048.

Figure 16-6. SPIBR–SPI Baud Rate Register

Example 16-1

Using BusFreq = 2 MHz, find the value for the SPI Baud Rate (SPIBR) register for the following baud rates: a) 1 MHz, b) 500 kHz c) 7,812.5 Hz

Solution:

(a) For 2 MHz/1 MHz = 2, we have SPIBR = 00000000 = 00 hex.
(b) For 2 MHz/500 kHz = 4, we have SPIBR = 0000001= 01 hex.
(c) For 2 MHz/7,812.5 Hz = 256, we have SPIBR = 01110100 = 74 hex.

SPI pins in HCS12

There are four pins associated with the SPI of the HCS12. They are MOSI (Master Out Slave In), MISO (Master In Slave out), SCK, and SS (Slave Select). Each HCS12 comes with two sets of SPI. They are called SPI0 and SPI1. The SPI modules use the PORTP pins, and therefore share pins with the PWM modules. The HCS12 SPI modules can be configured as master or slave. We use the HCS12 SPI as master in applications presented in this book. We must connect the the SS pin to VCC when the HCS SPI is used as master. The main difference between the SPI master and SPI slave is the fact that only a master SPI can initiate transmission. In that case, the MOSI plays the role of the SDO pin and MISO is the SDI. See Figure 16-7. It must be noted that HCS12 requires the internal pull-up resistors for PTP to be enabled, if no external pull-up resistors are connected. We do that by placing 00 in the WOMS register. To transfer a byte of data, we place it in the SPI data Register (SPIDR). The SPIDR register also holds the byte received via the SPI bus. In the HCS12, a transmission begins by writing to the SPI Data Register.

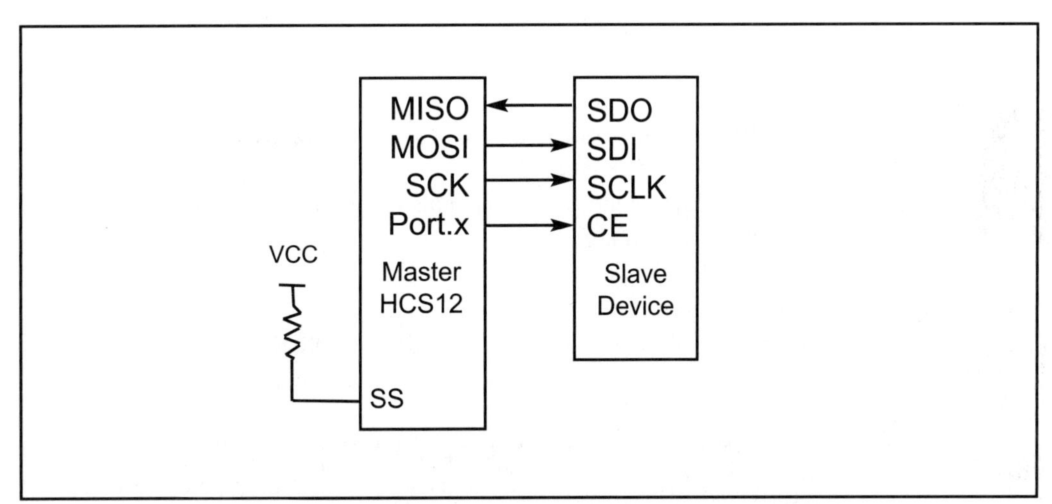

Figure 16-7. SPI Master Connection to Slave Device in HCS12

SPI control registers in HCS12

Figure 16-8 shows the SPICR1 (SPI Control Register 1) of the HCS12. We use SPICR1 to select the SPI mode operation of the HCS12. Notice that the SPE bit in the SPICR1 register must be set to HIGH to allow the use of the HCS12 pins for SPI data bus protocol. We choose the SPI Master mode by using the MSTR bit of SPICR1. The CPOL bit is used for selecting an inverted or noninverted SPI clock. In the active-HIGH (noninverted) SCK, it is low in the idle state. We must make sure that the SPI slave device has the same SCK polarity as the HCS12 master. We use the CPHA bit in the SPICR2 register to select the rising or falling edge of the SCK for sampling of data. We also have the option of sending out the LSB or the MSB first. See Example 16-2. With the options available in SPICR2 (SPI Control Register 2), many other features of the SPI module, such as bidirectional data transfer, can be used. See Figure 16-9 and the HCS12 manual.

Example 16-2

Using Figure 16-8, find SPICR1. Assume no interrupt, SPI enabled, SPI master, active-HIGH clock, sampling data on rising edge, and MSB first.

Solution:

SPICR1 = 01010000 or 50 in hex.

SPIE	SPE	SPTIE	MSTR	CPOL	CHPA	SSOE	LSBFE

SPIE D7 SPI Interrupt Enable. This bit enables SPI interrupt request, if SPIF = 1.
1 = SPI interrupts enabled
0 = SPI interrupts disabled

SPE D6 SPI System Enable bit
1 = Enables SPI port and configures pins as serial port pins
0 = Disables SPI port and configures these pins as I/O ports
(lower power consumption)

SPTIE D5 SPI Transmit Interrupt Enable. This bit enables the SPI interrupt request if SPTEF = 1.
1 = SPTEF interrupt enabled
0 = SPTEF interrupt disabled

MSTR D4 SPI Master/Slave mode Select bit. This bit selects master or slave mode.
1 = SPI in master mode
0 = SPI in slave mode

CPOL D3 SPI Clock Polarity bit
1 = Active-LOW clocks selected. In idle state SCK is high.
0 = Active-HIGH clocks selected. In idle state SCK is low.

CPHA D2 SPI Clock Phase bit
1 = Sampling of data occurs at even (falling) edges of the SCK clock.
0 = Sampling of data occurs at odd (rising) edges of the SCK clock.

SSOE D1 Slave select output enable. See the HCS12 manual.

LSBFE D0 LSB First Enable
1 = Data is transferred least significant bit first.
0 = Data is transferred most significant bit first.

Figure 16-8. SPICR1–SPI Control Register 1

0	0	0	MODFEN	BIDIROE	0	SPISWAI	SPC0

MODFEN D4 Mode Fault Enable Bit. See the HCS12 manual.
1 = SS port pin with MODF feature
0 = SS port pin is not used by the SPI (default)

BIDIROE D3 Bidirectional Output Enable. Used in bidirectional mode.
1 = Output Buffer enabled. See the HCS12 manual.
0 = Output Buffer enabled. (default)

SPC0 D0 Serial Pin Control bit 0. Used in bidirectional mode.
1 = Bidirectional. See the HCS12 manual.
0 = Normal (default)

Figure 16-9. SPICR2–SPI Control Register 2

SPI status register in HCS12

Figure 16-10 shows the SPISR (SPI status register) of the HCS12. We check the SPIF flag to see if a byte of data has been received. Before transferring a byte of data we use the SPTEF bit to make sure the SPI Data Register is empty. Tables 16-1 and 16-2 show the addresses for the SPI0 and SPI1 modules of the HCS12. Many of the HCS12 family members come with two on-chip SPI modules.

SPIF	0	SPTEF	MODF	0	0	0	0

SPIF D7 SPI Interrupt Flag. This bit is set after a received byte of data has been placed into the SPI Data Register. This bit is cleared by reading the SPI Status Register (SPISR) followed by a read from the SPI Data Register.
1 = New data has been received and placed in SPIDR.
0 = Transfer not yet complete

SPTEF D5 SPI Transmit Empty Interrupt Flag. If set, this bit indicates that the transmit data register is empty and ready for a new byte of data. To clear this and place new data into the transmit data register, SPISR has to be read with PTEF = 1, followed by a write to SPIDR. Any write to the SPI Data Register without reading SPTEF = 1 is effectively ignored.
1 = SPI Data Register empty
0 = SPI Data Register not empty

MODF D4 Mode Fault flag is used for mode selection error. See the HCS12 manual.
1 = Mode fault has occurred.
0 = Mode fault has not occurred.

Figure 16-10. SPISR–SPI Status Register

Table 16-1: SPI0 (Number 0) Registers and their Addresses

SCI #0 Register Name	Function	Addresses
SPI0CR1	SPI Control Reg. 1	$00D8
SPI0CR2	SPI Control Reg. 2	$00D9
SPI0BR	SPI Baud Rate Reg.	$00DA
SPI0SR	SPI Status Reg.	$00DB
SPI0DR	SPI Data Reg.	$00DD

Table 16-2: SPI1 (Number 1) Registers and their Addresses

SCI #1 Register Name	Function	Addresses
SPI1CR1	SPI Control Reg. 1	$00F0
SPI1CR2	SPI Control Reg. 2	$00F1
SPI1BR	SPI Baud Rate Reg.	$00F2
SPI1SR	SPI Status Reg.	$00F3
SPI1DR	SPI Data Reg.	$00F5

Steps in programming SPI

In programming the SPI of the HCS12 to transfer character bytes, the following steps must be taken:

1. The SPIBR register is loaded with a value to set the baud rate.
2. The SPICR1 register is loaded with the value $54, indicating SPI options.
3. The SPICR2 register is loaded with the value $00. This step is optional since the default value is 00. Also send the value 00 to the WOMS register to enable pull-up resistors on the PTP port.
4. Monitor the SPTEF bit of the SPI Status Register (SPISR) to make sure SPI is ready for a byte before writing another byte to SPIDR. If SPTEF = 1, then the SPI Data Register is empty and can go to the next step.
5. The character byte to be transmitted is written into the SPI Data Register (SPIDR).
6. Monitor the SPIF bit of the SPI Status Register (SPISR) to make sure the transfer is complete.
7. Read the SPIDR to clear the SPIF flag for the next round.
8. To transfer the next character, go to Step 4.

Example 16-3 shows the program to transfer data via the SPI pin.

Example 16-3

Write a program for the HCS12 to transfer the letter 'G' via the SPI pin at 100 kHz baud rate, continuously. Use SPI0 and assume XTAL = 4 MHz, BusFreq = 2 MHz, and CE is connected to ground.

Solution:

```
        LDAA  #$20           ;set the baud rate
        STAA  SPI0BR         ;
        LDAA  #$54           ;no interrupt, SPI Master, active-HIGH
        STAA  SPI0CR1        ;falling edge SCK, MSB first
        LDAA  #00            ;no bidirectional, no transfer during WAIT
        STAA  SPI0CR2        ;write to control reg 2
        STAA  WOMS           ;enable pull-up resistors
H1      BRCLR SPI0SR,mSPI0SR_SPTEF,H1 ;check to see SPI data reg empty
OVER    LDAA  #'G'           ;ASCII letter 'G' to be transferred
        STAA  SPI0DR         ;load it into SPI data reg
H2      BRCLR SPI0SR,mSPI0SR_SPIF,H2; check to see if transfer complete
        LDAA  SPI0DR         ;clear the SPIF flag bit for the next round
        BRA H1               ;keep sending letter 'G'
```

The C version is as follows:

```
unsigned char      x;
SPI0BR = 0x20;     //set the baud rate
SPI0CR1 = 0x54;    //no interrupt, SPI Master, active-HIGH
                   //falling edge SCK, MSB first
SPI0CR2 = 0x0;     //no bidirectional, no transfer during WAIT
WOMS = 0x0;        //enable pull-up resistors
while(1) {
  while (!(SPI0SR & SPI0SR_SPTEF_MASK)); //wait until empty
  SPI0DR = 'G';    //ASCII letter 'G' to be transferred
  while (!(SPI0SR & SPI0SR_SPIF_MASK))   //wait until received
  x=SPI0DR;        //read the Data reg to clear SPIF
  }
```

Receiving data by SPI

In programming the SPI of the HCS12 to receive character bytes, the following steps must be taken:

1. The SPIBR register is loaded with a value to set the baud rate.
2. The SPICR1 register is loaded with the value $54, indicating SPI options.
3. The SPICR2 register is loaded with the value $00. This step is optional since the default value is 00. Also send the value 00 to the WOMS register to enable pull-up resistors on the PTP port.
4. Monitor the SPTEF bit of the SPI Status Register (SPISR) to make sure SPI is ready to receive a byte. If SPTEF = 0, then the SPI Data Register is empty and can go to the next step.
5. The 0 is written into the SPI Data Register (SPIDR) to get the SPI to initiate the clocks.
6. Monitor the SPIF bit of the SPI Status Register (SPISR) to make sure a byte has been received.
7. Read the SPIDR to get the received data.
8. To receive the next character, go to Step 4.

Example 16-4 shows the program to receive data via the SPI pin.

Example 16-4

Write a program for the HCS12 to receive data via the SPI pin at a 100 kHz baud rate, continuously. Use the data in Example 16-3.

Solution:

```
MYBYTE      EQU   $2000
       ......             //see Example 16-3 for initialization
H1     BRCLR SPI0SR,mSPI0SR_SPTEF,H1 ;check to see SPI data reg empty
OVER   STAA  SPI0DR         ;send dummy value to SPI data reg
H2     BRCLR SPI0SR,mSPI0SR_SPIF,H2;check to see if receive complete
       LDAA  SPI0DR         ;get the character from SPI data reg
       STAA  MYBTYE         ;save the value
       BRA H1               ;keep receiving
```

The c version is as follow:

```
       unsigned char      x;
       ..............     //see Example 16-3 for initialization
       while(1) {
             while (!(SPI0SR & SPI0SR_SPTEF_MASK)); //wait until empty
             SPI0DR = 0;          //send dummy value to SPI data reg
             while (!(SPI0SR & SPI0SR_SPIF_MASK))//wait until received
             x=SPI0DR;            //reading the data reg clears SPIF
         }
```

Review Questions

1. True or false. In the HCS12, we have at least two sets of SPI modules.
2. True or false. The SPIxBR register is used for setting SCLK speed.
3. True or false. In the HCS12, the highest SCLK is BusFreq/2.
4. True or false. In transferring data, the use of SPISR is optional.

SECTION 16.3: DS1306 RTC INTERFACING WITH HCS12

The real-time clock (RTC) is a widely used device that provides accurate time and date information for many applications. Many systems such as the x86 IBM PC come with such a chip on the motherboard. The RTC chip in the IBM PC provides the time components of hour, minute, and second, in addition to the date/calendar components of year, month, and day. Many RTC chips use an internal battery, which keeps the time and date even when the power is off. Although some microcontrollers, such as the DS5000T, come with the RTC already embedded into the chip, we have to interface the vast majority of them to an external RTC chip. One of the most widely used RTC chips is the DS12887 from Dallas Semiconductor/Maxim Corp. This chip is found in the vast majority of x86 PCs. The original IBM PC/AT used the MC14618B RTC from Motorola. The DS12887 is the replacement for that chip. It uses an internal lithium battery to keep operating for over 10 years in the absence of external power. The DS12887 is a parallel RTC with 8 pins for the data bus. In this chapter, we interface and program the DS1306 RTC, which has an SPI bus. According to the DS1306 data sheet from Maxim, it keeps track of "seconds, minutes, hours, day of week, date, month, and year with leap-year compensation valid up to year 2099." The DS1306 RTC provides the above information in BCD format only. It supports both 12-hour and 24-hour clock modes, with AM and PM in the 12-hour mode. It does not support the Daylight Savings Time option. The DS1306 has a total of 128 bytes of nonvolatile RAM. It uses 28 bytes of RAM for clock/calendar and control registers, and the other 96 bytes of RAM are for general-purpose data storage. Next, we describe the pins of the DS1306. See Figure 16-11.

V_{CC2}

Pin 1 provides an external backup supply voltage to the chip. This pin is connected to an external rechargeable power source. This option is called *trickle charge*. If this pin is not used, it must be grounded.

V_{bat}

Pin 2 can be connected to an external +3 V lithium battery, thereby providing the power source to the chip externally as backup supply voltage. We must connect this pin to ground if it is not used.

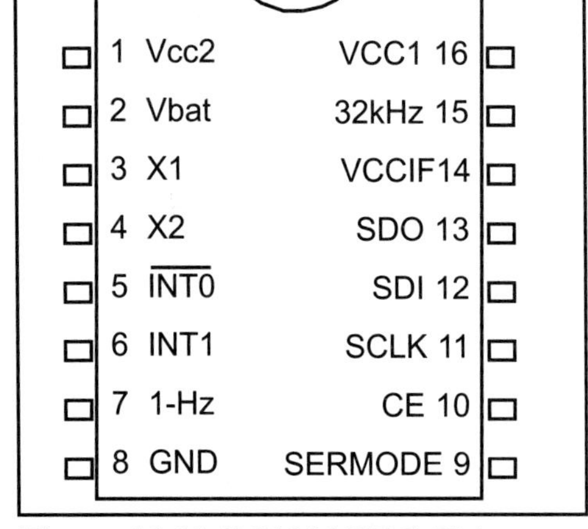

Figure 16-11. DS1306 RTC Chip (from Maxim/Dallas Semiconductor)

V_{CC1}

Pin 16 is used as the primary external voltage supply to the chip. This primary external voltage source is generally set to +5 V. When V_{cc1} falls below the V_{bat} voltage level, the DS1306 switches to Vbat and the external lithium battery provides power to the RTC. According to the DS1306 data sheet "upon power-up, the device switches from V_{bat} to V_{cc1} when V_{cc1} is greater than V_{bat} + 0.2 Volts." Because we can connect the standard 3 V lithium battery to the V_{bat} pin, the V_{cc1} voltage level must remain above 3.2 V in order for the V_{cc1} to remain as the primary voltage source to the chip. This nonvolatile capability of the RTC prevents any loss of data. See Figure 16-12.

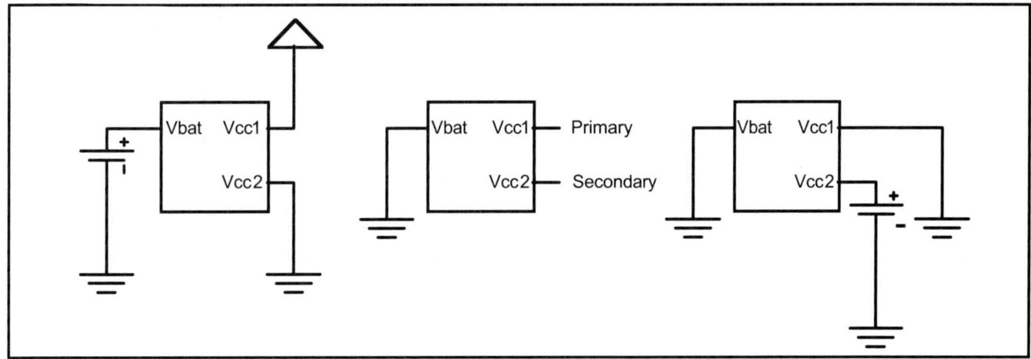

Figure 16-12. DS1306 Power Connection Options (Maxim/Dallas Semiconductor)

GND

Pin 8 is the ground.

SDI (Serial Din)

The SDI pin provides the path to bring data into the chip, one bit at a time.

SDO (Serial Dout)

The SDO pin provides the path to bring data out of the chip, one bit at a time.

32KHz

This is an output pin providing a 32.768 kHz frequency. This frequency is always present at the pin.

X1–X2

These are input pins that allow the DS1306 connection to an external crystal oscillator to provide the clock source to the chip. We must use the standard 32.768 kHz quartz crystal. The accuracy of the clock depends on the quality of this crystal oscillator. Heat can cause a drift on the oscillator. To avoid this, we use the DS32KHZ chip, which automatically adjusts for temperature variations. Note that when using the DS32KHZ or similar clock generators, we only need to connect X1 because the X2 loopback is not required.

SCLK (serial clock)

An input pin is used for the serial clock to synchronize the data transfer between the DS1306 and the microcontroller.

1-Hz

An output pin provides a 1-Hz square wave frequency. The DS1306 creates the 1-Hz square wave automatically. To get this 1-Hz frequency to show up on the pin, however, we must enable the associated bit in the DS1306 control register.

CE

Chip enable is an input pin and an active-HIGH signal. During the read and write cycle time, CE must be high.

INT0#

Interrupt request is an output pin and an active-LOW signal. To use INT0, the interrupt-enable bit in the RTC control register must be set HIGH. The interrupt feature of the DS1306 is discussed in Section 16.4.

INT1

Interrupt request is an output pin and an active-HIGH signal. To use INT1, the interrupt-enable bit in the RTC control register must be set HIGH. The interrupt feature of the DS1306 is discussed in Section 16.4.

SERMODE (serial mode selection)

Pin 9 is an input pin. If it is HIGH, then the SPI mode is selected. If it is connected to ground, the 3-wire mode is used. In our application, the SERMODE pin is connected to the V_{CC} pin because we program the 1306 chip using the SPI protocol.

V_{CCif}

Pin 14 is the interface logic power-supply input. This pin allows interfacing of the DS1306 with systems with 3 V logic in mixed supply systems. See the DS1306 data sheet if you are using a power source other than 5 V in your system.

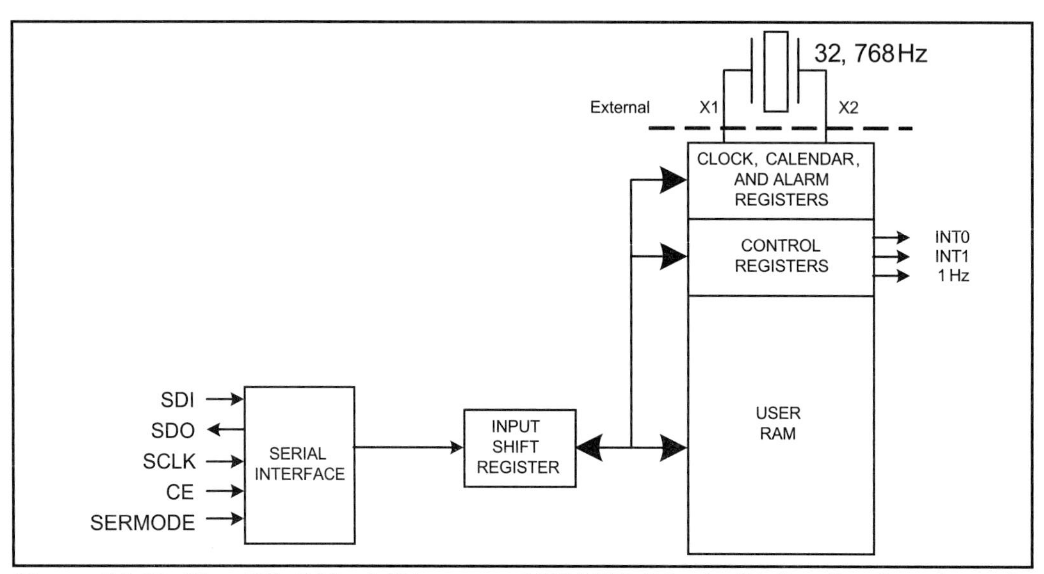

Figure 16-13. Simplified Block Diagram of DS1306 (Maxim/Dallas Semiconductor)

Importance of the WP bit in the Control register

As shown in Table 16-3, the Control register has an address of 8FH for write and 0FH for read. The most important bit in the Control register is the WP bit. The WP bit is undefined upon reset. In order to write to any of the registers of the DS1306, we must clear the WP bit first. See Figure 16-14. Upon powering up the DS1306, we have to clear the WP bit at least once. This means that after initializing the DS1306 we can write protect all the registers by making WP = 1.

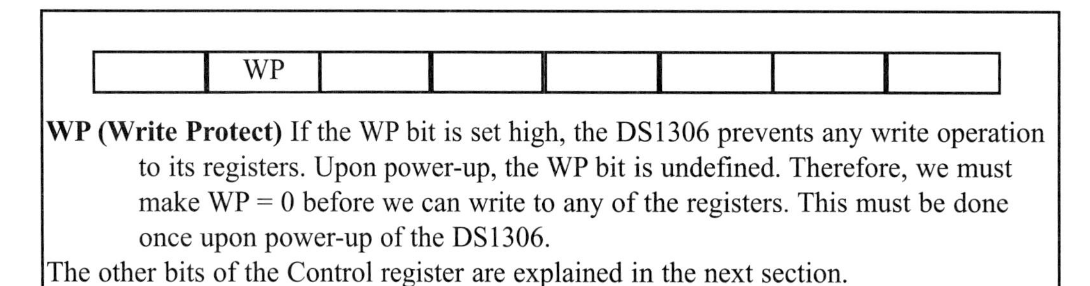

	WP						

WP (Write Protect) If the WP bit is set high, the DS1306 prevents any write operation to its registers. Upon power-up, the WP bit is undefined. Therefore, we must make WP = 0 before we can write to any of the registers. This must be done once upon power-up of the DS1306.

The other bits of the Control register are explained in the next section.

Figure 16-14. WP Bit of DS1306 Control Register (Write Location Address Is 8FH)

Address map of the DS1306

The DS1306 has a total of 128 bytes of RAM space with addresses 00–7FH. The first fifteen locations, 00–0E, are set aside for RTC values of time, date, and alarm data. The next three bytes are used for the control and status registers. They are located at addresses 0F–11 in hex. The next 14 bytes from addresses 12H to 1FH are reserved and cannot be used. That leaves 96 bytes, from addresses 20H to 7FH, available for general-purpose data storage. That means the entire 128 bytes of RAM except for addresses 12–1FH are accessible directly for read or write. Table 16-3 shows the address map of the DS1306. In this section we study the time and date. The alarm is examined in Section 16.5.

Table 16-3: Registers of the DS1306 (Modified from Data Sheet)

HEX ADDRESS READ	HEX ADDRESS WRITE	D7	D6	D5	D4	D3	D2	D1	D0	RANGE in HEX
0x00	0x80	0	10 SEC			SEC				00-59
0x01	0x81	0	10 MIN			MIN				00-59
0x02	0x82	0	24 / 12	20 HR P/A	10 HR	HOURS				00-23 01-12 P/A
0x03	0x83	0	0	0	0	0	DAY			01-07
0x04	0x84	0	0	10 DATE		DATE				01-31
0x05	0x85	0	0	10 MONTH		MONTH				01-12
0x06	0x86	0	10 YEAR			YEAR				00-99
0x07	0x87	M	10 SEC ALARM0			SEC ALARM0				00-59
0x08	0x88	M	10 MIN ALARM0			MIN ALARM0				00-59
0x09	0x89	M	24 / 12	20 HR P/A	10 HR	HOUR ALARM0				00-23 01-12 P/A
0x0A	0x08A	M	0	0	0	0	DAY ALARM0			01-07
0x0B	0x8B	M	10 SEC ALARM1			SEC ALARM1				00-59
0x0C	0x8C	M	10 MIN ALARM1			MIN ALARM1				00-59
0x0D	0x8D	M	24 / 12	20 HR P/A	10 HR	HOUR ALARM1				00-23 01-12 P/A
0x0E	0x8E	M	0	0	0	0	DAY ALARM1			01-07
0x0F	0x8F	CONTROL REGISTER								
0x10	0x90	STATUS REGISTER								
0x11	0x91	TRICKLE CHARGER REGISTER								
0x12-0x1F	0x92-0x9F	RESERVED								

Time and date address locations and modes

The byte addresses 0–6 are set aside for the time and date, as shown in Table 16-3. Table 16-4 is extracted from Table 16-3. It shows a summary of the address locations in read/write modes with data ranges for each location. The DS1306 provides data in BCD format only. Notice the data range for the hour mode. We can select 12-hour or 24-hour mode with bit 6 of hour location 02. When D6 = 1, the 12-hour mode is selected, and D6 = 0 provides us the 24-hour mode. In the 12-hour mode, we decide AM and PM with the bit 5. If D5 = 0, AM is selected and D5 = 1 is for PM. Example 16-5 shows how to get the range of the data acceptable for the hour location.

Table 16-4: DS1306 Address Locations for Time and Date (Extracted from Table 16-1)

Hex Address Location Read	Write	Function	Data Range BCD	Range in Hex
00	80	Seconds	00–59	00–59
01	81	Minutes	00–59	00–59
02	82	Hours, 12-Hour Mode	01–12	41–52 AM
		Hours, 12-Hour Mode	01–12	61–72 PM
		Hours, 24-Hour Mode	00–23	00–23
03	83	Day of the Week, Sun = 1	01–07	01–07
04	84	Day of the Month	01–31	01–31
05	85	Month	01–12	01–12
06	86	Year	00–99	00–99

Example 16-5

Using Table 16-3, verify the hour location values in Table 16-4.

Solution:

(a) For 24-hour mode, we have D6 = 0. Therefore, the range goes from 0000 **0000** to 0010 **0011,** which is 00–23 in BCD.
(b) For 12-hour mode, we have D6 = 1 and D5 = 0 for AM. Therefore, the range goes from 0100 **0001** to 0101 **0010,** which is 41–52 in BCD.
(c) For 12-hour mode, we have D6 = 1 and D5 = 1 for PM. Therefore, the range goes from 0110 **0001** to 0111 **0010,** which is 61–72 in BCD.

HCS12 interfacing to DS1306 using SPI

The DS1306 supports both SPI and 3-wire modes. In DS1306, we select the SPI mode by connecting the SERMODE pin to Vcc. If SERMODE = Gnd, then the 3-wire protocol is used. In this section, we use SPI mode only. In our application, we will use Fbus/20 speed (100 kBaud) for data transfer between the HCS12 and the DS1306 RTC. See Figure 16-15.

RTC setting, reading, and displaying time and date

Program 16-1 is the complete Assembly code for setting, reading, and displaying the time and date. The time and date are sent to the IBM PC screen via the HCS12 serial port after they are converted from packed BCD to ASCII.

```
;Program 16-1
DAY   DS.B 1              ;for day of the week
MON   DS.B 1              ;fileReg starting with month
DAT      DS.B 1           ;for day of the month
YR       DS.B 1           ;for year
HR       DS.B 1           ;for hour
MIN      DS.B 1           ;for minutes
SECS     DS.B 1           ;for seconds
```

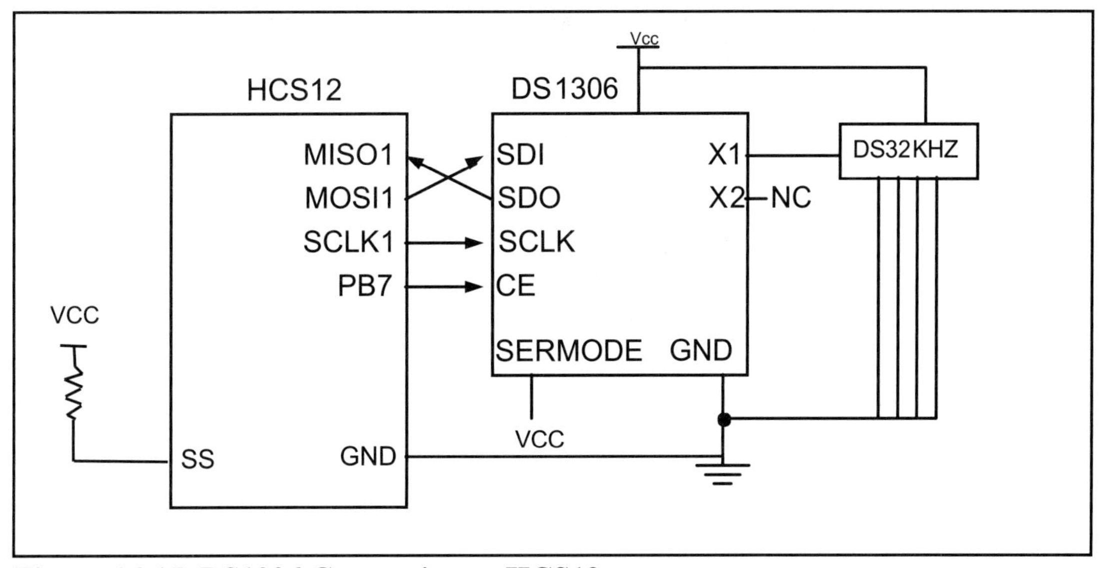

Figure 16-15. DS1306 Connection to HCS12
Note: For more accuracy, we use the DS32KHZ chip in place of a crystal.

```
CNT     DS.B 1              ;for counter
TMP     DS.B 1              ;for conversions

; code section
          ORG   ROMStart ;defined in CodeWarrior header file
Entry:
;-- initialize the stack pointer
      LDS    #RAMEnd+1
;-- initialize SPI
      LDAA #$20        ;set the baud rate
      STAA SPI1BR      ;
      LDAA #$54  ;no interrupt, SPI Master, active-HIGH
      STAA SPI1CR1     ;falling edge SCK, MSB first
      LDAA #00   ;no bidirectional, no transfer during WAIT
      STAA SPI1CR2     ;write to control reg 2
      STAA WOMS        ;enable PTP pull-up resistors
      BSET DDRB,mDDRB_BIT7 ; make PB7 output
;-- initialize UART
      LDAA #$0         ;set the baud rate
      STAA SCI0BDH ;we must write to High byte even if zero
      LDAA #$0D        ;set the baud rate to 9600
      STAA SCI0BDL     ;write to low byte
      LDAA #0         ;8-bit data, no parity and no interrupt
      STAA SCI0CR1     ;write to control reg 1
      LDAA #$0C        ;enable both transmit and receive
      STAA SCI0CR2     ;write to control reg 2
;-- initialize DS1306
      BSET PORTB,mPORTB_BIT7;make CE = 1 for single-byte
      JSR  SDELAY
      LDAA #$8F              ;DS1306 control register address
      JSR  SPI
      LDAA #$00              ;clear WP bit for write
```

```
     JSR  SPI
     BCLR PORTB,mPORTB_BIT7;CE = 0 (single-byte)
     JSR  SDELAY
;-- start a new line for UART communications
     LDAA #$0A            ;form feed
     JSR  TRANS
     LDAA #$0D            ;new line
     JSR  TRANS
;-- set the time
     BSET PORTB,mPORTB_BIT7;CE = 1 (start multibyte write)
     LDAA #$80       ;seconds register address for write
     JSR  SPI        ;send address
     LDAA #$55       ;55 seconds
     JSR  SPI        ;send seconds
     LDAA #$58       ;58 minutes
     JSR  SPI        ;send minutes
     LDAA #$16       ;24-hour clock at 16 hours
     JSR  SPI        ;send hour
;-- set the date
     LDAA #$03       ;Tuesday
     JSR  SPI        ;send address
     LDAA #$19       ;19th of the month
     JSR  SPI        ;send date
     LDAA #$10       ;October
     JSR  SPI        ;send month
     LDAA #$07       ;2007
     JSR  SPI        ;send year
     BCLR PORTB,mPORTB_BIT7;CE = 0 (end multibyte write)
;-- get the time from DS1306
RDA  BSET PORTB,mPORTB_BIT7;CE = 1 (start multibyte read)
     JSR  SDELAY
     LDAA #$00       ;seconds register address for read
     JSR  SPI        ;send address to DS1306
     JSR  SPI        ;start getting time/date
     STAA SECS       ;save the seconds
     JSR  SPI        ;get the minutes
     STAA MIN        ;save the minutes
     JSR  SPI        ;get the hour
     STAA HR         ;save the hour
;-- get the date from DS1306
     JSR  SPI        ;get the day
     STAA DAY        ;save the day
     JSR  SPI        ;get the date
     STAA DAT        ;save the date
     JSR  SPI        ;get the month
     STAA MON        ;save the month
     JSR  SPI        ;get the year
     STAA YR         ;save the year
     BCLR PORTB,mPORTB_BIT7;make CE=0 (end multibyte read)
;-- convert packed BCD to ASCII and display
     LDX  #RAMStart ;address of registers for time/date
```

```
        LDAA  #7           ;6 bytes of data to display
        STAA  CNT          ;set up the counter
B6      LDAA  0,X          ;get the BCD from RAM
        ANDA  #$F0         ;mask the lower nibble
        LSRA               ;shift right to get upper nibble
        LSRA
        LSRA
        LSRA
        ORAA  #$30         ;make it an ASCII
        JSR   TRANS        ;display the data
        LDAA  1,X+  ;get BCD data once more,increment pointer
        ANDA  #$0F         ;mask the upper nibble
        ORAA  #$30         ;make it an ASCII,
        JSR   TRANS        ;display the data
        LDAA  #':'
        JSR   TRANS
        DEC   CNT
        BNE   B6           ;keep looping
        LDAA  #$0D         ;line feed
        JSR   TRANS
        BRA   RDA   ;keep reading time/date and display them
;------------------------------------
SPI
        ;wait for SPI data register to empty
        BRCLR SPI1SR,mSPI1SR_SPTEF,SPI
        STAA  SPI1DR       ;load A into SPI data reg
H2      ;wait for the transfer to complete
        BRCLR SPI1SR,mSPI1SR_SPIF,H2
        LDAA  SPI1DR       ;store the received byte in A
        RTS
;------------------------------------
SDELAY
        LDAB  #$00
D1      DECB
        BNE D1
        RTS
;----serial data transfer subroutine
TRANS BRCLR SCI0SR1,mSCI0SR1_TDRE,TRANS ;wait until last byte
        STAA  SCI0DRL      ;load it into SCI data reg
        RTS                ;return to JSRer
```

Review Questions

1. True or false. All of the RAM contents of the DS1306 are nonvolatile.
2. How many bytes of RAM in the DS1306 are set aside for the clock and date?
3. How many bytes of RAM in the DS1306 are set aside for general-purpose applications?
4. True or false. The DS1306 has a single pin for Din.
5. Which pin of the DS1306 is used for clock in SPI connection?
6. True or false. To use the DS1306 in SPI mode, we make SERMODE = GND.

SECTION 16.4: DS1306 RTC PROGRAMMING IN C

In this section, we program the DS1306 in the C language. Before you embark on this section, make sure you understand the basic concepts of the DS1306 chip covered in the first section.

Setting, reading, and displaying time and date in C

Program 16-1C shows how to set the time and date for the DS1306 configuration in Figure 16-15.

```
//Program 16-1C : Setting and reading time and date
#include <mc9s12dp512.h>      /* derivative information */
unsigned char SPI(unsigned char);
void TRANS(unsigned char);
void BCDtoASCIIandSEND(unsigned char);
void SDELAY(int ms);
unsigned char data[ 7] ;      //holds date and time
unsigned char tmp;            //for BCD to ASCII conversion
int i;
void main()
{
  SPI1BR = 0x20;       //set the baud rate
  SPI1CR1 = 0x54;//no interrupt, SPI master, active-HIGH
  SPI1CR2 = 0x00;      //no bidirect., no trans. during WAIT
  WOMS = 0x00;         //enable PP pull-up resistors
  DDRB = 0x80;         //make PB7 output
//initialize UART
  SCI0BDH=0x0;         //choose
  SCI0BDL=0x0D;        //9600 baud rate
  SCI0CR1=0x0;         //8-bit data, no Int
  SCI0CR2=0X0C;//enable transmit and receive and no parity
//initialize DS1306
  PORTB = 0x80;             //CE = 1 single-byte write
  SDELAY(1);
  tmp = SPI(0x8F);          //control register address
  tmp = SPI(0x00);          //clear WP bit for write
  PORTB = 0;                //end of single-byte write
//start a new line for UART communications
  TRANS(0x0A);              //form feed
  TRANS(0x0D);              //new line
//set the time
  PORTB = 0x80;             //CE = 1 begin multibyte write
  tmp = SPI(0x80);          //seconds register address
  tmp = SPI(0x55);          //55 seconds
  tmp = SPI(0x58);          //58 minutes
  tmp = SPI(0x16);          //24-hour clock at 16 hours
  tmp = SPI(0x3);           //set the date, Tuesday
  tmp = SPI(0x19);          //19th of the month
  tmp = SPI(0x10);          //October
  tmp = SPI(0x07);          //2007
  PORTB = 0;                //end multibyte write
```

```
  for (;;)
  {
//-- get the time and date from RTC and save them
    PORTB = 0x80;          //CE = 1 begin multibyte read
    SDELAY(1);
    tmp = SPI(0x00);       //seconds register address
    data[6] = SPI(tmp);   //get the seconds
    data[5] = SPI(tmp);    //get the minutes
    data[4] = SPI(tmp);    //get the hour
    data[0] = SPI(tmp);    //get the day
    data[2] = SPI(tmp);    //get the date
    data[1] = SPI(tmp);    //get the month
    data[3] = SPI(tmp);    //get the year
    PORTB = 0;             //end of multibyte read
//-- convert time/date and display DW:MM:DM:YY:HH:MM:SS
    for(i=0;i<7;i++)
    {
      tmp = data[i];
      tmp = tmp & 0xF0;    //mask lower nibble
      tmp = tmp >> 4;      //swap it
      tmp = tmp | 0x30;    //make it ASCII
      TRANS(tmp);          //display
      tmp = data[i];       //for other digit
      tmp = tmp & 0x0F;    //mask upper nibble
      tmp = tmp | 0x30;    //make it ASCII
      TRANS(tmp);          //display
      TRANS(':');          //display separator
    }
    TRANS(0x0D);           //new line
  }
}
unsigned char SPI(unsigned char myByte)//-- SPI Write/Read
{
  while(!(SPI1SR & SPI1SR_SPTEF_MASK));//is SPI reg empty?
  SPI1DR = myByte;    //load into SPI data reg
  while(!(SPI1SR & SPI1SR_SPIF_MASK));//is tran complete?
  return SPI1DR;      //get from SPI data reg
}
//--------------------------------
void TRANS(unsigned char c)//-- UART transfer
{
  while(!(SCI0SR1 & SCI0SR1_TDRE_MASK));//wait until empty
  SCI0DRL=c;          //place value in buffer
}
//--------------------------------
void SDELAY(int ms) //--Delay routine
{
  unsigned int i, j;
  for(i=0;i<ms;i++)
    for(j=0;j<135;j++);
}
```

Review Questions

1. True or false. All the RAM contents of the DS1306 are volatile.
2. What locations of RAM in the DS1306 are set aside for the clock and date?
3. What locations of RAM in the DS1306 are set aside for general-purpose applications?
4. True or false. The DS1306 has a single pin for Dout.
5. True or false. CE is an output pin.
6. True or false. To use the DS1306 in SPI mode, we make SERMODE = VCC.

SECTION 16.5: ALARM AND INTERRUPT FEATURES OF THE DS1306

In this section, we program the alarm and interrupt features of the DS1306 chip using Assembly and C languages. These powerful features of the DS1306 can be very useful in many real-world applications. In the DS1306 there are two alarms, called Alarm0 and Alarm1, each with its own hardware interrupts. There is also a 1-Hz square wave output pin, which we discuss next. These features are accessed with the Control register shown in Figure 16-16.

Programming the 1-Hz feature

The 1-Hz pin of the DS1306 provides us a square wave output of 1-Hz frequency. Internally, the DS1306 generates the 1-Hz square wave automatically but it is blocked. We must enable the 1-Hz bit in the Control register to let it show up on the 1-Hz pin. This is shown below. Because we are writing to a single location, burst mode is not used.

```
;Program 16-2
     LDS   #RAMEnd+1
;-- initialize SPI
     LDAA #$20        ;set the baud rate
     STAA SPI1BR      ;
     LDAA #$54        ;no interrupt, SPI Master, act. hi
     STAA SPI1CR1     ;falling edge SCK, MSB first
     LDAA #00         ;no bidirect.,no trans during WAIT
     STAA SPI1CR2     ;write to control reg 2
     STAA WOMS
     BSET DDRB,mDDRB_BIT7       ;make PB7 output
;-- initialize DS1306
     BSET PORTB,mPORTB_BIT7;make CE=1 for single-byte
     JSR  SDELAY
     LDAA #$8F        ;DS1306 control register address
     JSR  SPI
     LDAA #$00        ;clear WP bit for write
     JSR  SPI
     BCLR PORTB,mPORTB_BIT7;make CE=0 end write
     JSR  SDELAY
```

```
;-- enable the 1 Hz pin
     BSET PORTB,mPORTB_BIT7;CE=1 (start single write)
     LDAA #$8F       ;DS1306 control register address
     JSR SPI         ;send address
     LDAA #$04       ;enable the 1 Hz pin
     JSR  SPI        ;send command
     BCLR PORTB,mPORTB_BIT7;CE = 0 (end single write)
;-- get the time and date from DS1306
     BRA $
```

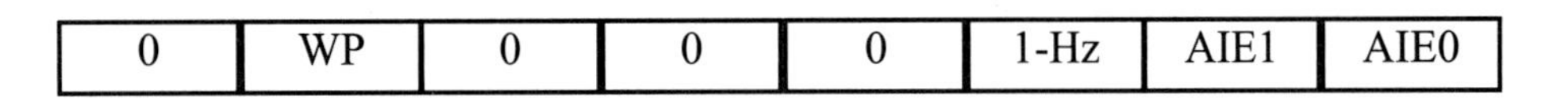

0	WP	0	0	0	1-Hz	AIE1	AIE0

WP (Write Protect) If the WP bit is set high, the DS1306 prevents any write operation to its registers. We must make WP = 0 before we can write to any of the registers. Upon power-up, the WP bit is undefined. Therefore, we must make WP = 0 before we can write to any of the registers.

1-Hz (1-Hz output enable) If this bit is set HIGH, it allows the 1-Hz frequency to come out of the 1-Hz pin of the DS1306. By making it LOW, we get High-Z on the 1-Hz pin. Notice that the 1-Hz frequency is automatically generated by the DS1306, but it will not show up at the 1-Hz pin unless we set this bit to HIGH.

AIE0 Alarm interrupt0 enable. If AIE 0 = 1, the INT0 pin will be asserted LOW when all three bytes of the real time (hh:mm:ss) are the same as the alarm bytes of hh:mm:ss. Also, if AIE0 = 1, the cases of once-per-second, once-per-minute, and once-per-hour will assert LOW the INT0 pin.

AIE1 Alarm interrupt1 enable. If AIE1 = 1, the INT1 pin will be asserted HIGH when all three bytes of the real time (hh:mm:ss) are the same as the alarm bytes of hh:mm:ss. Also, if AIE1 = 1, the cases of once-per-second, once-per-minute, and once-per-hour will assert HIGH the INT1 pin.

Figure 16-16. DS1306 Control Register (Write Location Address Is 8FH)

Alarm0, Alarm1, and interrupt

There are two time-of-day alarms in the DS1306 chip. They are referred to as Alarm0 and Alarm1. We can access Alarm0 by writing to its registers located at addresses 87H through 8AH, as shown in Table 16-5. Alarm1 is accessed by writing to its registers located at addresses 8BH through 8EH, as shown in Table 16-5. During each clock update, the RTC compares the clock registers and alarm registers. When the values stored in the time-keeping registers of 0, 1, and 2 match the values stored in the alarm registers, the corresponding alarm flag bit (IRQF0 or IRQF1) in the status register will go HIGH. See Figure 16-17. Because polling the IRQxF is too time-consuming, we can enable the AIEx bit in the Control register, and make it a hardware interrupt coming out of the INT0 and INT1 pins.

0	0	0	0	0	0	IRQF1	IRQF0

IRQF0 (Interrupt 0 Request Flag) The IRQF0 bit will go HIGH when all three bytes of the current real time (hh:mm:ss) are the same as the Alarm0 bytes of hh:mm:ss. Also, the cases of once-per-second, once-per-minute, and once-per-hour will assert HIGH the IRQ0 bit. We can use polling to see the status of IRQF0. However, in the Control register, if we make AIE0 = 1, IRQF0 will assert LOW the INT0 pin, making it a hardware interrupt. Any read or write of the Alarm0 registers will clear IRQF0.

IRQF1 (Interrupt 0 Request Flag) The IRQF1 bit will go HIGH when all three bytes of the current real time (hh:mm:ss) are the same as the Alarm1 bytes of hh:mm:ss. Also, the cases of once-per-second, once-per-minute, and once-per-hour will assert HIGH the IRQ1 bit. We can use polling to see the status of IRQF1. However, in the Control register, if we make AIE1 = 1, IRQF1 will assert HIGH the INT1 pin, making it a hardware interrupt. Any read or write of the Alarm1 registers will clear IRQF1.

Figure 16-17. Status Register (Read Location Address Is 10H)

Table 16-5: DS1306 Address Locations for Time, Calendar, and Alarm

Hex Address		Function		Data Range	
Read	**Write**		**D7**	**BCD**	**Possible Hex Range**
00H	80H	Seconds	0	00–59	00–59
01H	81H	Minute	0	00–59	00–59
02H	82H	Hours, 12-Hour Mode	0	01–12	41–52 AM
		Hours, 12-Hour Mode	0	01–12	61–72 PM
		Hours, 24-Hour Mode	0	00–23	00–23
03H	83H	Day of the Week, Sun = 1	0	01–07	01–07
04H	84H	Day of the Month	0	01–31	01–31
05H	85H	Month	0	01–12	01–12
06H	86H	Year	0	00–99	00–99
07H	87H	SEC Alarm**0**	0 or 1	00–59	00–59 or 89–A9
08H	88H	MIN Alarm**0**	0 or 1	00–59	00–59 or 89–A9
09H	89H	Hour Alarm**0**, 12-Hour	0 or 1	01–12	41–52 or C1–A2 AM
		Hour Alarm**0**, 12-Hour	0 or 1	01–12	61–72 or D1–F2 PM
		Hour Alarm**0**, 24-Hour	0 or 1	00–23	00–23 or 80–A3
0A	8A	Day Alarm**0**	0 or 1	1–7	01–07
0BH	8BH	SEC Alarm**1**	0 or 1	00–59	00–59 or 89–A9
0CH	8CH	MIN Alarm**1**	0 or 1	00–59	00–59 or 89–A9
0DH	8DH	Hour Alarm**1**, 12-Hour	0 or 1	01–12	41–52 or C1–A2 AM
		Hour Alarm**1**, 12-Hour	0 or 1	01–12	61–72 or D1–F2 PM
		Hour Alarm**1**, 24-Hour	0 or 1	00–23	00–23 or 80–A3
0EH	8EH	Day Alarm**1**	0 or 1	1–7	01–07
0FH	8FH	CONTROL REGISTER			
10H	90H	STATUS REGISTER			
11H	91H	TRICKLE REGISTER			
12–1FH	82–9FH	RESERVED			
20–7FH	A0–FFH	96-BYTE USER RAM			

Alarm and IRQ output pins

The alarm interrupts of INT0 and INT1 can be programmed to occur at rates of (a) once per week (b) once per day, (c) once per hour, (d) once per minute, and (e) once per second. Next, we look at each of these.

Once-per-day alarm

Table 16-5 shows the address locations belonging to the alarm seconds, alarm minutes, alarm hours, and alarm days. Notice the D7 bits of these locations. An alarm is generated every day when D7 of the day alarm location is set to HIGH. Therefore, to program the alarm for once-per-day, we must (a) write the desired time for the alarm into the hour, minute, and second of Alarm locations, and (b) set HIGH D7 of the alarm day. See Table 16-6. As the clock keeps the time, when all three bytes of hour, minute, and second for the real-time clock match the values in the alarm hour, minute, and second, the IRQxF flag bit in the Status register of the DS1306 will go high. We can poll the IRQxF bit in the Status register, which is a waste of microcontroller resources, or allow the hardware INTx pin to be activated upon matching the alarm time with the real time. It must be noted that in order to use the hardware INTx pin of the DS1306 for an alarm, the interrupt-enable bit for the alarm in the control register (AIEx) must be set HIGH. We will examine the process shortly.

Once-per-hour alarm

To program the alarm for once per hour, we must set HIGH D7 of both the day alarm and hour alarm registers. See Table 16-6.

Once-per-minute alarm

To program the alarm for once per minute, we must set HIGH D7 of all three, day alarm, hour alarm, and minute alarm locations. See Table 16-6.

Once-per-second alarm

To program the alarm for once per second, we must set HIGH D7 of all four locations of alarm day, alarm hour, alarm minute, and alarm second. See Table 16-6.

Once-per-week alarm

To program the alarm for once per week, we must clear D7 of all four locations of alarm day, alarm hour, alarm minute, and alarm second. See Table 16-6.

Table 16-6: DS1306 Time-of-day Alarm Mask Bits

Alarm Register Mask Bits (D7)

Seconds	Minutes	Hours	Days	Function
1	1	1	1	Alarm once per second
0	1	1	1	Alarm when seconds match (once-per-minute)
0	0	1	1	Alarm when minutes and seconds match (once-per-hour)
0	0	0	1	Alarm when hours, minutes, and seconds match (once-per-day)
0	0	0	0	Alarm when day, hours, minutes, and seconds match (once-per-week)

Using INT0 of DS1306 to activate the HCS12 interrupt

We can connect the INT0 bit of the DS1306 to the external interrupt pin of the HCS12 (IRQ). See Figure 16-18. This allows us to perform a task once per day, once per minute, and so on. Example 16-6 shows the values needed for the Alarm0 registers. One can write a program using the Alarm0 interrupt (INT0) to send the message "YES" to the serial port once per minute, at exactly 8 seconds past the minute. We can also use the 32 kHz output to sound an actual alarm. Because 32 kHz is too high a frequency for human ears, however, we must use multiple D flip-flops to bring down the frequency. See Figure 16-19. The development of these programs for Figures 16-18 and 16-19 are left to the reader.

Example 16-6

Using Table 16-6, find the values we must place in the Alarm1 register if we want to have an alarm activated at 16:05:07, and from then on once-per-minute at 7 seconds past the minute.

Solution:

Because we use 24-hour clock, we have D6 = 0 for the HR register. Therefore, we have 1001 0110 for 16 in BCD. This means that we must put value 96H into register location 8D of the DS1306. Notice that D7 is 1, according to Table 16-6.
For the MIN register, we have 1000 0101 for 05 in BCD. This means that we must put value 85H into register location 8C of the DS1306. Notice that D7 is 1, according to Table 16-6.
For the SEC register we have 0000 0111 for 07 in BCD. This means that we must put value 07H into register location 8B of the DS1306. Notice that D7 is 0 according to Table 16-6.
For once-per-minute to work, we must make sure that D7 of Alarm1 day is also set to 1. See Table 16-6.

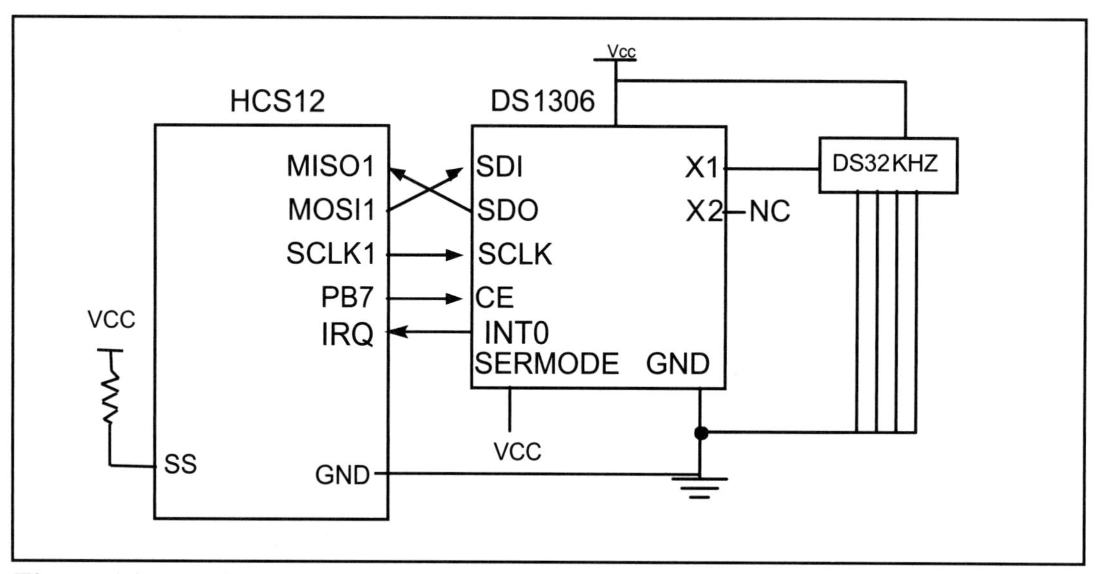

Figure 16-18. DS1306 Connection to HCS12 with Hardware INT0

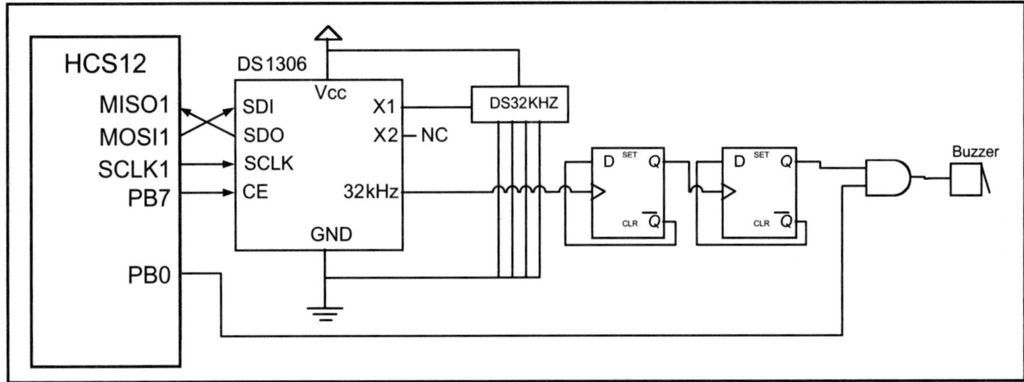

Figure 16-19. DS1306 Connection to HCS12 with Buzzer Control

Review Questions

1. Which bit of the Control register belongs to the 1-Hz pin?
2. True or false. The INT0 pin is an input for the DS1306.
3. True or false. The INT0 pin is active-LOW.
4. Which bit of the Control register belongs to the Alarm1 interrupt?
5. Give the address locations for Alarm1.

PROBLEMS

SECTION 16.1: SPI BUS PROTOCOL

1. True or false. The SPI bus needs an external clock.
2. True or false. The SPI CE is active-LOW.
3. True or false. The SPI bus has a single Din pin.
4. True or false. The SPI bus has multiple Dout pins.
5. True or false. When the SPI device is used as a slave, the SCLK is an input pin.
6. True or false. In SPI devices, data is transferred in 8-bit chunks.
7. True or false. In SPI devices, each bit of information (data, address) is transferred with a single clock pulse.
8. True or false. In SPI devices, the 8-bit data is followed by an 8-bit address.

SECTION 16.2: SPI MODULE IN THE HCS12

9. Assume BusFreq = 4 MHz (XTAL = 8 MHz). Find the value needed for the SPI Baud Rate (SPIBR) register for the following SCK frequencies.
   a) 500 kHz b) 125 kHz
   c) the highest allowed frequency d) the lowest allowed frequency
10. True or false. The SPI Data Register (SPIDR) is used for both send and receive.

SECTION 16.3: DS1306 RTC INTERFACING WITH HCS12

11. The DS1306 DIP package is a(n) ____-pin package.
12. Which pin is assigned as primary V_{cc}?

13. In the DS1306, how many pins are designated as address/data pins?
14. True or false. The DS1306 needs an external crystal oscillator.
15. True or false. The DS1306's crystal oscillator and heat affect the time-keeping accuracy.
16. What is the maximum year that the DS1306 can provide?
17. Describe the functions of pins SDI, SDO, and SCLK.
18 CE is an _______ (input, output) pin.
19. The CE pin is normally ______ (LOW, HIGH) and needs a _______ (LOW, HIGH) signal to be activated.
20. What keeps the contents of the DS1306 time and date registers if power to the primary V_{cc} pin is cut off?
21. The Vbat pin stands for ___________ and is an______ (input, output) pin.
22. For the DS1306 chip, pin V_{cc2} is connected to _____ (V_{cc}, GND).
23. SERMODE is an ______ (input, output) pin and it is connected to _____ for SPI mode.
24. V_{cc1} is an ________ (input, output) pin and is connected to _____ voltage.
25. 1-Hz is an ________ (input, output).
26. INT0 is an _______ (input, output) pin.
27. 32KHz is an _______ (input, output) pin.
28. INT1 is an _______ (input, output) pin.
29. DS1306 has a total of_____bytes of locations. Give the addresses for read and write operations.
30. What are the contents of the DS1306 time and date registers if power to the V_{cc} pin is lost?
31. What are the contents of the general-purpose RAM locations if power to the V_{cc1} is lost?
32. When does the DS1306 switch to a battery energy source?
33. What are the addresses assigned to the real-time clock (time) registers?
34. What are the addresses assigned to the calendar?
35. Which register is used to set the AM/PM mode? Give the bit location of that register.
36. Which register is used to set the 24-hour mode? Give the bit location of that register.
37. At what memory location does the DS1306 store the year 2007?
38. What is the address of the last location of RAM for the DS1306?
39. True or false. The DS1306 provides data in BCD format only.

SECTION 16.5: ALARM AND INTERRUPT FEATURES OF THE DS1306

40. INT0 is an _______ (input, output) pin and active-_______(LOW, HIGH).
41. 1-Hz is an _______ (input, output) pin.
42. Give the bit location of the Control register belonging to the alarm interrupt. Show how to enable it.
43. Give the bit location of the Control register belonging to the 1-Hz pin. Show how to enable it.
44. Give the bit location of the Status register belonging to the Alarm0 interrupt.

45. Give the bit location of the Status register belonging to the Alarm1 interrupt.
46. True or false. For the 32KHz output pin, the frequency is set and cannot be changed.
47. Give sources of interrupts that can activate the INT1 pin.
48. Why do we want to direct the AIE0 (Alarm0 flag) to an IRQ pin?
49. What is the difference between the IRQF0 and AIE0 bits?
50. What is the difference between the IRQF1 and AIE1 bits?
51. How do we allow the square wave to come out of the 1-Hz pin?
52. Which register is used to set the once-per-second Alarm1?
53. Explain how the IRQ1F pin is activated due to the once-per-minute alarm option.

ANSWERS TO REVIEW QUESTIONS

SECTION 16.1: SPI BUS PROTOCOL

1. True
2. True
3. False
4. False
5. In single-byte mode, after each byte, the CE pin must go LOW before the next cycle. In burst mode, the CE pin stays HIGH for the duration of the burst (multibyte) transfer.

SECTION 16.2: SPI MODULES IN THE HCS12

1. True
2. True
3. True
4. False

SECTION 16.3: DS1306 RTC INTERFACING WITH HCS12

1. True. Only if V_{bat} is connected to an external battery.
2. 7
3. 96
4. True
5. Pin 11 is SCLK.
6. False. SERMODE = V_{cc}

SECTION 16.4: DS1306 RTC PROGRAMMING IN C

1. True
2. 0–6
3. 20–7FH
4. True
5. False
6. False

SECTION 16.5: ALARM AND INTERRUPT FEATURES OF THE DS1306

1. Bit 2
2. False
3. True
4. Bit 1
5. Byte addresses of 0B–0E (in hex) for read and 8B–8E (in hex) for write

CHAPTER 17

PWM AND DC MOTOR CONTROL

OBJECTIVES

Upon completion of this chapter, you will be able to:

>> Describe the basic operation of a DC motor
>> Code HCS12 programs to control and operate a DC motor
>> Describe how PWM is used to control motor speed
>> Code PWM programs to control and operate a DC motor
>> Describe the PWM features of the HCS12
>> Code HCS12 Assembly and C programs to create PWM pulses

This chapter discusses the topic of PWM (pulse width modulation) and shows HCS12 interfacing with DC motors. The characteristics of DC motors are discussed along with their interfacing to the HCS12. We use both Assembly and C programming examples to create PWM pulses.

SECTION 17.1: DC MOTOR INTERFACING AND PWM

This section begins with an overview of the basic operation of DC motors. Then we describe how to interface a DC motor to the HCS12. Finally, we use Assembly and C language programs to demonstrate the concept of pulse width modulation (PWM) and show how to control the speed and direction of a DC motor.

DC motors

A direct current (DC) motor is a widely used device that translates electrical pulses into mechanical movement. In the DC motor we have only + and – leads. Connecting them to a DC voltage source moves the motor in one direction. By reversing the polarity, the DC motor will move in the opposite direction. One can easily experiment with the DC motor. For example, small fans used in many motherboards to cool the CPU are run by DC motors. By connecting their leads to the + and – voltage source, the DC motor moves. While a stepper motor moves in steps of 1 to 15 degrees, the DC motor moves continuously. In a stepper motor, if we know the starting position we can easily count the number of steps the motor has moved and calculate the final position of the motor. This is not possible in a DC motor. The maximum speed of a DC motor is indicated in rpm and is given in the data sheet. The DC motor has two types of RPM: no-load and loaded. The manufacturer's data sheet gives the no-load rpm. The no-load rpm can be from a few thousand to tens of thousands. The rpm is reduced when moving a load and it decreases as the load is increased. For example, a drill turning a screw has a much lower rpm speed than when it is in the no-load situation. DC motors also have voltage and current ratings. The nominal voltage is the voltage for that motor under normal conditions, and can be from 1 to 150 V, depending on the motor. As we increase the voltage, the rpm goes up. The current rating is the current consumption when the nominal voltage is applied with no load, and can be from 25 mA to a few amps. As the load increases, the rpm is decreased, unless the current or voltage provided to the motor is increased, which in turn increases the torque. With a fixed voltage, as the load increases, the current (power) consumption of a DC motor is increased. If we overload the motor it will stall, and that can damage the motor due to the heat generated by high current consumption.

Unidirectional control

Figure 17-1 shows the DC motor clockwise (CW) and counterclockwise (CCW) rotations. See Table 17-1 for selected DC motors.

Bidirectional control

With the help of relays or some specially designed chips we can change the direction of the DC motor rotation. Figures 17-2 through 17-4 show the basic con-

Table 17-1: Selected DC Motor Characteristics (http://www.Jameco.com)

Part No.	Nominal Volts	Volt Range	Current	RPM	Torque
154915CP	3 V	1.5–3 V	0.070 A	5,200	4.0 g-cm
154923CP	3 V	1.5–3 V	0.240 A	16,000	8.3 g-cm
177498CP	4.5 V	3–14 V	0.150 A	10,300	33.3 g-cm
181411CP	5 V	3–14 V	0.470 A	10,000	18.8 g-cm

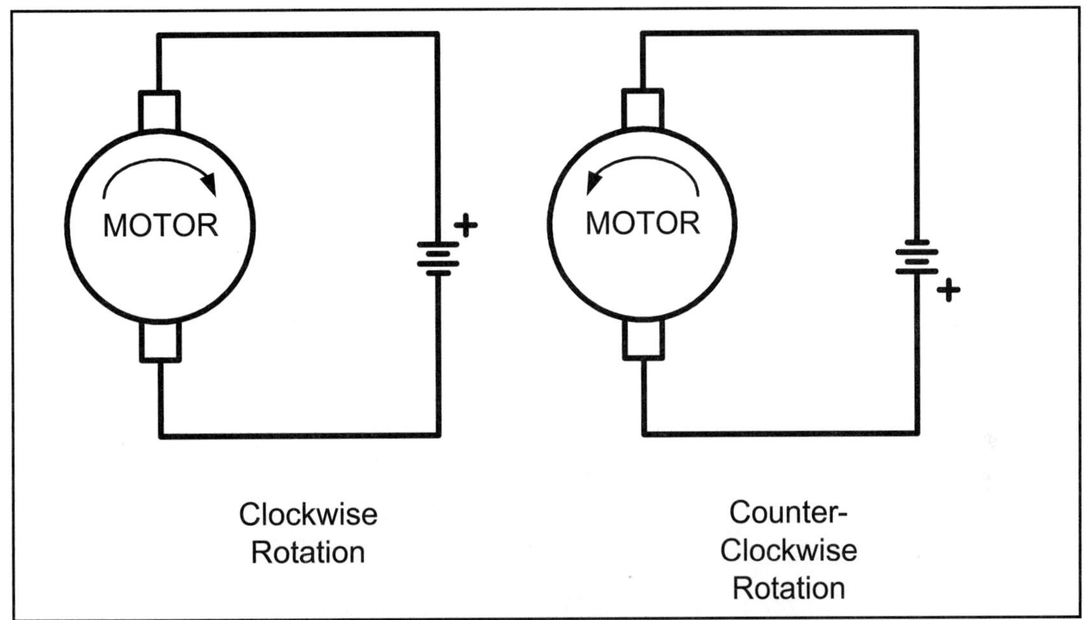

Figure 17-1. DC Motor Rotation (Permanent Magnet Field)

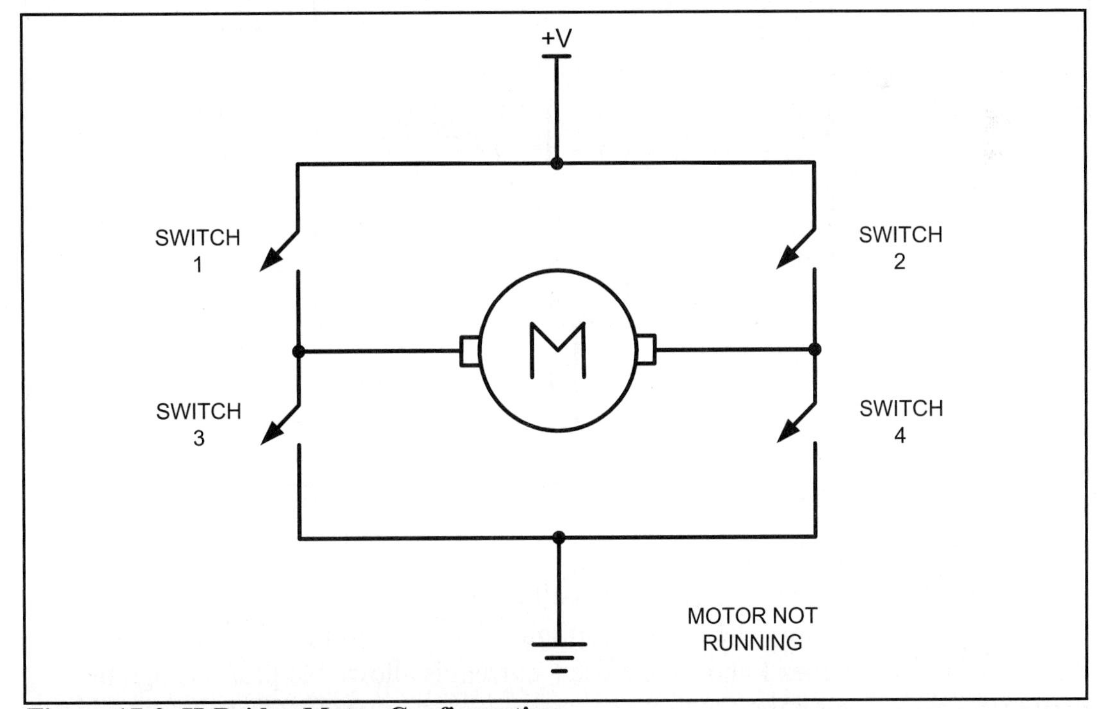

Figure 17-2. H-Bridge Motor Configuration

cepts of H-Bridge control of DC motors.

Figure 17-2 shows the connection of an H-Bridge using simple switches.

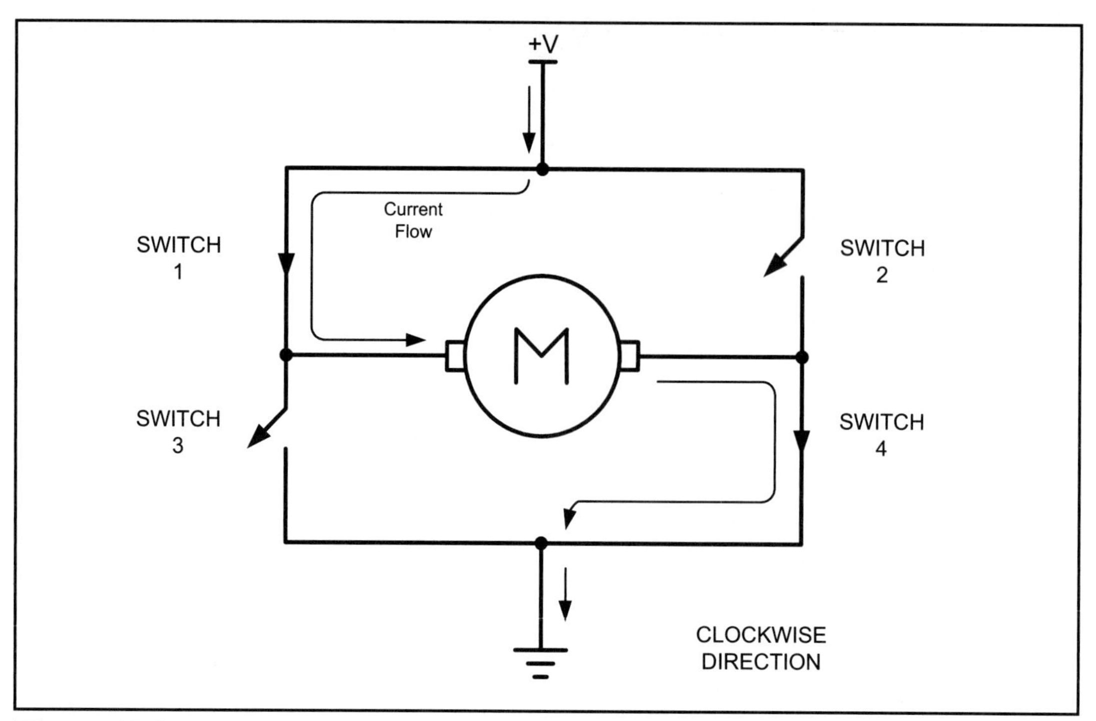

Figure 17-3. H-Bridge Motor Clockwise Configuration

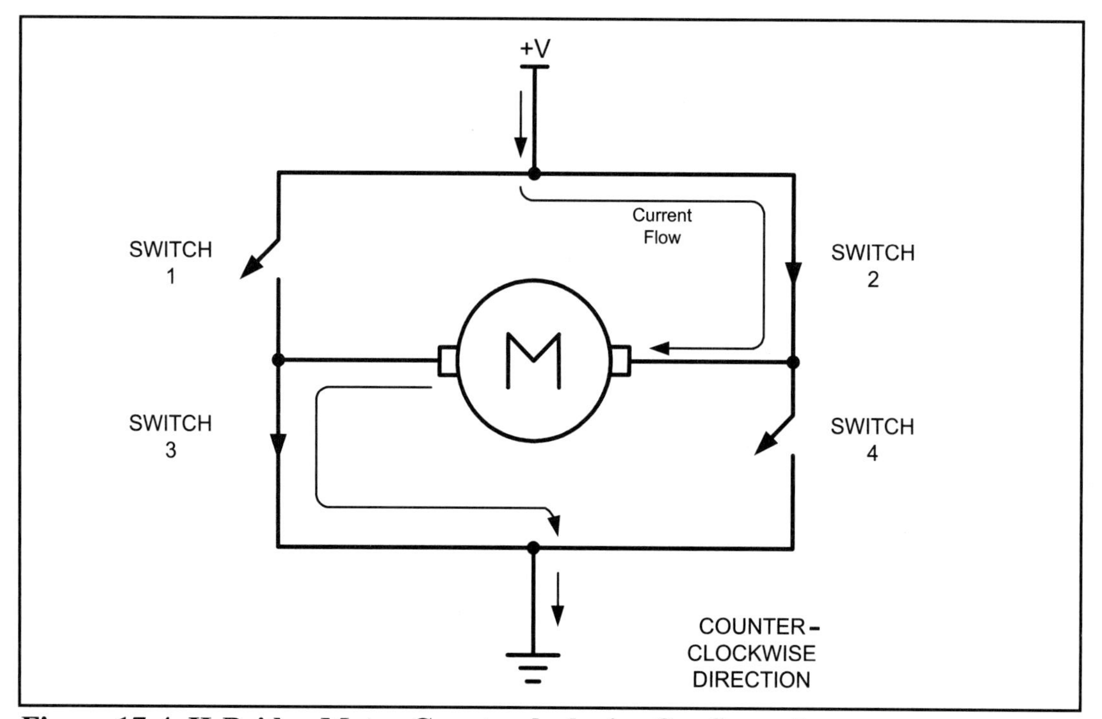

Figure 17-4. H-Bridge Motor Counterclockwise Configuration

All the switches are open, which does not allow the motor to turn.

Figure 17-3 shows the switch configuration for turning the motor in one direction. When switches 1 and 4 are closed, current is allowed to pass through the motor.

Figure 17-4 shows the switch configuration for turning the motor in the opposite direction from the configuration of Figure 17-3. When switches 2 and 3 are closed, current is allowed to pass through the motor.

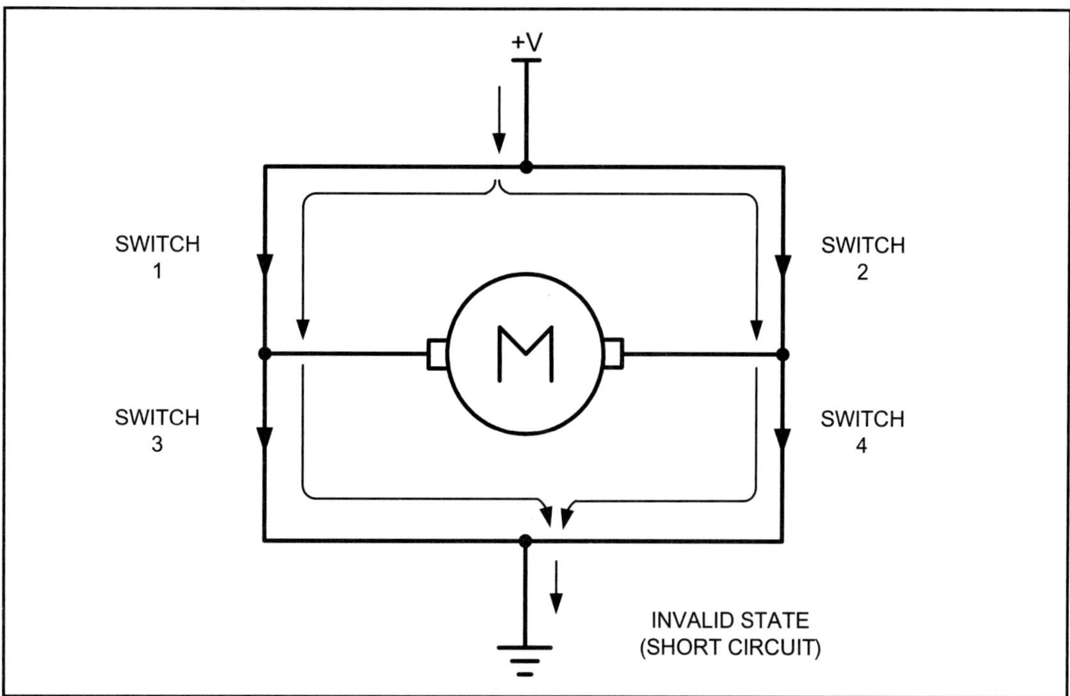

Figure 17-5. H-Bridge in an Invalid Configuration

Table 17-2: Some H-Bridge Logic Configurations for Figure 17-4

Motor Operation	SW1	SW2	SW3	SW4
Off	Open	Open	Open	Open
Clockwise	Closed	Open	Open	Closed
Counterclockwise	Open	Closed	Closed	Open
Invalid	Closed	Closed	Closed	Closed

Figure 17-5 shows an invalid configuration. Current flows directly to ground, creating a short circuit. The same effect occurs when switches 1 and 3 are closed or switches 2 and 4 are closed.

Table 17-2 shows some of the logic configurations for the H-Bridge design.

H-Bridge control can be created using relays, transistors, or a single IC solution such as the L293. When using relays and transistors, you must ensure that invalid configurations do not occur.

Although we do not show the relay control of an H-Bridge, Example 17-1 shows a simple program to operate a basic H-Bridge.

Figure 17-6 shows the connection of the L293 to an HCS12. Be aware that the L293 will generate heat during operation. For sustained operation of the motor, use a heat sink. Example 17-2 shows control of the L293.

Example 17-1

A switch is connected to pin PA7 (PORTA.7). Using a simulator, write a program to simulate the H-Bridge in Table 17-2. We must perform the following:
(a) If DIR = 0, the DC motor moves clockwise.
(b) If DIR = 1, the DC motor moves counterclockwise.

Solution:

```
     BSET DDRB,%00000001          ;PORTB.0 as output for switch 1
     BSET DDRB,%00000010          ;      .1  "            switch 2
     BSET DDRB,%00000100          ;      .2  "            switch 3
     BSET DDRB,%00001000          ;      .3  "            switch 4
     BCLR DDRA,%10000000          ;make PORTA.7 an input DIR
MONITOR:
     BRCLR PORTA,%10000000,CLKWISE
     BSET  PORTB,%00000001             ;switch 1
     BCLR  PORTB,%00000010             ;switch 2
     BCLR  PORTB,%00000100             ;switch 3
     BSET  PORTB,%00001000             ;switch 4
     BRA   MONITOR
CLKWISE:
     BCLR PORTB,%00000001              ;switch 1
     BSET PORTB,%00000010              ;switch 2
     BSET PORTB,%00000100              ;switch 3
     BCLR PORTB,%00001000              ;switch 4
     BRA  MONITOR
```

View the results on your simulator. This example is for simulation only and should not be used on a connected system.

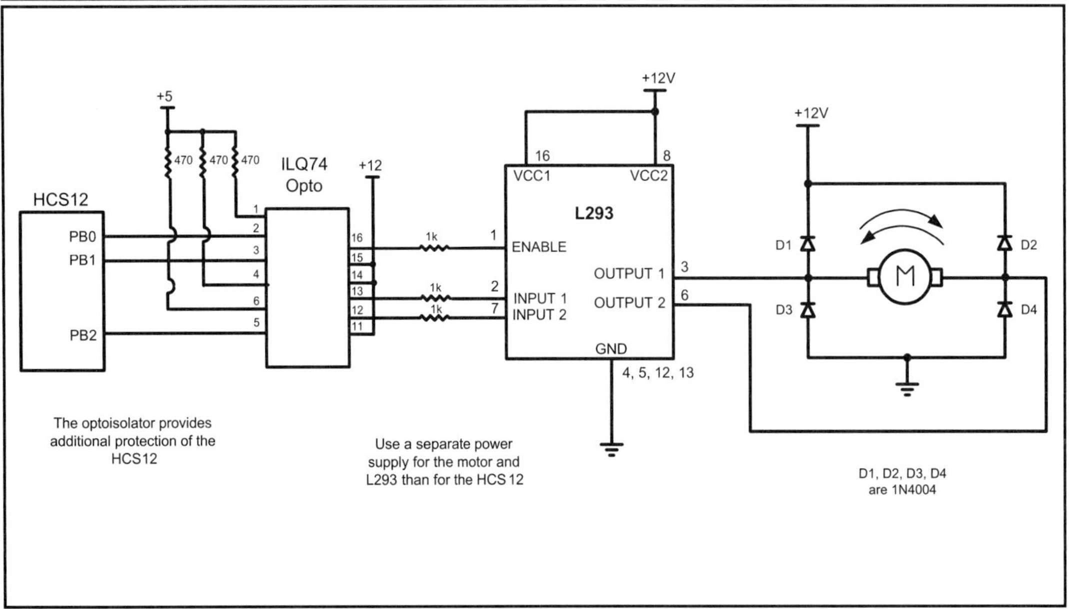

Figure 17-6. Bidirectional Motor Control Using an L293 Chip

Pulse width modulation (PWM)

The speed of the motor depends on three factors: (a) load, (b) voltage, and (c) current. For a given fixed load we can maintain a steady speed by using a

Example 17-2

Figure 17-6 shows the connection of an L293. Add a switch to pin PA7 (PORTA.7). Write a program to monitor the status of SW and perform the following:
(a) If SW = 0, the DC motor moves clockwise.
(b) If SW = 1, the DC motor moves counterclockwise.

Solution:

```
       BSET DDRB,%00000001
       BSET DDRB,%00000010
       BSET DDRB,%00000100
       BCLR DDRA,%10000000
       BSET PORTB,%00000001        ;enable the chip
CHK    BRCLR PORTA,%1000000,CWISE
       BCLR   PORTB,%00000001 ;turn the motor counterclockwise
       BSET   PORTB,%00000100
       BRA   CHK
CWISE BSET   PORTB,%00000001
       BCLR   PORTB,%00000010      ;turn motor clockwise
       BRA   CHK
```

method called *pulse width modulation* (PWM). By changing (modulating) the width of the pulse applied to the DC motor we can increase or decrease the amount of power provided to the motor, thereby increasing or decreasing the motor speed. Notice that, although the voltage has a fixed amplitude, it has a variable duty cycle. That means the wider the pulse, the higher the speed. PWM is so widely used in DC motor control that some microcontrollers come with the PWM circuitry embedded in the chip. In such microcontrollers all we have to do is load the proper registers with the values of the high and low portions of the desired pulse, and the rest is taken care of by the microcontroller. This allows the microcontroller to do other things. For microcontrollers without PWM circuitry, we must create the various duty cycle pulses using software, which prevents the microcontroller from doing other things. The ability to control the speed of the DC motor using PWM is one reason that DC motors are preferable over AC motors. AC motor speed is dictated by the AC frequency of the voltage applied to the motor and the frequency is generally fixed. As a result, we cannot control the speed of the AC motor when the load is increased. As will be shown later, we can also change the DC motor's direction and torque. See Figure 17-7 for PWM comparisons.

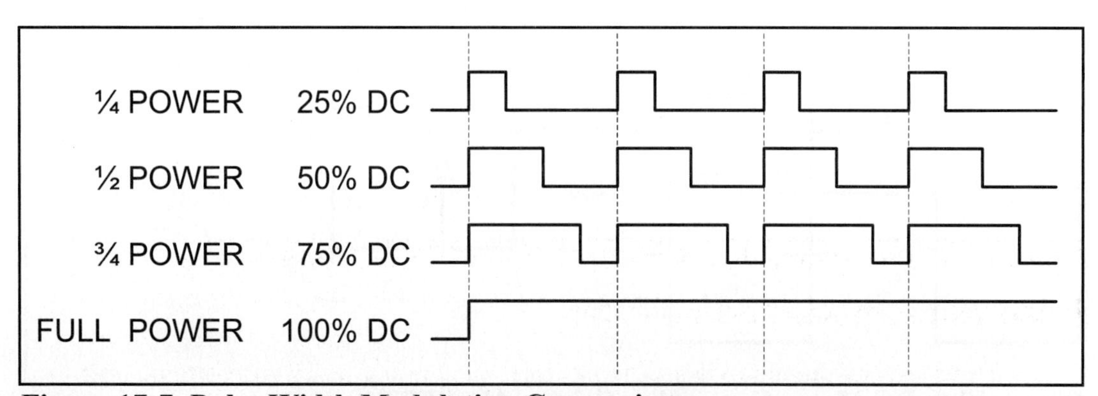

Figure 17-7. Pulse Width Modulation Comparison

DC motor control with optoisolator

As we discussed in Chapter 15, the optoisolator is indispensable in many motor control applications. Figures 17-8 and 17-9 show the connections to a simple DC motor using a bipolar and a MOSFET transistor. Notice that the HCS12 is protected from EMI created by motor brushes by using an optoisolator and a separate power supply.

Figures 17-8 and 17-9 show optoisolators for single directional motor control, and the same principle should be used for most motor applications. Separating the power supplies of the motor and logic will reduce the possibility of damage to the control circuity.

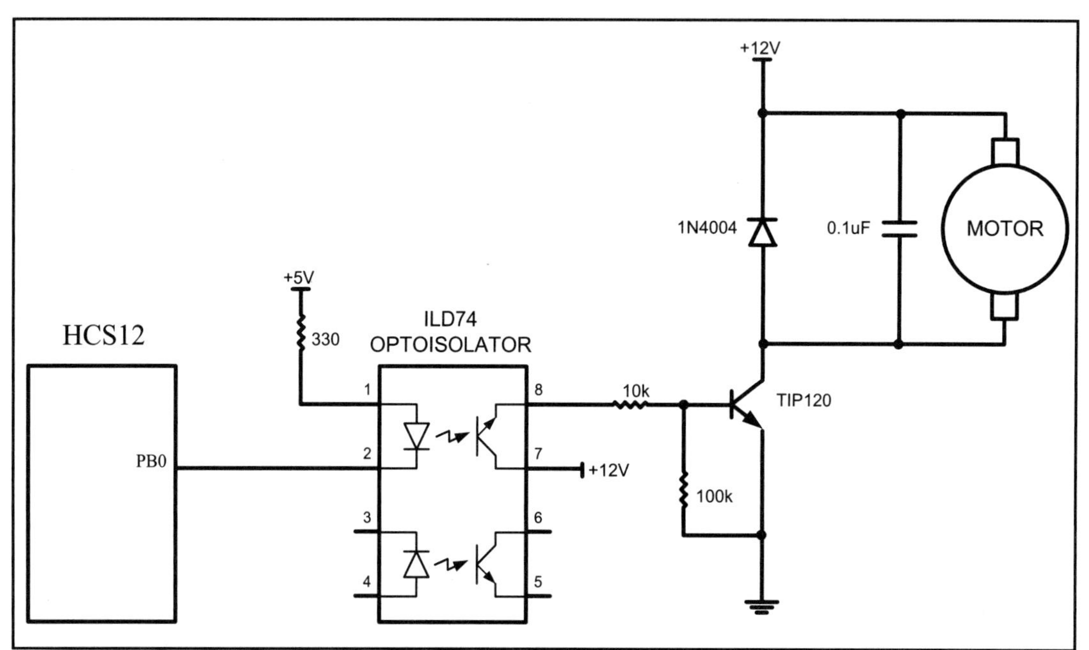

Figure 17-8. DC Motor Connection Using a Darlington Transistor

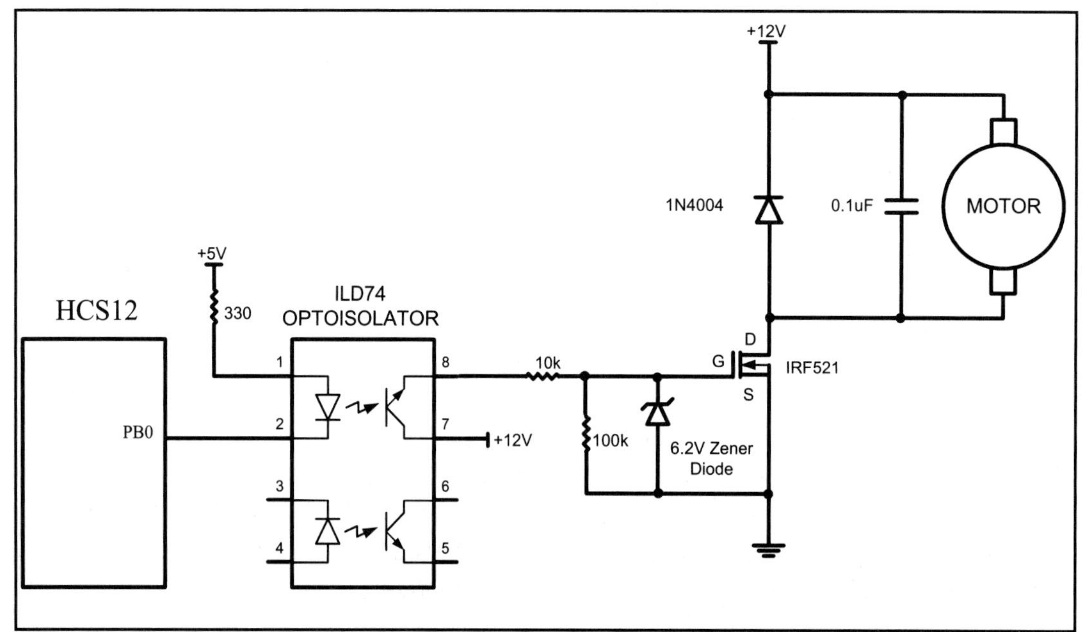

Figure 17-9. DC Motor Connection Using a MOSFET Transistor

Figure 17-8 shows the connection of a bipolar transistor to a motor. Protection of the control circuit is provided by the optoisolator. The motor and HCS12 use separate power supplies. The separation of power supplies also allows the use of high-voltage motors. Notice that we use a decoupling capacitor across the motor; this helps reduce the EMI created by the motor. The motor is switched on by clearing bit PB0.

Figure 17-9 shows the connection of a MOSFET transistor. The optoisolator protects the HCS12 from EMI. The zener diode is required for the transistor to reduce gate voltage below the rated maximum value. See Example 17-3.

Example 17-3

Refer to the figure in this example. Write a program to monitor the status of the switch and perform the following:
(a) If PORTA.7 = 1, the DC motor moves with 25% duty cycle pulse.
(b) If PORTA.7 = 0, the DC motor moves with 50% duty cycle pulse.

Solution:

```
          BSET DDRB,%00000001           ;PORTB.0 as output
          BCLR DDRA,%10000000           ;PORTA.7 as input
          BCLR PORTB,%00000001  ;turn on motor
CHK
          BRCLR PORTA,%10000000,PWM_50
          BCLR PORTB,%00000001  ;low portion of pulse (on)
          JSR DELAY
          BSET PORTB,%00000001  ;high portion of pulse (off)
          JSR DELAY
          JSR DELAY
          JSR DELAY
          BRA  CHK
PWM_50
          BSET PORTB,%00000001          ;high portion of pulse
          JSR DELAY
          JSR DELAY
          BCLR PORTB,%00000001          ;low portion of pulse
          JSR DELAY
          JSR DELAY
          BRA CHK
```

+5V
4.7k
PA7
HCS12
PB0
+5V
330
ILD74
OPTOISOLATOR
1 2 3 4 8 7 6 5
+12V
10k
100k
TIP120
+12V
1N4004
0.1uF
MOTOR

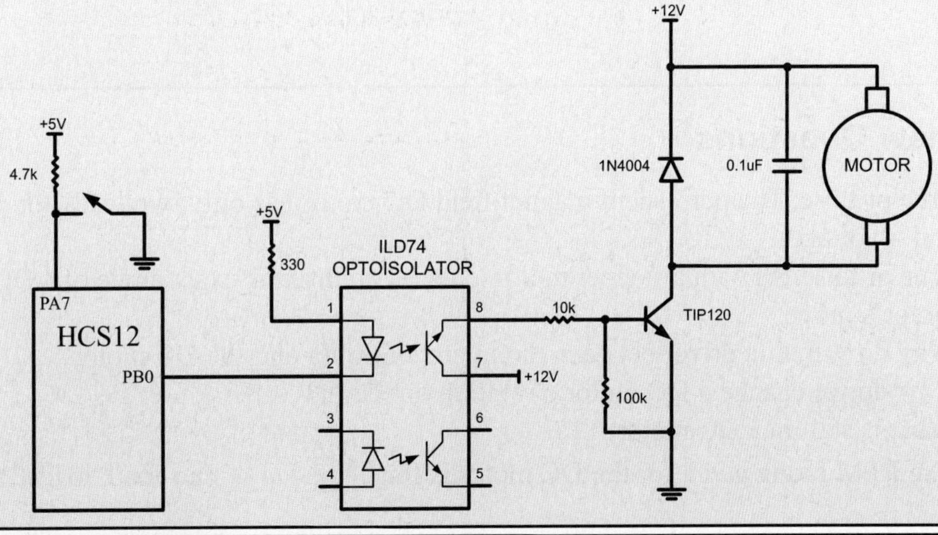

DC motor control and PWM using C

Examples 17-4 through 17-6 show the C versions of the earlier programs controlling the DC motor.

Example 17-4

Refer to Figure 17-6 for connection of the motor. A switch is connected to pin PA0. Write a C program to monitor the status of SW and perform the following:
(a) If SW = 0, the DC motor moves clockwise.
(b) If SW = 1, the DC motor moves counterclockwise.

Solution:

```
#include <mc9s12dp512.h>      /* derivative information */
#define SW PORTA_BIT7
#define ENABLE PORTB_BIT0
#define MTR_1 PORTB_BIT1
#define MTR_2 PORTB_BIT2

void main()
  {
  DDRA = DDRA & ~0x80;      //make PA7 input pin
  DDRB = DDRB | 0x07;      //make PORTB output
  SW = 1;
  ENABLE = 0;
  MTR_1 = 0;
  MTR_2 = 0;

  while(1)
    {
      ENABLE = 1;
      if(SW == 1)
        {
          MTR_1 = 1;
          MTR_2 = 0;
        }
      else
        {
          MTR_1 = 0;
          MTR_2 = 1;
        }
    }
  }
```

Review Questions

1. True or false. The permanent magnet field DC motor has only two leads for + and – voltages.
2. True or false. As with a stepper motor, one can control the exact angle of a DC motor's move.
3. Why do we put a driver between the microcontroller and the DC motor?
4. How do we change a DC motor's rotation direction?
5. What is stall in a DC motor?
6. The RPM rating given for the DC motor is for __________ (no-load, loaded).

Example 17-5

Refer to the figure in this example. Write a C program to monitor the status of SW and perform the following:

(a) If SW = 0, the DC motor moves with 50% duty cycle pulse.
(b) If SW = 1, the DC motor moves with 25% duty cycle pulse.

Solution:

```
#include <mc9s12dp512.h>       /* derivative information */
#define SW PORTA_BIT7
#define MTR  PORTB_BIT0
void MSDelay(unsigned int value);
void main()
  {
  DDRA=0x7F;           //make PA7 input pin
  DDRB=0x01;           //make PB0 output pin
  while(1)
    {
      if(SW == 1)
        {
          MTR = 0;
          MSDelay(25);
          MTR = 1;
          MSDelay(75);
        }
      else
        {
          MTR = 1;
          MSDelay(50);
          MTR = 0;
          MSDelay(50);
        }
    }
  }
void MSDelay(unsigned int value)
  {
    unsigned char x, y;
    for(x=0; x<1275; x++)
      for(y=0; y<value; y++);
  }
```

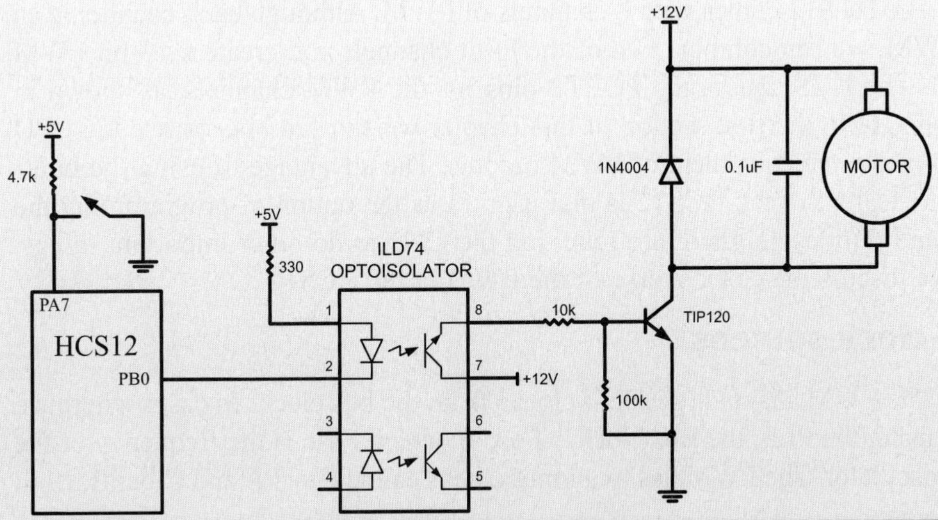

Example 17-6

Refer to Figure 17-8 for connection to the motor. Two switches are connected to pins PORTA0 and PORTA1. Write a C program to monitor the status of both switches and perform the following:

SW2 (PD.1)	SW1 (PD.0)	
0	0	DC motor moves slowly (25% duty cycle).
0	1	DC motor moves moderately (50% duty cycle).
1	0	DC motor moves fast (75% duty cycle).
1	1	DC motor moves very fast (100% duty cycle).

Solution:

```
#include <mc9s12dp512.h>      /* derivative information */
#define MTR   PORTB_BIT0
void MSDelay(unsigned int value);
void main()
  {
  unsigned int duty;
  DDRB = 0x01;
  DDRA = 0xFC;
  while(1)
    {
     duty = PORTA&0x03;     //mask off all but PA1 and PA0
     duty++;
     duty *= 25;
     MTR = 0;
     MSDelay(duty);         //see Example 17-5
     MTR = 1;
     MSDelay(100-duty);     //see Example 17-5
    }
  }
```

SECTION 17.2: PROGRAMMING PWM IN HCS12

This section discusses the PWM feature of the HCS12 and shows the programming of the PWM module of the HCS12 using Assembly and C.

Programming PWM in HCS12

The HCS12 comes with 8 channels of PWM. Although each channel is an 8-bit PWM, we can combine two of the 8-bit channels and create a 16-bit PWM channel. The HCS12 uses the PORTP pins for the PWM channels, as shown in Figure 17-10. In the first section of this chapter we showed how to use the CPU itself to create the equivalent of PWM outputs. The advantage of using the built-in PWM feature of the HCS12 is that it gives us the option of programming the period and duty cycle, therefore relieving the CPU to do other important things. Next, we discuss the clock source of the PWM in the HCS12.

PWM clock sources

The PWM channels get their clocks from the bus clock. In cases where we are not using the PLL, the Bus clock = Fosc/2, where Fosc is the frequency of the crystal oscillator. The PWM has two input clocks called ClockA and ClockB. Both

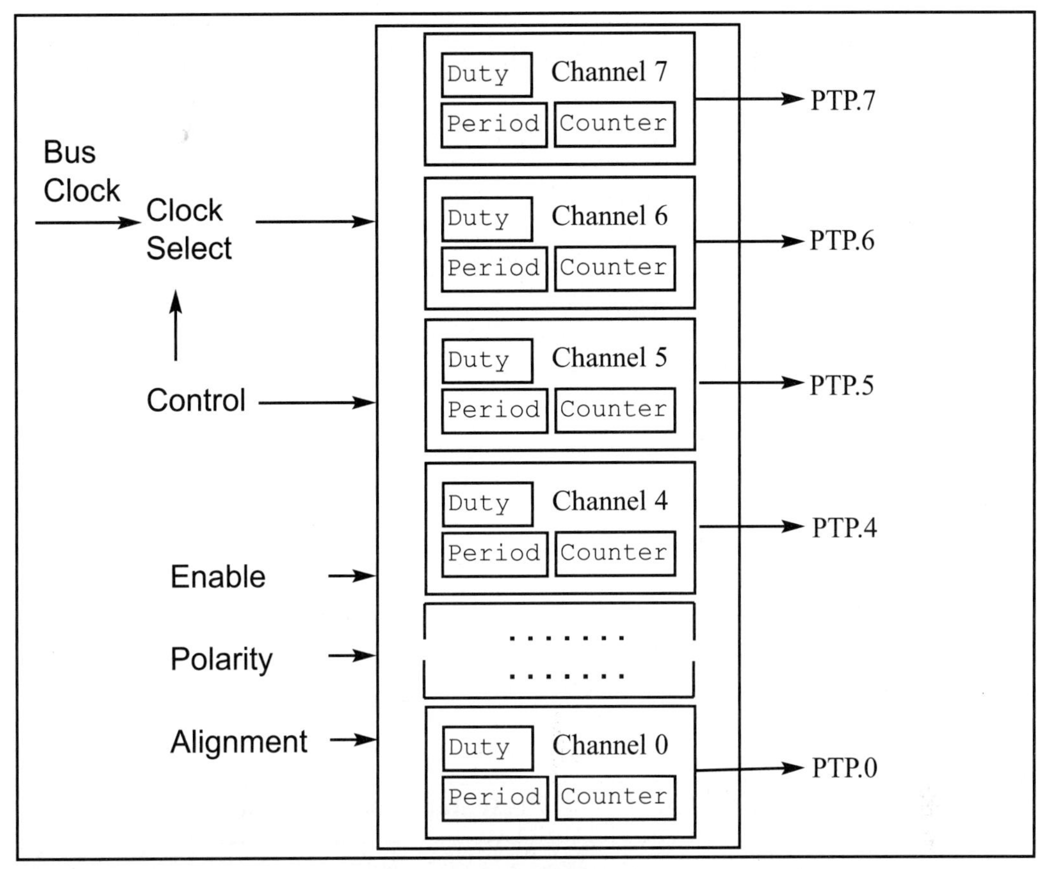

Figure 17-10. PWM Channels and PORTP

ClockA and ClockB come from the bus clock. ClockA is used for the input to Channels 0, 1, 4, and 5, while ClockB is used for the input to Channels 2, 3, 6, and 7. Using the prescaler, ClockA can be divided further. To do that, we use the PWMPRCLK register, as shown in Figure 17-11. The same is true for ClockB. Besides the use of the PWMPRCLK register, if we need to divide ClockA and ClockB further, we can use the SA and SB clocks.

ClockSA

To get a lower frequency for PWM channels, we can divide ClockA further by using the PWMSCLA register. See Figure 17-11a and 11b. Using the PWM-SCLA registers to divide ClockA, we get ClockSA (scaled A). Notice from Figure 17-12, the highest number we can divide ClockA by is 512 and the lowest number is 2. See Example 17-7.

ClockSB

We get ClockSB the same way we got ClockSA, except we use the PWM-SCLB register, as shown in Figure 17-13. That means similar rates are available for ClockSB. See Figure 17-14 and Example 17-8. In many applications we do not need to divide the clock.

0	PCKB2	PCKB1	PCKB0	0	PCKA2	PCKA1	PCKA0

PCKB2–PCKB0 Prescaler Select for ClockB

0	0	0	Bus clock = ClockB
0	0	1	Divide bus clock by 2 to get ClockB
0	1	0	Divide bus clock by 4
0	1	1	Divide bus clock by 8
1	0	0	Divide bus clock by 16
1	0	1	Divide bus clock by 32
1	1	0	Divide bus clock by 64
1	1	1	Divide bus clock by 128

PCKA2–PCKA0 Prescaler Select for ClockA

0	0	0	Bus clock = ClockA
0	0	1	Divide bus clock by 2 to get ClockA
0	1	0	Divide bus clock by 4
0	1	1	Divide bus clock by 8
1	0	0	Divide bus clock by 16
1	0	1	Divide bus clock by 32
1	1	0	Divide bus clock by 64
1	1	1	Divide bus clock by 128

Figure 17-11a. PWMPRCLK (PWM Prescaler Clock) Register

D7	D6	D5	D4	D3	D2	D1	D0

ClockSA = ClockA / (2 × D7–D0)

D7	**D6**	**D5**	**D4**	**D3**	**D2**	**D1**	**D0**	
0	0	0	0	0	0	0	1	Divide ClockA by 2
0	0	0	0	0	0	1	0	Divide ClockA by 4
0	0	0	0	0	0	1	1	Divide ClockA by 6
0	0	0	0	0	1	0	0	Divide ClockA by 8
0	0	0	0	0	1	0	1	Divide ClockA by 10
0	0	0	0	0	1	1	0	Divide ClockA by 12
0	0	0	0	0	1	1	1	Divide ClockA by 14
.	.	.	.	.	.	.	.	
.	.	.	.	.	.	.	.	
1	1	1	1	1	1	0	0	Divide ClockA by 504
1	1	1	1	1	1	0	1	Divide ClockA by 506
1	1	1	1	1	1	1	0	Divide ClockA by 508
1	1	1	1	1	1	1	1	Divide ClockA by 510
0	0	0	0	0	0	0	0	Divide ClockA by 512

Figure 17-11b. PWMSCLA (PWM Scale A) Register

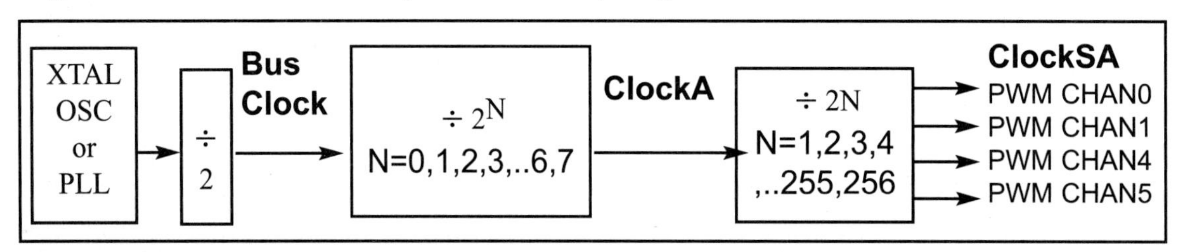

Figure 17-12. ClockA and ClockSA Frequencies

Example 17-7

Assume XTAL = 4 MHz and Bus Freq = 2 MHz. Find a) the lowest, and b) the highest frequency for ClockA. Give the values for the PWMPRCLK and PWMSCLA registers for the above options.

Solution:

a) The lowest frequency we can get for ClockA is 2 MHz/128 = 15,625 Hz. That means we have 15,625/512 = 30 Hz for the lowest ClockA. That gives us PWMPRCLK = 00000111 and PWMSCLA = 00000000.

b) The highest frequency we can get for ClockA is 2 MHz/1 = 2 MHz. That gives us 2 MHz/(1 × 2) = 2 MHz/2 = 1 MHz for the highest ClockA. That means PWMPRCLK = 00000000 and PWMSCLA = 00000001.

PWMCLK register and clock source selection

In programming the PWM channels 0, 1, 4, and 5, we have a choice of using ClockA or ClockSA. The same is true for ClockB and ClockSB when using the PWM channels 2, 3, 6, and 7. This is done using the PWMCLK register, as shown in Figure 17-15.

D7	D6	D5	D4	D3	D2	D1	D0

ClockSB = ClockB / (2 × D7–D0)

D7	D6	D5	D4	D3	D2	D1	D0	
0	0	0	0	0	0	0	1	Divide ClockB by 2
0	0	0	0	0	0	1	0	Divide ClockB by 4
0	0	0	0	0	0	1	1	Divide ClockB by 6
0	0	0	0	0	1	0	0	Divide ClockB by 8
0	0	0	0	0	1	0	1	Divide ClockB by 10
0	0	0	0	0	1	1	0	Divide ClockB by 12
0	0	0	0	0	1	1	1	Divide ClockB by 14
.	.	.	.	.	.	.	.	
.	.	.	.	.	.	.	.	
1	1	1	1	1	1	0	0	Divide ClockB by 504
1	1	1	1	1	1	0	1	Divide ClockB by 506
1	1	1	1	1	1	1	0	Divide ClockB by 508
1	1	1	1	1	1	1	1	Divide ClockB by 510
0	0	0	0	0	0	0	0	Divide ClockB by 512

Figure 17-13. PWMSCLB (PWM Scale B) Register

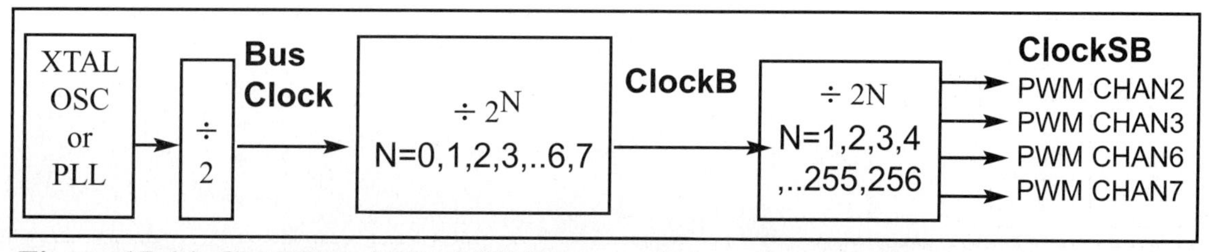

Figure 17-14. ClockB and ClockSB Frequencies

Example 17-8

Assume XTAL = 4 MHz and Bus Freq = 2 MHz. Find the values for PWMPRCLK and PWMSCLB if we want ClockSB = 2,500 Hz.

Solution:

Since the bus clock is 2 MHz, we get 2 MHz/8 = 250,000 Hz for PWMPRCLK = 00110000. Now, ClockSB = 250,000/(2 × 50) = 250,000/100 = 2,500 Hz. Therefore we need PWMSCLB = 00110010, since 00110010 = 50 decimal.

PCLK7	PCLK6	PCLK5	PCLK4	PCLK3	PCLK2	PCLK1	PCLK0

PCLK7–PCLK0 PWM Clock Selection

PCLK7: D7 PWM Channel 7 Clock Selection
1 = ClockSB is the clock source for PWM channel 7.
0 = ClockB is the clock source for PWM channel 7.

PCLK6: D6 PWM Channel 6 Clock Selection
1 = ClockSB is the clock source for PWM channel 6.
0 = ClockB is the clock source for PWM channel 6.

PCLK5: D5 PWM Channel 5 Clock Selection
1 = ClockSA is the clock source for PWM channel 5.
0 = ClockA is the clock source for PWM channel 5.

PCLK4: D4 PWM Channel 4 Clock Selection
1 = ClockSA is the clock source for PWM channel 4.
0 = ClockA is the clock source for PWM channel 4.

PCLK3: D3 PWM Channel 3 Clock Selection
1 = ClockSB is the clock source for PWM channel 3.
0 = ClockB is the clock source for PWM channel 3.

PCLK2: D2 PWM Channel 2 Clock Selection
1 = ClockSB is the clock source for PWM channel 2.
0 = ClockB is the clock source for PWM channel 2.

PCLK1: D1 PWM Channel 1 Clock Selection
1 = ClockSA is the clock source for PWM channel 1.
0 = ClockA is the clock source for PWM channel 1.

PCLK0: D0 PWM Channel 0 Clock Selection
1 = ClockSA is the clock source for PWM channel 0.
0 = ClockA is the clock source for PWM channel 0.

Figure 17-15. PWMCLK (PWM Clock Selection) Register

Selection channel polarity

We use the PWMPOL register to choose the polarity, as shown in Figure 17-16.

PPOL7	PPOL6	PPOL5	PPOL4	PPOL3	PPOL2	PPOL1	PPOL0

PPOL7–PPOL0 PWM Polarity Selection

PPOL7: PWM Channel 7 Polarity Selection
1 = PWM channel 7 output is high at the beginning of the period, then goes low when the duty count is reached.
0 = PWM channel 7 output is low at the beginning of the period, then goes high when the duty count is reached.

PPOL6: PWM Channel 6 Polarity Selection
1 = PWM channel 6 output is high at the beginning of the period, then goes low when the duty count is reached.
0 = PWM channel 6 output is low at the beginning of the period, then goes high when the duty count is reached.

PPOL5: PWM Channel 5 Polarity Selection
1 = PWM channel 5 output is high at the beginning of the period, then goes low when the duty count is reached.
0 = PWM channel 5 output is low at the beginning of the period, then goes high when the duty count is reached.

PPOL4: PWM Channel 4 Polarity Selection
1 = PWM channel 4 output is high at the beginning of the period, then goes low when the duty count is reached.
0 = PWM channel 4 output is low at the beginning of the period, then goes high when the duty count is reached.

PPOL3: PWM Channel 3 Polarity Selection
1 = PWM channel 3 output is high at the beginning of the period, then goes low when the duty count is reached.
0 = PWM channel 3 output is low at the beginning of the period, then goes high when the duty count is reached.

PPOL2: PWM Channel 2 Polarity Selection
1 = PWM channel 2 output is high at the beginning of the period, then goes low when the duty count is reached.
0 = PWM channel 2 output is low at the beginning of the period, then goes high when the duty count is reached.

PPOL1: PWM Channel 1 Polarity Selection
1 = PWM channel 1 output is high at the beginning of the period, then goes low when the duty count is reached.
0 = PWM channel 1 output is low at the beginning of the period, then goes high when the duty count is reached.

PPOL0: PWM Channel 0 Polarity Selection
1 = PWM channel 0 output is high at the beginning of the period, then goes low when the duty count is reached.
0 = PWM channel 0 output is low at the beginning of the period, then goes high when the duty count is reached.

Figure 17-16. PWMPOL (PWM Polarity Selection) Register

Left Aligned PWM output

To generate the pulses for the PWM output, we have the choice of making the pulse go high first and then low, or the other way around. We also have the choice of two types of outputs, Left Aligned or Center Aligned outputs. In the Left Aligned with positive polarity (PPOLx = 1), the output is high for the duration of the duty cycle and then goes low for the rest of the period. See Figure 17-17. See the HCS12 manual for the Center Aligned output. We use the PWMCAE register to choose the alignment, as shown in Figure 17-18.

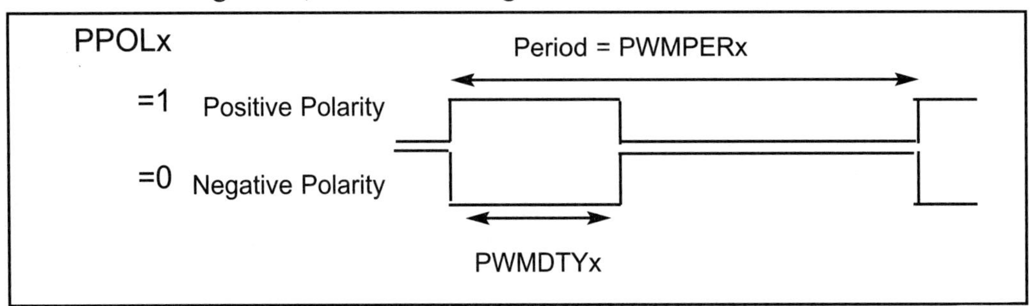

Figure 17-17. PWM Left Alignment Output with Polarity

CAE7	CAE6	CAE5	CAE4	CAE3	CAE2	CAE1	CAE0

CAE7–CAE0 PWM Center Aligned Enable

CAE7: D7 PWM Channel 7 Center Aligned output mode
1 = Channel 7 operates in Center Aligned output mode.
0 = Channel 7 operates in Left Aligned output mode.

CAE6: D6 PWM Channel 6 Center Aligned output mode
1 = Channel 6 operates in Center Aligned output mode.
0 = Channel 6 operates in Left Aligned output mode.

CAE5: D5 PWM Channel 5 Center Aligned output mode
1 = Channel 5 operates in Center Aligned output mode.
0 = Channel 5 operates in Left Aligned output mode.

CAE4: D4 PWM Channel 4 Center Aligned output mode
1 = Channel 4 operates in Center Aligned output mode.
0 = Channel 4 operates in Left Aligned output mode.

CAE3: D3 PWM Channel 3 Center Aligned output mode
1 = Channel 3 operates in Center Aligned output mode.
0 = Channel 3 operates in Left Aligned output mode.

CAE2: D2 PWM Channel 2 Center Aligned output mode
1 = Channel 2 operates in Center Aligned output mode.
0 = Channel 2 operates in Left Aligned output mode.

CAE1: D1 PWM Channel 1 Center Aligned output mode
1 = Channel 1 operates in Center Aligned output mode.
0 = Channel 1 operates in Left Aligned output mode.

CAE0: D0 PWM Channel 0 Center Aligned output mode
1 = Channel 0 operates in Center Aligned output mode.
0 = Channel 0 operates in Left Aligned output mode.

Figure 17-18. PWMCAE (PWM Center Aligned Enable) Register

Enabling (turning on) the PWM channel

Upon reset, the PWM channels are disabled, meaning no clock is provided to them. We use the PWME (PWM Enable) register to enable (turn on) a given channel. See Figure 17-19.

PWME7	PWME6	PWME5	PWME4	PWME3	PWME2	PWME1	PWME0

PWME7–PWME0 PWM Channel Enable

PWME7: D7 PWM Channel 7 Enable
1 = PWM channel 7 is enabled.
0 = PWM channel 7 is disabled.
PWME6: D6 PWM Channel 6 Enable
1 = PWM channel 6 is enabled.
0 = PWM channel 6 is disabled.
PWME5: D5 PWM Channel 5 Enable
1 = PWM channel 5 is enabled.
0 = PWM channel 5 is disabled.
PWME4: D4 PWM Channel 4 Enable
1 = PWM channel 4 is enabled.
0 = PWM channel 4 is disabled.
PWME3: D3 PWM Channel 3 Enable
1 = PWM channel 3 is enabled.
0 = PWM channel 3 is disabled.
PWME2: D2 PWM Channel 2 Enable
1 = PWM channel 2 is enabled.
0 = PWM channel 2 is disabled.
PWME1: D1 PWM Channel 1 Enable
1 = PWM channel 1 is enabled.
0 = PWM channel 1 is disabled.
PWME0: D0 PWM Channel 0 Enable
1 = PWM channel 0 is enabled.
0 = PWM channel 0 is disabled.

Notes:
1) The pulse-modulated output signal becomes available when its clock source begins its next cycle.
2) These bits are used for enabling the 8-bit channels only. When combining two 8-bit channels to create a 16-bit channel, these bits have no effect.

Figure 17-19. PWME (PWM Enable) Register

Counter, Period, and Duty Cycle registers for PWM

In the PWM, there are three registers associated with each channel. They are the Counter (PWMCNT) register, the Period (PWMPER) register, and the Duty Cycle (PWMDTY) register. The registers are 8 bits wide. The Counter registers are labeled as PWMCNT0, PWMCNT1, PWMCNT2, and so on. The Period registers are labeled as PWMPER0, PWMPER1, PWMPER2, and so on. The value in the period register (PWMPERx) determines the frequency of the pulses coming out of the PORTP pin. See Figure 17-20. The clock source (ClockA, ClockB, Clock SA, or Clock SB) is connected to the PWMCNTx registers. Upon enabling the PWM channel, the PWMCNTx register starts to count up from zero. As it counts up, its value is compared with the period register (PWMPERx) value and when they match, the output pin changes state. Assuming we choose PPOLx = 1 and left aligned output (CAEx = 0), we use the following equation to calculate the PWM frequency:

PWM_Freq = ClockFreq (A, B, SA, or SB) / PWMPERx

See Example 17-9. The Duty Cycle registers are called PWMDTY0, PWMDTY1, PWMDTY2, and so on. The value in the duty cycle register determines the duty cycle of the PWM pulses coming out of the PORTP pin. The counter register (PWMCNTx) is also used by the duty cycle register (PWMDTYx) to create the desired duty cycle. See Figure 17-20. The value for the duty cycle register is always a percentage of the value loaded into the period register. See Example 17-10.

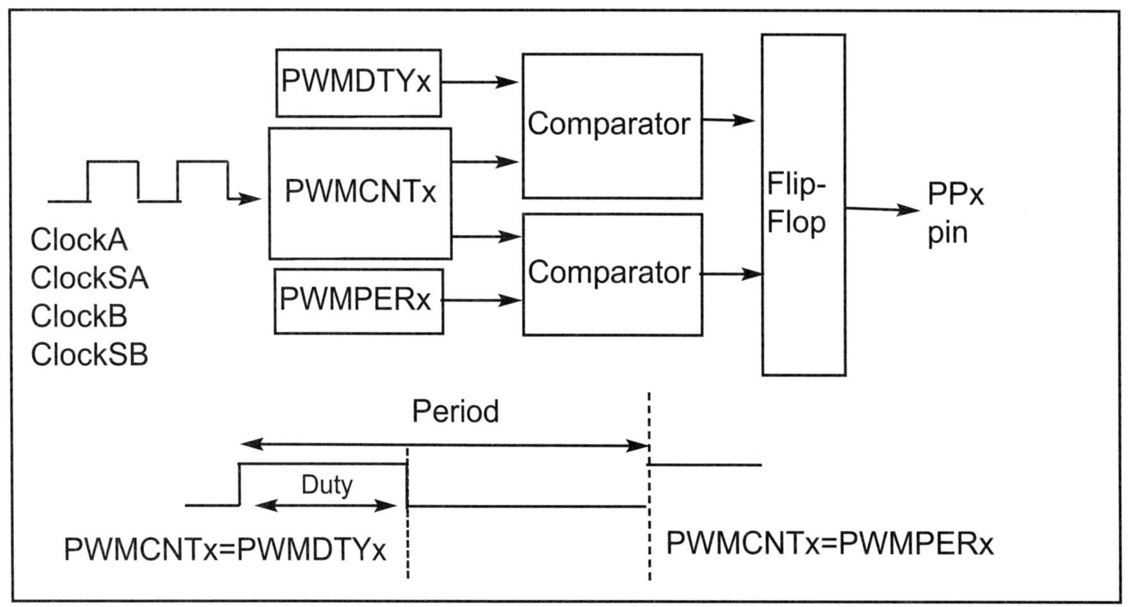

Figure 17-20. PWMCNT, PWMPER, and PWMDTY in Creating the Duty Cycle

Example 17-9

Assume ClockSB = 2,500 Hz. Find the value for the PWM frequency if PWMPERx has the value of 125.

Solution:

PWM_Freq = 2,500 Hz / 125 = 20 Hz

Example 17-10

Assume ClockSB = 2,500 Hz. Find the value for PWMPERx and PWMDTYx if we want a PWMx frequency of 50 Hz with 60% duty cycle.

Solution:

PWMPERx = 2,500 Hz / 50 Hz = 50. PWMDTYx = 60% × 50 = 30

PWM control register

The next register we need to discuss is the PWM Control (PWMCTL) register. The PWM control register can be used to select 8- or 16-bit PWM channels among other things. Upon reset, the channels are 8-bit by default. We also use PWMCTL to stop the PWM output during the Wait and Freeze modes of the CPU operation. See Figure 17-21.

CON67	CON45	CON23	CON01	PSWAI	PFRZ	0	0

CON67: D7 Concatenate channels 6 and 7 to create a 16-bit PWM channel.
1 = Channels 6 and 7 are concatenated to create a 16-bit PWM channel. Channel 6 is the high byte and channel 7 is the low byte of the 16 bits. All the options (output pin, clock source, and so on) for channel 7 are used for the 16-bit PWM.
0 = Channels 6 and 7 are separate 8-bit PWMs.

CON45: D6 Concatenate channels 4 and 5 to create a 16-bit PWM channel.
1 = Channels 4 and 5 are concatenated to create a 16-bit PWM channel. Channel 4 is the high byte and channel 5 is the low byte of the 16-bit. All the options (output pin, clock source, and so on) for channel 5 are used for the 16-bit PWM.
0 = Channels 4 and 5 are separate 8-bit PWMs.

CON23: D5 Concatenate channels 2 and 3 to create a 16-bit PWM channel.
1 = Channels 2 and 3 are concatenated to create a 16-bit PWM channel. Channel 2 is the high byte and channel 3 is the low byte of the 16-bit. All the options (output pin, clock source, and so on) for channel 3 are used for the 16-bit PWM.
0 = Channels 2 and 3 are separate 8-bit PWMs.

CON01: D4 Concatenate channels 0 and 1 to create a 16-bit PWM channel.
1 = Channels 0 and 1 are concatenated to create a 16-bit PWM channel. Channel 0 is the high byte and channel 1 is the low byte of the 16-bit. All the options (output pin, clock source, and so on) for channel 1 are used for the 16-bit PWM.
0 = Channels 0 and 1 are separate 8-bit PWMs.

PSWAI: D1 PWM stops in Wait mode.
1 = Stop the input clock to the prescaler whenever the CPU is in Wait mode.
0 = Allow the clock to the prescaler to continue while in Wait mode.

PFRZ: D0 PWM counter stops in Freeze mode.
1 = Disable PWM input clock to the prescaler whenever the part is in Freeze mode.
0 = Allow PWM to continue while in Freeze mode.

Note: Upon reset, the default value value for this register is 00000000.

Figure 17-21. PWMCTL (PWM Control) Register

Steps in programming the PWM 8-bit channels

To program the PWM of the HCS12, the following steps must be taken:

1. Use the PWMPRCLK register to set the prescaler value to bring down the Bus frequency (Fbus) to the desired value of ClockA (or ClockB).

2. If needed, use PWMSCLA (or PWMSCLB) to bring down ClockA (or ClockB) even further to get ClockSA (or ClockSB) if needed.
3. Use the PWMCLK register to select the clock source of ClockA or ClockB (or ClockSA or ClockSB) for the desired channel.
4. Use the PWMPOL register to select the polarity of the PWM output.
5. Use the PWMCAE register to select the Left or Center Aligned output.
6. Use the PWMCTL register to select the 8-bit channel and PWM operations during the wait and freeze modes of the CPU.
7. Load the value for the period of the PWM pulses into the PWMPERx register.
8. Load the value for the duty cycle of the PWM pulses into the PWMDTY register.
9. Clear the PWMCNTx register.
10. Enable (turn on) the desired PWM channel using the PWMEN register.

Example 17-11

Assume XTAL = 4 MHz (Fbus = 2 MHz). Using channel 2, write a program to create the PWM pulses with a frequency of 50 Hz and 60% duty cycle.

Solution:

We need to set PWMPRCLK = 30 for ClockB since ClockB = 2 MHz / 2^3 = 2 MHz/8 = 250 kHz. PWMSCLB = 50 since ClockSB = 250 kHz/(2 × 50) = 250 kHz/100 = 2,500 Hz. Since PWM_Freq = ClockSB/PWM_PER we need PWM_Freq = 2,500 Hz/50 = 50 Hz, which means we must have PWMPER2 = 50. For the duty cycle of 60%, we need PWMDTY2 = 30 since 50 × 60% = 30.

```
LDAA #$30                 ;
STAA PWMPRCLK             ;ClockB=Fbus/8
LDAA #50
STAA PWMSCLB              ;ClockSB=ClockB/2x50
LDAA #$04
STAA PWMCLK               ;use ClockSB for channel 2
LDAA #$04
STAA PWMPOL               ;high, then low for polarity
LDAA #$0
STAA PWMCAE               ;left aligned
LDAA #$0
STAA PWMCTL        ;8-bit chan,PWM during freeze and wait
LDAA #50
STAA PWMPER2              ;PWM_Freq=ClockSB/50
LDAA #30
STAA PWMDTY2              ;60% duty cycle
LDAA #$0
STAA PWMCNT2              ;clear PWMCNT2
BSET PWME,%00000100       ;turn on PWM chan.2(start PWMCNT)
BRA $
```

HCS12

PP2

Programming HCS12 PWM in Assembly

The Assembly language programming of the PWM channels is illustrated in Examples 17-11 through 17-13.

Example 17-12

Assume XTAL = 4 MHz (Fbus = 2 MHz). Using channel 0, write a program to create the PWM pulses with a frequency of 2 kHz and 80% duty cycle.

Solution:

We need to set PWMPRCLK = 02 for ClockA since 2 MHz / 2^2 = 2 MHz/4 = 500 kHz. We bypass the use of ClockA since we can get the PWM_Freq = ClockA/PWM_PER. PWM_Freq = 500 kHz/ 250 = 2 kHz, which means we must have PWMPER0 = 250. For the duty cycle of 80% we have PWMDTY0 = 200 since 250 × 80% = 200.

```
MOVB #$02,PWMPRCLK      ;ClockA=Fbus/4
MOVB #$0,PWMCLK         ;use ClockA for channel 0
MOVB #01,PWMPOL         ;high, then low for polarity
MOVB #0,PWMCAE          ;left aligned
MOVB #$0,PWMCTL ;8-bit chan,PWM during freeze and wait
MOVB #250,PWMPER0       ;PWM_Freq=ClockA/250
MOVB #200,PWMDTY0       ;80% duty cycle
MOVB #0,PWMCNT0         ;clear PWMCNT0
BSET PWME,%00000001     ;turn on PWM0 (start PWMCNT)
BRA $
```

HCS12
PP0

Example 17-13

Assume XTAL = 8 MHz (Fbus = 4 MHz). Using channel 5, write a program to create the PWM pulses with a frequency of 100 Hz and 50% duty cycle.

Solution:

We need to set PWMPRCLK = 04 for ClockA since ClockA = 4 MHz / 2^4 = 4 MHz/16 = 250 kHz. PWMSCLA = 125 since ClockA = 250 kHz/(2 × 125) = 250 kHz/250 = 1,000 Hz. Since PWM_Freq = ClockA/PWM_PER we need PWM_Freq = 1,000 Hz/10 = 100 Hz, which means we must have PWMPER5 = 10. For the duty cycle of 50%, we need PWMDTY5 = 5 since 10 × 50% = 5.

```
MOVB #$04, PWMPRCLK     ;ClockA=Fbus/16
MOVB #125, PWMSCLA      ;ClockA=ClockA/2x125
MOVB #$20, PWMCLK       ;use ClockA for channel 5
MOVB #$20,PWMPOL        ;high, then low for polarity
MOVB #$0,PWMCAE         ;left aligned
MOVB #$0, PWMCTL;8-bit chan,PWM during freeze and wait
MOVB #10, PWMPER5       ;PWM_Freq=ClockA/10
MOVB #5, PWMDTY5        ;50% duty cycle
MOVB #0,PWMCNT5         ;clear PWMCNT5
BSET PWME,%00100000     ;turn on PWM5 (start PWMCNT)
BRA $
```

Programming HCS12 PWM in C

The C versions of programming the PWM channels are shown in Examples 17-14 through 17-16.

Example 17-14

Assume XTAL = 4 MHz (Fbus = 2 MHz). Using channel 2, write a C program to create the PWM pulses with a frequency of 50 Hz and 60% duty cycle. This is a C version Example 17-11.

Solution:

We need to set PWMPRCLK = 30 for ClockB since ClockB = 2 MHz / 2^3 = 2 MHz/8 = 250 kHz. PWMSCLB = 50 since ClockSB = 250 kHz/(2 × 50) = 250 kHz/100 = 2,500 Hz. Since PWM_Freq = ClockSB/PWM_PER we need PWM_Freq = 2,500 Hz/50 = 50 Hz, which means we must have PWMPER2 = 50. For the duty cycle of 60%, we need PWMDTY2 = 30 since 50 × 60% = 30.

```
#include <mc9s12dp512.h>      /* derivative information */
void main()
  {
      PWMPRCLK=0x30;      //ClockSB=Fbus/8
      PWMSCLB=0x32;       //ClockSB/(2x50)
      PWMCLK=0x04;        //use ClockSB for channel 2
      PWMPOL=0x04;        //high, then low for polarity
      PWMCAE=0;           //left aligned
      PWMCTL=0;           //8-bit chan, PWM during freeze and wait
      PWMPER2=50;         //PWM_Freq=ClockSB/50
      PWMDTY2=30;         //60% duty cycle
      PWMCNT2=0;          //clear PWMCNT2
      PWME=0x04;          //turn on (start) PWM2
      while(1);
  }
```

16-bit PWM

In the HCS12, a 16-bit PWM can be formed by concatenating two 8-bit PWM channels into one. To do that we use the PWMCTL register. The PWMCTL register allows us to have 8 channels of 8-bit PWM or 4 channels of 16-bit, or a combination of both. We use the CON67 bit of PWMCTL to concatenate channels 6 and 7, CON45 for channels 4 and 5 and so on. See Table 17-3. Notice that in concatenating channels 6 and 7, channel 6 becomes the high byte of the 16-bit channel. This applies to channels 4 and 5 and so on. Assembly language programming of the PWM channels is illustrated in Examples 17-17 and 17-18.

Table 17-3: 16-Bit Concatenation Mode Summary

CONxx	PWMEx	PPOLx	PCLKx	CAEx	PWMx pin
CON67	PWME7	PPOL7	PCLK7	CAE7	PWM7 (PORTP.7)
CON45	PWME5	PPOL5	PCLK5	CAE5	PWM5 (PORTP.5)
CON23	PWME3	PPOL3	PCLK3	CAE3	PWM3 (PORTP.3)
CON01	PWME1	PPOL1	PCLK1	CAE1	PWM1 (PORTP.1)

Example 17-15

Assume XTAL = 4 MHz (Fbus = 2 MHz). Using channel 0, write a program to create the PWM pulses with a frequency of 2 kHz and 80% duty cycle. This is a C version of Example 17-12.

Solution:

We need to set PWMPRCLK = 02 for ClockA since 2 MHz / 2^2 = 2 MHz/4 = 500 kHz. We bypass the use of ClockA since we can get the PWM_Freq = ClockA/PWM_PER. PWM_Freq = 500 kHz/ 250 = 2 kHz, which means we must have PWMPER0 = 250. For the duty cycle of 80% we have PWMDTY0 = 200 since 250 × 80% = 200.

```
#include <mc9s12dp512.h>       /* derivative information */
void main()
  {
      PWMPRCLK=0x2;      //ClockA=Fbus/4
      PWMCLK=0x00;       //use ClockA for channel 0
      PWMPOL=0x01;       //high, then low for polarity
      PWMCAE=0;          //left aligned
      PWMCTL=0;          //8-bit chan,PWM during freeze and wait
      PWMPER0=250;       //PWM_Freq=ClockA/250
      PWMDTY0=200;       //80% duty cycle
      PWMCNT0=0;         //clear PWMCNT0
      PWME=0x01;         //turn on (start) PWM0
      while(1);
  }
```

Example 17-16

Assume XTAL = 8 MHz (Fbus = 4 MHz). Using channel 5, write a C program to create the PWM pulses with a frequency of 100 Hz and 50% duty cycle. This is a C version of Example 17-13.

Solution:

We need to set PWMPRCLK = 04 for ClockA since ClockA = 4 MHz / 2^4 = 4 MHz/16 = 250 kHz. PWMSCLA = 125 since ClockA = 250 kHz/(2 × 125) = 250 kHz/250 = 1,000 Hz. Since PWM_Freq = ClockA/PWM_PER we need PWM_Freq = 1,000 Hz/10 = 100 Hz, which means we must have PWMPER5 = 10. For the duty cycle of 50%, we need PWMDTY5 = 5 since 10 × 50% = 5.

```
#include <mc9s12dp512.h>       /* derivative information */
void main()
  {
      PWMPRCLK=0x04;    //ClockA=Fbus/16
      PWMSCLA=125;      //ClockA/(2x125)
      PWMCLK=0x20;      //use ClockA for channel 5
      PWMPOL=0x20;           //high, then low for polarity
      PWMCAE=0x0;            //left aligned
      PWMCTL=0x0;       //8-bit chan, PWM during freeze and wait
      PWMPER5=10;       //PWM_Freq=ClockA/10
      PWMDTY5=5;        //50% duty cycle
      PWMCNT5=0;        //clear PWMCNT5
      PWME=0x20;        //turn on (start) PWM5
      while(1);
  }
```

Example 17-17

Assume XTAL = 8 MHz (Fbus = 4 MHz). Using the 16-bit channels 4 and 5, write a program to create PWM pulses with a frequency of 2 Hz and 50% duty cycle.

Solution:

We need to set PWMPRCLK = 04 for ClockA since ClockA = 4 MHz / 2^4 = 4 MHz/16 = 250 kHz. PWMSCLA = 125 since ClockA = 250 kHz/(2 × 125) = 250 kHz/250 = 1,000 Hz. Since PWM_Freq = ClockA/PWM_PER we need PWM_Freq = 1,000 Hz/500 = 2 Hz, which means we must have PWMPER5 = 500. For the duty cycle of 50%, we need PWMDTY5 = 250.

```
MOVB #$04, PWMPRCLK         ;ClockA=Fbus/16
MOVB #125, PWMSCLA          ;ClockA=ClockA/2x125
MOVB #$20, PWMCLK           ;use ClockA for channels 4,5
MOVB #$20, PWMPOL           ;high, then low for polarity
MOVB #$0,  PWMCAE           ;left aligned
MOVB #$40, PWMCTL;16-bit chan,PWM during freeze and wait
MOVW #500, PWMPER45         ;PWM_Freq=ClockA/500
MOVW #250, PWMDTY45         ;50% duty cycle
MOVW #0,   PWMCNT45         ;clear PWMCNT45
BSET PWME,%00100000 ;turn on PWM5 (start PWMCNT)
BRA $                   ;Note: use an LED to see this
```

Example 17-18

Assume XTAL = 8 MHz (Fbus = 4 MHz). Using the 16-bit Channel 4 and 5, write a C program to create PWM pulses with a frequency of 2 Hz and 50% duty cycle.

Solution:

We need to set PWMPRCLK = 04 for ClockA since ClockA = 4 MHz / 2^4 = 4 MHz/16 = 250 kHz. PWMSCLA = 125 since ClockA = 250 kHz/(2 × 125) = 250 kHz/250 = 1,000 Hz. Since PWM_Freq = ClockA/PWM_PER we need PWM_Freq = 1,000 Hz/500 = 2 Hz, which means we must have PWMPER5 = 500. For the duty cycle of 50%, we need PWMDTY5 = 250.

```
#include <mc9s12dp512.h>      /* derivative information */
void main()
  {
     PWMPRCLK=0x04;   //ClockA=Fbus/8
     PWMSCLA=125;     //ClockA/(2x125)
     PWMCLK=0x20;     //use ClockA for channel 5
     PWMPOL=0x20;     //high, then low for polarity
     PWMCAE=0;        //left aligned
     PWMCTL=0x40;     //16-bit chan, PWM during freeze and wait
     PWMPER45=500;    //PWM_Freq=ClockA/500
     PWMDTY45=250;    //50% duty cycle
     PWMCNT45=0;      //clear PWMCNT45
     PWME=0x20;       //Note: turn on (start) both PWM4 and 5
     while(1); //Note: use an LED to see this on PORTP pin 5
}
```

Review Questions

1. True or false. For ClockA, we use the PWMPRCLK register.
2. True or false. For PWM, the PORTP pin must be configured as output.
3. In 8-bit channel 3 PWM, we use ________________ to set the period for PWM.
4. In 8-bit channel 3 PWM, we use ________________ to set the duty cycle for PWM.
5. In 16-bit channel 3,4 PWM, we use ________________ pin for the output.

PROBLEMS

SECTION 17.1: DC MOTOR INTERFACING AND PWM

1. Which motor is best for moving a wheel exactly 90 degrees?
2. True or false. Current dissipation of a DC motor is proportional to the load.
3. True or false. The RPM of a DC motor is the same for no-load and loaded.
4. The RPM given in data sheets is for ___________ (no-load, loaded).
5. What is the advantage of DC motors over AC motors?
6. What is the advantage of stepper motors over DC motors?
7. True or false. Higher load on a DC motor slows it down if the current and voltage supplied to the motor are fixed.
8. What is PWM, and how is it used in DC motor control?
9. A DC motor is moving a load. How do we keep the RPM constant?
10. What is the advantage of placing an optoisolator between the motor and the microcontroller?

SECTION 17.2: PWM PROGRAMMING IN HCS12

11. True or false. For ClockA, we use the PWMSCLA register.
12. True or false. For PWM, the PORTP pin must be configured as input.
13. In 8-bit channel 4 PWM, we use ________________ to set the period for PWM.
14. In 8-bit channel 4 PWM, we use ________________ to set the duty cycle for PWM.
15. In 16-bit chan 6,7 PWM, we use the ________________ pin for the output.
16. Assume XTAL = 8 MHz and Bus Freq = 4 MHz. Find the value for the PWMPRCLK and PWMSCLB registers if we want ClockSB = 2,500 Hz.
17. Assume XTAL = 8 MHz and Bus Freq = 4 MHz. Find a) the lowest, and b) the highest frequency for ClockA and ClockSA. Give the values for the PWMPRCLK and PWMSCLA registers for the above options.
18. Assume XTAL = 8 MHz and Bus Freq = 4 MHz. Find a) the value for the PWMPRCLK and PWMSCLB registers if we want ClockSB = 2,500 Hz.
19. Assume ClockSB = 5,000 Hz. Find the value for the PWM frequency if PWMPERx has the value of 250.
20. Assume ClockSB = 5,000 Hz. Find the value for the PWMPERx and

PWMDTYx registers if we want a PWMx frequency of 50 Hz with 60% duty cycle.

21. Assume XTAL = 8 MHz (Fbus = 4 MHz). Using channel 2, write a program to create the PWM pulses with frequency of 100 Hz and 60% duty cycle.
22. Assume XTAL = 8 MHz (Fbus = 4 MHz). Using channel 0, write a program to create the PWM pulses with frequency of 4 kHz and 80% duty cycle.

ANSWERS TO REVIEW QUESTIONS

SECTION 17.1: DC MOTOR INTERFACING AND PWM

1. True
2. False
3. Because microcontroller/digital outputs lack sufficient current to drive the DC motor, we need a driver.
4. By reversing the polarity of voltages connected to the leads
5. The DC motor is stalled if the load is beyond what it can handle.
6. No-load

SECTION 17.2: PWM PROGRAMMING IN HCS12

1. True
2. True
3. PWMRER3
4. PWMDTY3
5. PORTP.3

APPENDIX A

HCS12 INSTRUCTIONS EXPLAINED

OVERVIEW

In this appendix, we describe each intruction of the HCS12. In many cases, a simple code example is given to clarify the instruction.

ABA **Add B to A**

Flags: H, N, Z, V, C

This adds accumulator B to accumulator A, and places the result in A.

Example:

```
LDAA #$45               ;A=$45
LDAB #$4F               ;A=$4F
ABA                ;A=$94 ($45+$4F=$94)
                        ;C=0,H=1
```

ADCA **Add with Carry to A**

Flags: H, N, Z, V, C

This will add the source byte to A, in addition to the C flag (A = A + byte + C). If C = 1 prior to this instruction, 1 is also added to A. If C = 0 prior to the instruction, source is added to destination plus 0. This is used in multibyte additions. In the addition of $25F2 to $3189, for example, we use the ADCA instruction as shown below.

Example:

```
CLC             ;C=0
LDAA #$89       ;A=$89
ADCA #$F2       ;A=$89+$F2+0=$17B, A=7B, C=1
TAB             ;SAVE A
LDAA #$31
ADCA #$25       ;A=$31+$25+1=$57
```

Therefore the result is:

```
  $25F2
+ $3189
  $577B
```

The addressing modes for ADCA are the same as for ADDA.

ADCB **Add with Carry to B**

Flags: H, N, Z, V, C

This will add the source byte to B, in addition to the C flag (B = B + byte + C). If C = 1 prior to this instruction, 1 is also added to B. If C = 0 prior to the instruction, source is added to destination plus 0. This is used in multibyte additions. In the addition of $25F2 to $3187, for example, we use the ADCB instruction as shown below.

Example:

```
CLC             ;C=0
LDAB #$89       ;B=$87
ADCB #$F2       ;B=$87+$F2+0=$179, B=79, C=1
TBA             ;SAVE B
LDAB #$31
ADCB #$25       ;B=$31+$25+1=$57
```

Therefore the result is:

```
      $25F2
    + $3187
      $5779
```

The addressing modes for ADCB are the same as for ADDA.

ADDA Add byte to A

Flags: H, N, Z, V, C

This adds the source byte to the accumulator (A), and places the result in A. Since register A is one byte in size, the source operands must also be one byte.

The ADD instruction is used for both signed and unsigned numbers. Each one is discussed separately.

Unsigned addition

In the addition of unsigned numbers, the status of H, N, Z, V, and C may change. The most important of these flags is C. It becomes 1 when there is a carry from D7 out in 8-bit (D0–D7) operations.

Example:

```
        LDAA   $#45          ;A=$45
        ADDA   $#4F          ;A=$94 ($45+$4F=$94)
                             ;C=0,H=1
```

Example:

```
        LDAA   #$FE          ;A=$FE
        ADDA   #$75          ;A=FE+75=$73
                             ;C=1,H=1
```

Example:

```
        LDAA   #$25          ;A=$25
        ADDA   #$42          ;A=$67 ($25+$42=$67)
                             ;C=0,H=0
```

Addressing modes

The following addressing modes are supported for the ADDA instruction:

1. Immediate: `ADDA #data` Example: `ADDA #$25`
2. Direct: `ADDA PORTB`
3. Extended: `ADDA 16-bit memory address`
   Example: `ADDA $1550 ;add to A data in RAM loc. $1550`
4. Indexed: `ADDA 0,X ;add to A data pointed to by X`
   Examples:
   ```
   ADDA 0,X  ;add to A data pointed to by X+0
   ADDA 2,X  ;add to A data pointed to by X+2
   ```

In the following example, the contents of RAM locations $1500 to 155F are added together, and the sum is saved in RAM locations $1700 and $1701.

```
R2   EQU  $1600
     CLRA            ;A=0
```

```
      TAB               ;B=0
      LDX   #$1500      ;source pointer
      MOVB #16,R2       ;counter
A_1:  ADDA 0,X          ;ADD to A from source
      BCC   B_1         ;IF C=0 go to next byte
      INCB              ;otherwise keep carries
B_1:  INX               ;next location
      DEC   R2
      BNE A_1           ;repeat for all bytes
      STAA $1701        ;save low byte of sum
      STAB $1700        ;save high byte of sum
```

Notice in all the above examples that we ignored the status of the V flag. Although ADDA instructions do affect V, it is in the context of signed numbers that the V flag has any significance. This is discussed next.

Signed addition and negative numbers

In the addition of signed numbers, special attention should be given to the overflow flag (V) since this indicates if there is an error in the result of the addition. There are two rules for setting V in signed number operation. The overflow flag is set to 1:

1. If there is a carry from D6 to D7 and no carry from D7 out.
2. If there is a carry from D7 out and no carry from D6 to D7.

Notice that if there is a carry both from D7 out and from D6 to D7, V = 0.

Example:

```
        LDAA #+8          ;A=0000 1000
        ADDA #+4          ;A=0000 1000 + 0000 0100
                          ;A=0000 1100 V=0,C=0
```

Notice that D7 = 0 since the result is positive and V = 0 since there is neither a carry from D6 to D7 nor any carry beyond D7. Since V = 0, the result is correct [(+8) + (+4) = (+12)].

Example:

```
        LDAA #+66         ;A=0100 0010
        ADDA #+69         ;ADD 0100 0101 to A
                          ;A=1000 0111 = -121
                          ;(INCORRECT) C=0, D7=1, V=1
```

In the above example, the correct result is +135 [(+66) + (+69) = (+135)], but the result was -121. V = 1 is an indication of this error. Notice that D7 = 1 since the result is negative; V = 1 since there is a carry from D6 to D7 and C = 0.

Example:

```
        LDAA #-12         ;A=1111 0100
        ADDA #+18         ;ADDD 0001 0010 to A
                          ;A=0000 0110 (+6) correct
                          ;D7=0,V=0, and C=1
```

Notice that the result is correct (V = 0), since there is a carry from D6 to D7 and a carry from D7 out.

Example:

```
        LDAA #-30        ;A=1110 0010
        ADDA #+14        ;A=0000 1110 + 11100010
                         ;A=1111 0000 (-16, CORRECT)
                         ;D7=1,V=0, C=0
```

V = 0 since there is no carry from D7 out nor any carry from D6 to D7.

Example:

```
        LDAA   #-126          ;A=1000 0010
        ADDA   #-127          ;ADD 1000 0001 to A
                              ;A=0000 0011 (+3, WRONG)
                              ;D7=0, V=1
```

C = 1 since there is a carry from D7 out but no carry from D6 to D7.

From the above discussion we conclude that while the C flag is important in any addition, V is extremely important in signed number addition since it is used to indicate whether or not the result is valid. As we will see in instruction "DAA", the H flag is used in the addition of BCD numbers.See the description of these two instructions for further details.

ADDB Add byte to B

Flags: H, N, Z, V, C

This adds the source byte to the accumulator (B), and places the result in B. Since register B is one byte in size, the source operands must also be one byte.

Just like ADDA, the ADDB instruction is used for both signed and unsigned numbers.

Example:

```
        LDAB   $#45          ;B=$45
        ADDB   $#4F          ;B=$94 ($45+$4F=$94)
                             ;C=0,H=1
```

Example:

```
        LDAB   #$FE          ;B=$FE
        ADDB   #$75          ;B=FE+75=$73
                             ;C=1,H=1
```

Example:

```
        LDAB   #$25       ;B=$25
        ADDB   #$42       ;B=$67 ($25+$42=$67)
```

The signed addition of ADDB is the same as for ADDA.

The addressing modes for ADDB are the same as for ADDA.

```
R2    EQU  $1600

      CLRA              ;A=0
      TAB               ;B=0
      LDX  #$1500       ;source pointer
      MOVB #16,R2       ;counter
A_1:  ADDB 0,X          ;ADD to B from location X
      BCC  B_1          ;if C=0 go to next byte
      INCA              ;otherwise keep the carries
B_1:  INX               ;next location
      DEC  R2
      BNE A_1           ;repeat for all bytes
      STAB $1701        ;save low byte of sum
      STAA $1700        ;save high byte of sum
```

ADDD Add word to D

Flags: N, Z, V, C

This adds the 16-bit (2-byte) source word to the D (double register A:B) register, and places the result in D. Just like ADDA, the ADDD instruction is used for both signed and unsigned numbers. In the addition of unsigned numbers, the status of N, Z, V, and C may change. The most important of these flags is C. It becomes 1 when there is a carry from D15 out in 16-bit (D0–D15) operations.

Example:

```
        LDD     #$25F2          ;D=$25F2
        ADDD    #$3189          ;D=$25F2+$3189=$577B
                                ;C=0

                  $25F2
                + $3189
                  -----
                  $577B
```

The following addressing modes are supported for the ADDD instruction:

1. Immediate: `ADDD #data` Example: `ADDD #$2515`
2. Direct: `ADDD PORTB`
3. Extended: `ADDD 16-bit memory address`
   Example: `ADDD $1550 ;add to D data in RAM loc. $1550 and $1501. The content of loc $1501 is added to reg B and $1501 to A. The Big Endian convention.`
4. Indexed: `ADDD 0,X where X holds the address of data`
   Example:
   `ADDD 0,X ;add to D data pointed to by X+0 and X+1`

ANDA Logical AND A

Flags: N

This performs a logical AND on the operands, bit by bit, storing the result in the A register. Notice that both the source and destination values are byte-size only.

A	V	A AND V
0	0	0
0	1	0
1	0	0
1	1	1

Example:

```
LDAA #$39    ;A=$39
ANDA #$09    ;A=$39 ANDed with 09

39    0011 1001
09    0000 1001
09    0000 1001
```

Example:

```
LDAA #$32        ;A=32H     32    0011 0010
ANDA #$50    ;              50    0101 0000
                 ;(A=10H)   10    0001 0000
```

The following addressing modes are supported for the ANDA instruction:

1. Immediate: `ANDA #data` Example: `ANDA #$0F`
2. Direct: `ANDA PORTB`
3. Extended: `ANDA 16-bit memory address`
   Example: `ANDA $1550 ;AND A with data in RAM loc. $1550`
4. Indexed: `ANDA 0,X   ;AND A with data pointed to by X`

ANDB Logical AND B

Flags: N and Z

This performs a logical AND on the operands, bit by bit, storing the result in the B register. Notice that both the source and destination values are byte-size only.

Example:

```
LDAB #$39    ;B=$39
ANDB #$09    ;B=$39 ANDed with 09

39    0011 1001
09    0000 1001
09    0000 1001
```

Example:

```
LDAB #$32        ;B=32H     32    0011 0010
ANDB #$50    ;              50    0101 0000
                 ;(B=10H)   10    0001 0000
```

The addressing modes supported are the same as ANDA.

ANDCC **Logical AND CCR with mask**

Flags: H, N, Z, V, C

This performs a logical AND on the CCR register and mask byte, bit by bit, storing the result in the CCR register.

Example:

```
ANDCC #$01    ;CCR=XX ANDed with 01

83    1000 0011
01    0000 0001
01    0000 0001
```

ASL **Arithmetic Shift Left memory**

Flags: N, Z, V, C

This shifts all bits of memory location one bit position left. Bit 0 is loaded with a 0. The C bit is loaded from the MSB of memory location.

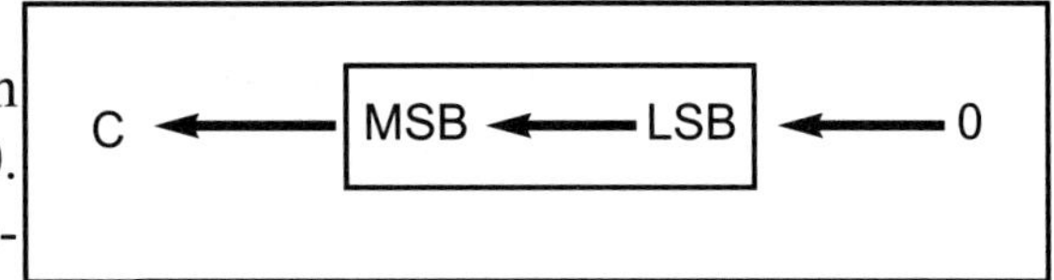

Example:

```
LDAA #$99        ;A=$99
STAA $1500       ;$1500=(10011001)
ASL $1500        ;Now $1500=(00110010) and C=1
ASL $1500        ;Now $1500=(01100100) and C=0
```

ASLA **Arithmetic Shift Left A**

Flags: N, Z, V, C

This shifts all bits of A one bit position left. Bit 0 is loaded with a 0. The C bit is loaded from the MSB of A.

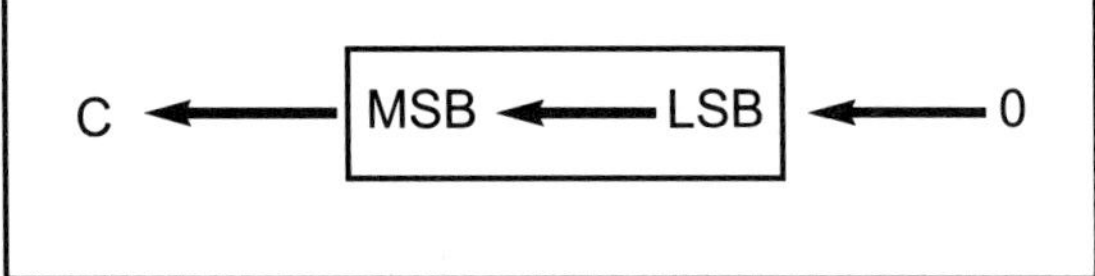

Example:

```
LDAA #$99        ;A=$99
ASLA             ;Now A=00110010 and C=1
ASLA             ;Now A=01100100 and C=0
```

ASLB **Arithmetic Shift Left B**

Flags: N, Z, V, C

This shifts all bits of B one bit position left. Bit 0 is loaded with a 0. The C bit is loaded from the MSB of B.

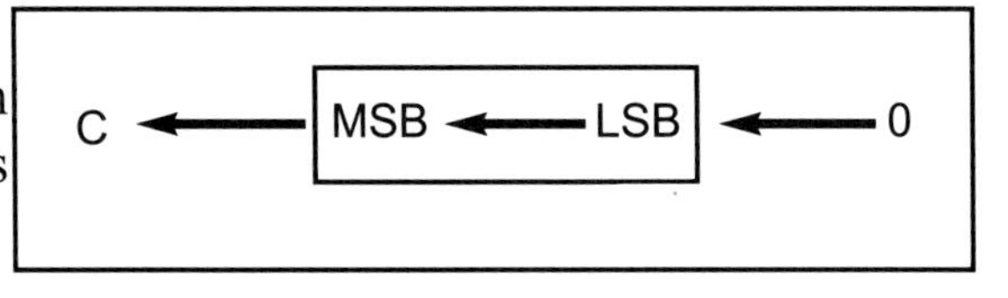

Example:

```
LDAB #$99        ;B=$99
ASLB             ;Now B=00110010 and C=1
ASLB             ;Now B=01100100 and C=0
```

ASLD **Arithmetic Shift Left D**

Flags: N, Z, V, C

This shifts all bits of D one bit position left. Bit 0 is loaded with a 0. The

C bit is loaded from the MSB of D.
Example:

```
LDD   #$9999      ;D=$9999
ASLD              ;Now D=$3332 and C=1
ASLD              ;Now D=$6664 and C=0
```

ASR Arithmetic Shift Right memory

Flags: N, Z, V, C

This shifts all bits of memory location one bit position right.

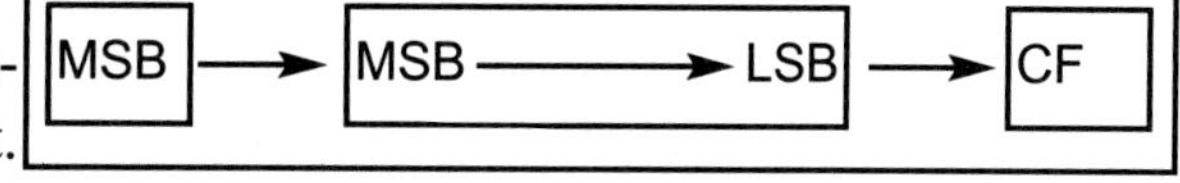

Bit 7 is held constant. Bit 0 is loaded into the C flag.
Example:

```
LDAA #$99        ;A=$99
STAA $1500 ;$1500=(10011001)
ASR $1500 ;Now $1500=(11001100) and C=1
ASR $1500 ;Now $1500=(11100110) and C=0
```

ASRA Arithmetic Shift Right A

Flags: N, Z, V, C

This shifts all bits of A one bit position right. Bit 7 is held constant. Bit 0 is loaded into the C flag.

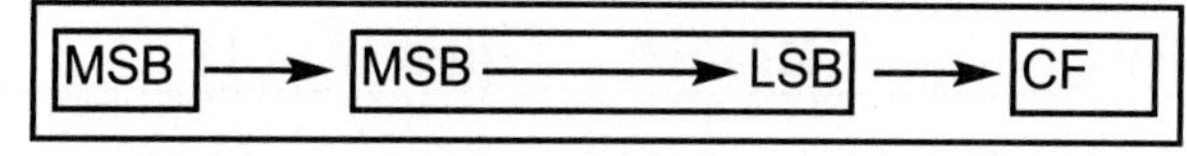

Example:

```
LDAA #$99         ;A=$99
ASRA              ;Now A=11001100 and C=1
ASRA              ;Now A=11100110 and C=0
```

ASRB Arithmetic Shift Right B

Flags: N, Z, V, C

This shifts all bits of B one bit position right. Bit 7 is held constant. Bit 0 is loaded into the C flag.

Example:

```
LDAB #$99         ;B=$99
ASRB              ;Now B=11001100 and C=1
ASRB              ;Now B=11100110 and C=0
```

BCC Branch if Carry Cleared (C = 0)

Flags: None

This instruction examines the C flag, and if it is zero it will jump (branch) to the target address. Notice that this is a 2-byte instruction and the target address cannot be farther than −128 to +127 bytes from the program counter. See Table A-1.

Table A-1: HCS12 Branch Instructions Using Flag Bits

Instruction	Action
BCS	Branch if C = 1
BCC	Branch if C = 0
BEQ	Branch if Z = 1
BNE	Branch if Z = 0
BMI	Branch if N = 1
BPL	Branch if N = 0
BVS	Branch if V = 1
BVC	Branch if V = 0

Note: The target address cannot be more than −128 to +127 bytes away from the current PC (program counter).

BCLR Bit Clear in memory

Flag: N, Z, V = 0

This instruction clears bits in a memory location. To clear a bit in memory, we set to high the corresponding bit in the mask byte. The 0 bits in the mask will leave the bits of memory unchanged.

Examples:

```
     BCLR PORTB,%00000001
;clear PORTB.0
     BCLR $1500,%10000000 ;clear bit 7 of Loc. $1500
     BCLR PORTB,%00001111     ;clear PORTB.3-PORTB.0
```

BCS Branch if Carry Set (C = 1)

Flags: None

This instruction examines the C flag, and if it is set it will jump to the target address. Notice that this is a 2-byte instruction and the target address cannot be farther than −128 to +127 bytes from the program counter.

BEQ Branch if Equal (Z = 1)

Flags: None

This instruction examines the Z flag and jumps if it is 1 (Z = 1).

Notice that this is a 2-byte instruction; therefore, the target address cannot be more than −128 to +127 bytes away from the program counter.

BGE Branch if Greater than or Equal

Flags: None

This instruction is used for branching after compare instructions such as CMPA, CMPB, and so on. It can also be used after subtract instructions such as SUBA, SUBB, and so on. Notice that this is a 2-byte instruction; therefore, the target address cannot be more than −128 to +127 bytes away from the program counter.

BGND Enter Background Debug Mode (BDM)

Flags: None

This instruction is used for entering the BDM. See the HCS12 manual for more details.

BGT **Branch if Greater than Zero**

Flags: None

This instruction is used for branching after compare instructions such as CMPA, CMPB, and so on. It can also be used after subtract instructions such as SUBA, SUBB, and so on. Notice that this is a 2-byte instruction; therefore, the target address cannot be more than −128 to +127 bytes away from the program counter.

BHI **Branch if Higher (if C + Z = 0)**

Flags: None

This instruction is used for branching after compare instructions such as CMPA, CMPB, and so on. It can also be used after subtract instructions such as SUBA, SUBB, and so on. BHI should not be used after instructions that do not affect the carry flag such as Increment, Decrement, and so on. Notice that this is a 2-byte instruction; therefore, the target address cannot be more than −128 to +127 bytes away from the program counter.

BHS **Branch if Higher or Same (if C = 0, which is same as BCC)**

Flags: None

This instruction is used for branching after compare instructions such as CMPA, CMPB, and so on. It can also be used after subtract instructions such as SUBA, SUBB, and so on. BHS should not be used after instructions that do not affect the carry flag such as Increment, Decrement, and so on. Notice that this is a 2-byte instruction; therefore, the target address cannot be more than −128 to +127 bytes away from the program counter.

BITA **Bit test A**

Flags: N, Z, V = 0

This performs a logical AND on the content of register A and the content of a memory location, bit by bit, and sets the flags in the CCR accordingly. Notice neither the content of A nor the content of the memory is affected.

Example:

```
;Assume Memory loc. $1500 has $39
LDAA #0    ;A=0
BITA $1500;A=00 ANDed with memory location $1500

        39    0011 1001
        00    0000 1001
        09    0000 1001        N=0, Z=0, V=0
```

BITB **Bit test B**

Flags: N, Z, V = 0

This performs a logical AND on the content of register B and the content of a memory location, bit by bit, and sets the flags in the CCR accordingly. Notice

neither the content of B nor the content of the memory is affected.
Example:

```
;Assume Memory loc. $1500 has 00
LDAB #$39    ;A=$39
BITB $1500;B=$39 ANDed with memory location $1500

          39    0011 1001
          09    0000 0000
          09    0000 0000        N=0, Z=1, V=0
```

BLE Branch if Less than or Equal

Flags: None

This instruction is used for branching after compare instructions such as CMPA, CMPB, and so on. It can also be used after subtract instructions such as SUBA, SUBB, and so on. Notice that this is a 2-byte instruction; therefore, the target address cannot be more than –128 to +127 bytes away from the program counter.

BLO Branch if Lower (if C = 0. This is the same as BCS.)

Flags: None

This instruction is used for branching after compare instructions such as CMPA, CMPB, and so on. It can also be used after subtract instructions such as SUBA, SUBB, and so on. BLO should not be used after instructions that do not affect the carry flag such as Increment, Decrement, and so on. Notice that this is a 2-byte instruction; therefore, the target address cannot be more than –128 to +127 bytes away from the program counter.

BLS Branch if Lower or Same (if C + Z = 1)

Flags: None

This instruction is used for branching after compare instructions such as CMPA, CMPB, and so on. It can also be used after subtract instructions such as SUBA, SUBB, and so on. BLS should not be used after instructions that do not affect the carry flag such as Increment, Decrement, and so on. Notice that this is a 2-byte instruction; therefore, the target address cannot be more than –128 to +127 bytes away from the program counter.

BLT Branch if Less than Zero

Flags: None

This instruction is used for branching after compare instructions such as CMPA, CMPB, and so on. It can also be used after subtract instructions such as SUBA, SUBB, and so on. Notice that this is a 2-byte instruction; therefore, the target address cannot be more than –128 to +127 bytes away from the program counter.

BMI Branch if Minus (if N = 1)

Flags: None

This instruction examines the N flag and jumps if it is 1 (N = 1).

Notice that this is a 2-byte instruction; therefore, the target address cannot be more than -128 to +127 bytes away from the program counter.

BNE Branch if Not Equal to Zero (if Z = 0)

Flags: None

This instruction examines the Z flag and jumps if it is 0 (Z = 0). Notice that this is a 2-byte instruction; therefore, the target address cannot be more than -128 to +127 bytes away from the program counter.

BPL Branch if Plus (if N = 0)

Flags: None

This instruction examines the N flag and jumps if it is 0 (N = 0). Notice that this is a 2-byte instruction; therefore, the target address cannot be more than -128 to +127 bytes away from the program counter.

BRA Branch Always

Flags: None

This is a 2-byte instruction. The first byte is the opcode and the second byte is the signed number displacement, which is added to the PC (program counter) of the instruction following the BRA to get the target address. Therefore, in this jump the target address must be within -128 to +127 bytes of the PC (program counter) of the instruction after the BRA since a single byte of address can take values of +127 to -128. This address is often referred to as a *relative address* since the target address is -128 to +127 bytes relative to the program counter (PC). In this Appendix, we have used the term target address in place of relative address only for the sake of simpilicity.

BRCLR Branch if Bits Cleared

Flags: None

This performs a logical AND on the content of a mask byte and the content of a memory location, bit by bit, then branches if and only if all bits with a value of 1 in the mask byte correspond to bits with a value of 0 in the tested byte.
Examples:

```
BRCLR PORTB,%00000100,OVER ;branch if PORTB.2 is 0
BRCLR PORTB,%10000000,NEXT ;branch if PORTB.7 is 0
```

BRN Branch Never

Flags: None

This is in reality a 2-byte NOP that is executed in one cycle. It is the complement of the BRA instruction. The following three instructions perform the same action.

```
        BRN
HERE    BRA   HERE
        BRA   $
```

BRSET Branch if bits Set

Flags: None

This performs a logical AND on the content of a mask byte and the content of a memory location, bit by bit, then branches if and only if all bits with a value of 1 in the mask byte correspond to bits with a value of 1 in the tested byte.

```
BRSET PORTB,%00000100,OVER ;branch if PORTB.2 is 1
BRSET PORTB,%10000000,NEXT ;branch if PORTB.7 is 1
```

BSET Bit Set in memory

Flag: N, Z, V = 0

This instruction sets to high the bits in a memory location. To set high a bit in memory, we set to high the corresponding bit in the mask byte. The 0 bits in mask will leave the bits of memory unchanged.

Examples:

```
BSET PORTB,%00000001      ;set PORTB.0
BSET $1500,%10000000      ;set bit 7 of Loc. $1500
BSET PORTB,%00001111      ;set PORTB.3-PORTB.0
```

BSR Branch to Subroutine

Flags: None

This instruction is like the JSR instruction except it is a 2-byte instruction; therefore, the target address cannot be more than –128 to +127 bytes away from the Program Counter. See the discussion on the CALL instruction.

BVC Branch if Overflow Cleared (V = 0)

Flags: None

This instruction examines the V flag, and if it is cleared it will jump to the target address. Notice that this is a 2-byte instruction; therefore, the target address cannot be more than –128 to +127 bytes away from the program counter.

BVS Branch if Overflow Set (V = 1)

Flags: None

This instruction examines the V flag, and if it is set it will jump to the target address. Notice that this is a 2-byte instruction; therefore, the target address cannot be more than –128 to +127 bytes away from the program counter.

CALL Call subroutine in Expanded memory (beyond 64 KB)

Flags: None

There are three types of calls: JSR, BSR, and CALL. In JSR (jump to subroutine), the target address is within 64K bytes of the current PC (program counter). In jumping to a subroutine with JSR, the PC register (which has the address of the instruction after the JSR) is pushed onto the stack, and the stack pointer (SP) is decremented by 2. Then the program counter is loaded with the new address and control is transferred to the subroutine. At the end of the procedure, when RTS (return from subroutine) is executed, PC is pulled off the stack, which returns control to the instruction after the JSR. Notice that JSR is a 3-byte instruction, in which one byte is the opcode, and the other two bytes are the 16-bit address of the target subroutine. Also notice that we must use RTS (return from subroutine) at the end of the subroutine if we use the JSR instruction.

BSR (branch to subroutine) is a 2-byte instruction, in which one byte is used for the opcode and the second byte is used for the target subroutine relative address. An 8-bit relative address limits it to the values in the range of –128 to +127 bytes away from the program counter. Notice that we must use RTS at the end with the BSR.

To reach the target address located in the expanded (memory space beyond 64 KB) pages of the HCS12, we must use CALL. CALL uses the PPAGE register to access the pages of expanded memory beyond the 64K bytes space. If calling a subroutine with CALL, the PC register is pushed onto the stack, and the stack pointer (SP) is decremented by 2. It also pushes onto the stack the current value of the PPAGE register and the stack pointer is decremented by 1. Then the program counter is loaded with the new address and the PPAGE register is loaded with a new value and control is transferred to the subroutine located in the expanded memory. At the end of the subroutine, when RTC (return from CALL) is executed, the PPAGE and PC are pulled off the stack, which returns control to the instruction after the CALL. Notice that CALL is a 4-byte instruction, in which one byte is the opcode, two bytes are the 16-bit extended address of the target subroutine, and the last byte is the 8-bit value of the PPAGE register. Also notice that we must use RTC (Return from Call) at the end of the subroutine if we use the CALL instruction.

CBA Compare B and A accumulators

Flag: N, V, Z, C

The magnitudes of B and A are compared and condition flags are set accordingly. This instruction is really A – B (B subtracted from A) and the flags are set as shown in table. The contents of A and B are not changed.

A < B	C = 1, Z = 0
A = B	C = 0, Z = 1
A > B	C = 0, Z = 0

CLC **Clear Carry (C = 0)**

Flag: C = 0

This instruction clears the carry flag. This instruction is the same as ANDCC #$FE.

CLI **Clear Interrupt mask**

Flag: I = 0

This instruction clears the I-bit mask in the CCR register. This instruction is the same as ANDCC #$EF.

CLR **Clear memory**

Flag: Z = 1, V = 0, C = 0, N = 0

This instruction clears the contents of memory location. All bits of the memory location are set to 0. The following addressing modes are supported for this instruction:

1. Extended: `CLR 16-bit memory address`
   Example: `CLR $1550 ;clear data in RAM loc. $1550`
2. Indexed: `CLR 0,X ;clear data in RAM pointed to by X`

CLRA **Clear A**

Flag: Z = 1, V = 0, C = 0, N = 0

This instruction clears register A. All bits of accumulator A are set to 0.

Example:

```
CLRA            ;A = 0
STAA PORTB      ;clear PORTB
```

CLRB **Clear B**

Flag: Z = 1, V = 0, C = 0, N = 0

This instruction clears register B. All bits of accumulator B are set to 0.

Example:

```
CLRB            ;B = 0
STAB PORTB      ;clear PORTB
```

CLV **Clear the V flag (V = 0)**

Flag: V = 0

This instruction clears the V flag. This instruction is the same as ANDCC #$FD.

CMPA **Compare A with memory**

Flag: N, V, Z, C

The magnitudes of the A and memory location contents are compared and condition flags are set accordingly. This instruction is really A – (M) and the flags are set as shown in the table. The contents of A and M are not changed.

A < M	C = 1, Z = 0
A = M	C = 0, Z = 1
A > M	C = 0, Z = 0

CMPB   Compare B with memory

Flag: N, V, Z, C

The magnitudes of the B and memory location contents are compared and condition flags are set accordingly. This instruction is really B – (M) and the flags are set as shown in the table. The contents of A and M are not changed.

B < M	C = 1, Z = 0
B = M	C = 0, Z = 1
B > M	C = 0, Z = 0

COM   Complement memory

Flags: N, Z, C = 1, V = 0

This complements the contents of a memory location. The result is the 1's complement of the memory. That is: 0s become 1s and 1s become 0s. The following addressing modes are supported for this instruction:

1. Extended: `COM 16-bit memory address`
   Example: `COM $1550 ;complement data in RAM loc. $1550`
2. Indexed: `COM 0,X   ;compl. data in RAM pointed to by X`

COMA   Complement A

Flags: N, Z, C = 1, V = 0

This complements the contents of register A, the accumulator. The result is the 1's complement of the accumulator. That is: 0s become 1s and 1s become 0s.
Example:

```
        LDAA #$55       ;A=01010101
AGAIN:  COMA            ;complement reg. A
        STAA PORTB      ;toggle bits of PORTB
        BRA  AGAIN      ;continuously
```

COMB   Complement B

Flags: N, Z, C = 1, V = 0

This complements the contents of accumulator B. The result is the 1's complement of accumulator B. That is: 0s become 1s and 1s become 0s.
Example:

```
        LDAB #$55       ;B=01010101
AGAIN:  COMB            ;complement reg. B
        STAB PORTB      ;toggle all bits of PORTB
        BRA  AGAIN      ;continuously
```

CPD   Compare D with memory

Flag: N, V, Z, C

The magnitudes of the D register (Double accumulator A:B) and a 16-bit word data are compared and condition flags are set accordingly. The 16-bit word data (W) can be an immediate value or 2 bytes of data located in memory locations M and M + 1. This instruction is really D – W and the flags are set as shown in the table. The contents of D and the contents of memory locations M and M + 1 are not changed.

D < Value	C = 1, Z = 0
D = Value	C = 0, Z = 1
D > Value	C = 0, Z = 0

CPS Compare Stack Pointer with memory

Flag: N, V, Z, C

The magnitudes of the SP (Stack Pointer) register and a 16-bit word data are compared and condition flags are set accordingly. The 16-bit word data (W) can be an immediate value or 2 bytes of data located in memory locations M and M + 1. This instruction is really SP – W and the flags are set accordingly. The contents of SP and the contents of memory locations M and M + 1 are not changed.

CPX Compare X with memory

Flag: N, V, Z, C

The magnitudes of the X register and a 16-bit word data (W) are compared and condition flags are set accordingly. The 16-bit word data (W) can be an immediate value or 2 bytes of data located in memory locations M and M + 1. This instruction is really X – W and the flags are set as shown in the table. The contents of X and the contents of memory locations M and M + 1 are not changed.

X < Value	C = 1, Z = 0
X = Value	C = 0, Z = 1
X > Value	C = 0, Z = 0

CPY Compare Y with memory

Flag: N, V, Z, C

The magnitudes of the Y register and a 16-bit word data (W) are compared and condition flags are set accordingly. The 16-bit word can be an immediate value or 2 bytes of data located in memory locations M and M + 1. This instruction is really Y – W and the flags are set as shown in the table. The contents of D and the contents of memory locations M and M + 1 are not changed.

Y < Value	C = 1, Z = 0
Y = Value	C = 0, Z = 1
Y > Value	C = 0, Z = 0

DAA Decimal Adjust A

Flags: C, N, Z

This instruction is used after addition of BCD numbers to convert the result back to BCD. The data is adjusted in the following two possible cases.

1. It adds 6 to the lower 4 bits of A if it is greater than 9 or if H = 1.
2. It also adds 6 to the upper 4 bits of A if it is greater than 9 or if C = 1.

Example:

```
LDAA #$47        ;A=0100 0111
ADDA #$38        ;A=$47+$38=$7F, invalid BCD
DAA              ;A=1000 0101=$85, valid BCD

  $47
+ $38
  $7F   (invalid BCD)
+   6   (after DA A)
  $85   (valid BCD)
```

In the above example, since the lower nibble was greater than 9, DAA added 6 to A. If the lower nibble is less than 9 but H = 1, it also adds 6 to the lower

nibble. See the following example.
Example:

```
LDAA #$29          ;A=0010 1001
ADDA #$18          ;A=0100 0001 INCORRECT
DAA                ;A=0100 0111 = $47 VALID BCD
```

```
   $29
 + $18
 -----
   $41   incorrect result in BCD
 +   6
 -----
  $ 47   correct result in BCD
```

The same thing can happen for the upper nibble. See the following example.

Example:

```
LDAA #$52          ;A=0101 0010
ADDA #$91          ;A=1110 0011 invalid BCD
DAA                ;A=0100 0011 and C=1
```

```
   $52
 + $91
 -----
   $E3   (invalid BCD)
 +   6   (after DAA, adding to upper nibble)
 -----
  $143   valid BCD
```

Similarly, if the upper nibble is less than 9 and C = 1, it must be corrected. See the following example.
Example:

```
LDAA #$94              ;A=1001 0100
ADDA #$91              ;A=0010 0101 incorrect
DAA                    ;A=1000 0101, valid BCD
                       ;for 85,C=1
```

It is possible that 6 is added to both the high and low nibbles. See the following example.

Example:

```
LDAA #$54              ;A=0101 0100
ADDA #$87              ;A=1101 1011 invalid BCD
DAA                ;A=0100 0001, C=1 (BCD 141)
```

DBEQ Decrement and Branch if Equal to Zero

Flags: None

In this instruction a register is decremented, and if the result is zero it will branch to the target address. The register can be A, B, D, X, Y, or SP.
Example:

```
DBEQ A    ;decrement A and branch if A=0
```

```
DBEQ B     ;decrement B and branch if B=0
DBEQ D     ;decrement D and branch if D=0
DBEQ X     ;decrement X and branch if X=0
DBEQ Y     ;decrement Y and branch if Y=0
```

Notice that the target address can be no more than 128 bytes backward or 127 bytes forward.

DBNE   Decrement and Branch if Not Equal to Zero

Flags: None

In this instruction a register is decremented, and if the result is not zero it will branch to the target address. The register can be A, B, D, X, Y, or SP.

Example:

```
DBNE A     ;decrement A and branch if A is not 0
DBNE B     ;decrement B and branch if B is not 0
DBNE D     ;decrement D and branch if D is not 0
DBNE X     ;decrement X and branch if X is not 0
DBNE Y     ;decrement Y and branch if Y is not 0
```

Notice that the target address can be no more than 128 bytes backward or 127 bytes forward.

DEC   Decrement memory

Flags: N, Z, V

This instruction subtracts 1 from the byte operand. Note that C (carry/ borrow) is unchanged even if a value 00 is decremented and becomes FF.

1. Extended: `DEC 16-bit memory address`

   Example: `DEC $1550 ;decrement data in RAM loc. $1550`
2. Indexed: `DEC 0,X ;decrement data in RAM pointed by X`

DECA   Decrement A

Flags: N, Z, V

This instruction subtracts 1 from accumulator A. Note that C (carry/ borrow) is unchanged even if a value 00 is decremented and becomes FF.

DECB   Decrement B

Flags: N, Z, V

This instruction subtracts 1 from accumulator B. Note that C (carry/ borrow) is unchanged even if a value 00 is decremented and becomes FF.

DES   Decrement Stack Pointer

Flags: None

This instruction subtracts 1 from the SP (Stack Pointer) register.

DEX **Decrement X**

Flags: Z

This instruction subtracts 1 from the X register.

DEY **Decrement Y**

Flags: Z

This instruction subtracts 1 from the Y register.

EDIV **Extended Divide 32-bit by 16-bit (Y:D/X, unsigned)**

Flags: N, Z, V, C

This instruction divides a 32-bit data in the Y:D registers by the 16-bit in register X. It is assumed that both registers Y:D and X contain an unsigned data. After the division, the 16-bit quotient will be in register Y and the 16-bit remainder in register D. Notice in this instruction that the carry flag is cleared, unless we divide Y:D by 0 (that is X = 0), in which case the result is invalid and C = 1 to indicate the invalid condition.

Example:

```
LDY #$0
LDD #65429
LDX #10
EDIV        ;Y=6542=$198E   D=9
```

Example:

```
LDY #$18
LDD #$8480       ;Y:D =2,000,000 =$188480
LDX #$1500            ;X=1500
EDIV        ;Y=1333=$535    D=500=$1F4
```

EDIVS **Extended Divide 32-bit by 16-bit (signed)**

Flags: N, Z, V, C

This instruction divides a 32-bit data in the Y:D registers by the 16-bit in register X. It is assumed that both registers Y:D and X contain signed data. After the division, the 16-bit quotient will be in register Y and the 16-bit remainder in register D. Notice in this instruction that the carry flag is cleared, unless we divide Y:D by 0 (that is X = 0), in which case the result is invalid and C = 1 to indicate the invalid condition.

EMACS **Extended Multiply and Accumulate (16-bit by 16-bit signed)**

Flags: N, Z, C

This multiplies an unsigned 16-bit value in memory by another unsigned 16-bit value in memory to produce a 32-bit intermediate result. This 32-bit intermediate result is then added to another 32-bit data. The EMACS is a signed integer operation. The first 16-bit value is located in memory locations pointed to by X and the second 16-bit value is located in memory locations pointed to by Y. The address location where the 32-bit result is stored is provided by the instruction itself. See the HCS12 manual for more information.

EMAXD Place Larger (Maximum) of two values in D

Flag: N, V, Z, C

The magnitudes of the D register (Double accumulator A:B) and a 16-bit word data are compared and the larger one is placed in D. The 16-bit word data (W) is located in memory locations M and M + 1. It is assumed that the data in both D and memory locations are unsigned. This instruction is really D –– W and the flags are set accordingly. The content of D is the larger value and the contents of memory locations M and M + 1 are unchanged.

EMAXM Place Larger (Maximum) of two values in Memory

Flag: N, V, Z, C

The magnitudes of the D register (Double accumulator A:B) and a 16-bit word data are compared and the larger one is placed in memory. The 16-bit word data (W) is located in memory locations M and M + 1. It is assumed that the data in both D and memory locations are unsigned. This instruction is really D – W and the flags are set accordingly. The content of D is unchanged and the contents of memory locations M and M + 1 have the larger value.

EMIND Place Smaller (Minimum) of two unsigned values in D

Flag: N, V, Z, C

The magnitudes of the D register (Double accumulator A:B) and a 16-bit word data are compared and the smaller one is placed in D. The 16-bit word data (W) is located in memory locations M and M + 1. It is assumed that the data in both D and the memory locations are unsigned. This instruction is really D – W and the flags are set accordingly. The content of D is the smaller value and the contents of memory locations M and M + 1 are unchnaged.

EMINM Place Smaller (Minimum) of two values in Memory

Flag: N, V, Z, C

The magnitudes of the D register (Double accumulator A:B) and a 16-bit word data are compared and the smaller one is placed in memory. The 16-bit word data (W) is located in memory locations M and M + 1. It is assumed that the data in both D and memory locations are unsigned. This instruction is really D – W and the flags are set accordingly. The content of D is unchanged and the contents of memory locations M and M + 1 have the smaller value.

EMUL Extended Multiply 16-bit by 16-bit (unsigned D × Y)

Flags: N, Z, C

This multiplies an unsigned 16-bit value in D register by an unsigned 16-bit value in register Y. The 32-bit result is placed in Y and D where D has the lower 16 bits and Y has the higher 16 bits.

Example:

```
LDD #5
LDY #7
EMUL        ;D=35=$23, Y=00
```

Example:

```
LDD #1000
LDY #150
EMUL        ;Y=02, D=$49F0
            (1000 x 150 = 150000=$000249F0
```

Example:

```
LDD #60000
LDY #$50000
EMUL        ;Y= $B2D0, D=$5E00
;(60000 x 50000 = 3000000000 =$B2D05E00)
```

Example:

```
LDD #100
LDY #200
EMUL ;D=$4E20, Y=0(100x200=20,000 = 4E20H)
```

EMULS Extended Multiply 16-bit by 16-bit (signed D × Y)

Flags: N, Z, C

This multiplies a 16-bit signed value in the D register by a signed 16-bit value in register Y. The 32-bit result is placed in Y and D where D has the lower 16 bits and Y has the higher 16 bits.

EORA Exclusive OR A

Flags: N, Z, V = 0

This performs a logical exclusive-OR on the operands, bit by bit, storing the result in the A register. Notice that both the source and destination values are byte-size only.

A	V	A XOR V
0	0	0
0	1	1
1	0	1
1	1	0

Example:

```
        LDAA #$39        ;A=$39
        EORA #$09        ;A=$39 ORed
with 09

       $39    0011 1001
       $09    0000 1001
       $30    0011 0000
```

Example:

```
LDAA   #$32       ;A=$32
EXORA #$50        ;A=$62

$32    0011 0010
$50    0101 0000
$62    0110 0010
```

The following addressing modes are supported for the EORA instruction:

1. Immediate: `EORA #data` Example: `EORA #$30`
2. Direct: `EORA PORTB`
3. Extended: `EORA 16-bit memory address`
   Example: `EORA $1550 ;OR A with data in RAM loc. $1550`
4. Indexed: `EORA 0,X   ;OR A with data pointed to by X`

EORB Exclusive OR B

Flags: N, Z, V = 0

This performs a logical exclusive-OR on the operands, bit by bit, storing the result in the B register. Notice that both the source and destination values are byte-size only.

Example:

```
        LDAB #$30        ;B=$30
        EORB #$09        ;B=$30 ORed with 09

       $30    0011 0000
       $09    0000 1001
       -----------------
       $39    0011 1001
```

Example:

```
        LDAB   #$32      ;B=$32
        EORB   #$5F      ;B=$6F

        $32    0011 0010
        $5F    0101 1111
        -----------------
        $62    0110 1111
```

The addressing modes supported are the same as for EORA.

ETBL Extended Table Look-up and Interpolate

See the HCS12 manual.

EXG Exchange Register Contents

See the HCS12 manual.

FDIV Fractional Divide

See the HCS12 manual.

IBEQ Increment and Branch if Equal to Zero

Flags: None

This instruction increments a register, and if the result is zero it will branch to the target address. The register can be A, B, D, X, Y, or SP.

Example:

```
        IBEQ A    ;increment A and branch if A=0
        IBEQ B    ;increment B and branch if B=0
        IBEQ D    ;increment D and branch if D=0
        IBEQ X    ;decrement X and branch if X=0
        IBEQ Y    ;decrement Y and branch if Y=0
```

Notice that the target address can be no more than 128 bytes backward or 127 bytes forward.

IBNE Increment and Branch if Not Equal to Zero

Flags: None

This instruction increments a register, and if the result is not zero it will branch to the target address. The register can be A, B, D, X, Y, or SP.

```
IBNE A     ;increment A and branch if A is not 0
IBNE B     ;increment B and branch if B is not 0
IBNE D     ;increment D and branch if D is not 0
IBNE X     ;increment X and branch if X is not 0
IBNE Y     ;increment Y and branch if Y is not 0
```

Notice that the target address can be no more than 128 bytes backward or 127 bytes forward.

IDIV Integer Divide (D/X, unsigned)

Flags: V = 0, Z, C

This instruction divides a 16-bit data in the D register by a 16-bit data in register X. It is assumed that both registers D and X contain unsigned data. After the division, the 16-bit quotient will be in register X and the 16-bit remainder in register D. Notice in this instruction that the carry flag is cleared, unless we divide D by 0 (that is X = 0), in which case the result is invalid and C = 1 to indicate the invalid condition.

Example:

```
LDD #65428
LDX #654
IDIV          ;X=10=$000A     D=28=$1C
```

Example:

```
LDD #$97
LDX #$12
IDIV          ;X=8 and D=7
```

IDIVS Integer Divide Signed (D/X, signed)

Flags: N, Z, V, C

This instruction divides a 16-bit data in the D register by the 16-bit data in register X. It is assumed that both registers D and X contain signed data. After the division, the 16-bit quotient will be in register Y and the 16-bit remainder in register D. Notice in this instruction that the carry flag is cleared, unless we divide D by 0 (that is X = 0), in which case the result is invalid and C = 1 to indicate the invalid condition.

INC Increment Memory

Flags: N, Z, V

This instruction adds 1 to the memory location specified by the operand. Note that C is not affected even if value FF is incremented to 00. This instruction supports the following addressing modes:

1. Extended: `INC 16-bit memory address`
   Example: `INC $1550 ;increment data in RAM loc. $1550`
2. Indexed: `INC 0,X ;increment data in RAM pointed by X`

INCA Increment A (A = A + 1)

Flags: N, Z, V

This instruction adds 1 to accumulator A. Note that C is unchanged even if a value $FF is incremented and becomes 00.

INCB Increment B (B = B + 1)

Flags: N, Z, V

This instruction adds 1 to accumulator B. Note that C (carry) is unchanged even if a value $FF is incremented and becomes 00.

INS Increment Stack Pointer (SP = SP + 1)

Flags: None

This instruction adds 1 to the SP (Stack Pointer) register.

INX Increment X (X = X + 1)

Flags: Z

This instruction adds 1 to the X register.

INY Increment Y (Y = Y + 1)

Flags: Z

This instruction adds 1 to the Y register.

JMP Jump

Flags: None

This is an unconditionnal jump. This instruction jumps to an effective address. The effective address is either extended address or indexed addressing mode. See the HCS12 manual.

1. Extended: `JMP 16-bit memory address`
   Example: `JMP $1550 ;jump to  address $1550`
2. Indexed: `JMP 0,X ;jump to an address pointed to by X`

JSR Jump to Subroutine

Flags: None

There are three types of calls: JSR, BSR, and CALL. In JSR (jump to subroutine), the target address is within 64K bytes of the current PC (program counter). If jumping to a subroutine using JSR, the PC register (which has the address of the instruction after the JSR) is pushed onto the stack, and the stack pointer (SP) is decremented by 2. Then the program counter is loaded with the new address and control is transferred to the subroutine. At the end of the procedure, when RTS (return from subroutine) is executed, PC is pulled off the stack, which returns control to the instruction after the JSR. Notice that JSR is a 3-byte instruction, in which one byte is the opcode, and the other two bytes are the 16-bit address of the target subroutine. For more information, see the CALL instruction.

LBCC Long Branch if Carry Cleared (C = 0)

Flags: None

This instruction examines the C flag, and if it is zero it will jump (branch) to the target address. Notice that in this instruction the target address cannot be farther than –32,768 to +32,767 bytes from the program counter.

See the table for the Long Branch instruction using flag bits.

Table A-3: HCS12 Long Branch Instructions Using Flag Bits

Instruction	Action
LBCS	Branch if C = 1
LBCC	Branch if C = 0
LBEQ	Branch if Z = 1
LBNE	Branch if Z = 0
LBMI	Branch if N = 1
LBPL	Branch if N = 0
LBVS	Branch if V = 1
LBVC	Branch if V = 0

Note: The target address cannot be more than –32,768 to +32,767 bytes away from the current PC (program counter).

LBCS Long Branch if Carry Set (C = 1)

Flags: None

This instruction examines the C flag, and if it is set it will jump to the target address. Notice that in this instruction the target address cannot be farther than –32,768 to +32,767 bytes from the program counter.

LBEQ Long Branch if Equal (Z = 1)

Flags: None

This instruction examines the Z flag and jumps if it is 1 (Z = 1). Notice that in this instruction the target address cannot be farther than –32,768 to +32,767 bytes from the program counter.

LBGE Long Branch if Greater than or Equal

Flags: None

This instruction is used for branching after compare instructions such as CMPA, CMPB, and so on. It can also be used after subtract instructions such as SUBA, SUBB, and so on. Notice that in this instruction the target address cannot be farther than –32,768 to +32,767 bytes from the program counter.

LBGT Long Branch if Greater than Zero

Flags: None

This instruction is used for branching after compare instructions such as CMPA, CMPB, and so on. It can also be used after subtract instructions such as SUBA, SUBB, and so on. Notice that in this instruction the target address cannot be farther than –32,768 to +32,767 bytes from the program counter.

LBHI Long Branch if Higher (if C + Z = 0)

Flags: None

This instruction is used for branching after compare instructions such as CMPA, CMPB, and so on. It can also be used after subtract instructions such as SUBA, SUBB, and so on. BHI should not be used after instructions that do not affect the carry flag such as Increment, Decrement, and so on. Notice that in this instruction the target address cannot be farther than –32,768 to +32,767 bytes from the program counter.

LBHS Long Branch if Higher or Same (if C = 0)

Flags: None

This instruction is used for branching after compare instructions such as CMPA, CMPB, and so on. It can also be used after subtract instructions such as SUBA, SUBB, and so on. BHS should not be used after instructions that do not affect the carry flag such as Increment, Decrement, and so on. Notice that in this instruction the target address cannot be farther than –32,768 to +32,767 bytes from the program counter.

LBLE Long Branch if Less than or Equal

Flags: None

This instruction is used for branching after compare instructions such as CMPA, CMPB, and so on. It can also be used after subtract instructions such as SUBA, SUBB, and so on. Notice that in this instruction the target address cannot be farther than –32,768 to +32,767 bytes from the program counter.

LBLO Long Branch if Lower (if C = 0)

Flags: None

This instruction is used for branching after compare instructions such as CMPA, CMPB, and so on. It can also be used after subtract instructions such as SUBA, SUBB, and so on. BLO should not be used after instructions that do not affect the carry flag such as Increment, Decrement, and so on. Notice that in this instruction the target address cannot be farther than –32,768 to +32,767 bytes from the program counter.

LBLS Long Branch if Lower or Same (if C + Z = 1)

Flags: None

This instruction is used for branching after compare instructions such as CMPA, CMPB, and so on. It can also be used after subtract instructions such as SUBA, SUBB, and so on. BLS should not be used after instructions that do not affect the carry flag such as Increment, Decrement, and so on. Notice that in this instruction the target address cannot be farther than –32,768 to +32,767 bytes from the program counter.

LBLT Long Branch if Less than Zero

Flags: None

This instruction is used for branching after compare instructions such as CMPA, CMPB, and so on. It can also be used after subtract instructions such as SUBA, SUBB, and so on. Notice that in this instruction the target address cannot be farther than -32,768 to +32,767 bytes from the program counter.

LBMI Long Branch if Minus (if N = 1)

Flags: None

This instruction examines the N flag and jumps if it is 1 (N = 1). Notice that in this instruction the target address cannot be farther than -32,768 to +32,767 bytes from the program counter.

LBNE Long Branch if Not Equal to Zero (if Z = 0)

Flags: None

This instruction examines the Z flag and jumps if it is 0 (Z = 0). Notice that in this instruction the target address cannot be farther than -32,768 to +32,767 bytes from the program counter.

LBPL Long Branch if Plus (if N = 0)

Flags: None

This instruction examines the N flag and jumps if it is 0 (N = 0). Notice that in this instruction the target address cannot be farther than -32,768 to +32,767 bytes from the program counter.

LBRA Long Branch Always

Flags: None

This is a 4-byte instruction. The 2 bytes are the opcode and the second 2 bytes are the signed number displacement, which is added to the PC (program counter) of the instruction following the LBRA to get the target address. Therefore, in this jump the target address must be within -32,768 to +32,767 bytes of the PC (program counter) of the instruction after the LBRA since 2 bytes of address can take values of -32,768 to +32,767 bytes. This address is often referred to as a *relative address* since the target address is -32,768 to +32,767 bytes relative to the program counter (PC).

LBVC Long Branch if Overflow Cleared (V = 0)

Flags: None

This instruction examines the V flag, and if it is cleared it will jump to the target address. Notice that in this instruction the target address cannot be farther than -32,768 to +32,767 bytes from the program counter.

LBVS **Long Branch if Overflow Set (V = 1)**

Flags: None

This instruction examines the V flag, and if it is set it will jump to the target address. Notice that in this instruction the target address cannot be farther than –32,768 to +32,767 bytes from the program counter.

LBRN **Long Branch Never**

Flags: None

This is in reality a 4-byte NOP that is executed in three cycles. It is the complement of the LBRA instruction.

The following three instructions perform the same action:

```
     LBRN

HERE LBRA HERE

     LBRA $
```

LDAA **Load Accumulator A**

Flags: N, V, Z

This loads a byte into register A. The value can be an immediate value or in some memory location. The following are some examples of how it is used for various addressing modes:

```
LDAA #$5F        ;A=$5F
LDAA #99         ;A=99
LDAA PORTB       ;load into A from PORTB
LDAA $1500 ;load into A from location $1500
LDAA 0,X ;load into A from location point to by X
LDAA 0,Y ;load into A from location point to by Y
```

LDAB **Load Accumulator B**

Flags: N, V, Z

This loads a byte into register B. The value can be an immediate value or in some memory location. The following are some examples of how it is used:

```
LDAB #$5F        ;B=$5F
LDAB #99         ;B=99
LDAB PORTB       ;load into B from PORTB
LDAB $1500 ;load into B from location $1500
LDAA 0,X ;load into A from location point to by X
LDAA 0,Y ;load into A from location point to by Y
```

LDD **Load Double Accumulator D**

Flags: N, V, Z

This loads a 16-bit word into register D. The value can be an immediate value or in some memory location. The following are some examples of how it is used:

```
LDD  #$5F77     ;D=$5F77
LDD  #9920      ;D=9920
LDD  $1500 ;load into D from loc $1500 and $1501
LDD  0,X ;load into D from location X and X+1
LDD  0,Y ;load into D from location Y and Y+1
```

LDS Load Stack Pointer

Flags: N, V, Z

This loads a 16-bit word into the SP register. The value can be an immediate value or in some memory location. The following are some examples of how it is used:

```
LDS  #$5F77     ;SP=$5F77
LDS  #9920      ;SP=9920
LDS  $1500;load into SP from loc $1500 and $1501
LDS  0,X ;load into SP from location X and X+1
LDS  0,Y ;load into SP from location Y and Y+1
```

LDX Load index register X

Flags: N, V, Z

This loads a 16-bit word into register X. The value can be an immediate value or in some memory location. The following are some examples of how it is used:

```
LDX  #$5F77     ;X=$5F77
LDX  #9920      ;X=9920
LDX  $1500 ;load into X from loc $1500 and $1501
LDX  0,Y ;load into X from location Y and Y+1
```

LDY Load index register Y

Flags: N, V, Z

This loads a 16-bit word into register Y. The value can be an immediate value or in some memory location. The following are some examples of how it is used:

```
LDY  #$5F77     ;Y=$5F77
LDY  #9920      ;Y=9920
LDY  $1500 ;load into Y from loc $1500 and $1501
LDY  0,X ;load into Y from location X and X+1
```

LEAS Load Stack Pointer with Effective Address

Flags: None

This loads an effective address into the SP register. The effective address is formed by combining the base index registers of X, Y, SP, or PC and an offset value, which can be an immediate value or the content of a register such as A, B, and so on.

LEAX Load X with Effective Address

Flags: None

This loads an effective address into register X. The effective address is formed by combining the base index registers of X, Y, SP, or PC and an offset value, which can be an immediate value or the content of a register such as A, B, and so on.

LEAY Load Y with Effective Address

Flags: None

This loads an effective address into register Y. The effective address is formed by combining the base index registers of X, Y, SP, or PC and an offset value, which can be an immediate value or the content of a register such as A, B, and so on.

LSL Logical Shift Left Memory

Flags: N, Z, V, C

This shifts all bits of a memory location one bit position left. Bit 0 is loaded with a 0. The C bit is loaded from the MSB of the memory location.

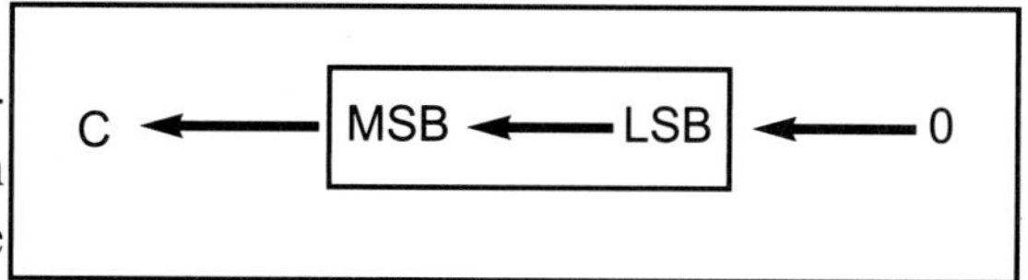

Example:

```
LDAA #$99        ;A=$99
STAA $1500       ;$1500=(10011001)
LSL $1500        ;Now $1500=(00110010) and C=1
LSL $1500        ;Now $1500=(01100100) and C=0
```

LSLA Logical Shift Left A

Flags: N, Z, V, C

This shifts all bits of A one bit position left. Bit 0 is loaded with a 0. The C bit is loaded from the MSB of A.

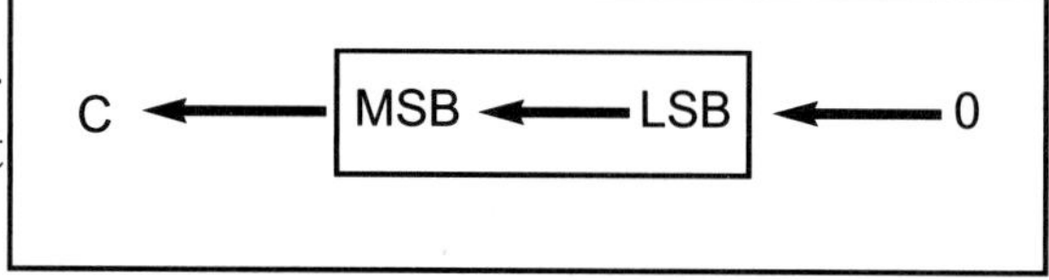

Example:

```
LDAA #$99        ;A=$99
LSLA             ;Now A=00110010 and C=1
LSLA             ;Now A=01100100 and C=0
```

LSLB Logical Shift Left B

Flags: N, Z, V, C

This shifts all bits of B one bit position left. Bit 0 is loaded with a 0. The C bit is loaded from the MSB of B.

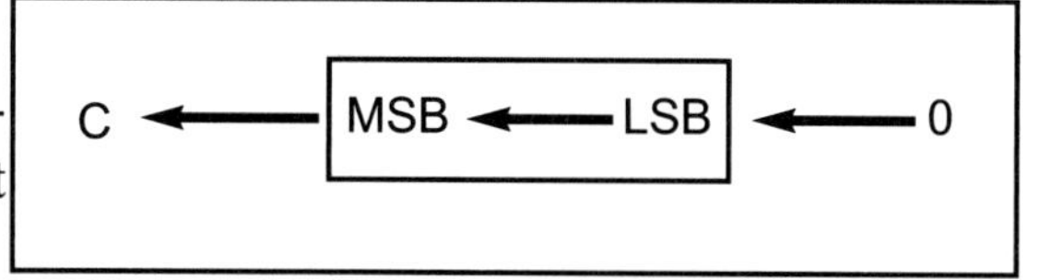

Example:

```
LDAB #$99        ;B=$99
LSRB             ;Now B=11001100 and C=1
LSRB             ;Now B=11100110 and C=0
```

LSLD Logical Shift Left D

Flags: N, Z, V, C

This shifts all bits of D one bit position left. Bit 0 is loaded with a 0. The C bit is loaded from the MSB of D.

Example:

```
LDD  #$9999     ;D=$9999
LSLD       ;Now D=$3332 and C=1
LSLD       ;Now D=$6664 and C=0
```

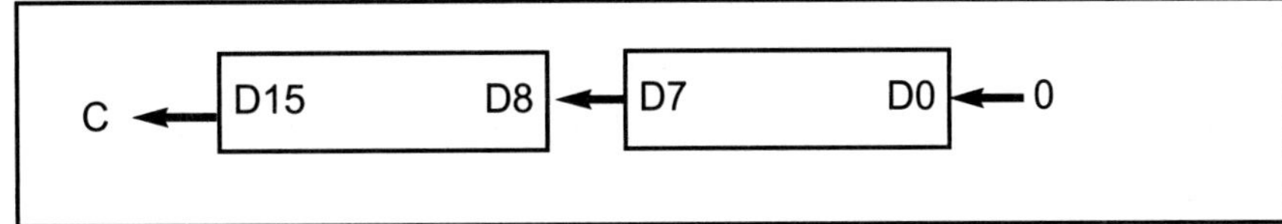

LSR Logical Shift Right Memory

Flags: N, Z, V, C

This shifts all bits of memory location one bit position right. Bit 7 is loaded with 0. The C bit is loaded with the LSB of memory.

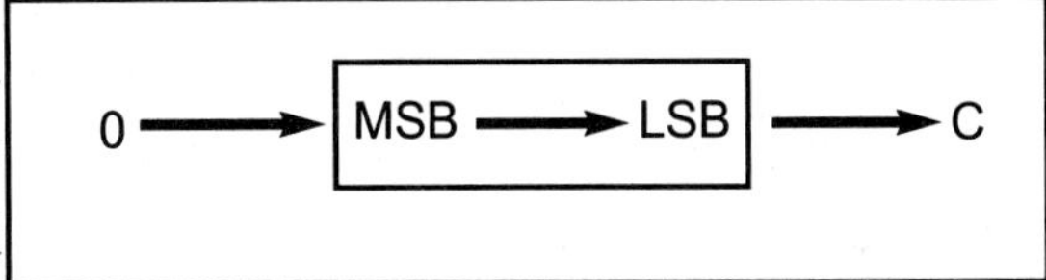

Example:

```
LDAA #$99       ;A=$99
STAA $1500 ;$1500=(10011001)
LSR $1500 ;Now $1500=(01001100) and C=1
LSR $1500 ;Now $1500=(00100110) and C=0
```

LSRA Logical Shift Right A

Flags: N, Z, V, C

This shifts all bits of A one bit position right. Bit 7 is loaded with 0. The C bit is loaded from the LSB of register A.

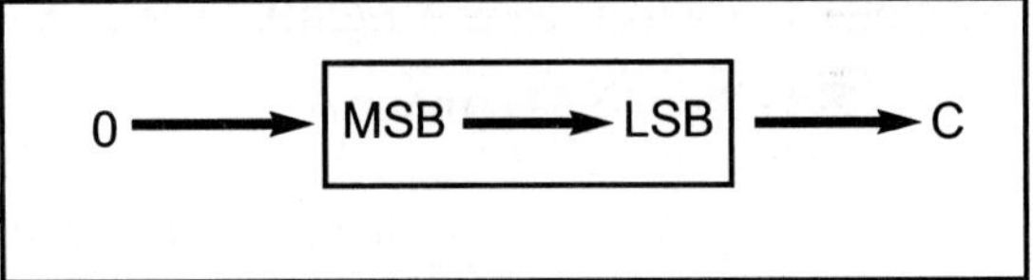

Example:

```
LDAA #$99       ;A=$99
LSRA            ;Now A=01001100 and C=1
LSRA            ;Now A=00100110 and C=0
```

LSRB Logical Shift Right B

Flags: N, Z, V, C

This shifts all bits of B one bit position right. Bit 7 is loaded with 0. The C bit is loaded from the LSB of register B.

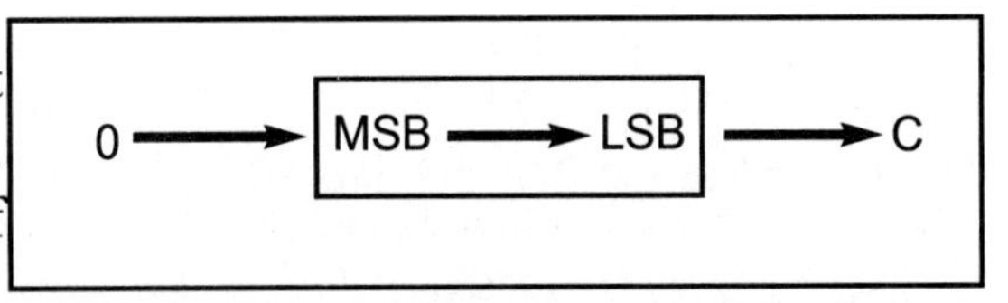

Example:

```
LDAB #$99       ;B=$99
LSRB            ;Now B=011001100 and C=1
LSRB            ;Now B=001100110 and C=0
```

LSRD **Logical Shift Right D**

Flags: N, Z, V, C

This shifts all bits of D one bit position right. Bit 15 is loaded with a 0. The C bit is loaded from the LSB of D.

Example:

```
LDD #$9999 ;D=$9999
LSRD       ;D=$6666 and C=1
LSRD       ;Now D=$3333 and C=0
```

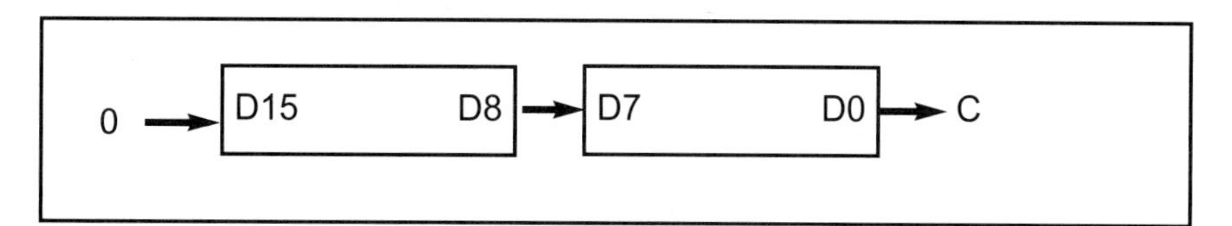

MAXA **Place Larger (Maximum) of two values in A**

Flags: N, V, Z, C

The magnitudes of the A register and a byte of data in memory are compared and the larger one is placed in A. The 8-bit data is located in memory location M. It is assumed that the data in both A and the memory location are unsigned.

MAXM **Place Larger (Maximum) of two values in Memory**

Flags: N, V, Z, C

The magnitudes of the A register and a byte of data are compared and the larger one is placed in memory. The byte data is located in memory location M. It is assumed that the data in both A and the memory location are unsigned.

MEM **Determine grade of Membership**

See the HCS12 manual.

MINA **Place Smaller (Minimum) of two unsigned values in A**

Flags: N, V, Z, C

The magnitudes of the A register and a byte of data are compared and the smaller one is placed in A. The byte of data is located in memory location M. It is assumed that the data in both A and the memory location are unsigned.

MINM **Place Smaller (Minimum) of two unsigned values in M**

Flags: N, V, Z, C

The magnitudes of the A register and a byte of data are compared and the smaller one is placed in memory. The byte of data is located in memory location M. It is assumed that the data in both A and the memory location are unsigned.

MOVB **Move a Byte of data from source to destination**

Flags: None

This copies a byte from the source location to the destination. See the following examples:

```
MOVB #$99,$1500 ;load $99 into location $1500
MOVB 0,X,0,Y ; move from location X to location Y
```

MOVW Move a Word of data from source to destination

Flags: None

This copies a 16-bit word from the source locations to the destinations. See the following examples:

```
MOVW #$9972,$1500    ;load $9972 into locations
                     ;$1500 and 1501
                     ; $1500=($99) and 1501=($72)

MOVW 0,X,0,Y      ;move word from locations X and X+1
                  ;to locations Y and Y+1
```

MUL Multiply byte by byte (D = A × B unsigned)

Flags: C

This multiplies an unsigned byte value in register A by an unsigned byte value in register B. The 16-bit result is placed in D (A and B) where B has the lower 8 bits and A has the higher 8 bits.

Example:

```
LDAA #5
LDDB #7
MUL          ;D=7x5=35=$23. A=0, B=$23
```

Example:

```
LDAA #10
LDAB #15
MUL          ;D=10x15=150=$96, A=0, B=$96
```

Example:

```
LDAA #$25
LDAB #$78
MUL          ;D=$1158  A=$11, B=$58
             ;($25 x $78 = $1158)
```

Example:

```
LDAA #100
LDAB #200
MUL          ;D=$4E20, A=$4E, B=$20
             ;(100 x 200 = 20,000 = $4E20)
```

NEG Negate memory

Flags: N, Z, C, V

This replaces the contents of a memory location with its 2’s complement. The result is the 2's complement of the memory.

Example:

```
LDAA #$58        ;A=01010000
STAA $1500       ;$1500($58)
NEG  $1500       ;$1500($B0)
```

1. Extended: `NEG 16-bit memory address`
   Example: `NEG $1550 ;2's comp of data in RAM loc. $1550`
2. Indexed: `NEG 0,X ;2's comp data in RAM pointed by X`

NEGA Negate A

Flags: N, Z, C, V

This takes the 2's complement of the contents of register A, the accumulator. The result is the 2's complement of the accumulator sitting in A.

Example:

```
LDAA #$58        ;A=01010000
NEGA             ;A=10110000
```

NEGB Negate B

Flags: N, Z, C, V

This takes the 2's complement of the contents of register B, the accumulator. The result is the 2's complement of the accumulator sitting in B.

Example:

```
LDAB #$74        ;B=01110100
NEGB             ;B=10001100
```

NOP No Operation

Flags: None

This performs no operation and execution continues with the next instruction. It is sometimes used for timing delays to waste clock cycles. This instruction only updates the PC (program counter) to point to the next instruction following NOP.

ORAA Logical OR A

Flags: N, Z, V = 0

This performs a logical OR on the operands, bit by bit, storing the result in register A. Notice that both the source and destination values are byte-size only.

A	B	A OR B
0	0	0
0	1	1
1	0	1
1	1	1

Example:

```
LDAA #$30     ;A=$30
ORAA #$09     ;A=$30 OR with 09 (A=$39)

$39    0011 0000
$09    0000 1001
$39    0011 1001
```

Example:

```
LDAA #$32       ;A=$32
ORAA #$50       ;A=$72

$32    0011 0010
$50    0101 0000
$72    0111 0010
```

The following addressing modes are supported for the ORAA instruction:

1. Immediate: `ORAA #data` Example: `ORAA #$30`
2. Direct: `ORAA PORTB`
3. Extended: `ORAA 16-bit memory address`
   Example: `ORAA $1550 ;OR A with data in RAM loc. $1550`
4. Indexed: `ORAA 0,X   ;OR A with data pointed to by X`

ORAB Logical OR B

Flags: N, Z, V = 0

This performs a logical OR on the operands, bit by bit, storing the result in register B. Notice that both the source and destination values are byte-size only.

Example:

```
LDAB #$39     ;B=$39
ORAB #$07     ;B=$39 OR with 07 (B=$3F)

$39    0011 1001
$09    0000 0111
$39    0011 1111
```

Example:

```
LDAB #$32       ;B=$32
ORAB #$50       ;B=$72

$32    0011 0010
$50    0101 0000
$72    0111 0010
```

The addressing modes supported are the same as for ORAA.

ORCC Logical OR CCR with mask

Flags: H, N, Z, V, C

This performs a logical OR on the CCR register and a mask byte, bit by bit, storing the result in the CCR register.

Example:

```
ORCC #$01     ;CCR=XX ORed with 01

XX     xxxx xxxx
00     0000 0001
X1     xxxx xxx1
```

PSHA Push A onto Stack

Flags: None

The SP (Stack Pointer) is decremented by 1, and the content of register A is stored on the stack memory location pointed to by SP.

PSHB Push B onto Stack

Flags: None

The SP (Stack Pointer) is decremented by 1, and the content of register B is stored on the stack memory location pointed to by SP.

PSHC Push CCR onto Stack

Flags: None

The SP (Stack Pointer) is decremented by 1, and the content of register CCR is stored on the stack memory location pointed to by SP.

PSHD Push D onto Stack

Flags: None

The SP (Stack Pointer) is decremented by 2, and the content of register D is stored on the stack memory location pointed to by SP. Notice that the high byte (register A) is stored in the low address location and the low byte (register B) is stored in the high address location. This means SP points to the high byte of register D. This is called *big endian*.

PSHX Push X onto Stack

Flags: None

The SP (Stack Pointer) is decremented by 2, and the content of register X is stored on the stack memory location pointed to by SP. Notice that the high byte is stored in the low address location and the low byte is stored in the high address location. This means SP points to the high byte of register X.

PSHY Push Y onto Stack

Flags: None

The SP (Stack Pointer) is decremented by 2, and the content of register Y is stored on the stack memory location pointed to by SP. Notice that the high byte is stored in the low address location and the low byte is stored in the high address location. This means SP points to the high byte of register Y.

PULA Pull A from Stack

Flags: None

This loads A with a byte pointed to by SP (stack pointer), and SP is incremented by 1.

PULB Pull B from stack

Flags: None

This loads B with a byte pointed to by SP (stack pointer), and SP is incremented by 1.

PULC Pull CCR from stack

Flags: None

This loads the CCR with a byte pointed to by SP (stack pointer), and SP is incremented by 1.

PULD Pull D from stack

Flags: None

This loads D with a word pointed to by SP (stack pointer), and SP is incremented by 2. Notice that the high byte (register A) is loaded with a byte pointed to by the low address location and the low byte (register B) is loaded with a byte pointed to by the high address location.

PULX Pull X from stack

Flags: None

This loads X with a word pointed to by SP (stack pointer), and SP is incremented by 2. Notice that the high byte of register X is loaded with a byte pointed to by the low address location and the low byte is loaded with a byte pointed to by the high address location.

PULY Pull Y from stack

Flags: None

This loads Y with a word pointed to by SP (stack pointer), and SP is incremented by 2. Notice that the high byte of register Y is loaded with a byte pointed to by the low address location and the low byte is loaded with a byte pointed to by the high address location.

REV Fuzzy Logic Rule Evaluation

See the HCS12 manual.

REVW Fuzzy Logic Rule Evaluation Weighted

See the HCS12 manual.

ROL Rotate Left memory

Flags: C, V, Z, N

This rotates the bits of a memory location left. The bits rotated out are rotated into C, and the C bit is rotated into the opposite end of the byte.

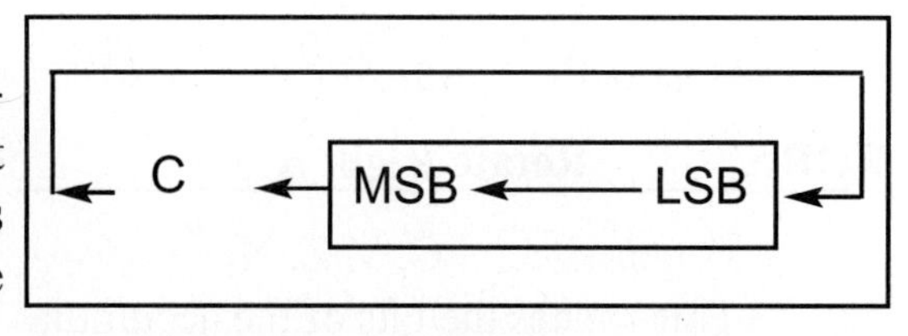

Example:

```
          CLC                     ;C=0
          LDAA #$99               ;A=10011001
          STAA $1500              ;$1500=(10011001)
          ROL  $1500 ;Now $1500=(00110010) and C=1
          ROL  $1500 ;Now $1500=(01100101) and C=0
```

ROLA **Rotate Left A**

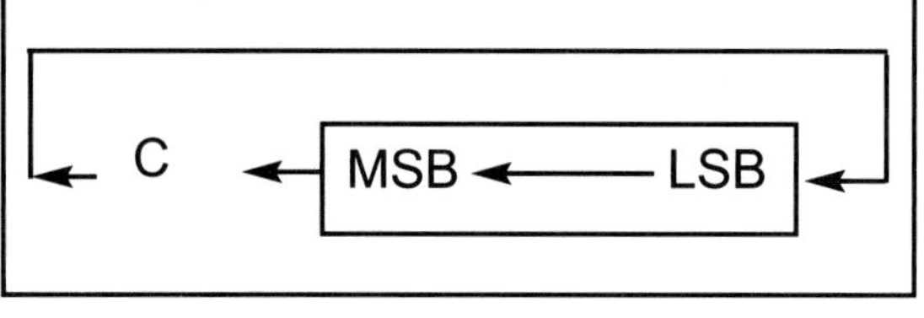

Flags: C, V, Z, N

This rotates the bits of the accumulator A left. The bits rotated out of register A are rotated into C, and the C bit is rotated into the opposite end of the accumulator.

```
          CLC             ;C=0
          LDAA #$99               ;A=10011001
          ROLA                    ;Now A=00110010 and C=1
          ROLA                    ;Now A=01100101 and C=0
```

ROLB **Rotate Left B**

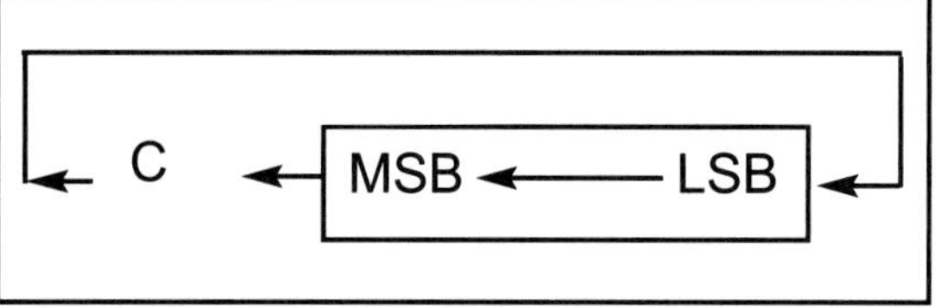

Flags: C, V, Z, N

This rotates the bits of the accumulator B left. The bits rotated out of register B are rotated into C, and the C bit is rotated into the opposite end of the accumulator.

```
          CLC             ;C=0
          LDAB #$99               ;A=10011001
          ROLB                    ;Now B=00110010 and C=1
          ROLB                    ;Now B=01100101 and C=0
```

ROR **Rotate Right memory**

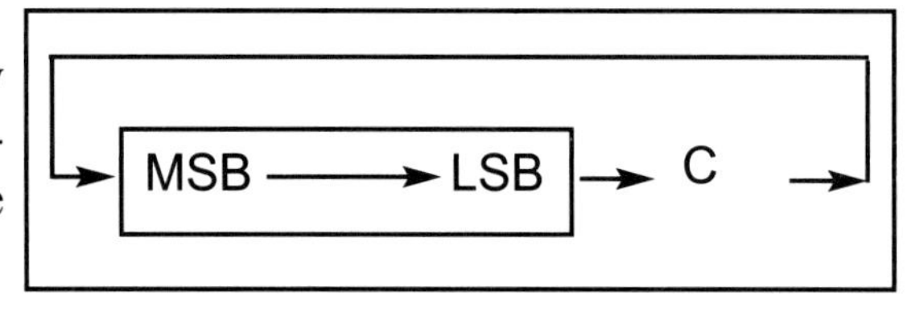

Flags: C, V, Z, N

This rotates the bits of a memory location right. The bits rotated out are rotated into C, and the C bit is rotated into the opposite end of the byte.

Example:

```
          SEC                     ;C=1
          LDAA #$99               ;A=10011001
          STAA $1500              ;$1500 = (10011001) and C=1
          ROR  $1500              ;Now $1500=(11001100) and C=1
          ROR  $1500              ;Now $1500=(11100110) and C=0
```

RORA **Rotate Right A**

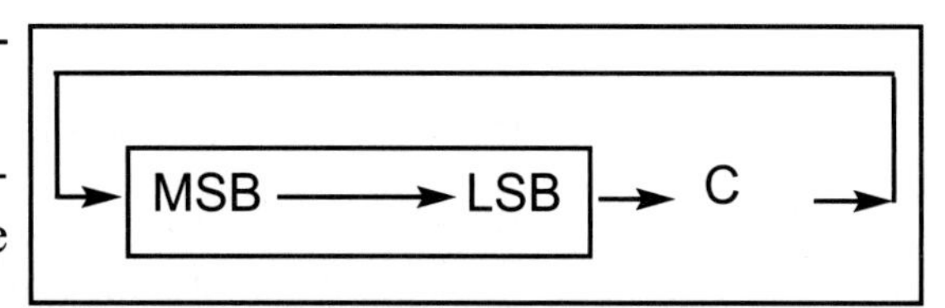

Flags: C, V, Z, N

This rotates the bits of the accumulator right. The bits rotated out of register A are rotated into C, and the C bit is rotated into the opposite end of the accumulator.

Example:

```
        SEC             ;C=1
        LDAA #$99       ;A=10011001
        RORA            ;Now A=11001100 and C=1
        RORA            ;Now A=11100110 and C=0
```

RORB   Rotate Right B

Flags: C, V, Z, N

This rotates the bits of B right. The bits rotated out of register B are rotated into C, and the C bit is rotated into the opposite end of B.

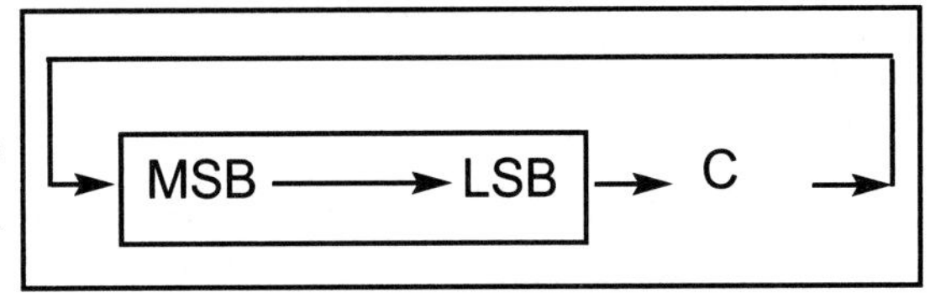

Example:

```
        SEC             ;C=1
        LDAB #$99       ;B=10011001
        RORB            ;Now B=11001100 and C=1
        RORB            ;Now B=11100110 and C=0
```

RTC   Return from Call

Flags: None

This instruction is used to return from a subroutine previously entered by CALL instructions. The top byte of the stack is pulled into the PPAGE register and SP is incremented by 1. Then, the top two bytes of the stack are pulled into the program counter (PC) and program execution continues at this new address. After pulling the top two bytes off the stack into the program counter, the stack pointer (SP) is incremented by 2.

RTI   Return from Interrupt

Flags: None

This is used at the end of an interrupt service routine (interrupt handler) to restore the CPU context to the state before the interrupt was executed. The top nine bytes of the stack are pulled into the CCR, D(A:B), X, PC, and Y registers, and program execution continues at the new address pointed to by PC. After pulling the top 9 bytes off the stack, the stack pointer (SP) is incremented by 9.

Notice that while the RTC instruction is used at the end of a subroutine associated with the CALL instruction, RTI must be used for the interrupt service subroutine (ISR).

RTS   Return from Subroutine

Flags: None

This instruction is used to return from a subroutine previously entered by JSR (Jump to Subroutine) instructions. The top two bytes of the stack are pulled into the program counter (PC), and program execution continues at this new address. After pulling the top two bytes off the stack into the program counter, the stack pointer (SP) is incremented by 2.

SBA Subtract B from A (A = A – B)

Flags: N, Z, V, C

This subtracts B from accumulator A and puts the result in accumulator A. B remains unchnaged. The steps for subtraction performed by the internal hardware of the CPU are as follows:

1. Take the 2's complement of B.
2. Add this to register A.
3. Invert the carry.

This instruction sets the carry and Z flags according to the following:

	C	Z	
A > B	0	0	The result is positive (notice C = 0).
A = B	0	1	The result is 0.
A < B	1	0	The result is negative in 2's complement.

Example:

```
LDAA #$3F
LDAB #$23
SBA         ;$3F-$23=$1C C=0,  Z=0

 $3F    0011 1111                  0011 1111
-$23    0010 0011       2's comp +1101 1101
+$1C                             1 0001 1100
                        C=0 (step 3, positive))
```

Example:

```
LDAA #$3F
LDAB #$45
SBA         ;$3F-$45=$1C C=0,  Z=0

 $35    0011 0101                  0011 0101
-$39    0011 1001       2's comp +1100 0111
-$04                             0 1111 1100
                        C=1 (step 3, negative)
```

SBCA Subtract memory with carry from A (A = A – M – C)

Flags: N, Z, V, C

This subtracts the byte in memory M and the carry flag from the accumulator A and puts the result in the accumulator A. M remains unchanged. The steps for subtraction performed by the internal hardware of the CPU are as follows:

1. Take the 2's complement of the memory byte.
2. Add this to register A.
3. Invert the carry.

This instruction sets the carry and Z flags according to the following:

	C	Z	
A > M	0	0	The result is positive (notice C = 0).
A = M	0	1	The result is 0.
A < M	1	0	The result is negative in 2's complement.

Example:

```
LDAA #$23
STAA $1500        ;$1500 = ($23)
CLC               ;C=0
LDAA #$45         ;A=$45
SBCA $1500        ;$45-$23-0=$22
```

SBCB **Subtract memory with Carry from B (B = B – M – C)**

Flags: N, Z, V, C

This subtracts the byte in memory M and the carry flag from the accumulator B and puts the result in B. M remains unchanged.

This instruction sets the carry and Z flags according to the following:

	C	Z	
B > M	0	0	The result is positive.
B = M	0	1	The result is 0.
B < M	1	0	The result is negative in 2's complement.

Example:

```
LDAA #$23
STAA $1500        ;$1500 = ($23)
SEC               ;C=1
LDAB #$45         ;B=$45
SBCB $1500        ;$45-$23-1=$21
```

SEC **Set Carry (C = 1)**

Flag: C = 1

This instruction sets the carry flag to high. This instruction is the same as ORCC #$01.

SEI **Set Interrupt bit (I = 1)**

Flag: I = 1

This instruction sets the I mask bit in the CCR to high. This instruction is the same as ORCC #$10.

SEV **Set V (V = 1)**

Flag: V = 1

This instruction sets the overflow (V) flag to high. This instruction is the same as ORCC #$02.

SEX **Sign EXtend into 16-bit register**

Flags: None

This copies the sign bit to the bits of the register.

Example:

```
        LDAA  #+9         ;A=$9
        SEX   D           ;D=0009
```

See the HCS12 manual for more information.

STAA **Store A**

Flags: N, Z, V = 0

This copies a byte from register A to a memory location. A is unchanged. The following are some examples of how it is used for various addressing modes:

```
           LDAA #$59      ;A=$59
DIR:       STAA PORTB     ;send A to PORTB
EXT:       STAA $1500     ;Memory loc. $1500 has $59
INDEXED:   STAA 0,X       ;store it in RAM loc pointed
                          ;to by X+0
           STAA 2,Y       ;store it in RAM loc pointed
                          ;to by Y+2
```

Example:

```
     LDAB #10                 ;B=10 counter
L1:  LDAA 0,X ;bring in from memory loc pointed by X
     STAA 0,Y           ;store it in loc pointed to by Y
     INX                ;increment X
     INY                ;increment Y
     DECB
     BNE  L1
```

STAB **Store B**

Flags: N, Z, V = 0

This copies a byte from register B to a memory location. B is unchanged.

Example:

```
        LDAB #$59      ;B=$59
        STAB $1500     ;Memory loc. $1500 has $95
```

STD **Store D**

Flags: N, Z, V = 0

This copies a word from register D to memory locations M and M + 1. D is unchanged.

Example:

```
        LDD #$2059     ;D=$2059
        STD  $1500     ;Memory loc. $1500 has $20
                       ;and loc $1501 has $59
```

STOP Stop Processing (standby mode stops all clocks)

Flags: None

If S = 1 (S bit in CCR), the STOP instruction acts as a 2-cycle NOP. If S = 0, the STOP instruction places the CPU context (registers PC, CCR, A, B, X, and Y) onto stack and stops all system clocks. That puts the CPU in standby mode, therefore minimizing the CPU power dissipation to an absolute minimum. The contents of registers and I/O remain unchanged. Activating any of the RSET, XIRQ, or IRQ signals ends the standby mode. These interrupts will force the CPU to fetch the program counter value from the interrupt vector table and start the execution of the interrupt service routine (ISR). At the end of the ISR, the RTI instruction pulls the original (before the execution of STOP instruction) CPU context from the stack and continues to execute the instructions after the STOP. The standby is also referred to as *sleep mode*.

STS Store SP

Flags: N, Z, V = 0

This copies a 16-bit value from the SP register to memory locations M and M + 1. SP is unchanged.
Example:

```
LDS #$2000      ;D=$2000
STS  $1500      ;Memory loc. $1500 has $20
                ;and loc $1501 has $00
```

STX Store X

Flags: N, Z, V = 0

This copies a 16-bit value from the X register to memory locations M and M + 1. X is unchanged.
Example:

```
LDX #$2059      ;X=$2059
STX  $1500      ;Memory loc. $1500 has $20
                ;and loc $1501 has $59
```

STY Store Y

Flags: N, Z, V = 0

This copies a 16-bit value from the Y register to memory locations M and M + 1. Y is unchanged.
Example:

```
LDY #$2059      ;y=$2059
STY  $1500      ;Memory loc. $1500 has $20
                ;and loc $1501 has $59
```

SUBA **Subtract M from A (A = A – M)**

Flags: N, Z, V, C

This subtracts M from accumulator A and puts the result in accumulator A. M remains unchanged.

This instruction sets the carry and Z flags according to the following:

	C	Z	
A > M	0	0	The result is positive (notice C = 0).
A = M	0	1	The result is 0.
A < M	1	0	The result is negative in 2's complement.

Example:

```
        LDAA #$23
        STAA $1500      ;$1500=($23)
        LDAA #$45       ;A=$45
        SUBA       ;$45-$23=$22 C=0, Z=0

 $45   0100 0101                    0100 0101
-$23   0010 0011         2's comp +1101 1101
+$22                              1 0010 0010
                           C=0 (step 3, positive)
```

Example:

```
        LDAA #$39
        STAA $1500      ;$1500=($39)
        LDAA #$35       ;A=$35
        SUBA       ;$35-$39=-$04 C=1, Z=0

 $35   0011 0101                    0011 0101
-$39   0011 1001         2's comp +1100 0111
-$04                              0 1111 1100
                           C=1 (step 3, negative)
```

SUBB **Subtract M from B (B = B – M)**

Flags: N, Z, V, C

This subtracts M from accumulator B and puts the result in B. M remains unchanged.

This instruction sets the carry and Z flags according to the following:

	C	Z	
B > M	0	0	The result is positive (notice C = 0).
B = M	0	1	The result is 0.
B < M	1	0	The result is negative in 2's complement.

Example:

```
        LDAA #$23
        STAA $1500      ;$1500=($23)
```

```
        LDAB #$45       ;B=$45
        SUBB      ;$45-$23=$22 C=0, Z=0
```

```
  $45    0100 0101                    0100 0101
 -$23    0010 0011        2's comp  +1101 1101
 +$22                               1 0010 0010
                          C=0 (step 3, positive)
```

Example:

```
        LDAA #$39
        STAA $1500      ;$1500=($39)
        LDAB #$35       ;B=$35
        SUBB            ;$35-$39=-$04 C=1, Z=0
```

```
  $35    0011 0101                    0011 0101
 -$39    0011 1001        2's comp  +1100 0111
 -$04                               0 1111 1100
                          C=1 (step 3, negative)
```

SUBD **Subtract M from D (D = D – M)**

Flags: N, Z, V, C

This subtracts M and M + 1 from D and puts the result in D. The contents of locations M and M + 1 remain unchanged.

This instruction sets the carry and Z flags according to the following:

	C	Z	
D > M	0	0	The result is positive (notice C = 0).
D = M	0	1	The result is 0.
D < M	1	0	The result is negative in 2's complement.

```
Assume D=$345 and M=($223)
 $345    0011 0100 0101                 0011 0100 0101
-$223    0010 0010 0011    2's comp   +1101 1101 1101
+$122                                 1 0001 0010 0010
                             C=0 (step 3, positive)

Assume D=$345 and M=($349)
 $345    0011 0100 0101                 0011 0100 0101
-$349    0011 0100 1001    2's comp   +1011 1011 0111
-$004                                 1  1111 1111 1100
                             C=1 (step 3, negative)
```

SWI **Software Interrupt**

Flags: I bit = 1

This is a software interrupt. It causes an interrupt without using an external hardware pin. Upon execution of the SWI, registers SP, CCR, A, B, X, and Y are pushed onto the stack and SP is decremented accordingly.

It also makes I bit = 1, therefore preventing another interrupt. The execution of SWI causes the CPU to load the PC (program counter) with fixed values at memory locations $FFF6 and $FFF7 and as a result the CPU starts to execute the ISR program associated with the SWI.

TAB Transfer from A to B

Flags: N, Z, V = 0

This instruction copies the byte in register A to register B. Register A remains unchanged. Look at the following examples.

```
LDAA #$99        ;A=$99
TAB              ;A=$99, B=$99
```

TAP Transfer from A to CCR

Flags: S, H, I, N, Z, V, C

This instruction copies the byte in register A to the CCR register. Register A remains unchanged. The X mask can be changed from 0 to 1 using this instruction. Look at the following examples:

```
LDAA #$99        ;A=$99
TAP              ;A=$99, CCR=10011001
```

TBA Transfer from B to A

Flags: N, Z, V = 0

This instruction copies the byte in register B to register A. Register B remains unchanged. Look at the following example:

```
LDAB #$99        ;B=$99
TBA              ;A=$99, B=$99
```

TBEQ Test and Branch if Equal to Zero

Flags: None

In this instruction a register is tested, and if the register's content is zero it will branch to the target address. The register can be A, B, D, X, Y, or SP. Notice that the target address can be no more than 128 bytes backward or 127 bytes forward.

TBL Table Lookup and Interpolate

See the HCS manual.

TBNE Test and Branch if Not Equal to Zero

Flags: None

In this instruction a register is tested, and if the register's content is not zero it will branch to the target address. The register can be A, B, D, X, Y, or SP. Notice that the target address can be no more than 128 bytes backward or 127 bytes forward.

TFR Transfer from source register to destination register

Flags: None unless the destination register is CCR

This copies a byte from the source register to the destination register.

TPA Transfer from CCR to A

Flags: None

This instruction copies the byte in the CCR register to register A. Register CCR remains unchanged.

TRAP Unimplemented opcode Trap

Flags: None except I

This instruction traps the unimplemented opcode, meaning that if we try to execute the nonexistent opcode it will cause an interrupt. The interrupt associated with this instruction is the address $FFF8 and $FFF9 in the interrupt vector table. See the HCS12 manual.

TST Test memory (M – 00)

Flags: N, Z, V = 0, C = 0

In this instruction, a memory location M is tested by subtracting 00 from the memory content and the flags are set accordingly. The memory remains unchanged.

TSTA Test A (A – 00)

Flags: N, Z, V = 0, C = 0

In this instruction, accumulator A is tested by subtracting 00 from A and the flags are set accordingly. The A remains unchanged.

TSTB Test B (B – 00)

Flags: N, Z, V = 0, C = 0

In this instruction, register B is tested by subtracting 00 from B and the flags are set accordingly. The B remains unchanged.

TSX Transfer from SP to X

Flags: None

This instruction copies the word in the SP register to register X. Register SP remains unchanged.

TSY Transfer from SP to Y

Flags: None

This instruction copies the word in the SP register to register Y. Register SP remains unchanged.

TXS Transfer from X to SP

Flags: None

This instruction copies the word in the X register to the SP register.

Register X remains unchanged.

TYS Transfer from Y to SP

Flags: None

This instruction copies the word in the Y register to the X register. Register Y remains unchanged.

WAI Wait for Interrupt (stop clock to CPU)

Flags: None

This instruction puts the CPU in wait state mode. Upon the execution of the WAI, the CPU context registers of PC, CCR, A, B, X, and Y are pushed onto the stack and SP is decremented accordingly. Then, the CPU enters a wait state for an integer number of bus clocks. Notice that during the wait state, all clocks to the CPU are stopped while the other parts of the MCU chip (UART, ATD, and so on) can continuc to run, if we choose the option for the peripherals. The CPU leaves the wait state upon activation of an interrupt that has not been masked. The interrupt will force the CPU to fetch the program counter value from the interrupt vector table and start to execute the interrupt service routine (ISR). At the end of the ISR, the RTI instruction pulls the original (before the execution of WAI instruction) CPU context from the stack and continues to execute the instructions after the WAI. Notice the fact that the WAI instruction is different from the STOP instruction. In the STOP instruction, clocks to the system (CPU and peripherals) are stopped and the entire chip is put in standby mode, therefore minimizing the CPU power dissipation to an absolute minimum. The WAI stops the clock to the CPU section of the chip, and not the peripherals.

WAV Weighted Average

See the HCS12 manual.

XGDX Exchange D and X registers

Flags: None

The XGDX instruction exchanges the D register with the X register.

XGDY Exchange D and Y registers

Flags: None

The XGDY instruction exchanges the D register with the Y register.

APPENDIX B

AsmIDE, ImageCraft C COMPILER, AND D-BUG12

OVERVIEW

This appendix shows how to use the AsmIDE assembler and ImageCraft C compiler with D-BUG12.

CodeWarrior is the industry standard IDE for the development of Freescale microcontroller products. It is a widely used GUI-based product. It is a complex and powerful tool for writing and debugging both Assembly and C programs. However, due to its complexity, some educators prefer to use a command-line assembler such as AsmIDE and D-BUG12. In this appendix, we will examine the AsmIDE assembler, D-BUG12, and the ImageCraft C Compiler for the HCS12 chip.

SECTION B-1: AsmIDE AND D-BUG12

In this section we examine AsmIDE and D-BUG12.

AsmIDE

Create the MyHCS12 directory and download the AsmIDE software from the following website:

http://www.ericengler.com/AsmIDE.aspx

Now extract AsmIDE into the MyHCS12 folder. Connect your HCS12 Trainer to the PC and configure the COM port for D-BUG12. Figure B-1 shows AsmIDE after configuration for the Dragon12 trainer. See www.MicroDigitalEd.com for details of the configuration for Dragon12 and trainers from other suppliers.

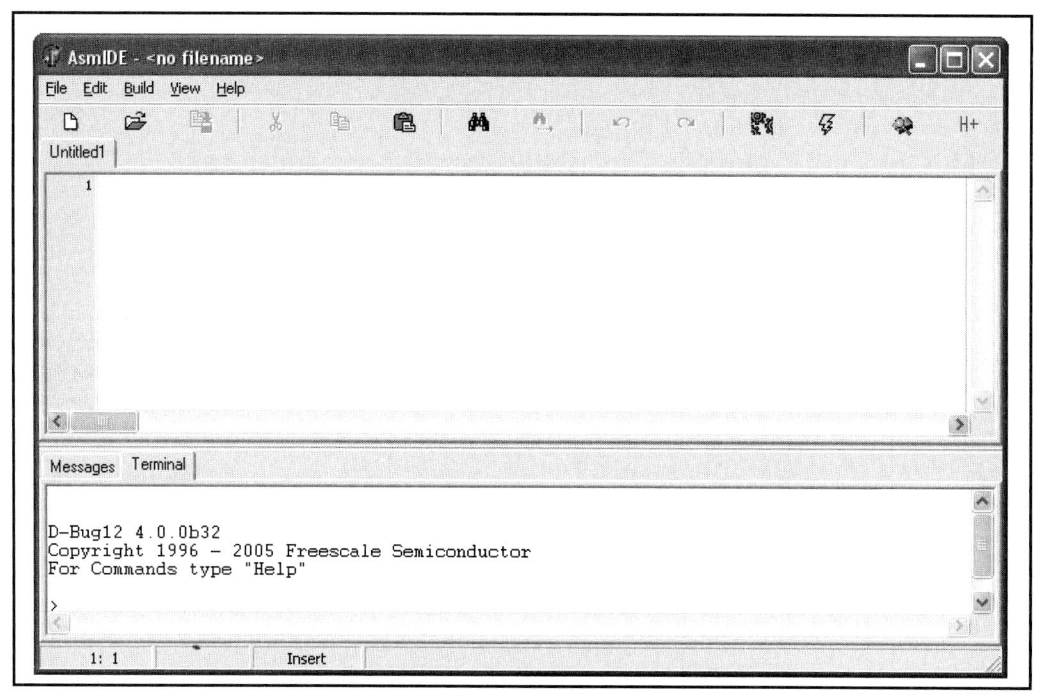

Figure B-1. AsmIDE Configured for Dragon12 Trainer

For your trainer, you need to know the memory map before you program the trainer. For example, on Dragon12 the address range for data RAM is \$1000–\$1FFF and for program code is \$2000–\$3FFF. You can modify the progam in Figure 2-9 of Chapter 2 for your trainer using AsmIDE. Figure B-2 shows the program assembled for the Dragon12 trainer. In Figure B-2 notice the use of "ORG \$2000" to put the program in the code space for the Dragon12 trainer. See www.MicroDigitalEd.com for more details.

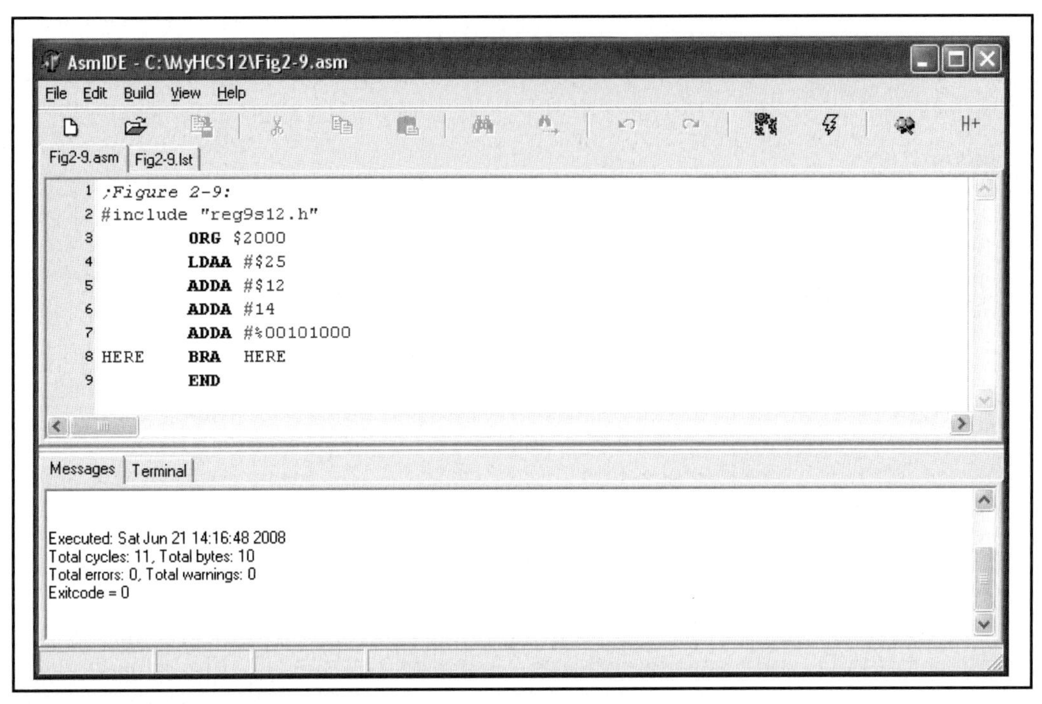

Figure B-2. Assembling and Building in AsmIDE

D-BUG12

After you have assembled and built the program, you can download it to your trainer using D-BUG12. Click on the Terminal tab in AsmIDE and press the reset button on your trainer. Figure B-3 shows the command line returned by D-BUG12 after you press the reset button. Notice the ">" prompt indicating that D-BUG12 is communicating with your PC. Download the s19 file into your trainer after typing the load command in the Terminal and pressing the Download icon. This is shown in Figure B-4.

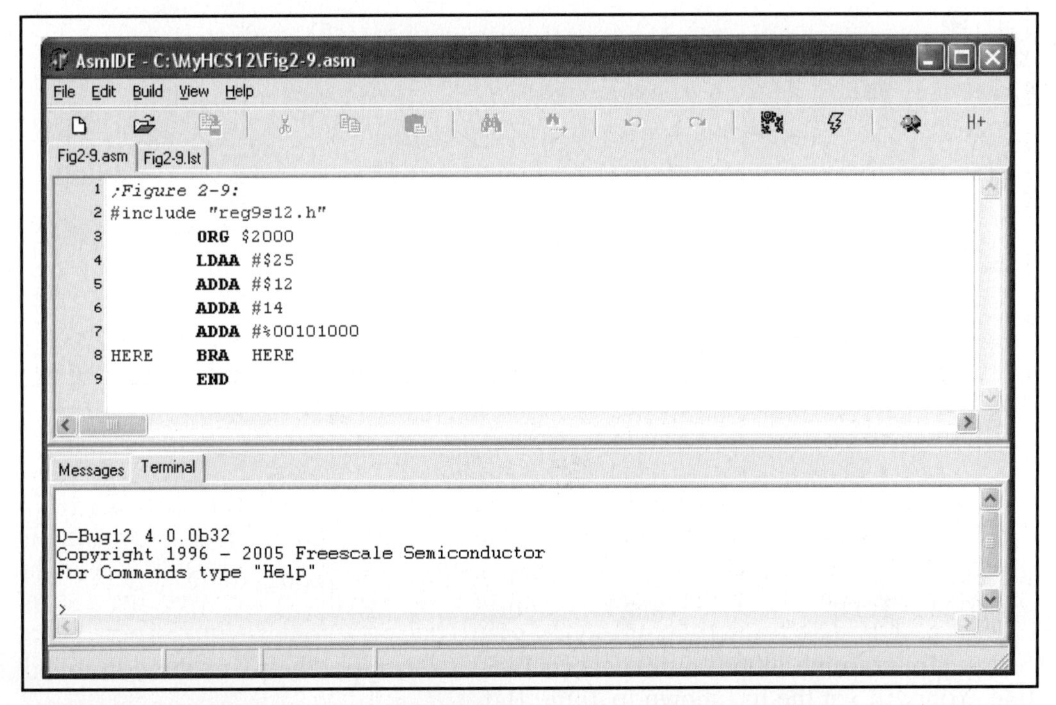

Figure B-3. D-BUG12 Command Line

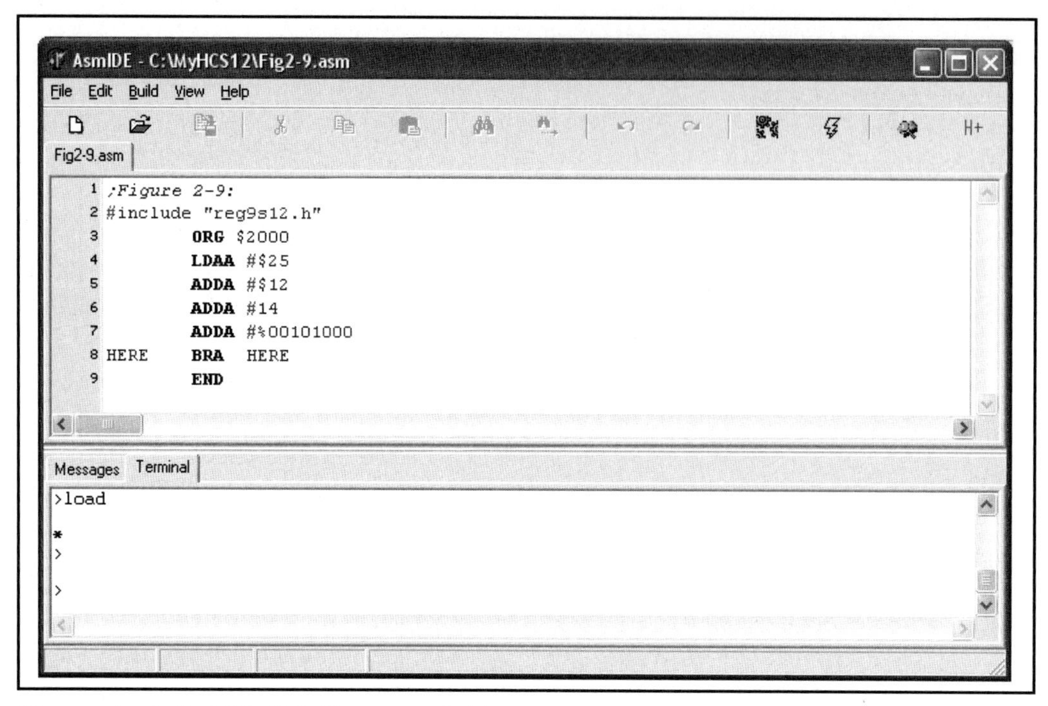

Figure B-4. D-BUG12 Command Line

To single-step the program, you must set the Program Counter (PC) to the ORG address, in this case $2000, and type "T" to trace the program as shown below:

```
>PC 2000

PP  PC     SP     X      Y        D = A:B    CCR = SXHI NZVC
38 2000   3C00   0000   0000        00:00          1001 0000
xx:2000   8625               LDAA   #$25

>T

PP  PC     SP     X      Y        D = A:B    CCR = SXHI NZVC
38 2002   3C00   0000   0000        25:00          1001 0000
xx:2002   8B12               ADDA   #$12

>T

PP  PC     SP     X      Y        D = A:B    CCR = SXHI NZVC
38 2004   3C00   0000   0000        37:00          1001 0000
xx:2004   8B0E               ADDA   #$0E

>T

PP  PC     SP     X      Y        D = A:B    CCR = SXHI NZVC
38 2006   3C00   0000   0000        45:00          1011 0000
xx:2006   8B28               ADDA   #$28

>T

PP  PC     SP     X      Y        D = A:B    CCR = SXHI NZVC
38 2008   3C00   0000   0000        6D:00          1001 0000
xx:2008   20FE               BRA    $2008
```

To examine other commands in D-BUG12, type "help" at the command line. You will get the list shown in Table B-1.

Table B-1: D-BUG12 Commands

<table>
<tr><th>Command</th><th>Description</th></tr>
<tr><td>ALTCLK [<BusClk>]</td><td>Specify an alternate BDM communications rate</td></tr>
<tr><td>ASM <Address></td><td>Single line assembler/disassembler</td></tr>
<tr><td><CR></td><td>Disassemble next instruction</td></tr>
<tr><td><.></td><td>Exit assembly/disassembly</td></tr>
<tr><td>BAUD <baudrate> [;t]</td><td>Set communications rate for the terminal</td></tr>
<tr><td>BDMDB</td><td>Enter the BDM command debugger</td></tr>
<tr><td>BDMPGMR</td><td>Load BDM Programmer Firmware Image Into MCU Flash</td></tr>
<tr><td>BF <StartAddress> <EndAddress> [<data>] [;nv]</td><td>Fill memory with data</td></tr>
<tr><td>BR [<Address>]</td><td>Set/Display breakpoints</td></tr>
<tr><td>BS <StartAddress> <EndAddress> '<String>' | <Data8> [<Data8>]</td><td>Block Search</td></tr>
<tr><td>BULK</td><td>Erase entire on-chip EEPROM contents</td></tr>
<tr><td>CALL [<Address>]</td><td>Call user subroutine at <Address></td></tr>
<tr><td>DEVICE [<DevName>]</td><td>display/select target device</td></tr>
<tr><td>EEBASE <Address></td><td>Set base address of on-chip EEPROM</td></tr>
<tr><td>FBULK [;np] [;<SecByteVal>]</td><td>Erase entire target FLASH contents</td></tr>
<tr><td>FLOAD [<AddressOffset> | ;b] [;np] [;nf]</td><td>Load S-Records into target Flash</td></tr>
<tr><td>FSERASE <StartAddress> [<EndAddress>]</td><td>Erase one or more sectors of Flash</td></tr>
<tr><td>G [<Address>]</td><td>Begin/continue execution of user code</td></tr>
<tr><td>GT <Address></td><td>Set temporary breakpoint at <Address> & execute user code</td></tr>
<tr><td>HELP</td><td>Display D-Bug12 command summary</td></tr>
<tr><td>LOAD [[<AddressOffset>] [;f]] | [;b]</td><td>Load S-Records into memory</td></tr>
<tr><td>MD <StartAddress> [<EndAddress>]</td><td>Memory Display Bytes</td></tr>
<tr><td>MDW <StartAddress> [<EndAddress>]</td><td>Memory Display Words</td></tr>
<tr><td>MM <StartAddress></td><td>Modify Memory Bytes</td></tr>
<tr><td><CR></td><td>Examine/Modify next location</td></tr>
<tr><td></> or <=></td><td>Examine/Modify same location</td></tr>
<tr><td><^> or <-></td><td>Examine/Modify previous location</td></tr>
<tr><td><.></td><td>Exit Modify Memory command</td></tr>
<tr><td>MMW <StartAddress></td><td>Modify Memory Words (same subcommands as MM)</td></tr>
<tr><td>MOVE <StartAddress> <EndAddress> <DestAddress></td><td>Move a block of memory</td></tr>
<tr><td>NOBR [<address>]</td><td>Remove One/All Breakpoint(s)</td></tr>
<tr><td>PCALL [<Address>]</td><td>Call user subroutine in expanded memory at <Address></td></tr>
<tr><td>RD</td><td>Display CPU registers</td></tr>
<tr><td>REGBASE <Address></td><td>Set base address of I/O registers</td></tr>
<tr><td>RESET</td><td>Reset target CPU</td></tr>
<tr><td>RM</td><td>Modify CPU Register Contents</td></tr>
<tr><td>SETVFP 12.0 | 12.6</td><td>Sets nominal Vfp Voltage to 12.0 or 12.6 volts</td></tr>
<tr><td>SO</td><td>Step Over subroutine calls</td></tr>
<tr><td>STOP</td><td>Stop target CPU</td></tr>
<tr><td>T [<count>]</td><td>Trace <count> instructions</td></tr>
<tr><td>TCONFIG [<Address>=<Data8>] | [DLY=<mSDelay>] | NONE</td><td>Configure Target Device</td></tr>
<tr><td>UPLOAD <StartAddress> <EndAddress> [;f] [;<SRecSize>]</td><td>S-Record Memory display</td></tr>
<tr><td>USEHBR [ON | OFF]</td><td>Use Hardware/Software Breakpoints</td></tr>
<tr><td>VER</td><td>Display D-Bug12's Version Number</td></tr>
<tr><td>VERF [[<AddressOffset>] [;f]] | [;b]</td><td>Verify S-Records against memory contents</td></tr>
<tr><td><Register Name> <Register Value></td><td>Set register contents
Register Names: PC, SP, X, Y, A, B, D, PP
CCR Status Bits: S, XM, H, IM, N, Z, V, C</td></tr>
</table>

We recommend that you use some basic D-BUG12 commands to familiarize yourself with their features. The "RD" command will dump the registers, as shown below:

```
>RD

PP   PC      SP     X      Y       D = A:B    CCR = SXHI NZVC
38 2008    3C00   0000   0000       6D:00          1001 0000
xx:2008    20FE                BRA    $2008
```

The "BF" command (byte fill) will fill the memory with byte-size data as shown below:

```
>BF 1100 111F 55
```

The "MD" command (memory dump) will dump the content of the memory locations in byte-size chunks, as shown below:

```
>MD 1100 111F

1100 55 55 55 55 - 55 55 55 55 - 55 55 55 55 - 55 55 55 55 U...
1110 55 55 55 55 - 55 55 55 55 - 55 55 55 55 - 55 55 55 55 U...
```

The "MDW" command (memory dump word) will dump the contents of memory locations in word-size chunks, as shown below:

```
>MDW 1100 1120

1100   5555 5555 - 5555 5555 - 5555 5555 - 5555 5555     UUUUU...
1110   5555 5555 - 5555 5555 - 5555 5555 - 5555 5555     UUUUU...
```

The breakpoint command runs a program to a specific address, allowing you to examine the results. The "BR" command (breakpoint) will display the current breakpoints:

```
>BR
Breakpoints:
```

Or you can insert a breakpoint by providing an address:

```
>BR 2044
Breakpoints: 2044
```

The following shows the results after executing the program from 0x2000 to 0x2044:

```
>G 2000
User Bkpt Encountered

PP   PC      SP     X      Y       D = A:B    CCR = SXHI NZVC
38 2044    3BF1   3E5E   0011       20:44          1001 0000
xx:2044    1B9E                LEAS   -2,SP
```

The "NOBR" command (no breakpoint) will remove all breakpoints or the breakpoint at the address provided.

```
>NOBR 2044
Breakpoints:

>NOBR
All Breakpoints Removed
```

SECTION B-2: ImageCraft C COMPILER

In this section we examine the ImageCraft C compiler for HCS12.

ImageCraft C for HCS12

Create the MyHCS12_C directory cand download the ImageCraft C compiler from the following website:

http://www.imagecraft.com/pub/iccv712_demo.exe

Run iccv712_demo.exe to install the compiler. Connect your HCS12 Trainer to the PC and configure the COM port for this compiler. Enter the C program in Example 7-7 from Chapter 7 and compile it. Figure B-5 shows the screen shot of the compiled program in the ImageCraft C compiler. Now, download the s19 file into the Dragon12 board and run it using D-BUG12. See Figure B-6. Notice that when running the C program in D-BUG12 you must use "g 1000". See www.MicroDigitalEd.com for details of the configuration for Dragon12 and trainers from other suppliers.

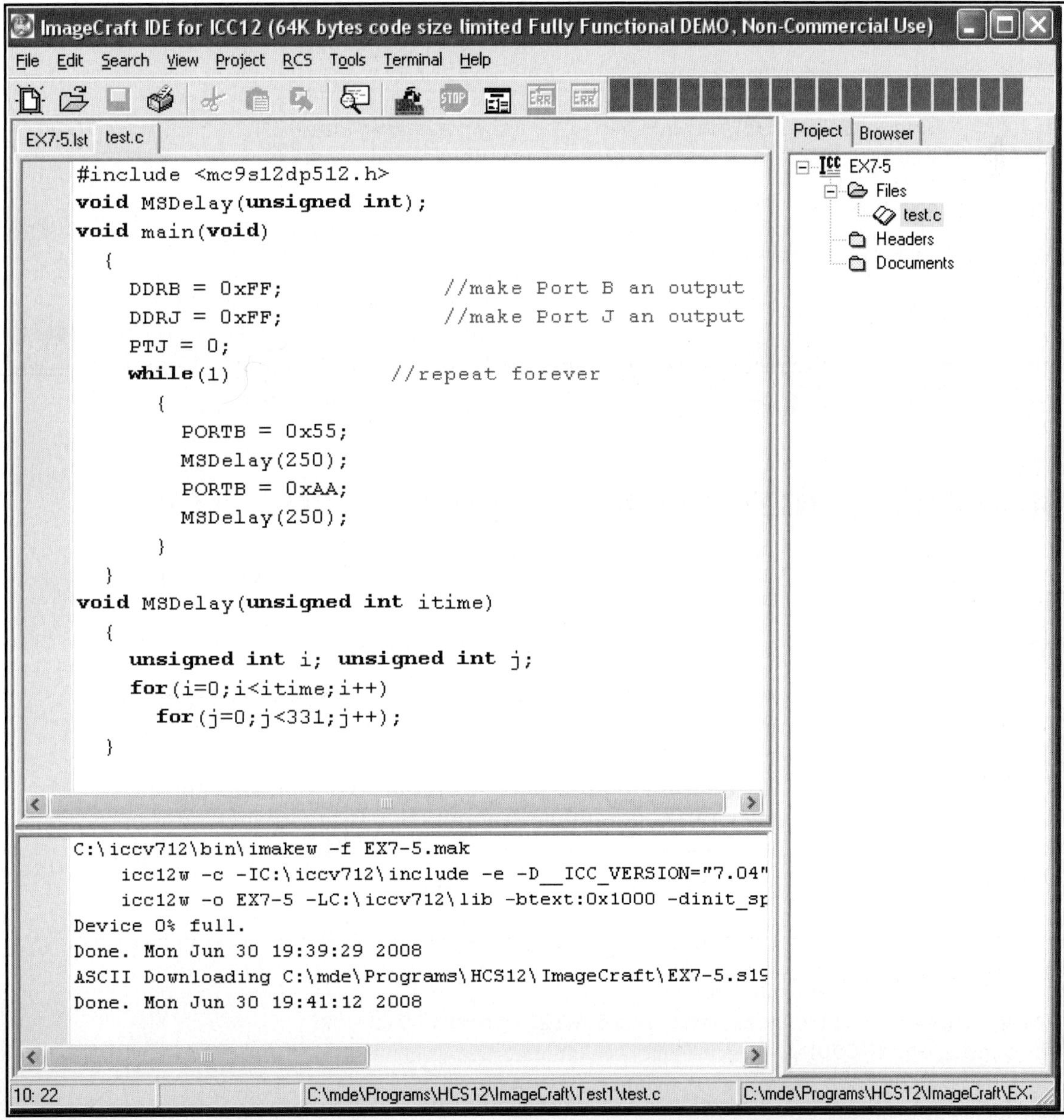

Figure B-5. Compiling a C Program in ImageCraft for Dragon12 Board

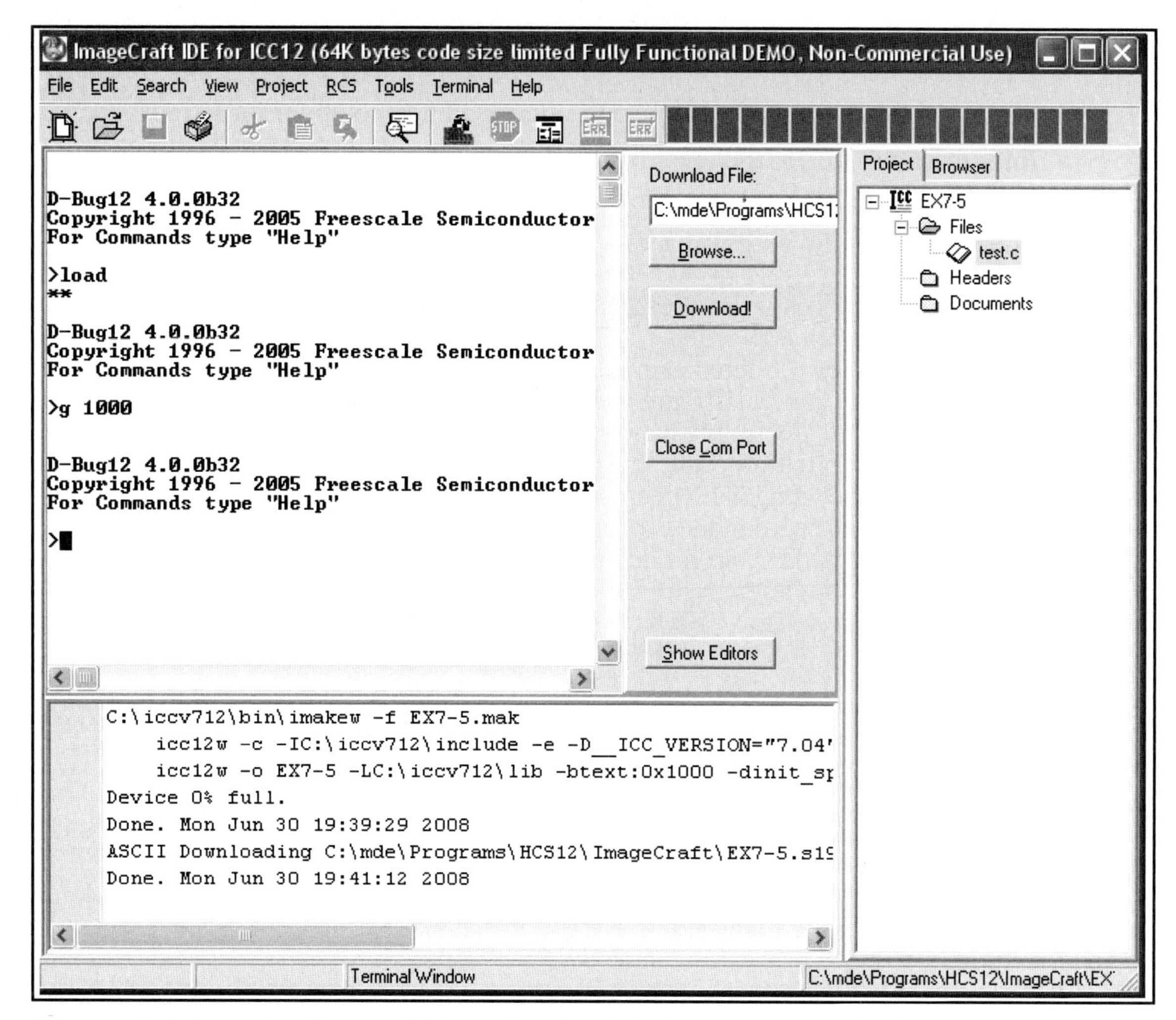

Figure B-6. Downloading and Running the C Code on Dragon12 Board

SECTION B-3: INTERRUPTS IN ImageCraft C COMPILER

In this section we examine interrupts in the ImageCraft C compiler for HCS12.

Interrupt program for D-BUG12 using ImageCraft C compiler

In Chapter 11 we showed how to compile an interrupt program in CodeWarrior. Programming the interrupt for D-BUG12 in the ImageCraft C compiler is tedious and involves several steps. The website www.MicroDigitalEd.com shows a tutorial on how to go through these steps. Program B-1 is a working shell for an interrupt in the ImageCraft C compiler. Program B-1 is the ImageCraft version of Program 11-1C, in Chapter 11, modified for D-BUG12, running on the Dragon12 trainer board.

```
#include <mc9s12dp256.h>

#pragma abs_address:0x1000
// Define variable here for Dragon12 board
unsigned char COUNT;

#pragma abs_address:0x2000
// Define your program here for Dragon12 board
```

```
void main(void)
{
  /* put your own code here */
  DDRB = 0xFF; //make PortB an output
  DDRA = 0xFF; //make PortA an output
  DDRH = 0x00; //make PTH an input
  DDRJ = 0xFF;
  PTJ = 0;
  COUNT = 0;    //initialize the count
  TSCR1 = 0x80;//enable the timer
  TSCR2 = 0x82;//enable interrupt, prescaler=64
  TFLG2 |= TFLG2 | 0x80; //clear flag
  asm("CLI");  //enable interrupts globally
  for(;;)
  {
     PORTA = PTH;
 } /* wait forever */
}

// Define the ISR for timer overflow
#pragma interrupt_handler TMR_ISR()
// The keyword interrupt_handler tells the compiler
// to use RTI at the end of the ISR
void TMR_ISR(void)
{
  COUNT++; //increment COUNT
  PORTB = COUNT; //display on PortB
  TFLG2 |= TFLG2 | 0x80; //clear flag
}
// Define the location for the timer overflow vector in D-BUG12
#pragma abs_address:0x3E5E

// This will redirect TMR_ISR to the vector location 0x3E5E
void (*TOV_VEC[ ])(void) = {TMR_ISR};
#pragma end_abs_address
```

Notice the following points about Program B-1:

1) The keyword "interrupt_handler" is used for defining the body of the ISR.
2) The vector address 0x3E5E is assigned to TMR_ISR 5 using a pointer. Although TMR_ISR must match the definition given in the interrupt_handler keyword, the name of the pointer can be anything.
3) For other interrupts, use the addresses given in Table B-2.

Table B-2: Complete Listing of Interrupt Vector Table for D-BUG12

Interrupt Source	Vector Address	Interrupt Source	Vector Address
Reserved $FF80	$3E00	IIC Bus	$3E40
Reserved $FF82	$3E02	DLC	$3E42
Reserved $FF84	$3E04	SCME	$3E44
Reserved $FF86	$3E06	CRG Lock	$3E46
Reserved $FF88	$3E08	Pulse Accumulator B Overflow	$3E48
Reserved $FF8A	$3E0A	Mod Down Counter Underflow	$3E4A
PWM Emergency Shutdown	$3E0C	Port H Interrupt	$3E4C
Port P Interrupt	$3E0E	Port J Interrupt	$3E4E
MSCAN 4 Transmit	$3E10	ATD1	$3E50
MSCAN 4 Receive	$3E12	ATD0	$3E52
MSCAN 4 Errors	$3E14	SCI1	$3E54
MSCAN 4 Wake-up	$3E16	SCI0	$3E56
MSCAN 3 Transmit	$3E18	SPI0	$3E58
MSCAN 3 Receive	$3E1A	Pulse Accumulator A Input Edge	$3E5A
MSCAN 3 Errors	$3E1C	Pulse Accumulator A Overflow	$3E5C
MSCAN 3 Wake-up	$3E1E	Timer Overflow	$3E5E
MSCAN 2 Transmit	$3E20	Timer Channel 7	$3E60
MSCAN 2 Receive	$3E22	Timer Channel 6	$3E62
MSCAN 2 Errors	$3E24	Timer Channel 5	$3E64
MSCAN 2 Wake-up	$3E26	Timer Channel 4	$3E66
MSCAN 1 Transmit	$3E28	Timer Channel 3	$3E68
MSCAN 1 Receive	$3E2A	Timer Channel 2	$3E6A
MSCAN 1 Errors	$3E2C	Timer Channel 1	$3E6C
MSCAN 1 Wake-up	$3E2E	Timer Channel 0	$3E6E
MSCAN 0 Transmit	$3E30	Real Time Interrupt	$3E70
MSCAN 0 Receive	$3E32	IRQ	$3E72
MSCAN 0 Errors	$3E34	XIRQ	$3E74
MSCAN 0 Wake-up	$3E36	SWI	$3E76
Flash	$3E38	Unimplemented Instruction Trap	$3E78
EEPROM	$3E3A	N/A	$3E7A
SPI2	$3E3C	N/A	$3E7C
SPI1	$3E3E	N/A	$3E7E

APPENDIX C

IC INTERFACING, SYSTEM DESIGN ISSUES, AND WIRE WRAPPING

OVERVIEW

This appendix provides an overview of IC interfacing and HCS12 interfacing. In addition, we look at the microcontroller-based system as a whole and examine some general issues in system design.

First, in Section C.1, we provide an overview of IC interfacing. Then, in Section C.2, the fan-out of HCS12 I/O ports and interfacing are discussed. Section C.3 examines system design issues. Section C.4 shows wire wrapping.

SECTION C.1: OVERVIEW OF IC TECHNOLOGY

In this section we examine IC technology and discuss some major developments in advanced logic families. Because this is an overview, it is assumed that the reader is familiar with logic families on the level presented in basic digital electronics books.

Transistors

The transistor was invented in 1947 by three scientists at Bell Laboratory. In the 1950s, transistors replaced vacuum tubes in many electronics systems, including computers. It was not until 1959 that the first integrated circuit was successfully fabricated and tested by Jack Kilby of Texas Instruments. Prior to the invention of the IC, the use of transistors, along with other discrete components such as capacitors and resistors, was common in computer design. Early transistors were made of germanium, which was later abandoned in favor of silicon. This was because the slightest rise in temperature resulted in massive current flows in germanium-based transistors. In semiconductor terms, it is because the band gap of germanium is much smaller than that of silicon, resulting in a massive flow of electrons from the valence band to the conduction band when the temperature rises even slightly. By the late 1960s and early 1970s, the use of the silicon-based IC was widespread in mainframes and minicomputers. Transistors and ICs at first were based on P-type materials. Later on, because the speed of electrons is much faster (about two-and-a-half times) than the speed of holes, N-type devices replaced P-type devices. By the mid-1970s, NPN and NMOS transistors had replaced the slower PNP and PMOS transistors in every sector of the electronics industry, including in the design of microprocessors and computers. Since the early 1980s, CMOS (complementary MOS) has become the dominant technology of IC design. Next we provide an overview of differences between MOS and bipolar transistors. See Figure C-1.

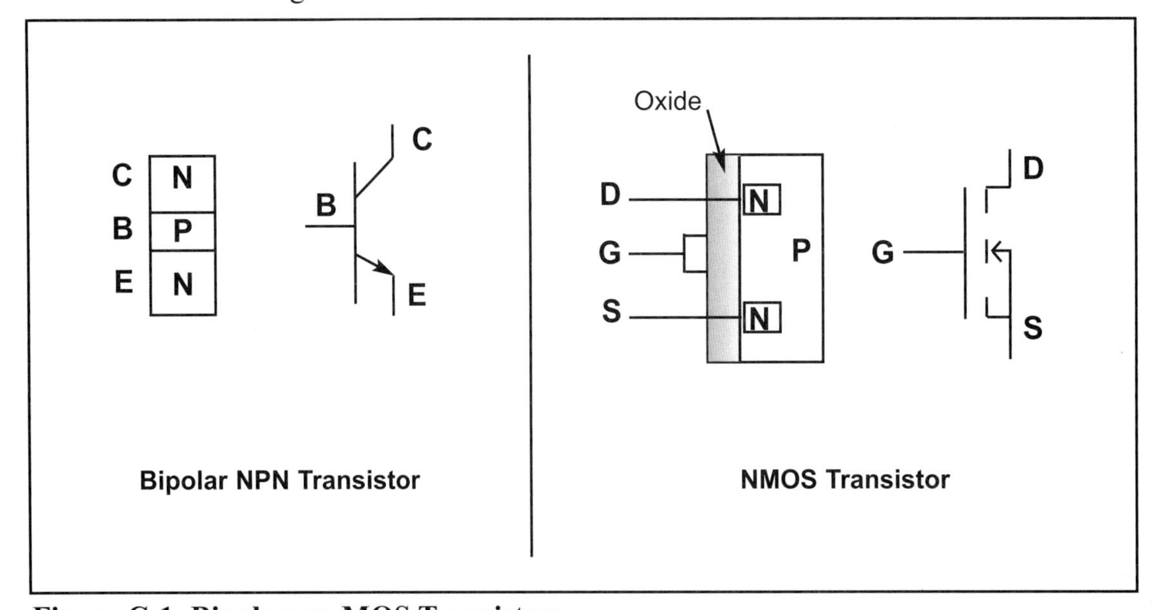

Figure C-1. Bipolar vs. MOS Transistors

MOS vs. bipolar transistors

There are two types of transistors: bipolar and MOS (metal-oxide semiconductor). Both have three leads. In bipolar transistors, the three leads are referred to as the *emitter*, *base*, and *collector*, while in MOS transistors they are named *source*, *gate*, and *drain*. In bipolar transistors, the carrier flows from the emitter to the collector, and the base is used as a flow controller. In MOS transistors, the carrier flows from the source to the drain, and the gate is used as a flow controller. In NPN-type bipolar transistors, the electron carrier leaving the emitter must overcome two voltage barriers before it reaches the collector (see Figure C-1). One is the N-P junction of the emitter-base and the other is the P-N junction of the base-collector. The voltage barrier of the base-collector is the most difficult one for the electrons to overcome (because it is reverse-biased) and it causes the most power dissipation. This led to the design of the unipolar type transistor called MOS. In N-channel MOS transistors, the electrons leave the source and reach the drain without going through any voltage barrier. The absence of any voltage barrier in the path of the carrier is one reason why MOS dissipates much less power than bipolar transistors. The low power dissipation of MOS allows millions of transistors to fit on a single IC chip. In today's technology, putting 10 million transistors into an IC is common, and it is all because of MOS technology. Without the MOS transistor, the advent of desktop personal computers would not have been possible, at least not so soon. The bipolar transistors in both the mainframes and minicomputers of the 1960s and 1970s were bulky and required expensive cooling systems and large rooms. MOS transistors do have one major drawback: They are slower than bipolar transistors. This is due partly to the gate capacitance of the MOS transistor. For a MOS to be turned on, the input capacitor of the gate takes time to charge up to the turn-on (threshold) voltage, leading to a longer propagation delay.

Overview of logic families

Logic families are judged according to (1) speed, (2) power dissipation, (3) noise immunity, (4) input/output interface compatibility, and (5) cost. Desirable qualities are high speed, low power dissipation, and high noise immunity (because it prevents the occurrence of false logic signals during switching transition). In interfacing logic families, the more inputs that can be driven by a single output, the better. This means that high-driving-capability outputs are desired. This, plus the fact that the input and output voltage levels of MOS and bipolar transistors are not compatible mean that one must be concerned with the ability of one logic family to drive the other one. In regard to the cost of a given logic family, it is high during the early years of its introduction but it declines as production and use rise.

The case of inverters

As an example of logic gates, we look at a simple inverter. In a one-transistor inverter, the transistor plays the role of a switch, and R_c is the pull-up resistor. See Figure C-2. For this inverter to work most effectively in digital circuits, however, the R value must be high when the transistor is "on" to limit the current flow from V_{CC} to ground in order to have low power dissipation (P = VI, where V

= 5 V). In other words, the lower I, the lower the power dissipation. On the other hand, when the transistor is "off", R_c must be a small value to limit the voltage drop across R_c, thereby making sure that V_{OUT} is close to V_{CC}. This is a contradictory demand on R_c. This is one reason that logic gate designers use active components (transistors) instead of passive components (resistors) to implement the pull-up resistor R_c.

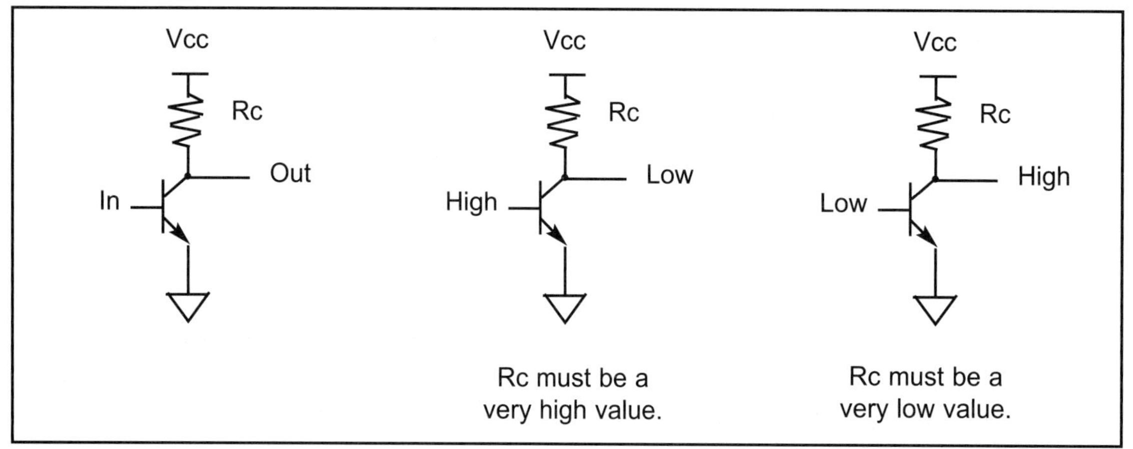

Figure C-2. One-Transistor Inverter with Pull-up Resistor

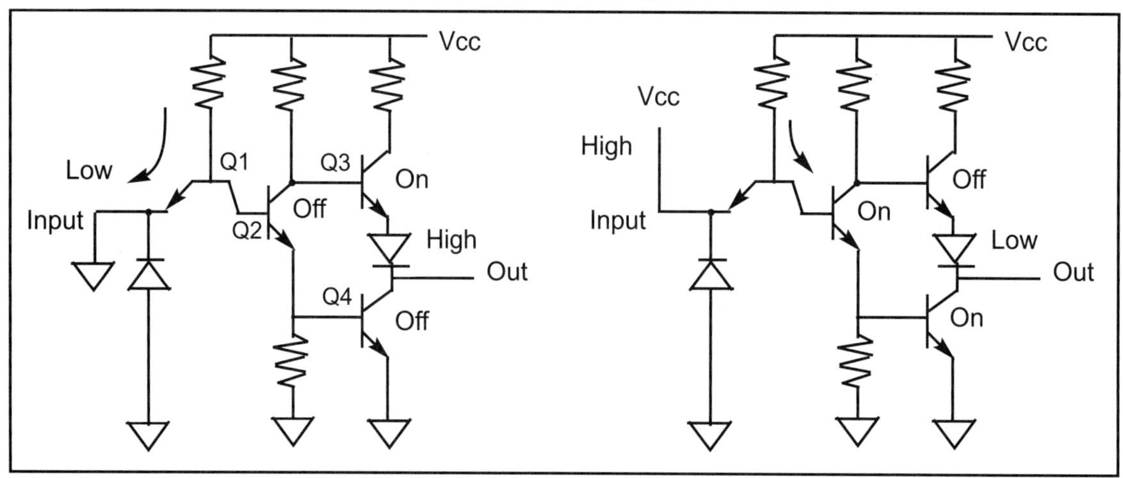

Figure C-3. TTL Inverter with Totem-Pole Output

The case of a TTL inverter with totem-pole output is shown in Figure C-3. In Figure C-3, Q3 plays the role of a pull-up resistor.

CMOS inverter

In the case of CMOS-based logic gates, PMOS and NMOS are used to construct a CMOS (complementary MOS) inverter as shown in Figure C-4. In CMOS inverters, when the PMOS transistor is off, it provides a very high impedance path, making leakage current almost zero (about 10 nA); when the PMOS is on, it provides a low resistance on the path of V_{DD} to output load. Because the speed of the hole is slower than that of the electron, the PMOS transistor is wider to compensate for this disparity; therefore, PMOS transistors take more space than NMOS transistors in the CMOS gates. At the end of this section we will see an open-collector gate in which the pull-up resistor is provided externally, thereby allowing system designers to choose the value of the pull-up resistor.

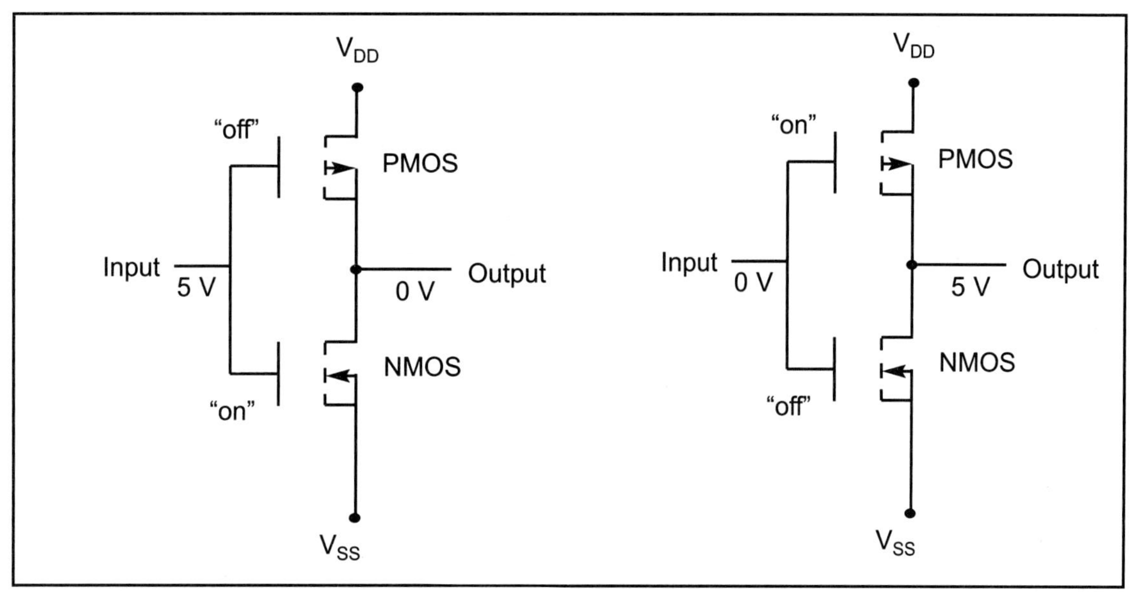

Figure C-4. CMOS Inverter

Input/output characteristics of some logic families

In 1968 the first logic family made of bipolar transistors was marketed. It was commonly referred to as the standard TTL (transistor-transistor logic) family. The first MOS-based logic family, the CD4000/74C series, was marketed in 1970. The addition of the Schottky diode to the base-collector of bipolar transistors in the early 1970s gave rise to the S family. The Schottky diode shortens the propagation delay of the TTL family by preventing the collector from going into what is called *deep saturation*. Table C-1 lists major characteristics of some logic families. In Table C-1, note that as the CMOS circuit's operating frequency rises, the power dissipation also increases. This is not the case for bipolar-based TTL.

Table C-1: Characteristics of Some Logic Families

Characteristic	STD TTL	LSTTL	ALSTTL	HCMOS
V_{CC}	5 V	5 V	5 V	5 V
V_{IH}	2.0 V	2.0 V	2.0 V	3.15 V
V_{IL}	0.8 V	0.8 V	0.8 V	1.1 V
V_{OH}	2.4 V	2.7 V	2.7 V	3.7 V
V_{OL}	0.4 V	0.5 V	0.4 V	0.4 V
I_{IL}	−1.6 mA	−0.36 mA	−0.2 mA	−1 μA
I_{IH}	40 μA	20 μA	20 μA	1 μA
I_{OL}	16 mA	8 mA	4 mA	4 mA
I_{OH}	−400 μA	−400 μA	−400 μA	4 mA
Propagation delay	10 ns	9.5 ns	4 ns	9 ns
Static power dissipation (f = 0)	10 mW	2 mW	1 mW	0.0025 nW
Dynamic power dissipation at f = 100 kHz	10 mW	2 mW	1 mW	0.17 mW

History of logic families

Early logic families and microprocessors required both positive and negative power voltages. In the mid-1970s, 5 V V_{CC} became standard. In the late 1970s, advances in IC technology allowed combining the speed and drive of the S family with the lower power of LS to form a new logic family called FAST (Fairchild Advanced Schottky TTL). In 1985, AC/ACT (Advanced CMOS Technology), a much higher speed version of HCMOS, was introduced. With the introduction of FCT (Fast CMOS Technology) in 1986, the speed gap between CMOS and TTL at last was closed. Because FCT is the CMOS version of FAST, it has the low power consumption of CMOS but the speed is comparable with TTL. Table C-2 provides an overview of logic families up to FCT.

Table C-2: Logic Family Overview

Product	Year Introduced	Speed (ns)	Static Supply Current (mA)	High/Low Family Drive (mA)
Std TTL	1968	40	30	-2/32
CD4K/74C	1970	70	0.3	-0.48/6.4
LS/S	1971	18	54	-15/24
HC/HCT	1977	25	0.08	-6/-6
FAST	1978	6.5	90	-15/64
AS	1980	6.2	90	-15/64
ALS	1980	10	27	-15/64
AC/ACT	1985	10	0.08	-24/24
FCT	1986	6.5	1.5	-15/64

Reprinted by permission of Electronic Design Magazine, c. 1991.

Recent advances in logic families

As the speed of high-performance microprocessors reached 25 MHz, it shortened the CPU's cycle time, leaving less time for the path delay. Designers normally allocate no more than 25% of a CPU's cycle time budget to path delay. Following this rule means that there must be a corresponding decline in the propagation delay of logic families used in the address and data path as the system frequency is increased. In recent years, many semiconductor manufacturers have responded to this need by providing logic families that have high speed, low noise, and high drive I/O. Table C-3 provides the characteristics of high-performance logic families introduced in recent years. ACQ/ACTQ are the second-generation advanced CMOS (ACMOS) with much lower noise. While ACQ has the CMOS input level, ACTQ is equipped with TTL-level input. The FCTx and FCTx-T are second-generation FCT with much higher speed. The "x" in the FCTx and FCTx-T refers to various speed grades, such as A, B, and C, where A means low speed and C means high speed. For designers who are well versed in using the FAST logic family, FASTr is an ideal choice because it is faster than FAST, has higher driving capability (I_{OL}, I_{OH}), and produces much lower noise than FAST. At the time of this writing, next to ECL and gallium arsenide logic gates, FASTr is the fastest logic family in the market (with the 5 V V_{CC}), but the power consumption is high relative to other logic families, as shown in Table C-3. The combining of high-speed bipolar TTL and the low power consumption of CMOS has given birth to what is called BICMOS. Although BICMOS seems to be the future trend in IC design, at this time it is expensive due to extra steps required in BICMOS IC fabrication, but in some cases there is no other choice. (For example, Intel's Pentium microprocessor, a BICMOS product, had to use high-speed bipolar transistors to speed up some of the internal functions.) Table C-3 provides advanced logic characteristics. The "x" is for different speeds designated as A, B, and C. A is the slowest one while C is the fastest one. The data is for the 74244 buffer.

Table C-3: Advanced Logic General Characteristics

Family	Year	Number Suppliers	Tech Base	I/O Level	Speed (ns)	Static Current	I_{OH}/I_{OL}
ACQ	1989	2	CMOS	CMOS/CMOS	6.0	80 μA	–24/24 mA
ACTQ	1989	2	CMOS	TTL/CMOS	7.5	80 μA	–24/24 mA
FCTx	1987	3	CMOS	TTL/CMOS	4.1–4.8	1.5 mA	–15/64 mA
FCTxT	1990	2	CMOS	TTL/TTL	4.1–4.8	1.5 mA	–15/64 mA
FASTr	1990	1	Bipolar	TTL/TTL	3.9	50 mA	–15/64 mA
BCT	1987	2	BICMOS	TTL/TTL	5.5	10 mA	–15/64 mA

Reprinted by permission of Electronic Design Magazine, c. 1991.

Since the late 70s, the use of a +5 V power supply has become standard in all microprocessors and microcontrollers. To reduce power consumption, 3.3 V V_{CC} is being embraced by many designers. The lowering of V_{CC} to 3.3 V has two major advantages: (1) it lowers the power consumption, prolonging the life of the battery in systems using a battery, and (2) it allows a further reduction of line size (design rule) to submicron dimensions. This reduction results in putting more transistors in a given die size. As fabrication processes improve, the decline in the line size is reaching submicron level and transistor densities are approaching 1 billion transistors.

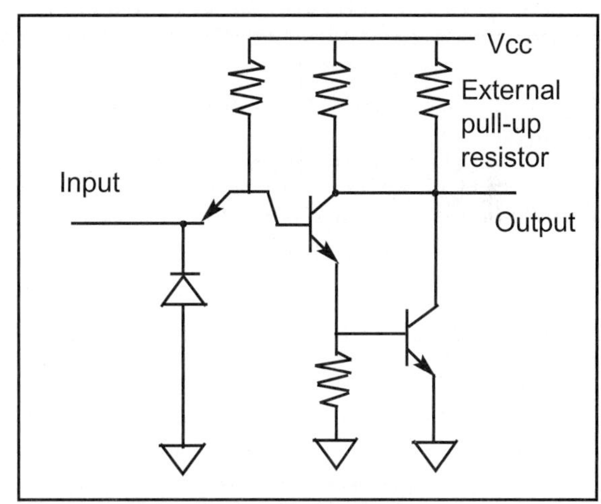

Figure C-5. Open Collector

Open-collector and open-drain gates

To allow multiple outputs to be connected together, we use open-collector logic gates. In such cases, an external resistor will serve as load. This is shown in Figures C-5 and C-6.

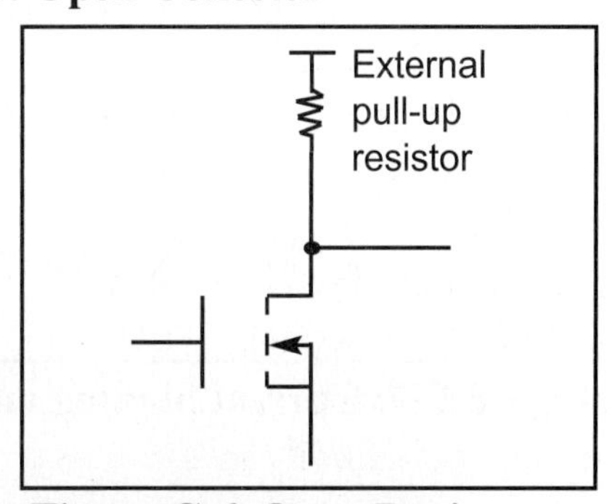

Figure C-6. Open Drain

SECTION C.2: HCS12 PORT FAN-OUT AND INTERFACING

In interfacing the HCS12 microcontroller with other IC chips or devices, fan-out is the most important issue. To understand the HCS12 fan-out we must first understand the port structure of the HCS12. This section provides a discussion of the HCS12 fan-out.

IC fan-out

When connecting IC chips together, we need to find out how many input pins can be driven by a single output pin. This is a very important issue and involves the discussion of what is called IC fan-out. The IC fan-out must be addressed for both logic "0" and logic "1" outputs. See Example C-1. Fan-out for logic LOW and fan-out for logic HIGH are defined as follows:

$$\text{fan-out (of LOW)} = \frac{I_{OL}}{I_{IL}} \qquad \text{fan-out (of HIGH)} = \frac{I_{OH}}{I_{IH}}$$

Of the above two values, the lower number is used to ensure the proper noise margin. Figure C-7 shows the sinking and sourcing of current when ICs are connected together.

Notice that in Figure C-7, as the number of input pins connected to a single output increases, I_{OL} rises, which causes V_{OL} to rise. If this continues, the rise of V_{OL} makes the noise margin smaller, and this results in the occurrence of false logic due to the slightest noise.

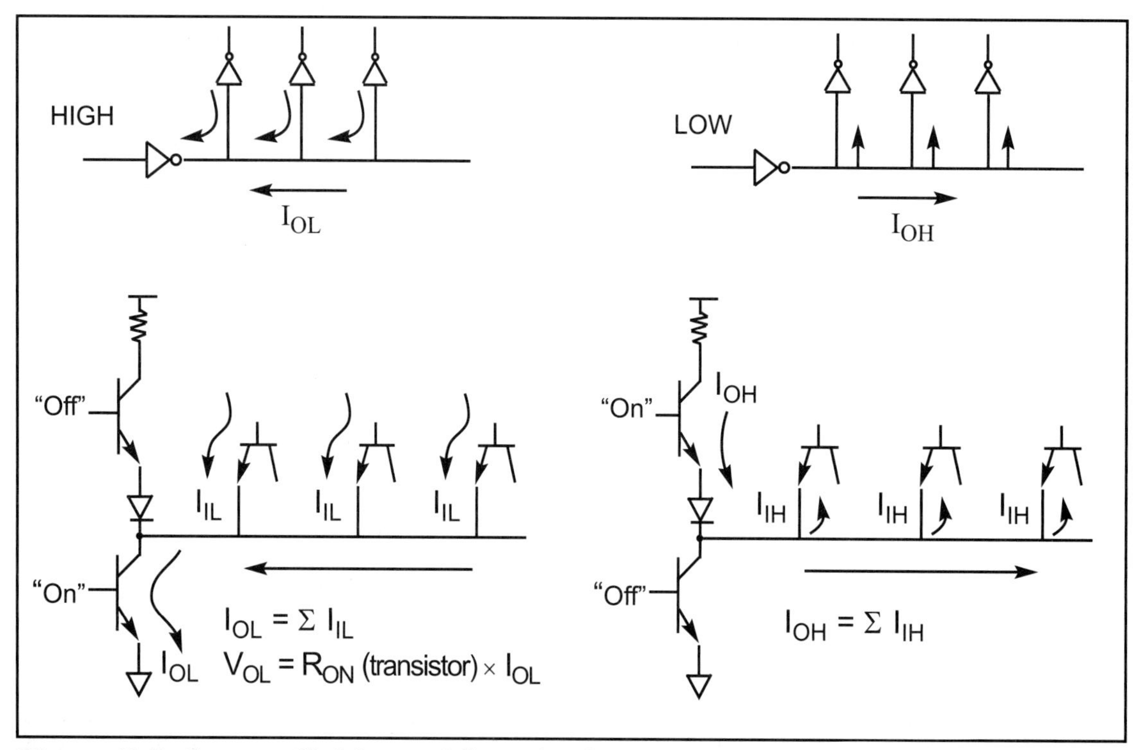

Figure C-7. Current Sinking and Sourcing in TTL

Example C-1

Find how many unit loads (UL) can be driven by the output of the LS logic family.

Solution:

The unit load is defined as I_{IL} = 1.6 mA and I_{IH} = 40 μA. Table C-1 shows I_{OH} = 400 μA and I_{OL} = 8 mA for the LS family. Therefore, we have

$$\text{fan-out (LOW)} = \frac{I_{OL}}{I_{IL}} = \frac{8\text{ mA}}{1.6\text{ mA}} = 5$$

$$\text{fan-out (HIGH)} = \frac{I_{OH}}{I_{IH}} = \frac{400\ \mu\text{A}}{40\ \mu\text{A}} = 10$$

This means that the fan-out is 5. In other words, the LS output must not be connected to more than 5 inputs with unit load characteristics.

74LS244 and 74LS245 buffers/drivers

In cases where the receiver current requirements exceed the driver's capability, we must use buffers/drivers such as the 74LS245 and 74LS244. Figure C-8 shows the internal gates for the 74LS244 and 74LS245. The 74LS245 is used for bidirectional data buses, and the 74LS244 is used for unidirectional address buses.

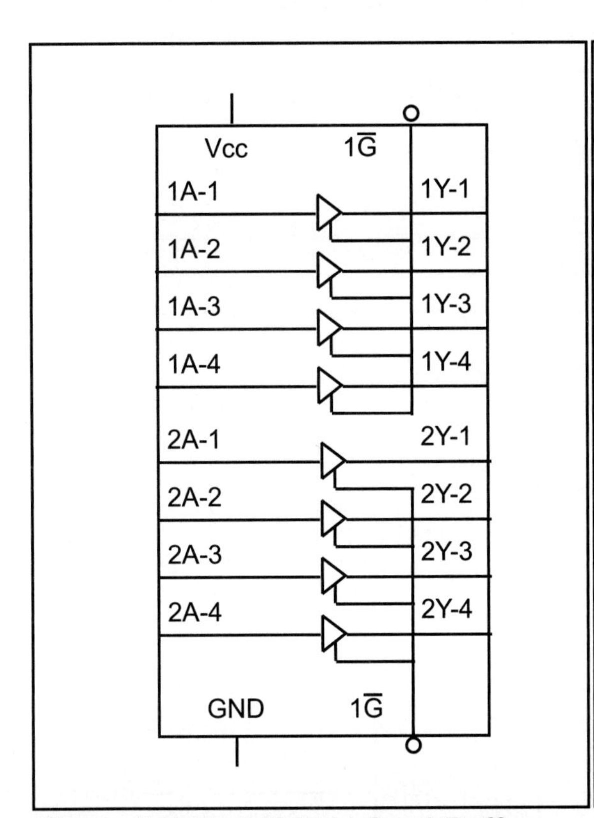

Figure C-8 (a). 74LS244 Octal Buffer

(Reprinted by permission of Texas Instruments, Copyright Texas Instruments, 1988)

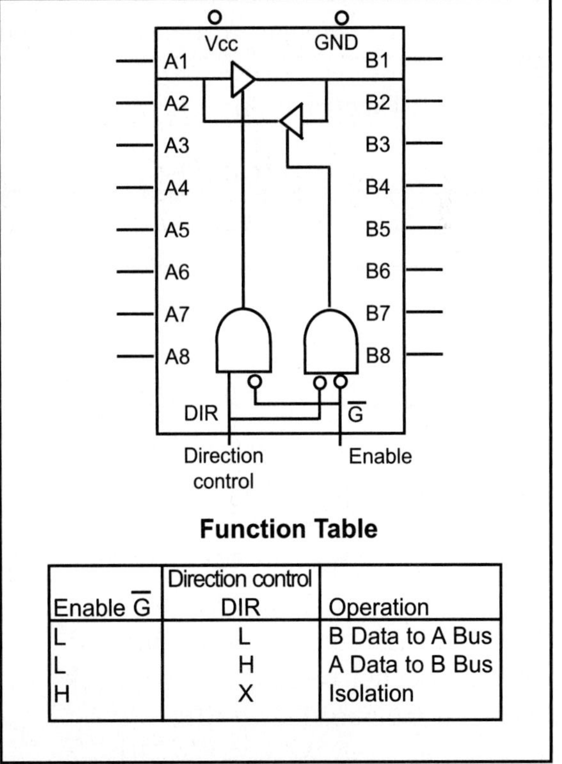

Enable $\overline{G}$	Direction control DIR	Operation
L	L	B Data to A Bus
L	H	A Data to B Bus
H	X	Isolation

Figure C-8 (b). 74LS245 Bidirectional Buffer

(Reprinted by permission of Texas Instruments, Copyright Texas Instruments, 1988)

Tri-state buffer

Notice that the 74LS244 is simply 8 tri-state buffers on a single chip. As shown in Figure C-9 a tri-state buffer has a single input, a single output, and the enable control input. By activating the enable, data at the input is transferred to the output. The enable can be an active-LOW or an active-HIGH. Notice that the enable input for the 74LS244 is active-LOW whereas the enable input pin for Figure C-9 is active-HIGH.

Figure C-9. Tri-State Buffer

74LS245 and 74LS244 fan-out

It must be noted that the output of the 74LS245 and 74LS244 can sink and source a much larger amount of current than that of other LS gates. See Table C-4. That is the reason we use these buffers for drivers when a signal is travelling a long distance through a cable or it has to drive many inputs.

Table C-4: Electrical Specifications for Buffers/Drivers

	I_{OH} (mA)	I_{OL} (mA)
74LS244	3	12
74LS245	3	12

74LS244 driving an output pin

In some cases, when a microcontroller port is driving multiple inputs, or driving a single input via a long wire or cable (e.g., printer cable), we can use the 74LS244 as a driver. When driving an off-board circuit, placing the 74LS244 buffer between your chip and the circuit is essential because the chip lacks sufficient current. See Figure C-10.

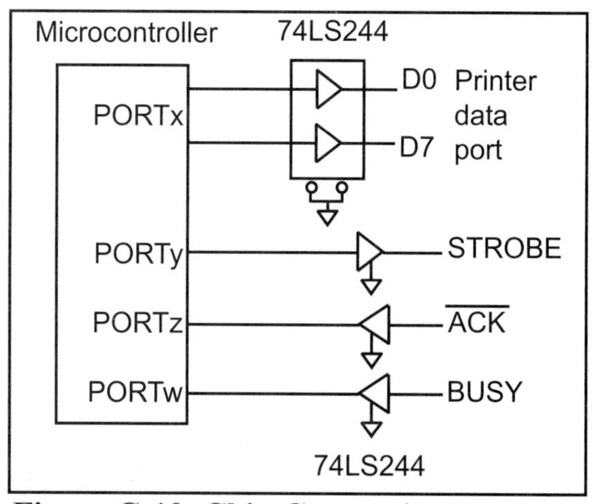

Figure C-10. Chip Connection to Printer Signals

HCS12 port fan-out and reduced-drive registers

Now that we are familiar with the fan-out of the chip, we need to examine the fan-out for the HCS12 microcontroller. While the early chips were based on NMOS IC technology, today's HCS12 microcontrollers are all based on CMOS technology. Note, however, that while the core of the HCS12 microcontroller is CMOS, the circuitry driving its pins is all TTL compatible. That is, the HCS12 is a CMOS-based product with TTL-compatible pins. All the ports of the HCS12 have the same I/O structure, and therefore the same fan-out. Table C-5 provides the I/O characteristics of HCS12 ports. Upon reset, all the ports are in full-drive mode. We can use the the RDRV (Reduced Drive) register to switch to reduced-drive mode for ports A, B, E, and K. In the case of these ports, we can only switch the entire port to reduced-drive mode and there is no option of selecting individual pins. That means all the pins are in either full-drive mode or reduced-drive mode. For all other ports, each one has its own RDRx (RDRH, RDRJ, RDRT, and so on) register. We can program the RDRx (reduced drive) register to switch any of the pins to reduced-drive mode we want since each RDRx register is an 8-bit register. See the HCS12 manual for the RDRx registers.

Table C-5: HCS12 Fan-out for Ports

Pin	Fan-out (full-drive, default on reset)	Fan-out (reduced-drive)
IOL	10 mA	2 mA
IOH	–10 mA	–2 mA
IIL	1 μA	1 μA
IIH	1 μA	1 μA

Note: Negative current is defined as current sourced by the pin.
Note: The HCS12 manual gives the details of Reduced Drive registers (RDRx) for each port. There is a single register for setting the fan-out for Ports A, B, E, and K. It is called RDRV. All other ports have their own RDRx registers.

WOMS register and wired-ORed option for PORTS (PTS)

In PORTS (PTS) we have the option of switching the pins to Wire-ORed mode. This is especially useful for connecting multiple SPI devices (see Chapter 16) to a single pin. In the wire-ORed option, the internal pull-up resistor is disengaged and the pin becomes an open-collector (or open-drained in the case of CMOS). Now, we can connect a single output pin to multiple input pins without causing damage to the pin. The HCS12 manual has the WOMS (Wired-ORed Mode S) register for the PORTS (PTS). No other port in HCS12 has this option.

SECTION C.3: SYSTEM DESIGN ISSUES

In addition to fan-out, the other issues related to system design are power dissipation, ground bounce, V_{CC} bounce, crosstalk, and transmission lines. In this section we provide an overview of these topics.

Power dissipation considerations

The power dissipation of a system is a major concern of system designers, especially for laptop and hand-held systems in which batteries provide the power. Power dissipation is a function of frequency and voltage as shown below:

$$Q = CV$$

$$\frac{Q}{T} = \frac{CV}{T}$$

since $F = \frac{1}{T}$ *and* $I = \frac{Q}{T}$

$$I = CVF$$

now $P = VI = CV^2F$

In the above equations, the effects of frequency and V_{CC} voltage should be noted. While the power dissipation goes up linearly with frequency, the impact of the power supply voltage is much more pronounced (squared). See Example C-2.

Example C-2

Compare the power consumption of two microcontroller-based systems. One uses 5 V and the other uses 3 V for V_{CC}.

Solution:

Because P = VI, by substituting I = V/R we have $P = V^2/R$. Assuming that R = 1, we have $P = 5^2 = 25$ W and $P = 3^2 = 9$ W. This results in using 16 W less power, which means power savings of 64% ($16/25 \times 100$) for systems using 3 V for power source.

Dynamic and static currents

There are two major types of currents flow through an IC: dynamic and static. A dynamic current is I = CVF. It is a function of the frequency under which the component is working. This means that as the frequency goes up, the dynamic current and power dissipation go up. The static current, also called DC, is the current consumption of the component when it is inactive (not selected). The dynamic current dissipation is much higher than the static current consumption. To reduce power consumption, many microcontrollers, including the HCS12, have a power-saving mode. The power-saving mode is called *sleep mode.* We describe the sleep mode next.

Sleep mode

In sleep mode the on-chip oscillator is frozen, which cuts off frequency to the CPU and peripheral functions, such as serial ports, interrupts, and timers. Notice that while this mode brings power consumption down to an absolute minimum, the contents of RAM and the peripheral registers (special function register) are saved and remain unchanged.

Ground bounce

One of the major issues that designers of high-frequency systems must grapple with is ground bounce. Before we define ground bounce, we will discuss lead inductance of IC pins. There is a certain amount of capacitance, resistance, and inductance associated with each pin of the IC. The size of these elements varies depending on many factors such as length, area, and so on.

The inductance of the pins is commonly referred to as *self-inductance* because there is also what is called *mutual inductance*, as we will show below. Of the three components of capacitor, resistor, and inductor, the property of self-inductance is the one that causes the most problems in high-frequency system design because it can result in ground bounce. Ground bounce occurs when a massive amount of current flows through the ground pin caused by many outputs changing from HIGH to LOW all at the same time. See Figure C-11(a). The induced voltage is related to the inductance of the ground lead as follows:

$$V = L\frac{di}{dt}$$

As we increase the system frequency, the rate of dynamic current, di/dt, is also increased, resulting in an increase in the inductance voltage L (di/dt) of the ground pin. Because the LOW state (ground) has a small noise margin, any extra voltage due to the inductance can cause a false signal. To reduce the effect of ground bounce, the following steps must be taken where possible:

1. The V_{CC} and ground pins of the chip must be located in the middle rather than at opposite ends of the IC chip (the 14-pin TTL logic IC uses pins 14 and 7 for ground and V_{CC}). This is exactly what we see in high-performance logic gates such as Texas Instruments' advanced logic AC11000 and ACT11000 families. For example, the ACT11013 is a 14-pin DIP chip in which pin numbers 4 and 11 are used for the ground and V_{CC}, instead of 7 and 14 as in the traditional TTL family. We can also use the SOIC packages instead of DIP.
2. Another solution is to use as many pins for ground and V_{CC} as possible to

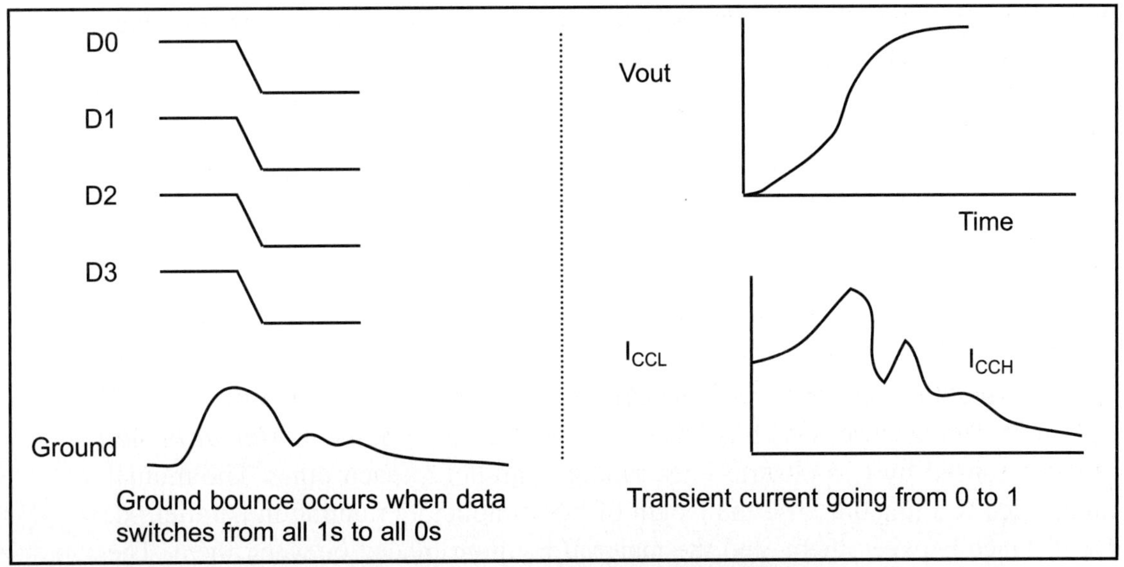

Figure C-11. (a) Ground Bounce (b) Transient Current

reduce the lead length. This is exactly why all high-performance microprocessors and logic families use many pins for V_{CC} and ground instead of the traditional single pin for V_{CC} and single pin for GND. For example, in the case of Intel's Pentium processor there are over 50 pins for ground, and another 50 pins for V_{CC}.

The above discussion of ground bounce is also applicable to V_{CC} when a large number of outputs changes from the LOW to the HIGH state; this is referred to as *V_{CC} bounce*. However, the effect of V_{CC} bounce is not as severe as ground bounce because the HIGH ("1") state has a wider noise margin than the LOW ("0") state.

Filtering the transient currents using decoupling capacitors

In the TTL family, the change of the output from LOW to HIGH can cause what is called *transient current*. In a totem-pole output in which the output is LOW, Q4 is on and saturated, whereas Q3 is off. By changing the output from the LOW to the HIGH state, Q3 turns on and Q4 turns off. This means that there is a time when both transistors are on and drawing current from V_{CC}. The amount of current depends on the R_{ON} values of the two transistors, which in turn depend on the internal parameters of the transistors. The net effect of this, however, is a large amount of current in the form of a spike for the output current, as shown in Figure C-11(b). To filter the transient current, a 0.01 μF or 0.1 μF ceramic disk capacitor can be placed between the V_{CC} and ground for each TTL IC. The lead for this capacitor, however, should be as small as possible because a long lead results in a large self-inductance, and that results in a spike on the V_{CC} line [$V = L\ (di/dt)$]. This spike is called V_{CC} bounce. The ceramic capacitor for each IC is referred to as a *decoupling capacitor*. There is also a bulk decoupling capacitor, as described next.

Bulk decoupling capacitor

If many IC chips change state at the same time, the combined currents drawn from the board's V_{CC} power supply can be massive and may cause a fluctuation of V_{CC} on the board where all the ICs are mounted. To eliminate this, a relatively large decoupling tantalum capacitor is placed between the V_{CC} and ground lines. The size and location of this tantalum capacitor varies depending on the number of ICs on the board and the amount of current drawn by each IC, but it is common to have a single 22 μF to 47 μF capacitor for each of the 16 devices, placed between the V_{CC} and ground lines.

Crosstalk

Crosstalk is due to mutual inductance. See Figure C-12. Previously, we discussed self-inductance, which is inherent in a piece of conductor. *Mutual inductance* is caused by two electric lines running parallel to each other. The mutual inductance is a function of l, the length of two conductors running in parallel, d, the distance between them, and the material medium placed between them. The effect of crosstalk can be reduced by increasing the distance between the parallel or adjacent lines (in printed circuit boards, they will be traces). In many cases, such as printer and disk drive cables, there is a dedicated ground for each signal. Placing ground lines (traces) between signal lines reduces the effect of crosstalk. This method is used even in some ACT logic families where a V_{CC} and a GND pin are next to each other. Crosstalk is also called *EMI* (electromagnetic interference). This is in contrast to *ESI* (electrostatic interference), which is caused by capacitive coupling between two adjacent conductors.

Transmission line ringing

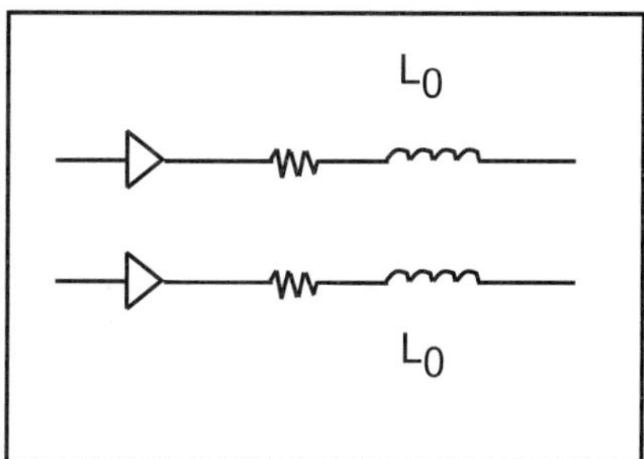

Figure C-12. Crosstalk (EMI)

The square wave used in digital circuits is in reality made of a single fundamental pulse and many harmonics of various amplitudes. When this signal travels on the line, not all the harmonics respond in the same way to the capacitance, inductance, and resistance of the line. This causes what is called *ringing*, which depends on the thickness and the length of the line driver, among other factors. To reduce the effect of ringing, the line drivers are terminated by putting a resistor at the end of the line. See Figure C-13. There are three major methods of line driver termination: parallel, serial, and Thevenin.

In serial termination, resistors of 30–50 ohms are used to terminate the line. The parallel and Thevenin methods are used in cases where there is a need to match the impedance of the line with the load impedance. This requires a detailed analysis of the signal traces and load impedance, which is beyond the scope of this book. In high-frequency systems, wire traces on the printed circuit board (PCB) behave like transmission lines, causing ringing. The severity of this ringing depends on the speed and the logic family used. Table C-6 provides the length of the traces, beyond which the traces must be regarded as transmission lines.

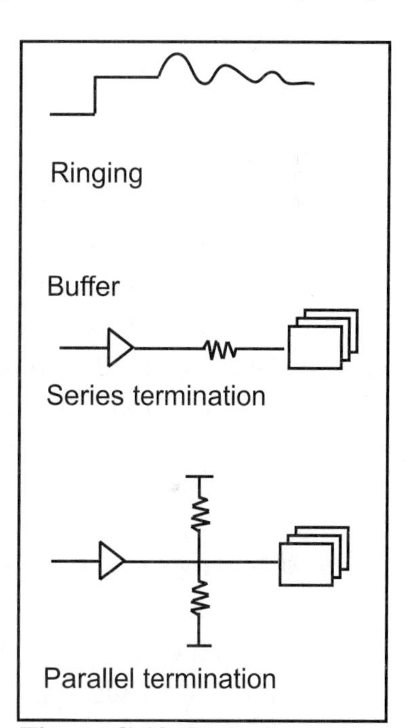

Figure C-13. Reducing Transmission Line Ringing

Table C-6: Line Length Beyond Which Traces Behave Like Transmission Lines

Logic Family	Line Length (in.)
LS	25
S, AS	11
F, ACT	8
AS, ECL	6
FCT, FCTA	5

(Reprinted by permission of Integrated Device Technology, copyright IDT 1991)

SECTION C.4: BASICS OF WIRE WRAPPING

Note: For this tutorial appendix, you will need the following:
Wire-wrapping tool (Radio Shack part number 276-1570)
30-gauge (30-AWG) wire for wire wrapping

The following describes the basics of wire wrapping:

1. There are several different types of wire-wrap tools available. The best one is available from Radio Shack for less than $10. The part number for the Radio Shack model is 276-1570. This tool combines the wrap and unwrap functions in the same end of the tool and includes a separate stripper. We found this to be much easier to use than the tools that combine all these features on one two-ended shaft. There are also wire-wrap guns, which are, of course, more expensive.
2. Wire-wrapping wire is available prestripped in various lengths or in bulk on a spool. The prestripped wire is usually more expensive and you are restricted to the different wire lengths you can afford to buy. Bulk wire can be cut to any length you wish, which allows each wire to be custom fit.
3. Several different types of wire-wrap boards are available. These are usually called *perfboards* or *wire-wrap boards*. These types of boards are sold at many electronics stores (such as Radio Shack). The best type of board has plating around the holes on the bottom of the board. These boards are better because the sockets and pins can be soldered to the board, which makes the circuit more mechanically stable.
4. Choose a board that is large enough to accommodate all the parts in your design with room to spare so that the wiring does not become too cluttered. If you wish to expand your project in the future, you should be sure to include enough room on the original board for the complete circuit. Also, if possible, the layout of the IC on the board needs to be such that signals go from left to right just like the schematic.
5. To make the wiring easier and to keep pressure off the pins, install one standoff on each corner of the board. You may also wish to put standoffs on the top of the board to add stability when the board is on its back.
6. For power hook-up, use some type of standard binding post. Solder a few single wire-wrap pins to each power post to make circuit connections (to at least one pin for each IC in the circuit).
7. To further reduce problems with power, each IC must have its own connection to the main power of the board. If your perfboard does not have built-in power buses, run a separate power and ground wire from each IC to the main power. In other words, DO NOT daisy chain power connections (make connections chip-to-chip), as each connection down the line will have more wire and more resistance to get power through. See Figure C-14. However, daisy chaining is acceptable for other connections such as data, address, and control buses.
8. You must use wire-wrap sockets. These sockets have long square pins whose edges will cut into the wire as it is wrapped around the pin.
9. Wire wrapping will not work on round legs. If you need to wrap to components, such as capacitors, that have round legs, you must also solder these con-

nections. The best way to connect single components is to install individual wire-wrap pins into the board and then solder the components to the pins. An alternate method is to use an empty IC socket to hold small components such as resistors and wrap them to the socket.

10. The wire should be stripped about 1 inch. This will allow 7 to 10 turns for each connection. The first turn or turn-and-a-half should be insulated. This prevents stripped wire from coming in contact with other pins. This can be accomplished by inserting the wire as far as it will go into the tool before making the connection.
11. Try to keep wire lengths to a minimum. This prevents the circuit from looking like a bird nest. Be neat and use color coding as much as possible. Use only red wires for V_{CC} and black wires for ground connections. Also use different colors for data, address, and control signal connections. These suggestions will make troubleshooting much easier.
12. It is standard practice to connect all power lines first and check them for continuity. This will eliminate trouble later on.
13. It's also a good idea to mark the pin orientation on the bottom of the board. Plastic templates are available with pin numbers preprinted on them specifically for this purpose, or you can make your own from paper. Forgetting to reverse pin order when looking at the bottom of the board is a very common mistake when wire wrapping circuits.
14. To prevent damage to your circuit, place a diode (such as IN5338) in reverse bias across the power supply. If the power gets hooked up backwards, the diode will be forward biased and will act as a short, keeping the reversed voltage from your circuit.
15. In digital circuits, there can be a problem with current demand on the power supply. To filter the noise on the power supply, a 100 μF electrolytic capacitor and a 0.1 μF monolithic capacitor are connected from V_{CC} to ground, in parallel with each other, at the entry point of the power supply to the board. These two together will filter both the high- and the low-frequency noise. Instead of using two capacitors in parallel, you can use a single 20–100 μF tantalum capacitor. Remember that the long lead is the positive one.
16. To filter the transient current, use a 0.1 μF monolithic capacitor for each IC. Place the 0.1 μF monolithic capacitor between V_{CC} and ground of each IC. Make sure the leads are as short as possible.

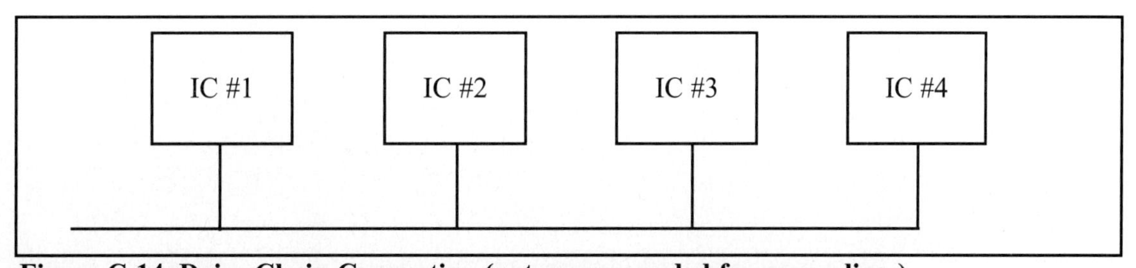

Figure C-14. Daisy Chain Connection (not recommended for power lines)

APPENDIX D

FLOWCHARTS AND PSEUDOCODE

OVERVIEW

This appendix provides an introduction to writing flowcharts and pseudocode.

Flowcharts

If you have taken any previous programming courses, you are probably familiar with flowcharting. Flowcharts use graphic symbols to represent different types of program operations. These symbols are connected together into a flowchart to show the flow of execution of a program. Figure D-1 shows some of the more commonly used symbols. Flowchart templates are available to help you draw the symbols quickly and neatly.

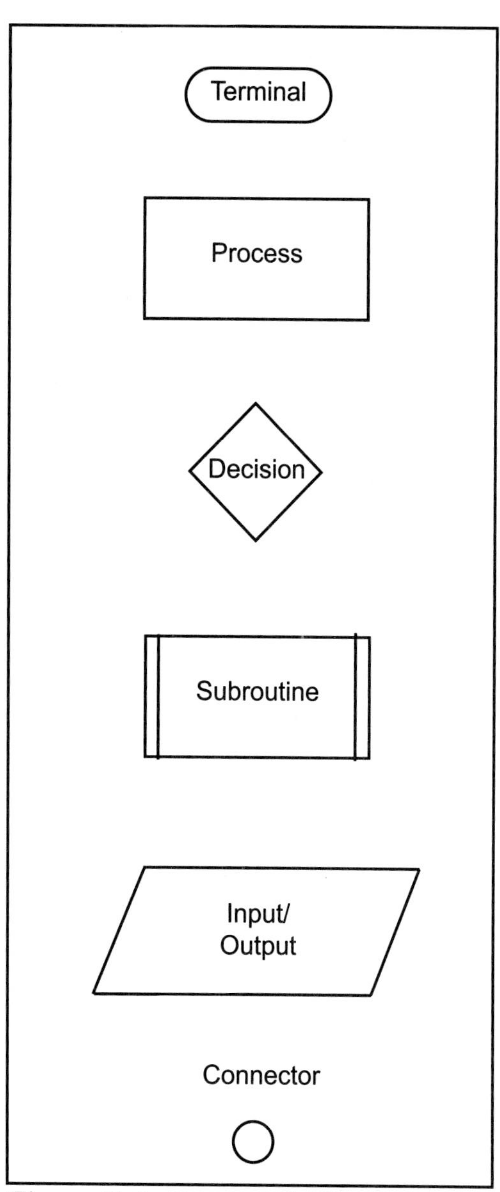

Figure D-1. Commonly Used Flowchart Symbols

Pseudocode

Flowcharting has been standard practice in industry for decades. However, some find limitations in using flowcharts, such as the fact that you can't write much in the little boxes, and it is hard to get the "big picture" of what the program does without getting bogged down in the details. An alternative to using flowcharts is pseudocode, which involves writing brief descriptions of the flow of the code. Figures D-2 through D-6 show flowcharts and pseudocode for commonly used control structures.

Structured programming uses three basic types of program control

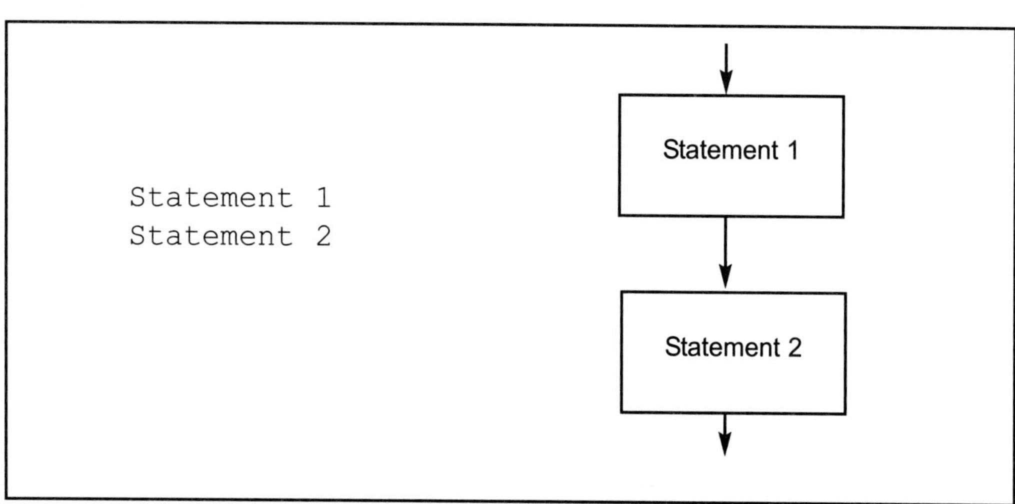

Figure D-2. SEQUENCE Pseudocode versus Flowchart

structures: sequence, control, and iteration. Sequence is simply executing instructions one after another. Figure D-2 shows how sequence can be represented in pseudocode and flowcharts.

Figures D-3 and D-4 show two control programming structures: IF-THEN-ELSE and IF-THEN in both pseudocode and flowcharts.

Note in Figures D-2 through D-6 that "statement" can indicate one statement or a group of statements.

Figures D-5 and D-6 show two iteration control structures: REPEAT UNTIL and WHILE DO. Both structures execute a statement or group of statements repeatedly. The difference between them is that the REPEAT UNTIL structure always executes the statement(s) at least once, and checks the condition after each iteration, whereas the WHILE DO may not execute the statement(s) at all because the condition is checked at the beginning of each iteration.

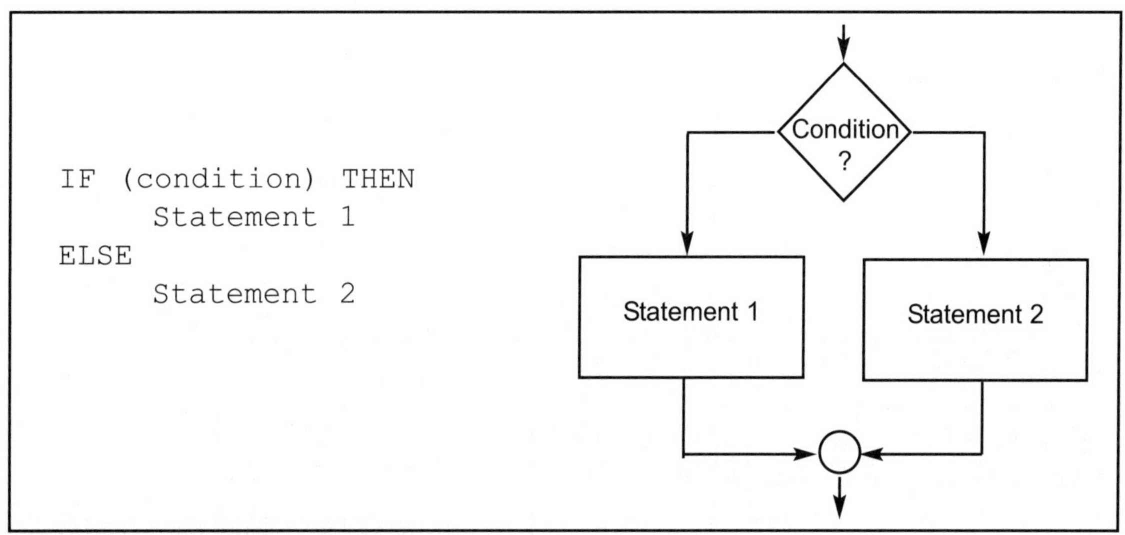

Figure D-3. IF THEN ELSE Pseudocode versus Flowchart

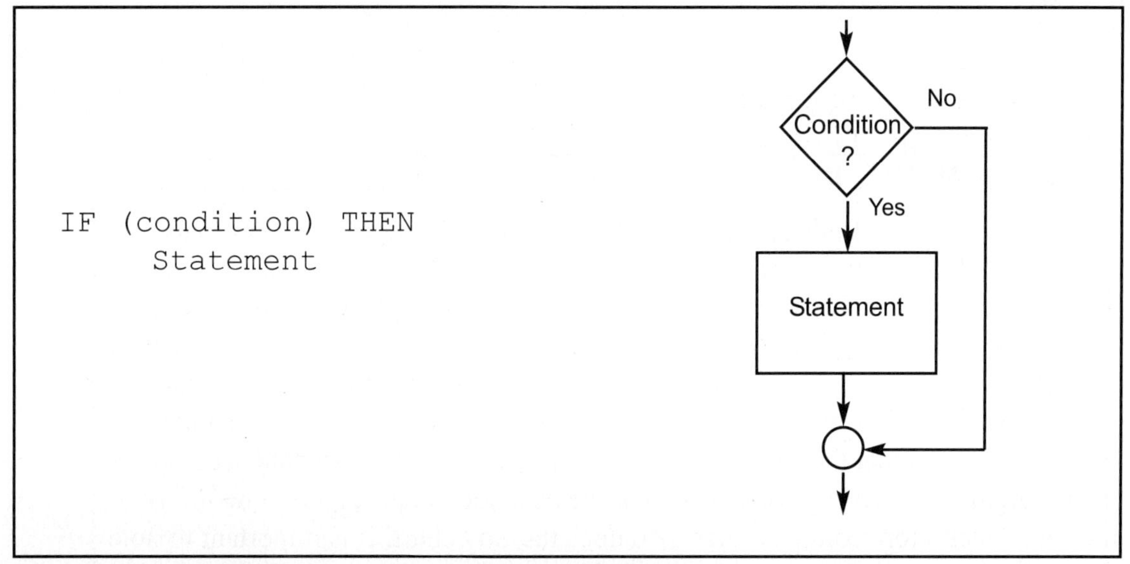

Figure D-4. IF THEN Pseudocode versus Flowchart

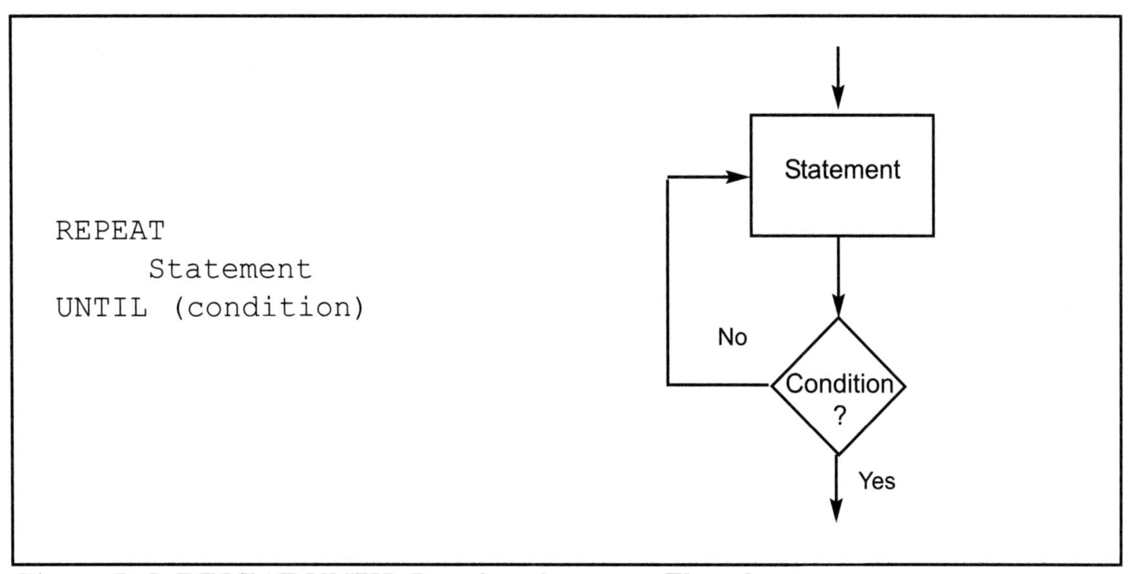

Figure D-5. REPEAT UNTIL Pseudocode versus Flowchart

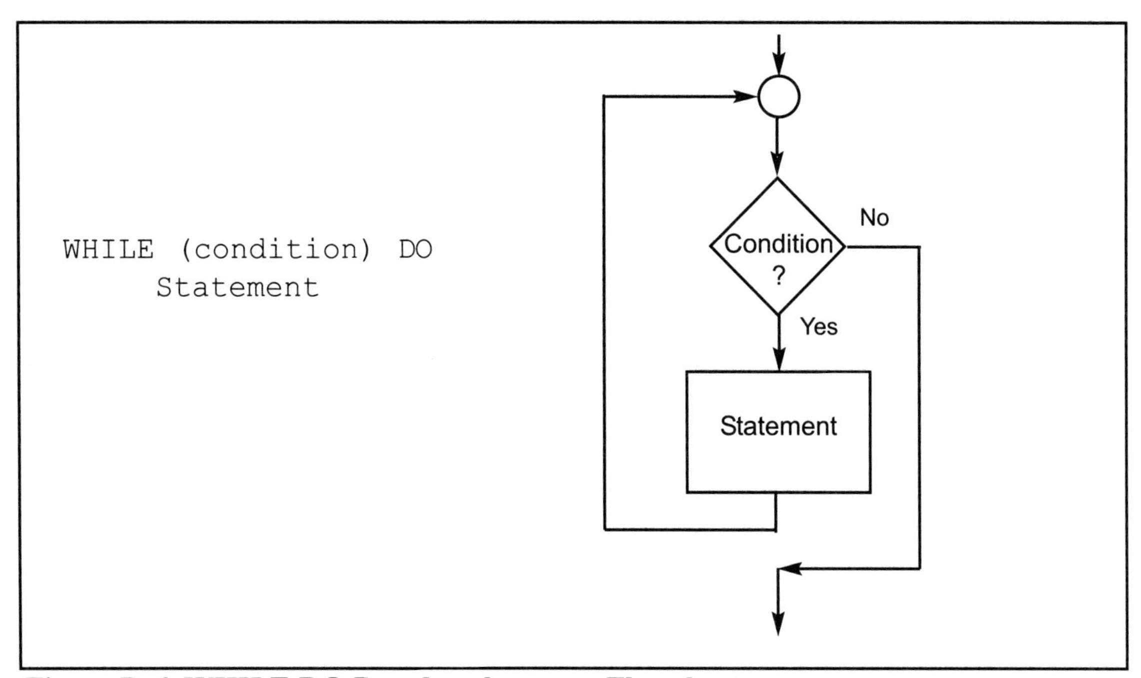

Figure D-6. WHILE DO Pseudocode versus Flowchart

Program D-1 finds the sum of a series of bytes. Compare the flowchart versus the pseudocode for Program D-1 (shown in Figure D-7). In this example, more program details are given than one usually finds. For example, this shows steps for initializing and decrementing counters. Another programmer may not include these steps in the flowchart or pseudocode. It is important to remember that the purpose of flowcharts or pseudocode is to show the flow of the program and what the program does, not the specific Assembly language instructions that accomplish the program's objectives. Notice also that the pseudocode gives the same information in a much more compact form than does the flowchart. It is important to note that sometimes pseudocode is written in layers, so that the outer level or layer shows the flow of the program and subsequent levels show more details of how the program accomplishes its assigned tasks.

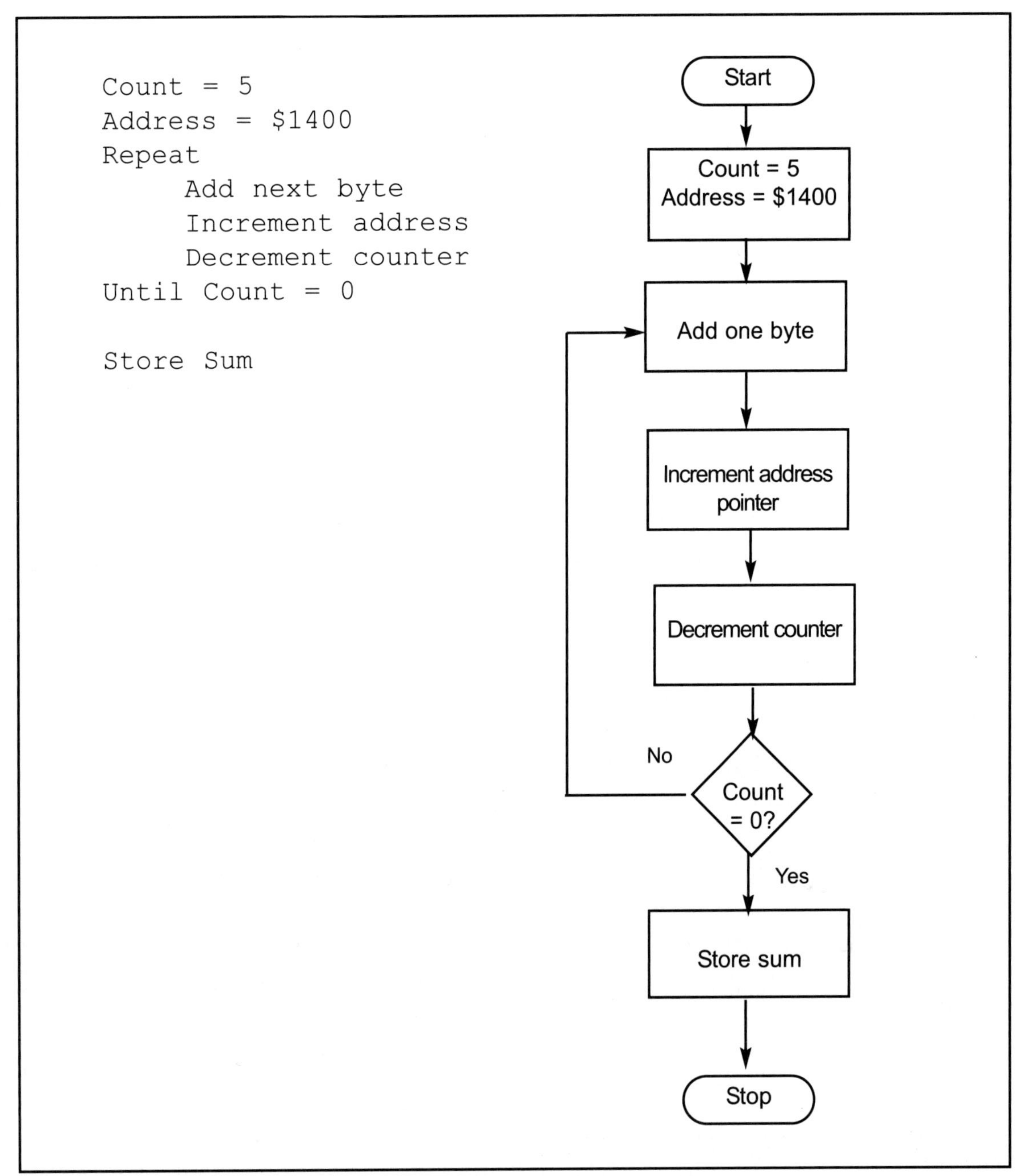

Figure D-7. Pseudocode versus Flowchart for Program D-1

```
COUNTVAL    EQU 5              ;COUNT = 5
COUNTREG    EQU $1500          ;set aside location for counter
SUM         EQU $1550          ;set aside location for sum
      LDAA   #COUNTVAL         ;A = 5
      STAA   COUNTREG          ;load the counter
      LDX    #$1400            ;load pointer to RAM data address
      CLRA                     ;clear A
B5    ADDA   0,X+              ;add RAM data to A and increment pointer
      DEC    COUNTREG          ;decrement counter
      BNE    B5                ;loop until counter = zero
      STAA   SUM               ;store the SUM
```

Program D-1

APPENDIX E

HCS12 PRIMER FOR x86 PROGRAMMERS

	x86	HCS12
8-bit registers:	AL, AH, BL, BH, CL, CH, DL, DH	A, B and up to several kilobytes of RAM locations
16-bit (data pointer):	BX, SI, DI	X, Y
Program Counter:	IP (16-bit)	PC (16-bit)
Input:	`MOV DX,port addr` `IN AL,DX`	`LDAA PORTx ;x = A,B,E,H,...`
Output:	`MOV DX,port addr` `OUT DX,AL`	`STAA PORTx ;x = A,B,E,H,...`
Loop:	`DEC CL` `JNZ TARGET`	`DEC MyReg` `BNE TARGET`
Stack Pointer:	SP (16-bit)	SP (16-bit)
	As we PUSH data onto the stack, it decrements the SP.	The same
	As we POP data from the stack, it increments the SP.	The same
Data movement:		
From the code segment:	`MOV AL,CS:[SI]`	`LDAA 0,X`
From the data segment:	`MOV AL,[SI]`	`LDAA 0,X`
From RAM:	`MOV AL,[SI]` (Use SI, DI, or BX only.)	`LDAA 0,X`
To RAM:	`MOV [SI],AL`	`STAA 0,X`

APPENDIX F

ASCII CODES

Ctrl	Dec	Hex	Ch	Code
^@	0	00		NUL
^A	1	01	☺	SOH
^B	2	02	☻	STX
^C	3	03	♥	ETX
^D	4	04	♦	EOT
^E	5	05	♣	ENQ
^F	6	06	♠	ACK
^G	7	07	•	BEL
^H	8	08	◘	BS
^I	9	09	○	HT
^J	10	0A	◙	LF
^K	11	0B	♂	VT
^L	12	0C	♀	FF
^M	13	0D	♪	CR
^N	14	0E	♫	SO
^O	15	0F	☼	SI
^P	16	10	►	DLE
^Q	17	11	◄	DC1
^R	18	12	↕	DC2
^S	19	13	‼	DC3
^T	20	14	¶	DC4
^U	21	15	§	NAK
^V	22	16	▬	SYN
^W	23	17	↨	ETB
^X	24	18	↑	CAN
^Y	25	19	↓	EM
^Z	26	1A	→	SUB
^[	27	1B	←	ESC
^\	28	1C	∟	FS
^]	29	1D	↔	GS
^^	30	1E	▲	RS
^_	31	1F	▼	US

Dec	Hex	Ch
32	20	
33	21	!
34	22	"
35	23	#
36	24	$
37	25	%
38	26	&
39	27	'
40	28	(
41	29	)
42	2A	*
43	2B	+
44	2C	,
45	2D	-
46	2E	.
47	2F	/
48	30	0
49	31	1
50	32	2
51	33	3
52	34	4
53	35	5
54	36	6
55	37	7
56	38	8
57	39	9
58	3A	:
59	3B	;
60	3C	<
61	3D	=
62	3E	>
63	3F	?

Dec	Hex	Ch
64	40	@
65	41	A
66	42	B
67	43	C
68	44	D
69	45	E
70	46	F
71	47	G
72	48	H
73	49	I
74	4A	J
75	4B	K
76	4C	L
77	4D	M
78	4E	N
79	4F	O
80	50	P
81	51	Q
82	52	R
83	53	S
84	54	T
85	55	U
86	56	V
87	57	W
88	58	X
89	59	Y
90	5A	Z
91	5B	[
92	5C	\
93	5D	]
94	5E	^
95	5F	_

Dec	Hex	Ch
96	60	`
97	61	a
98	62	b
99	63	c
100	64	d
101	65	e
102	66	f
103	67	g
104	68	h
105	69	i
106	6A	j
107	6B	k
108	6C	l
109	6D	m
110	6E	n
111	6F	o
112	70	p
113	71	q
114	72	r
115	73	s
116	74	t
117	75	u
118	76	v
119	77	w
120	78	x
121	79	y
122	7A	z
123	7B	{
124	7C	¦
125	7D	}
126	7E	~
127	7F	⌂

Dec	Hex	Ch	Dec	Hex	Ch	Dec	Hex	Ch	Dec	Hex	Ch
128	80	Ç	160	A0	á	192	C0	└	224	E0	α
129	81	ü	161	A1	í	193	C1	┴	225	E1	β
130	82	é	162	A2	ó	194	C2	┬	226	E2	Γ
131	83	â	163	A3	ú	195	C3	├	227	E3	π
132	84	ä	164	A4	ñ	196	C4	─	228	E4	Σ
133	85	à	165	A5	Ñ	197	C5	┼	229	E5	σ
134	86	å	166	A6	ª	198	C6	╞	230	E6	µ
135	87	ç	167	A7	º	199	C7	╟	231	E7	τ
136	88	ê	168	A8	¿	200	C8	╚	232	E8	Φ
137	89	ë	169	A9	⌐	201	C9	╔	233	E9	Θ
138	8A	è	170	AA	¬	202	CA	╩	234	EA	Ω
139	8B	ï	171	AB	½	203	CB	╦	235	EB	δ
140	8C	î	172	AC	¼	204	CC	╠	236	EC	∞
141	8D	ì	173	AD	¡	205	CD	═	237	ED	φ
142	8E	Ä	174	AE	«	206	CE	╬	238	EE	ε
143	8F	Å	175	AF	»	207	CF	╧	239	EF	∩
144	90	É	176	B0	░	208	D0	╨	240	F0	≡
145	91	æ	177	B1	▒	209	D1	╤	241	F1	±
146	92	Æ	178	B2	▓	210	D2	╥	242	F2	≥
147	93	ô	179	B3	│	211	D3	╙	243	F3	≤
148	94	ö	180	B4	┤	212	D4	?	244	F4	⌠
149	95	ò	181	B5	╡	213	D5	╒	245	F5	⌡
150	96	û	182	B6	╢	214	D6	╓	246	F6	÷
151	97	ù	183	B7	╖	215	D7	╫	247	F7	≈
152	98	ÿ	184	B8	╕	216	D8	╪	248	F8	≈
153	99	Ö	185	B9	╣	217	D9	┘	249	F9	∙
154	9A	Ü	186	BA	║	218	DA	┌	250	FA	·
155	9B	¢	187	BB	╗	219	DB	█	251	FB	√
156	9C	£	188	BC	╝	220	DC	▄	252	FC	^{n}
157	9D	¥	189	BD	╜	221	DD	▌	253	FD	^{2}
158	9E	₧	190	BE	╛	222	DE	▐	254	FE	■
159	9F	ƒ	191	BF	┐	223	DF	▀	255	FF	

APPENDIX G

ASSEMBLERS, DEVELOPMENT RESOURCES, AND SUPPLIERS

This appendix provides various sources for HCS12 assemblers, compilers, and trainers. In addition, it lists some suppliers for chips and other hardware needs. While these are all established products from well-known companies, neither the author nor the publisher assumes responsibility for any problem that may arise with any of them. You are neither encouraged nor discouraged from purchasing any of the products mentioned; you must make your own judgment in evaluating the products. This list is simply provided as a service to the reader. It also must be noted that the list of products is by no means complete or exhaustive.

The CodeWarrior from Freescale
http://www.freescale.com

ImageCraft
http://www.imagecraft.com

For AsmIDE and GNU C compiler and MiniIDE, see Eric Engler site
http://www.geocities.com/englere_geo
or
http://www.ericengler.com/EmbeddedGNU.aspx

Figure G-1. Suppliers of Assemblers and Compilers

HCS12 assemblers

The HCS12 assembler is provided by Freescale and other companies. Some of the companies provide shareware versions of their products, which you can download from their Web sites. However, the size of code for these shareware versions is limited to a few KB. Figure G-1 lists some suppliers of assemblers.

HCS12 trainers

There are many companies that produce and market HCS12 trainers. Figure G-2 provides a list of some of them.

Axiom Manufacturing Inc.
http://www.axman.com

Wytec Inc.
http://www.evbplus.com

Technological Arts Inc.
http://www.technologicalarts.com

PEMicro
http://www.pemicro.com

Freescale Corp.
http://www.freescale.com

Figure G-2. Trainer Suppliers

Parts Suppliers

Figure G-3 provides a list of suppliers for many electronics parts.

RSR Electronics
Electronix Express
365 Blair Road
Avenel, NJ 07001
Fax: (732) 381-1572
Mail Order: 1-800-972-2225
In New Jersey: (732) 381-8020
http://www.elexp.com

Altex Electronics
11342 IH-35 North
San Antonio, TX 78233
Fax: (210) 637-3264
Mail Order: 1-800-531-5369
http://www.altex.com

Digi-Key
1-800-344-4539 (1-800-DIGI-KEY)
Fax: (218) 681-3380
http://www.digikey.com

Radio Shack
http://www.radioshack.com

JDR Microdevices
1850 South 10th St.
San Jose, CA 95112-4108
Sales 1-800-538-5000
(408) 494-1400
Fax: 1-800-538-5005
Fax: (408) 494-1420
http://www.jdr.com

Mouser Electronics
958 N. Main St.
Mansfield, TX 76063
1-800-346-6873
http://www.mouser.com

Jameco Electronic
1355 Shoreway Road
Belmont, CA 94002-4100
1-800-831-4242
(415) 592-8097
Fax: 1-800-237-6948
Fax: (415) 592-2503
http://www.jameco.com

B. G. Micro
P. O. Box 280298
Dallas, TX 75228
1-800-276-2206 (orders only)
(972) 271-5546
Fax: (972) 271-2462
This is an excellent source of LCDs, ICs, keypads, etc.
http://www.bgmicro.com

Tanner Electronics
1100 Valwood Parkway, Suite #100
Carrollton, TX 75006
(972) 242-8702
http://www.tannerelectronics.com

Figure G-3. Electronics Suppliers

APPENDIX H

DATA SHEETS

SECTION H.1: HCS12 INSTRUCTION REFERENCE

Instruction Reference

A.1 Introduction

This appendix provides quick references for the instruction set, opcode map, and encoding.

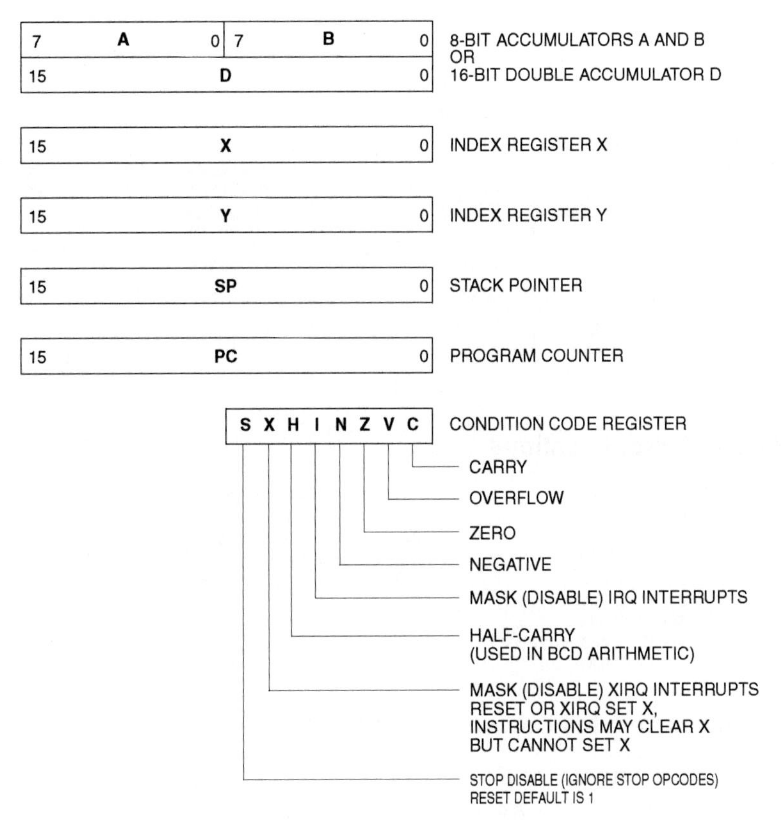

Figure A-1. Programming Model

A.2 Stack and Memory Layout

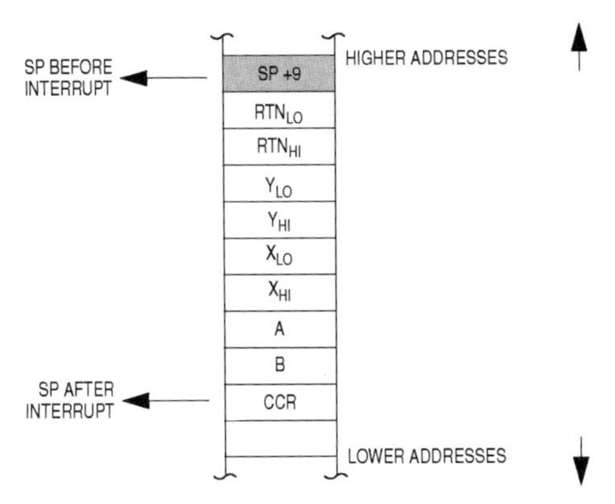

STACK UPON ENTRY TO SERVICE ROUTINE IF SP WAS ODD BEFORE INTERRUPT

SP +8	RTN$_{LO}$		SP +9
SP +6	Y$_{LO}$	RTN$_{HI}$	SP +7
SP +4	X$_{LO}$	Y$_{HI}$	SP +5
SP +2	A	X$_{HI}$	SP +3
SP	CCR	B	SP +1
SP –2			SP –1

STACK UPON ENTRY TO SERVICE ROUTINE IF SP WAS EVEN BEFORE INTERRUPT

SP +9			SP +10
SP +7	RTN$_{HI}$	RTN$_{LO}$	SP +8
SP +5	Y$_{HI}$	Y$_{LO}$	SP +6
SP +4	X$_{HI}$	X$_{LO}$	SP +4
SP +1	B	A	SP +2
SP –1		CCR	SP

A.3 Interrupt Vector Locations

$FFFE, $FFFF	Power-On (POR) or External Reset
$FFFC, $FFFD	Clock Monitor Reset
$FFFA, $FFFB	Computer Operating Properly (COP Watchdog Reset
$FFF8, $FFF9	Unimplemented Opcode Trap
$FFF6, $FFF7	Software Interrupt Instruction (SWI)
$FFF4, $FFF5	XIRQ
$FFF2, $FFF3	IRQ
$FFC0–$FFF1 (M68HC12)	Device-Specific Interrupt Sources
$FF00–$FFF1 (HCS12)	Device-Specific Interrupt Sources

A.4 Notation Used in Instruction Set Summary

CPU Register Notation

Accumulator A — A or a	Index Register Y — Y or y
Accumulator B — B or b	Stack Pointer — SP, sp, or s
Accumulator D — D or d	Program Counter — PC, pc, or p
Index Register X — X or x	Condition Code Register — CCR or c

Explanation of Italic Expressions in Source Form Column

abc — A or B or CCR

abcdxys — A or B or CCR or D or X or Y or SP. Some assemblers also allow T2 or T3.

abd — A or B or D

abdxys — A or B or D or X or Y or SP

dxys — D or X or Y or SP

msk8 — 8-bit mask, some assemblers require # symbol before value

opr8i — 8-bit immediate value

opr16i — 16-bit immediate value

opr8a — 8-bit address used with direct address mode

opr16a — 16-bit address value

oprx0_xysp — Indexed addressing postbyte code:
oprx3,–xys Predecrement X or Y or SP by 1 . . . 8
oprx3,+xys Preincrement X or Y or SP by 1 . . . 8
oprx3,xys– Postdecrement X or Y or SP by 1 . . . 8
oprx3,xys+ Postincrement X or Y or SP by 1 . . . 8
oprx5,xysp 5-bit constant offset from X or Y or SP or PC
abd,xysp Accumulator A or B or D offset from X or Y or SP or PC

oprx3 — Any positive integer 1 . . . 8 for pre/post increment/decrement

oprx5 — Any integer in the range –16 . . . +15

oprx9 — Any integer in the range –256 . . . +255

oprx16 — Any integer in the range –32,768 . . . 65,535

page — 8-bit value for PPAGE, some assemblers require # symbol before this value

rel8 — Label of branch destination within –128 to +127 locations

rel9 — Label of branch destination within –256 to +255 locations

rel16 — Any label within 64K memory space

trapnum — Any 8-bit integer in the range $30-$39 or $40-$FF

xys — X or Y or SP

xysp — X or Y or SP or PC

Operators

+ — Addition

− — Subtraction

• — Logical AND

+ — Logical OR (inclusive)

⊕ — Logical exclusive OR

× — Multiplication

÷ — Division

$\overline{M}$ — Negation. One's complement (invert each bit of M)

: — Concatenate
Example: A : B means the 16-bit value formed by concatenating 8-bit accumulator A with 8-bit accumulator B.
A is in the high-order position.

⇒ — Transfer
Example: (A) ⇒ M means the content of accumulator A is transferred to memory location M.

⇔ — Exchange
Example: D ⇔ X means exchange the contents of D with those of X.

Address Mode Notation

INH — Inherent; no operands in object code

IMM — Immediate; operand in object code

DIR — Direct; operand is the lower byte of an address from $0000 to $00FF

EXT — Operand is a 16-bit address

REL — Two's complement relative offset; for branch instructions

IDX — Indexed (no extension bytes); includes:
5-bit constant offset from X, Y, SP, or PC
Pre/post increment/decrement by 1 . . . 8
Accumulator A, B, or D offset

IDX1 — 9-bit signed offset from X, Y, SP, or PC; 1 extension byte

IDX2 — 16-bit signed offset from X, Y, SP, or PC; 2 extension bytes

[IDX2] — Indexed-indirect; 16-bit offset from X, Y, SP, or PC

[D, IDX] — Indexed-indirect; accumulator D offset from X, Y, SP, or PC

Machine Coding

dd — 8-bit direct address $0000 to $00FF. (High byte assumed to be $00).

ee — High-order byte of a 16-bit constant offset for indexed addressing.

eb — Exchange/Transfer post-byte. See **Table A-5** on page 369.

ff — Low-order eight bits of a 9-bit signed constant offset for indexed addressing, or low-order byte of a 16-bit constant offset for indexed addressing.

hh — High-order byte of a 16-bit extended address.

ii — 8-bit immediate data value.

jj — High-order byte of a 16-bit immediate data value.

kk — Low-order byte of a 16-bit immediate data value.

lb — Loop primitive (DBNE) post-byte. See **Table A-6** on page 370.

ll — Low-order byte of a 16-bit extended address.

mm — 8-bit immediate mask value for bit manipulation instructions. Set bits indicate bits to be affected.

pg — Program page (bank) number used in CALL instruction.

qq — High-order byte of a 16-bit relative offset for long branches.

tn — Trap number $30–$39 or $40–$FF.

rr — Signed relative offset $80 (–128) to $7F (+127). Offset relative to the byte following the relative offset byte, or low-order byte of a 16-bit relative offset for long branches.

xb — Indexed addressing post-byte. See **Table A-3** on page 367 and **Table A-4** on page 368.

Access Detail

Each code letter except (,), and comma equals one CPU cycle. Uppercase = 16-bit operation and lowercase = 8-bit operation.

f — Free cycle, CPU doesn't use bus
g — Read PPAGE internally
I — Read indirect pointer (indexed indirect)
i — Read indirect PPAGE value (CALL indirect only)
n — Write PPAGE internally
O — Optional program word fetch (P) if instruction is misaligned and has an odd number of bytes of object code — otherwise, appears as a free cycle (f); Page 2 prebyte treated as a separate 1-byte instruction
P — Program word fetch (always an aligned-word read)
r — 8-bit data read
R — 16-bit data read
s — 8-bit stack write
S — 16-bit stack write
w — 8-bit data write
W — 16-bit data write
u — 8-bit stack read
U — 16-bit stack read
V — 16-bit vector fetch (always an aligned-word read)
t — 8-bit conditional read (or free cycle)
T — 16-bit conditional read (or free cycle)
x — 8-bit conditional write (or free cycle)
() — Indicate a microcode loop
, — Indicates where an interrupt could be honored

Special Cases

PPP/P — Short branch, PPP if branch taken, P if not
OPPP/OPO — Long branch, OPPP if branch taken, OPO if not

Condition Codes Columns

– — Status bit not affected by operation.
0 — Status bit cleared by operation.
1 — Status bit set by operation.
Δ — Status bit affected by operation.
? — Status bit may be cleared or remain set, but is not set by operation.
⇑ — Status bit may be set or remain cleared, but is not cleared by operation.
? — Status bit may be changed by operation but the final state is not defined.
! — Status bit used for a special purpose.

Table A-1. Instruction Set Summary (Sheet 1 of 14)

Source Form	Operation	Addr. Mode	Machine Coding (hex)	Access Detail HCS12	Access Detail M68HC12	S X H I	N Z V C
ABA	(A) + (B) ⇒ A Add Accumulators A and B	INH	18 06	OO	OO	- - Δ -	Δ Δ Δ Δ
ABX	(B) + (X) ⇒ X *Translates to* LEAX B,X	IDX	1A E5	Pf	PP[1]	- - - -	- - - -
ABY	(B) + (Y) ⇒ Y *Translates to* LEAY B,Y	IDX	19 ED	Pf	PP[1]	- - - -	- - - -
ADCA #*opr8i*	(A) + (M) + C ⇒ A	IMM	89 ii	P	P	- - Δ -	Δ Δ Δ Δ
ADCA *opr8a*	Add with Carry to A	DIR	99 dd	rPf	rfP		
ADCA *opr16a*		EXT	B9 hh ll	rPO	rOP		
ADCA *oprx0_xysp*		IDX	A9 xb	rPf	rfP		
ADCA *oprx9,xysp*		IDX1	A9 xb ff	rPO	rPO		
ADCA *oprx16,xysp*		IDX2	A9 xb ee ff	frPP	frPP		
ADCA [D,*xysp*]		[D,IDX]	A9 xb	fIfrPf	flPrfP		
ADCA [*oprx16,xysp*]		[IDX2]	A9 xb ee ff	fIPrPf	fIPrfP		
ADCB #*opr8i*	(B) + (M) + C ⇒ B	IMM	C9 ii	P	P	- - Δ -	Δ Δ Δ Δ
ADCB *opr8a*	Add with Carry to B	DIR	D9 dd	rPf	rfP		
ADCB *opr16a*		EXT	F9 hh ll	rPO	rOP		
ADCB *oprx0_xysp*		IDX	E9 xb	rPf	rfP		
ADCB *oprx9,xysp*		IDX1	E9 xb ff	rPO	rPO		
ADCB *oprx16,xysp*		IDX2	E9 xb ee ff	frPP	frPP		
ADCB [D,*xysp*]		[D,IDX]	E9 xb	fIfrPf	fIfrfP		
ADCB [*oprx16,xysp*]		[IDX2]	E9 xb ee ff	fIPrPf	fIPrfP		
ADDA #*opr8i*	(A) + (M) ⇒ A	IMM	8B ii	P	P	- - Δ -	Δ Δ Δ Δ
ADDA *opr8a*	Add without Carry to A	DIR	9B dd	rPf	rfP		
ADDA *opr16a*		EXT	BB hh ll	rPO	rOP		
ADDA *oprx0_xysp*		IDX	AB xb	rPf	rfP		
ADDA *oprx9,xysp*		IDX1	AB xb ff	rPO	rPO		
ADDA *oprx16,xysp*		IDX2	AB xb ee ff	frPP	frPP		
ADDA [D,*xysp*]		[D,IDX]	AB xb	fIfrPf	fIfrfP		
ADDA [*oprx16,xysp*]		[IDX2]	AB xb ee ff	fIPrPf	fIPrfP		
ADDB #*opr8i*	(B) + (M) ⇒ B	IMM	CB ii	P	P	- - Δ -	Δ Δ Δ Δ
ADDB *opr8a*	Add without Carry to B	DIR	DB dd	rPf	rfP		
ADDB *opr16a*		EXT	FB hh ll	rPO	rOP		
ADDB *oprx0_xysp*		IDX	EB xb	rPf	rfP		
ADDB *oprx9,xysp*		IDX1	EB xb ff	rPO	rPO		
ADDB *oprx16,xysp*		IDX2	EB xb ee ff	frPP	frPP		
ADDB [D,*xysp*]		[D,IDX]	EB xb	fIfrPf	fIfrfP		
ADDB [*oprx16,xysp*]		[IDX2]	EB xb ee ff	fIPrPf	fIPrfP		
ADDD #*opr16i*	(A:B) + (M:M+1) ⇒ A:B	IMM	C3 jj kk	PO	OP	- - - -	Δ Δ Δ Δ
ADDD *opr8a*	Add 16-Bit to D (A:B)	DIR	D3 dd	RPf	RfP		
ADDD *opr16a*		EXT	F3 hh ll	RPO	ROP		
ADDD *oprx0_xysp*		IDX	E3 xb	RPf	RfP		
ADDD *oprx9,xysp*		IDX1	E3 xb ff	RPO	RPO		
ADDD *oprx16,xysp*		IDX2	E3 xb ee ff	fRPP	fRPP		
ADDD [D,*xysp*]		[D,IDX]	E3 xb	fIfRPf	fIfRfP		
ADDD [*oprx16,xysp*]		[IDX2]	E3 xb ee ff	fIPRPf	fIPRfP		
ANDA #*opr8i*	(A) • (M) ⇒ A	IMM	84 ii	P	P	- - - -	Δ Δ 0 -
ANDA *opr8a*	Logical AND A with Memory	DIR	94 dd	rPf	rfP		
ANDA *opr16a*		EXT	B4 hh ll	rPO	rOP		
ANDA *oprx0_xysp*		IDX	A4 xb	rPf	rfP		
ANDA *oprx9,xysp*		IDX1	A4 xb ff	rPO	rPO		
ANDA *oprx16,xysp*		IDX2	A4 xb ee ff	frPP	frPP		
ANDA [D,*xysp*]		[D,IDX]	A4 xb	fIfrPf	fIfrfP		
ANDA [*oprx16,xysp*]		[IDX2]	A4 xb ee ff	fIPrPf	fIPrfP		
ANDB #*opr8i*	(B) • (M) ⇒ B	IMM	C4 ii	P	P	- - - -	Δ Δ 0 -
ANDB *opr8a*	Logical AND B with Memory	DIR	D4 dd	rPf	rfP		
ANDB *opr16a*		EXT	F4 hh ll	rPO	rOP		
ANDB *oprx0_xysp*		IDX	E4 xb	rPf	rfP		
ANDB *oprx9,xysp*		IDX1	E4 xb ff	rPO	rPO		
ANDB *oprx16,xysp*		IDX2	E4 xb ee ff	frPP	frPP		
ANDB [D,*xysp*]		[D,IDX]	E4 xb	fIfrPf	fIfrfP		
ANDB [*oprx16,xysp*]		[IDX2]	E4 xb ee ff	fIPrPf	fIPrfP		
ANDCC #*opr8i*	(CCR) • (M) ⇒ CCR Logical AND CCR with Memory	IMM	10 ii	P	P	⇓ ⇓ ⇓ ⇓	⇓ ⇓ ⇓ ⇓

Note 1. Due to internal CPU requirements, the program word fetch is performed twice to the same address during this instruction.

Table A-1. Instruction Set Summary (Sheet 2 of 14)

Source Form	Operation	Addr. Mode	Machine Coding (hex)	Access Detail HCS12	Access Detail M68HC12	S X H I	N Z V C
ASL *opr16a* ASL *oprx0_xysp* ASL *oprx9,xysp* ASL *oprx16,xysp* ASL [D,*xysp*] ASL [*oprx16,xysp*] ASLA ASLB	C ← b7 … b0 ← 0 Arithmetic Shift Left Arithmetic Shift Left Accumulator A Arithmetic Shift Left Accumulator B	EXT IDX IDX1 IDX2 [D,IDX] [IDX2] INH INH	78 hh ll 68 xb 68 xb ff 68 xb ee ff 68 xb 68 xb ee ff 48 58	rPwO rPw rPwO frPwP fIfrPw fIPrPw O O	rOPw rPw rPOw frPPw fIfrPw fIPrPw O O	----	ΔΔΔΔ
ASLD	C ← b7 A b0 ← b7 B b0 ← 0 Arithmetic Shift Left Double	INH	59	O	O	----	ΔΔΔΔ
ASR *opr16a* ASR *oprx0_xysp* ASR *oprx9,xysp* ASR *oprx16,xysp* ASR [D,*xysp*] ASR [*oprx16,xysp*] ASRA ASRB	b7 … b0 → C Arithmetic Shift Right Arithmetic Shift Right Accumulator A Arithmetic Shift Right Accumulator B	EXT IDX IDX1 IDX2 [D,IDX] [IDX2] INH INH	77 hh ll 67 xb 67 xb ff 67 xb ee ff 67 xb 67 xb ee ff 47 57	rPwO rPw rPwO frPwP fIfrPw fIPrPw O O	rOPw rPw rPOw frPPw fIfrPw fIPrPw O O	----	ΔΔΔΔ
BCC *rel8*	Branch if Carry Clear (if C = 0)	REL	24 rr	PPP/P[1]	PPP/P[1]	----	----
BCLR *opr8a, msk8* BCLR *opr16a, msk8* BCLR *oprx0_xysp, msk8* BCLR *oprx9,xysp, msk8* BCLR *oprx16,xysp, msk8*	$(M) \bullet (\overline{mm}) \Rightarrow M$ Clear Bit(s) in Memory	DIR EXT IDX IDX1 IDX2	4D dd mm 1D hh ll mm 0D xb mm 0D xb ff mm 0D xb ee ff mm	rPwO rPwP rPwO rPwP frPwPO	rPOw rPPw rPOw rPwP frPwOP	----	ΔΔ0–
BCS *rel8*	Branch if Carry Set (if C = 1)	REL	25 rr	PPP/P[1]	PPP/P[1]	----	----
BEQ *rel8*	Branch if Equal (if Z = 1)	REL	27 rr	PPP/P[1]	PPP/P[1]	----	----
BGE *rel8*	Branch if Greater Than or Equal (if $N \oplus V = 0$) (signed)	REL	2C rr	PPP/P[1]	PPP/P[1]	----	----
BGND	Place CPU in Background Mode see *CPU12 Reference Manual*	INH	00	VfPPP	VfPPP	----	----
BGT *rel8*	Branch if Greater Than (if $Z + (N \oplus V) = 0$) (signed)	REL	2E rr	PPP/P[1]	PPP/P[1]	----	----
BHI *rel8*	Branch if Higher (if $C + Z = 0$) (unsigned)	REL	22 rr	PPP/P[1]	PPP/P[1]	----	----
BHS *rel8*	Branch if Higher or Same (if C = 0) (unsigned) same function as BCC	REL	24 rr	PPP/P[1]	PPP/P[1]	----	----
BITA #*opr8i* BITA *opr8a* BITA *opr16a* BITA *oprx0_xysp* BITA *oprx9,xysp* BITA *oprx16,xysp* BITA [D,*xysp*] BITA [*oprx16,xysp*]	$(A) \bullet (M)$ Logical AND A with Memory Does not change Accumulator or Memory	IMM DIR EXT IDX IDX1 IDX2 [D,IDX] [IDX2]	85 ii 95 dd B5 hh ll A5 xb A5 xb ff A5 xb ee ff A5 xb A5 xb ee ff	P rPf rPO rPf rPO frPP fIfrPf fIPrPf	P rfP rOP rfP rPO frPP fIfrfP fIPrfP	----	ΔΔ0–
BITB #*opr8i* BITB *opr8a* BITB *opr16a* BITB *oprx0_xysp* BITB *oprx9,xysp* BITB *oprx16,xysp* BITB [D,*xysp*] BITB [*oprx16,xysp*]	$(B) \bullet (M)$ Logical AND B with Memory Does not change Accumulator or Memory	IMM DIR EXT IDX IDX1 IDX2 [D,IDX] [IDX2]	C5 ii D5 dd F5 hh ll E5 xb E5 xb ff E5 xb ee ff E5 xb E5 xb ee ff	P rPf rPO rPf rPO frPP fIfrPf fIPrPf	P rfP rOP rfP rPO frPP fIfrfP fIPrfP	----	ΔΔ0–
BLE *rel8*	Branch if Less Than or Equal (if $Z + (N \oplus V) = 1$) (signed)	REL	2F rr	PPP/P[1]	PPP/P[1]	----	----
BLO *rel8*	Branch if Lower (if C = 1) (unsigned) same function as BCS	REL	25 rr	PPP/P[1]	PPP/P[1]	----	----

Note 1. PPP/P indicates this instruction takes three cycles to refill the instruction queue if the branch is taken and one program fetch cycle if the branch is not taken.

Table A-1. Instruction Set Summary (Sheet 3 of 14)

Source Form	Operation	Addr. Mode	Machine Coding (hex)	Access Detail		S X H I	N Z V C
				HCS12	M68HC12		
BLS *rel8*	Branch if Lower or Same (if C + Z = 1) (unsigned)	REL	23 rr	PPP/P[1]	PPP/P[1]	----	----
BLT *rel8*	Branch if Less Than (if N ⊕ V = 1) (signed)	REL	2D rr	PPP/P[1]	PPP/P[1]	----	----
BMI *rel8*	Branch if Minus (if N = 1)	REL	2B rr	PPP/P[1]	PPP/P[1]	----	----
BNE *rel8*	Branch if Not Equal (if Z = 0)	REL	26 rr	PPP/P[1]	PPP/P[1]	----	----
BPL *rel8*	Branch if Plus (if N = 0)	REL	2A rr	PPP/P[1]	PPP/P[1]	----	----
BRA *rel8*	Branch Always (if 1 = 1)	REL	20 rr	PPP	PPP	----	----
BRCLR *opr8a, msk8, rel8* BRCLR *opr16a, msk8, rel8* BRCLR *oprx0_xysp, msk8, rel8* BRCLR *oprx9,xysp, msk8, rel8* BRCLR *oprx16,xysp, msk8, rel8*	Branch if (M) • (mm) = 0 (if All Selected Bit(s) Clear)	DIR EXT IDX IDX1 IDX2	4F dd mm rr 1F hh ll mm rr 0F xb mm rr 0F xb ff mm rr 0F xb ee ff mm rr	rPPP rfPPP rPPP rfPPP PrfPPP	rPPP rfPPP rPPP rffPPP frPffPPP	----	----
BRN *rel8*	Branch Never (if 1 = 0)	REL	21 rr	P	P	----	----
BRSET *opr8, msk8, rel8* BRSET *opr16a, msk8, rel8* BRSET *oprx0_xysp, msk8, rel8* BRSET *oprx9,xysp, msk8, rel8* BRSET *oprx16,xysp, msk8, rel8*	Branch if $(\overline{M})$ • (mm) = 0 (if All Selected Bit(s) Set)	DIR EXT IDX IDX1 IDX2	4E dd mm rr 1E hh ll mm rr 0E xb mm rr 0E xb ff mm rr 0E xb ee ff mm rr	rPPP rfPPP rPPP rfPPP PrfPPP	rPPP rfPPP rPPP rffPPP frPffPPP	----	----
BSET *opr8, msk8* BSET *opr16a, msk8* BSET *oprx0_xysp, msk8* BSET *oprx9,xysp, msk8* BSET *oprx16,xysp, msk8*	(M) + (mm) ⇒ M Set Bit(s) in Memory	DIR EXT IDX IDX1 IDX2	4C dd mm 1C hh ll mm 0C xb mm 0C xb ff mm 0C xb ee ff mm	rPwO rPwP rPwO rPwP frPwPO	rPOw rPPw rPOw rPwP frPwOP	----	ΔΔ0-
BSR *rel8*	(SP) – 2 ⇒ SP; RTN_H:RTN_L ⇒ $M_{(SP)}$:$M_{(SP+1)}$ Subroutine address ⇒ PC Branch to Subroutine	REL	07 rr	SPPP	PPPS	----	----
BVC *rel8*	Branch if Overflow Bit Clear (if V = 0)	REL	28 rr	PPP/P[1]	PPP/P[1]	----	----
BVS *rel8*	Branch if Overflow Bit Set (if V = 1)	REL	29 rr	PPP/P[1]	PPP/P[1]	----	----
CALL *opr16a, page* CALL *oprx0_xysp, page* CALL *oprx9,xysp, page* CALL *oprx16,xysp, page* CALL [D,*xysp*] CALL [*oprx16, xysp*]	(SP) – 2 ⇒ SP; RTN_H:RTN_L ⇒ $M_{(SP)}$:$M_{(SP+1)}$ (SP) – 1 ⇒ SP; (PPG) ⇒ $M_{(SP)}$; pg ⇒ PPAGE register; Program address ⇒ PC Call subroutine in extended memory (Program may be located on another expansion memory page.) Indirect modes get program address and new pg value based on pointer.	EXT IDX IDX1 IDX2 [D,IDX] [IDX2]	4A hh ll pg 4B xb pg 4B xb ff pg 4B xb ee ff pg 4B xb 4B xb ee ff	gnSsPPP gnSsPPP gnSsPPP fgnSsPPP fIignSsPPP fIignSsPPP	gnfSsPPP gnfSsPPP gnfSsPPP fgnfSsPPP fIignSsPPP fIignSsPPP	----	----
CBA	(A) – (B) Compare 8-Bit Accumulators	INH	18 17	OO	OO	----	ΔΔΔΔ
CLC	0 ⇒ C *Translates to* ANDCC #$FE	IMM	10 FE	P	P	----	---0
CLI	0 ⇒ I *Translates to* ANDCC #$EF (enables I-bit interrupts)	IMM	10 EF	P	P	---0	----
CLR *opr16a* CLR *oprx0_xysp* CLR *oprx9,xysp* CLR *oprx16,xysp* CLR [D,*xysp*] CLR [*oprx16,xysp*] CLRA CLRB	0 ⇒ M Clear Memory Location 0 ⇒ A Clear Accumulator A 0 ⇒ B Clear Accumulator B	EXT IDX IDX1 IDX2 [D,IDX] [IDX2] INH INH	79 hh ll 69 xb 69 xb ff 69 xb ee ff 69 xb 69 xb ee ff 87 C7	PwO Pw PwO PwP PIfw PIPw O O	wOP Pw PwO PwP PIfPw PIPPw O O	----	0100
CLV	0 ⇒ V *Translates to* ANDCC #$FD	IMM	10 FD	P	P	----	--0-

Note 1. PPP/P indicates this instruction takes three cycles to refill the instruction queue if the branch is taken and one program fetch cycle if the branch is not taken.

Table A-1. Instruction Set Summary (Sheet 4 of 14)

Source Form	Operation	Addr. Mode	Machine Coding (hex)	Access Detail HCS12	Access Detail M68HC12	SXHI	NZVC

CMPA #*opr8i*
CMPA *opr8a*
CMPA *opr16a*
CMPA *oprx0_xysp*
CMPA *oprx9,xysp*
CMPA *oprx16,xysp*
CMPA [D,*xysp*]
CMPA [*oprx16,xysp*]
	(A) – (M)
	Compare Accumulator A with Memory
		IMM
		DIR
		EXT
		IDX
		IDX1
		IDX2
		[D,IDX]
		[IDX2]
			81 ii
			91 dd
			B1 hh ll
			A1 xb
			A1 xb ff
			A1 xb ee ff
			A1 xb
			A1 xb ee ff
				P
				rPf
				rPO
				rPf
				rPO
				frPP
				fIfrPf
				fIPrPf
					P
					rfP
					rOP
					rfP
					rPO
					frPP
					fIfrfP
					fIPrfP
						– – – –	Δ Δ Δ Δ

CMPB #*opr8i*
CMPB *opr8a*
CMPB *opr16a*
CMPB *oprx0_xysp*
CMPB *oprx9,xysp*
CMPB *oprx16,xysp*
CMPB [D,*xysp*]
CMPB [*oprx16,xysp*]
	(B) – (M)
	Compare Accumulator B with Memory
		IMM
		DIR
		EXT
		IDX
		IDX1
		IDX2
		[D,IDX]
		[IDX2]
			C1 ii
			D1 dd
			F1 hh ll
			E1 xb
			E1 xb ff
			E1 xb ee ff
			E1 xb
			E1 xb ee ff
				P
				rPf
				rPO
				rPf
				rPO
				frPP
				fIfrPf
				fIPrPf
					P
					rfP
					rOP
					rfP
					rPO
					frPP
					fIfrfP
					fIPrfP
						– – – –	Δ Δ Δ Δ

COM *opr16a* COM *oprx0_xysp* COM *oprx9,xysp* COM *oprx16,xysp* COM [D,*xysp*] COM [*oprx16,xysp*] COMA COMB	$(\overline{M}) \Rightarrow M$ *equivalent to* $FF – (M) ⇒ M 1's Complement Memory Location $(\overline{A}) \Rightarrow A$ Complement Accumulator A $(\overline{B}) \Rightarrow B$ Complement Accumulator B	EXT IDX IDX1 IDX2 [D,IDX] [IDX2] INH INH	71 hh ll 61 xb 61 xb ff 61 xb ee ff 61 xb 61 xb ee ff 41 51	rPwO rPw rPwO frPwP fIfrPw fIPrPw O O	rOPw rPw rPOw frPPw fIfrPw fIPrPw O O	– – – –	Δ Δ 0 1
CPD #*opr16i* CPD *opr8a* CPD *opr16a* CPD *oprx0_xysp* CPD *oprx9,xysp* CPD *oprx16,xysp* CPD [D,*xysp*] CPD [*oprx16,xysp*]	(A:B) – (M:M+1) Compare D to Memory (16-Bit)	IMM DIR EXT IDX IDX1 IDX2 [D,IDX] [IDX2]	8C jj kk 9C dd BC hh ll AC xb AC xb ff AC xb ee ff AC xb AC xb ee ff	PO RPf RPO RPf RPO fRPP fIfRPf fIPRPf	OP RfP ROP RfP RPO fRPP fIfRfP fIPRfP	– – – –	Δ Δ Δ Δ
CPS #*opr16i* CPS *opr8a* CPS *opr16a* CPS *oprx0_xysp* CPS *oprx9,xysp* CPS *oprx16,xysp* CPS [D,*xysp*] CPS [*oprx16,xysp*]	(SP) – (M:M+1) Compare SP to Memory (16-Bit)	IMM DIR EXT IDX IDX1 IDX2 [D,IDX] [IDX2]	8F jj kk 9F dd BF hh ll AF xb AF xb ff AF xb ee ff AF xb AF xb ee ff	PO RPf RPO RPf RPO fRPP fIfRPf fIPRPf	OP RfP ROP RfP RPO fRPP fIfRfP fIPRfP	– – – –	Δ Δ Δ Δ
CPX #*opr16i* CPX *opr8a* CPX *opr16a* CPX *oprx0_xysp* CPX *oprx9,xysp* CPX *oprx16,xysp* CPX [D,*xysp*] CPX [*oprx16,xysp*]	(X) – (M:M+1) Compare X to Memory (16-Bit)	IMM DIR EXT IDX IDX1 IDX2 [D,IDX] [IDX2]	8E jj kk 9E dd BE hh ll AE xb AE xb ff AE xb ee ff AE xb AE xb ee ff	PO RPf RPO RPf RPO fRPP fIfRPf fIPRPf	OP RfP ROP RfP RPO fRPP fIfRfP fIPRfP	– – – –	Δ Δ Δ Δ
CPY #*opr16i* CPY *opr8a* CPY *opr16a* CPY *oprx0_xysp* CPY *oprx9,xysp* CPY *oprx16,xysp* CPY [D,*xysp*] CPY [*oprx16,xysp*]	(Y) – (M:M+1) Compare Y to Memory (16-Bit)	IMM DIR EXT IDX IDX1 IDX2 [D,IDX] [IDX2]	8D jj kk 9D dd BD hh ll AD xb AD xb ff AD xb ee ff AD xb AD xb ee ff	PO RPf RPO RPf RPO fRPP fIfRPf fIPRPf	OP RfP ROP RfP RPO fRPP fIfRfP fIPRfP	– – – –	Δ Δ Δ Δ
DAA	Adjust Sum to BCD Decimal Adjust Accumulator A	INH	18 07	OfO	OfO	– – – –	Δ Δ ? Δ
DBEQ *abdxys, rel9*	(cntr) – 1⇒ cntr if (cntr) = 0, then Branch else Continue to next instruction Decrement Counter and Branch if = 0 (cntr = A, B, D, X, Y, or SP)	REL (9-bit)	04 lb rr	PPP (branch) PPO (no branch)	PPP	– – – –	– – – –

Table A-1. Instruction Set Summary (Sheet 5 of 14)

Source Form	Operation	Addr. Mode	Machine Coding (hex)	Access Detail HCS12	M68HC12	S X H I	N Z V C
DBNE *abdxys, rel9*	(cntr) – 1 ⇒ cntr If (cntr) not = 0, then Branch; else Continue to next instruction Decrement Counter and Branch if ≠ 0 (cntr = A, B, D, X, Y, or SP)	REL (9-bit)	04 lb rr	PPP (branch) PPO (no branch)	PPP	– – – –	– – – –
DEC *opr16a* DEC *oprx0_xysp* DEC *oprx9,xysp* DEC *oprx16,xysp* DEC [D,*xysp*] DEC [*oprx16,xysp*] DECA DECB	(M) – $01 ⇒ M Decrement Memory Location (A) – $01 ⇒ A Decrement A (B) – $01 ⇒ B Decrement B	EXT IDX IDX1 IDX2 [D,IDX] [IDX2] INH INH	73 hh ll 63 xb 63 xb ff 63 xb ee ff 63 xb 63 xb ee ff 43 53	rPwO rPw rPwO frPwP fIfrPw fIPrPw O O	rOPw rPw rPOw frPPw fIfrPw fIPrPw O O	– – – –	Δ Δ Δ –
DES	(SP) – $0001 ⇒ SP *Translates to* LEAS –1,SP	IDX	1B 9F	Pf	PP[1]	– – – –	– – – –
DEX	(X) – $0001 ⇒ X Decrement Index Register X	INH	09	O	O	– – – –	– Δ – –
DEY	(Y) – $0001 ⇒ Y Decrement Index Register Y	INH	03	O	O	– – – –	– Δ – –
EDIV	(Y:D) ÷ (X) ⇒ Y Remainder ⇒ D 32 by 16 Bit ⇒ 16 Bit Divide (unsigned)	INH	11	ffffffffffO	ffffffffffO	– – – –	Δ Δ Δ Δ
EDIVS	(Y:D) ÷ (X) ⇒ Y Remainder ⇒ D 32 by 16 Bit ⇒ 16 Bit Divide (signed)	INH	18 14	OffffffffffO	OffffffffffO	– – – –	Δ Δ Δ Δ
EMACS *opr16a* [2]	$(M_{(X)}:M_{(X+1)}) \times (M_{(Y)}:M_{(Y+1)}) + (M\sim M+3) \Rightarrow M\sim M+3$ 16 by 16 Bit ⇒ 32 Bit Multiply and Accumulate (signed)	Special	18 12 hh ll	ORROfffRRfW- WP	ORROfffRRfW- WP	– – – –	Δ Δ Δ Δ
EMAXD *oprx0_xysp* EMAXD *oprx9,xysp* EMAXD *oprx16,xysp* EMAXD [D,*xysp*] EMAXD [*oprx16,xysp*]	MAX((D), (M:M+1)) ⇒ D MAX of 2 Unsigned 16-Bit Values N, Z, V and C status bits reflect result of internal compare ((D) – (M:M+1))	IDX IDX1 IDX2 [D,IDX] [IDX2]	18 1A xb 18 1A xb ff 18 1A xb ee ff 18 1A xb 18 1A xb ee ff	ORPf ORPO OfRPP OfIfRPf OfIPRPf	ORfP ORPO OfRPP OfIfRfP OfIPRfP	– – – –	Δ Δ Δ Δ
EMAXM *oprx0_xysp* EMAXM *oprx9,xysp* EMAXM *oprx16,xysp* EMAXM [D,*xysp*] EMAXM [*oprx16,xysp*]	MAX((D), (M:M+1)) ⇒ M:M+1 MAX of 2 Unsigned 16-Bit Values N, Z, V and C status bits reflect result of internal compare ((D) – (M:M+1))	IDX IDX1 IDX2 [D,IDX] [IDX2]	18 1E xb 18 1E xb ff 18 1E xb ee ff 18 1E xb 18 1E xb ee ff	ORPW ORPWO OfRPWP OfIfRPW OfIPRPW	ORPW ORPWO OfRPWP OfIfRPW OfIPRPW	– – – –	Δ Δ Δ Δ
EMIND *oprx0_xysp* EMIND *oprx9,xysp* EMIND *oprx16,xysp* EMIND [D,*xysp*] EMIND [*oprx16,xysp*]	MIN((D), (M:M+1)) ⇒ D MIN of 2 Unsigned 16-Bit Values N, Z, V and C status bits reflect result of internal compare ((D) – (M:M+1))	IDX IDX1 IDX2 [D,IDX] [IDX2]	18 1B xb 18 1B xb ff 18 1B xb ee ff 18 1B xb 18 1B xb ee ff	ORPf ORPO OfRPP OfIfRPf OfIPRPf	ORfP ORPO OfRPP OfIfRfP OfIPRfP	– – – –	Δ Δ Δ Δ
EMINM *oprx0_xysp* EMINM *oprx9,xysp* EMINM *oprx16,xysp* EMINM [D,*xysp*] EMINM [*oprx16,xysp*]	MIN((D), (M:M+1)) ⇒ M:M+1 MIN of 2 Unsigned 16-Bit Values N, Z, V and C status bits reflect result of internal compare ((D) – (M:M+1))	IDX IDX1 IDX2 [D,IDX] [IDX2]	18 1F xb 18 1F xb ff 18 1F xb ee ff 18 1F xb 18 1F xb ee ff	ORPW ORPWO OfRPWP OfIfRPW OfIPRPW	ORPW ORPWO OfRPWP OfIfRPW OfIPRPW	– – – –	Δ Δ Δ Δ
EMUL	(D) × (Y) ⇒ Y:D 16 by 16 Bit Multiply (unsigned)	INH	13	ffO	ffO	– – – –	Δ Δ – Δ
EMULS	(D) × (Y) ⇒ Y:D 16 by 16 Bit Multiply (signed)	INH	18 13	OfO (if followed by page 2 instruction) OffO	OfO OfO	– – – –	Δ Δ – Δ
EORA #*opr8i* EORA *opr8a* EORA *opr16a* EORA *oprx0_xysp* EORA *oprx9,xysp* EORA *oprx16,xysp* EORA [D,*xysp*] EORA [*oprx16,xysp*]	(A) ⊕ (M) ⇒ A Exclusive-OR A with Memory	IMM DIR EXT IDX IDX1 IDX2 [D,IDX] [IDX2]	88 ii 98 dd B8 hh ll A8 xb A8 xb ff A8 xb ee ff A8 xb A8 xb ee ff	P rPf rPO rPf rPO frPP fIfrPf fIPrPf	P rfP rOP rfP rPO frPP fIfrfP fIPrfP	– – – –	Δ Δ 0 –

Notes:
1. Due to internal CPU requirements, the program word fetch is performed twice to the same address during this instruction.
2. *opr16a* is an extended address specification. Both X and Y point to source operands.

Table A-1. Instruction Set Summary (Sheet 6 of 14)

Source Form	Operation	Addr. Mode	Machine Coding (hex)	Access Detail HCS12	M68HC12	SXHI	NZVC
EORB #*opr8i* EORB *opr8a* EORB *opr16a* EORB *oprx0_xysp* EORB *oprx9,xysp* EORB *oprx16,xysp* EORB [D,*xysp*] EORB [*oprx16,xysp*]	(B) ⊕ (M) ⇒ B Exclusive-OR B with Memory	IMM DIR EXT IDX IDX1 IDX2 [D,IDX] [IDX2]	C8 ii D8 dd F8 hh ll E8 xb E8 xb ff E8 xb ee ff E8 xb E8 xb ee ff	P rPf rPO rPf rPO frPP fIfrPf fIPrPf	P rfP rOP rfP rPO frPP fIfrfP fIPrfP	----	ΔΔ0-
ETBL *oprx0_xysp*	(M:M+1)+ [(B)×((M+2:M+3) − (M:M+1))] ⇒ D 16-Bit Table Lookup and Interpolate Initialize B, and index before ETBL. <ea> points at first table entry (M:M+1) and B is fractional part of lookup value (no indirect addr. modes or extensions allowed)	IDX	18 3F xb	ORRffffffP	ORRffffffP	---- C Bit is undefined in HC12	ΔΔ-Δ ?
EXG *abcdxys,abcdxys*	(r1) ⇔ (r2) (if r1 and r2 same size) *or* $00:(r1) ⇒ r2 (if r1=8-bit; r2=16-bit) *or* ($r1_{low}$) ⇔ (r2) (if r1=16-bit; r2=8-bit) r1 and r2 may be A, B, CCR, D, X, Y, or SP	INH	B7 eb	P	P	----	----
FDIV	(D) ÷ (X) ⇒ X; Remainder ⇒ D 16 by 16 Bit Fractional Divide	INH	18 11	OffffffffffO	OffffffffffO	----	-ΔΔΔ
IBEQ *abdxys, rel9*	(cntr) + 1⇒ cntr If (cntr) = 0, then Branch else Continue to next instruction Increment Counter and Branch if = 0 (cntr = A, B, D, X, Y, or SP)	REL (9-bit)	04 lb rr	PPP (branch) PPO (no branch)	PPP	----	----
IBNE *abdxys, rel9*	(cntr) + 1⇒ cntr if (cntr) not = 0, then Branch; else Continue to next instruction Increment Counter and Branch if ≠ 0 (cntr = A, B, D, X, Y, or SP)	REL (9-bit)	04 lb rr	PPP (branch) PPO (no branch)	PPP	----	----
IDIV	(D) ÷ (X) ⇒ X; Remainder ⇒ D 16 by 16 Bit Integer Divide (unsigned)	INH	18 10	OffffffffffO	OffffffffffO	----	-Δ0Δ
IDIVS	(D) ÷ (X) ⇒ X; Remainder ⇒ D 16 by 16 Bit Integer Divide (signed)	INH	18 15	OffffffffffO	OffffffffffO	----	ΔΔΔΔ
INC *opr16a* INC *oprx0_xysp* INC *oprx9,xysp* INC *oprx16,xysp* INC [D,*xysp*] INC [*oprx16,xysp*] INCA INCB	(M) + $01 ⇒ M Increment Memory Byte (A) + $01 ⇒ A Increment Acc. A (B) + $01 ⇒ B Increment Acc. B	EXT IDX IDX1 IDX2 [D,IDX] [IDX2] INH INH	72 hh ll 62 xb 62 xb ff 62 xb ee ff 62 xb 62 xb ee ff 42 52	rPwO rPw rPwO frPwP fIfrPw fIPrPw O O	rOPw rPw rPOw frPPw fIfrPw fIPrPw O O	----	ΔΔΔ-
INS	(SP) + $0001 ⇒ SP *Translates to* LEAS 1,SP	IDX	1B 81	Pf	PP[1]	----	----
INX	(X) + $0001 ⇒ X Increment Index Register X	INH	08	O	O	----	-Δ--
INY	(Y) + $0001 ⇒ Y Increment Index Register Y	INH	02	O	O	----	-Δ--
JMP *opr16a* JMP *oprx0_xysp* JMP *oprx9,xysp* JMP *oprx16,xysp* JMP [D,*xysp*] JMP [*oprx16,xysp*]	Routine address ⇒ PC Jump	EXT IDX IDX1 IDX2 [D,IDX] [IDX2]	06 hh ll 05 xb 05 xb ff 05 xb ee ff 05 xb 05 xb ee ff	PPP PPP PPP fPPP fIfPPP fIfPPP	PPP PPP PPP fPPP fIfPPP fIfPPP	----	----

Note 1. Due to internal CPU requirements, the program word fetch is performed twice to the same address during this instruction.

Table A-1. Instruction Set Summary (Sheet 7 of 14)

Source Form	Operation	Addr. Mode	Machine Coding (hex)	Access Detail HCS12	Access Detail M68HC12	S X H I	N Z V C
JSR *opr8a*	$(SP) - 2 \Rightarrow SP$; $RTN_H:RTN_L \Rightarrow M_{(SP)}:M_{(SP+1)}$; Subroutine address ⇒ PC	DIR	17 dd	SPPP	PPPS	– – – –	– – – –
JSR *opr16a*		EXT	16 hh ll	SPPP	PPPS		
JSR *oprx0_xysp*		IDX	15 xb	PPPS	PPPS		
JSR *oprx9,xysp*		IDX1	15 xb ff	PPPS	PPPS		
JSR *oprx16,xysp*	Jump to Subroutine	IDX2	15 xb ee ff	fPPPS	fPPPS		
JSR [D,*xysp*]		[D,IDX]	15 xb	fIfPPPS	fIfPPPS		
JSR [*oprx16,xysp*]		[IDX2]	15 xb ee ff	fIfPPPS	fIfPPPS		
LBCC *rel16*	Long Branch if Carry Clear (if C = 0)	REL	18 24 qq rr	OPPP/OPO[1]	OPPP/OPO[1]	– – – –	– – – –
LBCS *rel16*	Long Branch if Carry Set (if C = 1)	REL	18 25 qq rr	OPPP/OPO[1]	OPPP/OPO[1]	– – – –	– – – –
LBEQ *rel16*	Long Branch if Equal (if Z = 1)	REL	18 27 qq rr	OPPP/OPO[1]	OPPP/OPO[1]	– – – –	– – – –
LBGE *rel16*	Long Branch Greater Than or Equal (if $N \oplus V = 0$) (signed)	REL	18 2C qq rr	OPPP/OPO[1]	OPPP/OPO[1]	– – – –	– – – –
LBGT *rel16*	Long Branch if Greater Than (if $Z + (N \oplus V) = 0$) (signed)	REL	18 2E qq rr	OPPP/OPO[1]	OPPP/OPO[1]	– – – –	– – – –
LBHI *rel16*	Long Branch if Higher (if $C + Z = 0$) (unsigned)	REL	18 22 qq rr	OPPP/OPO[1]	OPPP/OPO[1]	– – – –	– – – –
LBHS *rel16*	Long Branch if Higher or Same (if C = 0) (unsigned) same function as LBCC	REL	18 24 qq rr	OPPP/OPO[1]	OPPP/OPO[1]	– – – –	– – – –
LBLE *rel16*	Long Branch if Less Than or Equal (if $Z + (N \oplus V) = 1$) (signed)	REL	18 2F qq rr	OPPP/OPO[1]	OPPP/OPO[1]	– – – –	– – – –
LBLO *rel16*	Long Branch if Lower (if C = 1) (unsigned) *same function as LBCS*	REL	18 25 qq rr	OPPP/OPO[1]	OPPP/OPO[1]	– – – –	– – – –
LBLS *rel16*	Long Branch if Lower or Same (if $C + Z = 1$) (unsigned)	REL	18 23 qq rr	OPPP/OPO[1]	OPPP/OPO[1]	– – – –	– – – –
LBLT *rel16*	Long Branch if Less Than (if $N \oplus V = 1$) (signed)	REL	18 2D qq rr	OPPP/OPO[1]	OPPP/OPO[1]	– – – –	– – – –
LBMI *rel16*	Long Branch if Minus (if N = 1)	REL	18 2B qq rr	OPPP/OPO[1]	OPPP/OPO[1]	– – – –	– – – –
LBNE *rel16*	Long Branch if Not Equal (if Z = 0)	REL	18 26 qq rr	OPPP/OPO[1]	OPPP/OPO[1]	– – – –	– – – –
LBPL *rel16*	Long Branch if Plus (if N = 0)	REL	18 2A qq rr	OPPP/OPO[1]	OPPP/OPO[1]	– – – –	– – – –
LBRA *rel16*	Long Branch Always (if 1=1)	REL	18 20 qq rr	OPPP	OPPP	– – – –	– – – –
LBRN *rel16*	Long Branch Never (if 1 = 0)	REL	18 21 qq rr	OPO	OPO	– – – –	– – – –
LBVC *rel16*	Long Branch if Overflow Bit Clear (if V=0)	REL	18 28 qq rr	OPPP/OPO[1]	OPPP/OPO[1]	– – – –	– – – –
LBVS *rel16*	Long Branch if Overflow Bit Set (if V = 1)	REL	18 29 qq rr	OPPP/OPO[1]	OPPP/OPO[1]	– – – –	– – – –
LDAA #*opr8i*	$(M) \Rightarrow A$ Load Accumulator A	IMM	86 ii	P	P	– – – –	Δ Δ 0 –
LDAA *opr8a*		DIR	96 dd	rPf	rfP		
LDAA *opr16a*		EXT	B6 hh ll	rPO	rOP		
LDAA *oprx0_xysp*		IDX	A6 xb	rPf	rfP		
LDAA *oprx9,xysp*		IDX1	A6 xb ff	rPO	rPO		
LDAA *oprx16,xysp*		IDX2	A6 xb ee ff	frPP	frPP		
LDAA [D,*xysp*]		[D,IDX]	A6 xb	fIfrPf	fIfrfP		
LDAA [*oprx16,xysp*]		[IDX2]	A6 xb ee ff	fIPrPf	fIPrfP		
LDAB #*opr8i*	$(M) \Rightarrow B$ Load Accumulator B	IMM	C6 ii	P	P	– – – –	Δ Δ 0 –
LDAB *opr8a*		DIR	D6 dd	rPf	rfP		
LDAB *opr16a*		EXT	F6 hh ll	rPO	rOP		
LDAB *oprx0_xysp*		IDX	E6 xb	rPf	rfP		
LDAB *oprx9,xysp*		IDX1	E6 xb ff	rPO	rPO		
LDAB *oprx16,xysp*		IDX2	E6 xb ee ff	frPP	frPP		
LDAB [D,*xysp*]		[D,IDX]	E6 xb	fIfrPf	fIfrfP		
LDAB [*oprx16,xysp*]		[IDX2]	E6 xb ee ff	fIPrPf	fIPrfP		
LDD #*opr16i*	$(M:M+1) \Rightarrow A:B$ Load Double Accumulator D (A:B)	IMM	CC jj kk	PO	OP	– – – –	Δ Δ 0 –
LDD *opr8a*		DIR	DC dd	RPf	RfP		
LDD *opr16a*		EXT	FC hh ll	RPO	ROP		
LDD *oprx0_xysp*		IDX	EC xb	RPf	RfP		
LDD *oprx9,xysp*		IDX1	EC xb ff	RPO	RPO		
LDD *oprx16,xysp*		IDX2	EC xb ee ff	fRPP	fRPP		
LDD [D,*xysp*]		[D,IDX]	EC xb	fIfRPf	fIfRfP		
LDD [*oprx16,xysp*]		[IDX2]	EC xb ee ff	fIPRPf	fIPRfP		

Note 1. OPPP/OPO indicates this instruction takes four cycles to refill the instruction queue if the branch is taken and three cycles if the branch is not taken.

Table A-1. Instruction Set Summary (Sheet 8 of 14)

Source Form	Operation	Addr. Mode	Machine Coding (hex)	Access Detail HCS12	Access Detail M68HC12	SXHI	NZVC
LDS #*opr16i*	(M:M+1) ⇒ SP	IMM	CF jj kk	PO	OP	----	ΔΔ0–
LDS *opr8a*	Load Stack Pointer	DIR	DF dd	RPf	RfP		
LDS *opr16a*		EXT	FF hh ll	RPO	ROP		
LDS *oprx0_xysp*		IDX	EF xb	RPf	RfP		
LDS *oprx9,xysp*		IDX1	EF xb ff	RPO	RPO		
LDS *oprx16,xysp*		IDX2	EF xb ee ff	fRPP	fRPP		
LDS [D,*xysp*]		[D,IDX]	EF xb	fIfRPf	fIfRfP		
LDS [*oprx16,xysp*]		[IDX2]	EF xb ee ff	fIPRPf	fIPRfP		
LDX #*opr16i*	(M:M+1) ⇒ X	IMM	CE jj kk	PO	OP	----	ΔΔ0–
LDX *opr8a*	Load Index Register X	DIR	DE dd	RPf	RfP		
LDX *opr16a*		EXT	FE hh ll	RPO	ROP		
LDX *oprx0_xysp*		IDX	EE xb	RPf	RfP		
LDX *oprx9,xysp*		IDX1	EE xb ff	RPO	RPO		
LDX *oprx16,xysp*		IDX2	EE xb ee ff	fRPP	fRPP		
LDX [D,*xysp*]		[D,IDX]	EE xb	fIfRPf	fIfRfP		
LDX [*oprx16,xysp*]		[IDX2]	EE xb ee ff	fIPRPf	fIPRfP		
LDY #*opr16i*	(M:M+1) ⇒ Y	IMM	CD jj kk	PO	OP	----	ΔΔ0–
LDY *opr8a*	Load Index Register Y	DIR	DD dd	RPf	RfP		
LDY *opr16a*		EXT	FD hh ll	RPO	ROP		
LDY *oprx0_xysp*		IDX	ED xb	RPf	RfP		
LDY *oprx9,xysp*		IDX1	ED xb ff	RPO	RPO		
LDY *oprx16,xysp*		IDX2	ED xb ee ff	fRPP	fRPP		
LDY [D,*xysp*]		[D,IDX]	ED xb	fIfRPf	fIfRfP		
LDY [*oprx16,xysp*]		[IDX2]	ED xb ee ff	fIPRPf	fIPRfP		
LEAS *oprx0_xysp*	Effective Address ⇒ SP	IDX	1B xb	Pf	PP[1]	----	----
LEAS *oprx9,xysp*	Load Effective Address into SP	IDX1	1B xb ff	PO	PO		
LEAS *oprx16,xysp*		IDX2	1B xb ee ff	PP	PP		
LEAX *oprx0_xysp*	Effective Address ⇒ X	IDX	1A xb	Pf	PP[1]	----	----
LEAX *oprx9,xysp*	Load Effective Address into X	IDX1	1A xb ff	PO	PO		
LEAX *oprx16,xysp*		IDX2	1A xb ee ff	PP	PP		
LEAY *oprx0_xysp*	Effective Address ⇒ Y	IDX	19 xb	Pf	PP[1]	----	----
LEAY *oprx9,xysp*	Load Effective Address into Y	IDX1	19 xb ff	PO	PO		
LEAY *oprx16,xysp*		IDX2	19 xb ee ff	PP	PP		
LSL *opr16a*	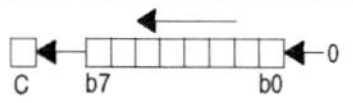 (diagram: C ◄ b7 … b0 ◄ 0)	EXT	78 hh ll	rPwO	rOPw	----	ΔΔΔΔ
LSL *oprx0_xysp*		IDX	68 xb	rPw	rPw		
LSL *oprx9,xysp*		IDX1	68 xb ff	rPwO	rPOw		
LSL *oprx16,xysp*	Logical Shift Left	IDX2	68 xb ee ff	frPPw	frPPw		
LSL [D,*xysp*]	same function as ASL	[D,IDX]	68 xb	fIfrPw	fIfrPw		
LSL [*oprx16,xysp*]		[IDX2]	68 xb ee ff	fIPrPw	fIPrPw		
LSLA	Logical Shift Accumulator A to Left	INH	48	O	O		
LSLB	Logical Shift Accumulator B to Left	INH	58	O	O		
LSLD	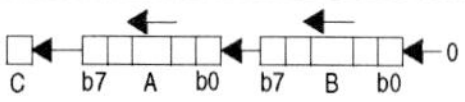 (diagram: C ◄ b7 A b0 ◄ b7 B b0 ◄ 0) Logical Shift Left D Accumulator same function as ASLD	INH	59	O	O	----	ΔΔΔΔ
LSR *opr16a*	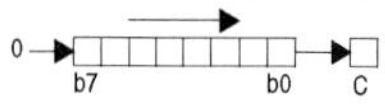 (diagram: 0 ► b7 … b0 ► C)	EXT	74 hh ll	rPwO	rOPw	----	0ΔΔΔ
LSR *oprx0_xysp*		IDX	64 xb	rPw	rPw		
LSR *oprx9,xysp*		IDX1	64 xb ff	rPwO	rPOw		
LSR *oprx16,xysp*	Logical Shift Right	IDX2	64 xb ee ff	frPwP	frPPw		
LSR [D,*xysp*]		[D,IDX]	64 xb	fIfrPw	fIfrPw		
LSR [*oprx16,xysp*]		[IDX2]	64 xb ee ff	fIPrPw	fIPrPw		
LSRA	Logical Shift Accumulator A to Right	INH	44	O	O		
LSRB	Logical Shift Accumulator B to Right	INH	54	O	O		
LSRD	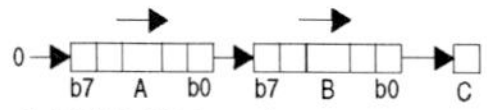 (diagram: 0 ► b7 A b0 ► b7 B b0 ► C) Logical Shift Right D Accumulator	INH	49	O	O	----	0ΔΔΔ
MAXA *oprx0_xysp*	MAX((A), (M)) ⇒ A	IDX	18 18 xb	OrPf	OrfP	----	ΔΔΔΔ
MAXA *oprx9,xysp*	MAX of 2 Unsigned 8-Bit Values	IDX1	18 18 xb ff	OrPO	OrPO		
MAXA *oprx16,xysp*		IDX2	18 18 xb ee ff	OfrPP	OfrPP		
MAXA [D,*xysp*]	N, Z, V and C status bits reflect result of	[D,IDX]	18 18 xb	OfIfrPf	OfIfrfP		
MAXA [*oprx16,xysp*]	internal compare ((A) – (M)).	[IDX2]	18 18 xb ee ff	OfIPrPf	OfIPrfP		

Note 1. Due to internal CPU requirements, the program word fetch is performed twice to the same address during this instruction.

Table A-1. Instruction Set Summary (Sheet 9 of 14)

Source Form	Operation	Addr. Mode	Machine Coding (hex)	Access Detail HCS12	Access Detail M68HC12	SXHI	NZVC
MAXM *oprx0_xysp*	MAX((A), (M)) $\Rightarrow$ M	IDX	18 1C xb	OrPw	OrPw	----	ΔΔΔΔ
MAXM *oprx9,xysp*	MAX of 2 Unsigned 8-Bit Values	IDX1	18 1C xb ff	OrPwO	OrPwO		
MAXM *oprx16,xysp*		IDX2	18 1C xb ee ff	OfrPwP	OfrPwP		
MAXM [D,*xysp*]	N, Z, V and C status bits reflect result of	[D,IDX]	18 1C xb	OfIfrPw	OfIfrPw		
MAXM [*oprx16,xysp*]	internal compare ((A) – (M)).	[IDX2]	18 1C xb ee ff	OfIPrPw	OfIPrPw		
MEM	μ (grade) $\Rightarrow M_{(Y)}$; (X) + 4 $\Rightarrow$ X; (Y) + 1 $\Rightarrow$ Y; A unchanged if (A) < P1 or (A) > P2 then μ = 0, else μ = MIN[((A) – P1)×S1, (P2 – (A))×S2, $FF] where: A = current crisp input value; X points at 4-byte data structure that describes a trapezoidal membership function (P1, P2, S1, S2); Y points at fuzzy input (RAM location). See *CPU12 Reference Manual* for special cases.	Special	01	RRfOw	RRfOw	--?-	????
MINA *oprx0_xysp*	MIN((A), (M)) $\Rightarrow$ A	IDX	18 19 xb	OrPf	OrfP	----	ΔΔΔΔ
MINA *oprx9,xysp*	MIN of 2 Unsigned 8-Bit Values	IDX1	18 19 xb ff	OrPO	OrPO		
MINA *oprx16,xysp*		IDX2	18 19 xb ee ff	OfrPP	OfrPP		
MINA [D,*xysp*]	N, Z, V and C status bits reflect result of	[D,IDX]	18 19 xb	OfIfrPf	OfIfrfP		
MINA [*oprx16,xysp*]	internal compare ((A) – (M)).	[IDX2]	18 19 xb ee ff	OfIPrPf	OfIPrfP		
MINM *oprx0_xysp*	MIN((A), (M)) $\Rightarrow$ M	IDX	18 1D xb	OrPw	OrPw	----	ΔΔΔΔ
MINM *oprx9,xysp*	MIN of 2 Unsigned 8-Bit Values	IDX1	18 1D xb ff	OrPwO	OrPwO		
MINM *oprx16,xysp*		IDX2	18 1D xb ee ff	OfrPwP	OfrPwP		
MINM [D,*xysp*]	N, Z, V and C status bits reflect result of	[D,IDX]	18 1D xb	OfIfrPw	OfIfrPw		
MINM [*oprx16,xysp*]	internal compare ((A) – (M)).	[IDX2]	18 1D xb ee ff	OfIPrPw	OfIPrPw		
MOVB *#opr8, opr16a*[1]	$(M_1) \Rightarrow M_2$	IMM-EXT	18 0B ii hh ll	OPwP	OPwP	----	----
MOVB *#opr8i, oprx0_xysp*[1]	Memory to Memory Byte-Move (8-Bit)	IMM-IDX	18 08 xb ii	OPwO	OPwO		
MOVB *opr16a, opr16a*[1]		EXT-EXT	18 0C hh ll hh ll	OrPwPO	OrPwPO		
MOVB *opr16a, oprx0_xysp*[1]		EXT-IDX	18 09 xb hh ll	OPrPw	OPrPw		
MOVB *oprx0_xysp, opr16a*[1]		IDX-EXT	18 0D xb hh ll	OrPwP	OrPwP		
MOVB *oprx0_xysp, oprx0_xysp*[1]		IDX-IDX	18 0A xb xb	OrPwO	OrPwO		
MOVW *#oprx16, opr16a*[1]	$(M:M+1_1) \Rightarrow M:M+1_2$	IMM-EXT	18 03 jj kk hh ll	OPWPO	OPWPO	----	----
MOVW *#opr16i, oprx0_xysp*[1]	Memory to Memory Word-Move (16-Bit)	IMM-IDX	18 00 xb jj kk	OPPW	OPPW		
MOVW *opr16a, opr16a*[1]		EXT-EXT	18 04 hh ll hh ll	ORPWPO	ORPWPO		
MOVW *opr16a, oprx0_xysp*[1]		EXT-IDX	18 01 xb hh ll	OPRPW	OPRPW		
MOVW *oprx0_xysp, opr16a*[1]		IDX-EXT	18 05 xb hh ll	ORPWP	ORPWP		
MOVW *oprx0_xysp, oprx0_xysp*[1]		IDX-IDX	18 02 xb xb	ORPWO	ORPWO		
MUL	(A) × (B) $\Rightarrow$ A:B 8 by 8 Unsigned Multiply	INH	12	O	ffO	----	---Δ
NEG *opr16a*		EXT	70 hh ll	rPwO	rOPw	----	ΔΔΔΔ
NEG *oprx0_xysp*	0 – (M) $\Rightarrow$ M *equivalent to* $(\overline{M})$ + 1 $\Rightarrow$ M	IDX	60 xb	rPw	rPw		
NEG *oprx9,xysp*	Two's Complement Negate	IDX1	60 xb ff	rPwO	rPOw		
NEG *oprx16,xysp*		IDX2	60 xb ee ff	frPwP	frPPw		
NEG [D,*xysp*]		[D,IDX]	60 xb	fIfrPw	fIfrPw		
NEG [*oprx16,xysp*]		[IDX2]	60 xb ee ff	fIPrPw	fIPrPw		
NEGA	0 – (A) $\Rightarrow$ A *equivalent to* $(\overline{A})$ + 1 $\Rightarrow$ A Negate Accumulator A	INH	40	O	O		
NEGB	0 – (B) $\Rightarrow$ B *equivalent to* $(\overline{B})$ + 1 $\Rightarrow$ B Negate Accumulator B	INH	50	O	O		
NOP	No Operation	INH	A7	O	O	----	----
ORAA *#opr8i*	(A) + (M) $\Rightarrow$ A	IMM	8A ii	P	P	----	ΔΔ0-
ORAA *opr8a*	Logical OR A with Memory	DIR	9A dd	rPf	rfP		
ORAA *opr16a*		EXT	BA hh ll	rPO	rOP		
ORAA *oprx0_xysp*		IDX	AA xb	rPf	rfP		
ORAA *oprx9,xysp*		IDX1	AA xb ff	rPO	rPO		
ORAA *oprx16,xysp*		IDX2	AA xb ee ff	frPP	frPP		
ORAA [D,*xysp*]		[D,IDX]	AA xb	fIfrPf	fIfrfP		
ORAA [*oprx16,xysp*]		[IDX2]	AA xb ee ff	fIPrPf	fIPrfP		

Note 1. The first operand in the source code statement specifies the source for the move.

Table A-1. Instruction Set Summary (Sheet 10 of 14)

Source Form	Operation	Addr. Mode	Machine Coding (hex)	Access Detail HCS12	Access Detail M68HC12	SXHI	NZVC
ORAB *#opr8i* ORAB *opr8a* ORAB *opr16a* ORAB *oprx0_xysp* ORAB *oprx9,xysp* ORAB *oprx16,xysp* ORAB [D,*xysp*] ORAB [*oprx16,xysp*]	$(B) + (M) \Rightarrow B$ Logical OR B with Memory	IMM DIR EXT IDX IDX1 IDX2 [D,IDX] [IDX2]	CA ii DA dd FA hh ll EA xb EA xb ff EA xb ee ff EA xb EA xb ee ff	P rPf rPO rPf rPO frPP fIfrPf fIPrPf	P rfP rOP rfP rPO frPP fIfrfP fIPrfP	----	ΔΔ0-
ORCC *#opr8i*	$(CCR) + M \Rightarrow CCR$ Logical OR CCR with Memory	IMM	14 ii	P	P	⇑-⇑⇑	⇑⇑⇑⇑
PSHA	$(SP) - 1 \Rightarrow SP$; $(A) \Rightarrow M_{(SP)}$ Push Accumulator A onto Stack	INH	36	Os	Os	----	----
PSHB	$(SP) - 1 \Rightarrow SP$; $(B) \Rightarrow M_{(SP)}$ Push Accumulator B onto Stack	INH	37	Os	Os	----	----
PSHC	$(SP) - 1 \Rightarrow SP$; $(CCR) \Rightarrow M_{(SP)}$ Push CCR onto Stack	INH	39	Os	Os	----	----
PSHD	$(SP) - 2 \Rightarrow SP$; $(A:B) \Rightarrow M_{(SP)}:M_{(SP+1)}$ Push D Accumulator onto Stack	INH	3B	OS	OS	----	----
PSHX	$(SP) - 2 \Rightarrow SP$; $(X_H:X_L) \Rightarrow M_{(SP)}:M_{(SP+1)}$ Push Index Register X onto Stack	INH	34	OS	OS	----	----
PSHY	$(SP) - 2 \Rightarrow SP$; $(Y_H:Y_L) \Rightarrow M_{(SP)}:M_{(SP+1)}$ Push Index Register Y onto Stack	INH	35	OS	OS	----	----
PULA	$(M_{(SP)}) \Rightarrow A$; $(SP) + 1 \Rightarrow SP$ Pull Accumulator A from Stack	INH	32	ufO	ufO	----	----
PULB	$(M_{(SP)}) \Rightarrow B$; $(SP) + 1 \Rightarrow SP$ Pull Accumulator B from Stack	INH	33	ufO	ufO	----	----
PULC	$(M_{(SP)}) \Rightarrow CCR$; $(SP) + 1 \Rightarrow SP$ Pull CCR from Stack	INH	38	ufO	ufO	Δ⇓ΔΔ	ΔΔΔΔ
PULD	$(M_{(SP)}:M_{(SP+1)}) \Rightarrow A:B$; $(SP) + 2 \Rightarrow SP$ Pull D from Stack	INH	3A	UfO	UfO	----	----
PULX	$(M_{(SP)}:M_{(SP+1)}) \Rightarrow X_H:X_L$; $(SP) + 2 \Rightarrow SP$ Pull Index Register X from Stack	INH	30	UfO	UfO	----	----
PULY	$(M_{(SP)}:M_{(SP+1)}) \Rightarrow Y_H:Y_L$; $(SP) + 2 \Rightarrow SP$ Pull Index Register Y from Stack	INH	31	UfO	UfO	----	----
REV	MIN-MAX rule evaluation Find smallest rule input (MIN). Store to rule outputs unless fuzzy output is already larger (MAX). For rule weights see REVW. Each rule input is an 8-bit offset from the base address in Y. Each rule output is an 8-bit offset from the base address in Y. $FE separates rule inputs from rule outputs. $FF terminates the rule list. REV may be interrupted.	Special	18 3A	Orf(t,tx)O (exit + re-entry replaces comma above if interrupted) ff + Orf(t,	Orf(t,tx)O ff + Orf(t,	--?-	??Δ?
REVW	MIN-MAX rule evaluation Find smallest rule input (MIN), Store to rule outputs unless fuzzy output is already larger (MAX). Rule weights supported, optional. Each rule input is the 16-bit address of a fuzzy input. Each rule output is the 16-bit address of a fuzzy output. The value $FFFE separates rule inputs from rule outputs. $FFFF terminates the rule list. REVW may be interrupted.	Special	18 3B	ORf(t,Tx)O (loop to read weight if enabled) (r,RfRf) (exit + re-entry replaces comma above if interrupted) ffff + ORf(t,	ORf(t,Tx)O (r,RfRf) fff + ORf(t,	--?-	??Δ!

Table A-1. Instruction Set Summary (Sheet 11 of 14)

Source Form	Operation	Addr. Mode	Machine Coding (hex)	Access Detail HCS12	Access Detail M68HC12	S X H I	N Z V C
ROL *opr16a* ROL *oprx0_xysp* ROL *oprx9,xysp* ROL *oprx16,xysp* ROL [D,*xysp*] ROL [*oprx16,xysp*] ROLA ROLB	C b7 b0 Rotate Memory Left through Carry Rotate A Left through Carry Rotate B Left through Carry	EXT IDX IDX1 IDX2 [D,IDX] [IDX2] INH INH	75 hh ll 65 xb 65 xb ff 65 xb ee ff 65 xb 65 xb ee ff 45 55	rPwO rPw rPwO frPwP fIfrPw fIPrPw O O	rOPw rPw rPOw frPPw fIfrPw fIPrPw O O	– – – –	Δ Δ Δ Δ
ROR *opr16a* ROR *oprx0_xysp* ROR *oprx9,xysp* ROR *oprx16,xysp* ROR [D,*xysp*] ROR [*oprx16,xysp*] RORA RORB	b7 b0 C Rotate Memory Right through Carry Rotate A Right through Carry Rotate B Right through Carry	EXT IDX IDX1 IDX2 [D,IDX] [IDX2] INH INH	76 hh ll 66 xb 66 xb ff 66 xb ee ff 66 xb 66 xb ee ff 46 56	rPwO rPw rPwO frPwP fIfrPw fIPrPw O O	rOPw rPw rPOw frPPw fIfrPw fIPrPw O O	– – – –	Δ Δ Δ Δ
RTC	$(M_{(SP)}) \Rightarrow PPAGE; (SP) + 1 \Rightarrow SP;$ $(M_{(SP)}:M_{(SP+1)}) \Rightarrow PC_H:PC_L;$ $(SP) + 2 \Rightarrow SP$ Return from Call	INH	0A	uUnfPPP	uUnPPP	– – – –	– – – –
RTI	$(M_{(SP)}) \Rightarrow CCR; (SP) + 1 \Rightarrow SP$ $(M_{(SP)}:M_{(SP+1)}) \Rightarrow B:A; (SP) + 2 \Rightarrow SP$ $(M_{(SP)}:M_{(SP+1)}) \Rightarrow X_H:X_L; (SP) + 4 \Rightarrow SP$ $(M_{(SP)}:M_{(SP+1)}) \Rightarrow PC_H:PC_L; (SP) - 2 \Rightarrow SP$ $(M_{(SP)}:M_{(SP+1)}) \Rightarrow Y_H:Y_L; (SP) + 4 \Rightarrow SP$ Return from Interrupt	INH	0B	uUUUUPPP (with interrupt pending) uUUUUVfPPP	uUUUUPPP uUUUUfVfPPP	Δ ⇓ Δ Δ	Δ Δ Δ Δ
RTS	$(M_{(SP)}:M_{(SP+1)}) \Rightarrow PC_H:PC_L;$ $(SP) + 2 \Rightarrow SP$ Return from Subroutine	INH	3D	UfPPP	UfPPP	– – – –	– – – –
SBA	$(A) - (B) \Rightarrow A$ Subtract B from A	INH	18 16	OO	OO	– – – –	Δ Δ Δ Δ
SBCA #*opr8i* SBCA *opr8a* SBCA *opr16a* SBCA *oprx0_xysp* SBCA *oprx9,xysp* SBCA *oprx16,xysp* SBCA [D,*xysp*] SBCA [*oprx16,xysp*]	$(A) - (M) - C \Rightarrow A$ Subtract with Borrow from A	IMM DIR EXT IDX IDX1 IDX2 [D,IDX] [IDX2]	82 ii 92 dd B2 hh ll A2 xb A2 xb ff A2 xb ee ff A2 xb A2 xb ee ff	P rPf rPO rPf rPO frPP fIfrPf fIPrPf	P rfP rOP rfP rPO frPP fIfrfP fIPrfP	– – – –	Δ Δ Δ Δ
SBCB #*opr8i* SBCB *opr8a* SBCB *opr16a* SBCB *oprx0_xysp* SBCB *oprx9,xysp* SBCB *oprx16,xysp* SBCB [D,*xysp*] SBCB [*oprx16,xysp*]	$(B) - (M) - C \Rightarrow B$ Subtract with Borrow from B	IMM DIR EXT IDX IDX1 IDX2 [D,IDX] [IDX2]	C2 ii D2 dd F2 hh ll E2 xb E2 xb ff E2 xb ee ff E2 xb E2 xb ee ff	P rPf rPO rPf rPO frPP fIfrPf fIPrPf	P rfP rOP rfP rPO frPP fIfrfP fIPrfP	– – – –	Δ Δ Δ Δ
SEC	$1 \Rightarrow C$ *Translates to* ORCC #$01	IMM	14 01	P	P	– – – –	– – – 1
SEI	$1 \Rightarrow I$; (inhibit I interrupts) *Translates to* ORCC #$10	IMM	14 10	P	P	– – – 1	– – – –
SEV	$1 \Rightarrow V$ *Translates to* ORCC #$02	IMM	14 02	P	P	– – – –	– – 1 –
SEX *abc,dxys*	$00:(r1) ⇒ r2 if r1, bit 7 is 0 *or* $FF:(r1) ⇒ r2 if r1, bit 7 is 1 Sign Extend 8-bit r1 to 16-bit r2 r1 may be A, B, or CCR r2 may be D, X, Y, or SP *Alternate mnemonic for* TFR r1, r2	INH	B7 eb	P	P	– – – –	– – – –

Table A-1. Instruction Set Summary (Sheet 12 of 14)

Source Form	Operation	Addr. Mode	Machine Coding (hex)	Access Detail HCS12	Access Detail M68HC12	SXHI	NZVC
STAA *opr8a*	$(A) \Rightarrow M$	DIR	5A dd	Pw	Pw	----	ΔΔ0–
STAA *opr16a*	Store Accumulator A to Memory	EXT	7A hh ll	PwO	wOP		
STAA *oprx0_xysp*		IDX	6A xb	Pw	Pw		
STAA *oprx9,xysp*		IDX1	6A xb ff	PwO	PwO		
STAA *oprx16,xysp*		IDX2	6A xb ee ff	PwP	PwP		
STAA [D,*xysp*]		[D,IDX]	6A xb	PIfw	PIfPw		
STAA [*oprx16,xysp*]		[IDX2]	6A xb ee ff	PIPw	PIPPw		
STAB *opr8a*	$(B) \Rightarrow M$	DIR	5B dd	Pw	Pw	----	ΔΔ0–
STAB *opr16a*	Store Accumulator B to Memory	EXT	7B hh ll	PwO	wOP		
STAB *oprx0_xysp*		IDX	6B xb	Pw	Pw		
STAB *oprx9,xysp*		IDX1	6B xb ff	PwO	PwO		
STAB *oprx16,xysp*		IDX2	6B xb ee ff	PwP	PwP		
STAB [D,*xysp*]		[D,IDX]	6B xb	PIfw	PIfPw		
STAB [*oprx16,xysp*]		[IDX2]	6B xb ee ff	PIPw	PIPPw		
STD *opr8a*	$(A) \Rightarrow M, (B) \Rightarrow M{+}1$	DIR	5C dd	PW	PW	----	ΔΔ0–
STD *opr16a*	Store Double Accumulator	EXT	7C hh ll	PWO	WOP		
STD *oprx0_xysp*		IDX	6C xb	PW	PW		
STD *oprx9,xysp*		IDX1	6C xb ff	PWO	PWO		
STD *oprx16,xysp*		IDX2	6C xb ee ff	PWP	PWP		
STD [D,*xysp*]		[D,IDX]	6C xb	PIfW	PIfPW		
STD [*oprx16,xysp*]		[IDX2]	6C xb ee ff	PIPW	PIPPW		
STOP	$(SP) - 2 \Rightarrow SP$; $RTN_H{:}RTN_L \Rightarrow M_{(SP)}{:}M_{(SP+1)}$; $(SP) - 2 \Rightarrow SP$; $(Y_H{:}Y_L) \Rightarrow M_{(SP)}{:}M_{(SP+1)}$; $(SP) - 2 \Rightarrow SP$; $(X_H{:}X_L) \Rightarrow M_{(SP)}{:}M_{(SP+1)}$; $(SP) - 2 \Rightarrow SP$; $(B{:}A) \Rightarrow M_{(SP)}{:}M_{(SP+1)}$; $(SP) - 1 \Rightarrow SP$; $(CCR) \Rightarrow M_{(SP)}$; STOP All Clocks Registers stacked to allow quicker recovery by interrupt. If S control bit = 1, the STOP instruction is disabled and acts like a two-cycle NOP.	INH	18 3E	(entering STOP) OOSSSSsf (exiting STOP) fVfPPP (continue) ff (if STOP disabled) OO	 OOSSSfSs fVfPPP fO OO	----	----
STS *opr8a*	$(SP_H{:}SP_L) \Rightarrow M{:}M{+}1$	DIR	5F dd	PW	PW	----	ΔΔ0–
STS *opr16a*	Store Stack Pointer	EXT	7F hh ll	PWO	WOP		
STS *oprx0_xysp*		IDX	6F xb	PW	PW		
STS *oprx9,xysp*		IDX1	6F xb ff	PWO	PWO		
STS *oprx16,xysp*		IDX2	6F xb ee ff	PWP	PWP		
STS [D,*xysp*]		[D,IDX]	6F xb	PIfW	PIfPW		
STS [*oprx16,xysp*]		[IDX2]	6F xb ee ff	PIPW	PIPPW		
STX *opr8a*	$(X_H{:}X_L) \Rightarrow M{:}M{+}1$	DIR	5E dd	PW	PW	----	ΔΔ0–
STX *opr16a*	Store Index Register X	EXT	7E hh ll	PWO	WOP		
STX *oprx0_xysp*		IDX	6E xb	PW	PW		
STX *oprx9,xysp*		IDX1	6E xb ff	PWO	PWO		
STX *oprx16,xysp*		IDX2	6E xb ee ff	PWP	PWP		
STX [D,*xysp*]		[D,IDX]	6E xb	PIfW	PIfPW		
STX [*oprx16,xysp*]		[IDX2]	6E xb ee ff	PIPW	PIPPW		
STY *opr8a*	$(Y_H{:}Y_L) \Rightarrow M{:}M{+}1$	DIR	5D dd	PW	PW	----	ΔΔ0–
STY *opr16a*	Store Index Register Y	EXT	7D hh ll	PWO	WOP		
STY *oprx0_xysp*		IDX	6D xb	PW	PW		
STY *oprx9,xysp*		IDX1	6D xb ff	PWO	PWO		
STY *oprx16,xysp*		IDX2	6D xb ee ff	PWP	PWP		
STY [D,*xysp*]		[D,IDX]	6D xb	PIfW	PIfPW		
STY [*oprx16,xysp*]		[IDX2]	6D xb ee ff	PIPW	PIPPW		
SUBA #*opr8i*	$(A) - (M) \Rightarrow A$	IMM	80 ii	P	P	----	ΔΔΔΔ
SUBA *opr8a*	Subtract Memory from Accumulator A	DIR	90 dd	rPf	rfP		
SUBA *opr16a*		EXT	B0 hh ll	rPO	rOP		
SUBA *oprx0_xysp*		IDX	A0 xb	rPf	rfP		
SUBA *oprx9,xysp*		IDX1	A0 xb ff	rPO	rPO		
SUBA *oprx16,xysp*		IDX2	A0 xb ee ff	frPP	frPP		
SUBA [D,*xysp*]		[D,IDX]	A0 xb	fIfrPf	fIfrfP		
SUBA [*oprx16,xysp*]		[IDX2]	A0 xb ee ff	fIPrPf	fIPrfP		

Table A-1. Instruction Set Summary (Sheet 13 of 14)

Source Form	Operation	Addr. Mode	Machine Coding (hex)	Access Detail HCS12	Access Detail M68HC12	S X H I	N Z V C

SUBB #*opr8i*
SUBB *opr8a*
SUBB *opr16a*
SUBB *oprx0_xysp*
SUBB *oprx9,xysp*
SUBB *oprx16,xysp*
SUBB [D,*xysp*]
SUBB [*oprx16,xysp*]
	(B) – (M) ⇒ B
	Subtract Memory from Accumulator B
		IMM
		DIR
		EXT
		IDX
		IDX1
		IDX2
		[D,IDX]
		[IDX2]
			C0 ii
			D0 dd
			F0 hh ll
			E0 xb
			E0 xb ff
			E0 xb ee ff
			E0 xb
			E0 xb ee ff
				P
				rPf
				rPO
				rPf
				rPO
				frPP
				fIfrPf
				fIPrPf
					P
					rfP
					rOP
					rfP
					rPO
					frPP
					fIfrfP
					fIPrfP
						– – – –	Δ Δ Δ Δ

SUBD #*opr16i*
SUBD *opr8a*
SUBD *opr16a*
SUBD *oprx0_xysp*
SUBD *oprx9,xysp*
SUBD *oprx16,xysp*
SUBD [D,*xysp*]
SUBD [*oprx16,xysp*]
	(D) – (M:M+1) ⇒ D
	Subtract Memory from D (A:B)
		IMM
		DIR
		EXT
		IDX
		IDX1
		IDX2
		[D,IDX]
		[IDX2]
			83 jj kk
			93 dd
			B3 hh ll
			A3 xb
			A3 xb ff
			A3 xb ee ff
			A3 xb
			A3 xb ee ff
				PO
				RPf
				RPO
				RPf
				RPO
				fRPP
				fIfRPf
				fIPRPf
					OP
					RfP
					ROP
					RfP
					RPO
					fRPP
					fIfRfP
					fIPRfP
						– – – –	Δ Δ Δ Δ

SWI	(SP) – 2 ⇒ SP; $RTN_H:RTN_L \Rightarrow M_{(SP)}:M_{(SP+1)}$; (SP) – 2 ⇒ SP; $(Y_H:Y_L) \Rightarrow M_{(SP)}:M_{(SP+1)}$; (SP) – 2 ⇒ SP; $(X_H:X_L) \Rightarrow M_{(SP)}:M_{(SP+1)}$; (SP) – 2 ⇒ SP; (B:A) ⇒ $M_{(SP)}:M_{(SP+1)}$; (SP) – 1 ⇒ SP; (CCR) ⇒ $M_{(SP)}$ 1 ⇒ I; (SWI Vector) ⇒ PC Software Interrupt	INH	3F	VSPSSPSsP* (for Reset) VfPPP	VSPSSPSsP* VfPPP	– – – 1 1 1 – 1	– – – – – – – –
*The CPU also uses the SWI microcode sequence for hardware interrupts and unimplemented opcode traps. Reset uses the VfPPP variation of this sequence.							
TAB	(A) ⇒ B Transfer A to B	INH	18 0E	OO	OO	– – – –	Δ Δ 0 –
TAP	(A) ⇒ CCR *Translates to* TFR A , CCR	INH	B7 02	P	P	Δ ⇓ Δ Δ	Δ Δ Δ Δ
TBA	(B) ⇒ A Transfer B to A	INH	18 0F	OO	OO	– – – –	Δ Δ 0 –
TBEQ *abdxys,rel9*	If (cntr) = 0, then Branch; else Continue to next instruction Test Counter and Branch if Zero (cntr = A, B, D, X,Y, or SP)	REL (9-bit)	04 lb rr	PPP (branch) PPO (no branch)	PPP	– – – –	– – – –
TBL *oprx0_xysp*	(M) + [(B) × ((M+1) – (M))] ⇒ A 8-Bit Table Lookup and Interpolate Initialize B, and index before TBL. <ea> points at first 8-bit table entry (M) and B is fractional part of lookup value. (no indirect addressing modes or extensions allowed)	IDX	18 3D xb	ORfffP	OrrffffP	– – – –	Δ Δ – Δ ? C Bit is undefined in HC12
TBNE *abdxys,rel9*	If (cntr) not = 0, then Branch; else Continue to next instruction Test Counter and Branch if Not Zero (cntr = A, B, D, X,Y, or SP)	REL (9-bit)	04 lb rr	PPP (branch) PPO (no branch)	PPP	– – – –	– – – –
TFR *abcdxys,abcdxys*	(r1) ⇒ r2 *or* \$00:(r1) ⇒ r2 *or* (r1[7:0]) ⇒ r2 Transfer Register to Register r1 and r2 may be A, B, CCR, D, X, Y, or SP	INH	B7 eb	P	P	– – – – or Δ ⇓ Δ Δ	– – – – Δ Δ Δ Δ
TPA	(CCR) ⇒ A *Translates to* TFR CCR ,A	INH	B7 20	P	P	– – – –	– – – –

Table A-1. Instruction Set Summary (Sheet 14 of 14)

Source Form	Operation	Addr. Mode	Machine Coding (hex)	Access Detail HCS12	Access Detail M68HC12	S X H I	N Z V C
TRAP *trapnum*	$(SP) - 2 \Rightarrow SP$; $RTN_H{:}RTN_L \Rightarrow M_{(SP)}{:}M_{(SP+1)}$; $(SP) - 2 \Rightarrow SP$; $(Y_H{:}Y_L) \Rightarrow M_{(SP)}{:}M_{(SP+1)}$; $(SP) - 2 \Rightarrow SP$; $(X_H{:}X_L) \Rightarrow M_{(SP)}{:}M_{(SP+1)}$; $(SP) - 2 \Rightarrow SP$; $(B{:}A) \Rightarrow M_{(SP)}{:}M_{(SP+1)}$; $(SP) - 1 \Rightarrow SP$; $(CCR) \Rightarrow M_{(SP)}$ $1 \Rightarrow I$; (TRAP Vector) $\Rightarrow$ PC Unimplemented opcode trap	INH	18 tn tn = $30-$39 or $40-$FF	OVSPSSPSsP	OfVSPSSPSsP	---1	----
TST *opr16a*	(M) – 0 Test Memory for Zero or Minus	EXT	F7 hh ll	rPO	rOP	----	ΔΔ00
TST *oprx0_xysp*		IDX	E7 xb	rPf	rfP		
TST *oprx9,xysp*		IDX1	E7 xb ff	rPO	rPO		
TST *oprx16,xysp*		IDX2	E7 xb ee ff	frPP	frPP		
TST [D,*xysp*]		[D,IDX]	E7 xb	fIfrPf	fIfrfP		
TST [*oprx16,xysp*]		[IDX2]	E7 xb ee ff	fIPrPf	fIPrfP		
TSTA	(A) – 0 Test A for Zero or Minus	INH	97	O	O		
TSTB	(B) – 0 Test B for Zero or Minus	INH	D7	O	O		
TSX	$(SP) \Rightarrow X$ *Translates to* TFR SP,X	INH	B7 75	P	P	----	----
TSY	$(SP) \Rightarrow Y$ *Translates to* TFR SP,Y	INH	B7 76	P	P	----	----
TXS	$(X) \Rightarrow SP$ *Translates to* TFR X,SP	INH	B7 57	P	P	----	----
TYS	$(Y) \Rightarrow SP$ *Translates to* TFR Y,SP	INH	B7 67	P	P	----	----
WAI	$(SP) - 2 \Rightarrow SP$; $RTN_H{:}RTN_L \Rightarrow M_{(SP)}{:}M_{(SP+1)}$; $(SP) - 2 \Rightarrow SP$; $(Y_H{:}Y_L) \Rightarrow M_{(SP)}{:}M_{(SP+1)}$; $(SP) - 2 \Rightarrow SP$; $(X_H{:}X_L) \Rightarrow M_{(SP)}{:}M_{(SP+1)}$; $(SP) - 2 \Rightarrow SP$; $(B{:}A) \Rightarrow M_{(SP)}{:}M_{(SP+1)}$; $(SP) - 1 \Rightarrow SP$; $(CCR) \Rightarrow M_{(SP)}$; WAIT for interrupt	INH	3E	OSSSSsf (after interrupt) fVfPPP	OSSSfSsf VfPPP	---- or ---1 or -1-1	---- ---- ----
WAV	$\sum_{i=1}^{B} S_i F_i \Rightarrow Y{:}D$ and $\sum_{i=1}^{B} F_i \Rightarrow X$ Calculate Sum of Products and Sum of Weights for Weighted Average Calculation Initialize B, X, and Y before WAV. B specifies number of elements. X points at first element in S_i list. Y points at first element in F_i list. All S_i and F_i elements are 8-bits. If interrupted, six extra bytes of stack used for intermediate values	Special	18 3C	Of(frr,ffff)O (add if interrupt) SSS + UUUrr,	Off(frr,fffff)O SSSf + UUUrr	--?-	?Δ??
wavr pseudo-instruction	*see* WAV Resume executing an interrupted WAV instruction (recover intermediate results from stack rather than initializing them to zero)	Special	3C	UUUrr,ffff (frr,ffff)O (exit + re-entry replaces comma above if interrupted) SSS + UUUrr,	UUUrrfffff (frr,fffff)O SSSf + UUUrr	--?-	?Δ??
XGDX	$(D) \Leftrightarrow (X)$ *Translates to* EXG D, X	INH	B7 C5	P	P	----	----
XGDY	$(D) \Leftrightarrow (Y)$ *Translates to* EXG D, Y	INH	B7 C6	P	P	----	----

1.5.1 Detailed Register Map

$0000 - $000F MEBI map 1 of 3 (HCS12 Multiplexed External Bus Interface)

Address	Name		Bit 7	Bit 6	Bit 5	Bit 4	Bit 3	Bit 2	Bit 1	Bit 0
$0000	PORTA	Read:	Bit 7	6	5	4	3	2	1	Bit 0
		Write:								
$0001	PORTB	Read:	Bit 7	6	5	4	3	2	1	Bit 0
		Write:								
$0002	DDRA	Read:	Bit 7	6	5	4	3	2	1	Bit 0
		Write:								
$0003	DDRB	Read:	Bit 7	6	5	4	3	2	1	Bit 0
		Write:								
$0004 - $0007	Reserved	Read:	0	0	0	0	0	0	0	0
		Write:								
$0008	PORTE	Read:	Bit 7	6	5	4	3	2	Bit 1	Bit 0
		Write:								
$0009	DDRE	Read:	Bit 7	6	5	4	3	Bit 2	0	0
		Write:								
$000A	PEAR	Read:	NOACCE	0	PIPOE	NECLK	LSTRE	RDWE	0	0
		Write:								
$000B	MODE	Read:	MODC	MODB	MODA	0	IVIS	0	EMK	EME
		Write:								
$000C	PUCR	Read:	PUPKE	0	0	PUPEE	0	0	PUPBE	PUPAE
		Write:								
$000D	RDRIV	Read:	RDPK	0	0	RDPE	0	0	RDPB	RDPA
		Write:								
$000E	EBICTL	Read:	0	0	0	0	0	0	0	ESTR
		Write:								
$000F	Reserved	Read:	0	0	0	0	0	0	0	0
		Write:								

$0010 - $0014 MMC map 1 of 4 (HCS12 Module Mapping Control)

Address	Name		Bit 7	Bit 6	Bit 5	Bit 4	Bit 3	Bit 2	Bit 1	Bit 0
$0010	INITRM	Read:	RAM15	RAM14	RAM13	RAM12	RAM11	0	0	RAMHAL
		Write:								
$0011	INITRG	Read:	0	REG14	REG13	REG12	REG11	0	0	0
		Write:								
$0012	INITEE	Read:	EE15	EE14	EE13	EE12	EE11	0	0	EEON
		Write:								
$0013	MISC	Read:	0	0	0	0	EXSTR1	EXSTR0	ROMHM	ROMON
		Write:								
$0014	Reserved	Read:	0	0	0	0	0	0	0	0
		Write:								

$0015 - $0016 INT map 1 of 2 (HCS12 Interrupt)

Address	Name		Bit 7	Bit 6	Bit 5	Bit 4	Bit 3	Bit 2	Bit 1	Bit 0
$0015	ITCR	Read:	0	0	0	WRINT	ADR3	ADR2	ADR1	ADR0
		Write:								
$0016	ITEST	Read: Write:	INTE	INTC	INTA	INT8	INT6	INT4	INT2	INT0

$0017 - $0019 Reserved

Address	Name		Bit 7	Bit 6	Bit 5	Bit 4	Bit 3	Bit 2	Bit 1	Bit 0
$0017- $0019	Reserved	Read:	0	0	0	0	0	0	0	0
		Write:								

$001A - $001B Device ID Register (Table 1-3)

Address	Name		Bit 7	Bit 6	Bit 5	Bit 4	Bit 3	Bit 2	Bit 1	Bit 0
$001A	PARTIDH	Read:	ID15	ID14	ID13	ID12	ID11	ID10	ID9	ID8
		Write:								
$001B	PARTIDL	Read:	ID7	ID6	ID5	ID4	ID3	ID2	ID1	ID0
		Write:								

$001C - $001D MMC map 3 of 4 (HCS12 Module Mapping Control, Table 1-4)

Address	Name		Bit 7	Bit 6	Bit 5	Bit 4	Bit 3	Bit 2	Bit 1	Bit 0
$001C	MEMSIZ0	Read:	reg_sw0	0	eep_sw1	eep_sw0	0	ram_sw2	ram_sw1	ram_sw0
		Write:								
$001D	MEMSIZ1	Read:	rom_sw1	rom_sw0	0	0	0	0	pag_sw1	pag_sw0
		Write:								

$001E - $001E MEBI map 2 of 3 (HCS12 Multiplexed External Bus Interface)

Address	Name		Bit 7	Bit 6	Bit 5	Bit 4	Bit 3	Bit 2	Bit 1	Bit 0
$001E	INTCR	Read:	IRQE	IRQEN	0	0	0	0	0	0
		Write:								

$001F - $001F INT map 2 of 2 (HCS12 Interrupt)

Address	Name		Bit 7	Bit 6	Bit 5	Bit 4	Bit 3	Bit 2	Bit 1	Bit 0
$001F	HPRIO	Read:	PSEL7	PSEL6	PSEL5	PSEL4	PSEL3	PSEL2	PSEL1	0
		Write:								

$0020 - $0027 Reserved

Address	Name		Bit 7	Bit 6	Bit 5	Bit 4	Bit 3	Bit 2	Bit 1	Bit 0
$0020 - $0027	Reserved	Read:	0	0	0	0	0	0	0	0
		Write:								

$0028 - $002F BKP (HCS12 Breakpoint)

Address	Name		Bit 7	Bit 6	Bit 5	Bit 4	Bit 3	Bit 2	Bit 1	Bit 0
$0028	BKPCT0	Read:	BKEN	BKFULL	BKBDM	BKTAG	0	0	0	0
		Write:								
$0029	BKPCT1	Read:	BK0MBH	BK0MBL	BK1MBH	BK1MBL	BK0RWE	BK0RW	BK1RWE	BK1RW
		Write:								
$002A	BKP0X	Read:	0	0	BK0V5	BK0V4	BK0V3	BK0V2	BK0V1	BK0V0
		Write:								
$002B	BKP0H	Read:	Bit 15	14	13	12	11	10	9	Bit 8
		Write:								
$002C	BKP0L	Read:	Bit 7	6	5	4	3	2	1	Bit 0
		Write:								
$002D	BKP1X	Read:	0	0	BK1V5	BK1V4	BK1V3	BK1V2	BK1V1	BK1V0
		Write:								
$002E	BKP1H	Read:	Bit 15	14	13	12	11	10	9	Bit 8
		Write:								
$002F	BKP1L	Read:	Bit 7	6	5	4	3	2	1	Bit 0
		Write:								

$0030 - $0031 MMC map 4 of 4 (HCS12 Module Mapping Control)

Address	Name		Bit 7	Bit 6	Bit 5	Bit 4	Bit 3	Bit 2	Bit 1	Bit 0
$0030	PPAGE	Read:	0	0	PIX5	PIX4	PIX3	PIX2	PIX1	PIX0
		Write:								
$0031	Reserved	Read:	0	0	0	0	0	0	0	0
		Write:								

$0032 - $0033 MEBI map 3 of 3 (HCS12 Multiplexed External Bus Interface)

Address	Name		Bit 7	Bit 6	Bit 5	Bit 4	Bit 3	Bit 2	Bit 1	Bit 0
$0032	PORTK	Read:	Bit 7	6	5	4	3	2	1	Bit 0
		Write:								
$0033	DDRK	Read:	Bit 7	6	5	4	3	2	1	Bit 0
		Write:								

$0034 - $003F CRG (Clock and Reset Generator)

Address	Name		Bit 7	Bit 6	Bit 5	Bit 4	Bit 3	Bit 2	Bit 1	Bit 0
$0034	SYNR	Read:	0	0	SYN5	SYN4	SYN3	SYN2	SYN1	SYN0
		Write:								
$0035	REFDV	Read:	0	0	0	0	REFDV3	REFDV2	REFDV1	REFDV0
		Write:								
$0036	CTFLG Test Only	Read:	TOUT7	TOUT6	TOUT5	TOUT4	TOUT3	TOUT2	TOUT1	TOUT0
		Write:								
$0037	CRGFLG	Read:	RTIF	PROF	0	LOCKIF	LOCK	TRACK	SCMIF	SCM
		Write:								
$0038	CRGINT	Read:	RTIE	0	0	LOCKIE	0	0	SCMIE	0
		Write:								

$0034 - $003F CRG (Clock and Reset Generator)

Address	Name		Bit 7	Bit 6	Bit 5	Bit 4	Bit 3	Bit 2	Bit 1	Bit 0
$0039	CLKSEL	Read:	PLLSEL	PSTP	SYSWAI	ROAWAI	PLLWAI	CWAI	RTIWAI	COPWAI
		Write:								
$003A	PLLCTL	Read:	CME	PLLON	AUTO	ACQ	0	PRE	PCE	SCME
		Write:								
$003B	RTICTL	Read:	0	RTR6	RTR5	RTR4	RTR3	RTR2	RTR1	RTR0
		Write:								
$003C	COPCTL	Read:	WCOP	RSBCK	0	0	0	CR2	CR1	CR0
		Write:								
$003D	FORBYP Test Only	Read:	RTIBYP	COPBYP	0	PLLBYP	0	0	FCM	0
		Write:								
$003E	CTCTL Test Only	Read:	TCTL7	TCTL6	TCTL5	TCTL4	TCLT3	TCTL2	TCTL1	TCTL0
		Write:								
$003F	ARMCOP	Read:	0	0	0	0	0	0	0	0
		Write:	Bit 7	6	5	4	3	2	1	Bit 0

$0040 - $007F ECT (Enhanced Capture Timer 16 Bit 8 Channels)

Address	Name		Bit 7	Bit 6	Bit 5	Bit 4	Bit 3	Bit 2	Bit 1	Bit 0
$0040	TIOS	Read:	IOS7	IOS6	IOS5	IOS4	IOS3	IOS2	IOS1	IOS0
		Write:								
$0041	CFORC	Read:	0	0	0	0	0	0	0	0
		Write:	FOC7	FOC6	FOC5	FOC4	FOC3	FOC2	FOC1	FOC0
$0042	OC7M	Read:	OC7M7	OC7M6	OC7M5	OC7M4	OC7M3	OC7M2	OC7M1	OC7M0
		Write:								
$0043	OC7D	Read:	OC7D7	OC7D6	OC7D5	OC7D4	OC7D3	OC7D2	OC7D1	OC7D0
		Write:								
$0044	TCNT (hi)	Read:	Bit 15	14	13	12	11	10	9	Bit 8
		Write:								
$0045	TCNT (lo)	Read:	Bit 7	6	5	4	3	2	1	Bit 0
		Write:								
$0046	TSCR1	Read:	TEN	TSWAI	TSFRZ	TFFCA	0	0	0	0
		Write:								
$0047	TTOV	Read:	TOV7	TOV6	TOV5	TOV4	TOV3	TOV2	TOV1	TOV0
		Write:								
$0048	TCTL1	Read:	OM7	OL7	OM6	OL6	OM5	OL5	OM4	OL4
		Write:								
$0049	TCTL2	Read:	OM3	OL3	OM2	OL2	OM1	OL1	OM0	OL0
		Write:								
$004A	TCTL3	Read:	EDG7B	EDG7A	EDG6B	EDG6A	EDG5B	EDG5A	EDG4B	EDG4A
		Write:								
$004B	TCTL4	Read:	EDG3B	EDG3A	EDG2B	EDG2A	EDG1B	EDG1A	EDG0B	EDG0A
		Write:								
$004C	TIE	Read:	C7I	C6I	C5I	C4I	C3I	C2I	C1I	C0I
		Write:								
$004D	TSCR2	Read:	TOI	0	0	0	TCRE	PR2	PR1	PR0
		Write:								
$004E	TFLG1	Read:	C7F	C6F	C5F	C4F	C3F	C2F	C1F	C0F
		Write:								

$0040 - $007F ECT (Enhanced Capture Timer 16 Bit 8 Channels)

Address	Name		Bit 7	Bit 6	Bit 5	Bit 4	Bit 3	Bit 2	Bit 1	Bit 0
$004F	TFLG2	Read:	TOF	0	0	0	0	0	0	0
		Write:								
$0050	TC0 (hi)	Read: Write:	Bit 15	14	13	12	11	10	9	Bit 8
$0051	TC0 (lo)	Read: Write:	Bit 7	6	5	4	3	2	1	Bit 0
$0052	TC1 (hi)	Read: Write:	Bit 15	14	13	12	11	10	9	Bit 8
$0053	TC1 (lo)	Read: Write:	Bit 7	6	5	4	3	2	1	Bit 0
$0054	TC2 (hi)	Read: Write:	Bit 15	14	13	12	11	10	9	Bit 8
$0055	TC2 (lo)	Read: Write:	Bit 7	6	5	4	3	2	1	Bit 0
$0056	TC3 (hi)	Read: Write:	Bit 15	14	13	12	11	10	9	Bit 8
$0057	TC3 (lo)	Read: Write:	Bit 7	6	5	4	3	2	1	Bit 0
$0058	TC4 (hi)	Read: Write:	Bit 15	14	13	12	11	10	9	Bit 8
$0059	TC4 (lo)	Read: Write:	Bit 7	6	5	4	3	2	1	Bit 0
$005A	TC5 (hi)	Read: Write:	Bit 15	14	13	12	11	10	9	Bit 8
$005B	TC5 (lo)	Read: Write:	Bit 7	6	5	4	3	2	1	Bit 0
$005C	TC6 (hi)	Read: Write:	Bit 15	14	13	12	11	10	9	Bit 8
$005D	TC6 (lo)	Read: Write:	Bit 7	6	5	4	3	2	1	Bit 0
$005E	TC7 (hi)	Read: Write:	Bit 15	14	13	12	11	10	9	Bit 8
$005F	TC7 (lo)	Read: Write:	Bit 7	6	5	4	3	2	1	Bit 0
$0060	PACTL	Read:	0	PAEN	PAMOD	PEDGE	CLK1	CLK0	PAOVI	PAI
		Write:								
$0061	PAFLG	Read:	0	0	0	0	0	0	PAOVF	PAIF
		Write:								
$0062	PACN3 (hi)	Read: Write:	Bit 7	6	5	4	3	2	1	Bit 0
$0063	PACN2 (lo)	Read: Write:	Bit 7	6	5	4	3	2	1	Bit 0
$0064	PACN1 (hi)	Read: Write:	Bit 7	6	5	4	3	2	1	Bit 0
$0065	PACN0 (lo)	Read: Write:	Bit 7	6	5	4	3	2	1	Bit 0
$0066	MCCTL	Read:	MCZI	MODMC	RDMCL	0	0	MCEN	MCPR1	MCPR0
		Write:				ICLAT	FLMC			
$0067	MCFLG	Read:	MCZF	0	0	0	POLF3	POLF2	POLF1	POLF0
		Write:								

$0040 - $007F ECT (Enhanced Capture Timer 16 Bit 8 Channels)

Address	Name		Bit 7	Bit 6	Bit 5	Bit 4	Bit 3	Bit 2	Bit 1	Bit 0
$0068	ICPAR	Read:	0	0	0	0	PA3EN	PA2EN	PA1EN	PA0EN
		Write:								
$0069	DLYCT	Read:	0	0	0	0	0	0	DLY1	DLY0
		Write:								
$006A	ICOVW	Read:	NOVW7	NOVW6	NOVW5	NOVW4	NOVW3	NOVW2	NOVW1	NOVW0
		Write:								
$006B	ICSYS	Read:	SH37	SH26	SH15	SH04	TFMOD	PACMX	BUFEN	LATQ
		Write:								
$006C	Reserved	Read:								
		Write:								
$006D	TIMTST Test Only	Read:	0	0	0	0	0	0	TCBYP	0
		Write:								
$006E - $006F	Reserved	Read:								
		Write:								
$0070	PBCTL	Read:	0	PBEN	0	0	0	0	PBOVI	0
		Write:								
$0071	PBFLG	Read:	0	0	0	0	0	0	PBOVF	0
		Write:								
$0072	PA3H	Read:	Bit 7	6	5	4	3	2	1	Bit 0
		Write:								
$0073	PA2H	Read:	Bit 7	6	5	4	3	2	1	Bit 0
		Write:								
$0074	PA1H	Read:	Bit 7	6	5	4	3	2	1	Bit 0
		Write:								
$0075	PA0H	Read:	Bit 7	6	5	4	3	2	1	Bit 0
		Write:								
$0076	MCCNT (hi)	Read:	Bit 15	14	13	12	11	10	9	Bit 8
		Write:								
$0077	MCCNT (lo)	Read:	Bit 7	6	5	4	3	2	1	Bit 0
		Write:								
$0078	TC0H (hi)	Read:	Bit 15	14	13	12	11	10	9	Bit 8
		Write:								
$0079	TC0H (lo)	Read:	Bit 7	6	5	4	3	2	1	Bit 0
		Write:								
$007A	TC1H (hi)	Read:	Bit 15	14	13	12	11	10	9	Bit 8
		Write:								
$007B	TC1H (lo)	Read:	Bit 7	6	5	4	3	2	1	Bit 0
		Write:								
$007C	TC2H (hi)	Read:	Bit 15	14	13	12	11	10	9	Bit 8
		Write:								
$007D	TC2H (lo)	Read:	Bit 7	6	5	4	3	2	1	Bit 0
		Write:								
$007E	TC3H (hi)	Read:	Bit 15	14	13	12	11	10	9	Bit 8
		Write:								
$007F	TC3H (lo)	Read:	Bit 7	6	5	4	3	2	1	Bit 0
		Write:								

$0080 - $009F ATD0 (Analog to Digital Converter 10 Bit 8 Channel)

Address	Name		Bit 7	Bit 6	Bit 5	Bit 4	Bit 3	Bit 2	Bit 1	Bit 0
$0080	ATD0CTL0	Read:	0	0	0	0	0	0	0	0
		Write:								
$0081	ATD0CTL1	Read:	0	0	0	0	0	0	0	0
		Write:								
$0082	ATD0CTL2	Read:	ADPU	AFFC	AWAI	ETRIGLE	ETRIGP	ETRIG	ASCIE	ASCIF
		Write:								
$0083	ATD0CTL3	Read:	0	S8C	S4C	S2C	S1C	FIFO	FRZ1	FRZ0
		Write:								
$0084	ATD0CTL4	Read:	SRES8	SMP1	SMP0	PRS4	PRS3	PRS2	PRS1	PRS0
		Write:								
$0085	ATD0CTL5	Read:	DJM	DSGN	SCAN	MULT	0	CC	CB	CA
		Write:								
$0086	ATD0STAT0	Read:	SCF	0	ETORF	FIFOR	0	CC2	CC1	CC0
		Write:								
$0087	Reserved	Read:	0	0	0	0	0	0	0	0
		Write:								
$0088	ATD0TEST0	Read:	0	0	0	0	0	0	0	0
		Write:								
$0089	ATD0TEST1	Read:	0	0	0	0	0	0	0	SC
		Write:								
$008A	Reserved	Read:	0	0	0	0	0	0	0	0
		Write:								
$008B	ATD0STAT1	Read:	CCF7	CCF6	CCF5	CCF4	CCF3	CCF2	CCF1	CCF0
		Write:								
$008C	Reserved	Read:	0	0	0	0	0	0	0	0
		Write:								
$008D	ATD0DIEN	Read:	Bit 7	6	5	4	3	2	1	Bit 0
		Write:								
$008E	Reserved	Read:	0	0	0	0	0	0	0	0
		Write:								
$008F	PORTAD0	Read:	Bit 7	6	5	4	3	2	1	Bit 0
		Write:								
$0090	ATD0DR0H	Read:	Bit 15	14	13	12	11	10	9	Bit 8
		Write:								
$0091	ATD0DR0L	Read:	Bit 7	6	5	4	3	2	1	Bit 0
		Write:								
$0092	ATD0DR1H	Read:	Bit 15	14	13	12	11	10	9	Bit 8
		Write:								
$0093	ATD0DR1L	Read:	Bit 7	6	5	4	3	2	1	Bit 0
		Write:								
$0094	ATD0DR2H	Read:	Bit 15	14	13	12	11	10	9	Bit 8
		Write:								
$0095	ATD0DR2L	Read:	Bit 7	6	5	4	3	2	1	Bit 0
		Write:								
$0096	ATD0DR3H	Read:	Bit 15	14	13	12	11	10	9	Bit 8
		Write:								
$0097	ATD0DR3L	Read:	Bit 7	6	5	4	3	2	1	Bit 0
		Write:								
$0098	ATD0DR4H	Read:	Bit 15	14	13	12	11	10	9	Bit 8
		Write:								

$0080 - $009F ATD0 (Analog to Digital Converter 10 Bit 8 Channel)

Address	Name		Bit 7	Bit 6	Bit 5	Bit 4	Bit 3	Bit 2	Bit 1	Bit 0
$0099	ATD0DR4L	Read:	Bit 7	6	5	4	3	2	1	Bit 0
		Write:								
$009A	ATD0DR5H	Read:	Bit 15	14	13	12	11	10	9	Bit 8
		Write:								
$009B	ATD0DR5L	Read:	Bit 7	6	5	4	3	2	1	Bit 0
		Write:								
$009C	ATD0DR6H	Read:	Bit 15	14	13	12	11	10	9	Bit 8
		Write:								
$009D	ATD0DR6L	Read:	Bit 7	6	5	4	3	2	1	Bit 0
		Write:								
$009E	ATD0DR7H	Read:	Bit 15	14	13	12	11	10	9	Bit 8
		Write:								
$009F	ATD0DR7L	Read:	Bit 7	6	5	4	3	2	1	Bit 0
		Write:								

$00A0 - $00C7 PWM (Pulse Width Modulator 8 Bit 8 Channel)

Address	Name		Bit 7	Bit 6	Bit 5	Bit 4	Bit 3	Bit 2	Bit 1	Bit 0
$00A0	PWME	Read: Write:	PWME7	PWME6	PWME5	PWME4	PWME3	PWME2	PWME1	PWME0
$00A1	PWMPOL	Read: Write:	PPOL7	PPOL6	PPOL5	PPOL4	PPOL3	PPOL2	PPOL1	PPOL0
$00A2	PWMCLK	Read: Write:	PCLK7	PCLK6	PCLK5	PCLK4	PCLK3	PCLK2	PCLK1	PCLK0
$00A3	PWMPRCLK	Read:	0	PCKB2	PCKB1	PCKB0	0	PCKA2	PCKA1	PCKA0
		Write:								
$00A4	PWMCAE	Read: Write:	CAE7	CAE6	CAE5	CAE4	CAE3	CAE2	CAE1	CAE0
$00A5	PWMCTL	Read:	CON67	CON45	CON23	CON01	PSWAI	PFRZ	0	0
		Write:								
$00A6	PWMTST Test Only	Read:	0	0	0	0	0	0	0	0
		Write:								
$00A7	PWMPRSC	Read:	0	0	0	0	0	0	0	0
		Write:								
$00A8	PWMSCLA	Read: Write:	Bit 7	6	5	4	3	2	1	Bit 0
$00A9	PWMSCLB	Read: Write:	Bit 7	6	5	4	3	2	1	Bit 0
$00AA	PWMSCNTA	Read:	0	0	0	0	0	0	0	0
		Write:								
$00AB	PWMSCNTB	Read:	0	0	0	0	0	0	0	0
		Write:								
$00AC	PWMCNT0	Read:	Bit 7	6	5	4	3	2	1	Bit 0
		Write:	0	0	0	0	0	0	0	0
$00AD	PWMCNT1	Read:	Bit 7	6	5	4	3	2	1	Bit 0
		Write:	0	0	0	0	0	0	0	0
$00AE	PWMCNT2	Read:	Bit 7	6	5	4	3	2	1	Bit 0
		Write:	0	0	0	0	0	0	0	0

$00A0 - $00C7 PWM (Pulse Width Modulator 8 Bit 8 Channel)

Address	Name		Bit 7	Bit 6	Bit 5	Bit 4	Bit 3	Bit 2	Bit 1	Bit 0
$00AF	PWMCNT3	Read:	Bit 7	6	5	4	3	2	1	Bit 0
		Write:	0	0	0	0	0	0	0	0
$00B0	PWMCNT4	Read:	Bit 7	6	5	4	3	2	1	Bit 0
		Write:	0	0	0	0	0	0	0	0
$00B1	PWMCNT5	Read:	Bit 7	6	5	4	3	2	1	Bit 0
		Write:	0	0	0	0	0	0	0	0
$00B2	PWMCNT6	Read:	Bit 7	6	5	4	3	2	1	Bit 0
		Write:	0	0	0	0	0	0	0	0
$00B3	PWMCNT7	Read:	Bit 7	6	5	4	3	2	1	Bit 0
		Write:	0	0	0	0	0	0	0	0
$00B4	PWMPER0	Read: Write:	Bit 7	6	5	4	3	2	1	Bit 0
$00B5	PWMPER1	Read: Write:	Bit 7	6	5	4	3	2	1	Bit 0
$00B6	PWMPER2	Read: Write:	Bit 7	6	5	4	3	2	1	Bit 0
$00B7	PWMPER3	Read: Write:	Bit 7	6	5	4	3	2	1	Bit 0
$00B8	PWMPER4	Read: Write:	Bit 7	6	5	4	3	2	1	Bit 0
$00B9	PWMPER5	Read: Write:	Bit 7	6	5	4	3	2	1	Bit 0
$00BA	PWMPER6	Read: Write:	Bit 7	6	5	4	3	2	1	Bit 0
$00BB	PWMPER7	Read: Write:	Bit 7	6	5	4	3	2	1	Bit 0
$00BC	PWMDTY0	Read: Write:	Bit 7	6	5	4	3	2	1	Bit 0
$00BD	PWMDTY1	Read: Write:	Bit 7	6	5	4	3	2	1	Bit 0
$00BE	PWMDTY2	Read: Write:	Bit 7	6	5	4	3	2	1	Bit 0
$00BF	PWMDTY3	Read: Write:	Bit 7	6	5	4	3	2	1	Bit 0
$00C0	PWMDTY4	Read: Write:	Bit 7	6	5	4	3	2	1	Bit 0
$00C1	PWMDTY5	Read: Write:	Bit 7	6	5	4	3	2	1	Bit 0
$00C2	PWMDTY6	Read: Write:	Bit 7	6	5	4	3	2	1	Bit 0
$00C3	PWMDTY7	Read: Write:	Bit 7	6	5	4	3	2	1	Bit 0
$00C4	PWMSDN	Read:	PWMIF	PWMIE	PWM RSTRT	PWMLVL	0	PWM7IN	PWM7 INL	PWM7 ENA
		Write:								
$00C5 - $00C7	Reserved	Read:	0	0	0	0	0	0	0	0
		Write:								

$00C8 - $00CF SCI0 (Asynchronous Serial Interface)

Address	Name		Bit 7	Bit 6	Bit 5	Bit 4	Bit 3	Bit 2	Bit 1	Bit 0
$00C8	SCI0BDH	Read:	0	0	0	SBR12	SBR11	SBR10	SBR9	SBR8
		Write:								
$00C9	SCI0BDL	Read:	SBR7	SBR6	SBR5	SBR4	SBR3	SBR2	SBR1	SBR0
		Write:								
$00CA	SC0CR1	Read:	LOOPS	SCISWAI	RSRC	M	WAKE	ILT	PE	PT
		Write:								
$00CB	SCI0CR2	Read:	TIE	TCIE	RIE	ILIE	TE	RE	RWU	SBK
		Write:								
$00CC	SCI0SR1	Read:	TDRE	TC	RDRF	IDLE	OR	NF	FE	PF
		Write:								
$00CD	SC0SR2	Read:	0	0	0	0	0	BRK13	TXDIR	RAF
		Write:								
$00CE	SCI0DRH	Read:	R8	T8	0	0	0	0	0	0
		Write:								
$00CF	SCI0DRL	Read:	R7	R6	R5	R4	R3	R2	R1	R0
		Write:	T7	T6	T5	T4	T3	T2	T1	T0

$00D0 - $00D7 SCI1 (Asynchronous Serial Interface)

Address	Name		Bit 7	Bit 6	Bit 5	Bit 4	Bit 3	Bit 2	Bit 1	Bit 0
$00D0	SCI1BDH	Read:	0	0	0	SBR12	SBR11	SBR10	SBR9	SBR8
		Write:								
$00D1	SCI1BDL	Read:	SBR7	SBR6	SBR5	SBR4	SBR3	SBR2	SBR1	SBR0
		Write:								
$00D2	SC1CR1	Read:	LOOPS	SCISWAI	RSRC	M	WAKE	ILT	PE	PT
		Write:								
$00D3	SCI1CR2	Read:	TIE	TCIE	RIE	ILIE	TE	RE	RWU	SBK
		Write:								
$00D4	SCI1SR1	Read:	TDRE	TC	RDRF	IDLE	OR	NF	FE	PF
		Write:								
$00D5	SC1SR2	Read:	0	0	0	0	0	BRK13	TXDIR	RAF
		Write:								
$00D6	SCI1DRH	Read:	R8	T8	0	0	0	0	0	0
		Write:								
$00D7	SCI1DRL	Read:	R7	R6	R5	R4	R3	R2	R1	R0
		Write:	T7	T6	T5	T4	T3	T2	T1	T0

$00D8 - $00DF SPI0 (Serial Peripheral Interface)

Address	Name		Bit 7	Bit 6	Bit 5	Bit 4	Bit 3	Bit 2	Bit 1	Bit 0
$00D8	SPI0CR1	Read:	SPIE	SPE	SPTIE	MSTR	CPOL	CPHA	SSOE	LSBFE
		Write:								
$00D9	SPI0CR2	Read:	0	0	0	MODFEN	BIDIROE	0	SPISWAI	SPC0
		Write:								
$00DA	SPI0BR	Read:	0	SPPR2	SPPR1	SPPR0	0	SPR2	SPR1	SPR0
		Write:								
$00DB	SPI0SR	Read:	SPIF	0	SPTEF	MODF	0	0	0	0
		Write:								

$00D8 - $00DF   SPI0 (Serial Peripheral Interface)

Address	Name		Bit 7	Bit 6	Bit 5	Bit 4	Bit 3	Bit 2	Bit 1	Bit 0
$00DC	Reserved	Read:	0	0	0	0	0	0	0	0
		Write:								
$00DD	SPI0DR	Read:	Bit 7	6	5	4	3	2	1	Bit 0
		Write:								
$00DE - $00DF	Reserved	Read:	0	0	0	0	0	0	0	0
		Write:								

$00E0 - $00E7   IIC (Inter IC Bus)

Address	Name		Bit 7	Bit 6	Bit 5	Bit 4	Bit 3	Bit 2	Bit 1	Bit 0
$00E0	IBAD	Read:	ADR7	ADR6	ADR5	ADR4	ADR3	ADR2	ADR1	0
		Write:								
$00E1	IBFD	Read:	IBC7	IBC6	IBC5	IBC4	IBC3	IBC2	IBC1	IBC0
		Write:								
$00E2	IBCR	Read:	IBEN	IBIE	MS/$\overline{SL}$	TX/$\overline{RX}$	TXAK	0	0	IBSWAI
		Write:						RSTA		
$00E3	IBSR	Read:	TCF	IAAS	IBB	IBAL	0	SRW	IBIF	RXAK
		Write:								
$00E4	IBDR	Read:	D7	D6	D5	D4	D3	D2	D1	D 0
		Write:								
$00E5 - $00E7	Reserved	Read:	0	0	0	0	0	0	0	0
		Write:								

$00E8 - $00EF   BDLC (Bytelevel Data Link Controller J1850)

Address	Name		Bit 7	Bit 6	Bit 5	Bit 4	Bit 3	Bit 2	Bit 1	Bit 0
$00E8	DLCBCR1	Read:	IMSG	CLKS	0	0	0	0	IE	WCM
		Write:								
$00E9	DLCBSVR	Read:	0	0	I3	I2	I1	I0	0	0
		Write:								
$00EA	DLCBCR2	Read:	SMRST	DLOOP	RX4XE	NBFS	TEOD	TSIFR	TMIFR1	TMIFR0
		Write:								
$00EB	DLCBDR	Read:	D7	D6	D5	D4	D3	D2	D1	D0
		Write:								
$00EC	DLCBARD	Read:	0	RXPOL	0	0	BO3	BO2	BO1	BO0
		Write:								
$00ED	DLCBRSR	Read:	0	0	R5	R4	R3	R2	R1	R0
		Write:								
$00EE	DLCSCR	Read:	0	0	0	BDLCE	0	0	0	0
		Write:								
$00EF	DLCBSTAT	Read:	0	0	0	0	0	0	0	IDLE
		Write:								

$00F0 - $00F7 SPI1 (Serial Peripheral Interface)

Address	Name		Bit 7	Bit 6	Bit 5	Bit 4	Bit 3	Bit 2	Bit 1	Bit 0
$00F0	SPI1CR1	Read:	SPIE	SPE	SPTIE	MSTR	CPOL	CPHA	SSOE	LSBFE
		Write:								
$00F1	SPI1CR2	Read:	0	0	0	MODFEN	BIDIROE	0	SPISWAI	SPC0
		Write:								
$00F2	SPI1BR	Read:	0	SPPR2	SPPR1	SPPR0	0	SPR2	SPR1	SPR0
		Write:								
$00F3	SPI1SR	Read:	SPIF	0	SPTEF	MODF	0	0	0	0
		Write:								
$00F4	Reserved	Read:	0	0	0	0	0	0	0	0
		Write:								
$00F5	SPI1DR	Read:	Bit 7	6	5	4	3	2	1	Bit 0
		Write:								
$00F6 - $00F7	Reserved	Read:	0	0	0	0	0	0	0	0
		Write:								

$00F8 - $00FF SPI2 (Serial Peripheral Interface)

Address	Name		Bit 7	Bit 6	Bit 5	Bit 4	Bit 3	Bit 2	Bit 1	Bit 0
$00F8	SPI2CR1	Read:	SPIE	SPE	SPTIE	MSTR	CPOL	CPHA	SSOE	LSBFE
		Write:								
$00F9	SPI2CR2	Read:	0	0	0	MODFEN	BIDIROE	0	SPISWAI	SPC0
		Write:								
$00FA	SPI2BR	Read:	0	SPPR2	SPPR1	SPPR0	0	SPR2	SPR1	SPR0
		Write:								
$00FB	SPI2SR	Read:	SPIF	0	SPTEF	MODF	0	0	0	0
		Write:								
$00FC	Reserved	Read:	0	0	0	0	0	0	0	0
		Write:								
$00FD	SPI2DR	Read:	Bit 7	6	5	4	3	2	1	Bit 0
		Write:								
$00FE - $00FF	Reserved	Read:	0	0	0	0	0	0	0	0
		Write:								

$0100 - $010F Flash Control Register (fts512k4)

Address	Name		Bit 7	Bit 6	Bit 5	Bit 4	Bit 3	Bit 2	Bit 1	Bit 0
$0100	FCLKDIV	Read:	FDIVLD	PRDIV8	FDIV5	FDIV4	FDIV3	FDIV2	FDIV1	FDIV0
		Write:								
$0101	FSEC	Read:	KEYEN1	KEYEN0	NV5	NV4	NV3	NV2	SEC1	SEC0
		Write:								
$0102	FTSTMOD	Read:	0	0	0	WRALL	0	0	0	0
		Write:								
$0103	FCNFG	Read:	CBEIE	CCIE	KEYACC	0	0	0	BKSEL1	BKSEL0
		Write:								
$0104	FPROT	Read:	FPOPEN	NV6	FPHDIS	FPHS1	FPHS0	FPLDIS	FPLS1	FPLS0
		Write:								
$0105	FSTAT	Read:	CBEIF	CCIF	PVIOL	ACCERR	0	BLANK	0	0
		Write:								

$0100 - $010F Flash Control Register (fts512k4)

Address	Name		Bit 7	Bit 6	Bit 5	Bit 4	Bit 3	Bit 2	Bit 1	Bit 0
$0106	FCMD	Read:	0	CMDB6	CMDB5	0	0	CMDB2	0	CMDB0
		Write:								
$0107	Reserved	Read:	0	0	0	0	0	0	0	0
		Write:								
$0108	FADDRHI	Read:	Bit 15	14	13	12	11	10	9	Bit 8
		Write:								
$0109	FADDRLO	Read:	Bit 7	6	5	4	3	2	1	Bit 0
		Write:								
$010A	FDATAHI	Read:	Bit 15	14	13	12	11	10	9	Bit 8
		Write:								
$010B	FDATALO	Read:	Bit 7	6	5	4	3	2	1	Bit 0
		Write:								
$010C - $010F	Reserved	Read:	0	0	0	0	0	0	0	0
		Write:								

$0110 - $011B EEPROM Control Register (eets4k)

Address	Name		Bit 7	Bit 6	Bit 5	Bit 4	Bit 3	Bit 2	Bit 1	Bit 0
$0110	ECLKDIV	Read:	EDIVLD	PRDIV8	EDIV5	EDIV4	EDIV3	EDIV2	EDIV1	EDIV0
		Write:								
$0111 - $0112	Reserved	Read:	0	0	0	0	0	0	0	0
		Write:								
$0113	ECNFG	Read:	CBEIE	CCIE	0	0	0	0	0	0
		Write:								
$0114	EPROT	Read:	EPOPEN	NV6	NV5	NV4	EPDIS	EP2	EP1	EP0
		Write:								
$0115	ESTAT	Read:	CBEIF	CCIF	PVIOL	ACCERR	0	BLANK	0	0
		Write:								
$0116	ECMD	Read:	0	CMDB6	CMDB5	0	0	CMDB2	0	CMDB0
		Write:								
$0117	Reserved	Read:	0	0	0	0	0	0	0	0
		Write:								
$0118	EADDRHI	Read:	0	0	0	0	0	10	9	Bit 8
		Write:								
$0119	EADDRLO	Read:	Bit 7	6	5	4	3	2	1	Bit 0
		Write:								
$011A	EDATAHI	Read:	Bit 15	14	13	12	11	10	9	Bit 8
		Write:								
$011B	EDATALO	Read:	Bit 7	6	5	4	3	2	1	Bit 0
		Write:								

$011C - $011F Reserved for RAM Control Register

Address	Name		Bit 7	Bit 6	Bit 5	Bit 4	Bit 3	Bit 2	Bit 1	Bit 0
$011C - $011F	Reserved	Read:	0	0	0	0	0	0	0	0
		Write:								

$0120 - $013F ATD1 (Analog to Digital Converter 10 Bit 8 Channel)

Address	Name		Bit 7	Bit 6	Bit 5	Bit 4	Bit 3	Bit 2	Bit 1	Bit 0
$0120	ATD1CTL0	Read:	0	0	0	0	0	0	0	0
		Write:								
$0121	ATD1CTL1	Read:	0	0	0	0	0	0	0	0
		Write:								
$0122	ATD1CTL2	Read:	ADPU	AFFC	AWAI	ETRIGLE	ETRIGP	ETRIG	ASCIE	ASCIF
		Write:								
$0123	ATD1CTL3	Read:	0	S8C	S4C	S2C	S1C	FIFO	FRZ1	FRZ0
		Write:								
$0124	ATD1CTL4	Read:	SRES8	SMP1	SMP0	PRS4	PRS3	PRS2	PRS1	PRS0
		Write:								
$0125	ATD1CTL5	Read:	DJM	DSGN	SCAN	MULT	0	CC	CB	CA
		Write:								
$0126	ATD1STAT0	Read:	SCF	0	ETORF	FIFOR	0	CC2	CC1	CC0
		Write:								
$0127	Reserved	Read:	0	0	0	0	0	0	0	0
		Write:								
$0128	ATD1TEST0	Read:	0	0	0	0	0	0	0	0
		Write:								
$0129	ATD1TEST1	Read:	0	0	0	0	0	0	0	SC
		Write:								
$012A	Reserved	Read:	0	0	0	0	0	0	0	0
		Write:								
$012B	ATD1STAT1	Read:	CCF7	CCF6	CCF5	CCF4	CCF3	CCF2	CCF1	CCF0
		Write:								
$012C	Reserved	Read:	0	0	0	0	0	0	0	0
		Write:								
$012D	ATD1DIEN	Read:	Bit 7	6	5	4	3	2	1	Bit 0
		Write:								
$012E	Reserved	Read:	0	0	0	0	0	0	0	0
		Write:								
$012F	PORTAD1	Read:	Bit 7	6	5	4	3	2	1	Bit 0
		Write:								
$0130	ATD1DR0H	Read:	Bit 15	14	13	12	11	10	9	Bit 8
		Write:								
$0131	ATD1DR0L	Read:	Bit 7	6	5	4	3	2	1	Bit 0
		Write:								
$0132	ATD1DR1H	Read:	Bit 15	14	13	12	11	10	9	Bit 8
		Write:								
$0133	ATD1DR1L	Read:	Bit 7	6	5	4	3	2	1	Bit 0
		Write:								
$0134	ATD1DR2H	Read:	Bit 15	14	13	12	11	10	9	Bit 8
		Write:								
$0135	ATD1DR2L	Read:	Bit 7	6	5	4	3	2	1	Bit 0
		Write:								
$0136	ATD1DR3H	Read:	Bit 15	14	13	12	11	10	9	Bit 8
		Write:								
$0137	ATD1DR3L	Read:	Bit 7	6	5	4	3	2	1	Bit 0
		Write:								
$0138	ATD1DR4H	Read:	Bit 15	14	13	12	11	10	9	Bit 8
		Write:								

$0120 - $013F **ATD1 (Analog to Digital Converter 10 Bit 8 Channel)**

Address	Name		Bit 7	Bit 6	Bit 5	Bit 4	Bit 3	Bit 2	Bit 1	Bit 0
$0139	ATD1DR4L	Read:	Bit 7	6	5	4	3	2	1	Bit 0
		Write:								
$013A	ATD1DR5H	Read:	Bit 15	14	13	12	11	10	9	Bit 8
		Write:								
$013B	ATD1DR5L	Read:	Bit 7	6	5	4	3	2	1	Bit 0
		Write:								
$013C	ATD1DR6H	Read:	Bit 15	14	13	12	11	10	9	Bit 8
		Write:								
$013D	ATD1DR6L	Read:	Bit 7	6	5	4	3	2	1	Bit 0
		Write:								
$013E	ATD1DR7H	Read:	Bit 15	14	13	12	11	10	9	Bit 8
		Write:								
$013F	ATD1DR7L	Read:	Bit 7	6	5	4	3	2	1	Bit 0
		Write:								

$0140 - $017F **CAN0 (Motorola Scalable CAN - MSCAN)**

Address	Name		Bit 7	Bit 6	Bit 5	Bit 4	Bit 3	Bit 2	Bit 1	Bit 0
$0140	CAN0CTL0	Read:	RXFRM	RXACT	CSWAI	SYNCH	TIME	WUPE	SLPRQ	INITRQ
		Write:								
$0141	CAN0CTL1	Read:	CANE	CLKSRC	LOOPB	LISTEN	0	WUPM	SLPAK	INITAK
		Write:								
$0142	CAN0BTR0	Read:	SJW1	SJW0	BRP5	BRP4	BRP3	BRP2	BRP1	BRP0
		Write:								
$0143	CAN0BTR1	Read:	SAMP	TSEG22	TSEG21	TSEG20	TSEG13	TSEG12	TSEG11	TSEG10
		Write:								
$0144	CAN0RFLG	Read:	WUPIF	CSCIF	RSTAT1	RSTAT0	TSTAT1	TSTAT0	OVRIF	RXF
		Write:								
$0145	CAN0RIER	Read:	WUPIE	CSCIE	RSTATE1	RSTATE0	TSTATE1	TSTATE0	OVRIE	RXFIE
		Write:								
$0146	CAN0TFLG	Read:	0	0	0	0	0	TXE2	TXE1	TXE0
		Write:								
$0147	CAN0TIER	Read:	0	0	0	0	0	TXEIE2	TXEIE1	TXEIE0
		Write:								
$0148	CAN0TARQ	Read:	0	0	0	0	0	ABTRQ2	ABTRQ1	ABTRQ0
		Write:								
$0149	CAN0TAAK	Read:	0	0	0	0	0	ABTAK2	ABTAK1	ABTAK0
		Write:								
$014A	CAN0TBSEL	Read:	0	0	0	0	0	TX2	TX1	TX0
		Write:								
$014B	CAN0IDAC	Read:	0	0	IDAM1	IDAM0	0	IDHIT2	IDHIT1	IDHIT0
		Write:								
$014C - $014D	Reserved	Read:	0	0	0	0	0	0	0	0
		Write:								
$014E	CAN0RXERR	Read:	RXERR7	RXERR6	RXERR5	RXERR4	RXERR3	RXERR2	RXERR1	RXERR0
		Write:								
$014F	CAN0TXERR	Read:	TXERR7	TXERR6	TXERR5	TXERR4	TXERR3	TXERR2	TXERR1	TXERR0
		Write:								
$0150 - $0153	CAN0IDAR0 - CAN0IDAR3	Read:	AC7	AC6	AC5	AC4	AC3	AC2	AC1	AC0
		Write:								

$0140 - $017F CAN0 (Motorola Scalable CAN - MSCAN)

Address	Name		Bit 7	Bit 6	Bit 5	Bit 4	Bit 3	Bit 2	Bit 1	Bit 0
$0154 - $0157	CAN0IDMR0 - CAN0IDMR3	Read: Write:	AM7	AM6	AM5	AM4	AM3	AM2	AM1	AM0
$0158 - $015B	CAN0IDAR4 - CAN0IDAR7	Read: Write:	AC7	AC6	AC5	AC4	AC3	AC2	AC1	AC0
$015C - $015F	CAN0IDMR4 - CAN0IDMR7	Read: Write:	AM7	AM6	AM5	AM4	AM3	AM2	AM1	AM0
$0160 - $016F	CAN0RXFG	Read:	FOREGROUND RECEIVE BUFFER see **Table 1-2**							
		Write:								
$0170 - $017F	CAN0TXFG	Read: Write:	FOREGROUND TRANSMIT BUFFER see **Table 1-2**							

Table 1-2 Detailed MSCAN Foreground Receive and Transmit Buffer Layout

Address	Name		Bit 7	Bit 6	Bit 5	Bit 4	Bit 3	Bit 2	Bit 1	Bit 0
	Extended ID	Read:	ID28	ID27	ID26	ID25	ID24	ID23	ID22	ID21
$xxx0	Standard ID	Read:	ID10	ID9	ID8	ID7	ID6	ID5	ID4	ID3
	CANxRIDR0	Write:								
	Extended ID	Read:	ID20	ID19	ID18	SRR=1	IDE=1	ID17	ID16	ID15
$xxx1	Standard ID	Read:	ID2	ID1	ID0	RTR	IDE=0			
	CANxRIDR1	Write:								
	Extended ID	Read:	ID14	ID13	ID12	ID11	ID10	ID9	ID8	ID7
$xxx2	Standard ID	Read:								
	CANxRIDR2	Write:								
	Extended ID	Read:	ID6	ID5	ID4	ID3	ID2	ID1	ID0	RTR
$xxx3	Standard ID	Read:								
	CANxRIDR3	Write:								
$xxx4 - $xxxB	CANxRDSR0 - CANxRDSR7	Read:	DB7	DB6	DB5	DB4	DB3	DB2	DB1	DB0
		Write:								
$xxxC	CANRxDLR	Read:					DLC3	DLC2	DLC1	DLC0
		Write:								
$xxxD	Reserved	Read:								
		Write:								
$xxxE	CANxRTSRH	Read:	TSR15	TSR14	TSR13	TSR12	TSR11	TSR10	TSR9	TSR8
		Write:								
$xxxF	CANxRTSRL	Read:	TSR7	TSR6	TSR5	TSR4	TSR3	TSR2	TSR1	TSR0
		Write:								
$xx10	Extended ID CANxTIDR0	Read: Write:	ID28	ID27	ID26	ID25	ID24	ID23	ID22	ID21
	Standard ID	Read: Write:	ID10	ID9	ID8	ID7	ID6	ID5	ID4	ID3
$xx11	Extended ID CANxTIDR1	Read: Write:	ID20	ID19	ID18	SRR=1	IDE=1	ID17	ID16	ID15
	Standard ID	Read: Write:	ID2	ID1	ID0	RTR	IDE=0			
$xx12	Extended ID CANxTIDR2	Read: Write:	ID14	ID13	ID12	ID11	ID10	ID9	ID8	ID7
	Standard ID	Read: Write:								

Table 1-2 Detailed MSCAN Foreground Receive and Transmit Buffer Layout

Address	Name		Bit 7	Bit 6	Bit 5	Bit 4	Bit 3	Bit 2	Bit 1	Bit 0
$xx13	Extended ID CANxTIDR3	Read:	ID6	ID5	ID4	ID3	ID2	ID1	ID0	RTR
		Write:	ID6	ID5	ID4	ID3	ID2	ID1	ID0	RTR
	Standard ID	Read:								
		Write:								
$xx14 - $xx1B	CANxTDSR0 - CANxTDSR7	Read:	DB7	DB6	DB5	DB4	DB3	DB2	DB1	DB0
		Write:	DB7	DB6	DB5	DB4	DB3	DB2	DB1	DB0
$xx1C	CANxTDLR	Read:					DLC3	DLC2	DLC1	DLC0
		Write:					DLC3	DLC2	DLC1	DLC0
$xx1D	CANxTTBPR	Read:	PRIO7	PRIO6	PRIO5	PRIO4	PRIO3	PRIO2	PRIO1	PRIO0
		Write:	PRIO7	PRIO6	PRIO5	PRIO4	PRIO3	PRIO2	PRIO1	PRIO0
$xx1E	CANxTTSRH	Read:	TSR15	TSR14	TSR13	TSR12	TSR11	TSR10	TSR9	TSR8
		Write:								
$xx1F	CANxTTSRL	Read:	TSR7	TSR6	TSR5	TSR4	TSR3	TSR2	TSR1	TSR0
		Write:								

$0180 - $01BF CAN1 (Motorola Scalable CAN - MSCAN)

Address	Name		Bit 7	Bit 6	Bit 5	Bit 4	Bit 3	Bit 2	Bit 1	Bit 0
$0180	CAN1CTL0	Read:	RXFRM	RXACT	CSWAI	SYNCH	TIME	WUPE	SLPRQ	INITRQ
		Write:	RXFRM		CSWAI		TIME	WUPE	SLPRQ	INITRQ
$0181	CAN1CTL1	Read:	CANE	CLKSRC	LOOPB	LISTEN	0	WUPM	SLPAK	INITAK
		Write:	CANE	CLKSRC	LOOPB	LISTEN		WUPM		
$0182	CAN1BTR0	Read:	SJW1	SJW0	BRP5	BRP4	BRP3	BRP2	BRP1	BRP0
		Write:	SJW1	SJW0	BRP5	BRP4	BRP3	BRP2	BRP1	BRP0
$0183	CAN1BTR1	Read:	SAMP	TSEG22	TSEG21	TSEG20	TSEG13	TSEG12	TSEG11	TSEG10
		Write:	SAMP	TSEG22	TSEG21	TSEG20	TSEG13	TSEG12	TSEG11	TSEG10
$0184	CAN1RFLG	Read:	WUPIF	CSCIF	RSTAT1	RSTAT0	TSTAT1	TSTAT0	OVRIF	RXF
		Write:	WUPIF	CSCIF					OVRIF	RXF
$0185	CAN1RIER	Read:	WUPIE	CSCIE	RSTATE1	RSTATE0	TSTATE1	TSTATE0	OVRIE	RXFIE
		Write:	WUPIE	CSCIE	RSTATE1	RSTATE0	TSTATE1	TSTATE0	OVRIE	RXFIE
$0186	CAN1TFLG	Read:	0	0	0	0	0	TXE2	TXE1	TXE0
		Write:						TXE2	TXE1	TXE0
$0187	CAN1TIER	Read:	0	0	0	0	0	TXEIE2	TXEIE1	TXEIE0
		Write:						TXEIE2	TXEIE1	TXEIE0
$0188	CAN1TARQ	Read:	0	0	0	0	0	ABTRQ2	ABTRQ1	ABTRQ0
		Write:						ABTRQ2	ABTRQ1	ABTRQ0
$0189	CAN1TAAK	Read:	0	0	0	0	0	ABTAK2	ABTAK1	ABTAK0
		Write:								
$018A	CAN1TBSEL	Read:	0	0	0	0	0	TX2	TX1	TX0
		Write:						TX2	TX1	TX0
$018B	CAN1IDAC	Read:	0	0	IDAM1	IDAM0	0	IDHIT2	IDHIT1	IDHIT0
		Write:			IDAM1	IDAM0				
$018C - $018D	Reserved	Read:	0	0	0	0	0	0	0	0
		Write:								
$018E	CAN1RXERR	Read:	RXERR7	RXERR6	RXERR5	RXERR4	RXERR3	RXERR2	RXERR1	RXERR0
		Write:								
$018F	CAN1TXERR	Read:	TXERR7	TXERR6	TXERR5	TXERR4	TXERR3	TXERR2	TXERR1	TXERR0
		Write:								
$0190 - $0193	CAN1IDAR0 - CAN1IDAR3	Read:	AC7	AC6	AC5	AC4	AC3	AC2	AC1	AC0
		Write:	AC7	AC6	AC5	AC4	AC3	AC2	AC1	AC0

$0180 - $01BF CAN1 (Motorola Scalable CAN - MSCAN)

Address	Name		Bit 7	Bit 6	Bit 5	Bit 4	Bit 3	Bit 2	Bit 1	Bit 0
$0194 - $0197	CAN1IDMR0 - CAN1IDMR3	Read:	AM7	AM6	AM5	AM4	AM3	AM2	AM1	AM0
		Write:	AM7	AM6	AM5	AM4	AM3	AM2	AM1	AM0
$0198 - $019B	CAN1IDAR4 - CAN1IDAR7	Read:	AC7	AC6	AC5	AC4	AC3	AC2	AC1	AC0
		Write:	AC7	AC6	AC5	AC4	AC3	AC2	AC1	AC0
$019C - $019F	CAN1IDMR4 - CAN1IDMR7	Read:	AM7	AM6	AM5	AM4	AM3	AM2	AM1	AM0
		Write:	AM7	AM6	AM5	AM4	AM3	AM2	AM1	AM0
$01A0 - $01AF	CAN1RXFG	Read:	FOREGROUND RECEIVE BUFFER see **Table 1-2**							
		Write:								
$01B0 - $01BF	CAN1TXFG	Read:	FOREGROUND TRANSMIT BUFFER see **Table 1-2**							
		Write:	FOREGROUND TRANSMIT BUFFER see **Table 1-2**							

$01C0 - $01FF CAN2 (Motorola Scalable CAN - MSCAN)

Address	Name		Bit 7	Bit 6	Bit 5	Bit 4	Bit 3	Bit 2	Bit 1	Bit 0
$01C0	CAN2CTL0	Read:	RXFRM	RXACT	CSWAI	SYNCH	TIME	WUPE	SLPRQ	INITRQ
		Write:	RXFRM		CSWAI		TIME	WUPE	SLPRQ	INITRQ
$01C1	CAN2CTL1	Read:	CANE	CLKSRC	LOOPB	LISTEN	0	WUPM	SLPAK	INITAK
		Write:	CANE	CLKSRC	LOOPB	LISTEN		WUPM		
$01C2	CAN2BTR0	Read:	SJW1	SJW0	BRP5	BRP4	BRP3	BRP2	BRP1	BRP0
		Write:	SJW1	SJW0	BRP5	BRP4	BRP3	BRP2	BRP1	BRP0
$01C3	CAN2BTR1	Read:	SAMP	TSEG22	TSEG21	TSEG20	TSEG13	TSEG12	TSEG11	TSEG10
		Write:	SAMP	TSEG22	TSEG21	TSEG20	TSEG13	TSEG12	TSEG11	TSEG10
$01C4	CAN2RFLG	Read:	WUPIF	CSCIF	RSTAT1	RSTAT0	TSTAT1	TSTAT0	OVRIF	RXF
		Write:	WUPIF	CSCIF					OVRIF	RXF
$01C5	CAN2RIER	Read:	WUPIE	CSCIE	RSTATE1	RSTATE0	TSTATE1	TSTATE0	OVRIE	RXFIE
		Write:	WUPIE	CSCIE	RSTATE1	RSTATE0	TSTATE1	TSTATE0	OVRIE	RXFIE
$01C6	CAN2TFLG	Read:	0	0	0	0	0	TXE2	TXE1	TXE0
		Write:						TXE2	TXE1	TXE0
$01C7	CAN2TIER	Read:	0	0	0	0	0	TXEIE2	TXEIE1	TXEIE0
		Write:						TXEIE2	TXEIE1	TXEIE0
$01C8	CAN2TARQ	Read:	0	0	0	0	0	ABTRQ2	ABTRQ1	ABTRQ0
		Write:						ABTRQ2	ABTRQ1	ABTRQ0
$01C9	CAN2TAAK	Read:	0	0	0	0	0	ABTAK2	ABTAK1	ABTAK0
		Write:								
$01CA	CAN2TBSEL	Read:	0	0	0	0	0	TX2	TX1	TX0
		Write:						TX2	TX1	TX0
$01CB	CAN2IDAC	Read:	0	0	IDAM1	IDAM0	0	IDHIT2	IDHIT1	IDHIT0
		Write:			IDAM1	IDAM0				
$01CC - $01CD	Reserved	Read:	0	0	0	0	0	0	0	0
		Write:								
$01CE	CAN2RXERR	Read:	RXERR7	RXERR6	RXERR5	RXERR4	RXERR3	RXERR2	RXERR1	RXERR0
		Write:								
$01CF	CAN2TXERR	Read:	TXERR7	TXERR6	TXERR5	TXERR4	TXERR3	TXERR2	TXERR1	TXERR0
		Write:								
$01D0 - $01D3	CAN2IDAR0 - CAN2IDAR3	Read:	AC7	AC6	AC5	AC4	AC3	AC2	AC1	AC0
		Write:	AC7	AC6	AC5	AC4	AC3	AC2	AC1	AC0
$01D4 - $01D7	CAN2IDMR0 - CAN2IDMR3	Read:	AM7	AM6	AM5	AM4	AM3	AM2	AM1	AM0
		Write:	AM7	AM6	AM5	AM4	AM3	AM2	AM1	AM0

$01C0 - $01FF **CAN2 (Motorola Scalable CAN - MSCAN)**

Address	Name		Bit 7	Bit 6	Bit 5	Bit 4	Bit 3	Bit 2	Bit 1	Bit 0
$01D8 - $01DB	CAN2IDAR4 - CAN2IDAR7	Read: Write:	AC7	AC6	AC5	AC4	AC3	AC2	AC1	AC0
$01DC - $01DF	CAN2IDMR4 - CAN2IDMR7	Read: Write:	AM7	AM6	AM5	AM4	AM3	AM2	AM1	AM0
$01E0 - $01EF	CAN2RXFG	Read:	FOREGROUND RECEIVE BUFFER see **Table 1-2**							
		Write:								
$01F0 - $01FF	CAN2TXFG	Read: Write:	FOREGROUND TRANSMIT BUFFER see **Table 1-2**							

$0200 - $023F **CAN3 (Motorola Scalable CAN - MSCAN)**

Address	Name		Bit 7	Bit 6	Bit 5	Bit 4	Bit 3	Bit 2	Bit 1	Bit 0
$0200	CAN3CTL0	Read:	RXFRM	RXACT	CSWAI	SYNCH	TIME	WUPE	SLPRQ	INITRQ
		Write:								
$0201	CAN3CTL1	Read:	CANE	CLKSRC	LOOPB	LISTEN	0	WUPM	SLPAK	INITAK
		Write:								
$0202	CAN3BTR0	Read: Write:	SJW1	SJW0	BRP5	BRP4	BRP3	BRP2	BRP1	BRP0
$0203	CAN3BTR1	Read: Write:	SAMP	TSEG22	TSEG21	TSEG20	TSEG13	TSEG12	TSEG11	TSEG10
$0204	CAN3RFLG	Read:	WUPIF	CSCIF	RSTAT1	RSTAT0	TSTAT1	TSTAT0	OVRIF	RXF
		Write:								
$0205	CAN3RIER	Read: Write:	WUPIE	CSCIE	RSTATE1	RSTATE0	TSTATE1	TSTATE0	OVRIE	RXFIE
$0206	CAN3TFLG	Read:	0	0	0	0	0	TXE2	TXE1	TXE0
		Write:								
$0207	CAN3TIER	Read:	0	0	0	0	0	TXEIE2	TXEIE1	TXEIE0
		Write:								
$0208	CAN3TARQ	Read:	0	0	0	0	0	ABTRQ2	ABTRQ1	ABTRQ0
		Write:								
$0209	CAN3TAAK	Read:	0	0	0	0	0	ABTAK2	ABTAK1	ABTAK0
		Write:								
$020A	CAN3TBSEL	Read:	0	0	0	0	0	TX2	TX1	TX0
		Write:								
$020B	CAN3IDAC	Read:	0	0	IDAM1	IDAM0	0	IDHIT2	IDHIT1	IDHIT0
		Write:								
$020C - $020D	Reserved	Read:	0	0	0	0	0	0	0	0
		Write:								
$020E	CAN3RXERR	Read:	RXERR7	RXERR6	RXERR5	RXERR4	RXERR3	RXERR2	RXERR1	RXERR0
		Write:								
$020F	CAN3TXERR	Read:	TXERR7	TXERR6	TXERR5	TXERR4	TXERR3	TXERR2	TXERR1	TXERR0
		Write:								
$0210 - $0213	CAN3IDAR0 - CAN3IDAR3	Read: Write:	AC7	AC6	AC5	AC4	AC3	AC2	AC1	AC0
$0214 - $0217	CAN3IDMR0 - CAN3IDMR3	Read: Write:	AM7	AM6	AM5	AM4	AM3	AM2	AM1	AM0
$0218 - $021B	CAN3IDAR4 - CAN3IDAR7	Read: Write:	AC7	AC6	AC5	AC4	AC3	AC2	AC1	AC0

$0200 - $023F CAN3 (Motorola Scalable CAN - MSCAN)

Address	Name		Bit 7	Bit 6	Bit 5	Bit 4	Bit 3	Bit 2	Bit 1	Bit 0
$021C - $021F	CAN3IDMR4 - CAN3IDMR7	Read: Write:	AM7	AM6	AM5	AM4	AM3	AM2	AM1	AM0
$0220 - $022F	CAN3RXFG	Read:	FOREGROUND RECEIVE BUFFER see **Table 1-2**							
		Write:								
$0230 - $023F	CAN3TXFG	Read: Write:	FOREGROUND TRANSMIT BUFFER see **Table 1-2**							

$0240 - $027F PIM (Port Integration Module PIM_9DP256)

Address	Name		Bit 7	Bit 6	Bit 5	Bit 4	Bit 3	Bit 2	Bit 1	Bit 0
$0240	PTT	Read: Write:	PTT7	PTT6	PTT5	PTT4	PTT3	PTT2	PTT1	PTT0
$0241	PTIT	Read:	PTIT7	PTIT6	PTIT5	PTIT4	PTIT3	PTIT2	PTIT1	PTIT0
		Write:								
$0242	DDRT	Read: Write:	DDRT7	DDRT7	DDRT5	DDRT4	DDRT3	DDRT2	DDRT1	DDRT0
$0243	RDRT	Read: Write:	RDRT7	RDRT6	RDRT5	RDRT4	RDRT3	RDRT2	RDRT1	RDRT0
$0244	PERT	Read: Write:	PERT7	PERT6	PERT5	PERT4	PERT3	PERT2	PERT1	PERT0
$0245	PPST	Read: Write:	PPST7	PPST6	PPST5	PPST4	PPST3	PPST2	PPST1	PPST0
$0246 - $0247	Reserved	Read:	0	0	0	0	0	0	0	0
		Write:								
$0248	PTS	Read: Write:	PTS7	PTS6	PTS5	PTS4	PTS3	PTS2	PTS1	PTS0
$0249	PTIS	Read:	PTIS7	PTIS6	PTIS5	PTIS4	PTIS3	PTIS2	PTIS1	PTIS0
		Write:								
$024A	DDRS	Read: Write:	DDRS7	DDRS7	DDRS5	DDRS4	DDRS3	DDRS2	DDRS1	DDRS0
$024B	RDRS	Read: Write:	RDRS7	RDRS6	RDRS5	RDRS4	RDRS3	RDRS2	RDRS1	RDRS0
$024C	PERS	Read: Write:	PERS7	PERS6	PERS5	PERS4	PERS3	PERS2	PERS1	PERS0
$024D	PPSS	Read: Write:	PPSS7	PPSS6	PPSS5	PPSS4	PPSS3	PPSS2	PPSS1	PPSS0
$024E	WOMS	Read: Write:	WOMS7	WOMS6	WOMS5	WOMS4	WOMS3	WOMS2	WOMS1	WOMS0
$024F	Reserved	Read:	0	0	0	0	0	0	0	0
		Write:								
$0250	PTM	Read: Write:	PTM7	PTM6	PTM5	PTM4	PTM3	PTM2	PTM1	PTM0
$0251	PTIM	Read:	PTIM7	PTIM6	PTIM5	PTIM4	PTIM3	PTIM2	PTIM1	PTIM0
		Write:								
$0252	DDRM	Read: Write:	DDRM7	DDRM7	DDRM5	DDRM4	DDRM3	DDRM2	DDRM1	DDRM0
$0253	RDRM	Read: Write:	RDRM7	RDRM6	RDRM5	RDRM4	RDRM3	RDRM2	RDRM1	RDRM0

$0240 - $027F PIM (Port Integration Module PIM_9DP256)

Address	Name		Bit 7	Bit 6	Bit 5	Bit 4	Bit 3	Bit 2	Bit 1	Bit 0
$0254	PERM	Read: Write:	PERM7	PERM6	PERM5	PERM4	PERM3	PERM2	PERM1	PERM0
$0255	PPSM	Read: Write:	PPSM7	PPSM6	PPSM5	PPSM4	PPSM3	PPSM2	PPSM1	PPSM0
$0256	WOMM	Read: Write:	WOMM7	WOMM6	WOMM5	WOMM4	WOMM3	WOMM2	WOMM1	WOMM0
$0257	MODRR	Read:	0	MODRR6	MODRR5	MODRR4	MODRR3	MODRR2	MODRR1	MODRR0
		Write:								
$0258	PTP	Read: Write:	PTP7	PTP6	PTP5	PTP4	PTP3	PTP2	PTP1	PTP0
$0259	PTIP	Read:	PTIP7	PTIP6	PTIP5	PTIP4	PTIP3	PTIP2	PTIP1	PTIP0
		Write:								
$025A	DDRP	Read: Write:	DDRP7	DDRP7	DDRP5	DDRP4	DDRP3	DDRP2	DDRP1	DDRP0
$025B	RDRP	Read: Write:	RDRP7	RDRP6	RDRP5	RDRP4	RDRP3	RDRP2	RDRP1	RDRP0
$025C	PERP	Read: Write:	PERP7	PERP6	PERP5	PERP4	PERP3	PERP2	PERP1	PERP0
$025D	PPSP	Read: Write:	PPSP7	PPSP6	PPSP5	PPSP4	PPSP3	PPSP2	PPSP1	PPSS0
$025E	PIEP	Read: Write:	PIEP7	PIEP6	PIEP5	PIEP4	PIEP3	PIEP2	PIEP1	PIEP0
$025F	PIFP	Read: Write:	PIFP7	PIFP6	PIFP5	PIFP4	PIFP3	PIFP2	PIFP1	PIFP0
$0260	PTH	Read: Write:	PTH7	PTH6	PTH5	PTH4	PTH3	PTH2	PTH1	PTH0
$0261	PTIH	Read:	PTIH7	PTIH6	PTIH5	PTIH4	PTIH3	PTIH2	PTIH1	PTIH0
		Write:								
$0262	DDRH	Read: Write:	DDRH7	DDRH7	DDRH5	DDRH4	DDRH3	DDRH2	DDRH1	DDRH0
$0263	RDRH	Read: Write:	RDRH7	RDRH6	RDRH5	RDRH4	RDRH3	RDRH2	RDRH1	RDRH0
$0264	PERH	Read: Write:	PERH7	PERH6	PERH5	PERH4	PERH3	PERH2	PERH1	PERH0
$0265	PPSH	Read: Write:	PPSH7	PPSH6	PPSH5	PPSH4	PPSH3	PPSH2	PPSH1	PPSH0
$0266	PIEH	Read: Write:	PIEH7	PIEH6	PIEH5	PIEH4	PIEH3	PIEH2	PIEH1	PIEH0
$0267	PIFH	Read: Write:	PIFH7	PIFH6	PIFH5	PIFH4	PIFH3	PIFH2	PIFH1	PIFH0
$0268	PTJ	Read:	PTJ7	PTJ6	0	0	0	0	PTJ1	PTJ0
		Write:								
$0269	PTIJ	Read:	PTIJ7	PTIJ6	0	0	0	0	PTIJ1	PTIJ0
		Write:								
$026A	DDRJ	Read:	DDRJ7	DDRJ7	0	0	0	0	DDRJ1	DDRJ0
		Write:								
$026B	RDRJ	Read:	RDRJ7	RDRJ6	0	0	0	0	RDRJ1	RDRJ0
		Write:								
$026C	PERJ	Read:	PERJ7	PERJ6	0	0	0	0	PERJ1	PERJ0
		Write:								

$0240 - $027F PIM (Port Integration Module PIM_9DP256)

Address	Name		Bit 7	Bit 6	Bit 5	Bit 4	Bit 3	Bit 2	Bit 1	Bit 0
$026D	PPSJ	Read: Write:	PPSJ7	PPSJ6	0	0	0	0	PPSJ1	PPSJ0
$026E	PIEJ	Read: Write:	PIEJ7	PIEJ6	0	0	0	0	PIEJ1	PIEJ0
$026F	PIFJ	Read: Write:	PIFJ7	PIFJ6	0	0	0	0	PIFJ1	PIFJ0
$0270 - $027F	Reserved	Read:								

$0280 - $02BF CAN4 (Motorola Scalable CAN - MSCAN)

Address	Name		Bit 7	Bit 6	Bit 5	Bit 4	Bit 3	Bit 2	Bit 1	Bit 0
$0280	CAN4CTL0	Read: Write:	RXFRM	RXACT	CSWAI	SYNCH	TIME	WUPE	SLPRQ	INITRQ
$0281	CAN4CTL1	Read: Write:	CANE	CLKSRC	LOOPB	LISTEN	0	WUPM	SLPAK	INITAK
$0282	CAN4BTR0	Read: Write:	SJW1	SJW0	BRP5	BRP4	BRP3	BRP2	BRP1	BRP0
$0283	CAN4BTR1	Read: Write:	SAMP	TSEG22	TSEG21	TSEG20	TSEG13	TSEG12	TSEG11	TSEG10
$0284	CAN4RFLG	Read: Write:	WUPIF	CSCIF	RSTAT1	RSTAT0	TSTAT1	TSTAT0	OVRIF	RXF
$0285	CAN4RIER	Read: Write:	WUPIE	CSCIE	RSTATE1	RSTATE0	TSTATE1	TSTATE0	OVRIE	RXFIE
$0286	CAN4TFLG	Read: Write:	0	0	0	0	0	TXE2	TXE1	TXE0
$0287	CAN4TIER	Read: Write:	0	0	0	0	0	TXEIE2	TXEIE1	TXEIE0
$0288	CAN4TARQ	Read: Write:	0	0	0	0	0	ABTRQ2	ABTRQ1	ABTRQ0
$0289	CAN4TAAK	Read: Write:	0	0	0	0	0	ABTAK2	ABTAK1	ABTAK0
$028A	CAN4TBSEL	Read: Write:	0	0	0	0	0	TX2	TX1	TX0
$028B	CAN4IDAC	Read: Write:	0	0	IDAM1	IDAM0	0	IDHIT2	IDHIT1	IDHIT0
$028C - $028D	Reserved	Read: Write:	0	0	0	0	0	0	0	0
$028E	CAN4RXERR	Read: Write:	RXERR7	RXERR6	RXERR5	RXERR4	RXERR3	RXERR2	RXERR1	RXERR0
$028F	CAN4TXERR	Read: Write:	TXERR7	TXERR6	TXERR5	TXERR4	TXERR3	TXERR2	TXERR1	TXERR0
$0290 - $0293	CAN4IDAR0 - CAN4IDAR3	Read: Write:	AC7	AC6	AC5	AC4	AC3	AC2	AC1	AC0
$0294 - $0297	CAN4IDMR0 - CAN4IDMR3	Read: Write:	AM7	AM6	AM5	AM4	AM3	AM2	AM1	AM0
$0298 - $029B	CAN4IDAR4 - CAN4IDAR7	Read: Write:	AC7	AC6	AC5	AC4	AC3	AC2	AC1	AC0

$0280 - $02BF **CAN4 (Motorola Scalable CAN - MSCAN)**

Address	Name		Bit 7	Bit 6	Bit 5	Bit 4	Bit 3	Bit 2	Bit 1	Bit 0
$029C - $029F	CAN4IDMR4 - CAN4IDMR7	Read: Write:	AM7	AM6	AM5	AM4	AM3	AM2	AM1	AM0
$02A0 - $02AF	CAN4RXFG	Read:	FOREGROUND RECEIVE BUFFER see **Table 1-2**							
		Write:								
$02B0 - $02BF	CAN4TXFG	Read: Write:	FOREGROUND TRANSMIT BUFFER see **Table 1-2**							

$02C0 - $03FF **Reserved**

Address	Name		Bit 7	Bit 6	Bit 5	Bit 4	Bit 3	Bit 2	Bit 1	Bit 0
$02C0 - $03FF	Reserved	Read:	0	0	0	0	0	0	0	0
		Write:								

1.6 Part ID Assignments

The part ID is located in two 8-bit registers PARTIDH and PARTIDL (addresses $001A and $001B after reset). The read-only value is a unique part ID for each revision of the chip. **Table 1-3** shows the assigned part ID number.

Table 1-3 Assigned Part ID Numbers

Device	Mask Set Number	Part ID[1]
MC9S12DP512	0L00M	$0400
MC9S12DP512	1L00M	$0401
MC9S12DP512	2L00M	$0402
MC9S12DP512	3L00M	$0403
MC9S12DP512	4L00M	$0404

NOTES:
1. The coding is as follows:
Bit 15 - 12: Major family identifier
Bit 11 - 8: Minor family identifier
Bit 7 - 4: Major mask set revision number including FAB transfers
Bit 3 - 0: Minor - non full - mask set revision

1.7 Memory Size Assignments

The device memory sizes are located in two 8-bit registers MEMSIZ0 and MEMSIZ1 (addresses $001C and $001D after reset). **Table 1-4** shows the read-only values of these registers. Refer to HCS12 Module Mapping Control (MMC) Block Guide for further details.

Table 1-4 Memory size registers

Register name	Value
MEMSIZ0	$26
MEMSIZ1	$82

SECTION H.3: MC9212D-Family

Freescale Semiconductor, Inc.

MC9S12D-FamilyPP
Rev 6.1, 23-Oct-02

MC9S12D-Family

Product Brief
16-Bit Microcontroller

Designed for automotive multiplexing applications, members of the MC9S12D-Family of 16 bit Flash-based microcontrollers are fully pin compatible and enable users to choose between different memory and peripheral options for scalable designs. All MC9S12D-Family members are composed of standard on-chip peripherals including a 16-bit central processing unit (CPU12), up to 512K bytes of Flash EEPROM, 14K bytes of RAM, 4K bytes of EEPROM, two asynchronous serial communications interfaces (SCI), three serial peripheral interfaces (SPI), IIC-bus, an enhanced capture timer (ECT), two 8-channel 10-bit analog-to-digital converters (ADC), an eight-channel pulse-width modulator (PWM), J1850 interface and up to five CAN 2.0 A, B software compatible modules (MSCAN12). System resource mapping, clock generation, interrupt control and bus interfacing are managed by the system integration module (SIM). The MC9S12D-Family has full 16-bit data paths throughout, however, the external bus can operate in an 8-bit narrow mode so single 8-bit wide memory can be interfaced for lower cost systems. The inclusion of a PLL circuit allows power consumption and performance to be adjusted to suit operational requirements. In addition to the I/O ports available in each module, up to 22 I/O ports are available with interrupt capability allowing Wake-Up from STOP or WAIT mode.

Features

NOTE

Not all features listed here are available in all configurations.
Additional information about D and B family inter-operability is given in:
EB386 "HCS12 D-Family Compatibility Considerations" and
EB388 "Using the HCS12 D-Family as a development platform for the HCS12 B family"

- **16-bit CPU12**
  - — Upward compatible with M68HC11 instruction set
  - — Interrupt stacking and programmer's model identical to M68HC11
  - — HCS12 Instruction queue
  - — Enhanced indexed addressing
- **Multiplexed bus**
  - — Single chip or expanded
  - — 16 address/16 data wide or 16 address/8 data narrow modes
  - — External address space 1MByte for Data and Program space (112 pin package only)
- **Wake-up interrupt inputs depending on the package option**
  - — 8-bit port H
  - — 2-bit port J1:0
  - — 2-bit port J7:6 shared with IIC, CAN4 and CAN0 module
  - — 8-bit port P shared with PWM or SPI1,2
- **Memory options**
  - — 32K, 64K, 128K, 256K, 512K Byte Flash EEPROM
  - — 1K, 2K, 4K Byte EEPROM
  - — 2K, 4K, 8K, 12K, 14K Byte RAM

- **Analog-to-Digital Converters**
  - — One or two 8-channel modules with 10-bit resolution depending on the package option
  - — External conversion trigger capability
- **Up to five 1M bit per second, CAN 2.0 A, B software compatible modules**
  - — Five receive and three transmit buffers
  - — Flexible identifier filter programmable as 2 x 32 bit, 4 x 16 bit or 8 x 8 bit
  - — Four separate interrupt channels for Receive, Transmit, Error and Wake-up
  - — Low-pass filter wake-up function in STOP mode
  - — Loop-back for self test operation
- **Enhanced Capture Timer (ECT)**
  - — 16-bit main counter with 7-bit prescaler
  - — 8 programmable input capture or output compare channels; 4 of the 8 input captures with buffer
  - — Input capture filters and buffers, three successive captures on four channels, or two captures on four channels with a capture/compare selectable on the remaining four
  - — Four 8-bit or two 16-bit pulse accumulators
  - — 16-bit modulus down-counter with 4-bit prescaler
  - — Four user-selectable delay counters for signal filtering
- **8 PWM channels with programmable period and duty cycle (7 channels on 80 Pin Packages)**
  - — 8-bit, 8-channel or 16-bit, 4-channel
  - — Separate control for each pulse width and duty cycle
  - — Center- or left-aligned outputs
  - — Programmable clock select logic with a wide range of frequencies
- **Serial interfaces**
  - — Two asynchronous serial communications interfaces (SCI)
  - — Up to three synchronous serial peripheral interfaces (SPI)
  - — IIC
- **SAE J1850 Compatible Module (BDLC)**
  - — 10.4 kbps Variable Pulse Width format
  - — Byte level receive and transmit
  - — 4x receive mode supported
- **SIM (System Integration Module)**
  - — CRG (windowed COP watchdog, real time interrupt, clock monitor, clock generation and reset)
  - — MEBI (multiplexed external bus interface)
  - — INT (interrupt control)
- **Clock generation**
  - — Phase-locked loop clock frequency multiplier
  - — Limp home mode in absence of external clock
  - — Clock Monitor
  - — Low power 0.5 to 16 MHz crystal oscillator reference clock
- **Operating frequency for ambient temperatures T_A -40°C <= T_A <= 125°C**
  - — 50MHz equivalent to 25MHz Bus Speed for single chip
    40MHz equivalent to 20MHz Bus Speed in expanded bus modes.
- **Internal 5V to 2.5V Regulator**
- **112-Pin LQFP or 80-Pin QFP package**
  - — I/O lines with 5V input and drive capability
  - — 5V A/D converter inputs and 5V I/O
  - — 2.5V logic supply
- **Development support**
  - — Single-wire background debug™ mode (BDM)
  - — On-chip hardware breakpoints

Table 1 List of MC9S12D-Family members

Flash	RAM	EEPROM	Package	Device	CAN	J1850	SCI	SPI	IIC	A/D	PWM	I/O
512K	14K	4K	112LQFP	DP512	5	1	2	3	1	2/16	8	91
				DT512	3	0	2	3	1	2/16	8	91
				DJ512	2	1	2	3	1	2/16	8	91
256K	12K	4K	112LQFP	DT256	3	0	2	3	1	2/16	8	91
				DJ256	2	1	2	3	1	2/16	8	91
				DG256	2	0	2	3	1	2/16	8	91
			80QFP	DJ256	2	1	2	3	1	1/8	7	59
				DG256	2	0	2	3	1	1/8	7	59
128K	8K	2K	112LQFP	DT128	3	0	2	2	1	2/16	8	91
				DJ128	2	1	2	2	1	2/16	8	91
				DG128	2	0	2	2	1	2/16	8	91
			80QFP	DJ128	2	1	2	2	1	1/8	7	59
				DG128	2	0	2	2	1	1/8	7	59
64K	4K	1K	112LQFP	DJ64	1	1	2	1	1	2/16	8	91
				D64	1	0	2	1	1	2/16	8	91
			80QFP	DJ64	1	1	2	1	1	1/8	7	59
				D64	1	0	2	1	1	1/8	7	59
32K	2K	1K	80QFP	D32	1	0	2	1	0	1/8	7	59

- **Pin out explanations:**
  - — A/D is the number of modules/total number of A/D channels.
  - — I/O is the sum of ports capable to act as digital input or output.
    - 112 Pin Packages:
      - Port A = 8, B = 8, E = 6 + 2 input only, H = 8, J = 4, K = 7, M = 8, P = 8, S = 8, T = 8, PAD = 16 input only.
      - 22 inputs provide Interrupt capability (H =8, P= 8, J = 4, IRQ, XIRQ)
    - 80 Pin Packages:
      - Port A = 8, B = 8, E = 6 + 2 input only, J = 2, M = 6, P = 7, S = 4, T = 8, PAD = 8 input only.
      - 11 inputs provide Interrupt capability (P= 7, J = 2, IRQ, XIRQ)
  - — CAN0 pins are shared between J1850 pins.
  - — CAN0 can be routed under software control from PM1:0 to pins PM3:2 or PM5:4 or PJ7:6.
  - — CAN4 pins are shared between IIC pins.
  - — CAN4 can be routed under software control from PJ7:6 to pins PM5:4 or PM7:6.
  - — Versions with 4 CAN modules will have CAN0, CAN1, CAN2 and CAN4.
  - — Versions with 3 CANs modules will have CAN0, CAN1 and CAN4.
  - — Versions with 2 CAN modules will have CAN0 and CAN4.
  - — Versions with one CAN module will have CAN0.
  - — Versions with 2 SPI modules will have SPI0 and SPI1.
  - — Versions with 1 SPI will have SPI0.
  - — SPI0 can be routed to either Ports PS7:4 or PM5:2.
  - — SPI2 pins are shared with $\overline{PWM}$7:4; In 112 pin versions SPI2 can be routed under software control to PH7:4. In 80 pin packages $\overline{SS}$-signal of SPI2 is not bonded out!

NOTE

CAN and SPI routing features are not available on the 1st PC9S12DP256 mask set 0K36N!

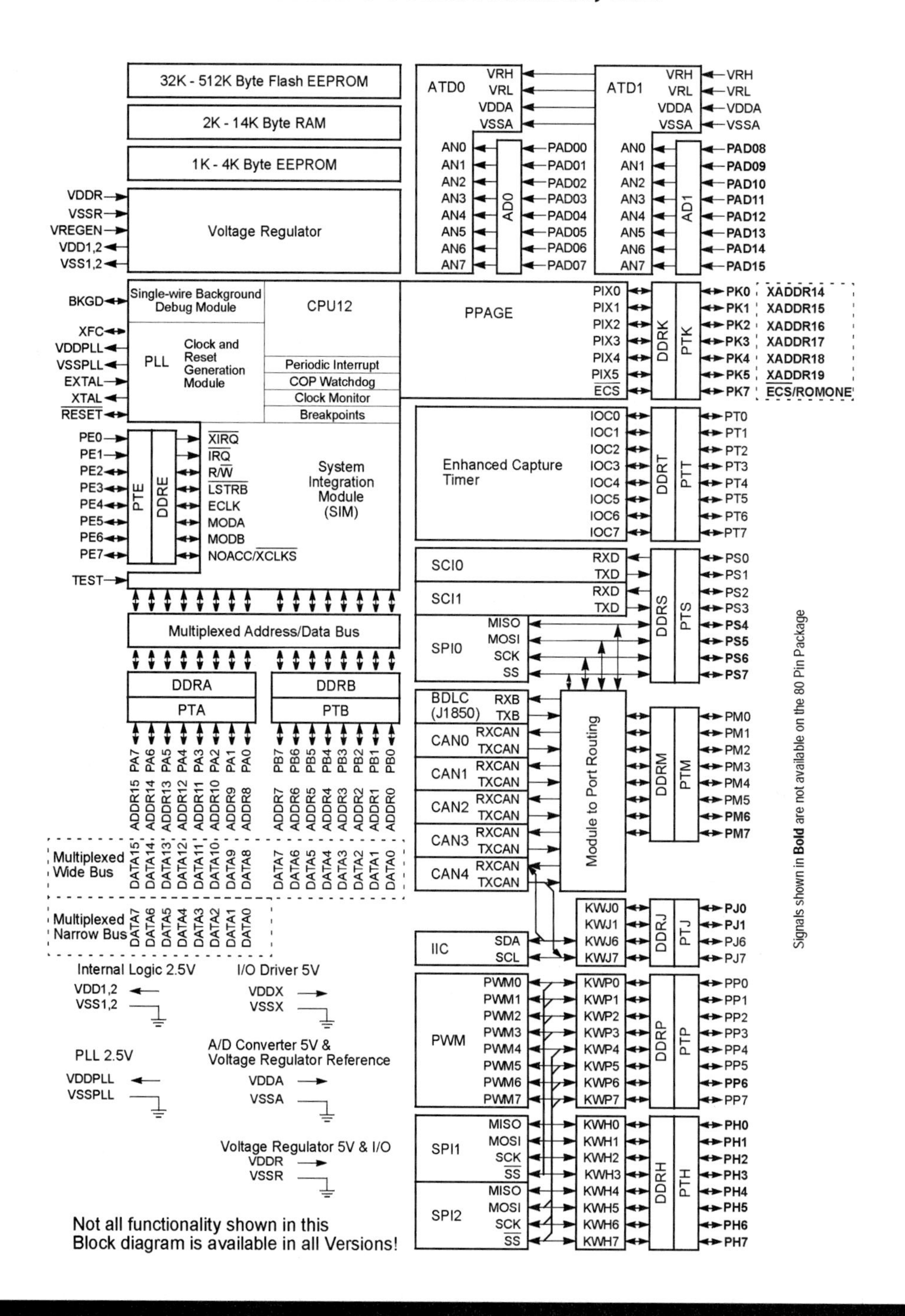

32K - 512K Byte Flash EEPROM
2K - 14K Byte RAM
1K - 4K Byte EEPROM
VDDR
VSSR
VREGEN
VDD1,2
VSS1,2
Voltage Regulator
BKGD
Single-wire Background Debug Module
CPU12
XFC
VDDPLL
VSSPLL
EXTAL
XTAL
RESET
PLL
Clock and Reset Generation Module
Periodic Interrupt
COP Watchdog
Clock Monitor
Breakpoints
PE0
PE1
PE2
PE3
PE4
PE5
PE6
PE7
PTE
DDRE
XIRQ
IRQ
R/W
LSTRB
ECLK
MODA
MODB
NOACC/XCLKS
System Integration Module (SIM)
TEST
Multiplexed Address/Data Bus
DDRA
PTA
DDRB
PTB
PA7 PA6 PA5 PA4 PA3 PA2 PA1 PA0
PB7 PB6 PB5 PB4 PB3 PB2 PB1 PB0
ADDR15 ADDR14 ADDR13 ADDR12 ADDR11 ADDR10 ADDR9 ADDR8
ADDR7 ADDR6 ADDR5 ADDR4 ADDR3 ADDR2 ADDR1 ADDR0
Multiplexed Wide Bus
DATA15 DATA14 DATA13 DATA12 DATA11 DATA10 DATA9 DATA8
DATA7 DATA6 DATA5 DATA4 DATA3 DATA2 DATA1 DATA0
Multiplexed Narrow Bus
DATA7 DATA6 DATA5 DATA4 DATA3 DATA2 DATA1 DATA0
Internal Logic 2.5V
VDD1,2
VSS1,2
I/O Driver 5V
VDDX
VSSX
PLL 2.5V
VDDPLL
VSSPLL
A/D Converter 5V & Voltage Regulator Reference
VDDA
VSSA
Voltage Regulator 5V & I/O
VDDR
VSSR
Not all functionality shown in this Block diagram is available in all Versions!
ATD0
VRH
VRL
VDDA
VSSA
AN0 AN1 AN2 AN3 AN4 AN5 AN6 AN7
AD0
PAD00 PAD01 PAD02 PAD03 PAD04 PAD05 PAD06 PAD07
ATD1
AD1
PAD08 PAD09 PAD10 PAD11 PAD12 PAD13 PAD14 PAD15
PPAGE
PIX0 PIX1 PIX2 PIX3 PIX4 PIX5 ECS
DDRK
PTK
PK0 PK1 PK2 PK3 PK4 PK5 PK7
XADDR14 XADDR15 XADDR16 XADDR17 XADDR18 XADDR19 ECS/ROMONE
Enhanced Capture Timer
IOC0 IOC1 IOC2 IOC3 IOC4 IOC5 IOC6 IOC7
DDRT
PTT
PT0 PT1 PT2 PT3 PT4 PT5 PT6 PT7
SCI0
RXD
TXD
SCI1
SPI0
MISO
MOSI
SCK
SS
DDRS
PTS
PS0 PS1 PS2 PS3 PS4 PS5 PS6 PS7
BDLC (J1850)
RXB
TXB
CAN0
CAN1
CAN2
CAN3
CAN4
RXCAN
TXCAN
Module to Port Routing
DDRM
PTM
PM0 PM1 PM2 PM3 PM4 PM5 PM6 PM7
IIC
SDA
SCL
KWJ0 KWJ1 KWJ6 KWJ7
DDRJ
PTJ
PJ0 PJ1 PJ6 PJ7
PWM
PWM0 PWM1 PWM2 PWM3 PWM4 PWM5 PWM6 PWM7
KWP0 KWP1 KWP2 KWP3 KWP4 KWP5 KWP6 KWP7
DDRP
PTP
PP0 PP1 PP2 PP3 PP4 PP5 PP6 PP7
SPI1
SPI2
KWH0 KWH1 KWH2 KWH3 KWH4 KWH5 KWH6 KWH7
DDRH
PTH
PH0 PH1 PH2 PH3 PH4 PH5 PH6 PH7
Signals shown in Bold are not available on the 80 Pin Package

Figure 1 Pin assignments 112 LQFP for MC9S12D-Family

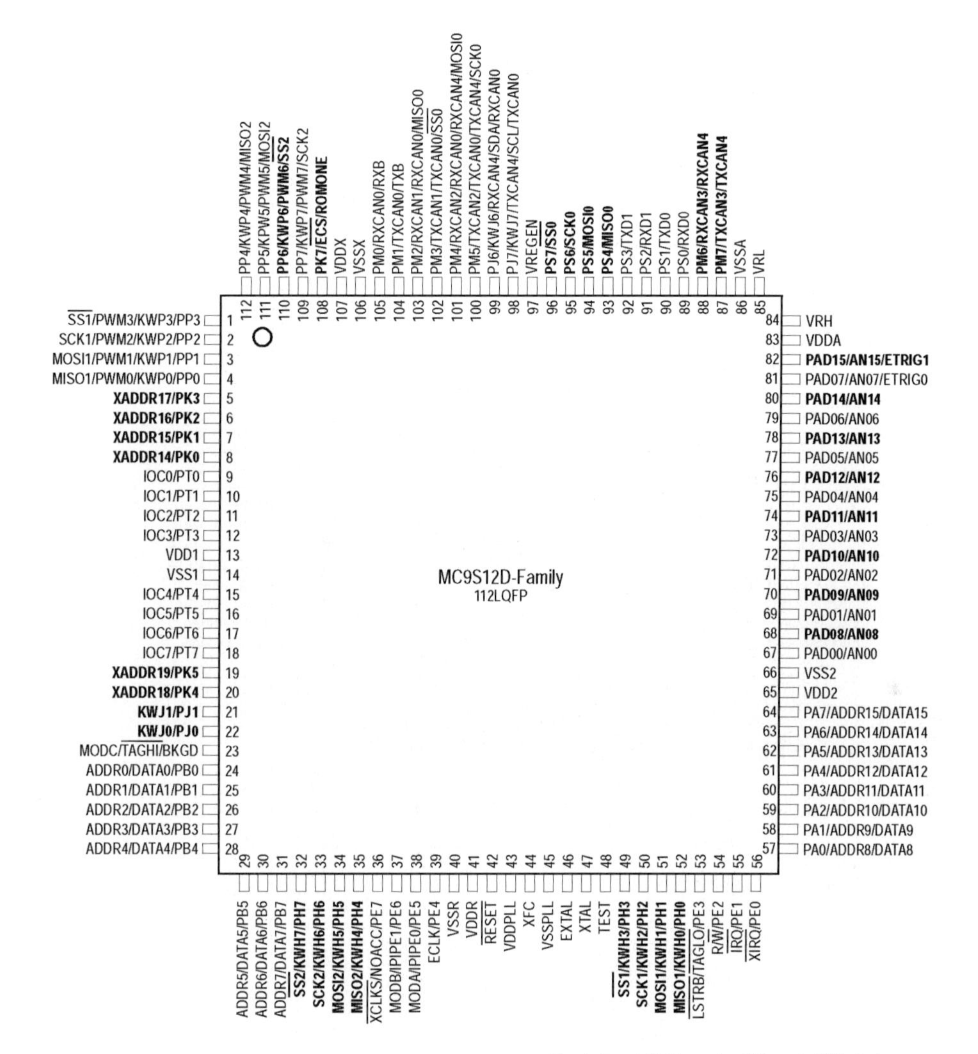

Figure 2 Pin Assignments in 80 QFP for MC9S12D-Family

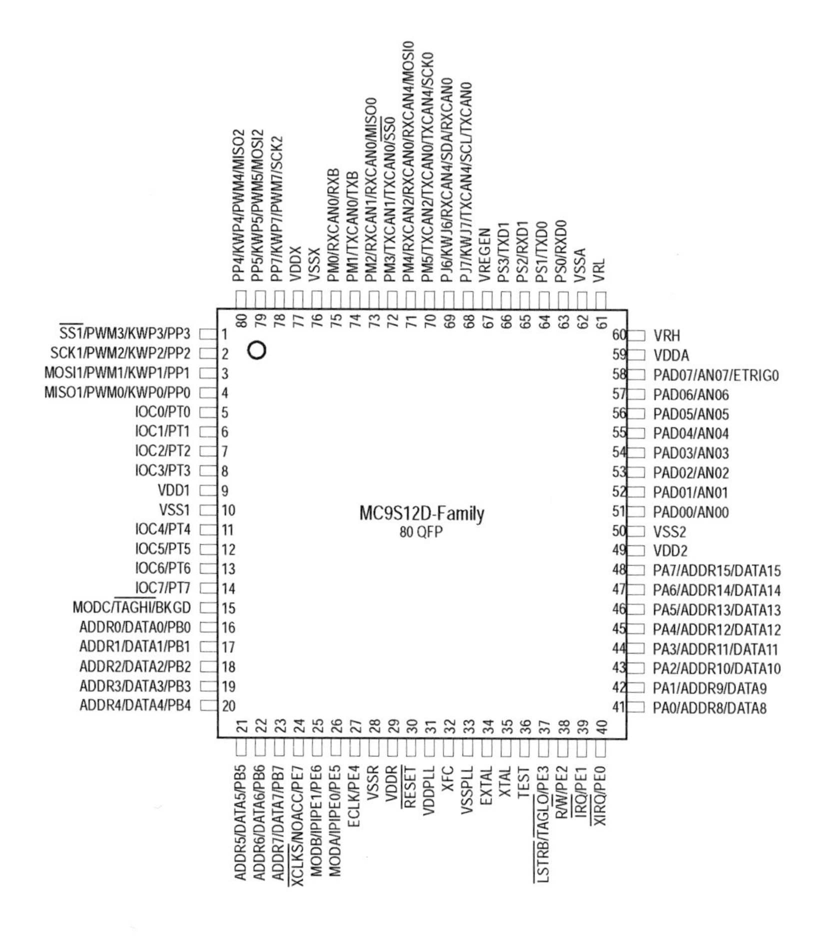

Figure 3 MC9S12Dx512 User Configurable Memory Map

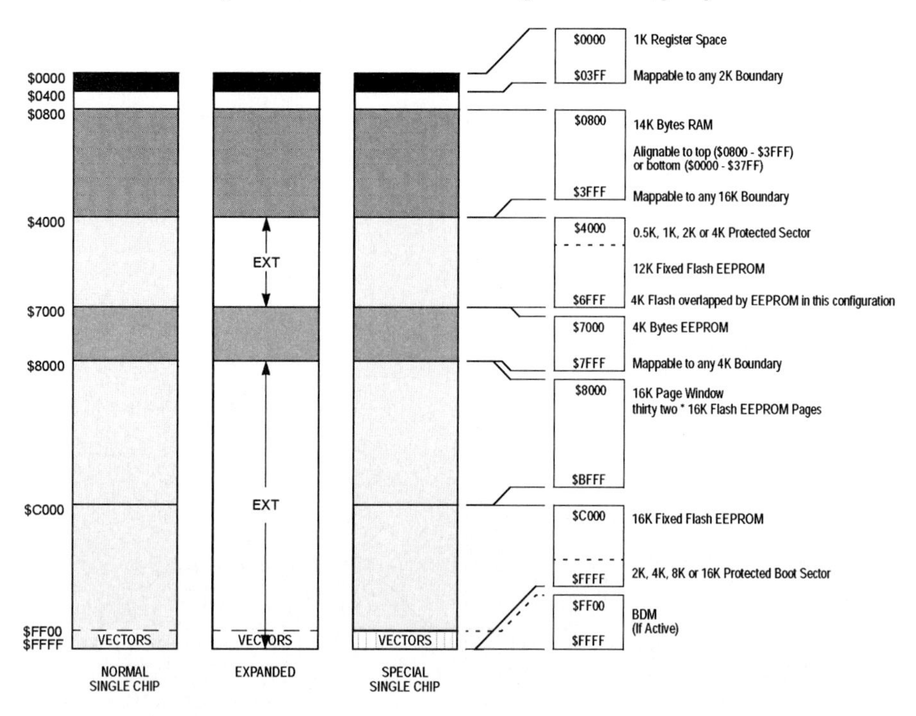

The figure shows a useful map, which is not the map out of reset. After reset the map is:

$0000 - $03FF: Register Space
$0800 - $3FFF: 14K RAM
$0000 - $0FFF: 4K EEPROM (1k $0400 - $07FF visible, $0000 - $03FF and $0800 - $0FFF are not visible)
Various possibilities to make more of the EEPROM fully visible are available, one of them is shown above

Figure 4 MC9S12Dx256 User Configurable Memory Map

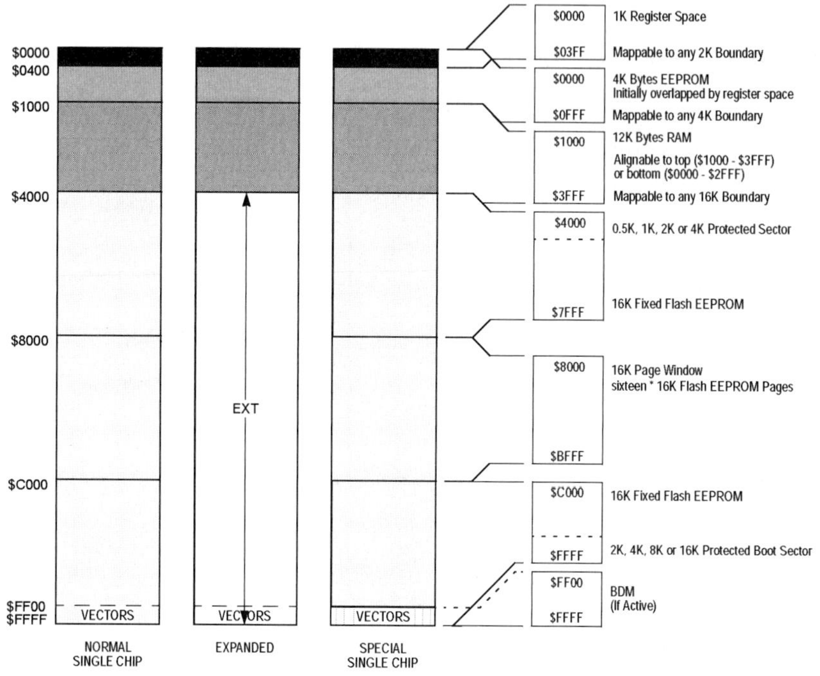

$0000
$0400
$1000
$4000
$8000
$C000
$FF00
$FFFF

VECTORS
NORMAL SINGLE CHIP

EXT
VECTORS
EXPANDED

VECTORS
SPECIAL SINGLE CHIP

$0000 – $03FF: 1K Register Space; Mappable to any 2K Boundary

$0000 – $0FFF: 4K Bytes EEPROM; Initially overlapped by register space; Mappable to any 4K Boundary

$1000 – $3FFF: 12K Bytes RAM; Alignable to top ($1000 - $3FFF) or bottom ($0000 - $2FFF); Mappable to any 16K Boundary

$4000 – $7FFF: 0.5K, 1K, 2K or 4K Protected Sector; 16K Fixed Flash EEPROM

$8000 – $BFFF: 16K Page Window; sixteen * 16K Flash EEPROM Pages

$C000 – $FFFF: 16K Fixed Flash EEPROM; 2K, 4K, 8K or 16K Protected Boot Sector

$FF00 – $FFFF: BDM (If Active)

Figure 5 MC9S12Dx128 User Configurable Memory Map

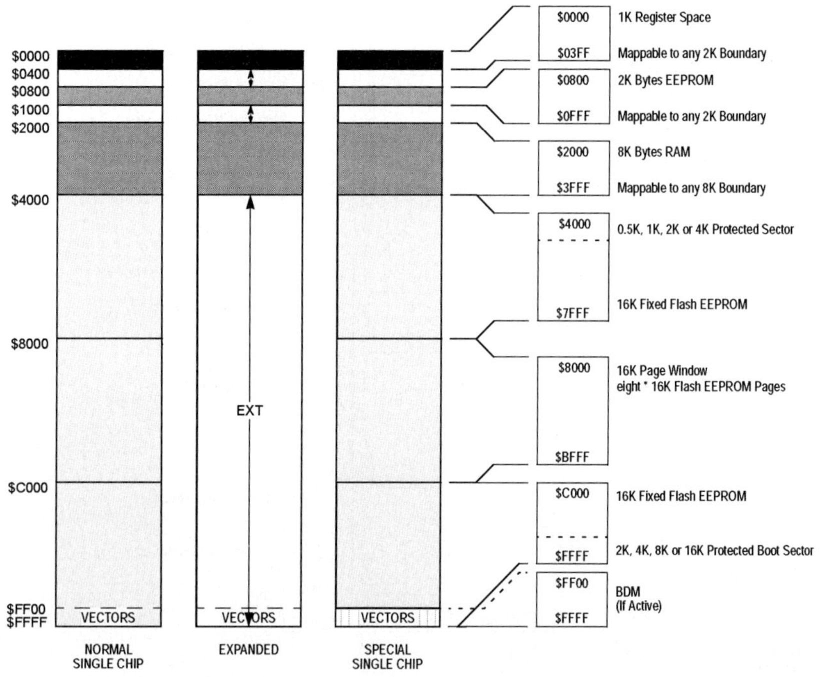

$0000
$0400
$0800
$1000
$2000
$4000
$8000
$C000
$FF00
$FFFF

VECTORS
NORMAL SINGLE CHIP

EXT
VECTORS
EXPANDED

VECTORS
SPECIAL SINGLE CHIP

$0000 – $03FF: 1K Register Space, Mappable to any 2K Boundary
$0800 – $0FFF: 2K Bytes EEPROM, Mappable to any 2K Boundary
$2000 – $3FFF: 8K Bytes RAM, Mappable to any 8K Boundary
$4000 – $7FFF: 0.5K, 1K, 2K or 4K Protected Sector; 16K Fixed Flash EEPROM
$8000 – $BFFF: 16K Page Window, eight * 16K Flash EEPROM Pages
$C000 – $FFFF: 16K Fixed Flash EEPROM; 2K, 4K, 8K or 16K Protected Boot Sector
$FF00 – $FFFF: BDM (If Active)

The figure shows a useful map, which is not the map out of reset. After reset the map is:

$0000 - $03FF: Register Space
$0000 - $1FFF: 8K RAM
$0000 - $07FF: 1K EEPROM (not visible)

Figure 6 MC9S12Dx64 User Configurable Memory Map

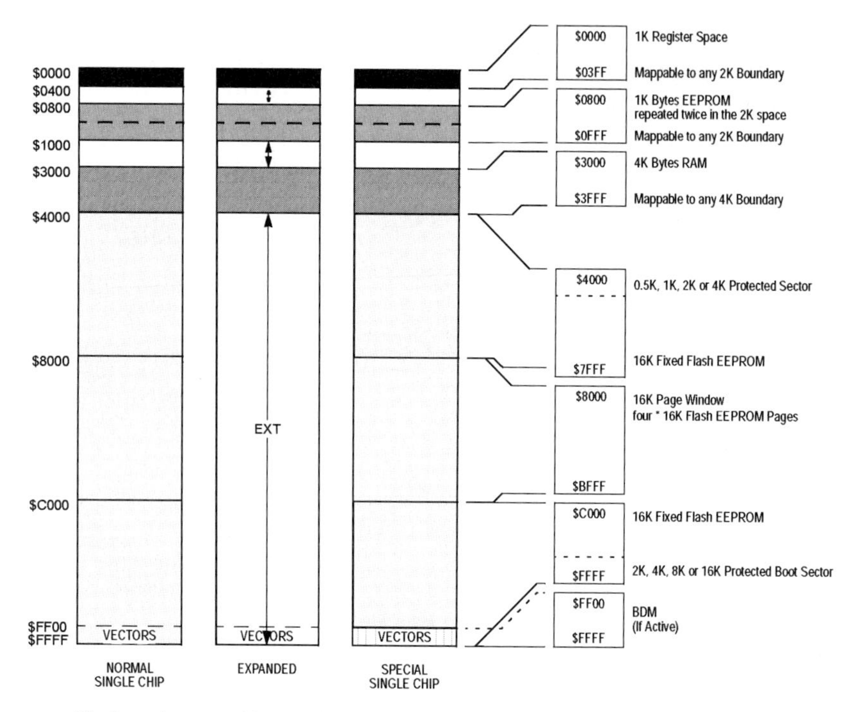

The figure shows a useful map, which is not the map out of reset. After reset the map is:

$0000 - $03FF: Register Space
$0000 - $0FFF: 4K RAM
$0000 - $07FF: 1K EEPROM (not visible)

Figure 7 MC9S12Dx32 User Configurable Memory Map

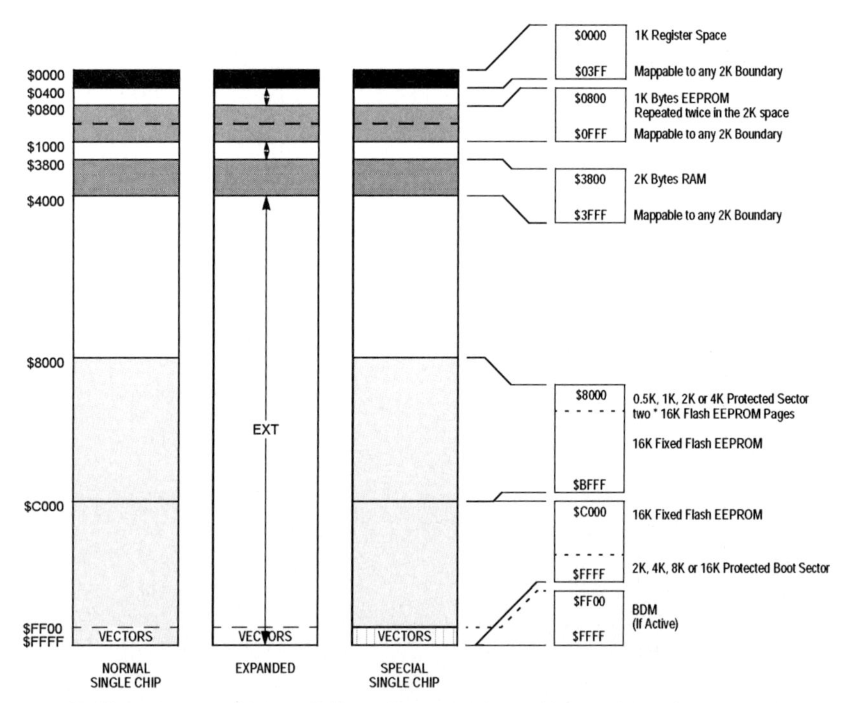

The figure shows a useful map, which is not the map out of reset. After reset the map is:

$0000 - $03FF: Register Space
$0800 - $0FFF: 2K RAM
$0000 - $07FF: 1K EEPROM (not visible)

INDEX

A

ADC interfacing, 434-438
  conversion time, 435
  data output, 436
  devices, 434-438
  input channels, 438
  parallel or serial, 436-437
  reference voltage, 435-436
  resolution, 435
  successive approximation, 438-470
add instructions, 63
  ADCA, 152-154
  ADDA, 152
  DAA, 156
  impact on CCR, 79-80
addressing, HCS12, 70-77
  direct mode, 74-75
  extended mode, 72
  immediate mode, 71-72
  indexed mode, 75-76
  inherent/register mode, 71
  relative mode, 75
Application-Specific Integrated Circuit(ASIC), 47
Arithmetic instructions
  ABA, 584
  ADCA, 584
  ADCB, 584-585
  ADDA, 585-587
  ADDB, 587-588
  ADDD, 588
  DAA, 600-601
  DEC, 602
  DECA, 602
  DECB, 602
  DECS, 602
  DEX, 603
  DEY, 603
  EDIV, 603
  EDIVS, 603
  EMACS, 603
  EMAXD, 604
  EMAXM, 604
  EMIND, 604
  EMINM, 604
  EMUL, 604-605
  EMULS, 605
  IDIV, 607
  IDIVS, 607
  INCA, 608
  INCB, 608
  INS, 608
  INX, 608
  INY, 608
  MAXA, 616
  MAXM, 616
  MINA, 616
  MINM, 616
  MUL, 617
  NEG, 617-618
  NEGA, 618
  NEGB, 618
  SBA, 624
  SBCA, 624-625
  SBCB, 625
  SUBA, 628
  SUBB, 628-629
  SUBD, 629
Arithmetic/Logic Unit (ALU), CPU, 30
ASCII code, 8
  BCD to ASCII conversion, 209-210
  binary to ASCII conversion, 210-211
ASCII codes, 668
AsmIDE, 634
assembler
  directives
    DC.x, 201-203
    INCLUDE, 214-215
    MLIST, 215
    XDEF, 218-219
    XREF, 217-218
  macros, 212-214
    definition, 212-213
    local labels, 213-214
    vs. subroutines, 217
  modules, 217
    linking, 220
assembler, HCS12, 81-86
  data format, 81-82
  directives, 82-83

DC.x, 85
DS.x, 85
END, 84
EQU, 82-83
FCB, 84
FCC, 85
ORG, 84
SET, 83-84
label naming, 85
Assembly language, 86-88
asm and object files, 89-90
debug, 90-92
linking, 88-92
simulators, 97
structure, 86-88
ATD interfacing, 439-458
ATDCTL2, 444
ATDCTL3, 444
ATDCTL4, 440
ATDCTL5, 445
conversion frequency, 440-441
conversion time, 442
features, 439-440
programming multiple channels, 456-458
programming multiple conversion, 454-456
programming using interrupts, 452-454
programming using polling, 450
in Assembly, 450-451
in C, 452
result registers, 446
status registers, 447
voltage connection, 448

B

Background Debug Mode (BDM), 56
BDM (Background Debugger Mode), 263
BGND instruction, 592
big endian, 73-74
bit programming, 142-150
checking an input pin, 144
instructions
BCLR
BRCLR
BRSET
BSET
branch instructions, 106-116
and $, 116
and looping, 106-111
BCC, 591-592
BCS, 592
BEQ, 592
BGE, 592
BGT, 593
BHI, 593
BHS, 593
BLE, 594
BLO, 594
BLS, 594
BLT, 594
BMI, 595
BNE, 595
BPL, 595
BRA, 595
BRCLR, 595
BRN, 595-596
BRSET, 596
BVC, 596
BVS, 596
DBEQ, 601-602
DBNE, 602
IBEQ, 606
IBNE, 607
JMP, 608
LBCC, 609
LBCS, 609
LBEQ, 609
LBGE, 609
LBGT, 609
LBHI, 610
LBHS, 610
LBLE, 610
LBLO, 610
LBLS, 610
LBLT, 611
LBMI, 611
LBNE, 611
LBPL, 611
LBRA, 611
LBRN, 612
LBVC, 611
LBVS, 612
long branch, 114
short branch, 113

BCC, 112-115
BEQ, 112
BNE, 106
BRA, 116
TBEQ, 630
TBNE, 630
unconditional branch, 115-116
JMP, 115-116

C

C programming
binary to decimal and ASCII, 242
bit-wise logic operators, 234
bit-wise shift operators, 234-235
checksum generation and testing, 239-240
code space allocation, 248-255
for data, 249
specific ROM address, 249-250
converting ASCII to packed BCD, 238
converting BCD to ASCII, 238
data types, 226-230
short long and long, 230
signed char, 228
signed int, 230
unsigned char, 227
unsigned int, 229
I/O bit-manipulattion, 243
serializing data, 246
time delay, 230-231
why C?, 226
call instructions, 117-126
CALL, 121-122
in the main program, 120-121
JSR, 117
Central Processing Unit (CPU), 13-15
choosing a microcontroller, 46-47
availability, 47
ease of development, 47
meeting the task, 46
compare instructions, 168-173
BCC, 169
BCS, 169
BHI, 169
BLO, 169
Complex Instruction Set Computer (CISC), 34-36
computer architecture, 13-15, 29-33
address bus, 14-15
data bus, 14
Harvard and von Neumann, 32-33
inside the CPU, 29-31
RISC and CISC, 34-36
condition code register (CCR), 77-81
C, the carry flag, 78-79
decision making, 81
H, the half-byte carry flag, 78-79
instruction, 79-80
N, the negative flag, 79
V, the overflow flag, 78-79
Z, the zero flag, 78-79
conversion of numbers.
See number systems, converting
COP Watchdog timer, 261
CRG (clock and reset generation), 264-265
operation in critical conditions, 265-266

D

DAC interfacing, 462-470
converting Iout to voltage, 464
generating a sine wave, 464-466
in Assembly, 466
in C, 466-467
MC1408, 463
data sheets, 672
data transfer instructions, 96
See also move instructions.
See also transfer instructions.
See also exchange instructions.
D-BUG12, 635-639
DC motors, 556-559
bidirectional control, 556-559
control with optoisolator, 562-563
pulse width modulation, 560-561
in C, 564
unidirectional control, 556
development tools, 670
division instructions
EDIV, 160-161
EDIVS, 189
IDIV, 160
IDIVS, 189
downloading the program, 263-264

Dynamic RAM (DRAM).
See memory, DRAM

E

EEPROM memory, 490-498
accessing the EEPROM in C, 496-498
registers, 490
setting the EEPROM clock, 498
writing to EEPROM, 492-496
embedded applications, 45-46
embedded processor, 46
embedded processors, high-end, 46
embedded system, 45
Erasable Programmable ROM (EPROM).
See memory, EPROM and UV-EPROM

F

Field-Programmable Gate Array (FPGA), 47
find the highest number, 170-171
flash memory.
See memory, Flash
Flash memory paging, 472-478
CALL instruction, 474-477
remapping registers, 477-479
Flash memory programming, 478-490
erasing Flash, 486-487, 489
registers, 479-480
setting the Flash clock, 498
writing to Flash, 482-486, 488-489
flowcharts, 662

G

general-purpose microprocessors, 44-45

H

Harvard architecture, 32-33
HCS12
serial communication interface, 335-354
baud rates, 337-338
control registers, 340
data register, 339
error calculation, 348
idle and break characters, 349
programming in Assembly, 337
programming in C, 351-354
RDRF flag, 346-347
RXD and TXD pins, 335
SCI0 and SCI1, 342-343
status registers, 341
TDRE flag, 344-345
HCS12 assemblers, 670
HCS12 peripherals
free-running timer, 278-281
control registers
(TSCR1 and TSCR2), 280-281
prescaler, 281-282
TOF flag, 279-280
input capture function, 290-293
output compare function, 283-290
programming, 284-289
registers, 284
pulse accumulator, 296-304
programming accumulator A, 298-300
programming accumulator B, 301-303
timer, 304-306
registers in C, 304
HCS12 pin connections, 258-261
BKGD, 263
Reset, 261
Vdd1,2 and Vss1,2, 261
VddPLL and VssPLL, 258-261
VddR (Vcc) and VssR (Gnd), 258-261
Vregen, 261
XTAL and EXTAL, 258-261
HCS12 Trainer, 266-267
troubleshooting tips, 267
HCS12 trainers, 670
hex to decimal conversion, 161

I

IC chips
74LS138 decoder, 27-28
ImageCraft, 639-642
immediate operand, 63
indexed addressing mode, 126-131, 196-200
auto-decrement, 198
auto-increment, 198
clearing RAM, 126-127

effective address, 199-200
transfer blocks of data, 127
instruction decoder, CPU, 30
interfacing devices, 650-653
buffers/drivers, 651
driving an output pin, 652
HCS12 port fan-out, 653
wired-ORed option, 653
IC fan-out, 650, 652
tri-state buffer, 652
interrupts, 360-366
and the flag register, 360-361
D-BUG12 vector table, 365-366
external, 373-379
IRQ, 373-375
Edge-triggered, 375
Level-triggered, 373-374
STOP and WAI instructions, 378
XIRQ, 376
fast context saving, 385-386
free-running timer, 366-372
TIE register, 368-372
global control, 365-366
interrupt service routines, 360
interrupt vector table, 362-363
latency, 386
major categories, 363-364
nested interrupt, 386-388
priority, 383-388
HPRIO register, 383-384
programming in C, 388
interrupt numbering, 389-399
real time, 398-405
calculating the period, 398-399
serial communication, 379-383
TDRE and RDRF, 379-380, 382-383
steps in executing, 361-362
vs. polling, 360
inverter gate. See logic gates, inverter

K

keyboard interfacing, 421-431
detecting the key, 422-429
using polling, 425-427
C version, 427-429
identifying the key, 421

L

LCD interfacing, 408-421
data sheet, 415
pin descriptions, 408-409
sending commands, 4-bit, 412-414
C version, 419-420
sending commands, 8-bit, 410-411
C version, 418-419
sending from a look-up table, 417-418
C Version, 420-421
little endian, 73-74
load (LD) instructions, 60-63, 76
load instructions
LDAA, 612
LDAB, 612
LDD, 612-613
LDS, 613
LDX, 613
LDY, 613
LEAS, 613
LEAX, 614
LEAY, 614
logic design, 10-12
decoder, 12
flip-flop, 12
full-adder, 11
half-adder, 11
logic families, 645-659
advances, 648-649
CMOS inverter, 646
history, 648
input/output characteristics, 647
inverters, 645-646
open-collector, 649
open-drain, 649
logic gates, 9-10
AND, 9-10
inverter, 10
NAND and NOR, 10
OR, 9-10
tri-state buffer, 9-10
XOR, 10
logic instructions
ANDA, 163

BCLR, 165
BRCLR, 167-168
BRSET, 167
BSET, 166
COM, 165
EORA, 164
NEG, 165
ORAA, 163
logical instructions
ANDA, 589
ANDB, 589
ANDCC, 590
BCLR, 592
BITA, 593
BITB, 593-594
BSET, 596
CBA, 597
CLC, 598
CLI, 598
CLR, 598
CLRA, 598
CLRB, 598
CLV, 598
CMPA, 598
CMPB, 599
COM, 599
COMA, 599
COMB, 599
CPD, 599
CPS, 600
CPX, 600
CPY, 600
EORA, 605
EORB, 606
ORAA, 618-619
ORAB, 619
ORCC, 619
SEC, 625
SEI, 625
SEV, 625
SEX, 625-626
TST, 631
TSTA, 631
TSTB, 631
TSX, 631
TSY, 631
TXS, 631-632
TYS, 632
logical vs. arithmetic shift, 175
lowercase to uppercase conversion, 172

M

Mask ROM. See memory, Mask ROM
mechatronics, 47-48
memory, 13-29
address decoding, 25-26
address decoder, 26. See also logic design, decoder
chip enable (CE), 25-26
chip select (CS), 25-26
programmable logic, 28
capacity, 15
DRAM, 23-24
organization, 25
RAS and CAS, 24
EEPROM, 19
EPROM and UV-EPROM, 17-18
Flash, 19-20
Mask ROM, 20
NV-RAM, 22-23
organization, 16
PROM and OTP, 16-17
RAM, 20
ROM, 16
speed, 16
SRAM, 21-23
width, 13
memory, HCS12, 64-70
accessing beyond 64KB, 68-69
checksum byte in ROM, 206
checksum generation and testing, 207-209
code ROM space, 65-66
where the HCS12 wakes up, 69
CodeWarrior simulator, 203
data in code space, 201-205
define constant byte, 201-203
look-up table, 204
data RAM space, 64-65, 66
EEPROM space, 65, 66
instruction size, 76-77
register space, 64
single-chip and expanded modes, 69-70

microcontroller, 44-48, 56
choosing.
See choosing a microcontroller
data RAM and EEPROM, 55
I/O pins, 56
in embedded systems, 45
in mechatronics, 47
peripherals, 54-55
program ROM, 53-54
microcontroller architecture
68xx, 48
8051, 48-49
ColdFire, 49
CPU08, 48-49
CPU12, 48-49
the Controller Continuum, 49
microprocessor, 44
migrating to the HCS12, 96-97
move instructions, 128
MOVB, 128, 616-617
MOVW, 128, 617
STAA, 626
STAB, 626
STD, 626
STS, 627
STX, 627
STY, 627
TAB, 630
TAP, 630
TBA, 630
TFR, 631
TPA, 631
XGDX, 632
XGDY, 632
multiplication instructions
EMUL, 160
EMULS, 188
MUL, 159-160

N

Nonvolatile RAM.
See memory, NV-RAM
NOP instruction, 618
number systems, 2
addition of bases, 6
ASCII, 180
BCD, 155-156
packed, 155-156
unpacked, 155
conversion programs in C.
See C programming, converting ...
converting ASCII to packed BCD, 181-182
converting BCD to ASCII, 209-210
converting between binary and hex, 4
converting binary to ASCII, 210-211
converting binary to decimal, 3
converting decimal to binary, 2
converting decimal to hex, 4
converting hex to decimal, 5
converting packed BCD to ASCII, 180
counting in bases, 6
decimal and binary, 2
hexadecimal (hex), 4
addition, 7
subtraction, 7

O

One-Time Programmable (OTP) memory.
See memory, PROM and OTP
optoisolator interfacing, 512-513

P

parts suppliers, 671
PLL (phase lock loop), 267-270
setting the clock, 267
port programming, 134-142
DDR registers, 138
in expanded mode, 141
PAD (analog-to-digital), 142
peripherals, 134
Port A, 140
Port B, 140-141
Port E, 141-142
Ports H through T, 142
primer for x86 programmers, 667
program counter, CPU, 30
program execution
branch penalty, 124
delay calculation, 122-123
instruction timing, 124-125
pipelining, 123

Programmable ROM (PROM). See memory, PROM and OTP
pseudocode, 662-664
PWM interfacing, 566-582
  16-bit PWM, 578-582
  channel alignment, 572-582
  channel polarity, 571-582
  clock sources, 566-567
    clock SA, 567
    clock SB, 567
    PWMCLK, 569-582
  control register, 575-582
  counter, period and duty cycle, 573-582
  steps in programming, 575-582
    in Assembly, 577-582
    in C, 578-582

R

Random Access Memory (RAM). See memory, RAM
Read-Only Memory (ROM). See memory, ROM
real-time clock(DS1306) interfacing, 536-543, 544-546, 546-553
  1-Hz feature, 546-547
  address map, 539
  Alarm and IRQ output pins, 549
  Alarm0, Alarm1, and interrupt, 547
  interfacing using SPI, 540
    in C, 544-546
  once-per-day alarm, 549
  once-per-hour alarm, 549
  once-per-minute alarm, 549
  once-per-second alarm, 549
  once-per-week alarm, 549
  pins, 536-538
  time and date, 539
  using INT0 to interrupt HCS12, 550
  WP bit, 538
Reduced Instruction Set Computer (RISC), 33-36
registers, CPU, 29
registers, HCS12, 60
relay interfacing, 508-511
  driving a relay, 510
  electromechanical, 508-509
  reed switch, 512
  solid-state, 511
rotate instructions
  ROL, 175, 621-622
  ROLA, 174-175, 622
  ROLB, 622
  ROR, 174, 622
  RORA, 174, 622-623
  RORB, 623

S

s19 file, 270-276
sensor interfacing, 458-462
  in Assembly, 461
  in C, 461-462
  LM34 and LM35, 459
  reading and converting temperature, 461-462
  signal conditioning, 459-461
  temperature sensors, 458
serial communication, 328-334
  COM ports, 334
  communication classification, 332
  data framing, 330
  half- and full-duplex, 329
  handshaking signals, 332-334
  MAX232, 335-336
  MAX233, 336
  RS232 pins, 331
  RS232 standards, 331
  start and stop bits, 330-331
  transfer rate, 331
serializing data, 176-177
shift instructions
  ASL, 590
  ASLA, 187, 590
  ASLB, 590
  ASLD, 590-591
  ASR, 187, 591
  ASRA, 187, 591
  ASRB, 591
  LSL, 176, 614
  LSLA, 175, 614
  LSLB, 614
  LSLD, 615
  LSR, 175, 615

LSRA, 175, 615
LSRB, 615
LSRD, 616
signed numbers, 182-193
arithmetic shift, 187. See also shift instructions
comparing, 187-188
multiplication and division, 188
V (overflow) flag, 184-186, 186-187
word-sized (16-bit), 186
SPI bus protocol, 526-529
read and write, 527
reading data, 528-529
multibyte, 528-549
single-byte, 528
writing data, 527-528
multibyte, 527-528
single-byte, 527
SPI interfacing, 529-535
baud rate selection, 530
control registers, 531
pins, 531
receiving data, 535-538
status register, 533
steps in programming, 534
stack, 92-104
access, 92
and call instructions, 95
instructions
pull, 93
push, 93
scratch pad conflict, 93-94
stack instructions
PSHA, 620
PSHB, 620
PSHC, 620
PSHD, 620
PSHX, 620
PSHY, 620
PULA, 620
PULB, 621
PULC, 621
PULD, 621
PULX, 621
PULY, 621
Static RAM (SRAM). See memory, SRAM
stepper motor interfacing, 514-524
4-step sequence, 517
8-step sequence, 518
controlling via optoisolator, 521
holding torque, 518
motor speed, 518
number of teeth on rotor, 517-518
step angle, 515-516
transistors as drivers, 519-520
unipolar versus bipolar, 519
wave drive sequence, 518
STOP instruction, 627
store (ST) instructions, 72-74, 76
subroutine instructions
BSR, 596
CALL, 597
JSR, 608-609
RTC, 623
RTI, 623
RTS, 623-624
subtract instructions
NEG, 157-158
SBCA, 158-193
SUBA, 157
SWI instruction, 629-630
system design issues, 653-657
bulk decoupling capacitor, 656
crosstalk, 656-657
decoupling capacitors, 656
dynamic and static currents, 654
ground bounce, 654-656
sleep mode, 654
transient currents, 656
transmission line ringing, 657-659

T

Trainers, 56
transistors, 644-645
MOS vs. bipolar, 645
TRAP instruction, 631
tri-state buffer. See logic gates, tri-state buffer

U

UltraViolet EPROM (UV-EPROM).
See memory, EPROM and UV-EPROM
unsigned numbers, 152
addition, 152
16-bit numbers, 152-154
individual bytes, 152
multi-byte numbers, 154-155
packed BCD, 155-156
division of unsigned numbers, 160
multiplication, 159-160
subtraction, 156-159

V

von Neumann (Princeton) architecture, 32-33

W

WAI instruction, 632